CANCER

CANCER
How Lifestyles May Impact Disease Development, Progression, and Treatment

Hymie Anisman
Canada Research Chair in Behavioral Neuroscience, Department of Neuroscience, Carleton University, Ottawa, ONT, Canada

Alexander W. Kusnecov
Professor, Behavioral and Systems Neuroscience Program, Department of Psychology, Rutgers University, Piscataway, NJ, United States

Academic Press is an imprint of Elsevier
125 London Wall, London EC2Y 5AS, United Kingdom
525 B Street, Suite 1650, San Diego, CA 92101, United States
50 Hampshire Street, 5th Floor, Cambridge, MA 02139, United States
The Boulevard, Langford Lane, Kidlington, Oxford OX5 1GB, United Kingdom

Copyright © 2022 Elsevier Inc. All rights reserved.

No part of this publication may be reproduced or transmitted in any form or by any means, electronic or mechanical, including photocopying, recording, or any information storage and retrieval system, without permission in writing from the publisher. Details on how to seek permission, further information about the Publisher's permissions policies and our arrangements with organizations such as the Copyright Clearance Center and the Copyright Licensing Agency, can be found at our website: www.elsevier.com/permissions.

This book and the individual contributions contained in it are protected under copyright by the Publisher (other than as may be noted herein).

Notices

Knowledge and best practice in this field are constantly changing. As new research and experience broaden our understanding, changes in research methods, professional practices, or medical treatment may become necessary.

Practitioners and researchers must always rely on their own experience and knowledge in evaluating and using any information, methods, compounds, or experiments described herein. In using such information or methods they should be mindful of their own safety and the safety of others, including parties for whom they have a professional responsibility.

To the fullest extent of the law, neither the Publisher nor the authors, contributors, or editors, assume any liability for any injury and/or damage to persons or property as a matter of products liability, negligence or otherwise, or from any use or operation of any methods, products, instructions, or ideas contained in the material herein.

Library of Congress Cataloging-in-Publication Data
A catalog record for this book is available from the Library of Congress

British Library Cataloguing-in-Publication Data
A catalogue record for this book is available from the British Library

ISBN 978-0-323-91904-3

For information on all Academic Press publications
visit our website at https://www.elsevier.com/books-and-journals

Publisher: Stacy Masucci
Acquisitions Editor: Rafael E. Teixeira
Editorial Project Manager: Kristi L. Anderson
Production Project Manager: Omer Mukthar
Cover Designer: Miles Hitchen

Typeset by STRAIVE, India

Dedication

These trees which he plants, and under whose shade he shall never sit, he loves them for themselves, and for the sake of his children and his children's children, who are to sit beneath the shadow of their spreading boughs.

Hyacinthe Loyson, Paris in 1866.

Contents

Preface xi
Acknowledgments xiii

1. Cancer biology and pathology

The distress of cancer 1
Extrinsic influences on cancer 3
Features of cancer development and progression 9
Cancer processes 10
Sources of cancer 13
Treatment resistance 20
Metastases 21
Ingredients for cancer growth and metastasis 23
Impact of hormones and hormone receptors on cancer processes 28
Inflammatory factors in the cancer process 29
Intrinsic vs extrinsic contributions to cancer occurrence 33
Summary and conclusions 34
References 34

2. Cancer and immunity

Changing times concerning diseases 39
Fundamentals of immunity 40
Innate and adaptive components of the immune system 43
Anatomy of the immune system: The lymphoid system 52
Adaptive (acquired) immunity 53
Cytokines 60
Inflammasome 63
Immune tolerance 65
Summary and conclusions 66
References 66

3. Microbiota and health

How it began 69
Features of microbiota 70
An evolutionary perspective 72
Microbiota development 75
Microbiota and immune functioning 80
Microbiota and illness comorbidities 83
Identifying good and bad microbes 87
The influence of environmental toxicants 88
Summary and conclusions 89
References 89

4. Genetic and epigenetic processes linked to cancer

Genetics and natural selection: The good and the bad 93
From Mendelian genetics to molecular biology 94
DNA and RNA 98
Epigenetic processes 110
Complex gene × environment interactions 119
Population-level genetics 121
Precision (personalized) medicine 125
Summary and conclusions 129
References 130

5. Stressors: Psychological and neurobiological processes

Attributions and misattributions of stressor effects on illness 135
Features of stressors 136
Chronic stressors and allostatic load 137
Stress sensitization 139
Identifying and responding to stressors 139
Resilience and vulnerability 145
Neurobiological actions of stressors 147
Growth factors 157
Cellular stress responses 164
Prenatal and early-life stressor effects 165
Summary and conclusions 171
References 172

6. Stress, immunity, and cancer

Blame it on stressor experiences 178
Brain and immune system interactions 178
Stressor influences on immunity 182
Cytokine variations associated with stressors 187
Stress and microbiota 191
Stressful events and cancer 195
Influence of psychosocial stressors on immunity and cancer 202
Prenatal and early-life stressors in relation to cancer 204
Influence of diverse life stressors 207
Psychological factors associated with cancer treatment 212
Linking neuropsychiatric and neurodegenerative disorders to cancer 217
Summary and conclusions 218
References 219

7. Eating and nutrition links to cancer

Then and now 225
The digestive process 227
Hormonal and brain processes underlying eating 229
Diets and weight loss 234
Diet and nutrition in relation to cancer 239
Summary and conclusions 248
References 249

8. Dietary components associated with being overweight, having obesity, and cancer

Being overweight or having obesity as a health risk 253
Linking the development of obesity to cancer 254
Relations between obesity, immunity, inflammation, and cancer 257
Genetic influences on obesity 260
Dietary components in relation to cancer and its treatment 262
Summary and conclusions 274
References 275

9. Microbiota in relation to cancer

Finding the right microbial mix 279
Microbiota in relation to cancer 282
Short-chain fatty acids 288
Influence of specific microbiota in diverse forms of cancer 292
Parasites and cancer 298
The inflammatory link between microbiota and cancer 298
Impact of the prenatal and perinatal diet 301
The other side of microbiota: Is there value to prebiotic and probiotic supplements? 304
Summary and conclusions 304
References 305

10. Exercise

The broad effects of exercise 311
Immune and cytokine changes associated with exercise 312
Cytokine changes associated with exercise 314
Exercise and cancer prevention 316
Impact of exercise on existent cancers 318
Impact of exercise on cancer features and side effects of treatments 322
Fueling cancer progression 327
Exercise and microbiota 329
Sedentary behaviors 332
The social element 334
Roadblocks to exercise and how to get around them 334
Summary and conclusions 336
References 337

11. Sleep and circadian rhythms

A brief history of thoughts and research related to sleep 341
Functions of sleep 343
Links between sleep and other lifestyle factors 345
Neurobiological aspects related to sleep and circadian rhythms 347
Circadian rhythms: Immune and cytokine changes 349
Sleep, circadian rhythms, and cancer progression 352
Occupations associated with altered sleep cycles and cancer occurrence 358
Sleep disruption and diurnal variations in relation to microbiota 361
Implications of circadian rhythmicity to cancer treatments 363

Summary and conclusions   364
References   365

12. Adopting healthy behaviors: Toward prevention and cures

Understanding counterproductive behaviors   369
A brief look back   370
Intervention approaches   372
Changing attitudes and behaviors and barriers to change   374
Psychosocial and cognitive approaches to enhance health behaviors   378
Positive psychology   390
Social support   392
Summary and conclusions   396
References   396

13. Cancer therapies: Caveats, concerns, and momentum

Promises, promises   401
Cancer screening   402
Cancer screening in common types of cancer   403
Caveats concerning screening and early cancer detection   406
Unintended consequences and unintended benefits   411
Progress in cancer treatment development   412
Complementary and alternative medicine   416
Precision treatment: Obstacles and challenges   419
Summary and conclusions   426
References   427

14. Traditional therapies and their moderation

The role of serendipity   431
Chemotherapy and radiation therapy   432

The challenge of treatment resistance   438
Nutrients related to treatment resistance   443
Manipulating energy and metabolism   444
Microbiota in relation to chemotherapy   451
Summary and conclusions   455
References   455

15. Immunotherapies and their moderation

Immunotherapeutic approaches   462
Stem cell therapy   464
Nonspecific immunotherapies   467
Treatment vaccines   471
Antibody-based therapies   474
Enhancing immunotherapy's effectiveness   479
Side effects of checkpoint inhibitors and CAR T therapy   482
Use of biomarkers   485
Influence of microbiota on treatment responses to cancer therapies   489
Prebiotics and probiotics in cancer treatment   495
Summary and conclusions   497
References   497

16. Moving forward—The science and the patient

Where we stand   503
Moving forward   504
Prevention vs treatment   506
Living with dignity   509
Dying with dignity   512
Summary and conclusions   514
References   515

Index   517

Preface

Leo Tolstoy informed us in *Anna Karenina* that "happy families are all alike; every unhappy family is unhappy in its own way." It can be said that all chronic illnesses are horrid, and each is horrid in its own way. Determining the processes and mechanisms responsible for chronic illnesses, such as cancer, has encountered multiple challenges, made much more difficult since the processes that are linked to a given type of cancer may vary appreciably across individuals. Several illnesses are determined by single genes or gene mutations, whereas others involve the confluence of multiple genetic, environmental, and experiential factors. Many chronic illnesses have their roots in prenatal experiences or those encountered during early development that can have repercussions throughout life. For that matter, even experiences of our forebears can have consequences that span generations, as in the case of cumulative, historical traumas.

Illnesses can be caused by environmental toxicants over which individuals have little control. Other chronic diseases, in contrast, can also be fomented by factors potentially within a person's control, such as lifestyle choices or how they cope with certain psychosocial challenges. In fact, to varying degrees the development and the progression of many chronic conditions is preventable. This shouldn't be taken to imply that risky behavior individuals adopt is solely their responsibility. The person's early life environment, including being raised in poverty, poor parental care, stressor experiences, and the diet to which they had been accustomed, can set the developmental trajectory, so that certain high-risk behaviors will be more likely to be adopted.

Of the many chronic diseases that can develop, the most dreaded is certainly cancer, which Siddhartha Mukherjee aptly described as *The Emperor of All Maladies*. Heart disease is responsible for more deaths than cancer, and mental illnesses, particularly depressive disorders, affect a larger portion of people and are comorbid with numerous illnesses. But cancer holds a special place in the collective psyche owing to the travails encountered by the therapies applied, often without providing a cure.

Researchers and oncologists have for generations attempted to devise better therapies, and to some extent these efforts have been successful, although their side effects are often brutal. Chemotherapy and radiation therapy continue to be the mainstays of cancer therapy, albeit in improved forms relative to that of a generation ago, and the introduction of immunotherapies has been changing the therapeutic landscape. As we'll describe throughout this book, it has become increasingly apparent that lifestyle factors (maintaining proper diets, replacing sedentary behaviors with moderate exercise, obtaining sufficient sleep, limiting stressful experiences, and coping effectively) may be fundamental in limiting cancer occurrence and may also enhance the effectiveness of therapies for many types of cancer. In fact, hope has been expressed that through novel therapeutic strategies and improved appreciation of the value of lifestyle factors (e.g., by altering microbiota, modifying immune functioning, and diminishing inflammatory responses), certain forms of cancer can eventually be treated more successfully, especially when therapies are tailored to individual patients.

Hymie Anisman
Alexander W. Kusnecov

Acknowledgments

As usual, we had considerable help in preparing this book. Zul Merali, Kim Matheson, Alfonso Abizaid, Robyn McQuaid, Marie-Claude Audet, Melissa Chee, Stephen Ferguson, and Richard Contrada provided needed insights into a wide range of specific topic areas. Tarek Benzouak and Ajani Asokumar were wonderfully kind in helping with references and in creating figures. We are exceptionally grateful to our editor Rafael Teixeira without whom this book would never have been created. We are also grateful to Kristi Anderson who shepherded the book through the many stages and who was patient with our many requests. Mohan Raj Rajendran is acknowledged for being very kind and helping us obtain permission to reuse numerous figures. As well, the help by Omer Mukthar Moosa throughout the editing process was invaluable. Our thanks to all.

The original research reported by both authors was supported by the Natural Sciences and Engineering Research Council of Canada and the Canadian Institutes of Health Research (H.A.) and the National Institutes of Health, the Rutgers University intramural Busch grants program, and The Brain Health Institute of Rutgers University (A.W.K.).

CHAPTER 1

Cancer biology and pathology

OUTLINE

The distress of cancer	1	Nutritional and energy support	23
Extrinsic influences on cancer	3	Glucose and the Warburg effect	24
The broad landscape	5	Impact of hormones and hormone receptors on cancer processes	28
Features of cancer development and progression	9	Inflammatory factors in the cancer process	29
Cancer processes	10	What inflammation implies	30
Classification	11	Two faces of inflammation	30
Sources of cancer	13	Mechanisms of inflammation	31
Immune surveillance	13	Intrinsic vs extrinsic contributions to cancer occurrence	33
Avoiding detection: The garden of good and evil (mostly evil)	15	Summary and conclusions	34
Treatment resistance	20	References	34
Metastases	21		
Ingredients for cancer growth and metastasis	23		

The distress of cancer

Henning Mankell, the Swedish author most notable for the creation of the fictional (and somewhat morose) detective, Kurt Wallander, offered the following reflection after being diagnosed with lung and neck cancer: "I remember that time as a fog, a shattering mental shudder that occasionally transmuted into an imagined fever. Brief, clear moments of despair. And all the resistance my willpower could muster. Looking back, I can now think of it all as a long drawn-out nightmare that paid no attention to whether I was asleep or awake."[a]

---

[a] https://www.theguardian.com/lifeandstyle/2014/feb/12/henning-mankell-diagnosed-cancer.

Cancer is a frightening illness. It is an intractable, often incurable disease with a history of treatments that are experienced as brutal, dehumanizing, and distressing. Even when patients are designated "cancer survivors," they may be acutely aware that if the disease recurs, further treatment may be worse and less tolerable. Indeed, patients typically undergo repeated testing to determine whether the cancer has recurred, experiences that can raise questions of treatment-related damage, and the uncertainty of waiting for the "other shoe to drop." To extend Mankell's thoughts, it is not only a nightmare to process the diagnosis, but one that can extend into the postdiagnosis treatment and recovery phase.

It is not uncommon for people to believe that they have been afflicted with cancer due to factors outside their control. Perhaps unwitting exposure to environmental toxicants, inheriting mutated genes, or eating the wrong foods and not being a better or calmer person have placed them at risk—any manner of theories will pass through a person's thoughts as they fathom the reasons for their condition. With the exception of environmental disasters—such as the Chernobyl nuclear meltdown or, more recently, the Fukushima nuclear disasters—it can be difficult to make causal links between any number of factors and the appearance of cancer. In effect, in many ways, cancer occurrence is outside an individual's control, most especially if it is inherited. Yet in other respects, individuals do have a say in determining their destiny and can take preventative measures: not smoking, limiting sun exposure, avoiding certain foods, not becoming overweight, engaging in exercise, not becoming dependent on alcohol, being vaccinated against carcinogenic viruses—all courses of action that can reduce cancer risk.

Once an individual becomes a patient—which in and of itself is a distinction that may be significant—he or she may have some control in the selection of treatments; but often this is illusory, as they likely know and understand little about the therapies being discussed with them and often simply follow the advice of the treating oncologist. In recent years, this blind trust on the part of patients has been replaced, to a modest extent, by patients wanting to understand more about the disease and its treatment, so that they can make informed choices. It is not unusual for patients to opt for unsubstantiated alternative treatments (e.g., herbal medicines gleaned from the internet) or complementary adjunct treatments (e.g., natural treatments to supplement standard medications) either because of their mistrust or fear of traditional treatments. It is not uncommon for people to display skepticism toward medical care, and in the face of multiple alternative points of view (legitimate and illegitimate), it can leave a patient confused and open to exploitation.[b]

When an individual suspects the presence of cancer, this triggers a cascade of perceived and actual threats. The anticipation and anxiety related to a cancer diagnosis are invariably distressing, most notably among individuals who are less adept at dealing with uncertainty, which only

---

[b] At the time of this writing, the historic repercussions of the COVID-19 pandemic have highlighted deep class, race, and political divisions, as well as a profound distrust of the biomedical scientific community, whose urgent calls for social isolation have promoted economic hardship due to business closures. With societies subjected to unrelenting stress, crackpot theories and treatments are being given equal weight with more traditional, tried, and true approaches (viz., safe use of effective vaccines) that have withstood the rigor of scientific scrutiny. To think that the range of options open to a cancer patient may not leave them bewildered and helpless is unrealistic. This would be especially the case when people are disadvantaged by a lack of adequate knowledge and/or insufficient social support networks that might help them make an informed decision.

leaves them more vulnerable to psychological disturbance. The distress is exacerbated with frequent delays in obtaining treatment, especially as this might allow the cancer to progress to stages of pathology that could predict undesirable treatment outcomes. These factors, together with the treatment itself, may result in patients experiencing cognitive disturbances as well as chronic anxiety, depression, and posttraumatic stress disorder (PTSD) (Cordova et al., 2017).

Overall, the diagnosis and treatment of cancer envelop the individual in a cloud of uncertainty, physical discomfort, and overwhelming psychological imbalance. The therapy itself may precipitate such problems, but it is important to note that the extreme distress of knowing, living, and even surviving cancer can contribute further to these problems. The individual's holistic mind-body state needs to be addressed, taking into account that the person's psychosocial relations with others also need to be considered if their battle with cancer is to have outcomes that positively impact both physical and psychological well-being.

Extrinsic influences on cancer

The etiology of disease now encompasses models that combine genetic factors with environmental chemicals, as described in Fig. 1.1. The figure illustrates the concept of an "exposome," an aggregation of numerous nongenomic factors that provoke biological changes that

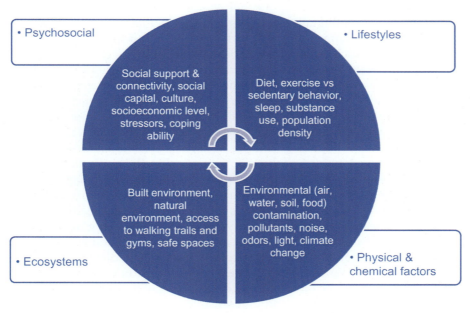

FIG. 1.1 Both good health and illness are determined by intrinsic factors (e.g., biological processes) and extrinsic factors. These extrinsic factors, sometimes referred to as exposome, comprise psychosocial factors, lifestyles endorsed, ecosystems, and encounters with physical-chemical stimuli. Some of the many components of each sector of the exposome are provided in the figure. These are not independent of one another, and their dynamic interactions influence one another and influence biological processes that affect quality of life and health. *Based on Vermeulen, R., Schymanski, E.L., Barabási, A.L., Miller, G.W., 2020. The exposome and health: where chemistry meets biology. Science 367, 392–396.*

create the circumstances for disease (Vermeulen et al., 2020). The development and progression of many types of cancer, as well as the response to cancer treatments, can be influenced by numerous environmental and lifestyle factors (diet, exercise, sleep). These may exert their effects through hormonal changes, inflammatory immune events, microbial colonization, and other biological mechanisms.

Cancer is often considered as a disease of cellular aging and prevention of this disease could be achieved, in a sense, by turning back the clock. One of the factors that may contribute to cellular aging is the accumulation of epigenetic changes in which the actions of genes are altered (suppressed or enhanced) without frank alterations of the genome. Thus, it is fascinating that a lifestyle manipulation that comprised an 8-week program in which individuals engaged in healthy eating, exercise, enhanced sleep, and relaxation training, supplemented by foods that provide probiotics and phytonutrients, was sufficient to diminish DNA-related epigenetic variations. Even this relatively short-term intervention seemed to turn back the DNA clock by about 3 years and thus could have profound implications for age-related diseases (Fitzgerald et al., 2021). Lifestyle and environmental influences are also intertwined with stress responses and behavioral and psychosocial functioning, which can affect immune and neurobiological processes. These varied factors may collectively impact cancer progression or recurrence and, importantly, may undermine the efficacy of cancer treatments.

Maintaining a healthy diet, engaging in exercise, avoiding sedentary behaviors, and diminishing stressful experiences may have benefits in preventing the occurrence of diseases, and there is reason to believe that these lifestyle choices can diminish the progression of some forms of cancer. For all its limitations standard cancer therapies have been effective in diminishing mortality stemming from many forms of cancer, but there has been considerable criticism of how slowly such advances have been made. This has encouraged the view that the answer to limit cancer deaths will largely come from modifications of lifestyles and the adoption of remedies; however, in many instances these remedies are simply too far out to be credible. Perhaps some of these approaches may have some value, but such perspective needs to be matched by convincing and reliable data that far too often haven't been provided. So, patients are often faced with the dilemma of following the science that hasn't uniformly provided cures on the one hand and adopting entirely untested and uncertain remedies on the other.

Within most developed countries, about 35% of people will be affected by some form of cancer, although rates in certain countries are appreciably lower (e.g., in Japan, Israel, Poland, Iceland), and nearly 30% of patients eventually develop a secondary metastatic tumor. According to the World Health Organization, cancer is the second leading cause of death globally (heart diseases are at the top) with 9.6 million deaths being attributable to some form of cancer. Of these deaths, about 70% occur in low- and middle-income countries, with various lifestyle factors (cigarette smoking, alcohol consumption, diet and obesity, poor food choices, and lack of exercise) being the greatest contributors. In ensuing chapters, we will consider these and other risk factors in greater detail, along with the possibility that avoiding or modifying these risks can be done through sensible prevention strategies.

The frequency of cancer occurrences coupled with the difficulties so often encountered during and after treatment led to altered patient care strategies. Among other things, this required continued surveillance regarding the individual's well-being and the recurrence of illness, and because of the protracted effects of cancer and its treatment, the development of other health problems needed to be considered. This entailed the involvement of specialties other than oncology (e.g., endocrinologists, dieticians, physiotherapists, occupational therapists, social workers, and specialists to deal with pain management), and greater patient engagement in selecting their treatment. There was also an obvious need for patient mental health to be considered in much greater depth than it had been previously, which encouraged the inclusion of clinical psychologists and/or psychiatrists in attending to patient needs. The availability of such broad treatment teams was thought to improve quality of life for patients, and there was the belief—albeit still debatable—that this would enhance the effectiveness of chemotherapy and radiation therapy.

The broad landscape

As depicted in Fig. 1.2, efforts to keep cancer patients alive longer have improved markedly over the past four decades, although for some cancers the advances have been limited. These data reflect the situation in the United States but are matched by data from the United Kingdom, Canada, Australia, New Zealand, and much of the European Union (EU) (Carioli et al., 2020). In developing countries, the situation has been improving, but still lags Western countries. What

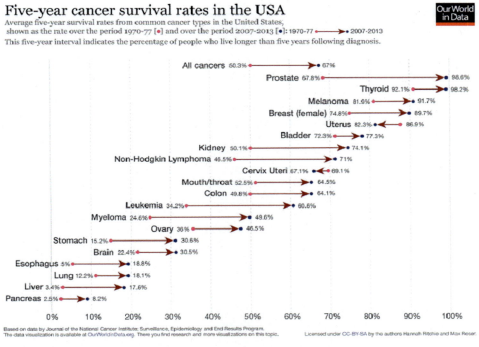

FIG. 1.2 Five-year survival rates determined over a 36–37-year period increased appreciably for several types of cancer (e.g., leukemia, myeloma) or have become highly treatable (prostate, thyroid cancer) or moderately treatable illnesses (melanoma, breast, uterine, bladder, and kidney cancer). Some improvements were also realized with esophageal, lung, liver, and pancreatic cancer, but the outlook for these forms of cancer have generally been grim. *Figure and caption from Roser, M., Ritchie, H., 2018. Our World in Data: Cancer. https://ourworldindata.org/cancer. Retrieved Jan 2020.*

this figure does not show is that no matter which country is examined, disparities exist with sex, ethnicity (race), and socioeconomic status (e.g., Ginsburg et al., 2017). Likewise, it does not portray the physical and mental cost of the illness and its treatment, nor what life is like for cancer survivors. This can vary with the treatments received, the individual's age, and a constellation of psychosocial factors. Typically, life span might not be the only or even the most important issue that preoccupies cancer patients. Instead, they may be more concerned with *health span* following treatment, which essentially amounts to the years lived without further illness or disability, defining the capacity to which they are happy and fulfilled.

The most recent update regarding the overall occurrence of cancer indicated that incidence rates for cancer among males have been stable, whereas that among women increased somewhat between 2013 and 2017, and this was also apparent among children, adolescents, and young adults (Islami et al., 2021). As depicted in Fig. 1.3, these statistics vary appreciably across different forms of cancer. In contrast, in both sexes, cancer-related deaths declined during this period, although again this depended on the nature of the cancer and varied between sexes. Among males, cancer-related death rates declined in 11 of the 19 forms of cancer, and in females, a decline was reported in 14 of the 20 types of cancer, including some of the deadliest cancers (e.g., melanoma, lung cancer).

As encouraging as it is to see increased survival among cancer patients, one should not assume that cancer elimination equates with full health restoration. In destroying cancer cells, standard cytotoxic treatments also kill healthy cells and may engender inflammation in many neighboring cells. Aside from the possibility of the cancer recurring, treatments may cause organ damage (including the heart, lung and airways, liver, and kidney), osteoporosis, persistent pain, sexual dysfunction and infertility, lymphedema (lymphatic fluid buildup in tissues, which produces painful inflammation, swelling, and restricted movement), and in response to some treatments, urinary or bowel (fecal) incontinence. Some of these consequences may not appear until years later. For instance, among individuals who survived childhood cancer (e.g., leukemia), it was common (80%) for a severe illness to occur before survivors reached 45 years of age. The protracted effects of therapies range broadly. Some were not life-threatening and were often being as "manageable," such as hearing loss (62%), memory problems (25%), male infertility (66%), and female infertility (12%). Others, however, comprised serious life-threatening conditions, including cardiac disturbances (in 63% of individuals), abnormal lung functioning (65%), and endocrine dysfunction (61%). In some instances, a second cancer appeared, possibly owing to genetic factors that favor the emergence of different types of cancer or perhaps due to the initial cancer treatment. Gentler treatments are thought to diminish these long-term side effects, and a personalized treatment approach may offer the hope of diminished proactive effects.

The cancer treatment landscape has been changing over the past decade, but it is certain that much more needs to be achieved. Historically, the focus of most research had been on factors related to cancer diagnosis and treatment with less attention devoted to cancer prevention. But it is now common knowledge that behavior is a common mode of facilitating the onset of some cancers—e.g., to avoid lung, throat, and skin cancer the prescription is don't smoke and avoid prolonged direct exposures to the sun. It would be rare to find people who would consider such advice as inappropriate, although it wasn't always accepted that smoking caused cancer or that certain rays of the sun can promote melanoma. Indeed, until a few decades ago children and teenagers spent as much time as possible in the sun, smothered in lotions that would

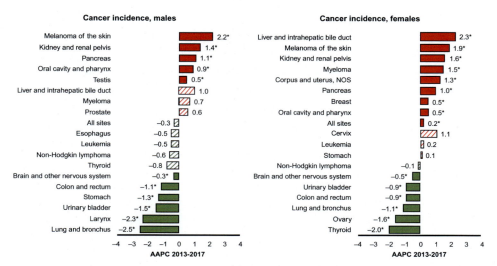

FIG. 1.3 Average annual percent change (AAPC) in age-standardized, delay-adjusted incidence rates for the period 2013–17 for all sites and the 18 most common cancers in men and women of all ages, all races/ethnicities combined. The AAPC was a weighted average of the annual percent change (APCs) over the fixed 5-year interval (incidence, 2013–17) using the underlying joinpoint regression model, which allowed up to three different APCs, for the 17-year period 2001–17. AAPCs were significantly different from zero (two-sided $P < .05$), using a t-test when the AAPC laid entirely within the last joinpoint segment and a z-test when the last joinpoint fell within the last 5 years of data, and are depicted as *solid-colored bars*; AAPCs with hash marks were not statistically significantly different from zero (stable). Abbreviations: *NOS*, not otherwise specified. *From Islami, F., Ward, E.M., Sung, H., Cronin, K.A., Tangka, F.K.L., et al., 2021. Annual report to the nation on the status of cancer, part 1: national cancer statistics. J. Natl. Cancer Inst. djab131.*

enhance their tans, rather than blocking the harmful ultraviolet rays. Equally, the influence of diet, exercise, stress, and sleep on cancer occurrence had frequently been ignored or simply neglected, despite these lifestyle factors receiving attention in the context of heart disease and type 2 diabetes. It ought to go without saying that preventive strategies are largely an individual's responsibility, but community-level preventive approaches might facilitate the adoption of methods to limit cancer development. Remarkably, however, until about a decade ago most people were unaware of the linkages between many lifestyle factors (e.g., diet, exercise) and cancer.[c] At that time it was not uncommon for people to believe that cancer was caused by injury, or that irreligiosity in some manner contributed to cancer occurrence. As well, there was limited awareness that diet or exercise was linked to cancer, and it was not unusual for people to believe that alternative medicines were as effective as standard therapies in cancer treatment

---

[c] It isn't our intention to stand on a soap box preaching about what should or shouldn't be done in relation to cancer prevention. Nobody likes the reformed sinner who takes every opportunity to hector others. Instead, as we deal with specific topics, we'll simply present some of the data and describe the suspected reasons for certain agents or behaviors having the effects that they do. Changing people's minds and behaviors in some instances is exceptionally difficult, witness for instance the ongoing debate between the antivaccine community and knowledgeable scientists who base their opinions on evidence (but who nonetheless seem to be banging their heads against a brick wall). Nevertheless, in Chapter 12, we will provide information relevant to behavior change strategies.

(Lord et al., 2012). Since then, there has been greater awareness of the sources of cancer occurrence, but a significant number of people are still unaware of the cancer risks associated with obesity or alcohol consumption (e.g., Meyer et al., 2019).

Meta-analyses and systematic reviews

The rate of scientific output has increased exponentially over the years, and as much as this ought to be advantageous, it is not unusual for inconsistent findings to be reported. This could occur owing to subtle or pronounced procedural differences between studies, or the findings of some studies may simply be unreliable for any number of reasons. Thus, it has become increasingly desirable to have cogent reviews of the literature that both synthesize the available literature and facilitate conclusions regarding where the bulk of evidence lies. Meta-analyses became a progressively more popular way of organizing and then reviewing large swaths of research while taking into account the effect size in each study (i.e., the strength of associations that exist between variables) and the number of participants in these studies (studies with a small number of participants are usually seen as being more prone to unreliable outcomes). Ordinarily, before undertaking the analyses, investigators specify inclusion and exclusion criteria for studies. This may entail the type of participants or research approach adopted in the studies, such as whether they involved retrospective or prospective approaches or whether they consisted of random controlled trials. By rigorously following sets of criteria, estimates can be determined as to how reliable and meaningful the results are, and what key variables could moderate the observed relationships.

Another approach of summarizing large amounts of data has entailed systematic reviews. These analyses follow rigorous guidelines concerning the specific issues to be addressed, the identification of the relevant work that should be included or excluded, and how to assess the quality of the research considered. Often, the data come from quantitative analyses, but qualitative analyses (e.g., individual narratives) can be included, and meta-analyses can be incorporated as part of a systematic review. It is of great importance that not all studies are given equivalent weight in that methodological issues might be considered by independent raters who determine the goodness of studies and hence the weight attributed to them. Numerous clinical conditions have been assessed using these approaches, including therapeutic methods to treat illnesses, the side effects of various treatments, and the effectiveness of social and public health interventions, as well as cost-benefit analyses related to the risks associated with particular treatments relative to what could be expected if treatments had not been applied.

Meta-analyses and systematic reviews have been remarkably effective in summarizing large sets of studies to bring order and understanding disparate findings. However, these analyses can also be misleading. These reviews are only as good as the inclusion and exclusion criteria that are used in the analyses. For instance, a given analysis might include both observational studies and randomized controlled studies whereas another analysis might only include the latter, thus they may provide different conclusions. Likewise, some analyses might ignore certain variables, such as sex differences, and hence biased or inexact conclusions may be derived. Another serious shortcoming of these analyses is that they can only include the available reports, which are typically those that have been published in peer-reviewed journals. Most often, however, published reports comprise those that showed significant outcomes, whereas studies in which significant effects are not obtained are often never published. As a result, one could easily be misled to believe that a given treatment is effective when, in fact, many more experiments indicated otherwise. It may unfortunately take many years of research before it becomes apparent that certain treatments weren't as effective as initially claimed.

Features of cancer development and progression

Cancers are characterized by uncontrolled cell growth, culminating in tumor development that can destroy surrounding tissue and spread promiscuously to other regions. Cancer cells are typically considered to be cells that can escape the main tumor mass (i.e., the primary tumor), migrate to remote sites via the vascular or lymphatic systems, and establish a new cancer colony (a metastasis or secondary tumor). For example, a biopsy on a lung tumor might reveal that the cells are actually from another organ (say, the pancreas), thereby establishing the tumor as a secondary tumor, and evidence of metastatic spread from the primary source (the pancreas). Such metastasizing cells present with a unique *gene signature*, which distinguishes them from more stable, nonmigrating cells that remain in the primary tumor. To a large degree cancer cells display unprogrammed and/or mutated behavior characterized by runaway mitosis, which becomes biologically threatening when unrestrained. The persistent cell division is a result of numerous cellular pathways being coopted to enable cells to avoid or disregard the inherent constraints on cell growth that are observed in healthy cells. In so doing, they modify their microenvironment to favor their survival and proliferation, drawing on energetic processes needed by normal cells and avoiding surveillance by cytotoxic immune cells. In short, cancer cells are capable of breaking through barriers that might otherwise restrict their growth and can spread to other organs. Such a situation is *malignant* because it impairs the function of colonized tissues and organs. In contrast, benign tumors do not invade or infiltrate into the cellular network of surrounding tissues or organs. These tumors grow as an isolated mass with their own capsule and, with some exceptions, lack metastatic capacity. Benign tumors do grow, but slowly, and in the best-case scenario, simply stop growing. And while they do not interfere with the physiological and biochemical functions of intrinsic cells in a tissue, they do compress nearby structures, thereby causing pain or other medical complications that will warrant their removal. Otherwise, benign tumors are sometimes judged to be better left alone.

In their analysis of the elements that make for cancer development and spread, Hanahan and Weinberg (2011) identified six characteristics that are common among many forms of cancer. These hallmarks of cancer are shown in Fig. 1.4. Cancer cells are seen as self-sufficient in being able to respond to cell growth signals and insensitive to signals that inhibit growth. Moreover, in addition to being able to evade apoptosis (programmed cell death to eliminate cells that are damaged or contain potentially dangerous mutations), they have a remarkable capacity to replicate and to promote and sustain their growth. They can stimulate blood flow to themselves (angiogenesis), evade attempts of the immune system to destroy them, and can invade local tissues and metastasize. Beyond these features, inflammatory processes, and the ensuing genetic dysregulation, might represent a seventh essential characteristic that supports cancer processes.

In addition to these fundamental processes, several biomechanical features of tumors may sustain and enhance their growth and spread. Specifically, owing to permeable blood vessels in tumors that leak blood plasma into tissues surrounding the tumor, together with limited drainage of lymphatic fluid, elevated interstitial fluid pressure may be created that causes edema, the release of growth factors, and the facilitation of cancer cell invasion of nearby and distal tissues. As well, the proliferating and migrating cancer cells compress blood and lymphatic vessels, thereby disturbing blood flow and hence diminish immune cell presence and limit oxygen to the region and may concurrently hinder drug presence that would otherwise limit tumor growth. At the same time, the immune microenvironment surrounding a tumor may limit the access of immune cells, and interactions between cancer cells and their

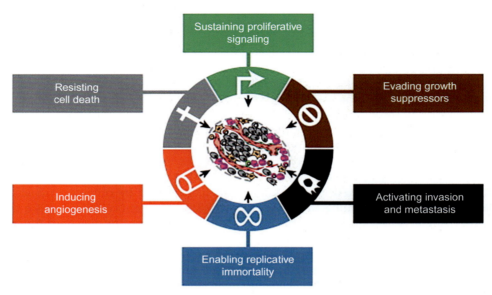

FIG. 1.4 The six hallmarks of cancer that facilitate their growth and survival. *From Hanahan, D., Weinberg, R.A., 2011. Hallmarks of cancer: the next generation. Cell 144, 646–674.*

microenvironment may be disturbed, consequently affecting signaling pathways that may affect cancer cell invasion and metastasis.

The tumor microenvironment is complex, typically being distinct from that of normal cells with respect to oxygen availability, metabolic processes, acidity, and interstitial fluid pressure. This environment, the battlefield between the cancer cells and a person's defenses, holds considerable sway over tumor progression and the efficacy of various treatments. Cancer cells can influence and be influenced by the lymphocytes that had infiltrated the area as well as by the presence of stromal cells that make up connective tissues, fibroblasts (a type of cell present in connective tissue), blood vessels, and the extracellular matrix. The latter, a noncellular component present in tissues and organs, comprises the scaffolding for cells. Accordingly, approaches could be adopted to take advantage of the microenvironment to enhance the efficacy of varied therapeutic strategies.

Cancer processes

More than 200 forms of cancer have been identified, involving different organs and different cell types. In general, these cancers fall into distinct classes (see Table 1.1) and are also described by their stage of progression (the size of the tumor and how extensively it has spread). Cancers can also be distinguished from each other based on their genetic and epigenetic signatures, as well as the speed of expansion (viz., slow vs fast growth). They are also differentiated according to sensitivity to immune and hormonal mechanisms, such as whether they are influenced by certain hormones or inflammatory processes. Each type of cancer may contain multiple subtypes. Breast cancer, for instance, may comprise ductal, lobular, tubular, invasive, infiltrating,

TABLE 1.1 Cancer classification.

Cancers are broadly classified within six main categories that affect different organs, such as breast, lung, prostate, or hematological cancers, or they may affect muscles or connective tissue (fat, cartilage, bone, fibrous tissue). Each of these broad classes can be broken down further into multiple subcategories.

- **Carcinoma**: involves the epithelial cells of tissues that line body surfaces and cavities. Cancers that affect skin and other organs, including the breast, prostate, colon, and lung are frequently carcinomas.
- **Sarcoma**: a cancer of connective tissues that includes cartilage, bone, tendons, adipose tissue, lymphatic tissue, and components of blood cells.
- **Myeloma**: a cancer that originates in the plasma cells of bone marrow.
- **Lymphoma**: These develop in the nodes or glands of the lymphatic system, the network of vessels, nodes, and organs, that serve to purify bodily fluids and produce white blood cells (lymphocytes) to fight infection. Lymphomas are broadly categorized as Hodgkin lymphoma and non-Hodgkin lymphoma, as well as multiple myeloma and immunoproliferative diseases.
- **Melanoma**: This cancer originates in cells that contain melanin (melanocytes) and are primarily found on the skin, but can be found elsewhere (pigmented tissue, including the eye).
- **Leukemia**: a diffuse type of cancer (as opposed to a solid cancer) that originates in the bone marrow, culminating in abnormal white blood cells appearing in excessively high numbers (e.g., leukemia cells)
- **Blastoma**: a type of cancer that involves "blasts," which are primitive and incompletely differentiated precursor cells. Several types of blastoma appear in children, and some forms—such as glioblastoma multiforme, which is the most aggressive brain tumor—appear in middle-aged adults.
- **Brain and spinal cord tumors**: these involve different types of cells and are named accordingly. These comprise astrocytic tumors, gliomas, oligodendroglial tumors, medulloblastomas, ependymal tumors, meningeal tumors, and craniopharyngioma.
- **Germ cell tumor**: a cancer involving germ cells (i.e., sperm or egg that unite during sexual reproduction). In essence, this type of cancer most often occurs within the ovaries or testis and can occur in babies and children.
- **Mixed type**: These may come from a single category or may span different categories (e.g., carcinosarcoma, adenosquamous carcinoma)

Modified from Anisman, H., 2021. Health Psychology: A Biopsychosocial Approach. SAGE, London.

mucinous, and medullary types, and can also be classified based on whether certain receptors are present, such as those for estrogen, progesterone, and HER2 (as in the case of triple-negative breast cancer). By virtue of this heterogeneity of cell types, each form of cancer requires a specific therapy, and even within a given cancer type, differences might prompt specialized treatments.

Classification

Cancer subtypes are classified based on the tissue affected, features of the cancer, and the extent to which the cancer has progressed (i.e., stage of the disease). These serve as prognostic indicators for the probability of successful treatment or life expectancy and guide the choice of treatment. For instance, when the cancer is localized, surgery may be selected, but chemotherapy may be adopted as an *adjuvant* therapy to eliminate cancer cells that might persist following surgery.

The general diagnosis of cancer involves the use of a staging system that ranges from abnormal but not immediately dangerous cellular states to those that have advanced to a life-threatening pathological condition. These are summarized as follows:

- Stage 0, also referred to as carcinoma in situ, refers to cancer cells having been detected that have the *potential* to spread.
- Stage I, or early-stage cancer, refers to the cancer being small and localized to a specific area.

- Stage II refers to a larger cancer that has infiltrated nearby tissues or lymph nodes.
- Stage III also describes the situation in which the cancer is larger and affects nearby tissues, as well as lymph nodes that are nearby as well as those that are more distant.
- Stage IV refers to cancer that has spread (metastasized) to other parts of the body.

Several biological processes and their interactions have been implicated in the provocation and progression of cancers, including diverse hormones, growth factors, a broad assortment of immune factors, and related genetic influences, as well as immune signaling molecules, most notably cytokines[d] (Thorsson et al., 2018). Given the links between cancer and immune functioning, appreciable efforts have focused on altering the immune system's capacity to eliminate unwanted cells. However, even with the best immunotherapeutic approaches, only some types of cancer can successfully be treated, and even then, only a proportion of individuals benefit from the treatment. Nonetheless, advances on this front have been impressive, and a continuous flow of immunotherapeutic agents and combinations of agents have been derived that can treat an ever-broader range of cancers, with moderate improvements realized in the number of patients effectively treated.

BFF

Many cancers, such as lymphoma, appear across species, including in our beloved pets, which are heavily interbred, while certain species rarely develop cancers, a list that includes whales, elephants, and mole rats. What protects certain species from developing cancer, and is this common across species? Provided that they are not hunted to extinction, cross-species comparisons of long-lived animals may offer important clues regarding the factors that promote cancer occurrence or conversely render them cancer-free. In the case of elephants, it seems likely that the absence of cancer may be related to their excellent ability to eliminate damaged cells, rather than simply repairing damaged DNA, or because they have an overabundance of genes that act as tumor suppressors (Vazquez and Lynch, 2021).

Cancer is the leading cause of death in certain dog breeds, such as the Golden and Labrador retrievers who are particularly prone to lymphoma, while the beautiful Bernese Mountain dogs are susceptible to histiocytic sarcoma characterized by an abnormal elevation of macrophages, dendritic cells, and monocytes. One type of cancer, canine transmissible venereal tumor, likely stemming from a Siberian dog strain that existed about 6000 years ago, is transmitted through sexual contact. By analyzing the genetics of these tumors, useful clues may arise as to why certain cancer-driving genes have been successful (Baez-Ortega et al., 2019). Humans and their canine friends may develop several similar cancers, such as sarcomas, lymphoma and leukemia, bladder cancer, glioma, and melanoma, whereas several others are not seen in animals, including lung, prostate, and testicular cancers. Why humans are unique in developing certain forms of cancer is uncertain, although

---

[d] Cytokines are molecules released by cells of the immune system and function in a signaling capacity between immune cells. They are like neurotransmitters that operate in the nervous system to allow neurons to "speak" to each other. We will consider cytokines in greater detail in Chapter 2 but suffice to say for now that some cytokines can promote inflammation (proinflammatory) and immune cell proliferation and functioning, whereas others inhibit inflammation (antiinflammatory) and immune activity.

they could be related to environmental triggers or may be secondary to unfortunate adaptations to environmental and social challenges.

Despite the differences between humans and their dogs, they share environments and consume some of the same foods, and therefore where dogs may develop the same cancers the reasons may be similar to those of humans. As it happens, the efficacy of some canine cancer vaccines that are in development may be effective in preventing a variety of different cancers, and programmed stem cell transfer methods could be used to diminish canine cancers. These methodologies may facilitate the search for treatments in humans. Analysis of particular tumor types and how they are passed across generations in dogs may also have implications for the evolution of human cancers (Maley and Shibata, 2019). Finally, it's no secret that having a dog is often good for heart health, possibly because it gets the dog's person out walking. If, as we'll see later, exercise can also have beneficial effects on cancer progression and treatment, it will be one more way in which dogs may be a human's best friend.

Many cat lovers know that these wonderful creatures may also develop a variety of different cancers, some of which also occur in humans. During their lifetime, one of four cats is diagnosed with cancer, most commonly lymphoma, squamous cell carcinoma, mammary tumors, and bone cancer. At one time, when our pets developed cancer, it was unlikely that considerable efforts would be expended to extend their lives, but this has been changing in recent years, and methods have been developed to deal with cancers in pets. Because of the similarities between cancer in cats and those observed in humans, treatments in cats might provide clues as to what might be beneficial in humans. Cats often develop oral squamous cell carcinoma that is reminiscent of head and neck cancer in humans, and cats also develop triple-negative breast cancer more often than humans, providing the opportunity to study this type of cancer when these pets are treated. Pet lovers would be aghast at the thought of having cats and dogs serve as experimental subjects to evaluate treatments of cancer much like mice are used in this regard. This would be impractical for a variety of reasons, but when veterinarians go about treating their patients, important information can be obtained for the benefit of other animals and humans provided that the findings are included in broad registries.

Sources of cancer

Cancer can be developed due to hereditary and/or genetic factors (that we will discuss in Chapter 4), random DNA mutations, or those brought about by carcinogens and other extrinsic factors. To a considerable extent, the immune system is reasonably proficient in protecting against foreign invaders, but the fact that we so often become ill attests to its limitations. Of course, the remarkable capacity of the immune system may be undermined by aging, as well as the adoption of poor lifestyles, culminating in increased disease risk. It also appears that with aging, chronic infection and inflammation associated with several diseases may undermine immune functioning, thereby increasing the risk of cancers as well as cardiovascular disease and stroke.

Immune surveillance

Once immune cells detect the presence of cancer cells and decide that they need to be eliminated, a battle ensues that can result in immune cells coming out on top, but all too often cancer cells are victorious. As stated in other contexts, the immune system can win many battles, but it only needs to lose once for catastrophe to ensue. In some instances, the two sides battle to a stalemate, in which case the tumor would not appear at that time (tumor dormancy), but a breakout may occur by cancer cells as long as decades later. This begs the question as to whether lifestyle factors or stressor experiences influence the emergence of cancer during the tumor dormancy period, but convincing data one way or another remains to be provided. It should be said that although a lengthy period may exist between the first cancer cells appearing and a frank cancer being apparent, there are instances in which this period can be greatly abbreviated. In some instances, such as childhood brain cancers, the culprit cells might originally appear during the embryonic stage of development when the immune system is in a rudimentary or immature state and then manifest after birth as a cancer during early life (Vladoiu et al., 2019).

According to an early view, the immune surveillance hypothesis (Burnet, 1970), immune cells located within secondary immune organs and circulating through the body were continuously on alert for the presence of newly transformed cells. Cancer would therefore be the result of mutated cells being, for whatever reason, undetected or not destroyed by the immune system. As tantalizing as this view was, experimental evidence was scant; but a revised version of this hypothesis, a cancer immunoediting perspective, was proposed and is illustrated in Fig. 1.5 (Dunn et al., 2004; Lussier and Schreiber, 2016). Like the earlier view, the cancer process was seen as involving several phases. During the first of these, the *elimination phase*, innate immune cells ought to recognize the presence of cancer, giving rise to several aspects of the immune system being activated. Most prominently, cytotoxic T cells and natural killer (NK) cells can directly confront tumor cells and destroy them, whereas others are responsible for carting away dead tumor cells and depositing them in draining lymph nodes. Supplementing these immune processes were tumor-disrupting chemical mediators and endogenous genetic processes that act to suppress tumor proliferation. Thereafter, during the ensuing *equilibrium phase*, various immune cells continue to be called upon to engage the cancerous cells, working to eliminate them, or at best keep them at bay. Finally, during the *escape phase*, tumor cells can infiltrate the epithelium—the cellular layer lining the outer surface of blood vessels, organs, and varied body cavities—ultimately overrunning the body's defenses and expanding into an uncontrolled mass.

The effectiveness of the immune system is often compromised in certain diseases (e.g., AIDS) or not quite up to the job, as in aging. This can permit tumor growth to occur more readily. Several other variables, such as the presence of inflammatory factors, can also moderate tumor development during these phases. As we will see, modest inflammation has multiple beneficial actions when present for brief periods, but left unchecked, inflammation becomes chronic and can precipitate numerous problems. Moreover, cancer cells can escape immune surveillance through changes in gene functioning associated with environmental or experiential variables. This can silence genes associated with immune functioning or inhibit protective tumor suppressor genes. It had long been considered that certain genes acted as tumor suppressors, but it now seems that some genes (e.g., a gene named GNA13) not only act in this capacity but can also help tumor cells evade detection (Martin et al., 2021). Chapter 4 will describe how genetic factors and interactions between genes, as well as genes and other processes, may promote cancer occurrence and exacerbate its progression.

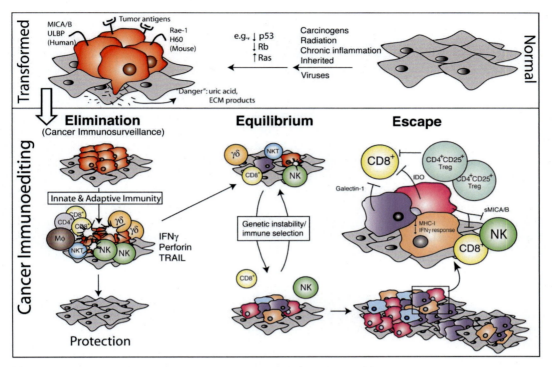

FIG. 1.5 The three phases of the cancer immunoediting process. Normal cells *(gray)* subject to common oncogenic stimuli ultimately undergo transformation and become tumor cells *(red)* (top). Even at early stages of tumorigenesis, these cells may express distinct tumor-specific markers and generate proinflammatory "danger" signals that initiate the cancer immunoediting process (bottom). In the first phase of elimination, cells and molecules of innate and adaptive immunity, which comprise the cancer immunosurveillance network, may eradicate the developing tumor and protect the host from tumor formation. However, if this process is not successful, the tumor cells may enter the equilibrium phase where they may be either maintained chronically or immunologically sculpted by immune "editors" to produce new populations of tumor variants. These variants may eventually evade the immune system by a variety of mechanisms and become clinically detectable in the escape phase. *Figure and caption from Dunn, G.P., Old, L.J., Schreiber, R.D., 2004. The immunobiology of cancer immunosurveillance and immunoediting. Immunity 21, 137–148.*

Avoiding detection: The garden of good and evil (mostly evil)

To devise effective treatment methods, it is necessary to understand how cancer develops, and why our immune system seems to fail us. Cancer may occur owing to random or inherited genetic mutations, those that are provoked by environmental factors or specific toxicants, and because the immune system failed to function effectively. However, rather than simply attributing cancer development to immunological missteps, it may be more profitable to also consider the talents of cancer cells in circumventing our defenses. In learning the repertoire of tricks that cancer cells use, one cannot help but marvel at how cunning they are in keeping scientists from uncovering the many ways they thwart our best efforts to eliminate them. Hopefully, by countering the ways tumor cells evade detection and survive attacks, coupled with ways of enhancing the capabilities of the immune system, progressively better methods of cancer treatment will be developed.

Circumventing defenses

Researchers attempting to devise methods to combat cancer have repeatedly encountered the devious ways by which cancer cells are able counter strategies to eliminate them. When one pathway for cancer growth was blocked, they were able to find other routes to carry out their mission, often leading to even more aggressive actions. For instance, a major stress pathway, mammalian target of rapamycin (mTOR), contributed to drug-provoked mutagenesis, but blocking this pathway did not provide beneficial actions, indicating that other pathways concurrently contributed to mutagenic outcomes. Some of these routes may be modulated by cancer-related genes, such as BRAF, MEK, and PAK, and targeting more than one of these at a time might optimize treatment responses, even if the benefits obtained were relatively transient. Similarly, targeting both BRAF and the gene that encodes insulin-like growth factor-binding protein 2 (IGFBP2), which predicts a poor prognosis among melanoma patients, provided a better treatment response than dealing with either gene separately (Strub et al., 2018). Another view has it that proinflammatory cytokines released from immune cells stimulate a protein, ITCH, which acts to modify BRAF so that it does not interact with inhibitory proteins, thereby allowing melanoma growth and metastasis (Yin et al., 2019). Immune cells are exceptionally adept at recognizing foreign invaders or the presence of cells that ought to be eliminated. Since cancer cells are a mutated form of the host's cells, they might still bear sufficient similarity to healthy cells. This enables them to evade attack, such as displaying a type of "exemption" or tolerance signal to other immune cells, such as macrophages (which detect, engulf, and destroy cellular pathogens). In acute myeloid leukemia (AML), for instance, many cancer cells overexpress the gene encoding CD47, a surface molecule known to serve as a "don't eat me" signal in healthy cells. This allows them to slip under the radar and evade ingestion by macrophages (Majeti et al., 2009).[e] Cancer cells can also be viewed as "cold" or "immune-privileged," such that neoepitopes (new markers on a foreign particle or cell) that ought to signal attack are absent or downregulated, as in the case of pancreatic cancer, in which the high mutation rate is accompanied by very few neoepitopes. Cold tumors also comprise those that are surrounded by cells that can suppress T cells so that the tumor cells are not attacked. Furthermore, when cancer cells are attacked (e.g., by chemotherapeutic agents) they can, as a group, go into a hibernation mode until the threat has gone.

As we will see in Chapter 15, coordination occurs so immune cells do not ordinarily attack healthy cells. Specifically, checkpoint proteins (e.g., PD-1) located on immune cells serve as an "off-switch" when they bind to a programmed death-ligand 1 (PD-L1) protein that is present on healthy cells. This type of suppression limits immune cell attacks on self-tissue, thereby reducing autoimmune disorders, tissue allografts (transplanted tissue from another person), and rejection of the fetus during pregnancy. As sophisticated and effective as this regulatory mechanism might be, it isn't always fully effective. Immune cells do, in some instances, turn

---

[e] Markers are present on the surface of immune cells (CD: clusters of differentiation) that allow us to recognize what they are and their presumed function. We will deal with these CDs in more detail when we delve into the immune system in Chapter 2. For now, it is only necessary to know that these CDs identify different immune cells and that they can serve as receptors for signaling molecules and can be used as targets to influence immune functioning.

on the self to produce autoimmune disorders, and certain cancer cells also display PD-L1, thereby taking advantage of the T cell, which fails to attack the cancer cell that is free to proliferate.

Aside from the few we have mentioned, cancer cells have many other ways of getting around immune defenses, making it much more difficult to thwart their pathological consequences. Some tumors downregulate aspects of the immune system, essentially making them invisible to T lymphocytes, while other tumors use structural defensive strategies—producing collagen and fibrin that serve as physical barriers. Tumor or infected cells ought to display features—like the presence of MICA and MICB proteins—that signal them for destruction by NK cells. However, some tumor cells acquire the ability to delete these markers, making an attack on them less likely (Ferrari de Andrade et al., 2018). Moreover, some genes, like the PARP1 gene, which encodes an enzyme involved in DNA repair, can make cancer cells less susceptible to NK cytotoxicity. By manipulating this gene, cancer stem cells were made more vulnerable to NK cells (Cerwenka and Lanier, 2018), in keeping with the notion that gene modification therapies can augment immunological tumor surveillance.

Duplicitous cancer cells engage in still other methods to avoid detection. A molecular coating on tumor cells can make them difficult to recognize by immune cells, or they may link with normal cells, thereby getting immune defenses (a friend of my friend, is also *my* friend). Alternatively, they might release protective factors to avoid proteins, such as metastasis suppressors, which normally would act against them. And transforming growth factor-β (TGF-β), which is produced by multiple healthy cells in the body, can be produced by lung cancer cells thereby limiting detection by NK cells (Donatelli et al., 2014). Cancer cells can also use substances within the body, such as neurotransmitters, to cloak their mutated nature when they reach particular milieus, such as the brain (essentially, hiding in plain sight).

In other cancer conditions, such as brain glioblastoma, there is a redistribution of immune cells such that circulating leukocytes are reduced, whereas an abundance of cells appears in the bone marrow, possibly through messages sent by the tumor (Chongsathidkiet et al., 2018). Glioblastomas and other types of cancer cells are highly plastic (changeable), altering their basic surface characteristics in response to a microenvironment consisting of tumor blood vessels and local immune cells and their signaling molecules, which makes the tumor more difficult to target (Dirkse et al., 2019). Indeed, gliomas are capable of adapting to the brain environment. This involves the formation of new functional synapses with nearby neurons, and glial cell release of the excitatory neurotransmitter, glutamate, which may drive the proliferation, survival, and overall invasiveness of glioma cells. In mouse models, modifying glutamate receptors on glioma cells can influence their proliferation and survival times (Venkataramani et al., 2019).

Cancer cells have still other dirty tricks to evade immune cells, deceiving them to think that they are "friends" rather than foes. Through their receptors, T cells produce a minute mechanical tug (a handshake in a sense) on other cells to determine whether it is a threat. Those that are not threatening maintain a weak handshake, but if they are threats, then stronger and longer handshakes occur, which leads to a cascade of immune factors to attack the foreign substance. However, unlike other foreign substances, some cancer cells have an extra molecule that can alter their handshake, causing T cells to become lethargic and less apt to function properly (Liu et al., 2016).

Cancer's active defenses

Aside from adopting passive methods to get by the immune system, cancer cells may take an active role to facilitate their own growth. Tumor cells can disturb immune functioning by producing toxic metabolites or by influencing inhibitory pathways. This can be through the secretion of particular growth factors by tumor cells (e.g., TGF-β), which inhibits metabolism in $CD4^+$ T cells and production of the cytokine interferon (IFN)-γ, resulting in diminished attacks on cancer cells and hence reduced patient survival. In addition, components of the innate immune system, specifically myeloid cells (i.e., granulocytes, monocytes, and tissue resident macrophages) that ought to be attacking cancerous cells can be coopted so that they become allied with cancer cells. Indeed, neutrophils that initially serve in a protective capacity may switch sides to become pro-tumorigenic (Magod et al., 2021).

Normally, the presence of integrin CD11b recruits myeloid cells to sites of damaged tissue so that a proinflammatory type of macrophage (categorized as M1) could attack tumor cells. However, tumor cells can develop the capacity to suppress CD11b functioning such that less aggressive, antiinflammatory macrophage types (called M2 macrophages) are recruited. These macrophages can act against T cells, thereby limiting their negative control over cancer cells and metastasis (Martinez and Gordon, 2014). In effect, innate immune cells may become polarized in the context of a tumorigenic environment, ultimately disrupting the cytotoxic actions of antigen-specific T cells, even producing an immunosuppressive milieu and undermining the plasticity of varied immune cells that are required to diminish cancer progression (Chang and Beatty, 2020). It should be said, however, that the plasticity which is a core characteristic of macrophages might turn out to be a good target in the treatment of varied inflammatory disorders, including cancer. Essentially, in the presence of IFN-γ, the M2 macrophage can be flipped to the M1 form, in the hope that this promotes antitumor cytotoxicity.

Macrophages and macrophage-stimulating genes are an inherent feature of the tumor environment, and while they could act against certain brain cancers, such as glioblastoma, they are able to reprogram these immune cells, thereby promoting tumor growth. Other types of macrophages have similarly been linked to enhanced tumor growth and poor outcomes, and the greater the macrophage density in breast and other cancers, the worse is the prognosis (Mantovani et al., 2017). These tumor-associated macrophages (TAMs) suppress the actions of other immune cells by releasing the inhibitory cytokine IL-10 (Ruffell et al., 2014), promoting angiogenesis (formation of new blood vessels in tumors), and facilitating metastasis. Further, in the face of some types of chemotherapy or radiation, TAMs release molecules (e.g., cathepsin) that support cancer progression. Fortunately, by inhibiting these processes, the effectiveness of treatments to detect and limit the presence of TAMs could be augmented to improve cancer therapy (Arlauckas et al., 2017) as shown for colon tumors in mice. Likewise, synergistic therapeutic actions have been achieved. For example, when an anti-CD47 (blocking the "don't eat me" signal) treatment was combined with antibodies against the lung cell marker, epidermal growth factor receptor (EGFR), the efficacy of the cancer treatment was enhanced (Weiskopf et al., 2013). There are also instances in which TAMs can augment the actions of some chemotherapies, making them an obvious target to improve treatment methods (Mantovani et al., 2017).

Cancer cells display a "social" element, influencing their neighbors by releasing exosomes (cell-derived vesicles) that contain proteins, fats, or genetic material into the extracellular space. It had long been thought that exosomes were involved in cellular debris being tossed out, but it is now understood that they have more complicated functions, even serving cell-to-cell signaling. Exosomes may create an inflammatory milieu that leads to more aggressive cancer cells (Nabet et al., 2017), or they can infiltrate adjacent healthy cells, causing them to become cancerous. Moreover, they can transplant tumor information (e.g., tainted RNA) and prepare cancer cells for metastatic travel to distant sites. As exosomes are associated with immune functioning and can influence the development of numerous diseases, including cancer, the notion arose that they could potentially be engineered to carry toxic payloads, such as immune modulators, chemotherapeutic, or immunotherapeutic agents, to destroy tumor cells (Kalluri and LeBleu, 2020). To be sure, this point hasn't been reached yet, but since exosomes (and their contents) can be detected in biological fluids, they may be useful as biomarkers that could operate as "multicomponent" readouts related to cancer diagnosis.

At this point, it's clear that cancer is a complex and slippery problem. Our reasons for providing this compendium of the different ways by which cancer cells can circumvent our defenses are twofold. One was simply to illustrate the powerful capacity of cancer cells to survive, which points to the challenges faced in developing ways to get rid of them. The second reason was to highlight the progress made to counter the nefarious efforts of cancer cells. Cancer still takes many lives and generates considerable fear. But through one step at a time, even if seemingly "baby steps," important advances have been made. To this point, we have primarily considered pharmacological treatment strategies that attempted to thwart the capabilities of cancer cells or that enhanced immune functioning. Little, in contrast, was said about how lifestyle factors could be used as an auxiliary method to facilitate or enhance the effects of other treatments, which could go a very long way in prophylactically stemming cancer occurrence. We will get to these prospects in later chapters. In the next and succeeding sections, we will address the recalcitrant nature of cancer and how it emerges and disseminates as part of metastatic spread.

Turning winners and losers

Tissues and organs are crowded with millions of cells that have to maintain the dictates of their genetic programming so that they support tissue function. But gene mutations occur and are actually quite common. Immune functioning undoubtedly plays a pivotal role in diminishing cancer occurrence, but some nonimmune cellular mechanisms also act in a protective capacity. Indeed, ordinary cells may compete with mutated cells for dominance, and in so doing may eliminate precancerous cells that could eventually evolve into cancer. They may do so through the promotion of apoptosis (programmed cell death), pushing cancer cells out of tissues, or by engulfing and cannibalizing them (autophagy). These cells act as a group, and when one of their number seems not to be doing what it was meant to do, it is banished by the group (considered a "black sheep"). Eliminating these cells—which we can call "loser" cells—is adaptive in that resources that would otherwise have been lost to them will be conserved. And yet, cancers grow and tumors form: these are the mutations that have won. "Winner" cells continue to proliferate, reflecting a natural selection process to maximize cellular well-being. Some of these cells are "supercompetitors," carrying mutations, such as the MYC or the Minute gene, or those that diminish p53 that ordinarily limits cell division, and that are especially effective in maintaining the fitness of the cell population (Dejosez et al., 2013). Such findings led to the obvious question as to whether groups of such cells could be harnessed to eliminate cancerous cells. This will require defining what constitutes an undesirable cell (or a "loser" cell) by the markers that it carries. Alternatively, if a cancer cell carries a "winner" marker, it could possibly be used as a target for destruction (Madan et al., 2019). Identifying these cells at very early stages, for example, by detecting mutations through RNA sequencing, may facilitate cancer prevention (Yizhak et al., 2019).

Treatment resistance

Even when cancer therapies seem to be having positive effects in reducing tumor size, it is not unusual for the efficacy of the treatment to diminish. The development of resistance to therapies is among the greatest problems encountered in effectively treating cancer. Drug resistance may involve processes in which individuals are inherently resistant to certain drugs (*intrinsic resistance*) or through *acquired resistance* that develops with continued exposure to specific treatments. Cancer cells are unstable so gene mutations occur frequently and become more common with the growth of the cancer. As the features of the cancer change, the effectiveness of treatments may wane, possibly owing to the activation of specific pathways that ordinarily act against toxicants, which can include drugs to fight cancer. The cancer cells may also carry mutations that limit the effects of therapeutics, and selection processes may also be at play so that those cancer cells that are inherently less sensitive to therapies will be more likely to survive the assault by anticancer agents, thus allowing for more such cells to multiply.

Acquired resistance to cancer treatments can develop through mutations of the genes within cancer cells, alterations of the tumor microenvironment that favors cancer cell survival, or the promotion of another oncogene that becomes the primary driver of the tumor. Early in the disease process, selection pressures placed on tumor cells by the attacking immune system may increase the propensity for tumor gene variants to occur. As a result, some cancer cells will have a greater capacity to escape and survive attacks by therapeutic agents and generating similarly resistant progeny (Russo et al., 2019). As cancer cells continue to multiply, they may lose some of their basic features, so it may become increasingly more difficult for the immune system to identify them as threats. In this context, considerable heterogeneity exists between cancer cells within a tumor (e.g., varying genetically and metabolically), making it difficult to destroy all of them with a single treatment. Once again, those cancer cells that survive may be key in the development of treatment resistance.

Changes can also occur in the expression of cancer-related genes (through epigenetic processes discussed in Chapter 4) so they become less responsive to therapeutic agents. Beyond the altered actions related to genetic processes, treatment resistance can emerge owing to increased drug inactivation (e.g., by endogenous substrates within the body) and it similarly appears that the efflux of anticancer drugs may increase (Vadlapatla et al., 2013). Moreover, enhanced DNA repair processes may occur so cancer cell damage created by chemotherapeutic agents can be reversed (Mansoori et al., 2017). Being voracious consumers of energy, it has also been maintained that processes related to variations of adenosine triphosphate (ATP), which is fundamental to generating energy within cells, are fundamental in cancer progression and the emergence of treatment resistance (Wang et al., 2019)—as we will see shortly when we discuss the Warburg effect.

Metastases

Each cancer is different in its own way, and some cancers are more treatable than others. Primary tumors are often manageable, although treatment resistance may develop, and disease recurrence is not uncommon. After treatment cancer stem cells may still be present, waiting for the opportunity to appear elsewhere in a more aggressive form. They are much more difficult to treat, and there are few ways to determine whether these stem cells are hanging about. Small numbers of these cells can be detected in culture but identifying them in the living organism is another issue entirely. Still, there have been indications that certain enzymes can be used to find stem cells that might develop into cancers.

Most cells in the body seem to know that they belong where they are and stay put. They are retained in the tissue or organ through genetic instructions and local intercellular signaling molecules, operating in a concerted manner to fulfill the functions of the organ. Cancer cells, in contrast, are not content to remain as part of the main tumor mass and can migrate to distant sites where a new colony of cells is established. Cancer cells also have an uncanny ability to get around so that cancer can spread to multiple organs. It had been thought at one time that cancer cells were oblivious to their environment, focusing instead on their multiplication. Based on analyses that combine evolutionary biology and artificial intelligence it was deduced that cancer cells are spatially aware of their surroundings so they can change their shape allowing them to get by obstacles they encounter in their travels and get through membranes and into the bloodstream. The more obstacles they encounter, the more likely they are to change their shape. After reaching a hospitable destination they return to their original form and continue on their quest to multiply prodigiously (Butler et al., 2020).

In addition to dealing with treatment resistance, the challenge in cancer therapy has been to limit metastasis, since the vagrant metastatic forms of the disease are far more difficult to contain than the primary tumor. Reasons for the difficulty of treating metastasized tumors could be due to tolerance to the initial treatment, or because the metastasized tumor has many more genetic mutations than did the primary cancer, possibly making them more aggressive. As well, epigenetic changes, in which the gene's actions are silenced or enhanced by nongenomic factors (see Chapter 4), can occur in circulating tumor cells and may be pivotal in metastatic seeding. It seems possible that drug treatments can alter these epigenetic transformations and promote improved therapeutic outcomes (Gkountela et al., 2019). Beyond these features, genetic aspects of the host that exist prior to cancer occurrence might contribute to metastasis or might be a marker for its occurrence. For example, the gene APOE, which makes the protein apolipoprotein E, and appears in several forms, APOE2, APOE3, and APOE4, has been linked to cancer. In particularl, the APOE2 form can create proteins that facilitate processes related to metastasis, including obtaining a blood supply, enhancing the ability of cancer cells to embed themselves more deeply in tissues, and defending against attacks by immune cells. In contrast, individuals carrying the APOE4 gene are best equipped to respond positively to agents that enhance immune functioning (Ostendorf et al., 2020). Thus, knowing a patient's APOE status may provide a warning of what to expect regarding the spread of the cancer.[f]

---

[f] Carrying the APOE4 variant may be beneficial in one context, but harmful in another, as carrying this variant is a risk factor for Alzheimer's disease.

The foregoing raises several questions. What is it about cancer cells that get them to break from the main tumor mass and promote their migration? How do they survive the journey and flourish when they find a new home? This multistep cascade for metastasis entails initial escape from the main tumor mass, transport through vessels, settling at the distal site, and obtaining an energy supply. Understandably, multiple mechanisms are involved in this complex process. Traveling to a distant site can be difficult for free-floating cancer cells so only a subset of cells that leaves the main tumor mass is successful in metastasizing. Moreover, not every site they occupy is a welcoming perfect host. Local factors may not be conducive to growth or retention, and what benefits the host, may not benefit the guest. Nonetheless, it seems apparent that originating metastatic primary tumor cells quite commonly migrate to lung, liver, and/or bone, although each type of primary tumor cell (e.g., breast, colonic, prostate) can also have additional distinct homing sites. For example, breast cancer is most likely to metastasize to bone, brain, liver, and lung, whereas melanoma is more likely to metastasize to these sites as well as the lung, muscle, and skin. Ultimately, the transition from one phase of cancer development to another (e.g., from tumor initiation, invasion, and eventual metastasis) can involve a variety of active processes and may require distinguishable microenvironments and gene regulatory networks.

One view of the metastatic process is that when the density of tumor cells becomes too great, several cytokines synergistically signal cancer cells to migrate (Jayatilaka et al., 2017). Yet, the disseminated cancer cells may appear relatively early in cancer progression, in some cases before the cancer is detectable. While many of these cells may die, some surviving cells may already have developed mutations that help them survive in new environments. The disseminated cells may remain in a dormant state until they receive signals for them to awaken. These signals may come about through a variety of challenges, such as wounding (e.g., surgery) or the presence of inflammation and can be enhanced by growth factors. While it is ordinarily thought that metastasized cancer cells adapt to their environment, it also appears that cancer cells can shape their microenvironment so that they obtain what they need to multiply.

It seems that before cancer cells have disseminated, chronic inflammatory responses secondary to chromosomal instability and leakage of DNA can promote processes that favor cancer spread (Bakhoum and Cantley, 2018). Thus, it might be expected that factors that promote inflammation might increase the potential for successful metastasis. Once cells have left the main tumor mass, they can travel through the bloodstream, and then kill cells in the vascular wall, allowing them to get through and lodge nearby. For cancer cells to attach to the vascular wall, a receptor protein EPHA2 must be active, but if this protein is switched off, then the cancer cells can detach from the vascular wall and travel elsewhere. When a metastasized cell arrives at a new site, nutritional support may be provided through existing vasculature, as well as angiogenesis, and further encouraged by the sympathetic nervous system, the presence of stress hormones, and the stimulatory actions of the cytokine MCP-1 (Cole et al., 2015).

There are cases in which the functioning of the immune system that ought to be protective may inadvertently facilitate metastasis. By expelling their DNA, neutrophils can form a gauzy trap, referred to as neutrophil extracellular traps (NETs), laced with toxic enzymes that can destroy pathogens. In the presence of inflammation, however, enzymes that are present may affect the structure of the protein that can be recognized by dormant metastasized tumor cells so that they are spurred into action and allow them to proliferate (Albrengues et al., 2018).

Similarly, lymphocytes may also secrete a web of sticky DNA that can capture circulating cancer cells. But rather than killing these cells, it makes them stronger, more aggressive, and more adept at reaching distal organs. Of therapeutic relevance, nanoparticles that degrade NETs have been shown to attenuate metastasis (Park et al., 2016). Consequently, as we learn the many different ways that cancer cells establish conditions to promote their survival and colonization of secondary body locations, therapeutic approaches can be devised to render cancer cells more vulnerable to cytotoxic targeting.

The survival of cancer cells can also be enhanced by certain types of integrins, which together with another protein, c-Met, encourage the cells to resist death (Barrow-McGee et al., 2016). Integrins are molecules that traverse the cell membrane and allow for cells to "stick" or adhere to the extracellular matrix in a given tissue. For this reason, they are called adhesion molecules. Eliminating the ability of cancer cells to adhere or set up camp in a given location is a useful goal in reducing metastatic tumor growth. By targeting specific integrins, cancer cells could potentially be eliminated, with side effects kept to a minimum.

It will be recalled that cancer cells may remain in a dormant state for a considerable time before emerging, likely being linked to the microenvironment in which they reside. This can also be said of cancer cells that have metastasized. During the initial phase of breast cancer, macrophages may be attracted to the site of breast ducts, but rather than destroying the cells located there, they may be coopted so that their ability to exit the breast is enhanced. In essence, metastasis may have occurred even before the primary tumor was detected (Linde et al., 2018). Metastasis may stem from a single primary tumor cell or could descend from multiple clones. The chronology of the spread has not been fully worked out, although it seems that this differs across cancer types, beginning early in tumor growth, with clonal mutations being vulnerable to a variety of treatments.

Ingredients for cancer growth and metastasis

Nutritional and energy support

In so far as cancer cells represent renegade cells with unrestrained, exponential growth cycles, their expansion is largely dependent on the receipt of sufficient nutritional support. We mentioned earlier that tumors grow and develop new blood vessels. In the absence of angiogenesis the supply of blood-derived growth factors, oxygen, glucose, and other important factors would be limited. For this reason, a branch of cancer research focuses on finding ways to limit tumor angiogenesis. So, what are some of the key ingredients in a cancer cell's "meal" that allow it to thrive? It is likely that whatever is good for the rest of the body is likely to appeal to cancer cells. But some elements of a typical cell's diet might be more important for cancer cells than is typically normal. That is, the mutated state of a cancer cell may render it more dependent on certain nutritional factors.

Fat and fatty acids could be an important ingredient for cancer spread as its utilization is needed by lymphatic vessels, which serve as pathways by which cancer can spread. Beyond this, located within the membrane of several types of cancer cells is a protein (CD36), a cell surface molecule that can facilitate fatty acid transport into the cytoplasm (among many other functions) that can take up fatty acids. In mice transplanted with a human oral tumor that had been treated with palmitic acid (a component of animal and vegetable fat), metastasis was elevated, whereas blocking CD36 prevented this outcome (Pascual et al., 2017). Such studies were consistent with the possibility of preventing metastases by diet or other processes that diminish the cellular intake of fat.

Glucose and the Warburg effect

Not to give the impression of an antifat bias (in the sugar vs fat corporate wars played out in the public sphere), it should be said that sugar had earlier been implicated as a major culprit in the development and metastasis of various cancers (Jiang et al., 2016). Ordinarily, through a process known as oxidative phosphorylation, healthy cells take glucose and convert it into pyruvate and in the presence of oxygen it is converted into adenosine triphosphate (ATP), which is the primary source of energy for cells. This occurs primarily in the mitochondria, but also in the cytoplasm through a process referred to as glycolysis, which does not require oxygen. When oxygen is less available (e.g., after a long-distance run), a shift occurs so that energy is made through glycolysis rather than oxidative phosphorylation, which is far less efficient in the production of energy.

In evaluating cancer cells, principally during the early 1920s, Otto Warburg had noted that while cancer cells exhibited increased rates of glycolysis (aerobic glycolysis) they favor fermentation, an anaerobic process in which pyruvate (the last product of glycolysis) is converted into lactate. In essence, cancer cells can efficiently convert glucose and glutamine to meet their energetic needs. This phenomenon, which came to be known as the Warburg effect (Warburg, 1956), has received increasing attention over the years and spawned varied approaches to diminish tumor cells. It has been demonstrated that oxidative phosphorylation is largely reduced in cancer cells and hence ATP is not produced through this route. However, cancer cells are particularly enamored of sugars (glucose) as an energy source, taking up large amounts of glucose that are converted to pyruvate. The pyruvate in turn is converted to lactate and lactate itself can produce pyruvate. Therefore, cancer cells appear to recycle lactate more actively than healthy cells, using this as fuel for their increased energy needs (Boroughs and DeBerardinis, 2015). The increased lactate also produces lactic acid, resulting in the reduction of pH levels, which favors tumor growth and invasion and serves to diminish immune functioning that might otherwise attack cancer cells (see Fig. 1.6).

This process is not unique to cancer cells being evident in rapidly dividing cells, such as in embryonic tissue. However, in normal cells, this process halts when they are no longer growing, whereas in cancer cells this process of aerobic glycolysis does not stop, instead the process seems to be stuck in place. In the rapidly multiplying cancer cells, there might be the possibility that oxygen supply would not meet the needs of these growing cells and thus having an anaerobic source of energy would be highly adaptive for them. Besides, during this process the raw material necessary for the creation of building blocks for cells are available, thereby allowing for rapid cell multiplication.

Understanding how and why cancer cells choose to use anaerobic metabolism may allow for the development of methods to deter their growth. One way this might occur is because of the overexpression of hypoxia-inducible factor-1 (HIF-1) in response to low oxygen levels so that the transcription of multiple genes that code for proteins that facilitate cancer growth is promoted. These genes may be important for angiogenesis, resistance to apoptosis in which damaged cells die (cell suicide), and very importantly, glucose metabolism is greatly

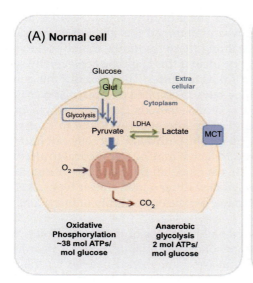
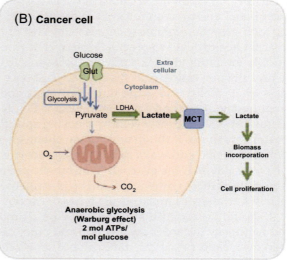

FIG. 1.6 Differences in glycolysis pathways between normal cells and cancer cells. (A) In the presence of oxygen, normal cells produce carbon dioxide up to 38 ATPs per glucose molecule through glycolysis, the tricarboxylic acid cycle (TCA or Krebs cycle), and an electron transport system. In a hypoxic environment, pyruvates are accumulated without going through the TCA cycle. These accumulated pyruvates in the muscle tissue are converted to lactic acid and only produce two ATPs. (B) Cancer cells only use the glycolysis process, regardless of the presence or absence of oxygen; two ATPs are produced per glucose molecule and, therefore, compared to normal cells, more glucose is required to obtain energy. *Figure and caption from Kim, S.H., Baek, K.H., 2021. Regulation of cancer metabolism by deubiquitinating enzymes: the Warburg effect. Int. J. Mol. Sci. 22, 6173.*

increased. Interestingly, certain aggressive cancer cells make and store enormous amounts of glycogen from glucose, and this can be converted back to glucose should the need arise during rapid cell multiplication (Altemus et al., 2019). As cancer cells are voracious consumers of glucose, this can be exploited for cancer diagnosis. Specifically, a form of radioactive glucose is administered to patients, which is rapidly taken up by cancer cells. These cells can then be imaged through positron emission tomography (PET) so that hot spots can be readily detected. Furthermore, glycolytic inhibitors can be used to inhibit cancer growth.

The Warburg effect had been seen as a possible breakthrough in cancer treatment and Warburg was awarded the Nobel Prize in Physiology in 1931, but attention to this area of research was eclipsed when DNA was unraveled by Watson and Crick (and arguably Rosalind Franklin). Over the past two decades, there has been a great resurgence in the interest of the Warburg effect and it has been implicated as an early step in the production of oncogenic mutations as well as being a key component in signaling tumor cells so that multiple cellular processes can be affected (Liberti and Locasale, 2016). As discussed in ensuing chapters, the implications of the Warburg effect to the connection between lifestyles and cancer are very substantial.

As mentioned, an end product of glycolysis is the formation of pyruvate that can be converted into energy through the formation of lactate by the enzyme pyruvate kinase. As it happens, one form of pyruvate kinase (M1) is present in healthy cells, whereas another form,

M2, is primarily found in all cancer cells (as well as in some healthy cells). Because of the exceptional energy requirements of cancer cells, it was maintained that the key to restraining their growth might be to increase the cost of obtaining these resources, thereby allowing adjacent healthy cells to prevail. Moreover, it would be ideal if this could be achieved by specifically inhibiting the M2 pathway that seems to be linked to cancer. Theoretically, this could be useful when cancer cells first appear, but once they have multiplied and reached a critical mass, their ability to escape these restraints and use other methods to foster growth may predominate (Oña and Lachmann, 2020). Pointedly, when attempts are made to diminish energy supplies to cancer cells by blocking specific metabolic pathways, the cells can find (or create) new routes through which they obtain energy. In fact, it may be impractical to attempt to inhibit single pathways to limit tumor growth. This is made all the more complicated by the possibility that the alternative metabolic pathways may differ across cancer types. At the same time, if these pathways could be identified, it may offer the opportunity of using a more selective strategy to treat specific types of cancer.

It should be added that lactate may be a fundamental regulator of macrophage activity, and hence key in cancer progression and other diseases of aging and autoimmune disorders. Moreover, cancer cells can diminish nutrients (e.g., enolase 1, an enzyme needed for the use of glucose) required for ideal immune functioning so that cytotoxic cells may become too sluggish to be effective. To a considerable extent immune cells and cancer cells compete for needed resources but as cancer cells win out in this competition, the resources needed by immune cells, including glucose and amino acids (e.g., arginine, tryptophan, and cysteine), decline so these cancer-fighting immune cells lose their potency. This is exacerbated by tumor cells producing metabolic waste products that also reduce immune functioning.

As described in Fig. 1.7, cancer cells can gain access to both conventional and unconventional nutrient sources that contribute to the development of a new biomass. Cancer cells are also adept at taking advantage of select metabolites to promote their development (Pavlova and Thompson, 2016). In theory, repairing metabolic disturbances can convert weak killer cells into tougher cancer fighters. This could be achieved by altering amino acid metabolism and eliminating harmful waste products, and therapeutic strategies might do well by targeting metabolic processes that are not shared by pro- and antitumor cells (Renner et al., 2017). Since lifestyle factors, including diet, sleep, exercise, and psychological stressors, can affect metabolic processes, addressing their actions on tumorigenesis and immune functioning could potentially affect cancer processes and the response to treatments.

Being unneighborly

Cancer cells can also thrive by stealing nutrients from nearby cells that might be going through the process of autophagy—in which old and damaged cells are broken down and thus release sugars and amino acids (Katheder et al., 2017). Accordingly, the common medication used to treat type 2 diabetes, metformin, which reduces glucose production (and hence a primary cellular energy source), has antitumor actions in a variety of cancer types (Jia et al., 2015), increasing susceptibility to autophagy-related cell death. Although autophagy has often been related to cancer provocation and progression, it might act against the development and growth of cancer cells and could thus be a potential therapeutic target, even at different stages of the cancer process. Since autophagy is stimulated in response to cellular stress, such as nutrient deprivation, the possibility had been advanced that certain foods could affect tumor growth by affecting this process.

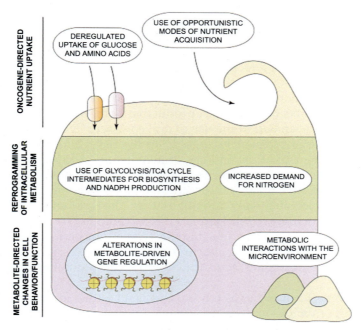

FIG. 1.7 Cancer cells accumulate metabolic alterations that allow them to gain access to both conventional and unconventional nutrient sources. They utilize these nutrients toward the creation of a new biomass to sustain deregulated proliferation, and take advantage of the ability of select metabolites to affect the fate of cancer cells themselves as well as a variety of normal cell types within the tumor microenvironment. Three layers of cell-metabolite interaction are depicted, all of which become reprogrammed in cancer. On top are the adaptations that involve nutrient uptake, followed by alterations to intracellular metabolic pathways in the middle. Finally, long-range effects of metabolic reprogramming on the cancer cell itself, as well as on other cells within its microenvironment, are depicted at the bottom. *From Pavlova, N.N., Thompson, C.B., 2016. The emerging hallmarks of cancer metabolism. Cell Metab. 23, 27–47.*

A common view of metastases is the "seed and soil" hypothesis, in which the tumor cells—the "seeds"—propagate most readily in the right "soil" (specific locations in the body). By way of example, cancer frequently spreads to the liver, possibly because liver cells respond to inflammation by activating the transcription factor STAT3 that is fundamental in the production of proteins that are involved in signaling pathways within cells. This includes the production of serum amyloid A (SAA) proteins, which then promotes remodeling of the liver, creating just the right soil for cancer cells to prosper and grow (Lee et al., 2019). It would be ideal if such seeds could be attacked while they were in a dormant state, but this has been difficult to accomplish in mouse models designed to produce rapidly growing cancers. Instead of focusing on the seeds, attention has been directed toward the soil to determine the characteristics of certain organs that make them particularly fertile ground for seeding. For instance, cancer cells communicate with normal tissues, perhaps, through exosomes; this can stimulate further changes that enhance or prepare the bedding and overall fertility of the soil for the seed's growth (e.g., Nakata et al., 2017). Thinking in metaphors can be useful and, in this case, given that metastasis is an example of pulling up stumps and relocating to a new site, it is apt to think in such terms to identify the molecular conditions that would allow such a move to be successful. And ultimately, how to parry and block such a venture.

Impact of hormones and hormone receptors on cancer processes

Diverse forms of cancer not only differ in their basic characteristics and genetic makeup, but they are also differentially affected by a variety of endogenous substrates. In the ensuing chapters, we will learn that numerous growth factors and hormones, as well as treatments that influence hormonal functioning, can markedly influence tumor growth. Breast cancer, for instance, has been linked to reproductive hormones, and the development of this form of cancer has been tied to the use of oral contraceptives (see text box), as well as the number and timing of pregnancies, age when the first child was born, breastfeeding practices, age at first menstruation, and age of menopause. Accordingly, it is common for breast cancer therapy to include treatments that affect estrogen or testosterone functioning. In the case of triple-negative breast cancer, hormone manipulations have much less effect, making this form of cancer relatively difficult to treat. The fact that the triple negative form of breast cancer does not involve estrogen receptors should not be misconstrued to suggest that estrogen does not play a role in affecting this type of cancer. Estrogen is present within the tumor's microenvironment, being able to stimulate astrocytes, which release the growth factor brain-derived neurotrophic factor (BDNF) that can promote metastasis (Contreras-Zárate et al., 2019).

Triple-negative breast cancer has also been associated with a mutation of the tumor suppressor TP53 gene. For various reasons, targeting this mutation was not deemed to be a viable option, but using nanotechnology, a nearby gene (POLR2A) that selectively acted on cancerous cells could be targeted (Xu et al., 2019). Other genes that may be important for this form of cancer have also been identified and targeting them could be more effective in dealing with both triple-negative breast cancer and prostate cancer.

Influence of hormone manipulations in women

For years, women had been using hormone replacement therapy (HRT) to diminish physical discomforts and to enhance bone strength following menopause. Despite HRT having several positive effects (e.g., in diminishing risk of osteoporosis and cardiovascular disease), these were outweighed by an elevated risk of breast and ovarian cancer, as well as gallbladder disease, stroke, and the development of pulmonary embolism (blood clots in the lungs). Indeed, if the HRT comprised a combination of estrogen and progesterone, breast cancer risk increased by 2% after 5 years of use and doubles with 10 years of use.

Shortly after the introduction of oral contraceptives, widespread concern developed that the hormonal changes elicited by the pill could favor cancer development. Large-scale prospective studies (Hunter et al., 2010) indicated that despite successive iterations of the pill comprising lower estrogen and progesterone concentrations, women using oral contraceptives were still at somewhat elevated risk of developing breast cancer, particularly those women using the "triphasic pill" in which the dose of the hormone changed in three stages during each monthly cycle (Hunter et al., 2010). However, subsequent studies revealed that while oral contraceptives increased breast cancer risk, especially with longer use, the increased frequency of cancer occurrences was relatively small. A recent report comprising several databases within the United Kingdom that amounted to more than 550,000 women indicated that estrogen-based contraceptives were associated with a 15% increase in the risk of breast cancer, whereas the combination of estrogen and progesterone was accompanied by a substantially greater increase (Vinogradova et al., 2020).

Cervical cancer was also elevated among women using oral contraceptives, and as in the case of breast cancer, the risk declined after their use was curtailed. A recent meta-analysis, however, reported that oral contraceptive use was not accompanied by increased cervical cancer incidence (Peng et al., 2017). In contrast to the potential harms attached to some hormone treatments, they can also have benefits in relation to certain forms of cancer, depending on genetic factors. Indeed, unlike the cancer risk, oral contraceptive use was accompanied by diminished risk for ovarian and endometrial cancer. These protective effects were evident even 35 years after discontinuing the use of an oral contraceptive (Karlsson et al., 2021).

Like the involvement of estrogen in breast cancer, testosterone may contribute to prostate cancer, possibly through actions on immune functioning and pro- and antiinflammatory cytokines and has also been associated with increased risk for colorectal cancer (Roshan et al., 2016). While the level of testosterone in early prostate cancer did not predict overall survival, higher testosterone levels in advanced prostate cancer were predictive of reduced death rates. Based on a systematic review and meta-analysis it was suggested that androgen deprivation therapy, particularly when coupled with the chemotherapy medication, docetaxel, was associated with increased survival among patients with metastatic hormone-naive prostate cancer (Botrel et al., 2016).

The links to cancer are certainly not restricted to estrogen and testosterone, with many other hormones showing direct or indirect relations. For example, insulin-like growth factor has been associated with osteosarcoma and its metastasis, and insulin (as well as type 2 diabetes) has been related not only to pancreatic cancer but also to leukemia. In line with these findings, the type 2 diabetes medication metformin is useful as a therapeutic agent in leukemia, lymphomas, and multiple myeloma (Cunha Júnior et al., 2018). The actions of metformin are in line with reports that linked glucose availability and cancer progression, as well as the ties to diabetes (and obesity) and inflammatory processes. There is also good reason to believe that leptin and ghrelin, two feeding-related hormones, are linked to breast cancer, possibly owing to effects on adipokines (fat-storing cells) as well as their actions on cytokines and insulin-like growth factor (Andò et al., 2019). Beyond these ties, several hormones that are associated with stressors (e.g., cortisol, norepinephrine, oxytocin) have also been linked to cancer development. We shall discuss the involvement of these hormones in cancer in later chapters and indicate how they may be related to stressor experiences and the adoption of specific lifestyle variables.

Inflammatory factors in the cancer process

It has been more than a hundred years since Rudolf Virchow made the link between inflammation and cancer, and at about the same time, surgeon William Coley induced a therapeutic inflammatory response using heat-killed bacteria to treat sarcomas. Since those early days, it has become apparent that inflammation, which can be brought about by a remarkably wide array of factors, may be associated with organ decline and can be fundamental in the provocation and maintenance of several diseases.

What inflammation implies

Inflammation largely comprises a rapidly mobilized nonspecific immunological defensive strategy. Inflammatory triggers range from microbial infection, noninfectious challenges—such as physical tissue injury or radiation, chemical exposure, metals or trace elements—and intrinsic disease states or other challenges that may induce inflammation. Regardless of the cause, in each instance, disruption of homeostasis occurs, and the main role of the inflammatory response is to correct this. Hence, inflammation is a fundamentally important response, which can facilitate the clearance of pathogens and foster the recovery and/or regeneration of injured tissues. But, when unchecked, inflammation can have devastating consequences, favoring the occurrence of a variety of disease conditions. For instance, overly prolonged or an especially robust inflammatory response can provoke autoimmune and allergic responses or possibly even neuronal degeneration, and was implicated in the emergence of type 2 diabetes and heart disease. Excessive or prolonged inflammatory activation has also been implicated as a key component in the onset and progression of several forms of cancer. An important caveat in this respect is that different leukocytes can exert divergent effects on growth, progression, and spread, such that infiltrating immune cells can either promote or inhibit cancer development, depending on their microenvironment.

As shown in Fig. 1.8, inflammation is intimately related to cancer progression, being key in the orchestration of processes responsible for the survival, proliferation, and migration of cancer cells. It has been suggested that over 20% of human cancers have an inflammatory component likely owing to DNA damage (Soria-Valles et al., 2017), although in many cases, inflammation probably is indirectly or secondarily involved in the disease.

Two faces of inflammation

At the cellular level, inflammation itself can largely be boiled down to the infiltration of various leukocytes (white blood cells) into a particular area of the body. Of the various innate immune cells, neutrophils, macrophages, and NK cells are first on the scene, followed by the appearance of further specialized cells, such as adaptive T or B lymphocytes. Each area or organ also has its own patrolling sentinel cells that form the local inflammatory milieu. For instance, microglia are the resident immune cells that help kick-start neuroinflammatory responses within the brain, which may or may not be exacerbated by infiltrating immune cells from the periphery. Similarly, specialized liver and lung macrophages, Kupffer cells, and alveolar macrophages help form a core feature of inflammation in these organs and, once again, infiltrating immune cells typically join these when prompted by an inflammatory stimulus. While innate immune cells are the primary movers of inflammation, when the condition persists, other specialized cells are also recruited and participate in sculpting the microenvironment.

Diseases that promote inflammation can morph into serious cancers of the affected organ. In this regard, microbial disturbances (gut bacteria, for instance) can lead to chronic inflammation, which can affect genomic processes that instigate gastrointestinal cancers, which are among the leading causes of mortality worldwide. Inflammatory bowel disease was also linked to the occurrence of colon cancer, and ways to regulate cellular RNA were effective in stemming ulcerative colitis and bowel cancer development (Polytarchou et al.,

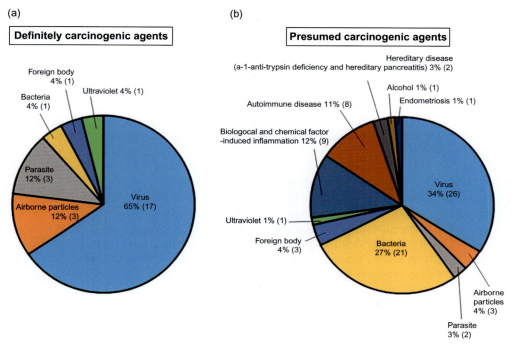

FIG. 1.8 Causes of inflammation-related carcinogenesis. The proportion of (A) definitely carcinogenic causes or (B) presumed carcinogenic causes attributed to inflammation. *Figure and caption from Kanda, Y., Osaki, M., Okada, F., 2017. Chemopreventive strategies for inflammation-related carcinogenesis: current status and future direction. Int. J. Mol. Sci. 18, 867.*

2015). Of course, noninfectious environmental insults, such as cigarette smoke and asbestos, have been directly linked to cancer, but in these cases, the inflammatory response may be secondarily recruited to modify the disease state. Whatever the stimulus, improper regulation of immune cells and their signaling agents (e.g., cytokines), together with various cancer-related genes (oncogenes) or other potentially DNA damaging factors, can lead to cancer development.

Mechanisms of inflammation

Many types of cancer share common features, particularly several fundamental inflammatory pathways. Among these, the NFκB and JAK-STAT pathways, which underlie many of the effects of the classic proinflammatory cytokines (IL-1β, TNF-α, IL-2, IL-6), can also affect cancer-related processes through their actions on apoptotic regulators, most notably the tumor suppressor p53 (Blaylock, 2015). It was also proposed that chronic inflammation at sites of infection causes the production of reactive oxygen and nitrogen species that can induce DNA damage. For that matter, the presence of an inflammatory microenvironment for prolonged periods may promote DNA damage (Anuja et al., 2017) that can be accelerated through a positive feedback loop in which these oxygen species amplify inflammatory responses leading to tumor cell multiplication (Klaunig, 2018). The presence of proinflammatory cytokines and other factors associated with inflammation (e.g., cyclooxygenase-2 and prostaglandins) within the tumor microenvironment and surrounding cells may ultimately promote tumor angiogenesis, invasion, and metastasis.

Commensurate with the involvement of inflammatory factors in cancer, elevated levels of C-reactive protein (CRP) released from the liver, together with other markers that signal the presence of inflammation, are predictive of the occurrence of diverse types of cancer (Hart et al., 2020). A meta-analysis of 12 studies indicated that the presence of circulating CRP was linearly related to the occurrence of breast cancer. This association was diminished when controlling for lifestyles, suggesting that these factors mediated the link between CRP and breast cancer occurrence (Chan et al., 2015). The effectiveness of these markers in predicting cancer is enhanced in the presence of other environmental contributors (e.g., smoking), possibly being related to several inflammatory cytokines, such as IL-1β, IL-6, and TNF-α and activation of NFκB, which influences DNA transcription related to immune functioning. Inflammatory factors could also promote cancer progression by encouraging cancer cell division, fostering immunosuppressive lymphocytes, and suppression of immunosurveillance, thereby protecting these cells from death. Under conditions in which inflammatory cytokines are released prematurely, certain types of T cells might be inhibited, which could act against autoimmune conditions but could concurrently undermine the ability to deal with tumors.

In response to strong environmental challenges, many cells respond by becoming senescent, which includes reduced cell division. However, as damaged, dying, and senescent cells build up, chronic inflammation may be instigated, which can promote age-related illnesses, including cancer. Moreover, when inflammation is present within a tumor's microenvironment, the proliferation and survival of cancer cells are enhanced, angiogenesis is promoted (thereby providing an energy source for the malignant cells), the odds of metastasis are elevated, and the effectiveness of chemotherapeutic agents and hormones are undermined. More than this, inflammation can promote genetic instability that may foster further tumor growth. Later chapters show how inflammation also gives rise to variations in gut microbes and further immune changes, which together favor DNA disturbances and hence mutations that promote cancer. Finding a way of restricting the "always-on" inflammatory system might be useful in limiting age-related diseases (Dou et al., 2017).

Up to this point, we have highlighted how inflammatory processes are most often associated with cancer development and progression. However, inflammation and its consequences are notorious for being difficult to reconcile with the ostensibly adaptive and/or protective functions of the immune system, a primary source of the inflammatory response. Perhaps inflammation is a reaction to the internally distressing changes taking place as a cancer grows. Nutrients are being consumed more heavily, new blood vessels are growing, the "winner" and "loser" cells are in a battle within microenvironments attempting to maintain biological integrity, with autophagic processes ramped up. Whatever the reason, an inflammatory response may in the first instance be an effort to restore some normality of function. However, negative effects may ultimately ensue as cancers grow and the inflammatory response persists and may even become more severe. To this point, the negative effects of inflammation may appear during different phases of cancer development. As previously stated, disseminated cancer cells can reside for an extended period in a dormant state within various organs, before breaking out to cause cancer recurrence, major tumor growth, and/or metastasis. In hormone-dependent cancers (e.g., prostate cancer and estrogen-positive breast cancer), the dormancy period may exceed a decade. Identifying the factors that promote the breakout—or reawakening—can be especially significant as this might allow for interventions to be undertaken (Aguirre-Ghiso, 2018). Indeed, it has been posited that dormant cancer cells can be stimulated to grow by the presence of inflammatory factors. We will consider this to a greater extent when we discuss stress processes. For the present, it is worth noting that surgical procedures can rouse dormant tumors under control by the immune system to emerge and start to grow. This may be instrumental in promoting metastasis, possibly owing to the activation of inflammatory processes (Krall et al., 2018).

Intrinsic vs extrinsic contributions to cancer occurrence

It has been suggested that 60% of all cancer occurrences are attributable to "bad luck" related to intrinsic factors, such as inherited gene mutations or those that occur on a random basis (Tomasetti et al., 2017). Others have maintained that this is an overestimate, and a sizeable number of cancer occurrences may be due to a wide variety of nonintrinsic influences. Even from the brief discussion provided to this point, it should be clear that numerous cancer occurrences can be due to a diverse range of extrinsic factors (see Table 1.2). The case was made that cancer risk was frequently due to several extrinsic factors, including those over which individuals have little control (exposure to carcinogens) as well as factors that undermined the adoption of healthy lifestyles that could prevent cancer occurrence. Often, cancer risk occurs owing to interactions between intrinsic and extrinsic influences (e.g., between genetic processes and environmental and experiential factors). Indeed, many cancer-related biological factors (including hormones and growth factors, metabolic effects, reactive oxygen species, immune responses, cytokines, inflammation) come about through variations in exogenous factors and lifestyles and may determine the actions of genes on cancer provocation.

Having said this, what we intend to explore in much of this book are the pathways that exist between the actions of different lifestyle behaviors and the cells that may or may not grow

TABLE 1.2 Multiple factors that affect cancer risk.

Genetic factors	Exogenous risk factors	Modifiable biological factors
Genetic vulnerability	Carcinogens	Epigenetic changes
Random mutations	Cancer-producing viruses	DNA repair mechanisms
Epigenetic changes	Lifestyles	Hormones
	Diet	Growth factors
	Obesity	Immune processes Inflammatory
	Smoking	factors Microbial processes
	Alcohol	Tumor microenvironment
	Sun exposure	
	Sedentary behaviors	
	Stressor experiences	

Modified from Wu, S., Zhu, W., Thompson, P., Hannun, Y.A., 2018. Evaluating intrinsic and non-intrinsic cancer risk factors. Nat. Commun. 9, 3490.

into a full-blown tumor. What we hope to achieve is a marriage of reductionistic observation and analysis with the more holistic aspects of what lifestyle means: ingesting foods, hiking trails, and enjoying the company of others—in short, doing stuff. How does the accrual of the myriad behaviors adopted by the individual sum up to a body whole being well or ill?

Summary and conclusions

The growth of tumors is shaped by their immediate cellular and vascular environment, their structural characteristics, immune and microbial factors, and, as we will repeatedly see a constellation of lifestyle factors. In response to their microenvironment, tumors undergo marked genetic alterations that result in modifications that enable them to evade host immunity. The cross-talk between tumor cells and immune cells plays a critical role in producing the microenvironment and hence the tumor status.

Chronic inflammation, such as that which accompanies infections, can precede tumor development, and there is ample reason to believe that chronic inflammation is causally linked to cancer occurrence and progression. These actions might stem from genomic instability, the provocation of oncogenic mutations, and enhanced angiogenesis. Given that obesity/diet, stressors, sleep, and exercise also affect inflammatory processes, raises the question as to their impact on the production or exacerbation of some types of cancer. The data in the case of obesity being linked to a variety of cancer types is very well established, whereas the involvement of other lifestyles in cancer processes is not as well documented. Yet, stressors, exercise, and sleep, as well as multiple exogenous and endogenous factors can instigate a cytokine-mediated inflammatory milieu. Thus, these factors could potentially favor cancer occurrence, progression, reemergence, and treatment resistance.

References

Aguirre-Ghiso, J.A., 2018. How dormant cancer persists and reawakens. Science 361, 1314–1315.

Albrengues, J., Shields, M.A., Ng, D., Park, C.G., Ambrico, A., et al., 2018. Neutrophil extracellular traps produced during inflammation awaken dormant cancer cells in mice. Science 361, eaao4227.

Altemus, M.A., Goo, L.E., Little, A.C., Yates, J.A., Cheriyan, H.G., et al., 2019. Breast cancers utilize hypoxic glycogen stores via PYGB, the brain isoform of glycogen phosphorylase, to promote metastatic phenotypes. PLoS One 14, e0220973.

Andò, S., Gelsomino, L., Panza, S., Giordano, C., Bonofiglio, D., et al., 2019. Obesity, leptin and breast cancer: epidemiological evidence and proposed mechanisms. Cancer 11, 62.

Anuja, K., Roy, S., Ghosh, C., Gupta, P., Bhattacharjee, S., et al., 2017. Prolonged inflammatory microenvironment is crucial for pro-neoplastic growth and genome instability: a detailed review. Inflamm. Res. 66, 119–128.

Arlauckas, S.P., Garris, C.S., Kohler, R.H., Kitaoka, M., Cuccarese, M.F., et al., 2017. In vivo imaging reveals a tumor-associated macrophage-mediated resistance pathway in anti-PD-1 therapy. Sci. Transl. Med. 9, eaal3604.

Baez-Ortega, A., Gori, K., Strakova, A., Allen, J.L., Allum, K.M., et al., 2019. Somatic evolution and global expansion of an ancient transmissible cancer lineage. Science 365, eaau9923.

Bakhoum, S.F., Cantley, L.C., 2018. The multifaceted role of chromosomal instability in cancer and its microenvironment. Cell 174, 1347–1360.

Barrow-McGee, R., Kishi, N., Joffre, C., Ménard, L., Hervieu, A., et al., 2016. Beta 1-integrin–c-Met cooperation reveals an inside-in survival signalling on autophagy-related endomembranes. Nat. Commun. 7, 11942.

Blaylock, R.L., 2015. Cancer microenvironment, inflammation and cancer stem cells: a hypothesis for a paradigm change and new targets in cancer control. Surg. Neurol. Int. 6, 92.

Boroughs, L.K., DeBerardinis, R.J., 2015. Metabolic pathways promoting cancer cell survival and growth. Nat. Cell Biol. 17, 351–359.

Botrel, T.E., Clark, O., Lima Pompeo, A.C., Horta Bretas, F.F., Sadi, M.V., et al., 2016. Efficacy and safety of combined androgen deprivation therapy (ADT) and docetaxel compared with adt alone for metastatic hormone-naive prostate cancer: a systematic review and meta-analysis. PLoS One 11, e0157660.

Burnet, F.M., 1970. The concept of immunological surveillance. Prog. Exp. Tumor Res. 13, 1–27.

Butler, G., Keeton, S.J., Johnson, L., Dash, P.R., 2020. A phenotypic switch in the dispersal strategy of breast cancer cells selected for metastatic colonization. Proc. R. Soc. B Biol. Sci. 287, 20202523.

Carioli, G., Bertuccio, P., Boffetta, P., Levi, F., La Vecchia, C., et al., 2020. European cancer mortality predictions for the year 2020 with a focus on prostate cancer. Ann. Oncol. 31, 650–658.

Cerwenka, A., Lanier, L.L., 2018. Natural killers join the fight against cancer. Science 359, 1460–1461.

Chan, D.S., Bandera, E.V., Greenwood, D.C., Norat, T., 2015. Circulating C-reactive protein and breast cancer risk-systematic literature review and meta-analysis of prospective cohort studies. Cancer Epidemiol. Biomark. Prev. 24, 1439–1449.

Chang, R.B., Beatty, G.L., 2020. The interplay between innate and adaptive immunity in cancer shapes the productivity of cancer immunosurveillance. J. Leukoc. Biol. 108, 363–376.

Chongsathidkiet, P., Jackson, C., Koyama, S., Loebel, F., Cui, X., et al., 2018. Sequestration of T cells in bone marrow in the setting of glioblastoma and other intracranial tumors. Nat. Med. 24, 1459–1468.

Cole, S.W., Nagaraja, A.S., Lutgendorf, S.K., Green, P.A., Sood, A.K., 2015. Sympathetic nervous system regulation of the tumour microenvironment. Nat. Rev. Cancer 15, 563–572.

Contreras-Zárate, M.J., Day, N.L., Ormond, D.R., Borges, V.F., Tobet, S., et al., 2019. Estradiol induces BDNF/TrkB signaling in triple-negative breast cancer to promote brain metastases. Oncogene 38, 4685–4699.

Cordova, M.J., Riba, M.B., Spiegel, D., 2017. Post-traumatic stress disorder and cancer. Lancet Psychiatry 4, 330–338.

Cunha Júnior, A.D., Pericole, F.V., Carvalheira, J.B.C., 2018. Metformin and blood cancers. Clinics 73, e412s.

Dejosez, M., Ura, H., Brandt, V.L., Zwaka, T.P., 2013. Safeguards for cell cooperation in mouse embryogenesis shown by genome-wide cheater screen. Science 341, 1511–1514.

Dirkse, A., Golebiewska, A., Buder, T., Nazarov, P.V., Muller, A., et al., 2019. Stem cell-associated heterogeneity in glioblastoma results from intrinsic tumor plasticity shaped by the microenvironment. Nat. Commun. 10, 1787.

Donatelli, S.S., Zhou, J.M., Gilvary, D.L., Eksioglu, E.A., Chen, X., et al., 2014. TGF-β-inducible microRNA-183 silences tumor-associated natural killer cells. Proc. Natl. Acad. Sci. U. S. A. 111, 4203–4208.

Dou, Z., Ghosh, K., Vizioli, M.G., Zhu, J., Sen, P., et al., 2017. Cytoplasmic chromatin triggers inflammation in senescence and cancer. Nature 550, 402–406.

Dunn, G.P., Old, L.J., Schreiber, R.D., 2004. The immunobiology of cancer immunosurveillance and immunoediting. Immunity 21, 137–148.

Ferrari de Andrade, L., Tay, R.E., Pan, D., Luoma, A.M., Ito, Y., et al., 2018. Antibody-mediated inhibition of MICA and MICB shedding promotes NK cell-driven tumor immunity. Science 359, 1537–1542.

Fitzgerald, K.N., Hodges, R., Hanes, D., Stack, E., Cheishvili, D., et al., 2021. Potential reversal of epigenetic age using a diet and lifestyle intervention: a pilot randomized clinical trial. Aging (Albany NY) 13, 9419–9432.

Ginsburg, O., Bray, F., Coleman, M.P., Vanderpuye, V., Eniu, A., et al., 2017. The global burden of women's cancers: a grand challenge in global health. Lancet 389, 847–860.

Gkountela, S., Castro-Giner, F., Szczerba, B.M., Vetter, M., Landin, J., et al., 2019. Circulating tumor cell clustering shapes DNA methylation to enable metastasis seeding. Cell 176, 98–112.e114.

Hanahan, D., Weinberg, R.A., 2011. Hallmarks of cancer: the next generation. Cell 144, 646–674.

Hart, P.C., Rajab, I.M., Alebraheem, M., Potempa, L.A., 2020. C-reactive protein and cancer-diagnostic and therapeutic insights. Front. Immunol. 11, 595835.

Hunter, D.J., Colditz, G.A., Hankinson, S.E., Malspeis, S., Spiegelman, D., et al., 2010. Oral contraceptive use and breast cancer: a prospective study of young women. Cancer Epidemiol. Biomark. Prev. 19, 2496–2502.

Islami, F., Ward, E.M., Sung, H., Cronin, K.A., Tangka, F.K.L., et al., 2021. Annual report to the nation on the status of cancer, part 1: national cancer statistics. J. Natl. Cancer Inst., djab131.

Jayatilaka, H., Tyle, P., Chen, J.J., Kwak, M., Ju, J., et al., 2017. Synergistic IL-6 and IL-8 paracrine signalling pathway infers a strategy to inhibit tumour cell migration. Nat. Commun. 8, 15584.

Jia, Y., Ma, Z., Liu, X., Zhou, W., He, S., et al., 2015. Metformin prevents DMH-induced colorectal cancer in diabetic rats by reversing the Warburg effect. Cancer Med. 4, 1730–1741.

Jiang, Y., Pan, Y., Rhea, P.R., Tan, L., Gagea, M., et al., 2016. A sucrose-enriched diet promotes tumorigenesis in mammary gland in part through the 12-lipoxygenase pathway. Cancer Res. 76, 24–29.

Kalluri, R., LeBleu, V.S., 2020. The biology, function, and biomedical applications of exosomes. Science 367, eaau6977.

Karlsson, T., Johansson, T., Höglund, J., Ek, W.E., Johansson, Å., 2021. Time-dependent effects of oral contraceptive use on breast, ovarian, and endometrial cancers. Cancer Res. 81, 1153–1162.

Katheder, N.S., Khezri, R., O'Farrell, F., Schultz, S.W., Jain, A., et al., 2017. Microenvironmental autophagy promotes tumour growth. Nature 541, 417–420.

Klaunig, J.E., 2018. Oxidative stress and cancer. Curr. Pharm. Des. 24, 4771–4778.

Krall, J.A., Reinhardt, F., Mercury, O.A., Pattabiraman, D.R., Brooks, M.W., et al., 2018. The systemic response to surgery triggers the outgrowth of distant immune-controlled tumors in mouse models of dormancy. Sci. Transl. Med. 10, eaan3464.

Lee, J.W., Stone, M.L., Porrett, P.M., Thomas, S.K., Komar, C.A., et al., 2019. Hepatocytes direct the formation of a pro-metastatic niche in the liver. Nature 567, 249–252.

Liberti, M.V., Locasale, J.W., 2016. The Warburg effect: how does it benefit cancer cells? Trends Biochem. Sci. 41, 211–218.

Linde, N., Casanova-Acebes, M., Sosa, M.S., Mortha, A., Rahman, A., et al., 2018. Macrophages orchestrate breast cancer early dissemination and metastasis. Nat. Commun. 9, 21.

Liu, Y., Blanchfield, L., Ma, V.P.Y., Andargachew, R., Galior, K., et al., 2016. DNA-based nanoparticle tension sensors reveal that T-cell receptors transmit defined pN forces to their antigens for enhanced fidelity. Proc. Natl. Acad. Sci. 113, 5610–5615.

Lord, K., Mitchell, A.J., Ibrahim, K., Kumar, S., Rudd, N., et al., 2012. The beliefs and knowledge of patients newly diagnosed with cancer in a UK ethnically diverse population. Clin. Oncol. 24, 4–12.

Lussier, D.M., Schreiber, R.D., 2016. Cancer immunosurveillance: immunoediting. In: Immunity to Pathogens and Tumors. Elsevier, pp. 396–405.

Madan, E., Pelham, C.J., Nagane, M., Parker, T.M., Canas-Marques, R., et al., 2019. Flower isoforms promote competitive growth incancer. Nature 572, 260–264.

Magod, P., Mastandrea, I., Rousso-Noori, L., Agemy, L., Shapira, G., et al., 2021. Exploring the longitudinal glioma microenvironment landscape uncovers reprogrammed pro-tumorigenic neutrophils in the bone marrow. Cell Rep. 36, 109480.

Majeti, R., Chao, M.P., Alizadeh, A.A., Pang, W.W., Jaiswal, S., et al., 2009. CD47 is an adverse prognostic factor and therapeutic antibody target on human acute myeloid leukemia stem cells. Cell 138, 286–299.

Maley, C.C., Shibata, D., 2019. Cancer cell evolution through the ages. Science 365, 440–441.

Mansoori, B., Mohammadi, A., Davudian, S., Shirjang, S., Baradaran, B., 2017. The different mechanisms of cancer drug resistance: a brief review. Adv. Pharm. Bull. 7, 339–348.

Mantovani, A., Marchesi, F., Malesci, A., Laghi, L., Allavena, P., 2017. Tumour-associated macrophages as treatment targets in oncology. Nat. Rev. Clin. Oncol. 14, 399–416.

Martin, T.D., Patel, R.S., Cook, D.R., Choi, M.Y., Patil, A., et al., 2021. The adaptive immune system is a major driver of selection for tumor suppressor gene inactivation. Science 373, 1327–1335.

Martinez, F.O., Gordon, S., 2014. The M1 and M2 paradigm of macrophage activation: time for reassessment. F1000Prime Rep. 6, 13.

Meyer, S.B., Foley, K., Olver, I., Ward, P.R., McNaughton, D., et al., 2019. Alcohol and breast cancer risk: middle-aged women's logic and recommendations for reducing consumption in Australia. PLoS One 14, e0211293.

Nabet, B.Y., Qiu, Y., Shabason, J.E., Wu, T.J., Yoon, T., et al., 2017. Exosome RNA unshielding couples stromal activation to pattern recognition receptor signaling in cancer. Cell 170, 352–366.e313.

Nakata, R., Shimada, H., Fernandez, G.E., Fanter, R., Fabbri, M., et al., 2017. Contribution of neuroblastoma-derived exosomes to the production of pro-tumorigenic signals by bone marrow mesenchymal stromal cells. J. Extracell. Vesicles 6, 1332941.

Oña, L., Lachmann, M., 2020. Signalling architectures can prevent cancer evolution. Sci. Rep. 10, 674.

Ostendorf, B.N., Bilanovic, J., Adaku, N., Tafreshian, K.N., Tavora, B., et al., 2020. Common germline variants of the human APOE gene modulate melanoma progression and survival. Nat. Med. 26, 1048–1053.

Park, J., Wysocki, R.W., Amoozgar, Z., Maiorino, L., Fein, M.R., et al., 2016. Cancer cells induce metastasis-supporting neutrophil extracellular DNA traps. Sci. Transl. Med. 8, 361ra138.

Pascual, G., Avgustinova, A., Mejetta, S., Martín, M., Castellanos, A., et al., 2017. Targeting metastasis-initiating cells through the fatty acid receptor CD36. Nature 541, 41–45.

References

Pavlova, N.N., Thompson, C.B., 2016. The emerging hallmarks of cancer metabolism. Cell Metab. 23, 27–47.

Peng, Y., Wang, X., Feng, H., Yan, G., 2017. Is oral contraceptive use associated with an increased risk of cervical cancer? An evidence-based meta-analysis. J. Obstet. Gynaecol. Res. 43, 913–922.

Polytarchou, C., Hommes, D.W., Palumbo, T., Hatziapostolou, M., Koutsioumpa, M., et al., 2015. MicroRNA214 is associated with progression of ulcerative colitis, and inhibition reduces development of colitis and colitis-associated cancer in mice. Gastroenterology 149, 981–992.e911.

Renner, K., Singer, K., Koehl, G.E., Geissler, E.K., Peter, K., et al., 2017. Metabolic hallmarks of tumor and immune cells in the tumor microenvironment. Front. Immunol. 8, 248.

Roshan, M.H.K., Tambo, A., Pace, N.P., 2016. The role of testosterone in colorectal carcinoma: pathomechanisms and open questions. EPMA J. 7, 22.

Ruffell, B., Chang-Strachan, D., Chan, V., Rosenbusch, A., Ho, C.M., et al., 2014. Macrophage IL-10 blocks CD8+ T cell-dependent responses to chemotherapy by suppressing IL-12 expression in intratumoral dendritic cells. Cancer Cell 26, 623–637.

Russo, M., Crisafulli, G., Sogari, A., Reilly, N.M., Arena, S., et al., 2019. Adaptive mutability of colorectal cancers in response to targeted therapies. Science 366, 1473–1480.

Soria-Valles, C., López-Soto, A., Osorio, F.G., López-Otín, C., 2017. Immune and inflammatory responses to DNA damage in cancer and aging. Mech. Ageing Dev. 165, 10–16.

Strub, T., Ghiraldini, F.G., Carcamo, S., Li, M., Wroblewska, A., et al., 2018. SIRT6 haploinsufficiency induces BRAFV600E melanoma cell resistance to MAPK inhibitors via IGF signalling. Nat. Commun. 9, 3440.

Thorsson, V., Gibbs, D.L., Brown, S.D., Wolf, D., Bortone, D.S., et al., 2018. The immune landscape of cancer. Immunity 48, 812–830.e814.

Tomasetti, C., Li, L., Vogelstein, B., 2017. Stem cell divisions, somatic mutations, cancer etiology, and cancer prevention. Science 355, 1330–1334.

Vadlapatla, R.K., Vadlapudi, A.D., Pal, D., Mitra, A.K., 2013. Mechanisms of drug resistance in cancer chemotherapy: coordinated role and regulation of efflux transporters and metabolizing enzymes. Curr. Pharm. Des. 19, 7126–7140.

Vazquez, J.M., Lynch, V.J., 2021. Pervasive duplication of tumor suppressors in Afrotherians during the evolution of large bodies and reduced cancer risk. eLife 10.

Venkataramani, V., Tanev, D.I., Strahle, C., Studier-Fischer, A., Fankhauser, L., et al., 2019. Glutamatergic synaptic input to glioma cells drives brain tumour progression. Nature 573, 532–538.

Vermeulen, R., Schymanski, E.L., Barabási, A.L., Miller, G.W., 2020. The exposome and health: where chemistry meets biology. Science 367, 392–396.

Vinogradova, Y., Coupland, C., Hippisley-Cox, J., 2020. Use of hormone replacement therapy and risk of breast cancer: nested case-control studies using the QResearch and CPRD databases. BMJ 371, m3873.

Vladoiu, M.C., El-Hamamy, I., Donovan, L.K., Farooq, H., Holgado, B.L., et al., 2019. Childhood cerebellar tumours mirror conserved fetal transcriptional programs. Nature 572, 67–73.

Wang, X., Zhang, H., Chen, X., 2019. Drug resistance and combating drug resistance in cancer. Cancer Drug Resist. 2, 141–160.

Warburg, O., 1956. On the origin of cancer cells. Science 123, 309–314.

Weiskopf, K., Ring, A.M., Ho, C.C., Volkmer, J.P., Levin, A.M., et al., 2013. Engineered SIRPα variants as immunotherapeutic adjuvants to anticancer antibodies. Science 341, 88–91.

Xu, J., Liu, Y., Li, Y., Wang, H., Stewart, S., et al., 2019. Precise targeting of POLR2A as a therapeutic strategy for human triple negative breast cancer. Nat. Nanotechnol. 14, 388–397.

Yin, Q., Han, T., Fang, B., Zhang, G., Zhang, C., et al., 2019. K27-linked ubiquitination of BRAF by ITCH engages cytokine response to maintain MEK-ERK signaling. Nat. Commun. 10, 1870.

Yizhak, K., Aguet, F., Kim, J., Hess, J.M., Kübler, K., et al., 2019. RNA sequence analysis reveals macroscopic somatic clonal expansion across normal tissues. Science 364, eaaw0726.

C H A P T E R

2

Cancer and immunity

OUTLINE

Changing times concerning diseases	39	Adaptive (acquired) immunity	53
Fundamentals of immunity	40	*Development of B cells*	55
Functions of the immune system	41	*T lymphocytes*	57
Innate and adaptive components of the immune system	43	Cytokines	60
		Chemokines	63
Innate immunity	44	Inflammasome	63
Innate immune function and cancer	46	Immune tolerance	65
Sterile inflammation	48	Summary and conclusions	66
Anatomy of the immune system: The lymphoid system	52	References	66

Changing times concerning diseases

Life span has more than doubled during the past two centuries, with the greatest increase occurring during the last 100 years. To a considerable extent, this came about with an increased understanding of how to deal with bacteria, viruses, and a constellation of noncommunicable diseases. The introduction of vaccines spared humans from pandemics, such as smallpox, and we can only imagine what the toll of SARS-CoV-2 might have been in the absence of new vaccines that were developed. Childhood illness and deaths declined with improved prenatal care, newborn screening of diseases, enhanced nutrition, diminished frequency of accidental deaths, and vaccines for multiple viral illnesses (e.g., the vaccine for measles, mumps, and rubella). The introduction of antibiotics diminished the odds of bacterial pandemics, almost eradicating tuberculosis, and spared us from the risks of everyday infections that might otherwise have been lethal (viz., the incidence of infection and death following virtually any form of surgery). Pasteurization of milk, and chlorine inclusion in water supplies, were breakthroughs in diminishing health risks, although this is hardly acknowledged in recent years.

At times the protection we've been afforded is taken for granted. Parents who hadn't witnessed the consequences of childhood measles may be more likely to choose not to have their children vaccinated against multiple diseases, and despite the frequent warnings, people frequently engage in self-destructive behaviors. The cavalier attitude toward antibiotic use and the absence of proper stewardship have also resulted in various bacteria developing resistance to these treatments, placing us at elevated risks of illnesses.

The fact is that it's a nasty world out there, and threats to our well-being are often ignored even when it has been abundantly clear that kicking the can down the road is not a viable way of dealing with global issues. Overpopulation has become unmanageable, natural resources are becoming depleted, ecosystems are collapsing, climate change has been relentlessly moving along, pollutants have become more prominent, food and water security have become endangered, and as we've seen, viral and bacterial threats are ever-present. These challenges aren't independent of one another. Eliminating the natural habitats of various species has increased the odds of viruses jumping from animals to humans, and inappropriate housing and breeding of poultry have similarly increased the odds of avian flu that can be transmitted to humans, which can with a few mutations be transmitted between humans. Beyond these interrelations, overpopulation has stretched natural resources and contributed to climate change, and alterations of ecosystems and the presence of pollutants affect how our immune system behaves. Global issues need to be dealt with at a societal level, but individual efforts are needed for this to occur.

The capacity of the immune system to deal with viral and bacterial challenges developed through eons of natural selection. But to what extent is the immune system capable of dealing with the multitude of novel threats that are being encountered? Immune system functioning is affected by numerous intrinsic factors (e.g., hormonal influences) that are readily affected by diverse environmental and experiential factors that also evolved through natural selection. With the rapid changes that have been experienced during the past century, human biological systems might simply not be prepared to tolerate or deal with the hazards that currently besiege us. Fortunately, with the understanding of how the immune system operates, novel approaches to deal with diseases followed, often attaining remarkable success. In this chapter, we will look into various components of the immune system and how they interact with one another in preventing and ameliorating certain diseases.

Fundamentals of immunity

Chapter 1 discussed how immune processes play a fundamental role in the prevention and amelioration of diseases, exerting a prominent influence during varied phases of the cancer process. Now we need to outline more deeply how the immune system operates. Much of the available information regarding immune functioning has been gleaned from studies in animals, and as much as we might wish otherwise, it has become apparent that these data are not uniformly relevant or translatable to what occurs in humans. Regardless of species, however, mammalian immune responses involve a heavily intertwined dialogue between numerous cellular and molecular elements that disarm potential threats to the host either by foreign microbial entities or endogenous factors within the body.

Functions of the immune system

Blood consists of two major categories of cells: red blood cells (erythrocytes) and white blood cells (leukocytes). The leukocytes belong to the immune system, present in blood as cells *en passage*, circulating throughout the body in search of foreign invaders (which we will more commonly refer to as *antigens*) or en route to a location where an invader has been spotted, sequestered, and assailed. A variety of cells make up our immune system, and while each has its own functions, they act in an organized and coordinated manner to deal with foreign particles. Their reaction to antigens can vary with the specific tissue environment being examined. Importantly, the immune response is modifiable by hormonal and neural factors, as well as intrinsically by immune-derived messenger molecules (cytokines) that guide the activity of the leukocytes.

Red blood cells were typically thought to primarily serve to shuttle oxygen around the body, but there is good evidence indicating that these cells may also contribute to immune processes. Specifically, red blood cells contain molecular sensors that signal the immune system of tissue damage and of the presence of foreign DNA. Unfortunately, when red cells encounter foreign DNA, it may result in these cells hiding a "don't eat me" message that ordinarily prevents their phagocytosis by macrophages (Anderson et al., 2018). As a result, in the presence of damaged red blood cells or infection, as in the case of COVID-19 patients, sepsis is more likely to occur, and if too many red blood cells are eaten by macrophages, anemia may develop (Lam et al., 2021).

Before microbial antigens engage immune cells, they must first breach physical barriers such as the skin and internal mucosal layers (such as the upper respiratory tract, the gastrointestinal system, and the anogenital areas). The skin (or derma) represents the largest defensive organ of the body and engages in frequent cycles of shedding and regeneration, thereby limiting prolonged microbial colonization. However, cuts and bruises are common, and so penetration of the skin by microbes can occur through wounds. At this point, subcutaneous or intradermal chemical and immune mechanisms "kick in" to contain the infectious microbe, stalling its growth and potential entry into the microvascular system of cutaneous tissue and ultimately the general circulation. Similar mobilization of defensive processes of containment readily occurs in other parts of the body that serve as portals of entry for infectious agents (mouth, nose, ears, eyes, genital surfaces).

Ordinarily, surveillance for the presence of foreign particles occurs by immune cells that are transported through a broad network of arteries and capillaries. Once discovered, leukocytes alert other immune cells of the danger, recruiting them to the site of invasion (i.e., infection) to aid in fending off the infectious agent. For example, should tissue damage occur (e.g., through a wound), particular types of signaling molecules (chemokines) inform and attract leukocytes circulating in the blood. These immune cells leave the blood (extravasate) and engage pathogens that may be present in damaged or infected tissue. In so doing, inflammation is produced that facilitates immune activity, while concurrently acting against pathogens. Once the pathogens are eliminated, components of the immune system facilitate the repair of damaged tissue, thereby limiting the risk of further infection.

Before delving further into the intricacies of immune function, it is important to underscore that the capacity of the immune system may be dictated by many factors to contend with microbes that threaten well-being. Specifically, when immune cells, such as T lymphocytes, encounter viruses or bacteria, they turn on a switch that allows them to detect oxygen (so-called oxygen sensors) and nutrient supply, as well as one that facilitates the transport of nutrients from the local environment into the immune cell. As such, the microenvironment comprising blood vessels, resident or recruited immune cells, and various signaling molecules has considerable sway in how effectively they deal with challenges (Howden et al., 2019). These diverse aspects of the immune response are affected by diet-related hormonal changes and by microbial variations within the gut. As well, lifestyle factors beyond diet, such as exercise and sleep, affect immune functioning and might have prophylactic actions in preventing the development of some forms of cancer and certainly with cancer progression (Harvie et al., 2015).

Good for the moment, but not necessarily for tomorrow

Over the course of evolution, humans developed protective mechanisms to prevent or mitigate a wide range of diseases. Once, when humans lived for three or four decades, the negative effects of chronic inflammation stemming from immune activation might not have become apparent and selection pressures against their occurrence would not have been notable. Now, with average age nearing eight decades, the nasty effects of chronic inflammation have become apparent. We are the unfortunate inheritors of immune functioning going awry, but the evolution of these systems is still ongoing. It's a reasonable bet that lifestyles and related factors will contribute to how the immune system ultimately adapts (Domínguez-Andrés and Netea, 2019), although with increased migration from rural to urban regions, as well as between countries, these adaptations may not always be advantageous.

Evolution may have affected our ability to deal with invaders in still other ways. Ordinarily, pathogens such as typhoid fever, cholera, mumps, whooping cough, and gonorrhea enter cells by manipulating sugar molecules (sialic acids) present on virtually every cell within the human body. With evolution, genetic changes may have occurred that enhanced defenses against pathogens by exploiting sialic acids. It is thought that about 2 million years ago, a mutation occurred so human cells were no longer coated with a sialic acid *N*-glycolylneuraminic acid (Neu5Gc) that is common in apes and other mammals. Instead, a small change appeared so human cells were coated with *N*-acetylneuraminic acid (Neu5Ac). This slight change might not appear to mean that much, but over the ensuing centuries, a variety of pathogens adapted to capitalize on Neu5Ac expression, latching on to it as an entry point into the cell. As a result, humans became vulnerable to diseases that don't appear in apes and other mammals whose cells express Neu5Gc on their surface.

The involvement of sugars might also have played an evolutionary role in the preeminence of humans over other species. The hominin family of prehistoric mammals, from which present-day humans emerged, included primarily Neanderthals, as well as a distinct cousin, the Denisovans. Analysis of DNA from 6 Neanderthals, 2 Denisovans, and 1000 humans revealed evolutionary changes linked to particular types of proteins, specifically, sialic acid-binding immunoglobulin-type lectins, or Siglecs (Khan et al., 2020). These Siglecs are not present in apes, but they were found in most Neanderthal, Denisovan, and human samples. It was suggested that the evolutionary change related to Siglecs occurred before our lineages diverged about 600,000 years ago, but well after the sialic acid mutation arose more than 2 million years ago, possibly in ancestors of humans and Neanderthals. Siglecs are located on human immune cells and recognize sialic acids that are damaged. This causes immune cell activation and an inflammatory response to eliminate damaged cells or invaders that fail to carry correct sialic acids. At some point during evolution, a mutation occurred within the SIGLEC12 gene so its protein SIGLEC-12 undermined the ability of the immune system to differentiate between the self and attacking microbes. This mutation is still present in a minority of individuals so the elevated levels of the SIGLEC-12 protein doubles the risk of some forms of cancer (Siddiqui et al., 2021).

As much as Siglecs may in some respects be beneficial, including protecting humans from pathogens, genetic changes in Siglecs have been associated with inflammation, autoimmune disorders (e.g., asthma), and meningitis. It seems that Siglecs may be useful insofar as they are continuously alert for invaders, but they may also favor immune reactions against self-tissue. Parenthetically, it was suggested that Siglecs may in some manner contribute to the excessive inflammation (cytokine storm) associated with death in COVID-19 patients. There are many other instances in which genetic changes have evolved to protect humans from particular diseases but may have unintended negative effects (sickle cell anemia being the classic example of this). Natural selection may have been effective in the development of phenotypes that were beneficial at a given time or place but may have been counterproductive in other environments or at other times.

Innate and adaptive components of the immune system

The classical notion of the immune system recognized two major arms of immune responsiveness, referred to as the *innate* and *adaptive* (acquired) immune components. This segregation has persisted for convenience of study and a recognition that each component plays an important role concerning particular pathogens, as well as the stepwise or phasic manner by which an immune response to most microbes evolves. The innate immune system—like the adaptive immune system— has a rudimentary ability to recognize a broad range of foreign matter—usually cellular in nature, as exemplified by bacteria and other parasites. The designation, "innate," arose based on observations that prior exposure to particular microbes was not necessary to solicit a response from certain immune cells. As we will discuss, this natural capacity to pounce on any invasive pathogen—even one never previously encountered—is based on a specialized recognition system involving molecules on the surface of innate immune cells and those expressed on microbial cells. By virtue of this seeming indiscriminate and broad range of surveillance, the innate immune system is often regarded as our first line of immune defense. As this innate response progresses, the adaptive immune system is recruited to partake in a more refined and selective range of responses that target specific proteins that are integral to the microbial pathogen. It is the adaptive immune response that will ultimately register the lethal insults against the pathogen. This will be in the form of antibodies and cytotoxic T cells that selectively target with considerable precision the viruses and bacteria that represent the major threats to our health. Throughout, there will be a careful calibration of the immune response to ensure that damage to self-tissue is minimized, and most importantly, that the immune response is not directed mistakenly against the cells and tissues of the host. This self/nonself discrimination is learned in utero so that immune cells know "what's me," and by extension, anything else that is "not me." To sum up, although the segregation of the immune system into these two different components is frequently used to explain the efficient workings of the immune system, this differentiation is, in a sense, artificial as they work together to eliminate unwanted microbial invaders. Fig. 2.1 shows some of

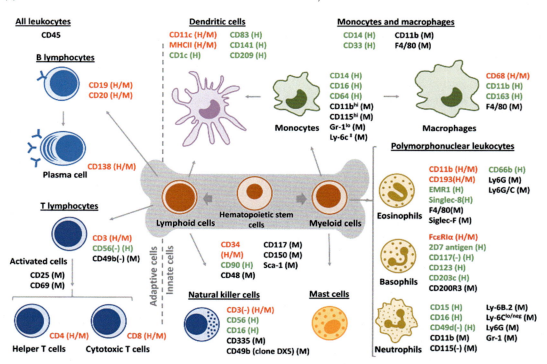

FIG. 2.1 Leukocytes originate from hematopoietic stem cells in the bone marrow. Stem cells give rise to lymphoid and myeloid cells, which are common progenitors of leukocytes and other blood cells. Monocytes (the precursor of macrophages), mast cells, and granulocytes (basophils, neutrophils, and eosinophils) originate from myeloid cells. Lymphoid cells produce B- and T-lymphocytes and natural killer cells. Dendritic cells can originate from monocytes or lymphoid precursors. All leukocytes exhibit the CD45 protein on their surface. Common receptors for each cell type in mice (M) and humans (H) are presented in *red* (both M/H), *green* (only H), or *black* (only M). According to their main function in the immune system, leukocytes can be subdivided into adaptive cells (left) or innate cells (right). *Figure and caption from Viana, I.M.O., Roussel, S., Defrêne, J., Lima, E.M., Barabé, F., Bertrand, N., 2021. Innate and adaptive immune responses toward nanomedicines. Acta Pharm. Sin. B 11, 852–870.*

the many innate and adaptive immune cells, the source of these cells, and the markers that are used to identify them.

Innate immunity

Innate immunity is a fundamental defense mechanism found in virtually all organisms from simple prokaryotic invertebrates to the most complex mammals. This component of the human immune system is less concerned with specific antigens as much as it is with general molecular signatures typically found on invaders, such as bacteria, viruses, or fungi. Of the many cells of the immune system, the two largest classes are *myeloid* and *lymphoid* cells. These cell groups emerge from stem cells in the bone marrow, a discussion we will defer until later (but see Fig. 2.1). The myeloid cells comprise granulocytes (primarily neutrophils), monocytes, and macrophages and are fundamental in mediating an inflammatory response. Cells of the lymphoid lineage are T and B lymphocytes (often referred to as T and B cells), as well as large granular lymphocytes (LGLs), which are commonly known as natural killer (NK) cells. As alluded to earlier, the innate immune cells operate in a fairly nonspecific manner using defensive strategies such as phagocytosis (literally, "eat up") or respiratory bursts (release of toxic free radicals) to attack invading microbes. Importantly, they liberate humoral mediators of immunity, proteins that include cytokines, lysozymes, and complement proteins. The release of cytokines is a common characteristic of all immune cells, whether innate or adaptive. These molecules exert a range of actions, including the functioning and proliferation of cells of the adaptive immune system, signaling circulating immune cells to redirect their movement to sites of infection, and enhance the cytotoxic (cell killing) abilities of innate (e.g., macrophages) and adaptive (e.g., LGLs and cytotoxic T cells) immune cells.

Tissue injury is a common stimulus for the recruitment of innate immune cells. Injury can be due to trauma, strain, or inherent metabolic or chemical deficiencies. In addition, tissue can be damaged by infection, in which case the role of the innate immune system is to react against the infectious agent and the damaged host cells. In such a scenario, restoration of healthy tissue status will involve removal of damaged or dead cells, eliminating the infection and establishing local conditions for rebuilding the tissue. Whatever the cause of tissue injury, circulating monocytes and neutrophils will infiltrate the site, while resident macrophages (those already present in the area of damage) will similarly be activated. These events will cause increased flow of plasma from the blood through the increased gaps between the cells that line the interior surface of blood vessels (i.e., endothelial cells). While this will expedite the movement of cells and blood-borne molecules, it will also result in edema (increased interstitial fluid volume), and hence swelling of the affected area. On cutaneous surfaces, such changes are characterized by redness and swelling, which is a reflection of the elevated leukocyte migration to the site of infection or damage. As such, it is a classic index of innate immune reactivity.

Inflammation is essentially a biologically useful phenomenon. As noted earlier, tissue repair and removal of infectious agents require the involvement of innate immune cells. And as we will learn, infiltrating monocytes and macrophages serve to pass on important molecular information to lymphocytes (the adaptive immune arm) that will expand the range of immune strategies for eliminating and protecting against microbial pathogens. However, as with most biological systems, when inflammation becomes excessive and long-lasting, the outcomes may be decidedly negative. For example, when allowed to proceed without a quick resolution—as when a wound fails to heal or when inflammatory events are occurring within critical organs like the pancreas, liver, or brain—pathology may result in the organs and/or tissues where inflammation occurred. As well, when inflammation occurs on a chronic basis, heart diseases, irritable bowel, or lung problems, as well as many other disturbances can be promoted. Importantly, the presence of circulating inflammatory factors can also signal subsequent downstream effects related to multiple diseases, even those of a psychological nature, such as anxiety and depression (Anisman et al., 2018).

A major cell that figures in most discussions of innate function is the macrophage. These cells develop from monocytes, which are myeloid cells that circulate in the blood. However, once circulating monocytes situate themselves in a particular tissue, such as the liver, they may continue to reside there undergoing morphological changes that convert them into what are referred to as "resident tissue macrophages." This sequestration and *differentiation* (a term used to denote a change in function and form) of the monocyte into a macrophage can occur anywhere in the body. As such, macrophages are ubiquitous, being present as resident tissue macrophages in most organs of the body, with some exceptions. Common organs that contain macrophages are the liver, lung, intestines, and spleen, along with the peritoneum (the cavity in which our intestines, liver, and other abdominal organs rest), where they are simply called peritoneal macrophages. From an immunological standpoint, macrophages comprise specialized cells that operate to detect and phagocytose (engulf and degrade) harmful microbes and infected host cells. On a more general level, they also serve housekeeping functions, "cleaning up" (through phagocytosis) necrotic tissue—host cells that have died as a result of injury, mysterious degenerative processes, or through programmed cell death (*apoptosis*). But macrophage function is not confined to just this first level sweep of unwanted material. Upon phagocytosis of foreign microbes, they present molecular components of the microbes to T cells, thereby expanding the scope of the immune response to include adaptive cellular mechanisms, of which we will speak in more detail. At the same time, macrophages are also fundamental in triggering inflammation through the release of signaling molecules (cytokines) that recruit other immune cells to enter the fray and engage the microbial intruders. In essence, macrophages are a key first line of immunological defense that not only offers offensive lethal actions against infectious agents, but also arms the remainder of the immune system for a more drawn-out but focused campaign that trounces them completely.

Innate immune function and cancer

Diversity of immune cells

On its face, the benefits of macrophages seem straightforward. However, the situation is actually far more complex, especially concerning cancer progression. Like many other cells, macrophages are malleable, their program of gene expression being subject to alteration over the course of tumor progression. Ordinarily, they display a phenotype that facilitates antitumor functions through their capacity to phagocytize foreign particles or release oxygen free radicals to kill tumor cells (typical of a macrophage class called the M1 phenotype). However, as we saw in Chapter 1, later in tumor development they can convert to a less tumoricidal M2 phenotype (Wang et al., 2019) and release factors, such as granulocyte-macrophage colony-stimulating factor (GM-CSF), which can promote tumor survival, possibly by inhibiting activation of many types of immune cells that normally attack pro-tumor cells (Orecchioni et al., 2019). Thus, the presence of macrophages can be seen as a double-edged sword, acting in a dual role as both protector and potentially toxic instigator of cancer progression. As addressed shortly, many other immune cells can have positive protective actions under most circumstances, but when dealing with cancer cells, their functioning may not only be compromised but may actually enhance disease progression. Indeed, cancer cells are as successful as they are owing to their frightening capacity to evade processes that ought to destroy them and in coopting various types of cells to ensure their survival.

Neutrophils, like macrophages, act in a protective capacity being among the first cells that migrate to sites of injury where they act to phagocytose microorganisms or release granules containing cytotoxic chemicals that serve to fight infection. They can also facilitate antitumor functions by altering transforming growth factor-β (TGF-β) signaling. Oddly, this cytokine can act as a tumor suppressor by inhibiting cell growth and division as well as apoptosis but can also act as a tumor promoter by influencing the local microenvironment to make it more conducive to tumor growth (Colak and Ten Dijke, 2017). As well, they can induce genetic instability through the release of destructive reactive oxygen radicals, thereby promoting tumor creation. These reactive oxygen species (ROS) comprise unstable oxygen-containing molecules that react readily with other molecules within a cell. While ROS is essential in the modulation of cell proliferation and programmed cell death, and is necessary for proper immune and cardiovascular functioning, this is possible only if ROS levels are maintained within circumscribed limits: not too high, not too low (the "Goldilocks principle").[a]

Most innate cells generate ROS. Prominent infiltrating innate inflammatory cells, neutrophils, macrophages, and eosinophils produce peroxide and superoxide, as well as nitric oxide, which can combine to produce the highly damaging radical, peroxynitrite, which normally acts to kill invading pathogens. Yet these factors can also induce DNA mutations, and oxidative stress-induced changes in genomic stability can fuel the transformation of healthy cells into tumors. Thus, while normal inflammatory cellular responses may be advantageous, chronic inflammation may give rise to cumulative damage wrought by oxygen-free radicals that were not neutralized by antioxidants. In fact, oxidative stress has been linked to aging processes, and aging-related diseases, such as cancer. For this reason, control of oxidative stress has become big business, with many foods and over-the-counter pharmacy labels exhorting the antioxidant value (or not) of everyday products. This topic is addressed later in the book.

In addition to myeloid cells like macrophages and neutrophils, NK cells are an essential element of the innate immune system that can attack cancer cells and viruses. Because of these abilities, NK cells have become important in the development of novel strategies to deal with several forms of cancer. Numerous factors can direct the actions of NK cells, including cytokines, such as interferon-α (IFN-α) and IL-15, with the latter serving to keep NK cells active and alive. Although NK cells are typically beneficial in dealing with certain cancers, such as those involving the liver and viral-associated cancers, they can also be deleterious because these same defenses are associated with tissue damage that can create a microenvironment that facilitates tumor growth (Abel et al., 2018). As well, certain NK cell subtypes that are associated with tumors can promote TGF-β release, which can suppress antitumor responses through its actions on cell proliferation and differentiation.

Pattern recognition receptors and pathogen-associated membrane patterns

An obvious question is how neutrophils and other granulocytes recognize foreign particles, such as bacterial cells, and know enough to eliminate them. Likewise, how are innate immune cells able to recognize dead (necrotic) tissue and again know the importance of removing them? Part of the answer, as already suggested, is that during prenatal development

---

[a] It is fairly common in the biological sciences to find that at a low dose a particular treatment may have only a modest beneficial effect, increasing with a somewhat higher dose, and then declining with doses that are still higher, even producing negative actions (an inverted U-shaped function). This not only occurs with pharmacological agents, but is also seen with lifestyle manipulations, such as exercise, dietary energy restrictions, and is apparent in relation to sleep durations. Pharmacologists and toxicologists refer to this phenomenon as "hormesis," but in other disciplines the "Goldilocks principle" is used to describe an environment that is "just right."

and the early postnatal period, the innate immune system experiences our cellular features and consequently should not attack these aspects of the self, but ought to act against those that had not previously been encountered. The apparatus necessary for this involves a variety of specialized detection receptors present on the surface or within immune cells and dubbed "pattern recognition receptors" (PRRs). These can recognize molecules that are broadly shared by pathogens, which collectively are referred to as "pathogen-associated molecular patterns" or "pathogen-associated membrane patterns" (PAMPs) (Riera Romo et al., 2016). Although innate immune cells evolved so PRRs could enable the destruction of foreign organisms bearing PAMPs, pernicious microbes have continued to coevolve clever mechanisms to either hide these PAMPs or alternatively to "turn off" the attacking immune cells. After all, on each side of the host:microbe divide, there is a struggle to survive, so each side acquired tricks as a function of natural selection: the host evolving ways to recognize microbes, while the latter evolved ways to evade potential assailants.

Substantial attention has been devoted to the function of toll-like receptors (TLRs), which are a critical type of PRR found on numerous immune cells. Toll-like receptors, which are the principal recognition receptors for PAMPs, comprise more than 10 members that fall into different groups, each of which responds to different exogenous stimuli. Extracellular TLRs (notably TLR1, 2, 4, 5, 6, and 10) recognize membrane components of microorganisms (i.e., those on the surface of microbes), whereas intracellular TLRs (TLR3, 7, 8, and 9) recognize microbial nucleic acids (DNA and RNA). TLRs are fundamental in the recognition and discrimination between self and nonself and are not only pivotal in the development of autoimmune disorders but may play a role in inflammatory diseases and cancer. Specifically, TLRs can promote effective immune responses that act against the tumor. Moreover, when TLRs on macrophages are activated, these highly adaptive immune cells quickly alter their metabolism and consume more glucose, presumably to facilitate actions against the challenges encountered (Lauterbach et al., 2019). Targeting TLRs has become a focus of research in cancer treatments, as they appear to enhance the effects of radiation and chemotherapeutic approaches. But once more we are warned of extremes, should TLR activation be excessive, the resulting inflammatory response or inhibition of immune cells that could suppress cancer progression can actually augment tumor cell proliferation (Cen et al., 2018). Calibrating how best to utilize our natural defenses against cancer remains a challenge.

Of the many TLRs expressed by immune cells, particular interest has been devoted to TLR3, which generally recognizes viruses, and TLR4, which recognizes signature bacterial molecules. TLR4 has also been implicated in cancer provocation by its ability to produce a chronic low-grade inflammatory state (see Fig. 2.2). In addition to its actions in the periphery, TLR4 is also highly expressed by microglial cells, specialized immunocompetent cells, which mediate the majority of innate immune responses in the central nervous system (CNS). As much as TLR4 activation of microglia helps deal with any wayward microbes that have found their way to the CNS, if sufficiently activated, damage to healthy cells may be induced (Anisman et al., 2018).

Sterile inflammation

For many decades the focus had been on how the immune system recognized and attacked foreign particles that could endanger animals and humans. However, even in the absence of microbes, inflammation can be engendered by noninfectious agents or events, such as tissue

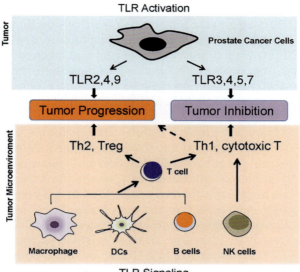

FIG. 2.2 Toll-like receptors and prostate cancer. TLR activation in tumor cells and its activation in the tumor microenvironment, such as in typical innate immune cells, lead to a complex scenario, which determines the role of TLRs in prostate cancer development. The activation of TLRs in antigen-presenting cells, such as DCs, macrophages, and B cells, can lead to either T_h1 and T cytotoxic responses or T_h2 and T_{reg} responses. The activation of TLR2, 4, and 9 in prostate cancer cells appear to promote tumor growth, but the activation of TLR3, 4, 5, and 7 might inhibit prostate cancer. *Figure and caption from Zhao, S., Zhang, Y., Zhang, Q., Wang, F., Zhang, D., 2014. Toll-x. Front. Immunol. 5, 352.*

damage, ischemia, toxicants, and even psychological stressors. This *sterile inflammation* has been implicated in the emergence of inflammatory-related diseases and could potentially influence tumor growth or spread. Aside from peripheral actions, inflammatory factors are also provoked within the brain through the release of cytokines and associated factors by glial cells, including astrocytes and microglia. Peripheral and central cytokine release is presumed to be adaptive (e.g., in facilitating healing), but as already indicated, if excessive, inflammatory factors may be neurodestructive and may promote psychological disturbances.

Danger-associated molecular patterns (DAMPs)

Sterile inflammation can be initiated by danger-associated molecular patterns (DAMPS) that are released by distressed cells, causing activation of PRRs located on immune cells, in a manner not unlike that induced by PAMPs on microbial cells. DAMPs can affect different neutrophils, monocytes, macrophages, dendritic cells, eosinophils, mast cells, NK cells, as well as T and B cells, which promote the release of proinflammatory factors. The macrophages, dendritic cells, and neutrophils present DAMP-derived peptides to T cells, and the NK cells and eosinophils activation produced by DAMPs have cytotoxic actions that can eliminate damaged cells or tumor cells. Several types of nonimmune cells, such as fibroblasts, epithelial cells, endothelial cells, can be activated by DAMPs, which may contribute to disease processes owing to chronic activation of inflammatory processes .

Danger may also be detected in the presence of necrotic cells, which in dying, ruptured their membrane and liberated intracellular molecules that function as DAMPs. These include heat shock proteins that can be promoted by cellular stress, molecules secreted by cells of the extracellular matrix that provide structural and biochemical support to the surrounding cells, and by fragments of enzymatically degraded proteins that are normally required for tissue repair. These *matrix metalloproteinases* are highly expressed within cancer cells, and it has been thought that their overproduction may promote further tumor development.

Additionally, the body's own primary cellular energy source, adenosine triphosphate (ATP), can act as a DAMP when released into the extracellular fluid in response to cell necrosis or strong cellular distress. Finally, DAMP secretion and subsequent sterile inflammation can also be induced by *pannexins*, glycoproteins (proteins containing lots of carbohydrate groups) which form transmembrane channels embedded in cell membranes to facilitate the passage of small molecules, such as ATP, in and out of cells.

Normally, ATP can signal through the purinergic system, which consists of receptors that respond to nucleotides, such as adenosine, or close relatives of nucleotides. Purinergic receptors such as P2X or P2Y are responsive to ATP and are found on immune cells. However, there are many more purinergic receptor subtypes that are present on most mammalian cells, and analogues have been identified in plants. In the nervous system, they serve as a fundamental communication mechanism between neurons and glial cells under resting conditions, but may be especially important in conveying relevant messages between these cells during times of cellular stress. For example, ATP released during cellular stress builds up in the extracellular environment and functionally becomes a DAMP that is detected by and activates microglia; this may then lead, for better or worse, to inflammatory changes in the brain. Therefore, increased DAMP levels, whether in the brain or the periphery, can activate processes that favor a variety of pathologies, including cancer.

Another nonmicrobial endogenous DAMP signal is the high-mobility group box 1 (HMGB1) protein. This has a multiplicity of regulatory roles in different cellular compartments (nucleus and cytoplasm), but can readily be secreted into the extracellular environment, an event that converts it into a DAMP. The extracellular accumulation of HMGB1 is recognized by PRRs on innate immune cells both in the periphery and in the brain by glial cells (Patel, 2018). As shown in Fig. 2.3, HMGB1 can be increased in response to tissue injury, which can result in processes that contribute to tissue repair. However, through actions on stem cells, HMGB1 may promote proinflammatory signaling and has been implicated in the development of hematopoietic malignancies, such as leukemia and multiple myeloma, and may contribute to chemotherapy resistance to these cancers (Yuan et al., 2020). As well, high levels of HMGB1 within the brain can provoke neuronal changes associated with neurological and psychiatric illnesses (Ratajczak et al., 2018), supporting the view that sterile inflammation may be a way by which stressors could contribute to such disorders. Indeed, elevated DAMPs, granulocytes, and monocytes were observed in depressed patients, as well as in those with neurodegenerative disorders, such as Alzheimer's and Parkinson's disease (Venegas and Heneka, 2017).

To briefly summarize the discussion so far, although immunity has most often been considered with respect to its actions against invading microbes, sterile inflammation triggered by physical, chemical, and metabolic stimuli, including hypoxic and nutrient stress, can have

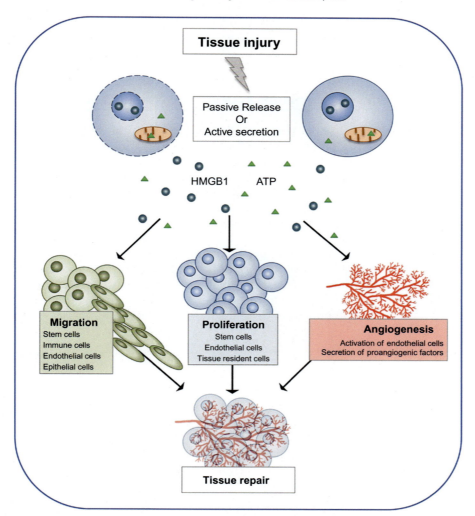

FIG. 2.3 HMGB1 and ATP in tissue repair. Following tissue injury, HMGB1 and ATP are passively released by dead cells or actively secreted by stressed cells. Then they recruit to the site of damage of the cell types required to heal the wound. First, immune cells are needed to clean the wound by engulfing dead cells and cellular debris. Then the stem cells and neighboring cells are induced to proliferate and build new tissue, together with its extracellular matrix. Endothelial cells are activated to form new blood vessels. *Figure and caption from Vénéreau, E., Ceriotti, C., Bianchi, M.E., 2015. DAMPs from cell death to new life. Front. Immunol. 6, 422.*

significant effects on well-being. As we have seen, chronic inflammation can engender multiple disease conditions related to immunity (cancer, autoimmune disorders), influence the course of viral illnesses, and affects metabolic disorders, type 2 diabetes, and heart disease (Rubartelli et al., 2013). Of particular relevance to the prevention of disease occurrence, engaging in exercise can have protective effects by the inhibition of sterile inflammation (Peeri and Amiri, 2015) as can particular diets and energy substrates (Ralston et al., 2017). It likewise appears that psychological stressors can not only activate the sterile inflammatory response potentially affecting disease occurrence, but may point to novel therapeutic targets (Fleshner et al., 2017).

Anatomy of the immune system: The lymphoid system

The bone marrow and thymus comprise the *primary* immune organs in which early stages of immune cell development occur. After they are fully differentiated and functional maturity has been reached, immune cells leave these sites and circulate throughout the body (called *cell trafficking*) monitoring the internal environment for potential invaders. Cell trafficking utilizes two major networks of transport: the cardiovascular and lymphatic vessels. When not in transit through the blood or lymphatics, many immune cells localize in secondary immune organs, such as the spleen, lymph nodes, and specialized areas in the small intestines. Immune reactions are needed where pathogens are likely to be most evident. This includes the upper respiratory tract, gut, and anogenital areas. As such, while lymph nodes are the predominant repository of lymphocytes and other immune cells, the cells are not sedentary, as they circulate from the lymph node to other parts of the body to conduct surveillance, and then return to the node via the blood and lymphatics. The lymphatic system and the lymph nodes are depicted in Fig. 2.4. When lymphocytes return to a

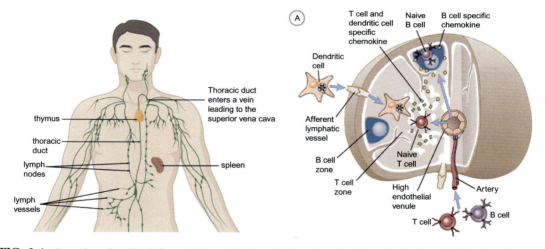

FIG. 2.4 Lymph nodes. The left panel shows the lymphatic system draining via the thoracic duct into the vena cava and into the general blood circulation. Lymphatics drain into and away from lymph nodes, which are shown distributed along the abdominal and upper regions in the cervical region. The right panel shows a typical lymph node. Note the afferent lymphatic vessel to the left through which lymphocytes and other cells can enter the node. Within the lymph node, B and T cells are segregated into distinct locations. Lymphocytes in blood can exit from arteries by reaching a high endothelial venule (HEV), which allows for easy extravasation, and then migrate to different areas of the node (e.g., note the B cell zones) as a result of attractive chemicals called chemokines, which are produced in different areas of the node. Note also the migration of dendritic cells (DCs) through the afferent lymph, which captures antigens that enter via the lymphatics. Once the DCs enter the node, they move to the T cell-rich areas of the node, where they present the antigen and induce T cell activation. *Left panel created using BioRender.com. The right panel is from Abbas, A.K., Lichtman, A.H., Pillai, S., 2015. Cellular and Molecular Immunology, eighth ed. Elsevier, Philadelphia, PA.*

node from a region where a pathogen was encountered, the cells can divide and multiply, while also activating other cells in the node to do the same. The node will rapidly swell with an expanded army of cells that can emigrate from the node to traffic back to the site of infection so the pathogen can be destroyed. This is why during upper respiratory tract infections, for instance, lymph nodes around the neck are easily palpable; they are bursting with cells that were activated and induced to multiply as a result of both encountering the pathogen in the throat (say, a cold virus or streptococcal bacteria) and draining of the antigen (the pathogen) to the nearest local lymph nodes. It is evident from this example that lymph nodes are strategically located, maximizing the speed at which an adaptive response can be generated against an infectious agent that enters via one of the more vulnerable regions of the body.

The entrance of cells into and out of a node is via *lymph*, which is essentially a mix of fluid and lymphocytes which fills the entire network of lymphatic vessels. These vessels will drain into a lymph node (called the *afferent* lymph), and a different section of the lymph node will have a lymphatic vessel that provides cells with egress from the node (called the *efferent* lymph). Of particular importance in understanding lymph nodes is that they are regions where information is gathered from the surrounding area to be passed on to immune cells residing in the node. This transmission of information, which comprises the nature of the antigens present (strictly speaking, the molecular makeup of a pathogen), is primarily conducted by dendritic cells (DCs), the major antigen-presenting cells (APCs) of the immune system. These cells can phagocytose microbes, digest them, and express peptide components of major microbial proteins on their surface. As we noted earlier, macrophages can perform a similar function. However, DCs appear to exist strictly for this purpose and can even receive the molecular fragments of a phagocytosed microbe from macrophages.

A closer look at lymph nodes reveals that they contain two types of lymphocytes: the T and B cells, which have distinct functions. Most importantly, these cells will focus their attention specifically and selectively on the various molecular components of the infectious agent that led to their activation. Only a subset of T and B cells will be activated, leaving many lymphocytes in a dormant state to await their recruitment at a later date, when their designated pathogen is encountered.

We have talked about lymph nodes and emphasized why they are important as an anatomical feature of the immune system. But we should also mention the spleen. The spleen is a major lymphoid organ that operates differently from that of lymph nodes. It has no connection to the lymphatic system, serving as a site for the deposition of lymphocytes circulating in the blood. In addition to cells, the spleen traps blood-borne antigens and microbial cells. Like lymph nodes, the spleen contains a full complement of macrophages, dendritic cells, and T and B lymphocytes. This stands in contrast to the liver which has similar filtration functions—being heavily vascularized—but contains mainly cells that belong to the innate component of immunity.

Adaptive (acquired) immunity

The second branch of immunity, the adaptive or acquired immune system, entails highly specific recognition of foreign entities and the formation of memory for antigens that they express. Execution of adaptive immunity is mediated by lymphocytes but requires the assistance of various accessory cells (such as DCs) and cells of the innate arm of the immune system. Immune memory ensures that the molecular features of encountered intruders are recalled easily and acted against quickly should they be met once more. Another feature of adaptive immunity includes mechanisms that enable immune cells to refrain from attacking self-tissues. This may be thought of as self-memory, something that is learned during the process of lymphocyte development and maturation. However, the frequent occurrence of diverse autoimmune disorders attests to the limitations of this self/nonself discriminative process.

As mentioned, lymphocytes (T and B cells) serve to implement adaptive immunity. The primary hallmarks of adaptive T and B lymphocyte responses comprise several features: (a) *Specificity*: any given B or T cell can respond to only a single antigen, although B and T cells can both respond to the same antigens. (b) *Diversity*: a foreign particle, especially a large protein, has multiple antigenic markers on its surface, and thus the enormous diversity of immune cells allows a single foreign molecule to be attacked at different target sites. (c) *Memory*: when T and B cells are prompted to multiply, a subset of their clones will become dormant and maintain a memory of the antigen. In future, these clones can respond to this antigen more rapidly and robustly, hence destroying it before a disease can develop. Parenthetically, although innate immune cells, including macrophages, neutrophils, and NK cells, are usually not considered to have "memory" abilities, this might not be fully accurate. A growing body of evidence has indicated that NK cells may have a short-lived "trained memory" that may occur through changes in the expression and function of genes but without alterations of the actual gene sequence (i.e., epigenetic modifications) of innate immune cells. As a result, these immune cells may exhibit a stronger and more efficient response to the same (or very similar) challenges (Nikzad et al., 2019). (d) *Self-limitation*: Just as memory cells should only attack antigens that signify a previously encountered foreign particle, they will not attack other molecules. Concurrently, regulatory mechanisms (notably the presence of regulatory T cells or T_{reg} cells) preclude cytotoxic T cells from attacking the host. (e) *Discrimination between self and nonself*: As described earlier, after developing into specific types of T cells, such as T helper or T cytotoxic cells, concurrent mechanisms are acquired that prevent aggressive or destructive reactivity to self-tissue molecules. This does not mean that cells do not recognize or process self, as this is needed to turn off any tendency that might be present to damage self-tissue. This ability is part of the T cell education that occurs in the thymus. This organ is a "primary" lymphoid organ since differentiation is ongoing and cells still need to acquire distinct functional states, including self/nonself discrimination which prevents them from turning against the organs and tissues they are meant to protect. Once this is achieved they are considered mature and leave the thymus to take up residence in secondary immune organs, such as the lymph nodes and spleen. (f) *Proliferation and elimination*: After an adaptive immune response is *induced* (known as the induction phase) and the intracellular biochemical machinery of the cell is activated, the *effector* arm of the immune response is engaged, which amounts to actions being promoted that eliminate the microbial antigens. This includes the production of antibodies and various cell-signaling molecules, such as cytokines, that facilitate the proliferation and differentiation of surrounding lymphocytes that bear the same antigenic specificity. As a result, the amplification of the number and range of lymphocyte functions provides a quantitative advantage over microbial infiltration and colonization.

Development of B cells

The development of B lymphocytes initially occurs within the bone marrow, moving through several stages before full maturation and competence. Thereafter, they enter circulation and take up residence within secondary lymphoid organs. The surface of B cells is replete with immunoglobulin (Ig) molecules that serve as antigen receptors. Upon activation with antigen, B cells produce enormous amounts of immunoglobulins (now functionally called "antibodies") that can neutralize viruses and destroy bacteria. These activated B cells, which are referred to as plasma cells (or antibody forming cells: AFC), are fundamental to antigen-specific memory. When B cells are activated, their release of Ig molecules into the extracellular environment serves to bind the antigen that stimulated the B cell in the first place. As such, the Ig molecules function as soluble antigen-binding receptors and can recognize the antigen that induced them on foreign particles or cells.

Antibodies come in several molecular forms, which have a universal antigen-binding region, that fall into the following classes: IgG, IgA, IgD, IgE, and IgM. Some of these classes tend to predominate in particular parts of the body as well as in particular types of clinical situations. IgM, which constitutes the first class of antibodies to be produced when the antigen is initially encountered, primarily appears in the blood and, to a lesser extent, in the lymph. IgA is common in saliva and the gastrointestinal tract, whereas IgE is a common class of antibody class produced in response to allergens. Because B cells tend to fine-tune IgG, it is found throughout the body. With repeated exposure to an antigen or as the antibody response to antigen develops, antibodies of the IgG class gain greater specificity and strength of binding to the antigen. IgG is present in the blood and extracellular fluid and appears shortly after IgM molecules have been produced. In time, IgG production increases, while IgM production begins to wane. Were the antigen to return and invade the organism on a second occasion, the immune response would more readily involve IgG production, the IgM response being much less important than it was during the initial or primary encounter with the antigen.

The usefulness of the IgG molecule is that it can efficiently *opsonize* pathogens. This process involves the antigen being bound by the IgG antibody, which marks it for engulfment by phagocytes, such as macrophages. Simultaneously, IgG binding of antigen activates a series or system of liver-derived proteins called *complement*, which then "punches" holes in the bacterial cell expressing the antigen-IgG complex, resulting in its death and removal through phagocytosis. So the better IgG can recognize antigens (show specificity or affinity), and the stronger that it can bind (avidity), the more efficiently it can sequester or neutralize the antigen for accessory functions like phagocytosis and complement activation.

As noted, IgA is typically found in secretions, particularly those of the mucus epithelium of the intestinal and respiratory tracts. Therefore, it is often the major Ig class encountered within the so-called mucosal immune system that monitors the interface between the world and the gastrointestinal, respiratory, and genital regions of the body. As can be imagined, these are the most likely portals of entry for most pathogens, and therefore mounting a strong IgA response is of considerable importance. As for IgE, this is considerably lower in concentration in blood or extracellular fluid than IgM and IgG. When IgE is produced, it can bind to receptors on mast cells present just beneath the skin and mucosa, as well as along blood vessels of connective tissue. Typical inducers of IgE are allergens that stimulate B cells to produce the IgE, which can then bind to a receptor for IgE called FcεRI. In the presence of the allergen, the IgE-FcεRI complex causes degranulation of the mast cells, which essentially involves the expulsion of histamine and other chemicals found in the granules. This degranulation and liberation of histamine induces the symptoms commonly associated with hay fever and other allergic reactions. This includes tearing and excessive secretion of nasal discharge (runny nose), as well as sneezing, coughing, and vomiting. One might think of these reactions as attempts to expel infectious agents. In some instances, an IgE overreaction can be elicited so that the allergic reaction can be exceptionally marked, as in the case of individuals with allergies to peanut butter or bee stings, leading to an *anaphylactic* reaction, characterized by throat swelling and low blood pressure, and possibly death.

Following the completion of the primary immune response, most of the responsive B cells die off but a subset reverts to a resting state. This population of dormant cells maintains a memory of the antigen, responding quickly to the antigen should it be encountered at a later time. This secondary response occurs more rapidly and is greater in magnitude, producing many more antibodies that have greater affinity and avidity for the antigen. During a secondary immune response, IgG is generally the most prominent antibody produced, but in mucosal regions, such as the gut, lungs, and respiratory tract, IgA production is predominant.

The B cells and secreted antibodies are abundant in the fluid portion of blood (called plasma) and circulate through lymphatic vessels. Accordingly, antibodies are present within lymphatic fluids and in lymph nodes (where B cells congregate in areas known as germinal centers), as well as in secretory fluids, such as the mucus of the upper respiratory and gastro-intestinal tract. Antibodies are also present in breast milk, thereby conferring passive protection to the neonate against pathogens to which mom had previously developed immunity. As a technical note, when blood is collected and allowed to clot, the fluid that oozes from the clot is called serum. This will contain the antibody we would normally find in plasma (which is what we collect from uncoagulated blood, such as that treated with heparin).

The antibody molecule

Immunoglobulin protein comprises two pairs of identical amino acid chains, as depicted in Fig. 2.5. One pair comprises *Heavy (H) chains*, whereas the second pair consists of *Light (L) chains*. The immunoglobulin type or class (i.e., IgG, IgA, IgM, IgD, and IgE) is determined by the physicochemical characteristics of the H chain. Although many regions of each H and L chain are constant across a particular type of Ig molecule, some regions are not shared, differing in the amino acid sequence present. These *variable regions* are fundamental to binding sites for antigenic determinants (the epitope, which comprises a portion of an antigen that can be recognized by the immune system). The sheer number of antibodies present allows for several different antigenic determinants being bound, and the specific amino acid sequences of the variable regions of an Ig molecule determine its affinity and avidity to antigenic determinants.

Antigen neutralization by antibodies

We have already mentioned the functional properties of antibodies. Here we introduce a few more concepts and expand on those already mentioned. When antigenic portions of a virus or bacteria circulate through the blood and lymph, they will eventually encounter an antibody that will bind to the soluble antigen. This antigen-antibody pairing can then be bound by free portions of an antibody that had already been paired with the same protein antigen. The resulting *immune complex* can be trapped in tissue where local macrophages can bind this

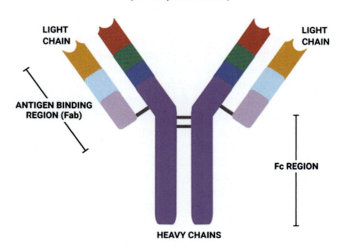

FIG. 2.5 The fundamental structure of the immunoglobulin molecule. The antigen-binding component (Fab) of the molecule comprises a variable amino acid sequence that exhibits high binding affinity and avidity to a specific antigen that is recognized. This variability in an amino acid sequence is dictated by sections of the light chain. The Fc region of the molecule also interacts with phagocytes and components of the complement pathway to eliminate antigens. *Figure created using BioRender.com.*

complex and then eliminate it by enzymatic breakdown. Alternatively, the *complement* system that comprises 20 proteins made by the liver (or by macrophages) can be engaged to eliminate foreign particles. These complement factors in particular combinations can kill foreign invaders on their own or they can do so in combination with an antibody. If the generation of antibodies is insufficient, illness can be provoked, and infection may be prolonged. After symptoms of a viral illness have resolved, the associated antibodies will remain in circulation, although a decline of the antibody will occur. Nonetheless, should appreciable antibodies be present, this can be used to determine whether an individual had experienced the illness and perhaps whether they might have protection against illness recurrence. For example, as we have seen in the case of the COVID-19 pandemic, antibodies may not be present permanently. However, this does not necessarily imply the absence of continued protection as T cells can also retain a memory of the infection and can act against the virus upon its reintroduction. Moreover, since the presence of antibodies for the virus implies prior exposure, it is also likely that memory B cells are present. Consequently, with the help of memory T cells, a B cell memory response is a promising possibility that can protect the individual should there be reexposure to the virus.

T lymphocytes

When immature T cells leave the bone marrow, they migrate to the thymus gland where they mature into different T cell subtypes with diverse functions. As intimated earlier, T cells come in several varieties that are involved in the recognition and response to foreign particles. The major histocompatibility complex (MHC), which comes in two forms, is fundamental to this process. Ordinarily, protein molecules are synthesized within a cell and then transported to the cell surface where they display a small peptide fragment (epitope) that can be recognized by T cells as belonging to the self and hence will not attack them.

After a virus enters a healthy cell, it uses the cell's machinery to multiply and eventually leaves the cell to infect other healthy cells nearby. At the same time, fragments of the invader can be loaded onto a protein assembly referred to as MHC I (or simply the MHC class I protein), after which this complex is transported to the cell surface for presentation of antigen to other immune cells. Should particular immune cells, specifically cytotoxic T cells (T_c; $CD8^+$) come by and recognize the displayed particle as being foreign, a process will be initiated whereby the T cell *directly* destroys the infected cell (referred to as a cell-mediated response). By destroying the cell, a whole factory of viruses is eliminated. In effect, T cells are taught in the thymus to leave the host tissue undisturbed, but at the same time ought to be able to recognize and attack foreign molecules rapidly. The cytotoxic T cells therefore will only kill virus-infected cells constituting self-tissue if it has been marked as infected by the antigen present on the MHC I molecule. Such collateral damage is the "cost of doing business" but hopefully is stopped rapidly once other immune components—such as antibody—are mounted to limit the spread and impact of infectious virus. Finally, because viruses can potentially infect any cell in the body, the MHC class I gene is active in most cells of the body, including the nervous system.

MHC class II molecules, in contrast to the MHC I molecules, are present on a restricted number of cell types, primarily comprising the antigen presenting cells (APCs), such as dendritic cells (e.g., monocytes) and macrophages, as well as B cells. In addition, T cells are known to express MHC II molecules, but only when they are activated. Of all these MHC II^+ cells, the dendritic cells are considered true professional APCs, whose role is exclusively to capture, process, and then present antigens to adaptive immune cells. The dendritic cells, which are prominent in all lymphoid organs, possess multiple dendritic arborizations that maximize the surface expression of MHC class II molecules loaded with antigenic peptides. The macrophages act similarly, ordinarily engulfing a foreign particle, say a bacteria or a virus, partially digest it, and then display fragments of the microbe bound to MHC II molecules on its surface. This will be recognized as foreign by the T cell receptor present on helper T (T_h; $CD4^+$) cells. Thereafter, T_h1 cells (a category of all T_h cells) will alert *Tc* cells, which will proliferate and act against the virus mediating the cytotoxic cell-mediated response. The second category of T_h cells, the T_h2 cell, also plays an important role in immune functioning, responding against extracellular pathogens (e.g., parasites), enhancing wound-healing, affecting chronic inflammatory disease, and ought to act against cancer. Fig. 2.6 shows some of the immune steps that occur throughout cancer progression.

An essential component of immune functioning is that the activity of different immune cells is orchestrated or regulated. On encountering a pathogen, a sufficiently powerful immune response must be mounted to ensure its elimination, but this has to be carefully calibrated against the need to spare healthy tissue. Examples where this has failed include autoimmune disorders or septic shock due to bacterial infection. So just as there needs to be a carefully regulated response against the foreign infectious agent, there also needs to be a mechanism for downregulating the response, especially once the offending pathogen has been stopped. In this case, the release of specific signaling molecules from T_h2 cells may act against further immune activity by T_c cells, B cells, or any other cell that selectively responded to the antigen. A major player here is the T regulatory (T_{reg}) cell, which in former times had been referred to as a T suppressor cell. While this term is no longer in use, T_{reg} cells serve in the maintenance

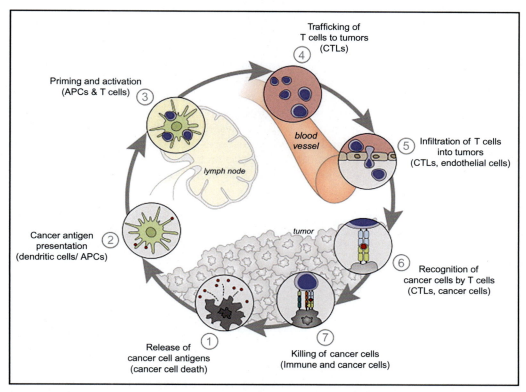

FIG. 2.6 The generation of immunity to cancer is a cyclic process that can be self-propagating, leading to an accumulation of immune-stimulatory factors that in principle should amplify and broaden T cell responses. The cycle is also characterized by inhibitory factors that lead to immune regulatory feedback mechanisms, which can halt the development or limit the immunity. This cycle can be divided into seven major steps, starting with the release of antigens from the cancer cell and ending with the killing of cancer cells. Abbreviations are as follows: *APCs*, antigen presenting cells; *CTLs*, cytotoxic T lymphocytes. *Figure and caption from Chen, D.S., Mellman, I., 2013. Oncology meets immunology: the cancer-immunity cycle. Immunity 39, 1–10.*

of immune tolerance, so that cytotoxic cells do not attack self-tissue (Romano et al., 2019).[b] Immune cell activity also declines owing to metabolic changes that affect energy production and hence limit immune functioning. Finally, intercellular communication occurs between immune cells which takes account of their expanded numbers at sites of infection, thereby resulting in self-regulated avoidance of adverse overreactivity (Muldoon et al., 2020). As much as these processes operate well in preventing immune cell assaults on the self, cytotoxic T cells can attack healthy cells that produce excessively high numbers of antigens (Pettmann et al., 2021).

[b] The varied types of T cells express distinct molecules on their surface that distinguish them from one another.

Cytokines

Communication (signaling) between immune cells is accomplished by the synthesis and secretion of regulatory substances (*cytokines*) that can bind to T and B cells. This also promotes the growth and differentiation of immune cells. Cytokines are produced by cells of the innate immune system as well as T and B cells, and therefore are involved in both innate and adaptive immunity. Numerous different cytokines serve as messenger proteins that direct the proliferation and movement of immune cells. As described in Table 2.1, each of these cytokines has its own function (although in several respects these can be redundant). Activated monocytes and macrophages secrete tumor necrosis factor-α (TNF-α), interleukin (IL)-1, IL-6, IL-8, and IL-12. When T_h1 cells are activated, they secrete IL-1, IL-2, interferon-γ (IFN-γ),

TABLE 2.1 Common cytokines encountered in neural-immune investigations.

Cytokine	Main immune cell origin	Receptor types	Most common function	Target cells
Interleukin-1 family				
IL-1 (IL-1α, IL-1β)	Monocytes, Macrophages, and other innate immune cells (incl. DCs)	IL-1R1 IL-1R2	Inflammation Fever	Endothelium, CNS, Hepatic cells, T cells
IL-18	As above	IL-18α IL-18β	Inflammation Induces IFN-γ by T and NK cells	T cells, NK cells, Monocytes, Neutrophils
Tumor necrosis factor family				
TNF	Macrophages NK Cells T cells	TNFRI TNFRII	Inflammation Cachexia	Endothelium Neutrophils
Type I cytokine family				
IL-2	T cells	IL-2Rα IL-2Rβ	Mitotic (proliferation) activity; T cell effector and memory formation; T_{reg} generation; NK cytotoxicity	T, B, and NK cells
IL-4	CD4 T cells Mast cells	IL-4Rα	Isotype switching to IgE; Proliferation and differentiation of T_h2 cells	B cells T cells Macrophages Mast cells
IL-6	Macrophages T cells	IL-6Rα gp130	Inflammation Acute phase response B cell differentiation (in plasma cell)	Hepatic cells B cells
IL-12 [IL-12A (p35), IL-12B (p40)]	Macrophages Dendritic cells	IL-12Rβ1 IL-12Rβ2	T_h1 differentiation; IFN-γ induction; Promotes cytotoxicity	T cells NK cells

TABLE 2.1 Common cytokines encountered in neural-immune investigations—cont'd

Cytokine	Main immune cell origin	Receptor types	Most common function	Target cells
IL-17 (IL-17A, IL-17F)	CD4 T cells (T_h17) Group 3 Innate Lymphoid Cells	IL-17RA IL-17RC	Increase chemokine and cytokine production by macrophages and endothelial cells	Endothelium Macrophages Epithelial cells
Type II cytokine family				
IFN-γ	CD4 T cell (T_h1) CD8 T cell NK cell	IFNGR1 IFNGR2	Induces MHC II Augments NK cytotoxicity	Macrophages NK cells
IL-10	Macrophages T cells (esp. T_{reg})	IL-10Rα IL-10Rβ	Suppression (cytokines, MHC II)	Macrophages Dendritic cells T cells (esp. T_h1)
Ungrouped cytokines				
Transforming growth factor-β (TGFβ)	T cells (esp. T_{regs}), macrophages	TGFβR1 TGFβR2 TGFβR3	Suppression (B and T cells; macrophages); T_h17 and T_{reg} differentiation; Tolerance promotion	T cells B cells Macrophages
Monocyte chemoattractant Protein (MCP-1/CCL2)	Osteoblasts following cytokine stimulation	CCR2	Regulates the migration and infiltration of monocytes and macrophages.	Recruits dendritic cells, monocytes, and memory cells to sites of inflammation.
Granulocyte-macrophage colony-stimulating factor (GM-CSF)	Macrophages, T-cells, NK cells, mast cells, endothelial cells, fibroblasts	Receptors on monocytes, macrophages, lymphocytes, granulocytes	White cell growth factor. Acts as an immune adjuvant to increase dendritic cell maturation. Used to treat neutropenia produced by chemotherapy, and after bone marrow transplantation.	Recruits circulating neutrophils, monocytes, and macrophages.

Compilation based on details from Abbas, A.K., Lichtman, A.H., Pillai, S., 2015. Cellular and Molecular Immunology, eighth ed. Elsevier, Philadelphia, PA, and modified from Anisman, H., Hayley, S., Kusnecov, A., 2018. The Immune System and Mental Health. Academic Press, San Diego, CA.

and tumor necrosis factor-α (TNF-α), which act in a proinflammatory capacity, as does IL-17 secreted by another type of Th cell, the T_h17 cell. The cytokines secreted by T_h2 cells (such as IL-4 and IL-10) are generally antiinflammatory. As stated previously, when inflammation occurs acutely, it is generally beneficial (e.g., in attenuating infection, facilitating repair of tissue), but if it persists, the inflammation can be destructive, promoting a variety of illnesses, such as heart disease, autoimmune responses, type 2 diabetes, and cancers. Thus, the balance between these pro- and antiinflammatory cytokines is critical for homeostasis to be maintained, and disturbances of these balances can favor disease occurrence. By limiting the actions of other T cells, B cells, and dendritic cells, the T_{reg} cells limit prolonged inflammatory responses, thereby acting against the development of inflammatory-related illnesses. In addition, activation of certain types of receptors on immune cells, such as the *V-type immunoglobulin domain-containing suppressor of T cell activation* (VISTA), inhibits naïve T lymphocytes so autoimmune responses will not occur readily (ElTanbouly et al., 2020).

Cytokines with proinflammatory actions, particularly IL-1β and TNF-α, have especially dramatic effects within both major branches of the immune system. These actions largely stem from the impact of these cytokines on a protein complex that functions as a transcription factor, the *nuclear factor kappa-light-chain-enhancer of activated B cells* (NFκB). Activation of NFκB results in a variety of cytokine genes being activated and is integral to responses against a variety of challenges (including stressful events). Other cytokines, including IL-2, IL-6, and the interferons, act through alternate signaling pathways that involve janus kinases (JAK) and *signal transducer and activator of transcription proteins* (STATs) pathways (referred to as the JAK-STAT pathway). Both the NFκB and various JAK-STAT pathways activated by these cytokines are of considerable importance in several types of cancers and disorders of the immune system, as well as contributing to mental illnesses, such as depression (Miller and Raison, 2016). We will return to these molecular signals in later chapters.

Additional cytokines, such as monocyte chemoattractant protein-1 (MCP-1) or the granulocyte-macrophage colony-stimulating factor (GM-CSF), promote the recruitment of immune cells to deal with challenges (Deshmane et al., 2009). It had at one time been thought that cytokines, such as MCP-1, acted against tumor cells by directing the recruitment of macrophages and neutrophils to the tumor microenvironment. However, it seems that MCP-1, which acts as a growth factor, can also promote tumor cell proliferation and fosters cancer growth (Yoshimura, 2018), possibly by encouraging angiogenesis (thereby providing nutrients that maintain tumor viability). In effect, immune cells and the cytokines they release are neither inherently "bad" nor "good" in relation to tumors, but rather act to shape the environment within which the tumor is embedded. They also interact with the tumor itself, but whether these limit or augment tumor growth depends on a host of factors that will be discussed later.

Cytokines involved in immune regulation are often subdivided into superfamilies based on the shared structural elements of the receptors to which they bind. Based on this sort of grouping, Table 2.1 provides a few examples from the (i) 35 members of the interleukin-1 family, (ii) 19 members of the tumor necrosis factor superfamily, (iii) other families of Type I (e.g., IL-2, IL-4), and (iv) Type II (IFN-γ and IL-10) cytokines, as well as other cytokines (or chemokines) that do not fit into any particular group. The primary function of each of these cytokines is indicated in the table, but these cytokines have numerous other functions. Moreover, small cytokines (chemokines) are present that are responsible for regulating the migration of cells during tissue repair, and serve in a proinflammatory capacity, operating to recruit immune cells to sites of infection.

Paralleling reports that immune cells can act in favor of or against cancer development, pro- and antiinflammatory cytokines can have both permissive and antagonistic influences on cancer progression. For instance, IL-10 is important in quelling excessive or prolonged immune responses. However, while it exerts inhibitory effects on antigen-presenting and memory T cells, and modulates the actions of T_{reg} cells, such effects can also favor cancer progression. It similarly appears that GM-CSF may have immunosuppressive actions that can be exploited by cancer cells. One way in which such immunosuppressive cytokines act is by reprogramming local macrophages to limit their affinity for the tumor (Waghray et al., 2016). Such macrophages can also suppress $CD8^+$ T cell responses and promote resistance to standard chemotherapy drugs.

Chemokines

Chemokines comprise small cytokines that belong to one of four major families (CXC, CC, CX3C, and XC). Chemokines operate during pathological conditions, functioning to attract immune cells through their *chemoattractant* capacity to engage foreign antigens at sites of tissue infection. At least 19 different types of chemokine receptors bound by members of major chemokine families have been identified. These receptors are expressed on all leukocytes but are most prominent on T cells, which prompts these continuously circulating lymphocytes to be directed to sites of injury and inflammation. In addition to signaling circulating leukocytes, should infection be discovered in a given tissue location during the course of immune surveillance, chemokines direct lymphocytes to the lymph nodes where the presence of pathogens is screened through interactions with antigen-presenting cells. In this sense, chemokines serve in a homeostatic capacity essentially assuring that leukocyte migration (homing) occurs as it should. Aside from their role in immune functioning, chemokines also contribute to the development of the nervous system by promoting axonal guidance and may contribute to the growth of new blood vessels (angiogenesis), which might also be involved in cancer growth.

Inflammasome

Although the often-used term "inflammasome" implies a singular entity, there are actually four inflammasomes (NLRP1, NLRP3, NLRC4, and AIM2). These constitute multiprotein complexes that activate the IL-1 family of cytokines and attendant inflammatory responses and serve as a mechanism by which alterations of cellular integrity and biological threats to tissue well-being are detected. Inflammasomes can be activated by some of the same processes that are responsible for sterile inflammatory responses, including environmental irritants and toxicants, as well as molecules tied to cellular damage (Levy et al., 2015). Inflammasome activation is promoted by PRRs (as mentioned in our discussion of sterile inflammation) that are stimulated by PAMPs and DAMPs released from distressed cells. After sensing an inflammatory response or DAMPs, the inflammasome engages adaptor proteins to form a complex cellular platform. The enzyme, caspase-1, is then recruited to the complex, resulting in the activation of IL-1 and IL-18.

NOD-like receptors (NLRs), a component of the PRRs, detect a variety of signals, including fatty acids, β-amyloid, and changes in the energy-carrying molecule ATP. Following a danger signal being detected, different types of NLRP (a form of NOD-like receptor) are activated, which can engender inflammation and the production of varied IL-1 family cytokines (Iwata et al., 2013). Together with toll-like receptors, they contribute to inflammation and apoptosis, and hence numerous illnesses, including psychological disorders (e.g., depression and anxiety) and a constellation of physical illnesses, such as diabetes, heart disease, and some cancers. In the latter regard, inflammasomes have been implicated in multiple aspects of tumor pathology, acting to either promote or suppress tumor functions, depending on the chronicity of inflammation. Both IL-1β and IL-18 released from inflammasomes promote cellular proliferation via hormonal processes. Additionally, beyond its proinflammatory actions, IL-1β can promote tumorigenesis by indirectly promoting oncogene expression, and IL-18 may do so by inhibiting caspase enzymes that normally result in DNA damage and subsequent apoptosis (preprogrammed cell death). The inflammasome can also induce immunosuppression by inhibiting antitumor T lymphocytes and dendritic cells, which has been particularly well documented in pancreatic adenocarcinoma and lung cancer (Lasithiotaki et al., 2018). More aggressive growth of lung, breast, and gastric cancers could be instigated by inflammasomes through their actions of fueling growth factors that maintain the survival and spread of cancer cells. Other types of inflammasomes can alter immune checkpoint proteins, such as cell death ligand 1 and alarmin proteins, that collectively induce tumorigenesis.

Besides promoting tumor growth, the inflammasome can paradoxically also have tumor-suppressive effects. In cases of colitis-associated cancer induced by chemical irritants, the inflammasome-mediated release of IL-18 fosters the regeneration of epithelial cells and hence protects against colorectal cancer. However, there are limits to this protection, so eventually IL-18 loses its protective ability and colorectal cancer can ultimately emerge. It is also known that IL-18 can catalyze the antitumor actions of NK cells. Accordingly, early release of IL-18 within the colon may facilitate repair of the injured epithelial barrier and limit tumorigenesis, but chronic inflammation may have the opposite effect and promote tumor growth.

It seems that IL-1β may also play a dual role in tumorigenesis. In the case of melanoma, inflammasome-derived IL-1β promotes tumor spread among epithelial cells but has the opposite effect in keratinocytes (an epidermal cell that produces keratin). It similarly appears that this cytokine promotes colon tumor progression through its impact on epithelial and T cells; however, IL-1β may also adopt an opposing role that is mediated by its impact on neutrophils and monocytes. Finally, high levels of colonic IL-1β can modulate the actions of gut microbiota, thereby influencing the "education" and production of immune cells. Clearly, inflammasome signaling is complex and is dependent upon the particular cells involved and may change over time with the growth of the cancer and associated alterations in the microenvironment.

Several processes other than those already mentioned may also contribute to the pathological outcomes attributable to persistent inflammation associated with the continued presence of bacteria and other irritants. Among other things, reactive oxygen species, as well as proteases (enzymes that break down proteins and peptides) produced by neutrophils and macrophages, can instigate tissue damage. These processes also give rise to the proliferation of fibroblasts (immature fiber-producing cells of connective tissue), and hence the accumulation of collagen and the development of excess fibrous connective tissue (fibrosis). These actions have been associated with heart disease, neurodegenerative disorders, and autoimmune disorders and may contribute to the development of some types of cancer. For instance, inflammatory bowel disease, which is accompanied by chronic inflammation, may lead to colorectal cancer, likely involving activation of macrophages and neutrophils, as well as cyclooxygenase enzymes (involved in the production of prostaglandins that contribute to fever and pain) and resolvins (metabolic by-products of omega-3 fatty acids). Persistent inflammation may also give rise to an anaerobic microenvironment that promotes DNA damage to epithelial cells, which can facilitate tumor development (Mariani et al., 2014). Accordingly, treatments that limit inflammation, including monoclonal antibodies targeting TNF-α, proton pump inhibitors, nonsteroidal antiinflammatory agents, and some "natural" agents, ought to have at least some beneficial effects (Lee et al., 2016).

A final point that warrants mention regarding sterile inflammation is the connection between psychological stress and cancer. While it is unlikely that stressors alone are a major precipitant of cancer development, when combined with other "hits," such as genetic vulnerability or the presence of toxins, stressors could potentially shape the progression of some forms of cancer and may interfere with the effectiveness of therapeutic procedures. Furthermore, given that variations of the inflammasome have been seen across numerous illnesses, including those of a psychological nature (e.g., depressive disorders), these processes might be a common denominator for the frequent comorbidities that exist between these illnesses (Anisman et al., 2018; Levy et al., 2015).

Immune tolerance

Host cells normally guard against unregulated attacks on the self by using highly specific tolerance mechanisms. These comprise a series of T cells, most notably T_{reg} cells, which regularly patrol the body and defend against aggressive responses to self-antigens. As we have seen, T cells interact with MHC/self-antigen complexes, and since self-antigens have a greater binding avidity for MHC than do nonself antigens, a process called negative selection ensures the elimination of self-reactive lymphocytes (Gronski et al., 2004). Further, T_{reg} cells play an important role in preventing autoimmunity by inhibiting dendritic cell maturation, blocking antigen presentation, and suppressing cells through the release of IL-10 and TGF-β (Misra et al., 2004).

Various T cell subsets are critical for long-term specific immune responses against tumor-associated antigens. T cells comprise almost 10% of all tumor-infiltrating cells within and on the margins of the tumor mass. Moreover, the presence of $CD8^+$ memory T cells is fundamental to a positive prognosis in certain cancers, including colorectal and lung cancers (e.g., Schollbach et al., 2019). Another subset of T lymphocytes, namely γδ T cells, can be protective or deleterious in cancer immunity. This dual role is characterized by an antitumor effect pursuant to treatment with chemotherapy drugs. Additionally, γδ T cells can produce IL-17, which can create an inflammatory microenvironment phenotype within mammary tumors, and in aged mice can favor cancer development (Prinz and Sandrock, 2019).

Over 2000 different tumor antigen-specific antibodies have been located within the blood of cancer patients, possibly reflecting ongoing adaptive humoral immune responses against various parts of the tumor tissue. There are B cell subsets whose characteristics vary depending on their location and whether they are activated in a T-dependent or a T-independent manner. Intriguingly, aside from B cells becoming long-lived memory cells, they can also form B_{reg} cells, which are often localized to tumors and can promote antitumor complement or enzymatic attacks.

The B cell-derived antibodies can kill tumor cells by stimulating the complement cascade or through activation of NK cell-dependent cytotoxicity. Independently of antibodies, B cells can directly attack tumor cells and induce their demise through the expression of Fas, a classical pro-apoptotic factor (Xia et al., 2016). Moreover, like their T_{reg} counterparts, B_{reg} cells can promote antitumor immunity, in part via actions on TGF-β and IL-10 (Gorosito Serrán et al., 2015). However, through their release of hormones and cytokines, tumors can also promote B cell differentiation into immunosuppressive B_{reg} cells. Thus, a reciprocal dialogue between B cells and tumors is of considerable importance in regulating cancer progression. In this respect, hormone-like substances, such as prostaglandins acting together with leukotrienes (inflammatory mediators), may activate B cells and promote their differentiation into immunosuppressive B$_{reg}$ cells. Some cancers also express chemokines that promote tumor-specific B cell migration or express trophic factors (e.g., FGF-2 and IGF-1) which further facilitate tumor growth.

Hijacking immunity

As much as the processes of immune tolerance are sophisticated and effective, they can be hijacked in an effort to protect malignant tissue during the early stages of oncogenesis. Tumor cells have evolved clever means of exploiting T$_{reg}$ cells to their benefit by essentially tricking the immune system into treating it as if it was a self-antigen, thereby signaling that they are off-limits. Tumors can also produce antiinflammatory cytokines that induce T cell suppression, further limiting attacks on their growth. Some immune tolerance mechanisms can also be undermined by the induction of anergy in T cells (when immune cells become unresponsive and unable to mount a proper response), possibly stemming from the chronic inflammatory environment. Indeed, after repeated assault from the tumor environment, T cells might fail to adopt their proper effector or memory cell roles. Interestingly, well before these suboptimal cells die, they lose their ability to proliferate and produce IL-2 and IFN-γ (Min et al., 2001). This ensures that their ability to be roused into a responsive state directed against dangerous cell forms—such as cancer cells— is limited.

Summary and conclusions

The immune system and its actions in relation to cancer occurrence and progression are without a doubt exceptionally complex and moderated by a great number of variables. The foregoing has provided some examples of this variety and complexity. As we have seen repeatedly, aspects of the immune system may be either pro- or antitumorigenic depending on features of the cancer cells and their stage of development, characteristics of the tumor microenvironment, chronicity of inflammation, a constellation of genetic and epigenetic factors, and the ability of the tumor to subvert immune functioning. Moreover, numerous hormones and growth factors impinge on tumor cells and moderate the actions of immune cells. If this wasn't a sufficiently complex set of factors, relatively more recent findings over the past decade concerning microbiota have increased this by an order of magnitude. One advantage, however, is that it has opened up new treatment targets and strategies.

References

Abel, A.M., Yang, C., Thakar, M.S., Malarkannan, S., 2018. Natural killer cells: development, maturation, and clinical utilization. Front. Immunol. 9, 1869.

Anderson, H.L., Brodsky, I.E., Mangalmurti, N.S., 2018. The evolving erythrocyte: red blood cells as modulators of innate immunity. J. Immunol. 201, 1343–1351.

Anisman, H., Hayley, S., Kusnecov, A., 2018. The Immune System and Mental Health. Academic Press, San Diego, CA.

Cen, X., Liu, S., Cheng, K., 2018. The role of toll-like receptor in inflammation and tumor immunity. Front. Pharmacol. 9, 878.

Colak, S., Ten Dijke, P., 2017. Targeting TGF-β signaling in cancer. Trends Cancer 3, 56–71.

References

Deshmane, S.L., Kremlev, S., Amini, S., Sawaya, B.E., 2009. Monocyte chemoattractant protein-1 (MCP-1): an overview. J. Interf. Cytokine Res. 29, 313–326.

Domínguez-Andrés, J., Netea, M.G., 2019. Impact of historic migrations and evolutionary processes on human immunity. Trends Immunol. 40, 1105–1119.

ElTanbouly, M.A., Zhao, Y., Nowak, E., Li, J., Schaafsma, E., et al., 2020. VISTA is a checkpoint regulator for naïve T cell quiescence and peripheral tolerance. Science 367, eaay0524.

Fleshner, M., Frank, M., Maier, S.F., 2017. Danger signals and inflammasomes: stress-evoked sterile inflammation in mood disorders. Neuropsychopharmacology 42, 36–45.

Gorosito Serrán, M., Fiocca Vernengo, F., Beccaria, C.G., Acosta Rodriguez, E.V., Montes, C.L., et al., 2015. The regulatory role of B cells in autoimmunity, infections and cancer: perspectives beyond IL10 production. FEBS Lett. 589, 3362–3369.

Gronski, M.A., Boulter, J.M., Moskophidis, D., Nguyen, L.T., Holmberg, K., et al., 2004. TCR affinity and negative regulation limit autoimmunity. Nat. Med. 10, 1234–1239.

Harvie, M., Howell, A., Evans, D.G., 2015. Can diet and lifestyle prevent breast cancer: what is the evidence? Am. Soc. Clin. Oncol. Educ. Book, e66–e73.

Howden, A.J.M., Hukelmann, J.L., Brenes, A., Spinelli, L., Sinclair, L.V., et al., 2019. Quantitative analysis of T cell proteomes and environmental sensors during T cell differentiation. Nat. Immunol. 20, 1542–1554.

Iwata, M., Ota, K.T., Duman, R.S., 2013. The inflammasome: pathways linking psychological stress, depression, and systemic illnesses. Brain Behav. Immun. 31, 105–114.

Khan, N., de Manuel, M., Peyregne, S., Do, R., Prufer, K., et al., 2020. Multiple genomic events altering hominin SIGLEC biology and innate immunity predated the common ancestor of humans and archaic hominins. Genome Biol. Evol. 12, 1040–1050.

Lam, L.M., Reilly, J.P., Rux, A.H., Murphy, S.J., Kuri-Cervantes, L., et al., 2021. Erythrocytes identify complement activation in patients with COVID-19. Am. J. Physiol. Lung Cell. Mol. Physiol. 321, L485–L489.

Lasithiotaki, I., Tsitoura, E., Samara, K.D., Trachalaki, A., Charalambous, I., et al., 2018. NLRP3/Caspase-1 inflammasome activation is decreased in alveolar macrophages in patients with lung cancer. PLoS One 13, e0205242.

Lauterbach, M.A., Hanke, J.E., Serefidou, M., Mangan, M.S.J., Kolbe, C.C., et al., 2019. Toll-like receptor signaling rewires macrophage metabolism and promotes histone acetylation via ATP-citrate lyase. Immunity 51, 997–1011. e1017.

Lee, H.J., Park, J.M., Han, Y.M., Gil, H.K., Kim, J., et al., 2016. The role of chronic inflammation in the development of gastrointestinal cancers: reviewing cancer prevention with natural anti-inflammatory intervention. Expert Rev. Gastroenterol. Hepatol. 10, 129–139.

Levy, M., Thaiss, C.A., Zeevi, D., Dohnalová, L., Zilberman-Schapira, G., et al., 2015. Microbiota-modulated metabolites shape the intestinal microenvironment by regulating NLRP6 inflammasome signaling. Cell 163, 1428–1443.

Mariani, F., Sena, P., Roncucci, L., 2014. Inflammatory pathways in the early steps of colorectal cancer development. World J. Gastroenterol. 20, 9716–9731.

Miller, A.H., Raison, C.L., 2016. The role of inflammation in depression: from evolutionary imperative to modern treatment target. Nat. Rev. Immunol. 16, 22–34.

Min, B., Legge, K.L., Bell, J.J., Gregg, R.K., Li, L., et al., 2001. Neonatal exposure to antigen induces a defective CD40 ligand expression that undermines both IL-12 production by APC and IL-2 receptor up-regulation on splenic T cells and perpetuates IFN-gamma-dependent T cell anergy. J. Immunol. 166, 5594–5603.

Misra, N., Bayry, J., Lacroix-Desmazes, S., Kazatchkine, M.D., Kaveri, S.V., 2004. Cutting edge: human CD4+CD25+ T cells restrain the maturation and antigen-presenting function of dendritic cells. J. Immunol. 172, 4676–4680.

Muldoon, J.J., Chuang, Y., Bagheri, N., Leonard, J.N., 2020. Macrophages employ quorum licensing to regulate collective activation. Nat. Commun. 11, 878.

Nikzad, R., Angelo, L.S., Aviles-Padilla, K., Le, D.T., Singh, V.K., et al., 2019. Human natural killer cells mediate adaptive immunity to viral antigens. Sci. Immunol. 4, eaat8116.

Orecchioni, M., Ghosheh, Y., Pramod, A.B., Ley, K., 2019. Macrophage polarization: different gene signatures in M1(LPS+) vs. classically and M2(LPS−) vs. alternatively activated macrophages. Front. Immunol. 10, 1084.

Patel, S., 2018. Danger-associated molecular patterns (DAMPs): the derivatives and triggers of inflammation. Curr. Allergy Asthma Rep. 18, 63.

Peeri, M., Amiri, S., 2015. Protective effects of exercise in metabolic disorders are mediated by inhibition of mitochondrial-derived sterile inflammation. Med. Hypotheses 85, 707–709.

Pettmann, J., Huhn, A., Shah, E.A., Kutuzov, M.A., Wilson, D.B., et al., 2021. The discriminatory power of the T cell receptor. eLife 10, e67092.

Prinz, I., Sandrock, I., 2019. Dangerous γδ T cells in aged mice. EMBO Rep. 20, e48678.

Ralston, J.C., Lyons, C.L., Kennedy, E.B., Kirwan, A.M., Roche, H.M., 2017. Fatty acids and NLRP3 inflammasome-mediated inflammation in metabolic tissues. Annu. Rev. Nutr. 37, 77–102.

Ratajczak, M.Z., Pedziwiatr, D., Cymer, M., Kucia, M., Kucharska-Mazur, J., et al., 2018. Sterile inflammation of brain, due to activation of innate immunity, as a culprit in psychiatric disorders. Front. Psychiatry 9, 60.

Riera Romo, M., Pérez-Martínez, D., Castillo Ferrer, C., 2016. Innate immunity in vertebrates: an overview. Immunology 148, 125–139.

Romano, M., Fanelli, G., Albany, C.J., Giganti, G., Lombardi, G., 2019. Past, present, and future of regulatory T cell therapy in transplantation and autoimmunity. Front. Immunol. 10, 43.

Rubartelli, A., Lotze, M., Latz, E., Manfredi, A., 2013. Mechanisms of sterile inflammation. Front. Immunol. 4, 398.

Schollbach, J., Kircher, S., Wiegering, A., Seyfried, F., Klein, I., et al., 2019. Prognostic value of tumour-infiltrating CD8+ lymphocytes in rectal cancer after neoadjuvant chemoradiation: is indoleamine-2, 3-dioxygenase (IDO1) a friend or foe? Cancer Immunol. Immunother. 68, 563–575.

Siddiqui, S.S., Vaill, M., Do, R., Khan, N., Verhagen, A.L., et al., 2021. Human-specific polymorphic pseudogenization of SIGLEC12 protects against advanced cancer progression. FASEB BioAdvances 3, 69–82.

Venegas, C., Heneka, M.T., 2017. Danger-associated molecular patterns in Alzheimer's disease. J. Leukoc. Biol. 101, 87–98.

Waghray, M., Yalamanchili, M., Dziubinski, M., Zeinali, M., Erkkinen, M., et al., 2016. GM-CSF mediates mesenchymal-epithelial cross-talk in pancreatic cancer. Cancer Discov. 6, 886–899.

Wang, Y., Smith, W., Hao, D., He, B., Kong, L., 2019. M1 and M2 macrophage polarization and potentially therapeutic naturally occurring compounds. Int. Immunopharmacol. 70, 459–466.

Xia, Y., Tao, H., Hu, Y., Chen, Q., Chen, X., et al., 2016. IL-2 augments the therapeutic efficacy of adoptively transferred B cells which directly kill tumor cells via the CXCR4/CXCL12 and perforin pathways. Oncotarget 7, 60461–60474.

Yoshimura, T., 2018. The chemokine MCP-1 (CCL2) in the host interaction with cancer: a foe or ally? Cell Mol. Immunol. 15, 335–345.

Yuan, S., Liu, Z., Xu, Z., Liu, J., Zhang, J., 2020. High mobility group box 1 (HMGB1): a pivotal regulator of hematopoietic malignancies. J. Hematol. Oncol. 13, 91.

CHAPTER 3

Microbiota and health

OUTLINE

How it began	69	Microbiota and immune functioning	80
Features of microbiota	70	Microbiota and inflammation	81
Beyond gut bacteria	71	Microbiota and illness comorbidities	83
Gut viruses	71	Microbiota, anxiety, and depressive disorders	84
An evolutionary perspective	72		
Microbiota development	75	Identifying good and bad microbes	87
Prenatal influences	75	The influence of environmental toxicants	88
Early postnatal influences	76	Summary and conclusions	89
Microbial factors derived through breastfeeding	77	References	89

How it began

In 1908, the Nobel Prize committee recognized the value of researchers who had linked immunity to diseases, awarding the prize for Physiology and Medicine jointly to Ilya Ilyich (Élie) Mechnikov and Paul Ehrlich for their work in immunology. Mechnikov was honored for the discovery of phagocytes (macrophages) that led to the concept of cell-mediated immunity, whereas Ehrlich was honored for his research on antitoxins (antibodies) released by immune cells that protected against diseases, ultimately being referred to as humoral immunity.

Like so many other brilliant scientists, their contributions were much broader. Ehrlich was instrumental in the development of chemicals (magic bullets) that affected immunity and could be used in a therapeutic capacity (an early version of chemotherapy). In fact, in 1909, a year after receiving the Nobel, one of his "magic bullets" was found to be effective in treating syphilis. For his part, Mechnikov was later responsible for the notion that fermented foods could provide health benefits by offsetting the toxic actions of gut bacteria, thereby increasing longevity. In his later years, Mechnikov turned his attention to the involvement of macrophages and gut microbiota in human aging and the development of dementia. He believed that with age the defenders of health, the phagocytes, faltered so that what he referred to as *autotoxins* derived from fermentation and putrefaction of gut bacteria could destroy the nerve cells and could similarly diminish the functioning of other body parts. Accordingly, he suggested that these degenerative processes could be slowed or prevented by avoiding certain foods (notably alcohol, rich meats, and badly cooked foods that were replete with undesirable microorganisms) as well as by inhibiting the influence of putrefactive bacteria by ingestion of other bacteria that produce lactate. Studies he conducted (including on himself and a couple of acolytes) were consistent with his views, but his theorizing went by the wayside during subsequent years. Mechnikov's notion did not generate appreciable traction a hundred years ago but has been rejuvenated over the past two decades. Interest in the microbiome has increased exponentially and so has the focus on natural ways (diet, exercise) to alter gut microbiota. However, it is less certain whether microbiota supplements (i.e., prebiotics and probiotics containing live bacteria) are beneficial (Zmora et al., 2018), although it may be premature to dismiss their value entirely.

Features of microbiota

The human gut contains trillions of bacteria as well as viruses, fungi, archaea,[a] and protozoa that may be intricately involved in well-being. Commensal bacterial species (those that derive benefits from other bacteria or the host) live in harmony with other bacteria (*eubiosis*), but some bacteria compete with others to obtain nutrients, or they can interact with one another for their mutual well-being. When the balance between diverse bacteria is disturbed (*dysbiosis*), the risk for a broad variety of illnesses might be elevated. For this reason, increased efforts have been expended to identify how specific bacteria within the gut (as well as at other sites) come to influence health, and whether these bacteria can be used in a prophylactic or therapeutic capacity. This is an enormously difficult undertaking given the huge variety of microbes that are present, many of which have yet to be identified. Moreover, well-being may be influenced by microbial metabolites, which in many cases are bioactive and may even migrate out of the gut and have effects throughout the body.

With appreciable research conducted over the past decade, there is hardly a disease for which microbial factors have not been implicated. Longevity has been attributed to the richness and diversity of several microbial species, whereas the presence of microbiota that are thought to be harmful predicted earlier mortality even after accounting for the influence of obesity and smoking (Salosensaari et al., 2021). The balance between beneficial and harmful gut bacteria presumably contributes to inflammatory bowel diseases (IBD), obesity, metabolic syndrome, type 1 and type 2 diabetes, cardiovascular disease and may constitute risk factors for cancer occurrence, and influences the efficacy of cancer therapies (Gentile and

---

[a] Archaea are single-celled microorganisms that are evolutionarily distinct from bacteria and eukaryotes although they are structurally similar to bacteria. Archaea are found in anaerobic environments and most often die at normal oxygen levels.

Weir, 2018). Aside from these physical illnesses, growing evidence has pointed to imbalances between beneficial and harmful microbiota exerting an influence on mental health. This likely stems from the multiple neurobiological changes induced by microbial challenges (Cryan et al., 2019). There is also reason to believe that gut microbiota influence sociability and social behaviors (Sherwin et al., 2019) that are needed to cope with stressors.

Gut microbiota have been implicated in so many disorders that a natural skepticism regarding their primacy in maintaining health or causing disease can almost certainly arise. Key questions are which illnesses are linked to bacteria, which are secondary to an existing illness, and to what extent is dysbiosis a bystander or associative phenomenon to disease? Mechanistic approaches in which the contribution of gut bacteria to brain or physical health is carefully and systematically dissected are possible, but mainly in animal studies. These have included the effects of various challenges that affect the microbial community—such as antibiotics or special diets—or the transfer of fecal bacteria from one mouse to another. These manipulations are sufficient to profoundly affect health status, and it similarly appears that when gut microbes from an ill person are transferred to recipient mice, they develop some of the symptoms evident in the human donor. As we will see, the transfer of fecal microbiota from stressed organisms can affect inflammatory processes, thereby producing downstream outcomes in the recipient animal.

Beyond gut bacteria

Much of the relatively recent focus on links between the microbiome and health has focused on bacteria located within the gut. However, other physiologic compartments also contain bacteria that may affect human health. The blood, saliva, adipose tissue, and several organs (e.g., skin, lung, liver, mouth, mammary glands, placenta, uterus, seminal fluid, ovarian follicles) contain microorganisms so their presence at these sites represents multiple opportunities for bacterial interactions within and between individuals. These microorganisms could have ramifications for a variety of health conditions, such as obesity, type 2 diabetes, as well as other illnesses that involve inflammatory processes, including cancers. That being said, microbes at different sites are fostered by very different processes, making it difficult to determine what local conditions drive their presence and evolution. We know that bacteria interact with multiple hormones and growth factors, as well as being influenced by host and microbe genetics. Consequently, it ultimately will be necessary to adopt a broad approach to understand how and why microbiota affect health risks (and benefits), and what these can tell us about establishing strategies in the treatment of many disorders.

Gut viruses

Modest attention has been devoted to the role of intestinal viruses in modulating behavioral functions. This is despite their abundance, with an estimated 140,000 unique viral populations being present in the human gut (the gut virome), with many more still to be identified. At parturition viral particles are absent, but dramatically increase in steps beginning shortly after birth, largely being obtained through breast milk (Liang et al., 2020). The virome, which can differ markedly across individuals, remains relatively stable over time but can vary in parallel with gut bacteria. The viruses may interact with gut bacteria, with some specifically infecting host bacteria. These bacteriophages (viruses that infect bacteria, usually referred to as "phages") generally share a symbiotic relationship with gut bacteria, although they may also act as human pathogens (Tetz and Tetz, 2018), and can gain access to diverse areas of the body, including the brain.

Phages can have multiple effects on bacteria. They can alter bacteria within the body by inserting genetic material into the host bacterium and altering its metabolism. Additionally, they can be lytic (virulent) and replicate within bacterial hosts, releasing virus particles that prove bactericidal, while the lysogenic phage types passively replicate in the host and create and release phage-encoded toxins. In contrast to bacterial microbiota, few of these phage-mediated effects have been studied in the human gut even though lysogenic phages can contribute to pathogenesis by genetically altering bacteria and causing dysbiosis (Tetz and Tetz, 2018).

An evolutionary perspective

The genomes of many bacteria (microbiome) were subject to selection pressures shaping the suitability of microbiota to particular environments, including the gut of mammalian hosts (see Fig. 3.1). This symbiotic relationship was subject to further pressures as early hu-

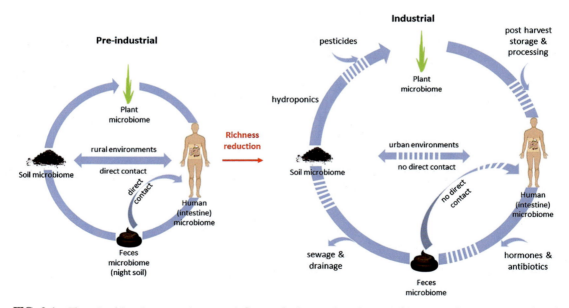

FIG. 3.1 The microbiota in our environment influence the human intestine microbiome, via direct contact with soil and feces as well as via food (quality). Our ancestors lived in close contact with the environment (A, a cycle for preindustrial microbiota). In contrast, human activities such as urbanization, industrialization of agriculture, and the modern lifestyle, including the use of pesticides and antibiotics as well as hormones (medication), together with the loss of direct contact with soil and feces has depleted the richness of microbiota (B, a cycle for industrial microbiota). This depletion of microbial richness in all compartments can substantially affect human health. *Figure and caption from Blum, W.E.H., Zechmeister-Boltenstern, S., Keiblinger, K.M., 2019. Does soil contribute to the human gut microbiome? Microorganisms, 7, 287.*

mans transitioned from small to large communities. In more recent times, such challenges were magnified when individuals migrated from rural to urban environments especially as this was accompanied by exposure to new foods (often those that were ultra-processed), environmental toxicants, and new medications (e.g., antibiotics that depleted gut microbiota). One result was a decline of certain commensal bacteria, whereas less beneficial or harmful bacteria expanded, contributing to an increased risk for various diseases (Sonnenburg and Sonnenburg, 2019a,b).[b] In fact, it has been argued that diet itself probably has more to say about the quality of microbiota than host genetics. This said, although it had been believed that ~10% of microbiota in humans were heritable this may have been due to analyses being restricted to only a single point in time. When fecal samples of 585 baboons were evaluated at a single time point, only 7% of the microbiome was deemed to be heritable. However, when the analysis was conducted several times over wet and dry seasons, which was accompanied by differences in diets, up to 97% of phenotypes related to the microbiome features were heritable (higher in the wet than a dry season) after controlling for age, diet, and socioecological factors (Grieneisen et al., 2021). Clearly, the heritability of the gut microbiome is considerable, but its expression is subject to variations attributable to season and diet that are accompanied by microbiota diversity.

As essential as antibiotics have been to treat bacterial infections, their use indiscriminately destroys bacteria that are presumed to be either good or bad, thereby increasing susceptibility to diseases, and may also diminish the efficacy of vaccines in vulnerable individuals (Hagan et al., 2019). There are, however, instances in which antibiotic-induced suppression of microbiota can have positive effects. Specifically, in line with the supposition that impaired arterial functioning associated with aging may be due to disturbed microbial presence, a broad-spectrum antibiotic reduced oxidative stress, inflammation, endothelial dysfunction, and arterial stiffening so that these features resembled that of younger animals (Brunt et al., 2019). Fig. 3.2 depicts some of the many factors that affect microbiota, which can promote multiple physiological phenotypes and the resulting enhanced disease susceptibility.

To some degree, gut microbiota are relatively stable. Perturbations are typically transient, but protracted alterations can arise after severe trauma, antibiotic treatment, or dietary changes. Moreover, as discussed later, early life experiences and diverse lifestyle factors exert a lasting impact on microbiota with consequences for well-being. However, there are vast interindividual differences in the composition of the gut microbiome of healthy people—even the consumption of similar foods may lead to differing microbial communities. It is consequently difficult to determine what comprises a "healthy" microbial community versus one that is counterproductive.

---

[b] The changes of microbiota associated with altered food intake didn't just occur with individuals moving from rural areas to cities. For instance, in some cases, as observed in Haiti and within Indigenous communities in Canada, colonization by white European settlers 300 years ago resulted in marked changes of diet. This occurred either because Indigenous people learned to be ashamed of the foods they had been eating or because the colonizers, having achieved their goal of taking the arable land, provided the population with the simplest foods (Steckley, 2016). Thus, the three sisters, corn, bean and squash, were replaced by the five white sins, flour, salt, sugar, milk, and lard (although for some reasons these are also referred to as "the five gifts").

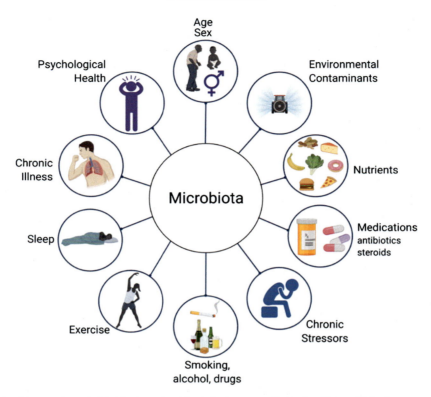

FIG. 3.2 A wide assortment of intrinsic and extrinsic factors can affect microbiota within the gut and numerous body organs, which can foster the emergence of numerous diseases. Many of these variables are readily modifiable, whereas others are unalterable, such as age. However, lifestyles that enhance the presence of beneficial microbiota can limit biological aging. *Created using BioRender.com.*

It may be germane that lateral transmission of microbiota, especially bacteria residing on the skin, occurs between people (especially within households), and is more notable among women than men (Brito et al., 2019). Interestingly, in response to an antibiotic subsequent microbiota colonization with *Escherichia coli* and *Salmonella* was more similar in pair-housed mice and they were more resistant to infection than were mice that had been housed singly. It seemed that the mice housed together shared a particular bacteria, *Klebsiella michiganensis*, that could outcompete bacterial pathogens, such as *E. coli* and *Salmonella* (Oliveira et al., 2020). This brings us to the important question of whether some apparently noncommunicable diseases are, in a sense, actually communicable. In ensuing chapters, we will show that microbiota are linked to obesity, which favors type 2 diabetes, heart disease, and cancer; if bacteria can be passed from one member of a household to another, then their risk for these conditions might be elevated (Finlay et al., 2020). By the same token, the transfer of beneficial bacteria, especially in early life, may have positive life-long effects even diminishing the risk of some illnesses developing.

Known unknowns and what we don't know

Given that we play host to trillions of bacteria, comprising more than 5000 species, it can be understood why we still don't have a handle on the multiple linkages to health and illness. To be sure, appreciable efforts have been made to tie specific bacterial genera or species with illnesses (or good health), but it seems that another layer of understanding is needed to account for these linkages, including how they come to interact with immune and hormonal processes. This is to say nothing of the fact that only a small portion of existing microbiota has been explored. And it's just as fair to plead ignorance and say that we don't know how much we don't know, although emerging technologies have opened up this field. Despite these limitations, detailed and convincing accounts of microbiota influences on multiple physiological processes have emerged that appeared to be relevant to the development of mental illness (Cryan et al., 2019). This report detailed the complexities inherent in tying microbiota to anxiety and depression, documenting the diverse neurochemical changes induced by different microbes, some of which could serve as biomarkers for the occurrence of affective disorders. Similar processes may underlie microbiota-related physical illnesses, and there is ample reason to think that these may be responsible for some comorbidities observed between mental and physical illnesses, such as depression and cancer.

Microbiota development

Prenatal influences

The placenta is often considered a sterile environment, although detection of bacterial DNA in placental tissue suggests the presence of bacteria (Aagaard et al., 2014). For instance, although a microbiome is not present within the placenta, streptococcal bacteria, such as *Streptococcus agalactiae*, has been detected (de Goffau et al., 2019). These observations may or may not arise from maternal infections, although placental bacteria can be found in the absence of intraamniotic infection, thus suggesting an intrauterine origin. In line with this thinking, amniotic fluid from women about to undergo C-sections revealed the presence of bacterial DNA, and the first feces of newborns contained bacteria (Stinson et al., 2019). It had been alleged that bacterial presence in newborns were due to contamination arising by the experimental procedures that had been used. Indeed, when prenatal stool (meconium) was collected from infants born by C-sections, thereby avoiding contamination by vaginal birth, bacteria were not found (Kennedy et al., 2021). However, following spontaneous preterm delivery, *Mycoplasma* and *Ureaplasma* were present, which may have contributed to preterm labor (Leon et al., 2018).

The makeup of a mother's microbiota may vary during pregnancy, including a substantial increase of *Bifidobacterium* during late pregnancy, possibly facilitated by progesterone (Nuriel-Ohayon et al., 2019). It is believed that microbiota present in utero could have potent consequences on postnatal microbiota in infants. When pregnant mice were colonized with a form of *E. coli* that does not survive postnatally, thereby allowing mice to be born germ free, pups showed multiple microbial and immune changes relative to those that had not received the prenatal bacterial colonization (Gomez de Aguero et al., 2016). It also seems that bacterial presence in humans could be detected in fetal tissue during the second trimester of pregnancy and could be involved in educating the immune system, particularly activation of memory T cells, which is already functional at this time (Mishra et al., 2021). Thus, while the microbial nature of the placenta is controversial, what happens prenatally has important implications for later immune development. In fact, among pregnant women who were overweight or obese, poor dietary quality was associated with reduced gut microbiota diversity, which could contribute to metabolic disturbances that may affect the fetus. No matter how it comes about, mom transmits significant aspects of her microbiome or the actions of microbiota to the fetus, which could have implications for the emergence of postnatal diseases.

Early postnatal influences

Microbial changes occur throughout life, varying with numerous lifestyles and drugs consumed. By the time a child is 1 year of age a distinct microbial profile is apparent, and by 2–5 years of age the microbiome is much like that of adults. This is true of gut microbiota as well as microbes present on the skin. Even at these early phases of life, the diversity of the gut microbiota varies subtly between males and females and may persist into adolescence and adulthood. These differences could sex-dependently influence hormonal processes and thereby could influence subsequent disease vulnerability (Jaggar et al., 2020). Thus, when the microbiota of young children is disturbed by antibiotic treatments, life-long repercussions may ensue. Children younger than 2 years of age who had received certain antibiotic treatment (particularly cephalosporins) were more likely to subsequently develop metabolic diseases (such as obesity) as well as a constellation of immune-related disorders, including food allergies and asthma (Aversa et al., 2021). As covered in later chapters, stressors experienced during adulthood affect microbiota diversity and abundance and this is also apparent among young children. For that matter, socioeconomic factors that have multiple knock-on effects owing to lifestyle and stressor experiences were accompanied by reduced levels of beneficial bacteria.

Much of the information available concerning microbiota and health have come from studies in rodents, and even though these bacteria are often similar to those found in humans, there are sufficient differences to make this translational step a bit tenuous. Nonetheless, the findings in mice can be particularly instructive. In two sets of mice with distinct microbial communities that were maintained in different colonies, the occurrence of colorectal tumors varied greatly, corresponding to the different gut microbiota that were present. When microbiota from each of these colonies were transplanted to other mice, the recipients developed tumors that aligned with those of the donors. Those mice that develop more tumors displayed elevated CD8+ T cells in gut mucous membranes prior to tumor occurrence but reduced levels of these T cells after tumorigenesis. These T cells may have become exhausted over the course of the battle so they were less able to eliminate the tumors.

It has become eminently clear that in humans the gut microbiota constitution can be shaped by multiple experiences, such as varied early life and psychosocial factors. There is likewise ample reason to believe that political and economic forces affect gut microbiota that affect health disparities. Understandably, because of the considerable differences that appear across humans as well as those that exist across countries and cultures, it is difficult to translate the findings from one group (culture) to another. An analysis of the gut microbiome in India, for instance, revealed 943,395 genes that were unique relative to those in an integrated gene catalog taken from people in North America (Dhakan et al., 2019). Whether foods consumed or geography (or both) were responsible for these differences is unknown. Nonetheless, as gut microbiota may play a significant role in the emergence of diseases, both culture and geography ought to be considered in disease emergence and progression as well as in treatment approaches.

The differences in microorganisms across cultures and regions may be related to evolutionary changes tied to selection pressures. The greatest incidence of colorectal cancer occurs among Indigenous people in Alaska, whereas the lowest incidence occurs in rural African residents. Numerous factors could certainly account for these differences, but it may be significant that these groups differ in their diets. This difference is particularly notable in the low-fiber, high-fat diet of Alaskan indigenous people, which could promote microbiota and their metabolites that have been implicated in colon cancers (Ocvirk et al., 2020). It is also intriguing that Greenland Inuit people, who consume large amounts of fatty acids through their seafood diet, were far less likely to develop heart disease. Selection pressures may have resulted in the evolution of genes that protected against the actions of high levels of cholesterol and triglycerides (Fumagalli et al., 2015). Similar adaptations may have occurred among Indigenous people within North America. Selection for a gene that controls fatty acid desaturases (FADS, a rate-limiting enzyme involved in unsaturated fatty acid synthesis) may have emerged when the first Indigenous people lived in arctic regions and was maintained as these people inhabited the remainder of the continent (Amorim et al., 2017). Such advantages, however, may have been lost over successive generations. For example, illnesses such as type 2 diabetes had been infrequent among Inuit people within Northern Canada owing to the presence of a gene that facilitated muscles being sustained by a diet rich in fat and protein but lacking in glucose. However, with continued changes in the Inuit diet, these well-adapted genes might no longer provide the benefits seen earlier, so the incidence of diabetes has been approaching that of the white Euro-Caucasian population.

Microbial factors derived through breastfeeding

Microbiota in infants varies as a function of whether they obtained milk directly from mom versus bottle-fed infants, including that obtained from pumped milk that had been stored. The latter was associated with lower microbiota diversity and richness of several beneficial bacteria. The actual contact between the infant's mouth and the mother's breast could provide the infant with skin bacteria and might introduce microorganisms that affect the milk constituency (Moossavi et al., 2019).

The specific bacteria obtained through breastfeeding that contribute to later well-being remain to be identified, and their determination may be complicated by individual differences in maternal diet, which affect bacterial composition and diversity, as can maternal weight and BMI, antibiotic use, medications, and smoking. Furthermore, studies in rodents indicated that exercise during pregnancy conferred health benefits to a mother's offspring. This may have been through breast milk, as the milk of pregnancy-exercised mothers that was provided to pups born to sedentary mothers, improved their health. It might be surmised that providing these ingredients to children whose mothers didn't exercise or who didn't breastfeed, might likewise augment their well-being (Harris et al., 2020). At the same time, it needs to be kept in mind that genetic factors can profoundly influence the nature of the microbial community.

In mice bred to be lean or obese, it was observed using a cross-fostering paradigm (where lean and obese female mice are assigned to raise each other's pups) that several important components of the pup's microbiota were determined by the nursing mother (Treichel et al., 2019). Still, as important as environmental factors are in determining the microbiota, after some time, the microbial profile may revert to its genetically determined phenotype.

A mother's milk contains several elements exceptionally beneficial for the infant. Key among these is betaine that mom receives through whole-grain food. The infant receives the betaine through breast milk, resulting in enhanced metabolic processes by promoting beneficial gut bacteria, thereby limiting the occurrence of the infant becoming overweight (Ribo et al., 2021). In addition, the benefits of breast milk come from simple and complex carbohydrates (glycans) that influence the neonate's immune functioning and have pronounced effects on commensal bacteria. Within a few hours of birth, *Lactobacillus* and *Bifidobacteria* appear in the infant's gut, having been obtained from mother's milk, and soon afterward other nonpathogenic bacteria appear. The bacterial profile of breast milk varies appreciably across mothers, although *Proteobacteria* and *Firmicutes* are predominant. These bacteria are higher in breast-fed infants than in those that had been bottle-fed, and pathways associated with fat and vitamin metabolism, and detoxification were reduced in the breast-fed infants (Ho et al., 2018). Finally, among infants with particular microorganisms that could break down human milk oligosaccharides, which serve to enhance specific gut bacteria that influence immune functioning, lower levels of inflammation were present in the gut and the blood (Henrick et al., 2021).

In addition to bacteria, *fungi* obtained from breast milk may also contribute to an infant's growth and development, and the availability of specific mycobiota (fungal population) varies with the mode of delivery and geographic location (Boix-Amoros et al., 2019). To date, the influence of early life mycobiota on later health has not been assessed as extensively as the influence of bacteria. Overall, it seemed that mom's diet during pregnancy and the perinatal period was key in the development of the infant's microbiome and could thereby markedly affect health.

Aside from the transfer of microorganisms, breastfeeding has been thought to have many other benefits (Jonas and Woodside, 2016). Specifically, breast milk contains complex sugars (oligosaccharides) that serve as prebiotics to promote the development of bacteria that are present in the infant, as well as immunoglobulins that can protect infants from disease-promoting microbes and predict their later growth and development. Additionally, infants that had been breast-fed for at least 3 months subsequently experienced fewer respiratory allergies and incidence of asthma (Bigman, 2020). These infants were also at lower risk of developing respiratory tract infections, problems associated with the gastrointestinal tract, obesity, diabetes, and childhood leukemia and lymphoma (e.g., Amitay and Keinan-Boker, 2015). Breast milk also contains nutrients, such as fat, protein, vitamins, minerals, and long-chain polyunsaturated fatty acids, which are believed to contribute to motor functioning and perhaps the development of cognitive processes.

Beyond the influence of nutrients, breastfeeding and the accompanying skin-to-skin contact may facilitate mother-infant bonding by causing the release of hormones, such as oxytocin, thereby promoting secondary benefits on the infant's emotional well-being. In support of breastfeeding, it was cited that childhood mortality could be reduced by 800,000 and since breastfeeding provides protection against breast cancer, this practice could reduce the incidence of breast cancer by 20,000 annually (Victora et al., 2016). Both the World Health

Organization and American Academy of Pediatrics recommended exclusive breastfeeding for the first 6 months of life, after which combined breast and formula feeding was recommended until the infant was 1–2 years of age. However, breastfeeding tends to be less common in industrialized countries than in developing nations. In fact, breastfeeding is adopted relatively infrequently in the United Kingdom, France, and the Netherlands (30%–40%), while it is more common in Scandinavian countries (70%–80%) and in the US breastfeeding is initially high (75%), but tails off to about 13% before infants are 6 months of age.

Microbial factors derived through vaginal birth

During the birth process (parturition) a diverse array of beneficial vaginal microbes are transferred to the neonate. Thus, suspicions arose that infants born through Caesarean section (C-section) might lack these advantageous microbes, leaving them more vulnerable to opportunistic bacteria encountered in hospitals and could increase susceptibility to subsequent inflammatory and metabolic diseases, such as allergies, asthma, and other chronic childhood immune disorders. For instance, in vaginally delivered newborns, beneficial gut *Bacteroides* and *Bifidobacteria* were abundant but were largely absent in babies born through C-section (Shao et al., 2019). In these infants, moreover, higher fecal concentrations of opportunistic bacteria were present (e.g., *Enterococcus* and *Klebsiella*) that are commonly found in hospital environments. While bacterial communities largely normalized within 9 months of birth, in most C-section infants the levels of *Bacteroides* were still relatively lower (Shao et al., 2019). The implications of these findings point to a vaginal "anointment" of protective bacteria that wards off opportunistic infectious agents.

A form of microbiome anointment was attempted through *vaginal seeding*, which involved vaginal swabs being applied to the neonate to avert potential health risks that could be incurred in C-section newborns. It was indeed observed that exposure to vaginal microbes at birth was associated with the normalization of the infants' microbial development over the ensuing year. Although such studies have offered support for the value of this procedure, the findings across studies were frequently inconsistent, and the possibility was considered that seeding could transmit pathogens to the infant. Ultimately, the American College of Obstetricians and Gynecologists indicated that the data were not sufficiently compelling to support a causal connection between vaginal seeding and later infant well-being.

Old microbial friends

It's not unusual for new parents to be a bit obsessed with germs encountered by their child and take steps to ensure her/his safety. This doesn't sound unreasonable, except that there is such a thing as being too extreme. Some years ago, the "hygiene hypothesis" was introduced, asserting that protecting children from every bug can lead to a relatively naïve and less well-versed immune system, which could diminish the ability to fight infection. Moreover, raising a child in an environment that was a bit too sterile could interfere with the development of immune tolerance and consequently later encounters with ordinarily harmless stimuli could promote exaggerated responses, manifested as the elevated occurrence of allergies and asthma (Strachan, 2000). This hypothesis had been developed based on observational studies and it is possible that the presumed increase of these disorders may have reflected improved illness detection rather than greater illness occurrence.

A modest variant of this notion, the "old friends" hypothesis asserted that "old microbial friends" were necessary for the development of functionally effective immunoregulatory processes. The absence of encounters with microorganisms during the early postnatal period might undermine immune processes, hence increasing vulnerability to physical diseases associated with excessive inflammatory immune responses (Rook et al., 2015) and also increasing the development of mental illnesses in response to psychosocial stressors. In a like fashion, diminishing the abundance of old friends (e.g., antibiotic use) during critical developmental windows might disturb T_{reg} cell functioning and thus might encourage the occurrence of some diseases in adolescence or adulthood and it was even maintained that these early experiences could influence susceptibility to autoimmune disorders, inflammatory bowel disease, and a constellation of psychological disorders.

As it happens, living in an environment enriched by microbial friends (e.g., living on farms) was accompanied by the reduced occurrence of microbiota-related illnesses, such as inflammatory bowel disorder, although encounters with pesticides have been implicated in the provocation of an exceptionally wide array of disorders, including cancer, neurodegenerative, and metabolic disorders (Mostafalou and Abdollahi, 2017). Essentially, we get to pick our poison.

Microbiota and immune functioning

Certain food constituents, particularly the fiber and starch present in grains, potatoes, and legumes, are not readily broken down in the gut and promote the availability of energy for microbiota, resulting in the formation of short-chain fatty acids (SCFAs), such as butyrate, propionate, and acetate. These SCFAs have multiple benefits, including the promotion of gut barrier integrity, together with their antiinflammatory and antioxidant actions. Microbial metabolites, such as butyrate, also influence immune activity and promote decreased tumor cell proliferation, arrest of the cell cycle, and apoptosis within some cancers. In addition, several small molecules, including amino acids, fatty acids, bile acids, and vitamins are effective in altering gut microbial processes, thereby affecting health. In contrast, members of the *Enterobacteriaceae* family (e.g., *Escherichia*, *Shigella*, *Proteus*, and *Klebsiella*) have been considered as nonbeneficial or harmful owing to their capacity to produce inflammation and inflammation-related illnesses (Huttenhower et al., 2014). It is important to underscore that while some microbiota are placed into the "bad" category, activation of genes for specific enzymes can be promoted by microbiota that affect bile acids in the gastrointestinal tract so they are instrumental in flipping these microbiota to the "good" side (Doden et al., 2021). By detoxifying bile acids that can provoke DNA damage, the occurrence of cancers of the colon, liver, and esophagus can be diminished. The functioning of these processes may be moderated by external factors, such as foods eaten, which may contribute to the ties between dietary factors and the development of some forms of cancer.

Many viral and bacterial challenges assail the human immune system, resulting in acquired protection, often aided by medications and particular lifestyles. Diverse microbiota communities also serve to support immunity, and hence disruption of the balance in beneficial and harmful microbiota can create inflammatory and endocrine variations that promote psychological and physical illnesses. Furthermore, within the gut, a mucosal barrier produced by epithelial cells segregates microbiota from innate immune cells, so intestinal permeability is limited. However, some microbiota populations can produce toxins that encourage inflammation and hence cancer occurrence (Vivarelli et al., 2019). These changes may promote further endocrine or immune alterations that aggravate the microbial state via feedback effects, creating the amplification of microbiota dysregulation and augmented inflammation that spirals out of control.

Changes in diet can also contribute to these escalating immune-related pathologies. It hardly needs saying that efficient immunity requires food-derived fatty acids, monounsaturated fats, as well as essential vitamins and minerals (e.g., zinc, selenium, iron, copper, and folic acid). The omission of fibers can markedly affect the microbial diversity of the gut, thereby affecting the capacity of the immune system to deal with pathogens. As described earlier, dietary factors in groups of people can cause the disappearance of certain gut microbiota, and the consumption of specific diets over generations can result in the microbial community becoming fixed. This may have some bearing on the selection processes that contribute to cultural differences in microbiota and disease, as well as the changes of microbiota that develop as individuals move from rural to urban environments (Sonnenburg et al., 2016) or simply living under conditions that act against beneficial microbiota.

Interplay occurs between microbial processes and immune functioning. Immunoglobulins, notably IgA released from specialized cells within the mucosa, can regulate gut bacteria (Rollenske et al., 2021). Conversely, commensal bacteria and their metabolites have impressive actions on numerous components of the immune system. Intestinal microbial composition and the timing and sequence of initial microbe exposure can affect an organism's B cell receptor diversity (Li et al., 2020), and can modulate T_h1, T_h2, T_h17, CD8+ T, and T_{reg} cell functioning, along with altered NK cell activity (Geva-Zatorsky et al., 2017). Indeed, *Bacteroides fragilis* influenced T_h1 development, whereas *Clostridium* promoted the production of T_{reg} cells. Still other species (e.g., segmented filamentous bacterium or *Porphyromonas uenonis*) mobilize T_h17 cells leading to further immune changes. By manipulating gut microbiota it may be possible to enhance the efficacy of vaccines that are often suboptimal, in part owing to age and the actions of lifestyles secondary to coming from impoverished areas (Lynn et al., 2021). Aside from the interaction between microbes and adaptive immune responses, as described in Fig. 3.3, innate lymphoid cells (ILCs) can be stimulated by microbiota and might thereby influence (enhance or suppress) the occurrence and development of cancer (Panda and Colonna, 2019). There is clearly ample opportunity for the microbiota to shape immune functioning and may do so within the tumor microenvironment.

Microbiota and inflammation

It is notable that shifts in bacterial species that lead to an increase of gut *Enterococcus faecium* can promote damaging inflammation and can act against cancer treatments. In addition, microbiota create an enormous number of very small proteins whose function is largely unknown but may have multiple roles, such as acting against unwanted bacteria, which keep the gut healthy. With such complexity being present, it is easy to see how either pro- or anti-tumor immunity could be modulated in the face of various environmental stimuli but finding how this comes about is exceptionally challenging.

Gut microbiota affect the release of cytokines by immune cells and can indirectly influence the gut epithelial barrier and alter mucus secretions. Furthermore, environmental factors may influence microbiota at sites where the integrity of the epithelial and associated protective

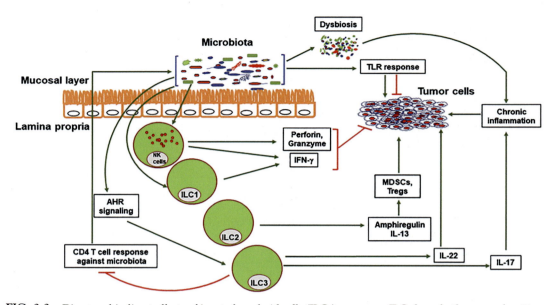

FIG. 3.3 Direct and indirect effects of innate lymphoid cells (ILCs) on cancer. ILCs have both pro- and antitumorigenic effect. ILC1s and NK cells block tumor cell growth through IFN-γ secretion. NK cells also lyse tumor cells through secretion of perforin and granzymes. IL-13 and amphiregulin secreted by ILC2s enhance myeloid-derived suppressor cells (MDSC) and T_{reg} functions, respectively, thereby facilitating immune evasion by tumors. ILC3 helps in maintaining diverse microbiota and induces CD4 T cell tolerance for commensal bacteria, preventing dysbiosis and chronic inflammation that create a pro-tumorigenic microenvironment. Species in a healthy microbiota can also augment anticancer therapy. However, ILC3s can promote tumor growth by excessive secretion of IL-22. Through inappropriate release of IL-17, ILC3s also contribute to chronic inflammation that promotes tumorigenesis. Products from microbiota enhance TLR activation, which can promote or inhibit tumors depending on the context. *Figure and caption from Panda, S.K., Colonna, M., 2019. Innate lymphoid cells: a potential link between microbiota and immune responses against cancer. Semin. Immunol. 41, 101271.*

tissue is compromised. If the coexistence of microbes and immune cells is disturbed at these sites pathogenic bacteria may expand and disturb the epithelial barrier to create a "leaky gut," so that intestinal bacteria can leak into the body, precipitating inflammatory-related pathologies (Yoo et al., 2020). Leaky gut syndrome has only recently received experimental or clinical attention, but has been implicated in the promotion of varied medical conditions, especially those in which inflammatory factors are prominent, such as Crohn's and celiac disease, and also appears in other conditions. The source for this is not fully understood, but may be related to food allergies and certain medical treatments (e.g., radiation) that cause gut bacterial disturbances.

With few exceptions, the processes by which microbiota affect immune functioning (and vice versa) are only now being elucidated, and in the main, the specific bacteria that are tied to particular diseases have yet to be identified, although certain bacteria are known to engender either positive or negative outcomes by affecting disparate systems. Some microbes can influence antiinflammatory components of the immune system (T_{reg} cells and IL-10), thereby maintaining self-tolerance and preventing autoimmune disorders. However, they can also interact with immune processes to promote inflammation, varying with specific contexts so that under some conditions pathogenic stimuli overwhelm the positive actions of beneficial bacteria and the immune system is affected accordingly (Wiles and Guillemin, 2019). Apart from promoting inflammation and suppressing anticancer immune functioning, the actions of microbiota in promoting diverse types of cancers might occur because they can produce DNA-damaging toxins and carcinogenic metabolites. Furthermore, should a cancer develop, microbiota can cause tumors to become resistant to chemotherapeutic agents.

As gut microbiota are suspected of being linked to cancer occurrence, it is reasonable to suppose that such outcomes would be particularly prominent within the digestive system and could even be tied to a cancer's progression and response to treatments. Commensurate with this perspective, colorectal cancer patients exhibited microbial alterations in which the prevalence of several pathogenic species was relatively high. Likewise, infection is strongly related to the development of colorectal cancer, as are other microbial species, including *B. fragilis*, *E. coli*, and *Fusobacterium nucleatum*. It also seems that microbial factors may promote cancers at distal locations, possibly by promoting inflammation at these sites (Buchta Rosean et al., 2019).

Since microbiota are present at many locations on and in the body, it might be counterproductive to focus solely on gut bacteria. For instance, the microenvironment associated with breast cancer may be distinct from that seen in the absence of a breast tumor. Developmental studies have also implicated in utero exposure to certain xenobiotics (those ordinarily foreign to the body) in programming mammary microbiota and consequently thought to promote tumorigenesis at much later times (Stiemsma and Michels, 2018).

In light of the actions of microbiota, it was suggested that they could be used in a therapeutic capacity to deliver specific bacteria or metabolites to particular niches, thus limiting off-target effects that are ordinarily provoked by standard medications. Furthermore, when combined with computational models or artificial intelligence (e.g., machine learning), it ought to be possible to devise individualized treatments (a "precision medicine" approach) to diminish disease occurrence and enhance treatment efficacy. This could potentially be enhanced by focusing on particular nutrients (e.g., foods that serve as prebiotics, probiotics, and a combination of two, i.e., synbiotics), and perhaps the use of synthetically engineered microorganisms. It will be shown later (Chapter 15), that probiotics could be beneficial for patients being treated with chemotherapy or immunotherapy, and once more this could potentially be enhanced based on the use of precision medicine. Significantly, there is ample reason to believe that the benefits of probiotics may occur more readily when obtained through diet than from supplements.

Microbiota and illness comorbidities

Some illnesses may come about on a seemingly random basis, unconnected to identifiable events or experiences. Other illnesses, in contrast, are often comorbid with one or more conditions, and may even signal increased risk for further illnesses. It shouldn't be surprising to find that heart disease, stroke, or cancer, are associated with the development of depressive symptoms. Importantly, however, depressive illness frequently precedes these conditions, often by many years. In fact, the prognosis for individuals who previously had been depressed (or experienced substance use disorders) was worse than for others and they died earlier as a result of cancer (and other illnesses) than did individuals who had not experienced depressive illness (Plana-Ripoll et al., 2019).

There may be many possible reasons for the appearance of comorbid conditions across multiple diseases. One illness can set in motion biological changes that promote the occurrence of a second condition. Alternatively, multiple illnesses may have common denominators, such as elevated levels of inflammatory factors that independently favor their occurrence. It is equally possible that illnesses share etiological factors, including the adoption of poor lifestyles, such as eating the wrong foods (e.g., refined carbohydrates) that lead to physiological changes that promote multiple illnesses.

Tied to these possibilities is that seemingly diverse illnesses are promoted by the presence of disturbed microbial balances that promote inflammation so their cooccurrence is more likely to evolve. In essence, factors like inflammation may create conditions that favor a set of illnesses, possibly being made more prominent owing to experiential or genetic factors, and in some instances, a second "hit" might be needed to move a benign condition to one that is malignant. The frequent occurrence of comorbid illnesses is important in yet another way. Specifically, the appearance of an illness, such as depression, might be a marker that could signal increased vulnerability to later disease occurrence (Anisman and Hayley, 2012), and family physicians might take note of this.

Microbiota, anxiety, and depressive disorders

In discussing the impact of lifestyles in relation to cancer as well as noncommunicable diseases, we will return repeatedly to discussions of the impact of various challenges on the development of psychological disorders, such as anxiety and depression. Indeed, of the many illnesses comorbid with cancer, the development of depression is most common. A review of this literature made a cogent argument for disturbed microbial functioning and altered microbial metabolites that can promote inflammation and various brain changes which likely foster mood disturbances (Cryan et al., 2019).

Consistent with the view that microbiota can affect psychological processes, germ-free mice (lacking a "normal" microbiota) exhibited lower anxiety than did mice that had been raised in a standard environment. Moreover, the transfer of gut microbiota from a mouse strain that was relatively passive to one that was rambunctious, or vice versa, resulted in the behavioral phenotypes going along with the microbiota (Bercik et al., 2011). Depressive-like consequences were similarly apparent following transplantation of bacteria from depressed humans to naïve mice (Kelly et al., 2015).

In line with the link between microbes and depression, certain gut microbiota species can produce or affect brain neurotransmitters, such as GABA, which has been implicated in anxiety and depressive disorders (Patterson et al., 2019). Moreover, microbiota-provoked depressive-like features in mice were paralleled by variations of serotonin functioning, which might contribute to depression (Yano et al., 2015). Microbiome variations were also accompanied by structural changes within the brain, including the number of newly born neurons within the hippocampus (Ogbonnaya et al., 2015), possibly implicating changes of growth factors (neurotrophins) in the microbiota-brain interface supporting depression.

As depicted in Fig. 3.4, the gut contains enteroendocrine cells that release neurotransmitters/hormones or their precursors (e.g., serotonin is synthesized from tryptophan) may come to affect CNS functioning. Serotonin is, in fact, a major player within the gut and could ostensibly affect brain functioning. The peripheral release of stress hormones, such as norepinephrine, can likewise affect bacterial gene expression or signaling between bacteria and may influence the microbial habitat. A large-scale investigation suggested that lower levels of two sets of microbiota, specifically *Coprococcus* and *Dialiste*, were associated with depressive mood as well as variations of dopamine metabolites (Valles-Colomer et al., 2019), although other bacteria, notably *Lactobacillus*, have also been implicated in mood states.

These relationships, as we'll see, are multidirectional. Variations of microbial diversity and particular bacterial strains can affect brain functioning and sympathetic activation (Muller et al., 2020) and sympathetic activation and release of norepinephrine can affect gut microbial communities, which can influence immune functioning and brain neurochemistry (Anisman et al., 2018; Levy et al., 2015). Considerable evidence has amassed showing that inflammatory mechanisms or the release of gut hormones, including serotonin secreted by enteroendocrine cells, can influence afferent neural pathways, such as the vagus nerve, thereby affecting mood states (Fülling et al., 2019). This may be a fundamental factor that contributes to the comorbidities that occur between anxiety and depressive disorders and other conditions involving immune and inflammatory processes, including cancer (Anisman et al., 2018).

Even though the blood-brain-barrier (BBB) limits brain infiltration by peripheral immune cells and large molecules, cytokines can gain entry to the brain at particular sites or when the BBB is partially compromised. To a considerable extent, however, the task of surveilling the brain and spinal cord to detect threats comprising infectious agents, plaque formation, damaged neurons, and unnecessary synapses, is largely left to microglia (Allen and Barres, 2009). In essence, to some extent microglia operate like peripheral macrophages, releasing certain cytokines in response to challenge, and the functioning of microglia can also be affected by gut microbiota. Mice lacking normal microbiota (having been raised in a germ-free environment or being exposed to antibiotic treatment) displayed a microglial phenotype that was consistent with an immature state, which could render the brain vulnerable to infectious agents and would reduce antitumor surveillance. The immature microglial state can be reversed with the transfer of SCFAs that are produced by gut microbiota, indicating the important influence of the microbiome on the brain. Other prominent glial cells in the brain are the astrocytes. These play a critical role in the regulation of BBB permeability, buffering of extracellular neurotransmitter levels, and neuronal excitability, which is influenced through the metabolism of nutrients. Microbial species and SCFAs can stimulate astrocytes, just as in the case of microglia, and as we will see, astrocytes are also affected by microbiota in ways that could potentially affect cancer development or progression in the brain.

Collectively, the available findings have convincingly linked microbiota and mood changes in animal models, and in later chapters we'll see that accumulating evidence has amassed showing similar relations in humans. The significance of these psychological disorders is underscored by the many reports pointing to the comorbidity that exists between them and a wide range of physical illnesses. These psychological conditions are also of importance for cancers since mood disorders can greatly influence lifestyle choices and adherence to cancer therapies. Given the vast microbiota differences that exist between individuals, it will be important to focus on how genetic, epigenetic, social, and cultural factors moderate the

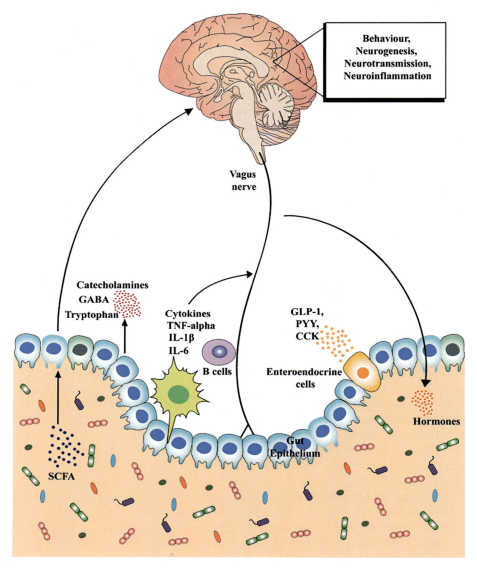

FIG. 3.4 Key communication pathways of the microbiota-gut-brain axis. There are numerous mechanisms through which the gut microbiota can signal the brain. These include activation of the vagus nerve, production of microbial antigens that recruit immune B cell responses, production of microbial metabolites (i.e., short-chain fatty acids [SCFAs]), and enteroendocrine signaling from gut epithelial cells (e.g., I-cells that release CCK, and L-cells that release GLP-1, PYY and other peptides). Through these routes of communication, the microbiota-gut-brain axis controls central physiological processes, such as neurotransmission, neurogenesis, neuroinflammation, and neuroendocrine signaling that are all implicated in stress-related responses. Dysregulation of the gut microbiota subsequently leads to alterations in all of these central processes and potentially contributes to stress-related disorders. *5-HT*, serotonin; *CCK*, cholecystokinin; *GABA*, γ-aminobutyric acid; *GLP*, glucagon-like peptide; *IL*, interleukin; *PYY*, peptide YY; *TNF*, tumor necrosis factor. *Figure and caption from Foster, J.A., Rinaman, L., Cryan, J.F., 2017. Stress & the gut-brain axis: regulation by the microbiome. Neurobiol. Stress 7, 124–136.*

influence of microbial manipulations in healthy individuals and, conversely, how they might promote poor health.

Identifying good and bad microbes

Although appropriate microbial balances are thought to favor good health, whereas imbalances may lead to poor outcomes, there remains the question of what actually constitutes a good balance and what comprises one that is counterproductive. As alluded to earlier, given the wide array of microbiota present in healthy individuals, varying across numerous factors (e.g., culture, sex, and age), there is a lack of clarity as to what a "normal" microbiome looks like, and it may not be productive to search for *the* healthy gut microbiota. Indeed, specific microbiota deemed to be healthy in one individual, might not act in the same way in a second person. Moreover, the "dys" in dysbiosis implies a negative state, but the changes of microbiota in response to challenges might actually reflect an effort to readjust microbiota in an adaptive way to meet ongoing needs. Because of these uncertainties, it was suggested that the term "dysbiosis" is being used inappropriately (Hooks and O'Malley, 2017). Simply suggesting that dysbiosis exists with a particular condition might not have any benefits for treatment any more than simply saying that schizophrenia, depression, or obsessive-compulsive disorder are due to some sort of brain chemical imbalance without specifying what such an imbalance comprises. This is not a matter of mere semantics but may have implications for directed research as well as treatment strategies.

Dysbiosis is typically considered as either a loss of microbial diversity, a loss of commensal bacteria, or the emergence of bacteria that could promote pathology (Levy et al., 2017). In addition, dysbiosis was described as comprising loss of primary (keystone) taxa that can have large effects even if they appear in low abundance, or it may entail a shift in metabolic functioning. These definitions are appropriate, but in practice, linkages to numerous pathologies likely reflect the interactions of multiple systems and while microbiota may play a role in many of these, the extent of their contribution is often unclear. Defining dysbiosis is made much more difficult by the very many different types of microorganisms, other than bacteria (e.g., archaea, viruses, fungi, protozoa), that are present within the gut, and which could contribute to diseases. As described earlier, viruses are present in very high numbers, and to a considerable extent their functions or effects are unknown, leading to them being dubbed "viral dark matter." Moreover, bacteriophages could alter the balance between ostensibly "good" and "bad" bacteria (Shkoporov and Hill, 2019), and depending upon the bacteria that the phages act upon, they may either promote inflammation and inflammatory-related illnesses or they may engender antiinflammatory actions (Huttenhower et al., 2014).

Despite the uncertainties about which bacteria are harmful and which might be beneficial, the *Lactobacillus* and *Bifidobacterium* genera are usually considered as the prototypical good microbes, and hence often appear in probiotic formulations. Among the many actions of good bacteria, often centering around effects related to immune functioning, disturbances of these lactic acid-producing bacteria have been implicated in gut-related disorders and cancer and may contribute to disturbed psychological functioning (Cryan et al., 2019). Within these genera, several species (e.g., *Eubacterium*, *Roseburia*, and *Faecalibacterium*) are viewed as being beneficial to health (Gibson et al., 2017), and it has been suggested that their inclusion in prebiotic supplements (in the form of galacto-oligosaccharides, fructo-oligosaccharides, and inulin) could foster well-being (Lordan et al., 2020).

Of the many bacterial phyla, *Firmicutes* and *Bacteroidetes* make up the largest bulk of gut bacteria, and through variations of vitamin A they affect intestinal immune functioning. Of particular importance to well-being is that alterations of these bacteria are linked to obesity, and hence may be related to cancer (as described in Chapter 7). In fact, the transfer of these and other microbiota species from an obese donor to germ-free mice resulted in an increase in total body fat relative to that observed when the donor was a lean mouse. Among germ-free mice that received fecal transplants from humans, their microbiota was altered so that it resembled that seen in the human donors, although features of many of these microbes had changed, pointing to difficulties that can be encountered using this procedure. There have, in fact, been questions as to what extent interspecies transplantation of microbiota is relevant to microbiota-related diseases in humans (Walter et al., 2020).

Ordinarily, bacterial diversity is lost with age so the appearance of pathobionts (e.g., *Fusobacteria*) increases, whereas beneficial butyrate-producing bacteria (*Faecalibacterium prausnitzii/Roseburia* vs *Eubacterium limosum*) may decline. In addition to these variations, coliforms that ordinarily reside within the gut increase with age and may exert a negative influence on beneficial microbes. Moreover, with age the microbes present in the gut shift so that those that survive without oxygen can become more plentiful, which can have implications on cancer growth (Leite et al., 2021).

Given the toll that aging extracts on microbial balances, it is informative that transferring gut microbiota from young to older mice could rejuvenate gut immune functioning (Stebegg et al., 2019) as well as immune processes within peripheral tissues and brain, and in parallel enhance cognitive functioning (Boehme et al., 2021). These data might thus have implications for diseases of aging, although to date this was assessed in mice only. Gut dysbiosis was likewise evident in human progeria patients (a rare genetic condition in which the body ages exceptionally quickly), and in mouse models of this condition, health and life-span were enhanced by transplantation of fecal microbiota from wild-type mice, possibly owing to specific microbial elevations and the subsequent increase of secondary bile acids (Barcena et al., 2019).

The influence of environmental toxicants

We have known for many years that numerous agents are hazardous for human health, although avoiding them entirely is virtually impossible, especially as we are often unaware of their presence. These agents can affect our well-being through diverse routes, but only relatively recently has it become apparent that effects on gut microbes may be involved in these processes.

The gut is essentially an open system that is integrated with the environment, making it particularly vulnerable to external insults that can promote microbiota imbalances and might thereby pose health risks. A detailed review of this topic outlined the microbial impact of some of the most common environmental toxins encountered by humans (Chiu et al., 2020). These included bisphenols, phthalates, several pesticides, organic pollutants, and heavy metals, which can affect various hormones and gut bacteria. Some of these chemicals are known to favor diseases, such as those related to immune dysfunction or inflammatory processes (e.g., type 2 diabetes, heart disease, liver and thyroid disorders, and obesity). Many of these compounds (e.g., phthalates) that act as endocrine disruptors are found in foods humans consume thereby affecting gut microbiota and metabolites essential for good health. Bisphenols that appear in numerous consumer goods (packaging, plastics) are commonly detected in human urine and by affecting gut microbiota may contribute to obesity. Persistent organic pollutants, such as flame retardants, pesticides, and herbicides, can affect microbiota throughout life, and polychlorinated biphenyls (PCBs) affect gut microbiota, gut permeability, and intestinal inflammation. We will discuss such agents further in later chapters when we consider the impact of carcinogens.

Summary and conclusions

The past two decades have seen remarkable attention devoted to the potential health benefits and harms associated with microbial processes. Even a cursory examination of the relevant literature attests to the many findings suggesting that gut microbiota are linked to a wide range of communicable and noncommunicable illnesses. Some of these conditions have been tied to microbiota-induced immune and inflammatory actions, whereas others may be related to peripheral and central neurochemical alterations. There have been indications of key microbiota species contributing to some illnesses, but in the main it is uncertain what combination of microbiota are fundamental in this regard even if it is widely thought that microbiota alterations through specific diets could ameliorate illnesses. Ultimately, it will be essential to determine the effectiveness of such treatments on an individual basis (i.e., through a precision medicine approach) rather than assuming that a given treatment will necessarily be effective for all people. Using innovative and exciting methodologies the feasibility of using a machine-learning algorithm was demonstrated based on blood parameters, dietary habits, anthropometric measures, physical activity, and gut microbiota indices that could predict individual glycemic responses to meals (Zeevi et al., 2015). Ultimately this could be enhanced by incorporating other essential biological data and measures of lifestyle factors, such as sleep quality and stressor experiences. The complexity of this approach is obvious, but this has not deterred charlatans from entering the marketplace and attempting to capitalize on the public's needs by selling simplistic approaches that are of little actual value. Go figure!

References

Aagaard, K., Ma, J., Antony, K.M., Ganu, R., Petrosino, J., et al., 2014. The placenta harbors a unique microbiome. Sci. Transl. Med. 6, 237ra265.

Allen, N.J., Barres, B.A., 2009. Neuroscience: glia—more than just brain glue. Nature 457, 675–677.

Amitay, E.L., Keinan-Boker, L., 2015. Breastfeeding and childhood leukemia incidence: a meta-analysis and systematic review. JAMA Pediatr. 169, e151025.

Amorim, C.E., Nunes, K., Meyer, D., Comas, D., Bortolini, M.C., et al., 2017. Genetic signature of natural selection in first Americans. Proc. Natl. Acad. Sci. 114, 2195–2199.

Anisman, H., Hayley, S., 2012. Illness comorbidity as a biomarker? J. Psychiatry Neurosci. 37, 221–223.

Anisman, H., Hayley, S., Kusnecov, A., 2018. The Immune System and Mental Health. Academic Press, United Kingdom.

Aversa, Z., Atkinson, E.J., Schafer, M.J., Theiler, R.N., Rocca, W.A., et al., 2021. Association of infant antibiotic exposure with childhood health outcomes. Mayo Clin. Proc. 96, 66–77.

Barcena, C., Valdes-Mas, R., Mayoral, P., Garabaya, C., Durand, S., et al., 2019. Healthspan and lifespan extension by fecal microbiota transplantation into progeroid mice. Nat. Med. 25, 1234–1242.

Bercik, P., Denou, E., Collins, J., Jackson, W., Lu, J., et al., 2011. The intestinal microbiota affect central levels of brain-derived neurotropic factor and behavior in mice. Gastroenterology 141, 599–609.

Bigman, G., 2020. Exclusive breastfeeding for the first 3 months of life may reduce the risk of respiratory allergies and some asthma in children at the age of 6 years. Acta Paediatr. 109, 1627–1633.

Boehme, M., Guzzetta, K.E., Bastiaanssen, T.F.S., van de Wouw, M., Moloney, G.M., et al., 2021. Microbiota from young mice counteracts selective age-associated behavioral deficits. Nat. Aging 1, 666–676.

Boix-Amoros, A., Puente-Sanchez, F., du Toit, E., Linderborg, K.M., Zhang, Y., et al., 2019. Mycobiome profiles in breast milk from healthy women depend on mode of delivery, geographic location, and interaction with bacteria. Appl. Environ. Microbiol. 85.

Brito, I.L., Gurry, T., Zhao, S., Huang, K., Young, S.K., et al., 2019. Transmission of human-associated microbiota along family and social networks. Nat. Microbiol. 4, 964–971.

Brunt, V.E., Gioscia-Ryan, R.A., Richey, J.J., Zigler, M.C., Cuevas, L.M., et al., 2019. Suppression of the gut microbiome ameliorates age-related arterial dysfunction and oxidative stress in mice. J. Physiol. 597, 2361–2378.

Buchta Rosean, C., Feng, T.Y., Azar, F.N., Rutkowski, M.R., 2019. Impact of the microbiome on cancer progression and response to anti-cancer therapies. Adv. Cancer Res. 143, 255–294.

Chiu, K., Warner, G., Nowak, R.A., Flaws, J.A., Mei, W., 2020. The impact of environmental chemicals on the gut microbiome. Toxicol. Sci. 176, 253–284.

Cryan, J.F., O'Riordan, K.J., Cowan, C.S.M., Sandhu, K.V., Bastiaanssen, T.F.S., et al., 2019. The microbiota-gut-brain axis. Physiol. Rev. 99, 1877–2013.

de Goffau, M.C., Lager, S., Sovio, U., Gaccioli, F., Cook, E., et al., 2019. Human placenta has no microbiome but can contain potential pathogens. Nature 572, 329–334.

Dhakan, D.B., Maji, A., Sharma, A.K., Saxena, R., Pulikkan, J., et al., 2019. The unique composition of Indian gut microbiome, gene catalogue, and associated fecal metabolome deciphered using multi-omics approaches. GigaScience 8, Giz004.

Doden, H.L., Wolf, P.G., Gaskins, H.R., Anantharaman, K., Alves, J.M.P., et al., 2021. Completion of the gut microbial epi-bile acid pathway. Gut Microbes 13, 1–20.

Finlay, B.B., Humans, C., Microbiome., 2020. Are noncommunicable diseases communicable? Science 367, 250–251.

Fülling, C., Dinan, T.G., Cryan, J.F., 2019. Gut microbe to brain signaling: what happens in vagus. Neuron 101, 998–1002.

Fumagalli, M., Moltke, I., Grarup, N., Racimo, F., Bjerregaard, P., et al., 2015. Greenlandic Inuit show genetic signatures of diet and climate adaptation. Science 349, 1343–1347.

Gentile, C.L., Weir, T.L., 2018. The gut microbiota at the intersection of diet and human health. Science 362, 776–780.

Geva-Zatorsky, N., Sefik, E., Kua, L., Pasman, L., Tan, T.G., et al., 2017. Mining the human gut microbiota for immunomodulatory organisms. Cell 168, 928–943 e911.

Gibson, G.R., Hutkins, R., Sanders, M.E., Prescott, S.L., Reimer, R.A., et al., 2017. Expert consensus document: the International Scientific Association for Probiotics and Prebiotics (ISAPP) consensus statement on the definition and scope of prebiotics. Nat. Rev. Gastroenterol. Hepatol. 14, 491–502.

Gomez de Aguero, M., Ganal-Vonarburg, S.C., Fuhrer, T., Rupp, S., Uchimura, Y., et al., 2016. The maternal microbiota drives early postnatal innate immune development. Science 351, 1296–1302.

Grieneisen, L., Dasari, M., Gould, T.J., Bjork, J.R., Grenier, J.C., et al., 2021. Gut microbiome heritability is nearly universal but environmentally contingent. Science 373, 181–186.

Hagan, T., Cortese, M., Rouphael, N., Boudreau, C., Linde, C., et al., 2019. Antibiotics-driven gut microbiome perturbation alters immunity to vaccines in humans. Cell 178, 1313–1328 e1313.

Harris, J.E., Pinckard, K.M., Wright, K.R., Baer, L.A., Arts, P.J., et al., 2020. Exercise-induced 3′-sialyllactose in breast milk is a critical mediator to improve metabolic health and cardiac function in mouse offspring. Nat. Metab. 2, 678–687.

Henrick, B.M., Rodriguez, L., Lakshmikanth, T., Pou, C., Henckel, E., et al., 2021. Bifidobacteria-mediated immune system imprinting early in life. Cell 184, 3884–3898.

Ho, N.T., Li, F., Lee-Sarwar, K.A., Tun, H.M., Brown, B.P., et al., 2018. Meta-analysis of effects of exclusive breastfeeding on infant gut microbiota across populations. Nat. Commun. 9, 4169.

Hooks, K.B., O'Malley, M.A., 2017. Dysbiosis and its discontents. mBio 8, e01492-17.

Huttenhower, C., Kostic, A.D., Xavier, R.J., 2014. Inflammatory bowel disease as a model for translating the microbiome. Immunity 40, 843–854.

References

Jaggar, M., Rea, K., Spichak, S., Dinan, T.G., Cryan, J.F., 2020. You've got male: sex and the microbiota-gut-brain axis across the lifespan. Front. Neuroendocrinol. 56, 100815.

Jonas, W., Woodside, B., 2016. Physiological mechanisms, behavioral and psychological factors influencing the transfer of milk from mothers to their young. Horm. Behav. 77, 167–181.

Kelly, J.R., Kennedy, P.J., Cryan, J.F., Dinan, T.G., Clarke, G., et al., 2015. Breaking down the barriers: the gut microbiome, intestinal permeability and stress-related psychiatric disorders. Front. Cell. Neurosci. 9, 392.

Kennedy, K.M., Gerlach, M.J., Adam, T., Heimesaat, M.M., Rossi, L., et al., 2021. Fetal meconium does not have a detectable microbiota before birth. Nat. Microbiol. 6, 865–873.

Leite, G., Pimentel, M., Barlow, G.M., Chang, C., Hosseini, A., et al., 2021. Age and the aging process significantly alter the small bowel microbiome. Cell Rep. 36, 109765.

Leon, L.J., Doyle, R., Diez-Benavente, E., Clark, T.G., Klein, N., et al., 2018. Enrichment of clinically relevant organisms in spontaneous preterm-delivered placentas and reagent contamination across all clinical groups in a large pregnancy cohort in the United Kingdom. Appl. Environ. Microbiol. 84, e00483-18.

Levy, M., Kolodziejczyk, A.A., Thaiss, C.A., Elinav, E., 2017. Dysbiosis and the immune system. Nat. Rev. Immunol. 17, 219–232.

Levy, M., Thaiss, C.A., Katz, M.N., Suez, J., Elinav, E., 2015. Inflammasomes and the microbiota—partners in the preservation of mucosal homeostasis. Semin. Immunopathol. 37, 39–46.

Li, H., Limenitakis, J.P., Greiff, V., Yilmaz, B., Scharen, O., et al., 2020. Mucosal or systemic microbiota exposures shape the B cell repertoire. Nature 584, 274–278.

Liang, G., Zhao, C., Zhang, H., Mattei, L., Sherrill-Mix, S., et al., 2020. The stepwise assembly of the neonatal virome is modulated by breastfeeding. Nature 581, 470–474.

Lordan, C., Thapa, D., Ross, R.P., Cotter, P.D., 2020. Potential for enriching next-generation health-promoting gut bacteria through prebiotics and other dietary components. Gut Microbes 11, 1–20.

Lynn, D.J., Benson, S.C., Lynn, M.A., Pulendran, B., 2021. Modulation of immune responses to vaccination by the microbiota: implications and potential mechanisms. Nat. Rev. Immunol. (May 17), 1–14. In press.

Mishra, A., Lai, G.C., Yao, L.J., Aung, T.T., Shental, N., et al., 2021. Microbial exposure during early human development primes fetal immune cells. Cell 184, 3394–3409.

Moossavi, S., Sepehri, S., Robertson, B., Bode, L., Goruk, S., et al., 2019. Composition and variation of the human milk microbiota are influenced by maternal and early-life factors. Cell Host Microbe 25, 324–335.

Mostafalou, S., Abdollahi, M., 2017. Pesticides: an update of human exposure and toxicity. Arch. Toxicol. 91, 549–599.

Muller, P.A., Schneeberger, M., Matheis, F., Wang, P., Kerner, Z., et al., 2020. Microbiota modulate sympathetic neurons via a gut-brain circuit. Nature 583, 441–446.

Nuriel-Ohayon, M., Neuman, H., Ziv, O., Belogolovski, A., Barsheshet, Y., et al., 2019. Progesterone increases Bifidobacterium relative abundance during late pregnancy. Cell Rep. 27, 730–736.

Ocvirk, S., Wilson, A.S., Posma, J.M., Li, J.V., Koller, K.R., et al., 2020. A prospective cohort analysis of gut microbial co-metabolism in Alaska Native and rural African people at high and low risk of colorectal cancer. Am. J. Clin. Nutr. 111, 406–419.

Ogbonnaya, E.S., Clarke, G., Shanahan, F., Dinan, T.G., Cryan, J.F., et al., 2015. Adult hippocampal neurogenesis is regulated by the microbiome. Biol. Psychiatry 78, e7–e9.

Oliveira, R.A., Ng, K.M., Correia, M.B., Cabral, V., Shi, H., et al., 2020. *Klebsiella michiganensis* transmission enhances resistance to Enterobacteriaceae gut invasion by nutrition competition. Nat. Microbiol. 5, 630–641.

Panda, S.K., Colonna, M., 2019. Innate lymphoid cells: a potential link between microbiota and immune responses against cancer. Semin. Immunol. 41, 101271.

Patterson, E., Ryan, P.M., Wiley, N., Carafa, I., Sherwin, E., et al., 2019. Gamma-aminobutyric acid-producing lactobacilli positively affect metabolism and depressive-like behaviour in a mouse model of metabolic syndrome. Sci. Rep. 9, 16323.

Plana-Ripoll, O., Pedersen, C.B., Agerbo, E., Holtz, Y., Erlangsen, A., et al., 2019. A comprehensive analysis of mortality-related health metrics associated with mental disorders: a nationwide, register-based cohort study. Lancet 394, 1827–1835.

Ribo, S., Sanchez-Infantes, D., Martinez-Guino, L., Garcia-Mantrana, I., Ramon-Krauel, M., et al., 2021. Increasing breast milk betaine modulates Akkermansia abundance in mammalian neonates and improves long-term metabolic health. Sci. Transl. Med. 13, eabb0322.

Rollenske, T., Burkhalter, S., Muerner, L., von Gunten, S., Lukasiewicz, Y., 2021. Parallelism of intestinal secretory IgA shapes functional microbial fitness. Nature 598, 657–661.

Rook, G.A., Lowry, C.A., Raison, C.L., 2015. Hygiene and other early childhood influences on the subsequent function of the immune system. Brain Res. 1617, 47–62.

Salosensaari, A., Laitinen, V., Havulinna, A.S., Meric, G., Cheng, S., et al., 2021. Taxonomic signatures of cause-specific mortality risk in human gut microbiome. Nat. Commun. 12, 2671.

Shao, Y., Forster, S.C., Tsaliki, E., Vervier, K., Strang, A., et al., 2019. Stunted microbiota and opportunistic pathogen colonization in caesarean-section birth. Nature 574, 117–121.

Sherwin, E., Bordenstein, S.R., Quinn, J.L., Dinan, T.G., Cryan, J.F., 2019. Microbiota and the social brain. Science 366, eaar2016.

Shkoporov, A.N., Hill, C., 2019. Bacteriophages of the human gut: the "Known Unknown" of the microbiome. Cell Host Microbe 25, 195–209.

Sonnenburg, E.D., Smits, S.A., Tikhonov, M., Higginbottom, S.K., Wingreen, N.S., et al., 2016. Diet-induced extinctions in the gut microbiota compound over generations. Nature 529, 212–215.

Sonnenburg, E.D., Sonnenburg, J.L., 2019a. The ancestral and industrialized gut microbiota and implications for human health. Nat. Rev. Microbiol. 17, 383–390.

Sonnenburg, J.L., Sonnenburg, E.D., 2019b. Vulnerability of the industrialized microbiota. Science 366, eaaw9255.

Stebegg, M., Silva-Cayetano, A., Innocentin, S., Jenkins, T.P., Cantacessi, C., et al., 2019. Heterochronic faecal transplantation boosts gut germinal centres in aged mice. Nat. Commun. 10, 2443.

Steckley, M., 2016. Eating up the social ladder: the problem of dietary aspirations for food sovereignty. Agric. Hum. Values 33, 549–562.

Stiemsma, L.T., Michels, K.B., 2018. The role of the microbiome in the developmental origins of health and disease. Pediatrics 141, e20172437.

Stinson, L.F., Boyce, M.C., Payne, M.S., Keelan, J.A., 2019. The not-so-sterile womb: evidence that the human fetus is exposed to bacteria prior to birth. Front. Microbiol. 10, 1124.

Strachan, D.P., 2000. Family size, infection and atopy: the first decade of the "hygiene hypothesis". Thorax 55, S2–S10.

Tetz, G., Tetz, V., 2018. Bacteriophages as new human viral pathogens. Microorganisms 6, 54.

Treichel, N.S., Prevorsek, Z., Mrak, V., Kostric, M., Vestergaard, G., et al., 2019. Effect of the nursing mother on the gut microbiome of the offspring during early mouse development. Microb. Ecol. 78, 517–527.

Valles-Colomer, M., Falony, G., Darzi, Y., Tigchelaar, E.F., Wang, J., et al., 2019. The neuroactive potential of the human gut microbiota in quality of life and depression. Nat. Microbiol. 4, 623–632.

Victora, C.G., Bahl, R., Barros, A.J., Franca, G.V., Horton, S., et al., 2016. Breastfeeding in the 21st century: epidemiology, mechanisms, and lifelong effect. Lancet 387, 475–490.

Vivarelli, S., Salemi, R., Candido, S., Falzone, L., Santagati, M., et al., 2019. Gut microbiota and cancer: from pathogenesis to therapy. Cancers 11, 38.

Walter, J., Armet, A.M., Finlay, B.B., Shanahan, F., 2020. Establishing or exaggerating causality for the gut microbiome: lessons from human microbiota-associated rodents. Cell 180, 221–232.

Wiles, T.J., Guillemin, K., 2019. The other side of the coin: what beneficial microbes can teach us about pathogenic potential. J. Mol. Biol. 431, 2946–2956.

Yano, J.M., Yu, K., Donaldson, G.P., Shastri, G.G., Ann, P., et al., 2015. Indigenous bacteria from the gut microbiota regulate host serotonin biosynthesis. Cell 161, 264–276.

Yoo, J.Y., Groer, M., Dutra, S.V.O., Sarkar, A., McSkimming, D.I., 2020. Gut microbiota and immune system interactions. Microorganisms 8, 1587.

Zeevi, D., Korem, T., Zmora, N., Israeli, D., Rothschild, D., et al., 2015. Personalized nutrition by prediction of glycemic responses. Cell 163, 1079–1094.

Zmora, N., Zilberman-Schapira, G., Suez, J., Mor, U., Dori-Bachash, M., et al., 2018. Personalized gut mucosal colonization resistance to empiric probiotics is associated with unique host and microbiome features. Cell 174, 1388–1405.e1321.

4

Genetic and epigenetic processes linked to cancer

OUTLINE

Genetics and natural selection: The good and the bad	93	Limitations of data linking epigenetic changes to specific phenotypes	118
From Mendelian genetics to molecular biology	94	Complex gene × environment interactions	119
DNA and RNA	98	Population-level genetics	121
Mutations	99	Sex variations in cancer	121
Mutations of specific genes related to cancer	102	Ethnic variations in relation to cancer	123
Carcinogens	109	Precision (personalized) medicine	125
Epigenetic processes	110	Multiple markers across domains	127
Predicting cancer based on epigenetic changes	112	Summary and conclusions	129
Intergenerational and transgenerational changes	115	References	130

Genetics and natural selection: The good and the bad

The notion that physical characteristics are inherited has been around for millennia. But the processes through which this occurred had been a mystery, with early theorizing, as seen through the modern lens, indistinct from magical thinking. During the late 17th and early 18th century, some theorists had proposed that individual traits were determined exclusively by semen, and others went so far as to suggest that an organism was actually *preformed* within the sperm or egg, growing during gestation, and emerging fully, complete with the soul. Despite centuries of this errant thinking, early theorists weren't oblivious to the importance of the environment, and some considered that the foods eaten by mom while pregnant contributed to the formation of an organism.

The influence of inherited contributions formally materialized with Gregor Johann Mendel's study of pea plants in the 1850s and 1860s, but not until the first quarter of the 20th century was genetics applied to studies of fruit flies (*Drosophila melanogaster*). This might be a bit of an overstatement as farmers and ranchers had long bred their animals so that they would exhibit desirable features. However, to meet their objectives it wasn't necessary to understand the genetic mechanisms by which a blue-ribbon bull and the right cow would produce ideal offspring for milk and meat production.

The post-Mendelian period went through a dark period owing to ethical and moral concerns before the emergence of more enlightened perspectives on genetics. Like the generations of scholars that preceded them, many theorists focused on inherited characteristics associated with race, attributing superior characteristics to the white "race" (whites of European descent) and less desirable behavioral features to a variety of nonwhite races. These and similar views were adopted to promote "inherited intelligence" and multiple forms of racism and eugenics (and forced sterilization) that culminated in the removal of inferior (or unwanted) groups of people and ultimately genocide. Eventually, these strong-held views and the pseudoscience behind them were dismissed through sound reasoning and evidence-based research (see the history and impressive critique of "scientific racism" provided by Gould, 1981).

Finally, no discussion of heritability should fail to mention Charles Darwin. A contemporary of Mendel, Darwin had developed his views on natural selection, introducing the notion that the appearance of the animal and human characteristics resulted from selection pressures created by the environment. Indeed, our understanding of illness-related processes has come to incorporate a greater appreciation of how natural selection and genetic processes operate to benefit human evolution. At the same time, it was determined that these same processes foster the survival of bacteria and viruses that plague us today and these mechanisms also operate in cancer cells as efforts were made to eliminate them.

From Mendelian genetics to molecular biology

Physical, biological, and behavioral characteristics (*phenotypes*) are to a considerable extent determined by genes (*genotype*) inherited from mom and dad. One gene component (*allele*) is inherited from dad and one from mom, which could appear in either a dominant or recessive form. When dominant, the presence of that allele would be sufficient to determine the phenotype, irrespective of whether the other allele was dominant or recessive. Only when both inherited alleles were recessive, would the alternative phenotype be expressed. If both parents were *heterozygous*, meaning that they both carried a dominant and recessive gene, they would similarly express the dominant phenotype. One in four of their offspring could potentially inherit the recessive gene from each parent, making them homozygous recessive, and would thus express the recessive phenotype.

Like pea plants that are either yellow or green (or round or wrinkled) owing to the influence of a single gene, some pathological conditions in humans are similarly determined by the presence of a single gene (e.g., Huntington's disease, Tay Sachs disease, sickle cell anemia, cystic fibrosis, and certain forms of cancer). Fig. 4.1 provides a simple Punnett square showing the gene inheritance for cystic fibrosis when mom and dad are both heterozygous, carrying both a dominant and recessive gene for this disorder. Since this condition is only evident when both

Cystic Fibrosis

	Mother	Mother
	F	f
Father F	FF	Ff
Father f	Ff	ff

Huntington's Disease

	Mother	Mother
	h	h
Father H	Hh	Hh
Father h	hh	hh

FIG. 4.1 Punnett square showing possible offspring genotypes when both parents are heterozygous for the cystic fibrosis gene (left panel). As cystic fibrosis is a recessive trait, one of four offspring may inherit the disease. The right-hand panel shows the possible genotypes related to Huntington's disease when one parent carries the dominant gene for this disorder and the other is homozygous recessive. In this instance, half the offspring may inherit the dominant gene and hence will develop the disease.

alleles are recessive, the disorder will not be present in either parent. As shown in this figure, one of their four offspring will likely carry only the recessive gene and hence will exhibit the cystic fibrosis phenotype. In another scenario, that of Huntington's chorea, the illness is apparent if individuals carry the dominant gene. So, if a male is heterozygous for Huntington's and the female carries two recessive genes, half their offspring will only carry the recessive genes and will not be affected, whereas half will be heterozygous and will thus develop Huntington's disease. As the illness does not appear until individuals are between 30 and 50 years of age, the affected person may pass on their gene before they are aware that they are a carrier. Testing can now be done to determine whether a person has inherited the dominant gene that invariably leads to the disease, and this also offers them the opportunity consider whether they will risk having children.

It is usually considered that the offspring of mating between someone who was homozygous dominant for a particular gene (e.g., AA) with someone who was homozygous recessive (aa) would be heterozygous (Aa) and their phenotype would be like that of the AA parent. This explanation is a bit simplistic as a heterozygous child might not fully resemble the dominant parent (intermediate inheritance). The term *penetrance* is used in reference to the probability of a phenotype being expressed when a dominant gene is present (i.e., not all individuals carrying a dominant gene will actually exhibit the expected phenotype). Relatedly, in the presence of penetrance, the degree to which a phenotype is manifested can vary, which is referred to as the gene's *expressivity*.

In some instances, inheritance may be a slight bit more complicated since alleles at a specific locus may come in different forms so multiple phenotypes can emerge (as in the case of the ABO blood type). The phenotype is determined by a single gene, but this gene comes from three alleles, I^A, I^B, and I, in which the I^A and I^B are codominant. As described in another Punnett square (Fig. 4.2), the combination of alleles inherited by offspring would result in blood types that comprise A, B, AB, or they may have inherited the I allele from both parents and thus have the relatively infrequent O blood type.

Unlike these relatively simple inheritance patterns, complex behaviors and most human illnesses are more often governed by the additive or interactive effects of many genes (*polygenic effects*) although core genes may play a particularly prominent role in disease occurrence. Furthermore, the proteins derived from several genes may interact with one another or with environmental influences (*epistatic interaction*) in determining the phenotypes that emerge. Accordingly, the appearance of specific phenotypes is much more difficult to predict.

	Blood type		
	I^A	I^B	i
I^A	A $\underline{I^A I^A}$	AB $\underline{I^A I^B}$	A $\underline{I^A i}$
I^B	AB $\underline{I^A I^B}$	B $\underline{I^B I^B}$	B $\underline{I^B i}$
i	A $\underline{I^A i}$	B $\underline{I^B i}$	O $\underline{i i}$

FIG. 4.2 Punnett square showing possible blood types (shown as large letter in the center of each cell of the square) inherited when parents carry the different alleles related to blood type. These comprise I^A, I^B, and I, in which the I^A and I^B are codominant. The designations in the corner of each cell denote the inherited alleles. In this instance three of nine individuals ought to inherit the Type A or Type B blood type, two should inherit the AB blood type, whereas only one of nine offspring should inherit the O blood type.

The term heritability refers to the extent to which variation of a phenotype can be attributed to genetic variation. Many studies have adopted the Mendelian approach to determine the heritability of diverse biological and behavioral outcomes, and how these might be related to the occurrence of a wide range of illnesses. A favorite method to evaluate heritability in humans had been that of comparing traits between monozygotic twins (genetically identical) relative to dizygotic twins (nonidentical twins who share 50% of their genome). If a trait were entirely determined by genetic factors, then the relationship observed between monozygotic twins should be high and ought to be about double that evident in dizygotic twins. However, environmental factors could influence phenotypes, and because monozygotic twins are often treated differently than dizygotic twins, the relationship in the latter sets of twins might have been influenced to a greater extent by social and experiential factors. To distinguish between the relative contribution of environmental versus genetic influences, studies were conducted in monozygotic twins who were reared together or apart (separated at birth). This was initially considered to be a powerful approach to determine whether certain diseases appeared in families and pointed to the heritability of the phenotype. However, the assumption that twins who were separated at birth experienced different environments was not accurate as they typically were both raised in good homes often in similar cultural environments. In some instances, the separated twins grew up living near to one another, even attending school together. Often, as well, twins weren't separated at birth but spent several years together, and as adults they frequently had close relations, even being reunited and living together (Joseph, 2014). The point here isn't whether monozygotic twins are behaviorally more like each other than are dizygotic twins, as they certainly are. The question is what factors are responsible for these differences and the studies that attempted to tease apart genetic versus environmental influences were in some instances influenced by confounding variables.

Numerous studies have been conducted on animals to determine the heritability of a diversity of traits. Most of these studies used inbred strains of mice that originated from repeated brother × sister matings so eventually all mice of a given strain were homozygous at each of their many alleles. In effect, all members of a given strain were genetically identical save for the occurrence of random mutations or epigenetic changes that might have occurred. By using various strains, it was possible to evaluate the presence of phenotypes (e.g., the inheritance of anxiety or the presence of certain cancers) and link these to specific biological characteristics. It could readily be determined whether the appearance of genotypes and multiple phenotypes of interest would be coinherited in crosses between strains of mice that differed from one another on these variables. Specifically, in a cross between two different inbred strains, the offspring (the first filial generation or F_1) might differ from a parent strain, but each member of the F_1 generation would still carry the same set of genes. As described earlier, if mom was dominant for the gene—i.e., *AA*—and dad was recessive in this respect (being *aa*), all offspring would be heterozygous (*Aa*). However, in *segregating generations*, such as the F_2 generation (a cross between an F_1 and another F_1) and in back-crosses in which an F_1 was crossed with mice of the original parent strains, homozygous dominant, homozygous recessive, and heterozygous offspring could be present. Thus, it could be determined whether specific genotypes would correspond to a particular phenotypes, such as a hormone or immune factor, and whether these were associated with the appearance of specific diseases, such as epilepsy or certain forms of cancer. This was instructive in many ways, but because the original parent mice had been inbred, they often developed unwanted characteristics that could confound the outcomes. Furthermore, although the mice were genetically identical, extraneous variables, such as the laboratory environment or the breeding farm from which mice were obtained, could affect phenotypes, attesting to the importance of environmental factors in determining these outcomes (Crabbe et al., 1999).

In addition to studies involving Mendelian approaches, efforts were made to select for specific phenotypes within laboratory settings (much like natural selection would occur in the wild) to determine the mechanisms responsible for specific pathological conditions. For instance, it might be of interest to determine whether the level of a particular hormone was linked to a particular form of cancer. In this case, animals from a randomly bred colony of mice that were not genetically identical could be selectively bred based on their hormone levels by mating those critters that were high for a particular phenotype. Concurrently, mice low in this same hormone could be interbred. This selection procedure would be repeated in offspring (avoiding brother × sister mating) and over successive generations, the phenotypes of these lines of mice ought to become progressively more different from one another and from that of randomly bred mice. These lines could then be assessed to determine whether the presence of exceptionally high or low levels of certain hormone levels was tied to certain forms of cancer.

With the later understanding of the molecular structure of DNA based on the work of Rosalind Franklin, Watson, and Crick, the mapping of the genome was independently achieved by Venter and by US and UK teams led by Collins and Sulson. This allowed for the identification of specific genes and gene subsets that were tied to certain phenotypes and led to attempts to alter wayward genes that ordinarily caused pathology. This included the development of genetically engineered mice (*transgenic* mice), in which a gene was knocked out or overexpressed, making it feasible to determine the contribution of specific genes that might influence cancer and other illnesses. Determining the genome and ways of manipulating it also made it possible to transfer a healthy fetal gene into an individual's blood stem cells to treat sickle cell disease (Telen et al., 2019), as well as the use of gene editing through CRISPR-Cas9 and variants of this procedure to add, remove, or alter genetic material at specific locations within the genome (e.g., Doudna, 2020). We will return to the use of CRISPR-Cas9 in relation to cancer treatments and ways to limit off-target effects in Chapter 14, but it is first necessary to consider the processes by which DNA comes to determine the proteins that make up phenotypes.

DNA and RNA

Genes in and of themselves do not perform the actions of a cell. Proteins do that. Walking, running, talking, thinking, laughing, getting hungry, and making antibodies during an immune response are all the results of protein actions within and between cells. However, protein production requires DNA and RNA, nucleic acids that are the first two steps in the classic sequence that defines molecular biology: DNA-RNA-Protein. In this basic sequence, the architect of the operation—DNA—codes for the delivery of instructions via messenger RNA (mRNA) to make protein. The code is in the genes. And the genes are in the DNA, which is contained in the 23 pairs of chromosomes that all animals package in the nucleus of their cells. Genetic information—or the genes that constitute segments of DNA—is composed of specific and long sequences of nucleotides. In DNA, these nucleotides are guanine (G), adenine (A), cytosine (C), and thymine (T), which in sets of three (a codon) code for specific amino acids that in lengthy chains form proteins and regulatory elements situated near these genes. In a sense, the nucleotides, when strung together like letters of the alphabet, form words that become paragraphs, providing the instructions (or blueprints) for the formation of the phenotypes expressed by each individual.

The DNA serves as a template for the formation of RNA through a process known as *transcription*, in which a gene's DNA is copied or transcribed to form an RNA molecule. To this end, a protein called RNA polymerase is responsible for making a single strand of RNA based on the DNA. This messenger RNA (mRNA) crosses from the nucleus into the cell's cytoplasm where a form of RNA (transfer RNA; tRNA), along with ribosomes, engages in *translation*. In essence, the nucleotide code is "read" so that lengthy chains of amino acids are formed to create the primary component of proteins, such as specific hormones or immune molecules (see Fig. 4.3). Once the primary sequence of a protein is made, subsequent modifications in other parts of the cell will allow it to fold into its tertiary structure, a three-dimensional shape that permits it to exert unique functions (e.g., if it is an antibody, it will assume a shape that can bind antigen—see Chapters 1 and 2). This process is known as *posttranslational modification*.

It was commonly thought that transcription of DNA to RNA was unidirectional, but RNA in the form of chromosome-associated regulatory RNAs (carRNAs) can influence how DNA is stored and transcribed (Liu et al., 2020). Be that as it may, alterations of DNA or mRNA can markedly influence proteins that affect the phenotypes that emerge. Beyond determining physical features, the formation of proteins also affects behavior and the responses that are emitted when challenges are encountered, and they play a considerable role in the development of illnesses.

Not all of the DNA in a cell is genetic—consisting only of genes. Many nucleotide sequences are noncoding and are interspersed between coding nucleotide sequences (genes) on a DNA strand. As these were deemed to have little function, these noncoding regions were considered as "junk DNA." However, some of this "junk" can serve useful functions, such as polymerase binding or binding of transcription factors (proteins that trigger the transcription of a gene). These noncoding sequences precede functional genes, acting as activators or repressors of gene transcription. These "promoters" or "promoter regions" essentially serve as an instruction manual for the gene that follows it (although the latter can sometimes be some distance away), telling them when to turn on or off, or even when to interact with other genes. It is especially relevant that these regulatory processes can be affected by experiences or by

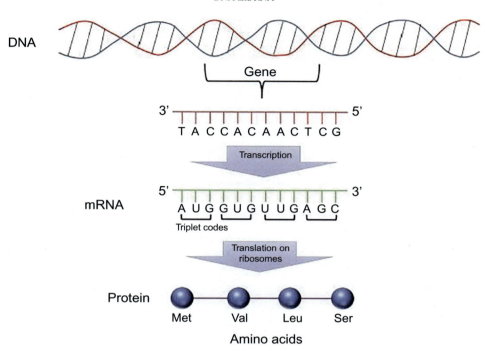

FIG. 4.3 A DNA strand, which comprises a double helix, is made up of a series of nucleotides that comprise guanine (G), adenine (A), cytosine (C), and thymine (T). The DNA is the template for the formation (transcription) of RNA, which is a single strand that corresponds to a DNA strand, except that uracil (U) is substituted for thymine. Sets of nucleotide triplets (codons) are translated to identify a given amino acid, and each amino acid will be connected to form a long string that becomes a particular protein. Specific codons also serve as "start" or a "stop" message located at the beginning or the end of an amino acid sequence that signals translation. *From Shen, H.C., 2019. Diagnostic Molecular Biology. Academic Press, New York.*

environmental stimuli, such as stressors or toxicants,[a] so that the genes' influence on other biological processes, such as hormone production, can be altered. In effect, features of the environment may influence or regulate gene transcription through actions on the noncoding regions of DNA, ultimately affecting health-related phenotypes.

Mutations

Random mutations

With continued cell duplication (and by exposure to toxicants), random mutations can arise within genes, which can have profound consequences. Even a small change in the nucleotide

---

[a] Toxins and toxicants both refer to substances that can be damaging to cells. Toxins are substances created in the body or those found naturally in the environment, such as in certain poisonous plants or mushrooms, as well as venom from insects or snakes. Toxicants, in contrast, refer to human-made products that appear in the environment. These include numerous pollutants, including pesticides and various industrial waste products within the air, water, and soil.

sequence (polymorphism) amounting to a substitution in a single nucleotide (referred to as a single nucleotide polymorphism: SNP), could potentially cause marked phenotypic alterations, including increased risk for a variety of diseases. This said mutation should not be viewed as a dirty word in genetics (the term "variant" might be a better alternative). The essence of natural selection and evolution is based firmly around the certainty that changes in genes can confer benefits that secure the survival of a species as these "mutated" genes are transmitted across generations (Nei, 2013). There has also been the view that some genetic features of cells have been retained for thousands of years because they have certain advantages, particularly during embryo development and wound healing. During the course of normal events, the actions of some genes are turned off, but in response to challenges that undermine regulatory control, these genes may be turned on, essentially reverting to their ancestral function (atavistic reversion or throwbacks) that can have beneficial consequences; but unfortunately, they may also allow for cancer cells to survive, proliferate, and metastasize (Lineweaver et al., 2021).

Studies comprising whole-genome analyses or involving targeted polymorphisms have attempted to link genes to specific health disturbances. These entailed finding a sample (cohort) of affected and nonaffected individuals (i.e., they have or do not have a particular phenotype or a family history for a particular phenotype), and then determining whether a match existed between the presence of certain genes or mutations and the appearance of the phenotype of interest. If the gene associated with an illness could be identified, then it might be possible to determine which proteins this gene is responsible for making (e.g., hormones and their receptors). If these turned out to be causally related to the illness, this would facilitate the development of treatments to attenuate or prevent pathology. Because of the very many polymorphisms that exist within everyone's genome, studies of this nature necessarily required analyses of a large number of participants. Many early studies involved relatively small cohorts, resulting in poor replicability. Moreover, when associations were observed with specific SNPs and illness, these were on many occasions interpreted as reflecting a causal connection, which of course, was inappropriate.

We all know better

There isn't a scientist in any discipline who isn't aware that correlation doesn't imply causality. Yet, it is not unusual to see findings described as if there was a causal connection, and even the direction of this relationship (If A preceded B, then A must have been the causal agent), even though the two might be independent of one another, having been promoted by a common third factor. We see these issues appear repeatedly, often encouraged by media looking for a good spin (e.g., violent video games are responsible for mass shootings; having certain genes cause people to adopt particular political views), but the language used in many research papers have been guilty of creating incorrect conclusions. In an interesting article published in the magazine Quillette, Blesske-Rechek (2019) described this as "The other crisis in psychology," but it's likely just as true in other scientific fields.

It's not just in this domain that serious errors of interpretation are encountered. Across numerous disciplines, the finding that a particular treatment leads to a biological change, as well as alleviation of symptoms, is on occasion taken (sometimes inappropriately) to suggest the identification of the mechanism responsible for the disease's occurrence or progression. For instance, increasing serotonin functioning using particular drug treatments (e.g., selective serotonin reuptake inhibitors; SSRIs) may alleviate depressive symptoms in a subset of individuals, but this doesn't necessarily mean that disturbed serotonin caused the initial depressive occurrence or their alleviation any more than the benefits of placing a feverish child in tepid water says anything about the relationship between water temperature and the mechanisms associated with fever production. This "affirming the consequent" (also referred to as "fallacy of the converse" or "Fallacy of Confirming the Consequence") is a common logical error. A randomized controlled trial can tell us that a particular treatment may or may not have a beneficial action, but this doesn't necessarily speak to the mechanism(s) responsible for a particular disease condition. After all, a treatment may seem to be effective simply because it masks symptoms.

Although we are fortunate in having cellular mechanisms that operate in a proofreading and editing capacity so that DNA integrity is maintained, some mutations still sneak through when the *spellchecker*, for whatever reason, fails to do the job adequately. A mutation may favor the occurrence of cancer by disturbing DNA repair processes, disrupting the functioning of tumor suppressor genes that ordinarily serve to prevent cancer-related processes, or by creating an oncogene in which a healthy gene is turned into one that is cancerous. A mutation could also appear in a section of a gene that interferes with apoptosis (programmed cell death). Ordinarily, apoptosis serves in a beneficial capacity for proper development or to eliminate the presence of damaged or unhealthy cells that use up resources but provide little benefit. However, when apoptosis is disturbed by a mutation, the loss of this fundamental regulatory mechanism could result in cells multiplying repeatedly, essentially becoming immortal, and possibly leading to the production of a tumor mass. A related process, ferroptosis, stems from reduced antioxidant functioning while concurrently increasing reactive oxygen species that can promote cell death. Ferroptosis has been implicated in the promotion of neuronally related illnesses, varied blood diseases, and cancer (Li et al., 2020a). The extent of the ferroptosis has also been predictive of the effectiveness of cancer therapy, potentially making it useful in predicting treatment outcomes.

Inherited mutations

Mutations are not only important in the occurrence of illnesses, such as cancer, but may have serious implications for the offspring of those carrying these mutations. In particular, mutations in germline cells (sperm and ova)—which are the vectors for transferring genetic information across generations—can be inherited from parents, creating a multigenerational problem. Indeed, across 33 cancer types, as many as 853 heritable germline variants were found. As the frequencies of mutations related to cancer can also increase with the age of the parents, presumably owing to more cell division as well as progressively greater exposure to environmental toxicants, the likelihood for cancer-relevant inherited mutations will similarly increase.

Inherited mutations may account for a significant proportion of the variance in the development of some types of cancer. By comparing germline mutations in 114 genes in healthy cells and affected cells of 12 different types of cancers it was deduced that these mutations occurred in 19% of ovarian cancer cases, 11% of stomach cancer cases, and somewhat fewer in other forms of cancer (e.g., 4% in acute myeloid leukemia cases).

Furthermore, by assessing more than 10,000 individuals with 33 different cancers, numerous germline variants were identified, some of which were related to tumor suppressors, oncogenes, and cancer drivers, and many others reflected variants that hadn't previously been extensively assessed. Of particular significance, many inherited genes could augment the ability to predict cancer treatment outcomes beyond that of usual clinical factors (Chatrath et al., 2020).

As much as specific mutations or sets of mutations may be instrumental in promoting the development of cancer, genome changes may occur that go beyond these relatively limited alterations. Specifically, chromosomal instability may be present in which genetic alterations occur. This may entail parts of chromosomes being duplicated or deleted, translocations from one part of a chromosome to another, and aneuploidy that comprises an abnormal number of chromosomes. Additionally, massive catastrophic mutations may occur through *chromothripsis* in which chromosomes shatter into multiple pieces and then reassemble but may do so in a manner that fosters cancer cell progression. Some pieces of the shattered chromosome may persist as extra-chromosomal DNA that can be reintegrated into the chromosome during radiation or chemotherapy. This may lead to more aggressive cancers and treatment resistance (Shoshani et al., 2021).

Ordinarily, in response to cellular challenges, immune system activation and localized inflammation should be triggered that facilitate immune cell migration to the site of damage or the presence of a tumor. This begs the question as to why these immune cells might be unable to destroy the cancer cells. When viral DNA or unstable cancer chromosomes appear within a cell's cytoplasm, a danger signal is activated in the form of a DNA sensing pathway cGAS-STING, in which cGAS binds to DNA, creating a compound molecule cGAMP that migrates to the cell's surface. This danger signal leads to activation of a unique immune response triggered by a protein called stimulator of interferon genes (STING) that ought to deal with the invader (Mackenzie et al., 2017). This works well for virally infected cells, but things may go awry in response to cancer cells. Among other things, using a protein, ENPP1, cancer cells can shred the cGAS-STING warning signal present on their surface, thereby averting attack by immune cells. As well, during the shredding of the danger signal, adenosine is released that diminishes inflammation, thereby limiting the accumulation of immune cells within the vicinity. This again illustrates the wiliness of cancer cells in overcoming our defenses; however, by using ENPP1 inhibitors it may be possible to outwit this aspect of cancer defense. As well, nanotechnologies have been developed that can influence different sites of the STING molecule, irrespective of the presence of cGAMP, making it a potentially effective strategy to deal with some forms of cancer (Li et al., 2021).

Mutations of specific genes related to cancer

It has been maintained that up to two-thirds of cancer risk may be attributable to random gene mutations that arise during DNA replication in normal dividing cells. The differential occurrence of cancer in diverse tissues is seen as being related to the rate of cell division that ordinarily occurs. Consistent with this perspective, an analysis of 31 tumor types showed that the lifetime risk for cancers was linked to the rate of total stem cell division (Tomasetti and Vogelstein, 2015). A greater number of mutations will occur in rapidly dividing cells than in slowly dividing, and cancer occurrence will vary accordingly.

The cancer cells present early in the progression of the disease are genetically unstable and mutate rapidly, placing them in a good position to avoid detection by the immune system. Selection pressures are also placed on tumor cells by the attacking immune system so that further tumor cell variants occur more frequently. Some of these gain the capacity to circumvent and survive attacks by the immune system so some will survive and produce similarly resistant progeny.

Several inherited mutations may cause individuals to become particularly vulnerable to multiple types of cancer. For instance, a particular gene, TP53, is responsible for the creation of the p53 protein, which serves as a tumor suppressor by detecting cellular stress or damage and initiating a process to stop cell division or promote cell death, and it is important for DNA repair. Thus, a mutation of this gene makes it more likely that cancer would develop (Yizhak et al., 2019). Indeed, TP53 mutations have been closely related to multiple forms of cancer (lung, colorectal, ovarian, esophageal, head, and neck), being present in up to 50% of affected individuals. In other forms of cancer (malignant melanoma, sarcoma, primary leukemia, testicular, and cervical cancer), the frequency is much lower, only reaching about 5%. In assessing tissue samples comprising 32 different cancers obtained from more than 10,000 individuals, TP53 mutations predicted poorer survival, and this gene could also act together with other genes in favoring cancer occurrence. Four upregulated genes were identified within mutant TP53-related tumors that also predicted a poorer prognosis, and a combination of other genes was identified that served either as tumor suppressors or pro-tumor oncogenes (Donehower et al., 2019). Findings such as these are not only informative in defining how cancers develop but also point to many cancers sharing common features, which can potentially be used as biomarkers to predict cancer development.

The prediction of breast cancer risk had been aided by the discovery that mutations of the BRCA1 and BRCA2 genes were heritable. These genes are ordinarily responsible for manufacturing a protein that facilitates the repair of damaged DNA so that inheriting a mutation of this gene results in progressively more DNA damage accumulating, rendering individuals at elevated risk of breast and ovarian cancer. Having a BRCA1 mutation was associated with a greater than 60% chance of developing breast cancer (somewhat less with a BRCA2 mutation), which is sufficiently threatening so that some women select double mastectomy (and removal of ovaries) to preclude cancer development. Clearly, however, the presence of this mutation doesn't guarantee that cancer will ultimately develop, and other genes may also contribute to breast cancer occurrence. In addition to mutations of these genes, the presence of PALB2 (partner and localizer of BRCA2) a familial-related mutation increases the risk for breast cancer by about sixfold, which has only recently been receiving significant attention (Breast Cancer Association Consortium et al., 2021). Smoking and alcohol consumption have been linked to cancer occurrence, and thus it might be expected that they would further increase cancer occurrence among individuals with BRCA1 and BRCA2 mutations, but this was not observed (Li et al., 2020c).

Several other specific genes have been identified that ordinarily suppress cell division, but in their aberrant form, they can promote cancer. The gene PTEN (phosphatase and tensin homolog) normally inhibits cell replication, but a mutation of this gene may allow glioblastoma, as well as lung, breast, and prostate cancer to grow relatively unimpeded. Ordinarily, macrophages can destroy cancer, but in PTEN deficient mice, the activation of a gene, YAP1, causes macrophages to disturb repair processes and encourage cancer progression. The macrophages may also secrete a growth factor, SPP1, which promotes the development of new blood vessels to feed the cancer cells, and concurrently protects them from programmed cell death (Chen et al., 2019).

Mutations can also occur in noncoding regions of DNA that primarily serve in a regulatory capacity. Indeed, tumor-specific antigens that uniquely appear on cancer cells were linked to junk DNA (Laumont et al., 2018). In essence, while some genes or gene mutations are associated with cancer emergence, behaving as oncogenes or inhibitors of tumor suppressors, alterations in noncoding portions of DNA may also contribute to cancer occurrence. For instance, a mutation within the noncoding region of DNA termed the U1-snRNA, was found in chronic lymphocytic leukemia, hepatocellular carcinoma, and within some brain tumors (Shuai et al., 2019). Moreover, the expression of these regulatory genes can differ between males and females, which may be significant in sexually dimorphic diseases as well as in responses to particular treatments.

As the lifetime risk of experiencing some form of cancer is relatively high—exceeding 40%—it might be expected that individuals could develop additional primary cancers. In some instances, however, the odds of developing a second type of cancer are far higher than expected probabilities. About 25% of people with cancer carry a particular genetic marker, the KRAS-variant. Among these individuals, more than 50% develop more than one type of cancer. It is thought that KRAS regulates cell division, and when it is mutated uncontrolled cell growth occurs, potentially leading to cancer. Thus, finding ways of turning off this action has been seen as critical in dealing with some forms of cancer, but this has been exceedingly difficult to accomplish.

Aside from possibly offering a potential target for cancer treatments, the KRAS mutation may also have value in signaling the occurrence and prognosis of specific cancers. Even though the presence of the BRCA1 and BRCA2 mutation has been a good predictor of breast and ovarian cancer, the presence of the KRAS mutation should also be an important consideration in helping women to make decisions as to whether they should undertake preventive surgery. This is particularly relevant as the presence of the KRAS mutation in women with a BRCA1 mutation also predicted the triple-negative form of breast cancer as well as ovarian cancer. The BRCA1 gene may also interact with other genes in determining DNA repair processes and may be important in the development of therapeutic strategies (Tarsounas and Sung, 2020).

The search for cancer-related genes has often focused on either inherited gene mutations or those influenced by environmental factors, but other DNA-related processes could potentially influence cancer occurrence and progression that have received far less attention. Specifically, beyond the DNA that makes up chromosomes, extrachromosomal DNA (ecDNA) has been identified, which may take the form of small circular DNA (described as resembling SpaghettiOs). These ecDNAs express exceptionally high levels of oncogenes across cancer types (Wu et al., 2019), including children with neuroblastoma. What causes the appearance of circular DNA is not known, nor is it entirely certain why their presence would promote tumors, although they might do so by disturbing normal DNA functioning. It is likewise possible that ecDNAs drive cancer by altered gene regulation and through greatly elevated oncogene transcription (Wu et al., 2021).

Understanding the ties between specific genes and an illness can guide the choice of treatment for some cancers, but it can be difficult to define these linkages, and challenging to modify their actions. Ultimately, through technologies to alter genes systematically, it might become possible to identify the main genetic culprits that drive even the deadliest cancers, such as glioblastoma, and perhaps point to feasible treatment strategies.

Driver genes

A high degree of tissue specificity exists in the appearance of many cancers, and genetic alterations (or altered gene functioning) were found in genes that promote tumorigenesis (Haigis et al., 2019). Such genes are considered to be cancer driver genes in that mutations are assumed to cause cancer occurrence by conferring a growth advantage to the cancer cell. An analysis based on whole-exome sequencing (i.e., the coding portions of the genome) from 11,873 tumor-normal pairs was able to identify 460 such driver genes that clustered at 21 cancer-related pathways (Dietlein et al., 2020).

Mutations of cancer driver genes within the tumor's microenvironment are thought to occur based on a natural selection process and tend to cluster with others that are cancer-promoting genes. Typically, several driver genes may be involved in promoting the development of cancers, varying with different types of cancer. Passenger mutations, in contrast, do not foster a growth advantage, but simply come along for the ride, and do not necessarily cluster in the vicinity of the cancer genes, and instead are randomly distributed. When certain genes have been associated with cancer occurrence, the next step would ideally be to turn off the regulatory processes that affect these genes to determine whether cancer occurrence is affected.

Driver genes are determined based on advanced algorithms of genome sequences within normal and cancer cells. It has been assumed that recurrence of mutations at specific positions is likely not a chance occurrence, and thus when the same mutations are seen across individuals, this signifies the presence of a driver gene, but this conclusion may not always be warranted. Some driver genes are not readily detected owing to the limits of gene sequencing that analyzes snippets rather than the whole chromosome. Thus, duplications or deletions within a DNA strand may be missed. Once again, algorithms have been developed that can deduce the presence of such mutations based on multiple snippets of DNA.

Gene signatures

It has been considered that whole-genome sequencing might make it possible to identify genetic mutations that are common across several types of cancer, which could have significant treatment implications. A large-scale whole-genome analysis of diverse metastatic solid tumors, comprising 2520 pairs of tumor and normal tissue, indicated that mutations varied broadly, although similar mutational profiles were apparent in specific cancer lesions. Features of these gene profiles were linked to therapy responsiveness and treatment resistance and thus such an approach could be used to stratify patients for specific treatments (Priestley et al., 2019).

In the same vein, a remarkably comprehensive set of studies coming from more than 1300 scientists across many countries sequenced the full genomes of 38 types of cancer obtained from 2658 samples. These studies, reported in several concurrently published reports, were broadly described in a summary paper (Campbell et al., 2020). This extensive work identified 705 mutations, including 100 or so that appeared outside of protein-coding regions that were evident in multiple cancer genomes. The signatures of these cancer-related mutations comprised base substitutions, small deletions and insertions, structural variations, and the downstream consequences of these mutations on RNA transcription. In general, individual cancers were determined by several driver mutations that often increased over the lifespan, encouraged by environmental contributions and genetic disposition. In some types of cancer (e.g., melanoma), multiple mutations appeared as parts of clusters of mutations likely stemming from genetic reconfiguration that occurred owing to a "single catastrophic" event. Significantly, several mutations may have preceded cancer occurrence by many years. These findings are all the more important since the data could be paired with individuals' family or medical history, the therapy they had received, and whether they responded to treatment.

Going a further step that comprised whole genome sequencing, whole-exome sequencing, and RNA sequencing, it was possible to identify variants that were relevant to diagnosis, prognosis, and the therapy that would be most beneficial for children with cancer. In fact, through this approach clinically significant mutations were apparent in about 20% of patients, which might not otherwise have been detected. This approach also facilitated the identification of effective therapeutic strategies in patients who had been nonresponsive to other treatments (Newman et al., 2021).

Impressive studies such as these tend to promote further questions that beg to be addressed. Although certain mutations existed prior to cancer developing, did their presence reliably predict subsequent cancer development? Further, as not all individuals who carried a given mutation developed cancer, which specific genes interacting with one another or with environmental or experiential factors foreshadowed cancer emergence? In the end, detailed prospective analyses will be needed to delineate the specific nonintrinsic factors that influence the actions of genetic actions on cancer development.

Aside from mutations that occur within cancer cells, the tumor microenvironment plays a significant regulatory role in the development of the tumor, since genetic factors have been linked to cells within the tumor's microenvironment. Elevated expression of factors that contribute to angiogenesis, the processes that provide a tumor with a blood supply and hence a source of energy might be influenced by genetic factors related to the microenvironment. Likewise, inflammation within a tumor's microenvironment can affect the growth of neoplastic cells, and immunosuppressive cells that overcome natural immunity appear in a tumor's microenvironment.

A multihit hypothesis

The occurrence of some cancers can be promoted by single mutations, but more commonly the development of cancer involves the combination of several inherited mutations as well as those caused by environmental factors. Ordinarily, proto-oncogenes may be present within cells, being aligned with the formation of proteins involved in cell division, cell growth, and regulation of programmed cell death. Under some conditions, a mutation may result in a normal gene becoming an oncogene that could favor cancer development (e.g., the presence of a germline mutation or disturbance of the actions of a tumor suppressor gene). This may comprise a first hit that puts cells in play to eventually become cancerous. With the accumulation of particular mutation(s), essentially being a second hit, the odds of cancer development increases. More than this, the influence of gene functioning may vary at different phases of an oncogene becoming a tumor. For instance, in the case of melanoma in which the BRAF mutation occurs in about one-half of cases, the "oncogenic competence" of cells occurs more during the early phases of cancer-forming processes (Baggiolini et al., 2021). The findings concerning multihits raise questions as to which factors cause a first hit that allows cells to potentially become cancerous. Are these random mutations or do environmental factors figure into this? Likewise, what are the characteristics of the second or third hit that causes these cells to become cancerous and multiply uncontrollably? To be sure, cancer occurrence might be linked to a great number of gene variants and their interactions, but lifestyles and other health conditions can also act as a second hit that favors the development of cancer, as in the case of obesity superimposed on a particular genetic backdrop.

Mammals are "mosaics" in which multiple cancer mutations are present in DNA (as well as in RNA) that can interact with one another to promote cancer occurrence. And sequential gene expression changes may occur that contribute to the provocation of some cancers. For instance, genetic factors interacting with an environmental trigger, such as smoking, can act in the provocation of some cancers, as in the case of lung cancer. Identifying critical genes for cancer occurrence may be complicated as they may be silent in the absence of a trigger stimulus, and some of these genes might also be developmentally regulated so that the passage of time (age) acts as the trigger for their expression. Evaluating these processes is made exponentially more difficult as the genetic features of cancer in one individual may differ from that evident in a second, and the genetic characteristics of cancer cells may change with continued tumor growth. Thus, despite the impressive data coming from whole-genome analyses across multiple cancer types, identifying the processes responsible for cancer in any given person can still be onerous.

These limitations notwithstanding, in attempting to identify genes associated with breast cancer subtypes, the data from multiple studies, including pooled data from 10 prospective reports, revealed that polygenic risk scores could be used to predict the occurrence of some forms of cancer. These polygenic scores were aligned with the risk for estrogen-related and unrelated breast cancer and could potentially be used in determining cancer prevention approaches (Mavaddat et al., 2019). Other studies similarly indicated that a variety of cancers involved multiple genes with each accounting for only a small portion of the variance. While polygenic risk scores could predict illness occurrence for some forms of cancer based on common variants (e.g., breast, prostate, colon), for others, this seemed less feasible (Zhang et al., 2020).

Mathematical modeling approaches indicated that cancer development generally stems from two to eight genetic "hits" (Tomasetti et al., 2015). A similar analysis based on 7500 tumors from 29 cancer types concluded that up to 10 mutations are needed for some cancers to develop (Martincorena et al., 2017). Intriguing new methods are being used to track multiple gene interactions and the development of cancer. For instance, using a mutation frequency approach to track combinations of genes in tumor samples, multihit combinations could be identified that were sensitive in distinguishing between normal and tumor tissue samples in 17 different types of cancer and could also distinguish between driver and passenger mutations (Dash et al., 2019).

In some instances, intrinsic or extrinsic challenges may alter the functioning of specific genes, thereby influencing cancer progression. We will repeatedly see that biological functioning must remain within a somewhat restricted range as variations either upward or downward can produce negative outcomes. For example, the activating transcription factor 3 (ATF3) ordinarily plays an important role in maintaining cellular health in that it may produce apoptosis if cells had been irretrievably damaged owing to lack of oxygen or irradiation. As well, ATF3 may be a regulator of immune and metabolic processes (e.g., glucose metabolism, adipocyte metabolism, immunoresponsiveness) and could thus affect various forms of cancer growth. In a sense, ATF3 has been seen as a master regulator so that variations of its functioning, either being diminished or elevated, can result in negative outcomes. Several stress-like conditions (e.g., elevated cytokines) can result in dysregulated ATF3 functioning, including disturbed immune cell functioning at the site of a tumor. In mice genetically engineered to lack this gene within immune cells, metastasis was less likely to occur. On the other side, overexpression of ATF3 was associated with worse outcomes among cancer patients. Thus, it was surmised that the ATF3 serves as a "stress gene" (stress being defined in terms of biological changes rather than psychogenic stressors) that could be turned on by diverse stress signals (Ku and Cheng, 2020), including radiation therapy and chemotherapeutic agents, as well as carcinogens, a high-fat diet, and perhaps chronic encounters with psychogenic and neurogenic stressors.

As much as polygenic risk scores can be used to predict the occurrence of specific cancers, using mathematical modeling it was demonstrated that genetic contributions account for only a fraction of the variance related to cancer occurrence. As expected, based on polygenic risk scores derived from a whole-genome analysis, cancer occurrence among individuals with high-risk scores were much more likely to develop cancer than were individuals with lower risk scores. More than this, among men and women in the high-risk polygenic category who maintained healthy lifestyles (related to diet, alcohol consumption, smoking, exercise, and body mass index), 5-year cancer incidence was appreciably lower than in high-risk individuals who maintained poor lifestyles. Indeed, men and women with the highest genetic risk and who maintained poor lifestyles were 2.99 and 2.38 times more likely to develop some form of cancer relative to individuals with low polygenic risk scores and who also maintained unhealthy lifestyles (Zhu et al., 2021). Thus, although polygenic risk scores may be an important indicator of cancer occurrence, this relation can be mitigated through the adoption of health behaviors.

When to look for factors that promote cancer development

There are, as we've seen, many instances in which a small number of critical genes are necessary for cancer occurrence. However, cancer cells may be present for many years before symptoms were noticeable (Körber et al., 2019). Among other things, this means that the search for factors that encourage cancer growth should be sought at times well before its frank appearance. Although attempts have been made to link lifestyle factors (as in the case of stressors) to cancer appearance, these have most often focused on the 12- or 24-month period preceding cancer appearance. In fact, early life experiences (stressor or diet) can have proactive hormonal, immune, and neurobiological sequelae that appear throughout adulthood. It is certainly possible that altered processes that occurred many years earlier might contribute to cancer development and progression.

Carcinogens

Aside from random and inherited mutations, as most people are aware, cancer frequently is associated with mutations promoted by environmental carcinogens, such as radiation and chemicals present in our water, food, and air. Besides causing cancer by way of promoting the appearance of oncogenes, carcinogens might also do so by interfering with DNA repair processes or through metabolic actions. Despite the induction of gene mutations that have been associated with cancer occurrence, carcinogenic exposure is not a guarantee for developing cancer—tobacco smoke does not cause lung cancer in all smokers, and ultraviolet rays do not cause malignant melanoma in all sun worshipers. Most individuals encounter numerous environmental contaminants, but only some people fall victim to illnesses that stem from these toxicants. It is likely that multiple environmental agents act additively or interactively to create gene mutations that increase the risk of developing cancer. Therefore, agents considered to be carcinogens may be sufficient to favor cancer occurrence, but their impact may also be influenced by inherited genetic variants, such as those that contribute to tumor suppression or DNA repair. For instance, in mice genetically engineered so that a fundamental DNA-repair protein was deleted, carcinogen exposure markedly increased biological aging processes, apparently being linked to the increased presence of free radicals. Ordinarily, the body has ways of mopping up free radicals, but with poor lifestyles, such as smoking and engaging in poor diets, these protective processes can be overwhelmed so in the presence of disturbed DNA repair age-related immune functioning can be undermined, thereby favoring cancer occurrence (Yousefzadeh et al., 2021). In essence, we might want to think of carcinogens as a second hit among individuals already carrying certain gene mutations, or they may promote mutations that disturb the ability of cells to correct errors or aggravate aging processes that lend themselves to the emergence of pathology (Pfeifer, 2015).

Most people would undoubtedly prefer to be the masters of their destiny, especially concerning disease prevention, but individuals frequently encounter disease triggers over which they have little control. Certain occupations may bring people into contact with toxicants or carcinogens. Farmers, for instance, are often exposed to pesticides, firefighters may be at risk owing to inhaled smoke, and house painters are exposed to high levels of chemicals, such as benzene and arsenic that can foster leukemia and lymphoma. Individuals might find themselves in a job with many smokers present so they are more likely to be affected by second-hand smoke. Beyond carcinogen exposure related to the workplace, residents in an industrial region, heavy in air pollutants or in which water supplies are contaminated, may be at increased risk for cancer development and/or other pathologies. The work of epidemiologists and those in other fields of preventive medicine has made it certain that the incidence of cancer was related to several environmental contributors comprising air and water quality, as well as built environments, and sociodemographic factors. Indeed, it is not unusual for individuals to incorrectly assume a genetic basis for an illness because it "ran in the family," when in fact, it may have reflected a shared environment, as in the case of families who live downwind from a carcinogen-spewing industrial plant or consumed water contaminated with carcinogens.

Climate change and cancer

When the issue of climate change comes up, the focus is often on warming temperatures, with inevitable comments about changes in levels of seawater, areas that are lush with vegetation turning into arid regions, and the mass movement of populations that will be created by food and water shortages. Less frequently is any mention made of diseases that are fueled by climate change, including those characteristically found in warmer regions moving northward (e.g., Lyme's disease, malaria, cholera, and dengue). With continued climate change we can count on bacteria in soil changing, as will chemicals that appear or disappear from foods.

It hardly needs to be said that with climate change, humans will be exposed to increasingly more carcinogens over which they have little influence. We've been warned for decades that various skin diseases and skin cancers will increase with climate change and altered radiation from sunlight, and with an increase of fine particulate matter in the air, there will likely be a commensurate rise of lung cancer. We should not have been surprised by the political intrigues that surrounded actions related to COVID-19 (maintain targeted lockdowns or open up to sustain the economy)—the very same considerations have long been key in the decisions related to whether and to what extent governments should place greater restraints on human-made environmental pollution.

The influence of carcinogens on cancer development has typically focused on the mutations produced by these agents together with failures of DNA repair processes. As well, factors not directly related to the genotoxic effects of carcinogens, such as oxidative stress and the production of inflammation, also play a critical role in cancer development and may act synergistically with the actions stemming from DNA damage. Indeed, numerous carcinogens (including alcohol and tobacco products as well as many other environmental toxicants) not only promote inflammation that exacerbates tumor growth, but also impair the resolution of inflammation. Conversely, diminishing inflammation may have benefits on cancer treatments by facilitating clearance of toxins and may also act against side effects attributable to therapies. These actions can be achieved through the adoption of healthy lifestyles (e.g., diet, exercise, and stress reduction procedures) and could thus limit cancer occurrence and its progression.

Epigenetic processes

Epigenetics refers to the regulation of gene expression within the chromosomal environment. To the extent that genes can be shut off or turned on by receptor-mediated processes that emerge from intercellular interactions, the second level of gene regulation operates "above" the genome itself (hence the prefix epi-). We have long known that many phenotypes reflect the conjoint actions of genetic factors or interactions between genes and either experiential or environmental variables. More than that, however, gene *expression* itself can be influenced by certain experiences, but without changes (or mutations) in the gene(s). Despite the absence of a change in code, gene expression may be turned off or turned on through several processes that are referred to as epigenetic changes. Physical and psychological challenges, including those experienced prenatally or during early life, can silence or activate the actions of some genes or their promoters (regulatory elements) (Szyf, 2019). In this section, we will examine the major mechanisms of epigenetic regulation of gene expression and consider how affecting these mechanisms can contribute to cancer occurrence.

Some of the epigenetic changes promoted by environmental challenges, much like inherited mutations, might help an organism adapt to its environment, and hence might favor the development of stable advantageous traits that are passed on across generations. Equally, epigenetic changes, even if they produce short-term advantages, may favor the occurrence of illnesses, including cancer development. Although many epigenetic changes are transient, others may be permanent, and could ostensibly be linked to health disturbances that emerge long after the initial environmental challenge and the accompanying epigenetic variations.

Epigenetic changes can come about through different processes, the most common being DNA methylation. This comprises the addition of a methyl group (one carbon atom connected to three hydrogen atoms, $-CH_3$) to DNA at specific sites, causing these genes to be less transcriptionally active. The second type of epigenetic alteration is referred to as histone modification. Histones are the primary proteins present in chromatin that facilitate the packaging of the lengthy (~ 2 m) DNA strand in the cell nucleus. Specifically, DNA is tightly wound around histones (like thread around a spool), thus allowing it to fit into the nucleus. But if the DNA is wound too tightly, regulatory regions are less accessible and hence gene expression is prevented. The process of "chromatin remodeling" solves this problem by modifying chromatin so that it is more accessible and hence transcriptionally active. This can occur through several processes, such as histone-modifying enzymes, ATP-dependent chromatin remodeling complexes, as well as methylation, phosphorylation, acetylation, and ubiquitylation (a tagging process in which the protein ubiquitin can mark proteins for degradation, alter cellular location and activity, and affect interactions between proteins).[b] Interference with chromatin remodeling serves to prevent gene expression and mutations within genes and has been implicated in the occurrence of several forms of cancer. Gene expression can also be influenced in other ways, which can include *posttranscriptional* regulation of genes (i.e., after the DNA has been transcribed to RNA). For example, a large number of microRNAs (small noncoding RNA) can influence gene expression, operating to silence RNA. Like epigenetic changes, circulating microRNAs have been related to the development and progression of some forms of cancer.

Epigenetic changes are particularly relevant to sensitive developmental periods before and shortly after birth. Numerous prenatal challenges, such as exposure to methyl mercury, diesel fumes, the androgenic fungicide vinclozolin, the estrogenic peptide methoxychlor, and the endocrine disruptor bisphenol-A can affect development. These chemical challenges induce epigenetic changes in genes coding for growth and several immune factors (IFN-γ and IL-4), some of which promote psychological and physical disturbances (e.g., Kundakovic and Champagne, 2011).

Lifestyles, including diet and exercise, have similarly been associated with variations of microRNAs that may be linked to regulatory processes that could affect disease resilience. Variations in food types and microbiota can affect microRNAs, ultimately affecting the appearance of psychological disorders, and conversely, altered brain functioning can influence microbiota (Moloney et al., 2019). Several microRNAs were also linked to bacteria within a tumor's microenvironment and were linked to colorectal cancer (Yuan et al., 2018), although it is uncertain whether these were causally related to cancer occurrence or progression.

---

[b] The 2004 Nobel Prize in chemistry was jointly awarded to Aaron Ciechanover, Avram Hershko, and Irwin Rose for their discovery of ubiquitin-mediated proteolysis in which an enzyme system tags unwanted proteins for their subsequent degradation. This process is involved in the regulation of the cell cycle, DNA repair and transcription, protein quality control, and the effectiveness of immune responses. As such, disturbances of this process may contribute to a variety of diseases, including some forms of cancer.

Predicting cancer based on epigenetic changes

Neither genes nor gene mutations fully account for the occurrence of many forms of cancer or the vast differences that occur among individuals with particular mutations. The demonstration of epigenetic changes with disease occurrence was a game-changer in analyses of how genetic processes and environmental factors may come to affect psychological and physical illnesses. Epigenetic changes reflected by either hypo- or hypermethylation were, in fact, apparent in various cancers, such as colon, bladder, and prostate cancer. An important way in which epigenetic changes influence cancer is by silencing genes that ordinarily act as tumor suppressors or by causing cells to grow and divide too quickly, as observed in the case of changes in the BRAF gene. And epigenetic dysregulation of immune cells in peripheral blood lymphocytes, raise the possibility that these changes contribute to cancer by modulating antitumor immune processes (Vogelstein et al., 2013).

Some epigenetic changes had long been suspected of provoking the occurrence of tumors. The epigenetic modifications may have been present prior to frank carcinogenesis (*somatic epitypes*) having formed in preneoplastic sites, as well as in apparently healthy tissue adjacent to the tumor. These epitypes increase with age so their accumulation over the lifetime could potentially increase cancer risk (Hitchins, 2015). As cancer progresses, further epigenetic changes can occur that were not apparent in their cells of origin, so a concatenation of epigenetic changes that develop over time shape the progression of the disease.

An analysis of epigenetic changes that occurred within single cells of a tumor indicated that these cells could appear in different states, which was taken to suggest that individual cancer cells go through different (or independent) evolutionary pathways. As epigenetic changes in premetastatic cells were sometimes similar to cells that had already metastasized, it was suggested that cells within the primary tumor develop epigenetic states that facilitate or allow them to become migratory, seeding cells at distant sites.

It would be valuable to determine whether specific epigenetic actions or gene mutations are causally linked to a specific form of cancer, which might allow for the development of drugs that target epigenetic processes. In some cases, epigenetic changes seemed to be drivers of the disease rather than bystanders, but in other instances, they may simply have been disease correlates. Even so, epigenetic alterations associated with cancer occurrence could serve as important biomarkers to predict the occurrence of the illness, the progression, and aggressiveness of the disease, as well as the most efficacious treatment strategies. In fact, epigenetic markers have been detected in early breast, colorectal, prostate, pancreatic, gastric, and lung cancers, and could often be observed in biological fluids, and as such, could serve as indicators of cancer presence and progression as well as the response to therapies. Regardless of how they came about, epigenetic signatures might prove useful in elucidating cancer risk and progression, although challenges exist in determining relevant markers in specific tissues, as opposed to more easily accessible biological fluids.

Aging

Cancer can affect individuals of all ages but is often seen as predominantly a disease of aging or one in which biological aging processes are exacerbated. Some epigenetic actions are subject to modification (e.g., through various lifestyle changes), whereas others are not (e.g., familial contributions, sex, age), but both can influence metabolic processes as well as cancer occurrence and growth. Moreover, socioeconomic status and exposure to environmental toxicants were accompanied by epigenetic aging that could favor the development of age-related diseases, and as discussed in Chapter 6, early-life stressors and cumulative stressful experiences can influence aging-related changes and vulnerability to illness (e.g., Palma-Gudiel et al., 2020).

The development of cancer with aging may occur for several reasons. With age, aspects of immune efficacy (e.g., T cell number and functioning) may diminish, and the opportunity for the occurrence of multiple mutations that favor cancer is increased, as are epigenetic changes within senescent cells. In this respect, during aging epigenetic changes were more frequent on promoters of various immune cells, including T and B lymphocytes (Jasiulionis, 2018). For example, early-stage breast cancer was associated with diminished NK cell activity, and reduced IFN-γ production, along with immune-related epigenetic effects that could be detected within peripheral blood mononuclear cells. Epigenetic regulation of immune functioning could also occur indirectly by affecting genes related to glucocorticoid functioning or a notch signaling pathway that is important in cell-cell communication, regulation of embryonic development, and the formation of the tumor microenvironment (Meurette and Mehlen, 2018). Of course, aging comes with many further hardships, ranging from depression and loneliness to the emergence of multiple physical illnesses, and these many challenges could generate further biological changes that favor cancer occurrence and poor responses to treatment.

Telomeres, aging, and cancer

It has been almost 40 years since the discovery of telomeres, and 30 years since the identification of telomerase, the enzyme involved in determining telomere length (Blackburn et al., 2015). Telomeres comprise bits of DNA at the tips of chromosomes, which prevent the DNA from unraveling, much like aglets on a shoelace serve in this capacity. With each replication of the DNA strand, these telomeres become shorter, and the length of the telomeres can, in a sense, serve as an index of cellular aging.

It wasn't very long after their discovery that shortened telomeres were associated with poor lifestyles, nutrition, and psychological stressor experiences, which predicted poorer health and earlier death. Indeed, a chronic stressor (caregiving among mothers with an autistic child) was associated with chronic inflammation, which might have contributed to the aging process reflected by telomere dysfunction in peripheral blood mononuclear cells (Lin et al., 2018). To a considerable extent, it wasn't (and still isn't) understood whether telomere length was causally linked to illnesses, or simply a marker of poor well-being.

As aging has been associated with both cancer development and telomere length, the possible causal connection between telomere length and disease occurrence couldn't be ignored. It had been expected that shorter telomere length would be linked to premature aging and cancer. However, it was discovered that longer telomeres in DNA obtained from lymphocytes were related to the increased likelihood of cancer occurrence (Rode et al., 2016), and most cancer cells contained elevated telomerase. Moreover, within families that were prone to developing cancers, it appeared that their telomeres were inordinately long. To be sure, the general notion had been that long telomeres reflect cellular machinery adept at repairing damaged cells, which is certainly advantageous. However, this reparatory ability might have counterproductive consequences if lengthened telomeres had the effect of sustaining cancer cells. Thus, cancer growth could conceivably be deterred by diminishing the ability of telomeres to lengthen.

When a gene was manipulated in mice to simulate the gene alteration in humans with long telomeres, namely the creation of mutations in the TINF2 gene, mice were born with long telomeres. It was maintained that this gene acts on a tumor suppressor pathway, thereby increasing vulnerability to breast, colorectal, thyroid cancers, and melanoma (Schmutz et al., 2020). Similarly, by attacking TRF1 protein, which serves as part of the telomere-protective complex, glioblastoma progression in a mouse model was reduced so survival was extended by 30%–80% (Bejarano et al., 2017). There have been efforts to target telomeres and the enzyme telomerase to deal with diseases of aging in a broader context, but so far, treatments following this approach haven't been able to extend aging or eliminate cancers. New technologies, however, have allowed for a more detailed analysis of the structure of telomerase, which may come with the rejuvenation of this approach in assessing cancer occurrence.

As with so many other discoveries relevant to health, charlatans have crawled out of their hideyholes and have been attempting to fleece customers who wish to lengthen their telomeres to turn back the clock. Carol Greider, who shared the 2009 Nobel Prize in Physiology and Medicine with Elizabeth Blackburn and Jack Szostak "for the discovery of how chromosomes are protected by telomeres and the enzyme telomerase" has weighed in strongly, criticizing the value of these efforts. She warned against these purveyors of pseudoscience and faulty treatments, advising that if a person wants a healthier life, then eating properly and exercising would be a far better approach.

Targeting epigenetic modifications in cancer therapy

Considerable efforts have been made to target DNA methylation and histone deacetylation enzymes to alter cancer progression, and promising data had been in the offing for more than a decade. This approach may be particularly effective in combination with already existing treatments, such as PARP inhibitors that can initiate a response to repair DNA breaks as well as to enhance immune functioning and attacks on cancer cells. In acute myeloid leukemia (AML), the effectiveness of a PARP inhibitor was enhanced when combined with the hypomethylating agent decitabine (Muvarak et al., 2016). No doubt, alternative manipulations will be necessary to deal with epigenetic variations in other forms of cancer.

In the end, treatment of cancers by targeting epigenetic processes will likely require a precision medicine approach, although this might prove difficult in hard-to-treat conditions, such as pancreatic cancer as well as in cancers in which multiple epigenetic markers have been detected. Moreover, it is likely that genetic and epigenetic processes act together in the development of some forms of cancer, necessitating combined targeting through therapies that involve immunotherapy and epitherapies (Topper et al., 2020).

The development of drugs that appropriately target epigenetic enzymes to treat varied cancers will not be an easy feat. Nevertheless, with epigenetic reference maps having become available (e.g., Roadmap Epigenetics, Kundaje et al., 2015), it has become feasible to predict improved abilities to identify epigenetic marks that could be used to predict cancer occurrence and may point to individualized treatment strategies. As described in Fig. 4.4, these efforts will likely entail multiple platforms being used for diagnostic and prognostic purposes, as well as treatment decisions (Nebbioso et al., 2018).

It is entirely understandable that efforts to modify cancer through epigenetic manipulations entailed the use of pharmacological treatments. Yet, various nutrients can affect the

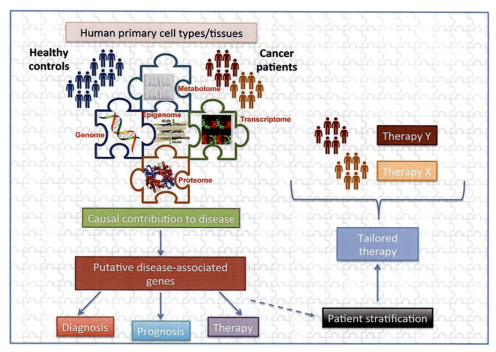

FIG. 4.4 Integrating and combining data from different platforms (genome-DNA sequence, transcriptome, proteome, metabolome, and epigenome) leads to a better understanding of the basis of cancer and paves the way toward personalized medicine. *Figure and caption from Nebbioso, A., Tambaro, F.P., Dell'Aversana, C., Altucci, L., 2018. Cancer epigenetics: moving forward. PLoS Genet. 14, e1007362.*

cancer process, possibly through effects on epigenetic changes. Early life nutrients as well as malnutrition can similarly engender epigenetic changes that were linked to some forms of cancer (Campisano et al., 2019). Thus, it might be possible to harness these actions on the epigenome in the treatment of some forms of cancer.

Further to this, as we'll discuss in greater detail in later chapters, depressive disorders have been aligned with cancer development. In some instances, cancer occurrence may predict later depression, and conversely, depressive illnesses can predict later cancer occurrence. In both cases, epigenetic factors may be a key player in mediating the comorbidity. For instance, epigenetic changes related to inflammatory cytokines have been linked to some forms of cancer (Ivashkiv, 2018), just as inflammation has been tied to depression among a subset of individuals. Once more, these findings raise the possibility that inflammatory processes are in some manner related to both types of illness.

Intergenerational and transgenerational changes

Epigenetic changes may promote intergenerational (two generations) and transgenerational (three or more generations) actions on disease processes (Cavalli and Heard, 2019). Toxicants and varied life events experienced at any time of life can induce epigenetic effects, and these outcomes can also occur owing to events that occurred prenatally, thereby engendering physical and psychological actions throughout an organism's life (Szyf, 2019). Should these epigenetic effects occur within germline cells (sperm or ova), then these changes could be recapitulated across multiple generations.

Aside from traumatic early life events or exposure to environmental toxicants, poor parenting in animals may have cascading epigenetic effects that can be transmitted across generations. In addition to epigenetic changes of glucocorticoid receptors, early life experiences were associated with DNA methylation of the gene coding for the serotonin transporter in monkeys, and in rats poor maternal care was accompanied by methylation of the gene promoter for estrogen receptor alpha (ERα) (Peña et al., 2013). This gene may be important for the establishment of good maternal behaviors, thus its modification in offspring may lead to them becoming poor mothers, which could be passed on across generations.

Transgenerational epigenetic changes in animals can be induced by many toxicants, as well as traumatic events and were accompanied by multiple behavioral phenotypes. Similarly, the offspring of mice that had been exposed to pollutants (e.g., dioxins) while pregnant exhibited impaired immunity against influenza A virus, which was also apparent in the ensuing generation, suggesting possible epigenetic transmission (Post et al., 2019). And pesticides, such as methoxychlor (which replaced the banned DDT), also promoted epigenetic actions that appeared over several generations (Manikkam et al., 2014). In essence, this type of research suggests that the substantial increase in modern-day diseases could have stemmed from toxicants encountered generations earlier.

Most studies assessed epigenetic links between women and their offspring, but what about dad? Can he contribute to epigenetic changes? Offspring of dams impregnated by a stressed male are epigenetically affected. But interestingly, when in vitro fertilization was undertaken using sperm from a stressed mouse, such effects were not apparent (Dietz and Nestler, 2012), raising questions about the factors necessary for epigenetic changes to occur. Still, paternal stress could influence the microRNA content of sperm and could cause the reprogramming of hypothalamic-pituitary-adrenal functioning, and it appeared that microinjections of sperm microRNAs directly into the zygote could reprogram gene expression in the offspring (Rodgers et al., 2015). Paternal stress may influence the sperm epigenome through histone modification, DNA methylation, and noncoding RNA expression (Xu et al., 2021). Moreover, the presence of methyl groups associated with histones in the sperm epigenome could be modified by feeding male mice a folate-deficient diet, resulting in effects on embryonic gene expression related to postnatal physical development (Lismer et al., 2021). In essence, the paternal transmission of epigenetic effects can occur through multiple processes and can be influenced by exercise, stressful events, dietary factors, and toxicants that may have exact consequences on offspring that are recapitulated over multiple generations. In the studies described here, the assumption was that stressors would promote epigenetic changes in male sperm, ignoring individual differences between mice. When male mice that had been exposed to chronic social defeat were categorized as being stress-sensitive versus stress-resilient, differential epigenetic actions were observed, pointing to the need to consider individual characteristics in tracing the transmission of stressor effects over generations (Cunningham et al., 2021).

Understandably, assessing intergenerational and transgenerational epigenetic actions in humans is much more difficult than in animals. Nonetheless, a small number of studies, typically involving a limited number of genes, have pointed to intergenerational effects stemming from epigenetic actions. Informative findings have been obtained through historical analyses of periods when whole societies or groups of people have encountered hardship. Among holocaust survivors, epigenetic changes were apparent within the FKBP5 gene, which is associated with glucocorticoid functioning, and was linked to PTSD and depression. A similar epigenetic change was also apparent in their children, which was associated with altered morning cortisol levels (Yehuda et al., 2016). In offspring born during the Great Chinese famine of 1959–61 (in the Suihua area of China), epigenetic changes were noted in genes related to glomerular filtration rate that is associated with chronic kidney disease, but these outcomes did not carry over to the next generation (Jiang et al., 2020). However, other phenotypes, such as the risk for type 2 diabetes in adults, were apparent across generations born after the Great Chinese famine (Li et al., 2017). Like the findings observed in China, famine-related intergenerational actions have been reported in other contexts. During the later portion of the Second World War, the Nazi regime cut off food and fuel to towns in western parts of the Netherlands, resulting in a severe famine, which came to be known as the Dutch Hunger Winter. Among individuals who were prenatally exposed to these distressing conditions, an epigenetic change related to IGF2 was detected decades later (Heijmans et al., 2008). There is the belief that epigenetic effects triggered by cumulative historical trauma (i.e., those experienced in multiple generations) could play a significant role in determining health outcomes that appear over successive generations (Bombay et al., 2014). Because these experiences are often accompanied by multiple individual and societal hardships, it has been difficult to identify the specific processes that lead to these health risks.[c]

There has been growing recognition that epigenetic changes experienced prenatally or during the early postnatal period (perhaps owing to poor diets and exposure to pollutants) may have broad intergenerational and transgenerational consequences. These epigenetic changes have been postulated to contribute to the etiology of numerous diseases including worldwide obesity and related health risks. With the increased prevalence of obesity, increased attention has focused on the transgenerational effects of this condition as well as the impact of dietary factors that affect adipogenesis, lipid balance, and obesity. It was similarly considered that exercise could promote epigenetic effects that are beneficial to health and that these could also be transmitted across generations (Denham, 2018).

It was proposed that prenatal epigenetic changes provoked by negative psychosocial experiences and exposure to toxicants may provide a first hit that increases vulnerability to pathology, and later experiential and environmental challenges comprise the second hit that promotes disease development. But there may be more to this than first and second hits. In many countries, the cumulative epigenetic changes that stemmed from food shortages and other challenges

---

[c] There is the belief that some health problems encountered by Indigenous people within Canada and elsewhere (high levels of type 2 diabetes, heart disease) might stem from generations of abuse experienced at the hands of European colonizers. These transgenerational epigenetic effects may have promoted these negative outcomes, but numerous related factors could also contributed in this regard, including continued rampant poverty, poor diet, water safety, social disturbances, or the ability of parents to care for their children adequately (Bombay et al., 2014). Whether the poor conditions stem from epigenetic effects or not, it's a fair bet that improving living conditions and medical assistance would go a long way in diminishing the health risks experienced.

during the 19th and 20th centuries may have increased vulnerability to diseases, including cancer, thereby contributing to its persistence over subsequent generations. Similarly, cross-generational exposure to fungicides and other pollutants may have been the source of epigenetic alterations that contributed to the progressive rise of some forms of cancer (Klukovich et al., 2019). Relatedly, chronic diseases may comprise a mismatch between the ancestral epigenome and second hits related to poor Western diets, excessive alcohol consumption, smoking, physical inactivity, and multiple environmental pollutants (Imam and Ismail, 2017).

The notion that cancers could be transmitted nongenetically across generations (epimutation) is often thought of as being a recent development. However, this perspective was proposed more than three decades ago. It had been surmised that these epigenetic actions could have effects indistinguishable from those associated with gene mutations. Moreover, it was maintained that while some cancer-related epitypes could be acquired, others were derived from the parental germline (or during embryogenesis) and were widely distributed within somatic tissues. These "constitutional epimutations" could be inherited much as any other genetic feature could be transmitted (Hitchins, 2015). Increasingly, these epimutations have been implicated in a variety of cancer types, although the intergenerational inheritance of epigenetic effects has only been demonstrated in a small number of studies. Support for this notion was found in mice that showed the presence of hypermethylation on many DNA regions which affected gene regulation and cancer incidence and progression across at least two generations (Lesch et al., 2019).

Analyses of intergenerational and transgenerational actions concerning cancer have most often been assessed in female rodents, but dietary factors and endocrine disruptors in either males or female parents led to increased risk of breast cancer transmission that could be recapitulated over generations. Likewise, paternal malnutrition can influence breast cancer risk in rodent offspring (da Cruz et al., 2020). The impact of dietary factors on intergenerational transmission of breast cancer risk also varied with the parental dietary intake, differing as a function of heavy lard versus corn oil intake, and these actions differed yet again as a function of whether the male or female parent consumed these foods (Fontelles et al., 2016).

Aside from a few studies of this nature, there are still limited data as to whether the cancer-related actions of specific epigenetic variations are transmitted across multiple generations and still less information concerning the specific mechanisms that underly the epigenetic-related risks for some forms of cancer. Yet, the simple fact that transgenerational epigenetic changes have been observed with respect to several biological mechanisms that can affect cancer processes makes this an issue that warrants continued investigation.

Limitations of data linking epigenetic changes to specific phenotypes

It had at one time been assumed that epigenetic changes were infrequent, but it quickly became clear that they were actually fairly common with an enormous number of epigenetic marks normally being present within the genome. Epigenetic modifications are part of normal development, being involved in the formation (cell differentiation) of the 200 or so different types of cells that humans possess. The number of epigenetic changes discovered has increased exponentially over the past two decades, and the NIH Roadmap Epigenomics project revealed epigenetic modifications associated with numerous regulatory elements that might be relevant for several disease conditions (Kundaje et al., 2015).

Given that most of the data collected concerning epigenetic modifications and specific phenotypes have been correlational, linking the vast number of epigenetic changes to specific phenotypes is onerous, and certainly can't be the basis for causal attributions. While some epigenetic changes might be linked to the later development of specific diseases, others may occur because of the illness or the accompanying distress, or they might simply be bystanders that aren't related to the illness in any meaningful way. Cumulative epigenetic changes (like the occurrence of mutations) vary with the age of an organism and with sex, which adds to the difficulties in linking epigenetic characteristics to specific physical or psychological phenotypes. With so many epigenetic changes appearing on the genome, many of which aren't permanent, the difficult task facing researchers has been to identify their phenotypic actions in promoting illnesses, and how these can be neutralized. Further to this, the cells that make up diverse tissues may have unique epigenetic profiles, so the information obtained from one type of tissue may not have predictive value about other tissues.

We have pointed to the view that complex illnesses typically involve multiple genetic factors, and it would be reasonable to suppose that this is true of epigenetic influences, making the significance of any single epigenetic mark questionable. Further, the pattern of methylation within cancer cells is in greater disarray than in healthy cells (Landau et al., 2014). While this may be an index of disease risk (or presence), it also works against the identification of specific epigenetic alterations linked to pathology. In addition, the functions of gene activity may vary at different points in the cancer process. For instance, a particular protein, AZH2, was reported to be an epigenetic regulator that might limit or delay the development of myeloid leukemia; however, once the disease is well entrenched, the protein related to this epigenetic action may facilitate cancer's progression (Basheer et al., 2019).

The difficulties in understanding genetic and epigenetic processes concerning cancer development are compounded by numerous other factors beyond the many already outlined. Not only can seemingly similar cancers differ from one another, but metastasized tumors may be genetically and epigenetically distinct from those present early, and even the treatments administered may change the characteristics of tumors so that they develop resistance to treatment. Despite these many difficulties, there is little question concerning the importance of epigenetic processes in cancer production, and it is ever more important to identify epigenetic markers as part of the effort to develop classification systems to determine the prognosis and treatment of specific types of cancer. As epigenetic processes are dynamic and vary over time, longitudinal analyses that track pathologies and epigenetic profiles would be essential in defining whether and how epigenetic changes are linked to cancer occurrence and its progression.

Complex gene × environment interactions

The development of many types of cancer generally involves multiple genes acting additively or interactively. But, even if certain mutations are prime predictors of cancer development, they don't necessarily guarantee cancer development. As indicated earlier, while carrying a BRCA gene mutation is associated with the increased occurrence of breast and ovarian cancer, it is not a sufficient condition to assure that cancer will develop. The view was advanced, as depicted in Fig. 4.5, that in addition to genomic processes, predictors of

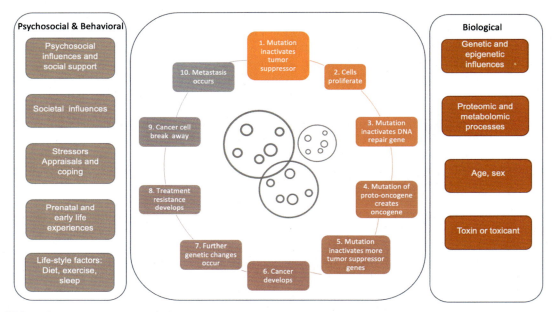

FIG. 4.5 The cancer process and its multiple moderating variables. Tumor suppressor genes ordinarily act against the development of cancer, but particular mutations can diminish their capacity to serve in this capacity. For instance, the gene coding for p53 is especially important in tumor suppression, and about 40% of cancers are accompanied by mutations of this gene. If a second mutation occurs, which limits DNA repair processes or favors a proto-oncogene becoming an oncogene, cancer cells may develop and prosper. Proto-oncogenes provide signals that promote cell division and may be involved in regulating programmed cell death (apoptosis). This can be followed by further mutations that favor tumor growth, and as still other changes occur in the tumor, it becomes more resistant to treatment, and as cells escape from their local milieus (which can occur earlier than depicted in the figure) metastasis may occur. A variety of factors, depicted in the outer portions of the figure, may contribute to cancer emergence or progression. These not only include biological processes, and exposure to toxicants, but also environmental and psychosocial influences, various lifestyles, prenatal experiences, stress reactions, and general societal challenges, especially if they limit or create roadblocks to preventive measures being undertaken.

cancer ought to include psychosocial factors together with experiential and environmental influences, such as diet (Letai, 2017). For that matter, variations of cancer occurrence across disparate groups could be due to differences in their exposure to carcinogens and pathogens, access to health care, and lifestyles adopted. Furthermore, in assessing the impact of stressful experiences it is important to consider the contribution of factors beyond the individual, particularly those challenges that entail whole communities. Society itself (with its myriad levels of social interaction and expression of—and reaction to—distinct attitudes and values) ought to be considered as an environment that can produce epigenetic variations that could influence emotional and physical health. From this perspective, having a positive social environment could act against the development of undesirable epigenetic variations, and it might be possible to modify epigenetic actions. Such effects can be produced by particular pharmacological interventions and can seemingly emerge through positive nurturing.

Population-level genetics

Sex variations in cancer

It was only recently that attention focused on differences that exist between males and females in the occurrence of a wide array of illnesses, as well as the efficacy of treatments in diminishing or eliminating these conditions. A machine reading approach that sampled data from 43,135 journal articles and 13,165 clinical trial records, indicated that females were underrepresented in the majority of illnesses assessed, being particularly notable in chronic kidney and cardiovascular diseases (Feldman et al., 2019), and women had historically been underrepresented in research related to pain. The situation was equally biased in Phase I and II clinical drug trials in which women had been explicitly excluded. In fact, it wasn't until 1993 that the FDA took steps to assure that women were included in all phases of drug evaluations. Of course, it's now well established that very marked genetic differences exist in numerous tissues of men and women and the relative frequency of many diseases varies as a function of sex, as do the effectiveness of some treatments. One needs to wonder what important treatment approaches could have been developed if earlier and greater attention had focused on the sex/gender biological disparities related to illnesses.

More than a third of all genes exhibit differential sex-related expression in one or more types of tissue (Oliva et al., 2020). As observed in so many other conditions, pronounced sex differences exist in cancer occurrence and mortality, the genetic factors tied to cancers, the involvement of certain biological processes (e.g., hormones, immune functioning, and cytokines), the efficacy of treatments, and the toxicity associated with therapies (Kim et al., 2018). In general, as shown in Fig. 4.6, the occurrence of tumors in nonreproductive tissues is more frequent in males than in females. The sex differences in cancer occurrence have also been apparent in several genetic features associated with cancer. This included differences in coding and noncoding drivers, the prevalence of mutations and mutational load, and mutational signatures (Li et al., 2020b). To an extent, sex hormones contribute to some of the differences that have been observed. Estrogen may play a protective role in limiting the occurrence of some forms of cancer, such as colorectal cancer but may increase the occurrence of other forms of cancer, such as thyroid cancer.

Genetic factors play a prominent role in sex dimorphisms and may contribute to the differential responses to treatment as might epigenetic factors that arise owing to environmental exposures. For example, the tumor suppressor protein, p53, was more likely to lose functioning in males than in females, allowing for greater tumor development (Sun et al., 2015). Likewise, tumor suppressor genes that escape X-chromosome inactivation (in which one copy of an X chromosome is inactivated) are more highly expressed in females than in males. Simply put, the two X chromosomes in females provide better protection than does the single X chromosome in males (Dunford et al., 2017).

Although males are more likely to develop cancer, the dimorphism related to therapeutic agents is very different in that chemotherapeutic drugs reliably produce greater adverse effects in women. The influence of various therapeutic agents on cancer survival, in contrast, was inconsistent. The sex differences that occur are dependent on the type of cancer being treated and the nature of the therapy used. An analysis of colorectal cancer therapy indicated that adjuvant chemotherapy resulted in longer overall and cancer-related survival among

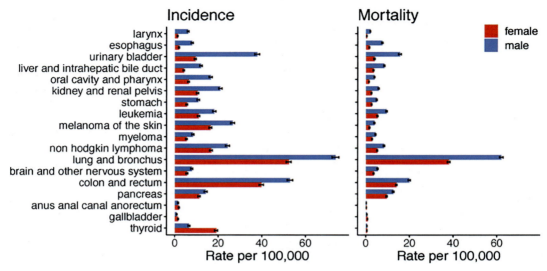

FIG. 4.6 Cancers with sex disparity in incidence and mortality rates. Age-adjusted incidence and mortality rates per 100,000 individuals in the United States were retrieved from the Surveillance, Epidemiology, and End Results explorer. Bars show the average rate from 2000 to 2017 and 95% confidence interval. Cancer sites are ordered according to the male-to-female incidence rate ratio; starting from cancer sites with higher incidence rates in males compared to females. *From Lopes-Ramos, C.M., Quackenbush, J., DeMeo, D.L., 2020. Genome-wide sex and gender differences in cancer. Front. Oncol. 10, 597788.*

women than in men (Quirt et al., 2017). In contrast, there was no sex difference regarding treatment efficacy for metastatic colorectal cancer, but females experienced greater toxicity of a hematologic or gastrointestinal nature. In other studies, chemotherapeutic agents yielded somewhat better outcomes in females, and toxicity was similarly greater. Unfortunately, chemotherapeutic agents have frequently been used as a treatment option without adequate consideration of sex-related differences in the efficacy or toxicity of the treatments.

The dimorphism reported in response to immunotherapy was somewhat different from that observed with chemotherapy. A meta-analysis that included 20 randomized control trials revealed that the response to a variety of immunotherapeutic agents was better in males than in females (Conforti et al., 2018), although a similar analysis that included 23 trials indicated that the effectiveness of checkpoint inhibitors for the treatment of advanced solid-organ cancers was comparable between the sexes (Wallis et al., 2019). The different outcomes may have come about owing to the specific types of cancer being considered, how treatment success was determined, and what was considered a meaningful difference in outcomes. For whatever reasons, tumors in males contain a greater number of tumor-relevant mutations and cancer antigens and might thus be more responsive to immunotherapy (Wang et al., 2019). It was also suggested that in female patients some cancer-causing genetic mutations that are present may comprise a type that is not readily presented to the immune system by MHC molecules. In effect, the female immune system may ordinarily be very effective in ridding the body of cells that present with mutant self-antigens, but those that remain might comprise cancer cells that have poorly presented mutations, thereby making them less accessible by treatments (Castro et al., 2020).

Microbiota in males and females

In addition to hormonal and cytokine-related sex differences, sex dimorphisms were apparent in the endogenous gut microbial response to stressors, as well as in the cross talk that occurs between microbiota and hormones. Relative to males, females exhibited lower intestinal permeability and microbiota diversity, as well as differential actions of hormones on gut microbiota (Org et al., 2016). Conversely, microbiota in females and males were differentially effective in modifying sex hormones (Audet, 2019). Supporting the cytokine-hormone-microbiota link, treating mice with progesterone increased *Lactobacillus* species and concurrently reduced IL-6 expression, thereby affecting a depressive-like condition (Sovijit et al., 2021). As described in Fig. 4.7, diverse factors may contribute to sex-dependent changes of a range of psychological disorders, and physical health may be affected by the microbiota and cytokine differences that exist.

Ethnic variations in relation to cancer

Marked racial and cultural differences occur in the occurrence of diverse types of cancer, varying both within and between countries. Based on genetic differences that exist across cultural and ethnic groups, it's hardly surprising that cancer morbidity and mortality would similarly vary with these factors. Many of the cultural differences may be related to genetic factors, but cultural differences have also been observed with respect to barriers to treatment, as well as the acceptance of prevention strategies.

Returning to our example of BRCA1 and BRCA2, relative to the general population this mutation occurs much more frequently (~ 8%) among Ashkenazi Jews (i.e., primarily those of European descent) and moderately more often among Hispanic American women (3.5%). The BRCA1 mutation occurs in about 1.3% of Black women, and still less frequently among women of Asian descent (0.5%). Differences in triple-negative breast cancer occurrence were also reported in African American women relative to women from East and Sub-Saharan Africa (Newman and Kaljee, 2017). These differences were seemingly unrelated to estrogen receptor status, and it was surmised that they might be due to differences in the genetic features of tumors that exist across ethnic or racial groups. While such differences may be based on genetic influences, epigenetic processes could promote the breast cancer disparities in the occurrence and therapy success that has been observed with ethnicity.

Several types of cancer, including cervical, colorectal, gastric, lung, and thyroid, also vary with ethnicity, culture, and geographical location. Moreover, the responses to drug treatments, including the emergence of adverse reactions, are influenced by polymorphisms that vary across cultures (Bachtiar et al., 2019). In some instances, the difference may be linked primarily to genetic factors, but they might also reflect Gene × Environment interactions involving diet or environmental toxicants. In this regard, early cancer detection varies with the socioeconomic status that might be linked to ethnic and cultural factors. For instance, diagnoses of breast cancer among Black and South Asian women occurred later than among Japanese women or whites of European descent, and mortality rates followed suit.

Considering the importance of culture in contributing to disease processes, it is curious that many of the genome-wide studies that have been conducted, which now include many millions of participants, typically involved limited diversity. Studies in China understandably involved participants from their own country, and the vast majority (78%) of studies conducted

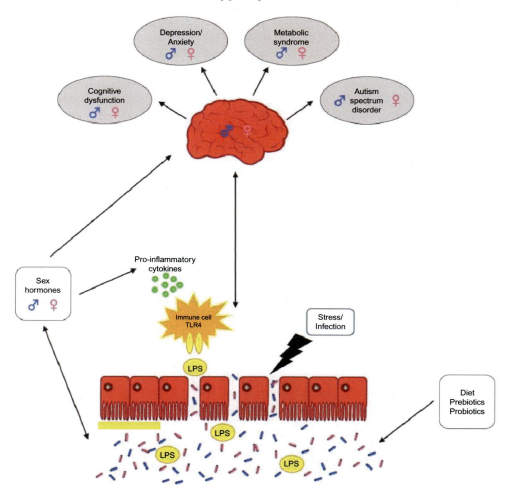

FIG. 4.7 Schematic representation of the potential influence of gut microbes and sex hormones on the immune-signaling routes mediating the cross-talk between the gut microbiota and the brain, ultimately leading to sexually dimorphic health outcomes. It is suggested that the reciprocal influence between sex hormones and the gut microbiota may come to modulate the gut microbiota-immune-brain axis. Elevations of the proinflammatory tone within the intestinal environment and damages to the intestinal barrier ordinarily elicited by stressor or immune challenges could be affected by these microbe-hormone relationships in a sex-dependent manner, with potential impacts on translocation of bacterial products (e.g., LPS) and subsequent activation of their respective signaling pathways (e.g., TLRs, proinflammatory cytokines). Inflammatory processes in the blood and the brain would remain sensitive to the effects of sex hormones. In addition, it is proposed that dietary or microbiota-targeted interventions (e.g., prebiotics, probiotics) may act on the gut microbiota in a sex-specific manner, potentially through their effects on sex hormones. Ultimately, sex-specific disturbances along the gut microbiota-immune-brain axis would promote sexually dimorphic features within the brain and distinct mental health phenotypes among men and women. *Figure and caption from Audet, M.C., 2019. Stress-induced disturbances along the gut microbiota-immune-brain axis and implications for mental health: does sex matter? Front. Neuroendocrinol. 54, 100772.*

in western countries primarily involved individuals of European descent (Gurdasani et al., 2019). Whole-genome analyses have been particularly infrequent across Africa (2%) despite it being the cradle for today's humans and may be the origin of numerous genetically determined diseases. Analyses that have been conducted within African countries have revealed 3 million previously unreported gene variants and multiple gene loci that are related to viral immunity, DNA repair, and metabolism (Choudhury et al., 2020). By identifying the genomic differences across cultures/ethnicity it might be possible to identify essential features that contribute to the development of specific illnesses and at the same might be instrumental in promoting equitable delivery of care across diverse populations. As we indicated in the context of sex/gender differences, one can't help but ponder about the information that has not been obtained owing to the restricted diversity in many studies, including the possibility of identifying better targets for treatment.

Finally, there is the issue of examining a specific individual's genome and then attempting to discern whether the array of genes is disturbed in some way. In some instances, the presence of a particular mutation could readily be identified, but in other instances, this can be challenging. Ordinarily, when parts of an individual's genome are evaluated, this is done in relation to the reference gene that is obtained from a control sample. However, the reference genes used for the original Human Genome Project might not have been representative of the population. In fact, the genome of a random individual may differ from that of the reference genome by as much as 16 million bases of the 3.1 billion or so that comprise the genetic code. Moreover, if the individual's ancestry differs from that comprising the original reference genome, the differences may be still more profound (Levy-Sakin et al., 2019). An analysis among 1000 Swedish individuals identified 61,000 novel genetic sequences that had been absent in the usual human reference genome (Eisfeldt et al., 2020). Evidently, better reference genomes are needed, especially to capture differences that exist across ancestries.

Precision (personalized) medicine

We have mentioned precision medicine on several occasions in this and earlier chapters without fully describing its therapeutic importance. It had long been recognized that significant problems were inherent in the diagnosis and treatment of varied disorders. For instance, in the case of clinical depression, two individuals could present with very different symptoms yet receive the same diagnosis. Conversely, two patients could present with the same symptoms, but when treated with a particular drug, they frequently showed very different responses. These disconcerting situations aren't restricted to mental illnesses, as they are also encountered in other conditions. Cancer patients can similarly present with many of the same illness features but might require very different treatments. With the realization that one-size-fits-all was simply inadequate in treating many patients, it wasn't long before it was widely accepted that individualized treatment strategies might be more effective. Rather than simply considering illnesses at face value, an approach was adopted in which specific genes were linked to measurable aspects of the disease condition or to specific features (symptoms) of an illness that could inform treatment methods.

Precision medicine approaches most often focus on identifying specific genes (or sets of genes) that are associated with particular forms of cancer, and then determining whether these genes were predictive of therapies that are optimal for individual patients. Many forms of cancer that are seemingly similar can be distinguished from one another genetically and might thus require different types of treatments. For instance, gene expression profiles might be useful in predicting the efficacy of chemotherapies, including in HER2 and estrogen receptor-negative tumors. The use of blood samples to detect difficult-to-reach tumors (e.g., small-cell lung cancer) has also been effective in identifying the "profile" of genetic faults that can be used to predict optimal treatments (Carter et al., 2017). Based on genetic characteristics, glioblastoma in children can be divided into 10 different subtypes, and it may be possible to target specific mutations using drugs currently available to treat (sub) forms of this cancer (Mackay et al., 2017). Consideration of acute myeloid leukemia (AML) may be another illustrative case in point. There are 11 variants of AML that are treated by chemotherapy or stem cell transplants. Although the stem cell approach could produce successful outcomes, about half the patients encounter long-term negative effects, and complications may cause death. By using a knowledge bank relevant to AML, together with clinical experiences, it may be possible to determine the most efficacious treatments for an individual, while taking into account the odds of negative outcomes (Gerstung et al., 2017). Commensurate with this, small cell lung cancer in mice could be subdivided into types that differed in their aggressiveness. The C-MYC subtype could be treated fairly well by a drug and chemotherapy combination, which is significant since this subtype accounts for about 20% of patients with small-cell lung cancer, making it possible that this could be used to decide therapy options in patients.

Some genes may play a particularly prominent role in the promotion or progression of certain forms of cancer. For example, in a study conducted in mice, 23 genes were identified that contributed to cancer spread from the skin to the lungs. But one of these (Spns2) seemed to be a core gene so its modification reduced tumor spread by 75%, making it a potential target for treatment (van der Weyden et al., 2017). In effect, even if a full gene profile isn't obtained, beneficial effects can be derived by identifying specific genetic attributes of cancer. As we'll see, many other genes serve in this capacity, and these genes or the proteins for which they code have frequently been targeted to treat cancers.

The presence of epigenetic marks can also be significant for the early detection of breast cancer and could serve to direct treatment strategies and microRNA methylation was predictive of early-stage pancreatic and gastrointestinal cancer (Konno et al., 2019). It also appears that in triple-negative breast cancer and certain prostate cancers, targeting specific microRNAs can have positive effects and could potentially limit metastasis (Gilam et al., 2016). It was similarly reported that the presence of cancer checkpoints (PD-L1) on melanoma cells, which are important for cancer cells to evade detection by the immune system, was accompanied by global DNA hypomethylation, which may have important implications for the choice of immunotherapy that might be administered (Chatterjee et al., 2018). A detailed analysis that assessed epigenetic marks in three different tissues (thyroid, heart, and brain) was used to create a map that could be used to detect "correlated regions of systemic interindividual variation" in which methylation in one tissue could predict the expression of related genes in other tissues (Gunasekara et al., 2019). This could facilitate the identification of fundamental processes tied to several diseases and could theoretically be used as a screen to determine the presence of epigenetic changes at distal sites and might eventually be a viable approach in determining the treatments patients receive.

Recent renditions of the precision medicine approach have broadened the physiological predictors of illnesses to include endocrine, neurotransmitter, immune, microbial, neuroanatomical, and neuropsychological factors that may or may not be tied to genetic and epigenetic processes. By linking these multiple processes to symptoms or features of the illness, it might be possible to identify more precisely those treatments that would be optimal for a given individual. In some instances, particular symptoms of an illness might fall into clusters, likely because they involve the same underlying processes. Thus, rather than focusing on specific individual symptoms, various symptom clusters could be included in this approach. Unfortunately, because illness comorbidities are common, it can be challenging to determine which features are linked to the original condition being considered and which are related to a second illness.

Related to this, illnesses can develop owing to pleiotropic gene actions in which two or more phenotypic traits can be influenced by a single gene. Pleiotropy can occur when a gene induces a series of biological changes that lead to single or multiple pathologies (or symptoms). Accordingly, targeting one or more steps on this biological path could potentially modify an illness condition. However, parallel pleiotropy can also occur so a gene directly or indirectly influences different biological pathways and might thereby promote more than a single pathology, but each of these might be independent of one another. Thus, manipulating elements that promote a particular phenotype may not affect the second phenotype. In addition to pleiotropy and epistatic interactions, the promotion of illnesses may be governed by gene networks that are orchestrated by a common element or set of interconnected elements, which can make it difficult to determine the specific process that ought to be targeted to enact a positive therapeutic response. These caveats should not be taken to imply that precision medicine approaches be abandoned, but they point to some of the difficulties that will need to be addressed. In a way, however, these might only represent the proverbial tip of the iceberg since a precision medicine approach can be procedurally demanding and costly, raising the specter that this approach attempted on a broad scale may not be feasible.

Multiple markers across domains

Rather than the choice of therapies being predicated based on a single marker, it may be necessary for multiple biomarkers to be identified to select treatment strategies. As the characteristics of a tumor may change over time or with treatment, these biomarkers and the preferred treatment options would necessarily vary accordingly. A study that tracked the genetic factors associated with the effectiveness of immunotherapy identified more than 100 genes (and presumably their associated proteins) that could play some role in resistance to treatment (Patel et al., 2017). It has similarly been maintained that the identification of epigenetic marks may also be important in deciding treatment options. This also raises the question concerning which factors were responsible for the appearance of epigenetic changes that were linked to cancer occurrence, including the social, experiential, and environmental factors that could interact with the actions of genes.

For decades psychiatry had faced a crisis regarding the efficacy of treatment strategies for diverse mental disorders. It was argued that, among other things, diagnoses of illness were too narrowly defined, and appropriate markers relevant to treatments had not been adequately formulated. As a step to deal with this, the Research Domain Criteria (RDoC) were established in which illnesses were described across several levels of analysis, ranging from the gene, molecule, cells, neural circuits, and behavioral characteristics (Cuthbert and Insel, 2013). Concurrently, illness symptoms were distinguished from one another across multiple domains (e.g., the nature of the symptoms presented, cognitive and social functioning, among others). Based on this matrix of biological factors and behavioral phenotypes it was hoped that it might ultimately be possible to link specific symptoms or symptoms profiles (clusters) to biological processes that could, in turn, determine the most efficacious treatment strategies. This represented a marked departure from traditional approaches that began by defining an illness as falling into a specific syndrome, which informed the treatment. The RDoC perspective certainly had its detractors for a variety of reasons, including the implementation difficulties that would certainly be encountered, as well as the inattention that was devoted to cultural differences that fed into illness appearance (Kirmayer and Crafa, 2014).

Cancer researchers had been ahead of their mental health colleagues in using biomarkers of illness to determine treatment strategies. But, given the cancer links to psychosocial factors and lifestyles, it may be time to extend the predictors of cancer occurrence, progression, and treatment efficacy to include factors beyond the presence of genetic, epigenetic, and particular hormonal and growth factor markers. Fig. 4.8 provides a perspective in which a constel-

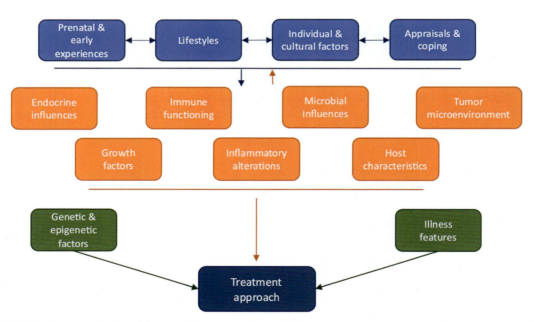

FIG. 4.8 An expanded precision medicine approach to predict cancer occurrence, progression, metastases, and choice of treatment, including the side effects that may emerge with treatment. The model suggests that features of a particular form of cancer, as well as the genes related to the individual, and those of the cancerous cells, and their microenvironment are important in dictating symptoms. Moreover, various hormonal factors (including those related to energy processing), growth factors, immune and inflammatory mechanisms, and microbial constituencies, may be fundamental in determining cancer occurrence and recurrence, and the effects of treatments. Many of these ingredients are linked to a variety of lifestyles factors and experiences, such as eating (the specific diet, and the presence of obesity), exercise, sleep, and circadian factors, stressor experience (especially those encountered during early life and stressors that are experienced chronically), and environmental toxicants.

lation of experiential factors, characteristics of the individual, lifestyles endorsed, and cancer itself (and the tumor microenvironment), might be linked to illness occurrence and progression, and might similarly be predictive of the response to treatments. In the ensuing chapters, we'll consider these lifestyle influences in some detail, but for the moment this model is only presented as a provisional template from which to work.

In theory, a precision medicine approach could be effective in defining new treatment targets, but in practice very large population studies would be essential to identify specific associations between genotypes and cancer-related phenotypes. Even then, such association studies may have only a limited capacity to predict specific phenotypes in any given individual (Khoury and Galea, 2016). As progressively more information accumulated tying genes to specific diseases, the realization dawned that the initial hopes might have been somewhat too optimistic and more clever approaches will be needed to make the connections between the presence of certain phenotypes (endophenotypes) and the effectiveness of specific treatments. This has been happening in several respects. The use of very large commercial databases (e.g., 23andMe) could potentially be instrumental in making links between genetic factors, disease vulnerability, and treatment efficacy. Likewise, large-scale analyses involving whole genome and epigenome sequencing were undertaken that linked gene variants to illnesses based on the information available on electronic health records (e.g., Dewey et al., 2016). With the evolution of machine learning approaches, connecting the dots, even very faint dots, will become accurate and feasible.

Summary and conclusions

Multiple factors play into the development and progression of most forms of cancer, including individual genetic factors and those associated with the cancer cells themselves, and the mutations that occur as cancer growth progresses. At the same time, multiple epigenetic changes occur that vary across cancer types, but also with individual experiences. Importantly, complex illnesses, such as cancer, are not only influenced by epigenetic processes or by Gene × Environment interactions but might also be connected to multiple psychosocial determinants (including stressful encounters, the availability of social support networks), lifestyles (diet, exercise, sleep), and cultural factors that additively or interactively operate to determine disease occurrence and the treatment strategies that should be adopted. To be sure, precision medicine has taken on increasing allure, but broadening this approach may be called for to provide optimal cancer treatments. Rather than relying exclusively on specific genes or biochemical markers, this could be achieved by including the psychosocial-biological signature presented by individual patients.

With the array of dynamic processes associated with cancer, including their remarkable ability to avoid detection and coopt immune cell functioning, finding the critical links that can be used as treatment targets has been exceptionally challenging. Indeed, the constellation of factors that are tied to specific cancers might be more a reflection of disharmony or discoordination between multiple processes that ordinarily ought to work in a synchronized manner (the orchestra has essentially lost its conductor). In the face of this discoordination, the normal processes that are essential for protection against cancers are lost. What this also implies, however, is that this quasirandomness could result in cancers differing subtly from one another and between individuals afflicted with the same cancer. Through the increased development of sophisticated detection approaches and machine learning and related algorithms, it may be possible to take advantage of cancer diversity to initiate individually tailored treatments. While many machine learning approaches have focused on identifying multiple genes and gene mutations that predict cancer occurrence, it was also demonstrated that mapping the multiple mutations situated on a cancer gene within a specific type of tissue may allow the identification of the optimal treatment strategy.

As more information is acquired concerning the actions of psychosocial and lifestyle factors, and biomarkers that signal the actions of genes, still better treatments might be developed, including the incorporation of psychosocial interventions. Even if they don't add to longevity, these psychological treatments can diminish distress and enhance the quality of life. At the same time, it is important to recognize that the cancer burden carried by underserved populations is inordinately high, stemming from various health care disparities together with greater exposure to diverse carcinogens, cultural barriers, social stressors, and poor lifestyles. Some of these same variables also represent roadblocks for obtaining access to diagnosis and treatment of cancers. Ultimately, much more will need to be accomplished in cancer prevention efforts, facilitating equity for cancer care, and adopting a broader perspective regarding the ingredients that makeup precision medicine approaches.

References

Audet, M.C., 2019. Stress-induced disturbances along the gut microbiota-immune-brain axis and implications for mental health: does sex matter? Front. Neuroendocrinol. 54, 100772.

Bachtiar, M., Ooi, B.N.S., Wang, J., Jin, Y., Tan, T.W., et al., 2019. Towards precision medicine: interrogating the human genome to identify drug pathways associated with potentially functional, population-differentiated polymorphisms. Pharmacogenomics J. 19, 516–527.

Baggiolini, A., Callahan, S.J., Montal, E., Weiss, J.M., Trieu, T., et al., 2021. Developmental chromatin programs determine oncogenic competence in melanoma. Science 373, eabc1048.

Basheer, F., Giotopoulos, G., Meduri, E., Yun, H., Mazan, M., et al., 2019. Contrasting requirements during disease evolution identify EZH2 as a therapeutic target in AML. J. Exp. Med. 216, 966–981.

Bejarano, L., Schuhmacher, A.J., Méndez, M., Megías, D., Blanco-Aparicio, C., et al., 2017. Inhibition of TRF1 telomere protein impairs tumor initiation and progression in glioblastoma mouse models and patient-derived xenografts. Cancer Cell 32, 590–607.

Blackburn, E.H., Epel, E.S., Lin, J., 2015. Human telomere biology: a contributory and interactive factor in aging, disease risks, and protection. Science 350, 1193–1198.

Blesske-Rechek, A.L., 2019. The Other Crisis in Psychology. Quillette. https://quillette.com/2019/07/30/the-other-crisis-in-psychology/. Accessed June 2020.

Bombay, A., Matheson, K., Anisman, H., 2014. The intergenerational effects of Indian Residential Schools: implications for the concept of historical trauma. Transcult. Psychiatry 51, 320–338.

Breast Cancer Association Consortium, Dorling, L., Carvalho, S., Allen, J., González-Neira, A., Luccarini, C., 2021. Breast cancer risk genes—association analysis in more than 113,000 women. N. Engl. J. Med. 384, 428–439.

Campbell, P.J., Getz, G., Korbel, J.O., Stuart, J.M., Jennings, J.L., et al., 2020. Pan-cancer analysis of whole genomes. Nature 578, 82–93.

Campisano, S., La Colla, A., Echarte, S.M., Chisari, A.N., 2019. Interplay between early-life malnutrition, epigenetic modulation of the immune function and liver diseases. Nutr. Res. Rev. 32, 128–145.

Carter, L., Rothwell, D.G., Mesquita, B., Smowton, C., Leong, H.S., et al., 2017. Molecular analysis of circulating tumor cells identifies distinct copy-number profiles in patients with chemosensitive and chemorefractory small-cell lung cancer. Nat. Med. 23, 114–119.

Castro, A., Pyke, R.M., Zhang, X., Thompson, W.K., Day, C.P., et al., 2020. Strength of immune selection in tumors varies with sex and age. Nat. Commun. 11, 4128.

Cavalli, G., Heard, E., 2019. Advances in epigenetics link genetics to the environment and disease. Nature 571, 489–499.

Chatrath, A., Przanowska, R., Kiran, S., Su, Z., Saha, S., et al., 2020. The pan-cancer landscape of prognostic germline variants in 10,582 patients. Genome Med. 12, 15.

Chatterjee, A., Rodger, E.J., Ahn, A., Stockwell, P.A., Parry, M., et al., 2018. Marked global DNA hypomethylation is associated with constitutive PD-L1 expression in melanoma. iScience 4, 312–325.

Chen, P., Zhao, D., Li, J., Liang, X., Li, J., et al., 2019. Symbiotic macrophage-glioma cell interactions reveal synthetic lethality in PTEN-null glioma. Cancer Cell 35, 868–884.e866.

Choudhury, A., Aron, S., Botigué, L.R., Sengupta, D., Botha, G., et al., 2020. High-depth African genomes inform human migration and health. Nature 586, 741–748.

Conforti, F., Pala, L., Bagnardi, V., De Pas, T., Martinetti, M., et al., 2018. Cancer immunotherapy efficacy and patients' sex: a systematic review and meta-analysis. Lancet Oncol. 19, 737–746.

Crabbe, J.C., Wahlsten, D., Dudek, B.C., 1999. Genetics of mouse behavior: interactions with laboratory environment. Science 284, 1670–1672.

Cunningham, A.M., Walker, D.M., Nestler, E.J., 2021. Paternal transgenerational epigenetic mechanisms mediating stress phenotypes of offspring. Eur. J. Neurosci. 53, 271–280.

Cuthbert, B.N., Insel, T.R., 2013. Toward the future of psychiatric diagnosis: the seven pillars of RDoC. BMC Med. 11, 126.

da Cruz, R.S., Chen, E., Smith, M., Bates, J., de Assis, S., 2020. Diet and transgenerational epigenetic inheritance of breast cancer: the role of the paternal germline. Front. Nutr. 7.

Dash, S., Kinney, N.A., Varghese, R.T., Garner, H.R., Feng, W.C., et al., 2019. Differentiating between cancer and normal tissue samples using multi-hit combinations of genetic mutations. Sci. Rep. 9, 1005.

Denham, J., 2018. Exercise and epigenetic inheritance of disease risk. Acta Physiol. 222.

Dewey, F.E., Murray, M.F., Overton, J.D., Habegger, L., Leader, J.B., et al., 2016. Distribution and clinical impact of functional variants in 50,726 whole-exome sequences from the DiscovEHR study. Science 354.

Dietlein, F., Weghorn, D., Taylor-Weiner, A., Richters, A., Reardon, B., et al., 2020. Identification of cancer driver genes based on nucleotide context. Nat. Genet. 52, 208–218.

Dietz, D.M., Nestler, E.J., 2012. From father to offspring: paternal transmission of depressive-like behaviors. Neuropsychopharmacology 37, 311–312.

Donehower, L.A., Soussi, T., Korkut, A., Liu, Y., Schultz, A., et al., 2019. Integrated analysis of TP53 gene and pathway alterations in The Cancer Genome Atlas. Cell Rep. 28, 1370–1384.e1375.

Doudna, J.A., 2020. The promise and challenge of therapeutic genome editing. Nature 578, 229–236.

Dunford, A., Weinstock, D.M., Savova, V., Schumacher, S.E., Cleary, J.P., et al., 2017. Tumor-suppressor genes that escape from X-inactivation contribute to cancer sex bias. Nat. Genet. 49, 10–16.

Eisfeldt, J., Mårtensson, G., Ameur, A., Nilsson, D., Lindstrand, A., 2020. Discovery of novel sequences in 1,000 Swedish genomes. Mol. Biol. Evol. 37, 18–30.

Feldman, S., Ammar, W., Lo, K., Trepman, E., van Zuylen, M., et al., 2019. Quantifying sex bias in clinical studies at scale with automated data extraction. JAMA Netw. Open 2, e196700.

Fontelles, C.C., Guido, L.N., Rosim, M.P., Andrade Fde, O., Jin, L., et al., 2016. Paternal programming of breast cancer risk in daughters in a rat model: opposing effects of animal- and plant-based high-fat diets. Breast Cancer Res. 18, 71.

Gerstung, M., Papaemmanuil, E., Martincorena, I., Bullinger, L., Gaidzik, V.I., et al., 2017. Precision oncology for acute myeloid leukemia using a knowledge bank approach. Nat. Genet. 49, 332–340.

Gilam, A., Conde, J., Weissglas-Volkov, D., Oliva, N., Friedman, E., et al., 2016. Local microRNA delivery targets Palladin and prevents metastatic breast cancer. Nat. Commun. 7, 12868.

Gould, S.J., 1981. The Mismeasure of Man. W.W. Norton & Company, New York.

Gunasekara, C.J., Scott, C.A., Laritsky, E., Baker, M.S., MacKay, H., et al., 2019. A genomic atlas of systemic interindividual epigenetic variation in humans. Genome Biol. 20, 105.

Gurdasani, D., Barroso, I., Zeggini, E., Sandhu, M.S., 2019. Genomics of disease risk in globally diverse populations. Nat. Rev. Genet. 20, 520–535.

Haigis, K.M., Cichowski, K., Elledge, S.J., 2019. Tissue-specificity in cancer: the rule, not the exception. Science 363, 1150–1151.

Heijmans, B.T., Tobi, E.W., Stein, A.D., Putter, H., Blauw, G.J., et al., 2008. Persistent epigenetic differences associated with prenatal exposure to famine in humans. Proc. Natl. Acad. Sci. U. S. A. 105, 17046–17049.

Hitchins, M.P., 2015. Constitutional epimutation as a mechanism for cancer causality and heritability? Nat. Rev. Cancer 15, 625–634.

Imam, M.U., Ismail, M., 2017. The impact of traditional food and lifestyle behavior on epigenetic burden of chronic disease. Global Chall. 1, 1700043.

Ivashkiv, L.B., 2018. IFNγ: signalling, epigenetics and roles in immunity, metabolism, disease and cancer immunotherapy. Nat. Rev. Immunol. 18, 545–558.

Jasiulionis, M.G., 2018. Abnormal epigenetic regulation of immune system during aging. Front. Immunol. 9, 197.

Jiang, W., Han, T., Duan, W., Dong, Q., Hou, W., Wu, H., et al., 2020. Prenatal famine exposure and estimated glomerular filtration rate across consecutive generations: association and epigenetic mediation in a population-based cohort study in Suihua China. Aging 12, 12206–12221.

Joseph, J., 2014. The Trouble With Twin Studies: A Reassessment of Twin Research in the Social and Behavioral Sciences. Routledge, United Kingdom.

Khoury, M.J., Galea, S., 2016. Will precision medicine improve population health? JAMA 316, 1357–1358.

Kim, H.I., Lim, H., Moon, A., 2018. Sex differences in cancer: epidemiology, genetics and therapy. Biomol. Ther. 26, 335–342.

Kirmayer, L.J., Crafa, D., 2014. What kind of science for psychiatry? Front. Hum. Neurosci. 8, 435.

Klukovich, R., Nilsson, E., Sadler-Riggleman, I., Beck, D., Xie, Y., et al., 2019. Environmental toxicant induced epigenetic transgenerational inheritance of prostate pathology and stromal-epithelial cell epigenome and transcriptome alterations: ancestral origins of prostate disease. Sci. Rep. 9, 2209.

Konno, M., Koseki, J., Asai, A., Yamagata, A., Shimamura, T., et al., 2019. Distinct methylation levels of mature microRNAs in gastrointestinal cancers. Nat. Commun. 10, 3888.

Körber, V., Yang, J., Barah, P., Wu, Y., Stichel, D., et al., 2019. Evolutionary trajectories of IDH(WT) glioblastomas reveal a common path of early tumorigenesis instigated years ahead of initial diagnosis. Cancer Cell 35, 692–704. e612.

Ku, H.-C., Cheng, C.-F., 2020. Master regulator activating transcription factor 3 (ATF3) in metabolic homeostasis and cancer. Front. Endocrinol. 11, 556.

Kundaje, A., Meuleman, W., Ernst, J., Bilenky, M., Yen, A., et al., 2015. Integrative analysis of 111 reference human epigenomes. Nature 518, 317–330.

Kundakovic, M., Champagne, F.A., 2011. Epigenetic perspective on the developmental effects of bisphenol A. Brain Behav. Immun. 25, 1084–1093.

Landau, D.A., Clement, K., Ziller, M.J., Boyle, P., Fan, J., et al., 2014. Locally disordered methylation forms the basis of intratumor methylome variation in chronic lymphocytic leukemia. Cancer Cell 26, 813–825.

Laumont, C.M., Vincent, K., Hesnard, L., Audemard, É., Bonneil, É., et al., 2018. Noncoding regions are the main source of targetable tumor-specific antigens. Sci. Transl. Med. 10, eaau5516.

Lesch, B.J., Tothova, Z., Morgan, E.A., Liao, Z., Bronson, R.T., et al., 2019. Intergenerational epigenetic inheritance of cancer susceptibility in mammals. eLife 8, e39380.

Letai, A., 2017. Functional precision cancer medicine-moving beyond pure genomics. Nat. Med. 23, 1028–1035.

Levy-Sakin, M., Pastor, S., Mostovoy, Y., Li, L., Leung, A.K.Y., et al., 2019. Genome maps across 26 human populations reveal population-specific patterns of structural variation. Nat. Commun. 10, 1025.

Li, J., Liu, S., Li, S., Feng, R., Na, L., et al., 2017. Prenatal exposure to famine and the development of hyperglycemia and type 2 diabetes in adulthood across consecutive generations: a population-based cohort study of families in Suihua, China. Am. J. Clin. Nutr. 105, 221–227.

Li, J., Cao, F., Yin, H.-L., Huang, Z.J., Lin, Z.T., et al., 2020a. Ferroptosis: past, present and future. Cell Death Dis. 11, 88.

Li, C.H., Prokopec, S.D., Sun, R.X., et al., 2020b. Sex differences in oncogenic mutational processes. Nat. Commun. 11, 4330.

Li, H., Terry, M.B., Antoniou, A.C., Phillips, K.A., Kast, K., et al., 2020c. Alcohol consumption, cigarette smoking, and risk of breast cancer for BRCA1 and BRCA2 mutation carriers: results from the BRCA1 and BRCA2 cohort consortium. Cancer Epidemiol. Biomark. Prev. 29, 368–378.

Li, S., Luo, M., Wang, Z., Feng, Q., Wilhelm, J., et al., 2021. Prolonged activation of innate immune pathways by a polyvalent STING agonist. Nat. Biomed. Eng. 5, 455–466.

Lin, J., Sun, J., Wang, S., Milush, J.M., Baker, C.A.R., et al., 2018. In vitro proinflammatory gene expression predicts in vivo telomere shortening: a preliminary study. Psychoneuroendocrinology 96, 179–187.

Lineweaver, C.H., Bussey, K.J., Blackburn, A.C., Davies, P.C.W., 2021. Cancer progression as a sequence of atavistic reversions. BioEssays 43, e2000305.

Lismer, A., Dumeaux, V., Lafleur, C., Lambrot, R., Brind'Amour, J., et al., 2021. Histone H3 lysine 4 trimethylation in sperm is transmitted to the embryo and associated with diet-induced phenotypes in the offspring. Dev. Cell 56, 671–686.e676.

Liu, J., Dou, X., Chen, C., Chen, C., Liu, C., et al., 2020. N (6)-methyladenosine of chromosome-associated regulatory RNA regulates chromatin state and transcription. Science 367, 580–586.

Mackay, A., Burford, A., Carvalho, D., Izquierdo, E., Fazal-Salom, J., et al., 2017. Integrated molecular meta-analysis of 1,000 pediatric high-grade and diffuse intrinsic pontine glioma. Cancer Cell 32, 520–537.e525.

Mackenzie, K.J., Carroll, P., Martin, C.A., Murina, O., Fluteau, A., et al., 2017. cGAS surveillance of micronuclei links genome instability to innate immunity. Nature 548, 461–465.

Manikkam, M., Haque, M.M., Guerrero-Bosagna, C., Nilsson, E.E., Skinner, M.K., 2014. Pesticide methoxychlor promotes the epigenetic transgenerational inheritance of adult-onset disease through the female germline. PLoS One 9, e102091.

Martincorena, I., Raine, K.M., Gerstung, M., Dawson, K.J., Haase, K., et al., 2017. Universal patterns of selection in cancer and somatic tissues. Cell 171, 1029–1041.e1021.

Mavaddat, N., Michailidou, K., Dennis, J., Lush, M., Fachal, L., et al., 2019. Polygenic risk scores for prediction of breast cancer and breast cancer subtypes. Am. J. Hum. Genet. 104, 21–34.

Meurette, O., Mehlen, P., 2018. Notch signaling in the tumor microenvironment. Cancer Cell 34, 536–548.

Moloney, G.M., Dinan, T.G., Clarke, G., Cryan, J.F., 2019. Microbial regulation of microRNA expression in the brain-gut axis. Curr. Opin. Pharmacol. 48, 120–126.

Muvarak, N.E., Chowdhury, K., Xia, L., Robert, C., Choi, E.Y., et al., 2016. Enhancing the cytotoxic effects of PARP inhibitors with DNA demethylating agents—a potential therapy for cancer. Cancer Cell 30, 637–650.

Nebbioso, A., Tambaro, F.P., Dell'Aversana, C., Altucci, L., 2018. Cancer epigenetics: moving forward. PLoS Genet. 14, e1007362.

Nei, M., 2013. Mutation-Driven Evolution. OUP, Oxford University Press, United Kingdom.

Newman, L.A., Kaljee, L.M., 2017. Health disparities and triple-negative breast cancer in African American women: a review. JAMA Surg. 152, 485–493.

Newman, S., Nakitandwe, J., Kesserwan, C.A., Azzato, E.M., Wheeler, D.A., 2021. Genomes for kids: the scope of pathogenic mutations in pediatric cancer revealed by comprehensive DNA and RNA sequencing. Cancer Discov. 11, 3008–3027.

Oliva, M., Muñoz-Aguirre, M., Kim-Hellmuth, S., Wucher, V., Gewirtz, A.D.H., et al., 2020. The impact of sex on gene expression across human tissues. Science 369, eaba3066.

Org, E., Mehrabian, M., Parks, B.W., Shipkova, P., Liu, X., et al., 2016. Sex differences and hormonal effects on gut microbiota composition in mice. Gut Microbes 7, 313–322.

Palma-Gudiel, H., Fañanás, L., Horvath, S., Zannas, A.S., 2020. Psychosocial stress and epigenetic aging. Int. Rev. Neurobiol. 150, 107–128.

Patel, S.J., Sanjana, N.E., Kishton, R.J., Eidizadeh, A., Vodnala, S.K., et al., 2017. Identification of essential genes for cancer immunotherapy. Nature 548, 537–542.

Peña, C.J., Neugut, Y.D., Champagne, F.A., 2013. Developmental timing of the effects of maternal care on gene expression and epigenetic regulation of hormone receptor levels in female rats. Endocrinology 154, 4340–4351.

Pfeifer, G.P., 2015. How the environment shapes cancer genomes. Curr. Opin. Oncol. 27, 71–77.

Post, C.M., Boule, L.A., Burke, C.G., O'Dell, C.T., Winans, B., et al., 2019. The ancestral environment shapes antiviral CD8+ T cell responses across generations. iScience 20, 168–183.

Priestley, P., Baber, J., Lolkema, M.P., Steeghs, N., de Bruijn, E., et al., 2019. Pan-cancer whole-genome analyses of metastatic solid tumours. Nature 575, 210–216.

Quirt, J.S., Nanji, S., Wei, X., Flemming, J.A., Booth, C.M., 2017. Is there a sex effect in colon cancer? Disease characteristics, management, and outcomes in routine clinical practice. Curr. Oncol. 24, e15–e23.

Rode, L., Nordestgaard, B.G., Bojesen, S.E., 2016. Long telomeres and cancer risk among 95568 individuals from the general population. Int. J. Epidemiol. 45, 1634–1643.

Rodgers, A.B., Morgan, C.P., Leu, N.A., Bale, T.L., 2015. Transgenerational epigenetic programming via sperm microRNA recapitulates effects of paternal stress. Proc. Natl. Acad. Sci. U. S. A 112, 13699–13704.

Schmutz, I., Mensenkamp, A.R., Takai, K.K., Haadsma, M., Spruijt, L., et al., 2020. TINF2 is a haploinsufficient tumor suppressor that limits telomere length. eLife 9, e61235.

Shoshani, O., Brunner, S.F., Yaeger, R., Ly, P., Nechemia-Arbely, Y., et al., 2021. Chromothripsis drives the evolution of gene amplification in cancer. Nature 591, 137–141.

Shuai, S., Suzuki, H., Diaz-Navarro, A., Nadeu, F., Kumar, S.A., et al., 2019. The U1 spliceosomal RNA is recurrently mutated in multiple cancers. Nature 574, 712–716.

Sovijit, W.N., Sovijit, W.E., Pu, S., Usuda, K., Inoue, R., et al., 2021. Ovarian progesterone suppresses depression and anxiety-like behaviors by increasing the Lactobacillus population of gut microbiota in ovariectomized mice. Neurosci. Res. 168, 76–82.

Sun, T., Plutynski, A., Ward, S., Rubin, J.B., 2015. An integrative view on sex differences in brain tumors. Cell. Mol. Life Sci. 72, 3323–3342.

Szyf, M., 2019. The epigenetics of perinatal stress. Dialogues Clin. Neurosci. 21, 369–378.

Tarsounas, M., Sung, P., 2020. The antitumorigenic roles of BRCA1–BARD1 in DNA repair and replication. Nat. Rev. Mol. Cell Biol. 21, 284–299.

Telen, M.J., Malik, P., Vercellotti, G.M., 2019. Therapeutic strategies for sickle cell disease: towards a multi-agent approach. Nat. Rev. Drug Discov. 18, 139–158.

Tomasetti, C., Vogelstein, B., 2015. Cancer etiology. Variation in cancer risk among tissues can be explained by the number of stem cell divisions. Science 347, 78–81.

Tomasetti, C., Marchionni, L., Nowak, M.A., Parmigiani, G., Vogelstein, B., 2015. Only three driver gene mutations are required for the development of lung and colorectal cancers. Proc. Natl. Acad. Sci. 112, 118–123.

Topper, M.J., Vaz, M., Marrone, K.A., Brahmer, J.R., Baylin, S.B., 2020. The emerging role of epigenetic therapeutics in immuno-oncology. Nat. Rev. Clin. Oncol. 17, 75–90.

van der Weyden, L., Arends, M.J., Campbell, A.D., Bald, T., Wardle-Jones, H., et al., 2017. Genome-wide in vivo screen identifies novel host regulators of metastatic colonization. Nature 541, 233–236.

Vogelstein, B., Papadopoulos, N., Velculescu, V.E., Zhou, S., Diaz Jr., L.A., et al., 2013. Cancer genome landscapes. Science 339, 1546–1558.

Wallis, C.J.D., Butaney, M., Satkunasivam, R., Freedland, S.J., Patel, S.P., et al., 2019. Association of patient sex with efficacy of immune checkpoint inhibitors and overall survival in advanced cancers: a systematic review and meta-analysis. JAMA Oncol. 5, 529–536.

Wang, S., Cowley, L.A., Liu, X.S., 2019. Sex differences in cancer immunotherapy efficacy, biomarkers, and therapeutic strategy. Molecules 24, 3214.

Wu, S., Turner, K.M., Nguyen, N., Raviram, R., Erb, M., et al., 2019. Circular ecDNA promotes accessible chromatin and high oncogene expression. Nature 575, 699–703.

Wu, S., Bafna, V., Mischel, P.S., 2021. Extrachromosomal DNA (ecDNA) in cancer pathogenesis. Curr. Opin. Genet. Dev. 66, 78–82.

Xu, X., Miao, Z., Sun, M., Wan, B., 2021. Epigenetic mechanisms of paternal stress in offspring development and diseases. Int. J. Genomics 2021, 6632719.

Yehuda, R., Daskalakis, N.P., Bierer, L.M., Bader, H.N., Klengel, T., et al., 2016. Holocaust exposure induced intergenerational effects on FKBP5 methylation. Biol. Psychiatry 80, 372–380.

Yizhak, K., Aguet, F., Kim, J., Hess, J.M., Kübler, K., et al., 2019. RNA sequence analysis reveals macroscopic somatic clonal expansion across normal tissues. Science 364, eaaw0726.

Yousefzadeh, M.J., Flores, R.R., Zhu, Y., Schmiechen, Z.C., Brooks, R.W., et al., 2021. An aged immune system drives senescence and ageing of solid organs. Nature 594, 100–105.

Yuan, C., Burns, M.B., Subramanian, S., Blekhman, R., 2018. Interaction between host micrornas and the gut microbiota in colorectal cancer. mSystems 3, e00205-17.

Zhang, Y.D., Hurson, A.N., Zhang, H., Choudhury, P.P., Easton, D.F., et al., 2020. Assessment of polygenic architecture and risk prediction based on common variants across fourteen cancers. Nat. Commun. 11, 3353.

Zhu, M., Wang, T., Huang, Y., Zhao, X., Ding, Y., et al., 2021. Genetic risk for overall cancer and the benefit of adherence to a healthy lifestyle. Cancer Res. 81, 4618–4627.

Stressors: Psychological and neurobiological processes

OUTLINE

Attributions and misattributions of stressor effects on illness	135	Growth factors	157
		Neurotrophins and brain functioning	158
Features of stressors	136	Growth factors linked to cancer	160
		Influence of stressors on growth factors	163
Chronic stressors and allostatic load	137	Cellular stress responses	164
Stress sensitization	139	Prenatal and early-life stressor effects	165
Identifying and responding to stressors	139	Promotion of premature birth	165
Appraisals	139	Hormonal consequences of prenatal stress	166
Heuristics, decision-making, and misappraisals	140	Epigenetic changes related to prenatal stress	166
Coping	142	Impact of prenatal infection	168
		Influence of early-life stressors	169
Resilience and vulnerability	145		
Neurobiological actions of stressors	147	Summary and conclusions	171
A systems perspective of stressor actions	147	References	172
Hormonal processes	149		

Attributions and misattributions of stressor effects on illness

Sadly, sometimes "bad things happen to good people" (Kushner, 1981). Surely, this shouldn't occur, but when it does, individuals might seek reasons for the distressing situation in which they find themselves, asking "Why me?" Depending on their spiritual orientation they might attribute these events to "God's will" or simply ascribe these experiences to fate. Others might look for specific causes for their situation, and it is not uncommon for individuals to attribute their illness to stressors that they had experienced, after which they might seek ways of diminishing their distress and enhance their psychological well-being.

But can stressors actually provoke the development or progression of illnesses? Conversely, can stress-reducing interventions inhibit tumor progression, or enhance the effects of cancer treatments? To understand these relationships, it is necessary to appreciate the cognitive and behavioral responses to stressors, as well as the stressor-related features that are apt to promote changes of hormone, immune, inflammatory, and microbial functioning. Recognition of these processes may also be relevant to diminish mental health problems and might be instrumental in limiting difficulties that could otherwise undermine health-promoting behaviors related to physical illnesses, including cancer and its treatment. Determining the biological changes that accompany stressor experiences could also provide insights concerning the processes by which stressors could lead to the frequent comorbidities that are so often apparent across seemingly different illnesses.

Features of stressors

It's probably safe to say that if an individual perceives an event or a stimulus as being a stressor, then it probably is. However, if the individual does not perceive a stressor to be present, it doesn't necessarily mean that a stressor isn't affecting them. Many stressors that we encounter are physical or psychological, termed neurogenic and psychogenic stressors, respectively, and are most often readily recognized. But some challenges comprise systemic stressors, such as bacterial or viral challenges, increased presence of inflammatory factors, or elevated levels of blood glucose, which may not be sensed by individuals. Yet, these systemic challenges stimulate many of the same biological circuits activated by physical and psychological stressors, which could potentially contribute to the development of illnesses (Anisman et al., 2018).

Severe stressors typically have more pronounced effects than do mild stressors, although the accumulation of day-to-day hassles can also be damaging, especially if these are superimposed on ongoing strong stressors. It's not just the severity of stressors that can lead to adverse outcomes. In general, events that are *unpredictable*, *uncertain*, or *ambiguous* promote greater negative consequences than do those that are predictable, certain, or unambiguous. Most of us are certainly aware of the anxiety elicited by situations that are vague and unpredictable and these events can have particularly pronounced consequences among individuals who are unable to deal with uncertainty.

Of the many aspects of stressor situations that can cause health problems, the ability to maintain control over outcomes is among the most important. Early studies had indicated that relative to animals that could control the offset of stressors through their behavior (a controllable stressor), those that were exposed to an identical stressor that was uncontrollable showed depressive-like behaviors and various other pathologies. These outcomes were initially attributed to the development of cognitive disturbances, such as "learned helplessness" (Maier and Seligman, 2016), but alternative positions attributed the behavioral impairments to several brain neurochemical changes (Anisman et al., 2018). Uncontrollable events in humans could likewise instigate behavioral disturbances, although control in humans involves elements that could not be assessed in animal studies. These comprise several distinct types of control (Cohen et al., 1986) that may be especially important to cancer and its treatments.

Control may be related to being able to determine outcomes through mental or action-based strategies (cognitive and behavioral control), and the ability or the opportunity to select a particular coping strategy (decisional control). It is also important for individuals to be able to determine the extent to which events can be predicted, allowing for preparatory behaviors to be adopted (information control). It was also maintained that attributions concerning control over events, specifically whether individuals believe that their actions have bearing on outcomes that are outside of their personal control (locus of control), are fundamental in determining well-being.

The importance of some forms of control has been documented in numerous studies that assessed psychological well-being among individuals diagnosed or being treated for cancer. In general, cognitive control rather than behavioral control was most closely aligned with good adjustment to a breast cancer diagnosis. In contrast, cancer occurrence that was attributed to features of the self, another person, chance, or external factors, were linked to poor adjustment. The individual's belief that they have the ability and capacity to adopt behaviors essential for their well-being (self-efficacy) was particularly important and has been included in various therapies to improve adjustment to illnesses (see Chapter 13). Likewise, adopting active coping efforts contributed to individuals' resilience, thereby influencing their general quality of life. Among women with breast cancer, acceptance and positive reappraisal strategies were likewise associated with better well-being and health, whereas avoidant and disengagement coping methods were accompanied by diminished well-being (Kvillemo and Bränström, 2014). As expected, beliefs related to being able to control outcomes or that others could do so (e.g., their physician) were linked to enhanced adjustment, which may be tied to brain neurochemical, hormonal, and immune processes. Given the impact of coping in dealing with stress related to breast cancer interventions and treatments, ways to manage distress have often been recommended, although these need to be considered on an individual basis (e.g., Borgi et al., 2020).

Chronic stressors and allostatic load

Arguably, the most important feature of stressors in the development of psychological and physical pathologies concerns the chronicity of the experience. Relative to acute threats, stressors that persist for extended periods, particularly if these are uncontrollable and occur unpredictably, are more likely to promote adverse effects. The nature of the stressors that lead to pathology vary widely and may include physical and psychological challenges, such as family issues, job strain, being repeatedly bullied in the schoolyard or at work, as well as acting as a caregiver for an ailing child, parent, or spouse.

Chronic illnesses and chronic pain, which have become progressively more common, are powerful stressors that are endured often without suitable medications. A 2009 *Lancet* editorial had estimated that within the United States, 133 million people suffered a chronic illness, and this number was expected to reach 157 million by 2020, with almost half these individuals being affected by multiple conditions. This turned out to be a vast underestimate. A Rand Corporation report indicated that by 2014 more than 150 million people within the US were living with a chronic illness, and about 100 million people had more than a single chronic condition. Furthermore, according to the CDC, in 2019 approximately 60% of US adults were experiencing a chronic illness, and 40% had more than a single chronic health condition.

The primary illnesses comprised diabetes, heart disease, and cancer, with the most common nongenetic risk factors being smoking, excessive use of alcohol, poor nutrition, and a lack of exercise. There is little doubt that with an aging population, chronic conditions will need to receive more attention, especially to deal with the multiple comorbid illnesses that will likely emerge.

The concepts of allostasis and allostatic overload were introduced to account for the impact of chronic stressors. In response to stressors, a series of biological changes occur that may facilitate coping and ought to preserve well-being. In a sense, this is like homeostasis that reflects adaptations that occur in response to moderate environmental changes. However, in response to strong threats, this process necessarily must occur more quickly and often entails much stronger biological responses (allostasis). Behavioral, cognitive, and emotional responses are typically sufficient to deal with most negative events that humans meet on a day-to-day basis. However, as described in Fig. 5.1, with chronic stressor experiences that are relatively severe, including multiple psychosocial threats, biological resources may be overly taxed so that their availability and capacity to deal with further negative events may be undermined (allostatic overload). For instance, if a stressor is mild, then the neurobiological responses elicited may facilitate well-being. However, if the stressor is severe, unpredictable, and uncontrollable, and especially if it occurs on a chronic basis, then the altered functioning (or "misfunctioning") of certain biological systems may eventually give rise to negative emotional and cognitive outcomes, as well as psychological and physical illnesses (McEwen and Akil, 2020).

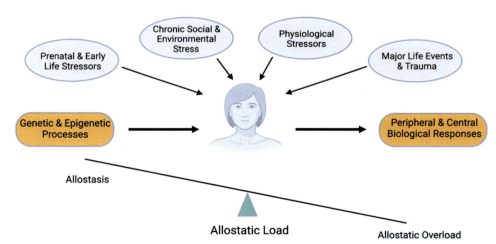

FIG. 5.1 Stressful events give rise to multiple neurobiological changes that ought to have adaptive value so psychological and neurological pathology will not occur (allostasis). However, in response to chronic stressors, especially if these are uncontrollable and unpredictable, neurobiological systems may be overly taxed or certain receptors may be excessively stimulated (allostatic overload). While this model was initially adopted to account for brain changes that governed psychological functioning, it is equally applicable to immune- and inflammatory-related disorders thereby promoting or accelerating disease processes that foster cardiovascular disease, diabetes, and cancer progression.

Stress sensitization

It's commonly thought that once a stressor has passed, it can be left behind and forgotten. Indeed, many of the neurobiological effects of stressors persist for a relatively brief period, measured in minutes or hours, and occasionally over days. But stressor experiences can have proactive effects so the response to later negative encounters may be markedly exaggerated, favoring the occurrence of pathology. These *sensitized* responses are not unique to specific stressors and can be elicited in response to later challenges that differed from those initially experienced. As well, exaggerated responses could also be provoked by drugs, such as amphetamine or cocaine, which affect neurobiological systems ordinarily activated by stressors. The increased response to the second stressor experience may occur because individuals have become more emotionally sensitive to events or triggers (cues) related to these experiences, which may be due to biological systems that had been primed (sensitized) so more pronounced neurochemical responses are elicited when further stressors are experienced (Anisman et al., 2018). These lasting negative outcomes can emerge irrespective of when the initial stressor was encountered but are especially prominent if they occurred during early life, thereby promoting lasting susceptibility to health disturbances.

Identifying and responding to stressors

Appraisals

In response to stressors, individuals make assessments (appraisals) about these events, which contribute to the methods of coping that are subsequently adopted (Lazarus and Folkman, 1984), ultimately contributing to the occurrence of illnesses (see Fig. 5.2). Some of the common factors that figure into accurate or poor appraisals being made may be dictated by several personality factors. These individual difference characteristics include perceived personal resources, self-efficacy (the belief that the person can succeed in specific situations) together with an individual's beliefs and expectancies concerning stressor outcomes, as well as the demands and the constraints that an individual faces in attempting to deal with challenges. On the surface, making accurate appraisals of situations might seem straightforward. However, the ability of people to make appropriate appraisals may not always be particularly accurate. Previous experiences with both similar and dissimilar aversive events (including early-life stressor experiences) can affect appraisals, as can characteristics of the stressor itself and a constellation of personality variables.

Annoyingly, some individuals maintain a cognitive bias (*illusory superiority*) in which they overestimate their abilities and personal qualities, fail to acknowledge their limitations, lack of skill, and the mistakes that they had made, and they concurrently underestimate the expertise and skills of others (Dunning-Kruger effect). This contrasts with top performers who are more accurate in their self-appraisals or who even underestimate their abilities and performance (Kruger and Dunning, 1999). Those high in illusory superiority are so self-assured that they tend to respond more quickly in certain tasks, which is accompanied by distinct brain electrophysiological responses (Muller et al., 2021). Such biases may create vast differences

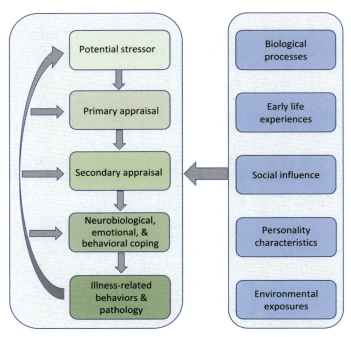

FIG. 5.2 Upon encountering a potential stressor, individuals consider whether this represents a threat or risk to well-being. Following this primary appraisal, individuals ordinarily make a secondary appraisal, assessing whether they have adequate coping resources to meet the demands of the situation. Based on these appraisals, specific coping responses to deal with the challenge will be adopted. Social heuristics, referring to ways of interpreting events based on similar or dissimilar experiences, can alter stressor appraisals, and social influences can likewise affect the secondary appraisals. The appraisals that are made and coping strategies used may also vary with numerous dispositional (personality) characteristics. Social buffers (social coping) can likewise affect other coping methods used. Accurate appraisals and the adoption of effective coping methods may limit stressor-related problems. *Modified from Anisman, H., 2021. Health Psychology: A Biopsychosocial Approach. Sage, London.*

in the accuracy of appraisals and could potentially influence strongly held beliefs concerning health risks, perhaps even contributing to outlandish views of events (conspiracy theories) that have increasingly been squirming out of the woodwork, which has been fairly obvious during the COVID-19 pandemic.

Heuristics, decision-making, and misappraisals

How decisions are made and why these decisions are at times inappropriate and counterproductive is affected by multiple factors, and too often decision errors occur under stressful situations. Of the theoretical perspectives offered in this regard, one of the best known is that offered by Tversky and Kahneman (Kahneman, 2011; Tversky and Kahneman, 1974) who maintained that several heuristics (cognitive shortcuts) are fundamental to decision-making. Their analyses and others that followed have had enormous implications as to why inappropriate and counterproductive decisions are frequently made.[a] Their research and theorizing hadn't been specifically concerned with performance under stressor conditions; nevertheless, their views have much to say about decision-making processes that are linked to stressor appraisals and the selection of specific coping methods.

The initial formulations that had been offered comprised several heuristics that influenced decision-making (Tversky and Kahneman, 1974). The "availability heuristics," which is observed fairly frequently, entails decisions based on how readily these come to mind. A second concerns "representativeness" in which decisions are based on their assumed similarity to other situations. A third major heuristic comprises "anchoring or adjustment" in which numerical predictions are based on information initially obtained and then modified as further information is garnered. For instance, in buying a used car, we might not have a clue as to its actual value. Thus, we might base our initial appraisals on the asking price, and then as additional information becomes available, we might alter our appraisals accordingly.

Since the introduction of this framework, dozens of other heuristics and cognitive biases were described that could affect decision-making. One of the more common is "attribute substitution" in which complex decisions are replaced by simpler approaches to form decisions, even if these are entirely unwarranted. Yet another frequently adopted heuristic, "associative coherence," is one in which decisions are based on preconceived or "primed intuitions" that had been established through previous experiences or perhaps second-hand information that was irrelevant to current decisions that need to be made. Individuals may also be affected by the "scarcity heuristic" in which it is assumed that the value of an object increases as it becomes more scarce. If it is more difficult to obtain, then it must be more valuable. It hardly needs saying that shortcuts in decision-making might have little bearing on the wisdom or correctness of these decisions; however, individuals may feel that these are "good enough" for the moment (Gigerenzer, 2008).

A decade ago, Kahneman (2011) suggested that dual systems operate in making decisions; this consisted of an automatic or fast thinking system (System 1) and a cognitively based, slow thinking system (System 2). The fast thinking response occurs based on already established heuristics so the automatic reactions occur in some decision-making situations, which may be important in stressful situations that call for rapid responses to be initiated. But as errors can occur under these circumstances, the cognitively oriented System 2 may be called upon to override the fast system, particularly when decisions need to be made in complex situations.

There are instances in which our dispositions, possibly cultivated over years, affect day-to-day behaviors (e.g., food brands selected, political parties being favored, and even discrimination and stigma directed at other groups), and based on brain neural responses decisions may be made even before individuals are consciously aware of having made them (Soon et al., 2013). These preconceived decisions may reflect strongly entrenched beliefs that aren't readily altered through logical arguments or even in dire situations. It's certainly unfortunate that individual appraisals may be inaccurate even under the best of circumstances,

---

[a] Daniel Kahneman won the 2002 Nobel Prize in Economics for his work applying psychological principles to economics. His research partner Amos Tversky sadly died in 1996, and so was not eligible for the prize. According to a statement released by the American Psychological Association, Kahneman stated "Certainly, we would have gotten this together….. There is that shadow over the joy I feel."

but when stressed, appraisals are more apt to be inexact and the decisions that follow will similarly be compromised. Indeed, it has been known for decades that under stressful circumstances individuals' decisional and response repertoire narrows, precluding the adoption of alternative modes of dealing with threats. As well, being in a poor mood may favor negative appraisals in numerous situations, whereas when an individual's spirits are elevated, appraisals will likely be more positive, and even patently negative situations may be perceived as being manageable. In essence, we might not always appraise situations accurately or behave rationally and we may be the victims of long-standing biases or affected by our current mood state.

In the context of making decisions concerning how to deter the likelihood of cancer development, most individuals fail to take appropriate steps to prevent illness (the "it won't happen to me syndrome") and may rely on heuristics that are counterproductive in guiding their behaviors. In contrast, individuals at high risk for certain types of cancer (e.g., based on family history) are more likely to engage in certain measures, such as more frequent cancer screening, to determine if cancer is present. Yet, many individuals fail to do so, and might not adopt behaviors that could potentially reduce cancer occurrence. Thus, increasing research attention has focused on shared decision-making (i.e., between the patient and her/his physician) to encourage people to take on a more active role in their health care. Even so, when cancer-related symptoms appear, individuals may, for better or worse, make decisions based on a variety of heuristics, including those that foster a delay in seeking health care (Kummer et al., 2019). Predictably, when faced with a cancer diagnosis, the resources that individuals have may be limited, and making well thought-out and considered appraisals that lead to optimal or satisfactory decisions may be questionable. As a result, they may revert to heuristics like those that are governed by intuitions, irrespective of data related to the efficacy of specific treatment strategies.

Coping

Assuming that proper appraisals have been made concerning a threat encountered, individuals will adopt varied means to deal with them. Some coping "styles" may be inherent to any given individual, essentially being part of their personality so particular coping methods predominate across situations. But coping may also comprise distinct "strategies" that vary across situations. As such, while individuals may have certain styles that are generally favored, a given strategy may be more apt to be adopted in specific contexts.

Dozens of ways are available to contend with the numerous stressors that individuals can encounter, with the majority falling into about 15 subcategories that, in turn, make up 3 broad classes that comprise problem-, emotion-, and avoidant-focused coping (see Table 5.1). Although most coping methods provided in the table were assigned to a specific category, it's fair to say that some of these coping methods don't sit comfortably within any single category, since many of these can serve multiple functions. Moreover, the coping methods adopted not only vary across stressor situations but may change with time as chronic stressors play out (e.g., chronic illness). For example, obtaining social support can operate in an emotion-focused capacity (a shoulder to cry on), as a problem-focused coping method (obtaining assistance to find a way to deal with a stressor), in an avoidant capacity (e.g., serving as a distractor), or as instrumental support (having friends help with chores).

TABLE 5.1 Coping methods.

Problem-focused strategies

Problem solving: Finding methods to eliminate the stressor or that might deter its impact.

Cognitive restructuring (positive reframing): Reassessing or placing a new spin on a situation (finding a silver lining to a black cloud) so that it may take on positive attributes.

Finding meaning (benefit finding): A form of cognitive restructuring that entails individuals finding some benefit or making sense of a traumatic experience. This might involve emotional or cognitive changes, or active efforts so that others will gain from the experience.

Avoidant or disengagement strategies

Active distraction: Using active behaviors (working out, going to movies) as a distraction from ongoing problems.

Cognitive distraction: Thinking about issues unrelated to the stressor, such as immersing oneself in work, or engaging in hobbies.

Denial/emotional containment: Not thinking about an issue or simply convincing oneself that it's not particularly serious.

Humor: Using laughter and amusement to diminish the distressing nature of a given situation.

Drug use: Using certain drugs to diminish the impact of stressors, including the physiological (physical discomfort) and emotional responses (anxiety) elicited by a stressor. Some individuals may engage in eating in a counterproductive effort to cope with challenges.

Emotion-focused strategies

Emotional expression: Using emotions such as crying, anger, and even aggressive behaviors to deal with stressors.

Blame (other): Blaming others for adverse events, possibly to avoid being blamed (or self-blame), or as a way to make sense of some situations.

Self-blame: Blaming ourselves for events that occurred. Often associated with feelings of guilt or anger directed at the self.

Rumination: Continued, sometimes unremitting thoughts about an issue or event, or replaying the events and the strategies that could have been used to deal with them.

Wishful thinking: Thinking what it would be like if the stressor were gone, or what it was like in happier times before the stressor had surfaced.

Passive resignation: Acceptance of a situation as it is, possibly reflecting feelings of helplessness, or simply accepting the situation without regret or malice ("it is what it is").

Social support

Social support seeking: Finding people or groups who may be beneficial in coping with stressors. This method is especially useful as it may buffer the impact of stressors as well as obtaining information as to how to best deal with a particular situation.

Religion and spirituality

Religiosity (internal): A belief in God to deal with adverse events. This may entail the simple belief in a better hereafter, a belief that a merciful God will help diminish a negative situation.

Religiosity (external): A social component of religion in which similar-minded people come together (congregate) and serve to support or buffer one another and facilitate coping.

Spirituality: Spirituality may be linked to religious worship, but it can also refer to other belief systems. It may reflect a search for meaning or purpose in life, or it might be obtained through connections to music, art, and nature, and of course, connections with others.

Modified from Anisman, H., 2021. *Health Psychology: A Biopsychosocial Approach.* Sage, London.

Problem-solving coping methods are often considered to be superior to other methods and are most likely to buffer against negative outcomes. In contrast, avoidant strategies are most often seen as being the worst coping method ("sticking your head in the sand won't make it go away"), whereas emotion-focused coping methods lie somewhere in the middle but are generally considered maladaptive. While this division might be intuitively appealing, the reality is that it is overly simplistic.

Other things being equal, problem-solving strategies are relatively effective in diminishing the actions of stressors if they are controllable or modifiable. In some instances, the stressor might not be controllable, but even if there is an *illusion* of control, problem-solving efforts may enhance emotional well-being. As in the case of placebo effects, simply believing or expecting positive changes might be sufficient to promote beneficial outcomes. There are, to be sure, limits as to how effective problem-focused coping strategies can be in preventing or reducing the adverse health effects of stressors. In response to a severe stressor, difficulties may be encountered in planning and initiating actions, and consequently, the benefits of this strategy might not readily be obtained. Further, as the stressor severity and chronicity increase, individuals might attempt excessively complex and unfeasible strategies.

As much as avoidant strategies are often ineffective and even harmful, when a situation is entirely out of an individual's control, and the outlook is bleak (as in the case of metastasized cancer), problem-solving likely won't be productive. In contrast, avoidant coping might be preferable insofar as it might preclude distress from being overwhelming. Emotion-focused coping might similarly be ineffective in many situations, especially if individuals endorse self-blame for things that should or shouldn't have been done, however incorrect and damaging this can be. There are, however, aspects of emotion-focused coping that can have positive attributes under certain circumstances. Of paramount importance is that emotion-focused coping may signal others that their help is needed, and it might serve as a way of helping a person come to terms with their feelings, and thus diminish the distress they might otherwise feel. Some stressors are exceptionally disturbing (e.g., loss of a child) and no amount of problem-solving might be effective in altering the situation, whereas social and emotional support can diminish the distress, especially when the support comes from those with whom the individual strongly identifies (Haslam et al., 2021).

In response to some extremely distressing situations, particularly in response to chronic illnesses or in the aftermath of such experiences (e.g., among cancer survivors), a form of cognitive restructuring may be adopted in which individuals attempt to find something positive coming from their traumatic experiences. This can emerge in the form of posttraumatic growth (finding meaning or benefit finding) or trying to make sense of their experience or attempting to help others through these experiences. In some instances, a person may come to believe that they have grown from the experience, as terrible as it had been (smelling the roses or coffee). Some people become more empathic and expend their energies so others might gain from their horrid, distressing encounters. This may appear in developing or joining organizations to raise funds for specific illnesses or engaging in efforts to promote public safety (e.g., eliminating gun violence or drunk driving). There is little doubt that some people gain appreciably from benefit-finding strategies, but this type of positive outcome is not uniformly attained. Engaging in a search for meaning doesn't ensure that meaning will be found.

Lamentably, some situations simply don't lend themselves to any benefits ever being found, and some individuals may be unable to gain from these efforts under any circumstances.

The notion that there is a single best strategy to deal with stressful events is contrary to what people actually do, as coping often involves a cocktail of strategies that vary from situation to situation. An individual may ruminate over an event, while simultaneously using self-blame, which is a very bad combination that predicts the development of depressive symptoms (Nolen-Hoeksema, 1998), or they might ruminate and concurrently use problem-focused coping, which is a much better combination (Kelly et al., 2007). Being flexible in the choice of coping methods and being able to shift between strategies as the situation evolves may be the most adaptive way to deal with stressors. This is particularly important given that aspects of stressors may vary over time as a chronic condition progresses; hence, optimal adaptability entails being able to shift coping methods accordingly. Cognitive flexibility is likewise important for stressor appraisals and may contribute to dealing with negative events effectively (Gabrys et al., 2018), and was indeed accompanied by less depression and anxiety in patients with breast cancer (Berrocal Montiel et al., 2016). To a considerable extent, the benefits of some forms of psychotherapy may come about because they foster a positive psychological state characterized by flexibility in being able to shift cognitive sets (Hinton and Kirmayer, 2017).

Resilience and vulnerability

Vulnerability refers to an individuals' or a groups' propensity to be at elevated risk for pathological outcomes in response to biological, environmental, or social threats, whereas resilience refers to the ability to recover from illness or for illness occurrence to be precluded. Vulnerability and resilience are often considered to be at opposite ends of a continuum, but this isn't fully accurate, and it might be more appropriate to view vulnerability and resilience as being intersecting factors. After all, the absence of vulnerability doesn't necessarily imply increased resilience. Even the healthiest individual, who ought to live a long and healthy life, can be brought down by the presence of even a single vulnerability factor (e.g., having a major gene that leads to a terrible outcome, such as Huntington's disease). Conversely, individuals who carry genes that ought to make them vulnerable to a pathology might also carry other genes that in some manner prevent negative outcomes or they may be part of social groups that are effective in supporting the person's general well-being.

Understandably, resilience may be tied to the availability of social support and behaviors (social connectedness) that can blunt the impact of stressors (Haslam et al., 2018), as well as the ability to adapt to distressing situations and be flexible in adopting effective coping strategies. In the aftermath of some forms of trauma (e.g., abuse, distress experienced by refugees), finding psychological or physical "safe places" is essential for healing, and to a significant extent, social supports are integral to this (Matheson et al., 2020). There has also been the view that dispositional factors—such as optimism, self-efficacy, self-esteem, self-empowerment, optimism, hardiness, and acceptance of illness—may be fundamental to resilience (Southwick and Charney, 2018); and for some individuals, religion or spirituality may be a driving force in dealing with challenging events (Ysseldyk et al., 2010). To an extent, resilience can be acquired through earlier stressor experiences. Unlike the impact of toxic stressors (e.g., psychological or physical abuse, neglect, maternal depression, parental substance abuse, family violence, poverty), children who encountered mild or tolerable stressors might have learned coping methods that prepared them to deal with later insults (McEwen, 2020).

It ought to be underscored that what defines resilience in one culture could be very different in a second culture, and these could differ across contexts (Gone and Kirmayer, 2020). For instance, collective historical trauma in some groups may have resulted in elevated illness vulnerability. But these experiences can also foster elevated resilience, especially when allies are available to promote their cause (witness the influence of the Black Lives Matter movement). Likewise, psychosocial and neuroendocrine processes that promote resilience in one age group, say young people, may be very different from those that foster resilience in older individuals. Appreciating these differences may be fundamental in recommending interventions among at-risk populations. Fig. 5.3 shows a smattering of key factors that contribute to resilience.

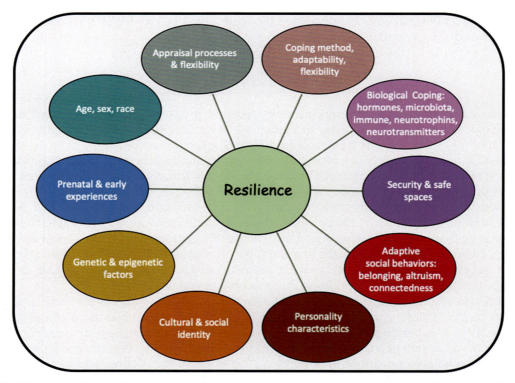

FIG. 5.3 Stress-related resilience can be influenced by multiple factors. Some are outside of the individual's control, but others (e.g., appraisal, coping) are skills that can be acquired and used to act against processes that increase vulnerability to stressor actions. The effectiveness of many factors that favor or act against resilience may be context-dependent, varying with different types of challenges and previous experiences, and can also vary over time as the stressor situation plays out. *Modified from Anisman, H., 2021. Health Psychology: A Biopsychosocial Approach. Sage, London.*

Neurobiological actions of stressors

A systems perspective of stressor actions

The multidimensional factors that contribute to the cognitive and emotional components of stressors may be mediated by the interactive actions of specific brain regions and several neurobiological systems. Although studies in animals and humans frequently assess the role of discrete brain regions and specific neurochemicals in mediating the effects of stressors, it is generally understood that they are components of complex systems that guide emotional, cognitive, and behavioral changes. Ordinarily, when a threat is encountered, diverse aspects of the prefrontal cortex are engaged to make sense of the situation. These regions are involved in appraisal and executive functioning, and the different aspects of the anterior cingulate cortex act as intermediaries (hubs) between multiple brain regions involved in cognitive, emotional, and motivational processes. Changes within these regions influence aspects of the amygdala and the bed nucleus of the stria terminalis that are fundamental in anxiety and fear responses and together with the hippocampus serve in memory processes related to stressor experiences. At the same time, being vigilant in the face of stressors may be mediated through these same systems as well as norepinephrine activity within the locus coeruleus that innervates large parts of the frontal cortex and the ventral tegmentum and affects sympathetic nervous system activity. Concurrent with these brain changes, dopamine functioning within the ventral tegmentum and nucleus accumbens are also part of a stress resilience circuit, being involved in denoting the significance or salience of events. These regions also underlie both reward processes that may be undermined in stressor situations. The diminished dopamine activity that is produced by stressors may be a component of depression often associated with strong stressors. Yet, when a stressor is first encountered, reducing the functioning of reward processes may be a highly adaptive response so that individual efforts are focused on methods of dealing with the stressor while concurrently diminishing actions (e.g., searching for food or obtaining pleasure) that can be a distractor or even place the organism at risk.

These brain regions and several others speak to one another directly or through intermediaries. The interconnected brain regions operating synchronously are fundamental in promoting psychological resilience so disturbances of any of these could result in increased risk of stress-related disorders. Accordingly, considering each of the brain regions independently might not provide a complete or accurate perspective of the stress and resilience processes. The complexity of such analyses is complicated by numerous features of the stressor. Among other things, the neural circuitry associated with different psychological challenges may be distinct from one another so some stressors are more apt to provoke particular hormonal, neurochemical, and inflammatory changes, and certain psychological insults may be more or less influential in the provocation of diverse pathologies (Finnell et al., 2017). It is also certain that the functioning of these neural circuits is subject to moderation by several characteristics of the organism and previous positive and negative experiences (e.g., genetic, epigenetic factors, age, earlier stressor encounters), which contribute to the vastly different responses displayed across individuals as well as their differential vulnerability to stressor-related pathologies.

Just as the actions of stressors may come to affect peripheral biological changes, such as sympathetic, immune, and inflammatory processes, the communication between these systems is multidirectional so several hormones, adaptive and innate immune responses, and inflammatory factors, as well as microbial processes, can affect the functioning of many stress-sensitive brain regions (Cathomas et al., 2019). Through the intersections between these systems, stressful experiences may contribute to the promotion and exacerbation of chronic physical illnesses, such as metabolic syndrome, type 2 diabetes, heart disease, and cancer. In fact, through manipulations of the inflammatory immune phenotype that is markedly affected by stressors, vulnerable individuals could be made more resilient, and conversely, resilient individuals could be made more vulnerable.

Stressful events influence a wide array of neurotransmitter alterations across many brain regions, often serving in an adaptive capacity. These neurotransmitter variations may be elicited by threats as soon as they appear, but the extent of the changes may be affected by appraisals of stressors, including their perceived controllability and predictability. In response to acute stressful experiences, neurochemical systems in collaboration with behavioral coping processes may limit the occurrence of adverse outcomes, although illness susceptibility may be instigated by acute stressors in vulnerable populations, such as older individuals, those with preexisting chronic medical conditions, or in the presence of specific genetic and epigenetic factors. We won't go into detail in documenting the broad array of neurotransmitter changes elicited by stressors as we recently described these in detail (Anisman et al., 2018). Instead, we will briefly describe some of the key neurotransmitters affected by stressful events, especially those thought to contribute to vulnerability and resilience.

Of the many neurotransmitter changes elicited by stressors, particular attention has been devoted to the impact on norepinephrine and serotonin variations that occur within cortical and several subcortical brain regions, as well as the dopamine changes that occur within the nucleus accumbens, which may account for the anhedonia that occurs in response to stressors. Additionally, stressors promote changes of GABA and glutamate in prefrontal cortical regions and within the hippocampus, which may affect cognitive functioning and depressive disorders, and may contribute to the development of substance use disorders. As we'll see shortly, changes of corticotrophin releasing hormone (CRH) in portions of the amygdala have been extensively assessed given its role in fear and anxiety and its functioning in the hypothalamus has received considerable attention because of its role in regulating hormonal processes. Many other hormones, which are discussed in the next section, also appear within the brain where they can profoundly influence behavioral and emotional responses elicited by stressors.

Stressors affect these diverse neurotransmitters in different ways, but we'll describe changes of norepinephrine and serotonin in somewhat greater detail as these are illustrative of some of the adaptations that occur in response to aversive events. Stressor experiences that are relatively short-lasting and are of moderate severity promote increased utilization of norepinephrine and serotonin within several cortical regions, possibly facilitating behavioral coping, and hence adverse outcomes are unlikely to occur. In contrast, when behavioral methods of coping with the stressor are unavailable, then the coping burden rests more heavily on these neurochemical systems. Under such conditions, the utilization (release) of neurotransmitters may be increased to meet the challenge experienced, and this is met by elevated neurotransmitter synthesis (production) to compensate for sustained utilization. However, if an uncontrollable stressor persists for a sufficiently long period, then the rate of neurotransmitter utilization will exceed its production so the availability of the transmitter is diminished. Once the stressor ends, the functioning of the neurotransmitters normalizes, typically within a short time (minutes or a few hours), although in older animals or highly stress-reactive animals these neurochemical changes can persist somewhat longer. Furthermore, beyond the immediate consequences of stressors, the sensitization of neuronal processes may be engendered, so later encounters with stressors or stressor cues (triggers) can promote rapid and exceptionally marked neurochemical changes that can have profound adverse consequences.

Despite the adaptive value of brain neurochemical changes induced by stressors, the elevated neurotransmitter activity associated with continued uncontrollable stressors can only go on for so long before these systems become overly taxed or certain brain regions are excessively activated resulting in neuronal loss (allostatic overload). In effect, with chronic stressor experiences, which include chronic pain and illnesses, central adaptive systems may be overwhelmed (McEwen and Akil, 2020). Together with variations of inflammatory and hormonal processes, altered neurotransmitter functioning favors the occurrence of anxiety, depression, PTSD, and diverse somatic pathologies. These actions may be moderated by sex hormones and genetic influences, accounting for the marked interindividual differences that are so often apparent and for the female bias commonly found in the appearance of disorders, such as depression (Duman et al., 2019).

Beyond neurotransmitter variations, several other biological processes (e.g., hormones, growth factors, inflammatory immune processes) are also affected, which may contribute to behavioral and somatic disturbances. We'll now briefly turn to several hormones that are involved in the stress process. Reiterating our earlier point, although these are presented as if they comprise independent changes, they are components of broad dynamic, interacting systems (the brain and body's "butterfly effect").

Hormonal processes

Neurotransmitters released from vesicles located at the terminal regions of neurons travel a short distance (across the synaptic cleft) to stimulate receptors on an adjacent neuron. Hormones, in contrast, can travel long distances through the blood and stimulate receptors on distal organs. Some of the peripheral hormones also appear within the brain where they can act as neurotransmitters and, conversely, several brain neurotransmitters may be present in the periphery where they behave like hormones. Several hormones whose release is instigated by stressors may interact with one another as well as with brain and immune processes. For simplicity, we will address different hormones independently, but we emphasize that in addition to their primary actions, each of these hormones operates as part of larger vulnerability and resilience networks.

Peripheral norepinephrine and acetylcholine

The autonomic nervous system (ANS) is fundamental in the promotion and regulation of involuntary responses associated with body organs (heart, gut). The ANS comprises two subsystems, wherein the activating effects of the sympathetic system are balanced to some degree by the inhibitory actions produced by parasympathetic functioning. The sympathetic nervous system is exquisitely sensitive to stressors, so marked changes of norepinephrine are provoked (mainly coming from sympathetic nerves that enmesh blood vessels), whereas epinephrine is released from the adrenal gland. These actions promote increased heart rate and hence greater oxygenation of the brain and organs and mobilization of glucose from storage sites. The sympathetic hormones also serve to inhibit the production of nitric oxide (NO) and neuropeptide Y (NPY) production by endothelial cells that line the interior surface of blood and lymphatic vessels that ordinarily produce vasoconstriction. Concurrently, immune cell trafficking is altered by norepinephrine, so immune cells are pushed from lymphoid organs and the spleen into circulation. To borrow a phrase from Dhabhar et al. (2012), immune cells are redistributed from the barracks where they might be comfortable, to the boulevards and battlefields where they are needed. Unfortunately, as described in Chapter 6, by activating norepinephrine receptors on tumor and stromal cells (connective tissue cells of various organs), sympathetic functioning can also contribute to cancer progression and may enhance tumor vascularization.

Acetylcholine (ACh), the primary neurotransmitter of the parasympathetic system, has multiple functions beyond the regulation of sympathetic activity, contributing to inflammatory processes, and may limit oxidative stress. The cholinergic inflammatory response is modulated by its two types of receptors (nicotinic and muscarinic), which contribute to innate and adaptive immune responses, and effector cell responses. Parasympathetic functioning may thus contribute to a range of immune-related disorders, such as allergic inflammatory responses. Given the links between sympathetic norepinephrine functioning and cancer promotion, it is tempting to assume that parasympathetic acetylcholine activity might have the opposite effect. However, the case for this has been weak, and there is only modest evidence that the effects of parasympathetic activation vary with the nature of the cancer and other related variables (Tibensky and Mravec, 2021).

Corticotropin releasing hormone (CRH) and cortisol

Neuronal activity involving CRH activity within the amygdala and hippocampus is involved in the acquisition and expression of fear as well as fear memories (LeDoux, 2014). Although fear and anxiety are intertwined, they can be distinguished from one another. Whereas the amygdala is associated with fear typically provoked by stressors, the bed nucleus of the stria terminalis (also referred to as the extended amygdala) may be aligned with the development and maintenance of anxiety elicited by uncertain stressor occurrences or stressor cues that are spatially or temporally distant (Klumpers et al., 2017). The influence of these brain regions on fear and anxiety may reflect the conjoint actions of CRH (and different CRH receptors) and that of the inhibitory neurotransmitter GABA. The actions of these brain areas can be moderated by frontocortical regions that are responsive to uncertain threats.

Communication between these brain regions may be integral to the appraisal and emotional responses provoked by acute dangers, and could potentially affect the way these regions behave in response to strong, uncertain life challenges. More than this, people who reported elevated levels of day-to-day annoyances and consequent negative mood, exhibited greater and more persistent left amygdala activation in response to negative images than did individuals who reported fewer such negative experiences. Harboring these unseen emotional biases could potentially have long-term health ramifications. If these relatively innocuous experiences can have such effects, then it can reasonably be expected that more significant threats, such as awaiting the results of a biopsy or uncertainties related to chronic illnesses, could have still more pernicious consequences.

Beyond its role in promoting fear and anxiety, CRH plays a role in the prototypical stress system that comprises the interplay between the hypothalamus, the pituitary gland, and the adrenal gland (HPA axis). As depicted in Fig. 5.4, activation of the amygdala results in the

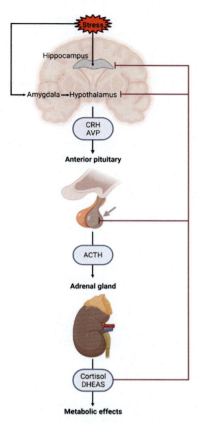

FIG. 5.4 Stressful events cause amygdala activation that provokes activation of the paraventricular nucleus of the hypothalamus. This promotes the release of CRH and arginine vasopressin from the median eminence situated at the base of the hypothalamus. These peptides promote the release of ACTH from the anterior pituitary, which reaches the adrenal gland through the bloodstream, causing the release of mineralocorticoids and glucocorticoids, such as cortisol. The cortisol enters the bloodstream, ultimately stimulating glucocorticoid receptors at the hypothalamus and mineralocorticoid and glucocorticoid receptors at the hippocampus, which serve to limit further HPA activation. *Black arrows* denote excitatory processes; *red lines* with blunted ends signify inhibitor processes. *Created using BioRender.com.*

paraventricular nucleus of the hypothalamus being stimulated, which results in the release of CRH from its terminal regions situated at the base of the hypothalamus (a region referred to as the median eminence). The CRH that is released then stimulates the anterior portion of the pituitary gland, causing the release of adrenocorticotropic hormone (ACTH). This hormone enters the bloodstream, and upon reaching the adrenal gland causes cortisol (corticosterone in rodents) to be released into circulation. The circulating cortisol reaches the brain where it stimulates receptors present within the hippocampus, which then informs the hypothalamus to cease the release of CRH. This cascade of changes, which is fundamental in regulating other stress responses, is not only provoked by frank stressors but may also occur in anticipation of such challenges, essentially serving in a preparatory capacity and may contribute to the selection of defensive responses (Daviu et al., 2020).

Cortisol release associated with stressor encounters has multiple beneficial actions, including metabolic regulation, and moderation of immune and inflammatory responses. This hormone similarly produces preparatory actions and facilitates coping to deal with impending stressors. In addition, cortisol has permissive effects allowing the actions of other hormones to be amplified or suppressed (Sapolsky et al., 2000). The cortisol response is highly adaptive, but contrary to common beliefs, its release may occur on a relatively selective basis. In humans, the cortisol rise is not pronounced or not evident in response to some stressors, such as anticipation of an academic exam, whereas a pronounced cortisol rise occurs in response to stressful events that entail a social-evaluative threat, such as public speaking and the emotions it elicits (Dickerson and Kemeny, 2004). Likewise, stressors that entail racial discrimination can produce cortisol changes, especially when the stressors promote strong emotions, such as anger (Matheson et al., 2021). Like so many other adaptive biological systems, the benefits of the HPA activation may be lost with chronic stressor experiences and can produce damaging effects when cortisol stimulates hippocampal receptors for too long. In this case, hippocampal cells may be damaged, leading to the shutdown process being disturbed and hence cortisol release will persist, leading to greater hippocampal damage and ensuing cognitive disturbances (McEwen and Akil, 2020).

Given the potentially damaging effects of cortisol being continuously released, it is particularly interesting that with chronic stressor experiences (as well as among individuals with PTSD) glucocorticoid receptor sensitivity may be diminished so some of the negative consequences that could arise with excessive cortisol release are precluded. Under these conditions, circulating cortisol levels may also be reduced, even falling below that of individuals who had not been stressed (Yehuda, 2002). This doesn't imply that a cortisol response can't be mounted, and indeed, in chronically stressed animals, the subsequent introduction of a novel aversive stimulus elicits an amplified corticosterone response. It is telling that among women who had experienced early-life trauma that the subsequent adult HPA response elicited by an exogenous CRH challenge was diminished. However, if women were exposed to a psychological challenge that could elicit shame and anger, an exaggerated response was elicited (Heim et al., 2008). Reduced cortisol was likewise evident among women who had been in an abusive relationship and who displayed PTSD features; however, when these women recounted these experiences, which were obviously meaningful to them, a marked cortisol response was triggered (Matheson and Anisman, 2012). It is likely that the diminished cortisol levels associated with a chronic or traumatic stressor experience reflect yet another adaptive change to limit the excessive hormone activation that could be damaging, but in response to potent psychosocial or meaningful threats, inhibitory processes can be overridden so a strong cortisol response can be initiated.

Cortisol, aging, and disease

Aging is obviously a key factor associated with the emergence of multiple pathologies, and stressful experiences may be especially pertinent in promoting cellular aging and age-related disorders. Among older individuals in whom a degree of HPA dysregulation may already exist owing to the natural loss of hippocampal glucocorticoid receptors, diurnal cortisol variations are altered, which may further disturb HPA feedback processes and perhaps prompt additional cell loss. These actions are exacerbated by chronic stressor experiences and may thereby contribute to the emergence of neurodegenerative disorders, such as Alzheimer's disease.

The full gamut of mechanisms that contribute to the impact of stressors in older individuals aren't fully understood, but chronic social stressors that occur with aging may induce varied epigenetic changes within the brain and in numerous other tissues (Palma-Gudiel et al., 2020). Furthermore, aging is accompanied by more pronounced stressor-provoked alterations of immune functioning or variations of sex hormones, as well as changes of insulin and insulin-like growth factor 1, thereby contributing to cellular aging that has been associated with cancer (Kruk et al., 2019). With chronic stressor experiences, allostatic overload can also emerge in the form of mitochondrial disturbances so the energy ordinarily provided to cells is disrupted, instigating cumulative damage that leads to diminished ability to deal with further challenges. In a metabolically stressed organism, mitochondrial DNA (as distinct from nuclear DNA) may be affected through elevated levels of reactive oxygen species and the presence of inflammatory factors, and apoptosis and impaired autophagy may be instigated so the damaged mitochondria are not removed as they should.

Aging and cancer share numerous features, such as genomic instability, increased occurrence of epigenetic alteration, elevated inflammation, and immune changes, as well as variations of damage-associated molecular patterns (DAMPS). Moreover, aging and cancer are both associated with altered metabolic functioning and mitochondrial metabolism, processes related to autophagy, and the degradation of systems that facilitate repair processes (Huang et al., 2015). Together, these actions may foster the development of age-related diseases. Of significance, lifestyle changes, such as obtaining adequate exercise (or otherwise obtaining mitochondria-targeted antioxidants) can prevent or even reverse the adverse effects that could otherwise accrue.

CRH and cortisol associated with food selection

Coordination between different behavioral processes is fundamental for survival. It would, for instance, be highly adaptive for eating to be suppressed in threatening situations, such as during attempts to evade a predator. In humans, intense stressors tend to be associated with anhedonia (reduced feelings of reward from normally pleasurable stimuli) and diminished eating, and a shift toward defensive and vigilant behaviors. This also frequently occurs in response to mild and moderate stressors, but in a subset of individuals, increased eating occurs ("I eat when I'm stressed"), just as this is evident in patients with a common subtype of depressive illness, that of atypical depression. The increased eating usually involves consumption of comfort foods that provide a fast caloric fix in the form of sugars and carbs (broccoli and celery sticks are rarely on the menu), perhaps reflecting an effort at self-medication. This response to stressors may be mediated by CRH stimulation of nucleus accumbens dopamine activity, which is part of a reward system, which might make comfort foods appear more salient or rewarding (Abizaid, 2019).

Cortisol functioning is also key in accounting for the links between stress, eating, and metabolic processes, ultimately promoting obesity. It is of particular significance that cortisol also contributes to the redistribution of stored fat so it preferentially appears as abdominal fat depots that contain proinflammatory cytokines, which may account for individuals with belly fat being at elevated risk for inflammatory-related diseases. We'll be considering this in greater detail in Chapters 7 and 8 where eating and obesity are discussed, and it will be clear that having obesity can be exceedingly harmful, but this depends on where fat is localized (belly versus thighs), to what extent individuals are metabolically fit, how long they had experienced obesity, and the contribution of several psychosocial factors, including the stigma associated with this condition.

Estrogen

Many inflammatory- and stress-related illnesses are far more common in women than in men, which may be linked to alterations of sex (most autoimmune disorders, as well as depression and anxiety), although other factors directly or indirectly contribute to the female biases. A great number of genes coding for hormones and growth factors within several brain regions are differentially influenced by acute and chronic stressors in the sexes (Brivio et al., 2020). These brain processes are intimately related to behavioral functioning and through the peripheral hormonal changes imparted, the development of physical illnesses may similarly be elevated.

Estrogen receptors are typically present on immune cells and might contribute to sex-dependent illnesses related to inflammatory diseases. Indeed, the sexes differ in their immune response to stressors, possibly being mediated by estrogen (Rainville et al., 2018). For that matter, the actions of estrogen may contribute to several other neuroendocrine changes elicited by stressors, including cortisol, which affect immune activity and might thereby influence the occurrence of physical illnesses that are more prominent in females than in males.

The impact of stressors in females varies over the course of the estrous cycle. Specifically, stressor-provoked cortisol responses are more pronounced during the luteal phase (the latter part of the menstrual cycle), during which progesterone is at its highest and estrogen is lower, and the stressor-elicited cortisol response is also altered after menopause. Predictably, the influence of some stressors (e.g., those that involve public speaking) on cortisol levels was diminished among women using oral contraceptives, again pointing to the importance of sex hormones in determining stress responses (Kudielka and Kirschbaum, 2005).

Like estrogen, testosterone may be affected by stressful events, leading to other biological and behavioral changes. To a considerable extent, interactions between testosterone and cortisol influence behavioral stress responses, including those elicited by social threats, and together with oxytocin may contribute to social anxiety. In addition to these actions, testosterone contributes to sex differences in immune responses and the appearance of certain immune-related disorders. This hormone promotes immunosuppression in both male and female rodents, but it is primarily in males that testosterone reduces circulating T_h cells. In essence, while elevated testosterone in males contributes to stress resilience, at least over the short run, this occurs at the risk of poorer longer term survival since the reduced T_h cells may impair humoral immunity, thereby rendering males vulnerable to a range of infectious diseases.

Prolactin

Prolactin, secreted from the pituitary, is best known for its involvement in milk production, nursing, ovulation, and may be important in parental nurturing behaviors. But, like so many other hormones, prolactin has multiple functions, being involved in metabolic homeostasis, adrenal stress responses, pancreatic functioning, effects on adipose tissue, and control of body weight. Stressors have long been known to increase circulating prolactin levels, which may affect biological systems that influence immune and inflammatory responses. For instance, this hormone serves as a mediator for T_{reg}-related intestinal inflammation that can be elicited by psychological stressors, thereby influencing autoimmune diseases that preferentially occur in women and are subject to change during pregnancy and the lactation period (Borba and Shoenfeld, 2019). There had been suspicions that prolactin may contribute to the emergence or progression of breast cancer, but in the main, this link has not been examined extensively. Nonetheless, there were indications that prolactin was linked to estrogen and progesterone receptor-positive breast cancer in postmenopausal women (Wang et al., 2016).

Oxytocin

For a while, oxytocin had been the darling of the hormone world, at least from the perspective of the media that dubbed it the "love hormone" based on its presumed role in social and pair bonding, as well as prosocial behaviors, such as generosity, altruism, and attention to positive cues. An early perspective was that together with opioid peptides and gonadal hormones, oxytocin in females promoted "tend-and-befriend" characteristics (Taylor et al., 2000). Beyond this form of altruism, wherein individuals' behaviors facilitate and encourage the well-being of group members, males display "parochial altruism" that entails both support of ingroup members and defense against the outgroup (De Dreu, 2012). In essence, among males, oxytocin serves to encourage an evolutionarily advantageous "tend-and-defend" characteristic.

This hormone was also implicated in coping, especially if the stressor involved social challenges, and oxytocin administration limited cortisol release elicited by a social stressor (McQuaid et al., 2014). As some behavioral changes provoked by stressors could primarily be attenuated by oxytocin in male mice, this hormone might be an essential player in determining sex-dependent mood processes. Aside from these actions, it seems that oxytocin could also enhance sensitivity to social stimuli, amplifying both positive and negative social interactions, possibly through interactions with dopamine-mediated reward systems (McQuaid et al., 2014). From this perspective, oxytocin could promote psychological well-being, yet because of the increased sensitivity to social cues associated with this hormone, it could also mediate poor mood and depression instigated by social stressors. In line with this, the presence of an oxytocin-related gene was associated with depression that stemmed from interpersonal stressors, but this was not apparent among individuals carrying a polymorphism on this gene (Tabak et al., 2016). The view that oxytocin serves to influence sensitivity to social cues and social events has several interesting implications that had not been broadly considered. While the diminished social sensitivity among individuals who carried a polymorphism of the oxytocin receptor could constrain the beneficial effects of positive and nurturing early life, the diminished reaction to social factors could also limit the negative effects of deficient early-life experiences, such as neglect (McQuaid et al., 2014).

Beyond these behavioral actions, oxytocin can moderate inflammation, promote wound healing and regeneration, attenuate oxidative stress and macrophage-related inflammation, probably mediated by regulation of the transcription factor NFκB that is involved in the generation of cytokines linked to innate immunity. As well, oxytocin could potentially suppress stressor-elicited immune disorders and could reduce TNF-α production by macrophages and that derived from adipose tissue of obese mice. Not unlike other processes that link microbial and inflammatory processes, oxytocin manipulations can also serve as a brake on intestinal motility and can reduce mucosal activation of enteric neurons, thereby acting against gut inflammation and related pathological conditions.

Neuropeptide Y

As we've seen, many hormones have multiple actions that are tied to well-being. Despite the remarkable array of actions attributable to neuropeptide Y (NPY), it has probably not received the attention that it should have. NPY is an important player in appetite regulation and storage of energy as fat and contributes to weight gain associated with stressor-induced eating in rodents. A high-fat diet induced marked increases in weight among animals that had been chronically stressed, but this outcome was curtailed when NPY functioning was switched off, possibly by affecting insulin variations that affected the amygdala and the emotional responses associated with activation of this brain region (Ip et al., 2019). It may be particularly significant that in obese mice, NPY regulates the function of adipose tissue macrophages and may thereby affect inflammation and could thus affect the occurrence of diseases related to obesity.

Aside from its role in eating processes, NPY has been touted as a major contributor to the promotion of stress resilience, serving to preclude the development of anxiety and PTSD. In rodents, behavioral resilience was accompanied by elevated NPY levels, whereas anxiety was generally high in mice genetically engineered to have low NPY levels. As expected, in rodents, administration of NPY prevented the adverse effects of strong stressors, including the HPA dysregulation characteristic of PTSD. Paralleling the animal studies, soldiers at war who were deemed resilient and least likely to develop PTSD had the highest levels of NPY (Sah et al., 2014), and this same profile was seen in other situations that were apt to promote PTSD. As expected, NPY administered intranasally was effective in diminishing symptoms of PTSD (Sayed et al., 2018).

In addition to these actions, NPY has been implicated in aging processes, such as stem cell exhaustion, cellular senescence, disturbed intercellular communication, dysregulated nutrient sensing, and mitochondrial dysfunction. Consistent with the changes in aging processes, NPY can have far-reaching immune actions, regulating adaptive and innate immunity, modulating phagocytosis, immune cell trafficking, NK cell activity, T_h cell differentiation, and cytokine secretion. The immune changes elicited by this peptide may consequently contribute to inflammation and inflammation-induced tumorigenesis (Jeppsson et al., 2017) and was implicated as a fundamental mediator between dietary restriction and cancer occurrence in mice.

Endocannabinoids

The endocannabinoid (eCB) system entails the actions of neurotransmitter-like substrates, endocannabinoids, which stimulate two types of receptors, CB_1 and CB_2. These receptors, which are present within the central nervous system and in other organs and tissues, are the primary targets of endogenous endocannabinoids, 2-arachidonoyl glycerol (2-AG), and arachidonoyl ethanolamide (anandamide). These substrates and their corresponding CB receptors contribute to the regulation of food consumption as well as neuronal processes tied to motivation and pleasure. The psychoactive component of cannabis, Δ^9-tetrahydrocannabinol (Δ^9-THC), binds to these receptors, thereby producing some of the effects of this compound (Hill et al., 2018). As well, Δ^9-THC along with cannabidiol, a primary phytocannabinoid obtained from cannabis plants but with lower binding to CB receptors, are responsible for many actions of cannabis, but cannabidiol does not produce the psychological high elicited by Δ^9-THC.

As activation of the eCB system and the CB_1 receptors promotes the release of several neurotransmitters that influence stress processes, it was suggested that this system might serve as a suitable target to attenuate anxiety and anxiety-related disorders. It seems that the amygdala may be intimately related to the antianxiety actions of cannabis as a CB_1 agonist applied directly to the basolateral amygdala limited stressor-elicited HPA activity, and the eCB system functioning might also be involved in the consolidation of emotional memories associated with stressor experiences (Hill et al., 2018). Considerable data have pointed to the benefits derived from cannabis in diminishing fear and anxiety and may be effective in the treatment of anxiety disorders and can diminish specific symptoms of PTSD, such as reducing nightmares. As well, there is reason to suppose that the effects of microbiota in diminishing anxiety in chronically stressed mice were tied to endocannabinoid variations (Chevalier et al., 2020).

There have been indications that cannabis might be effective in diminishing inflammation and has been used in treating inflammatory related conditions, such as early stages of osteoarthritis and there have been hints that cannabinoids might be useful for other rheumatic diseases. As some illnesses or the treatments used to treat diseases, particularly cancer chemotherapy, may produce nausea and vomiting, the ability of cannabis in diminishing these features makes it an important tool in cancer therapy, and by its effects on anxiety, it may also diminish the distress created by the illness (Parker, 2017). Cannabinoids may also influence several processes that are relevant to cancer progression. These have included the modulation of apoptosis, cell proliferation, angiogenesis, and tumor cell migration. As well, endocannabinoids may affect immune functioning and can influence inflammation that ordinarily fosters tumor progression (Braile et al., 2021). In vivo and in vitro studies have supported the supposition that cannabis can affect the progression of several forms of cancer, and it was suggested that its actions can be used to facilitate the actions of standard therapies (Nigro et al., 2021).

As much as the available data have been impressive, the purported medicinal value of cannabis has taken on mythic proportions despite the lack of data in humans fully supporting many claims that have been made (Parker, 2017). It was asserted that Δ^9-THC was involved in the control of a variety of gastrointestinal functions, including hunger signaling, gut motility and permeability, and dynamic interactions with gut microbiota (DiPatrizio, 2016). Although cannabis was alleged to have multiple benefits on mental health, a meta-analysis indicated that cannabis was ineffective in the treatment of most psychological disorders (Black et al., 2019). On the contrary, cannabis can instigate cognitive disturbances, persistent functional brain changes, and impaired neuronal plasticity and organization, especially in the adolescent brain. Moreover, in highly vulnerable individuals, powerful strains of cannabis could favor the development of schizophrenia or alter the disease trajectory. Still, the value of cannabis has continued to be advanced for the alleviation of symptoms associated with an ever-growing number of illnesses.

With the increasing legalization of marijuana in some countries, more research might be able to assess its value for diverse illnesses and might clarify the potential harms of chronic cannabis use, and under what conditions and in whom this is most likely to occur. So far, however, the barriers to cannabis-related health research have yet to be fully removed in several countries, and greater efforts are needed to provide health-care providers with correct information so that they can advise their cancer patients appropriately.

Growth factors

Growth factors comprise endogenous substances present in the brain and periphery that can stimulate cell proliferation and play a critical role in cellular organization. When growth factors bind to cell surface receptors, cellular proliferation and differentiation is provoked. A wide range of growth factors are present within the body and the brain where they behave much like hormones produced by a variety of tissues, and they also behave like cytokines released by immune cells. Once growth factors are released into the bloodstream, it carries them to target tissues where they can promote diverse actions. As we'll see, although growth factors serve essential functions to maintain the integrity of biological systems, in several instances they can facilitate adverse outcomes that result in cell death and can facilitate the growth of tumors. Several of the growth factors that appear in the brain and body, together with their functions, are provided in Table 5.2.

Neurotrophins and brain functioning

Some growth factors within the brain (termed neurotrophins), such as brain-derived neurotrophic factor (BDNF), fibroblast growth factor-2 (FGF-2), vascular endothelial growth factor (VEGF), and nerve growth factor (NGF) promote neuroplasticity and stimulate the growth of new blood vessels required for the vascularization of brain tissue. Owing to the high degree of plasticity and sensitivity of brain neurons, FGF-2 and BDNF might be fundamental in shaping stress reactivity and adaptation, so that reductions of these growth factors may have profound repercussions on psychological and cognitive functioning.

It was commonly observed that BDNF within the hippocampus was diminished by acute stressors or by reminders of strong stressors (Duman and Monteggia, 2006), and these changes were still more prominent following a chronic stressor regimen that comprised a series of different challenges. The impact of stressors on BDNF is not uniform across stress-sensitive brain regions, increasing rather than declining within portions of the prefrontal cortex. Unlike many other neurobiological changes associated with stressors, the elevated BDNF expression within the anterior cingulate cortex was more prominent in response to controllable than uncontrollable stressors, possibly reflecting the adoption of active coping efforts or to new learning related to the controllability of the stressful situation.

It has been maintained that the release of BDNF may contribute to the fast-acting antidepressant effects of compounds, such as the glutamate antagonist ketamine, among patients who hadn't responded to other treatments. It was further proposed that the effects of selective serotonin reuptake inhibitors (SSRIs) in the treatment of depression may operate through gradual changes of neurotrophins, which may account for the lengthy treatment period required for the positive effects of these drugs to appear (Simard et al., 2018). Other growth factors, such as insulin-like growth factor (IGF-1) and VEGF may similarly act against the behavioral disturbances that have been associated with stressors and might thereby affect depressive illnesses. As we'll see, some of these, such as NGF and epidermal growth factor (EGF), also affect cancer growth and proliferation and are subject to the actions of stressors.[b]

---

[b] For their remarkable work in discovering NGF and EGF, Rita Levi-Montalcini and her collaborator Stanley Cohen shared the 1986 Nobel Prize in Physiology and Medicine. The implications of their findings were clear at that time and have taken on still more importance in the ensuing years. Some of the early studies conducted by Levi-Montalcini began many years earlier when she was at the University of Turin in Italy. However, being a Jewish scientist, Mussolini's Fascist government forbade her from conducting her experiments after 1938. But her dedication to research would not be deterred and she continued the work from her bedroom that served as a makeshift secret laboratory. Of the many great women who have made lasting and significant contributions to science and health, Rita Levi-Montalcini stands at the front of the line of researchers (male or female) who have so successfully overcome obstacles that would have deterred most others.

TABLE 5.2 Growth factors and their functions.

Growth factor	Biological effect	Outcome
Brain-derived neurotrophic factor (BDNF)	Supports survival of neurons; encourages growth and differentiation of new neurons; promotes synaptic growth	Increased neuroplasticity; influences memory processes, stress responses, mood states
Basic fibroblast growth factor (bFGF or FGF-2)	Involved in the formation of new blood vessels; protective actions in relation to heart injury; essential for maintaining stem cell differentiation.	Increased neuroplasticity; contributes to wound healing; neuroprotective; diminishes tissue death (e.g., following heart attack); related to anxiety and depression
Glial cell line-derived neurotrophic factor (GDNF)	Encourages the survival of several types of neurons.	Acts strongly on dopaminergic neurons associated with motor functioning and thus was evaluated as a method of treating Parkinson's disease. Also influences proliferation and maturation of cells and may also play a role in cancer progression (e.g., neuroblastoma and glioblastoma).
Nerve growth factor (NGF) and family members Neurotrophin-3 (NT-3) and Neurotrophin-4 (NT-4)	Contributes to cell survival, growth, and differentiation of new neurons. Fundamental for maintenance and survival of sympathetic and sensory neurons; axonal growth.	Survival of neurons; new neuron formation from stem cells; related to neuron regeneration, myelin repair, and neurodegeneration. Implicated in cognitive functioning, inflammatory diseases, in several psychiatric disorders, addiction, dementia as well as in physical illness, such as heart disease, and diabetes.
Insulin-like growth factor 1 (IGF-1)	Secreted by the liver upon stimulation by growth hormone (GH). Promotes cell proliferation and inhibits cell death (apoptosis)	Primary mediator of insulin and growth hormone; activates AKT signaling that stimulates cell growth and proliferation; inhibits apoptosis; linked to signaling pathways (e.g., PI3K-AKT-mTOR) associated with some types of cancer.
Vascular endothelial growth factors (VEGF)	Signaling protein associated with the formation of the circulatory system (vasculogenesis) and the growth of blood vessels (angiogenesis)	Creates new blood vessels during embryonic development, encourages development of blood vessels following injury, and creates new blood vessels when some are blocked. Muscles stimulated following exercise. Implicated in various diseases, such as rheumatoid arthritis, and poor prognosis of breast cancer.
Epidermal growth factor (EGF)	Found in numerous tissues and is apparent in saliva, plasma, urine, milk, and tears. After binding to its receptor, EGFR promotes epidermal and endothelial cell proliferation, differentiation, and survival.	Pivotal in the healing of oral and gastric ulcers, is protective concerning injury that could be provoked by gastric acid, bile acids, trypsin, and pepsin, and could act against damage stemming from chemical and bacterial challenges.
Transforming growth factor (TGF)-β	Part of the TGF superfamily. A cytokine produced by macrophages and lymphocytes. Induces the	Plays a key role in gut inflammation, and has been implicated in cancer, autoimmune disorders, and infectious diseases.

Continued

TABLE 5.2 Growth factors and their functions—cont'd

Growth factor	Biological effect	Outcome
	transcription of several target genes that influence differentiation, proliferation, and regulation of several types of immune cells.	
Colony-stimulating factors (CSF)	Binds to specific receptors on hemopoietic stem cells, stimulates differentiation and proliferation of varied types of leukocytes. Comes in several forms. Such as macrophage colony-stimulating factor (M-CSF), granulocyte colony-stimulating factor (G-CSF), and granulocyte macrophage colony-stimulating factor (GM-CSF). Promotes proliferation, differentiation, and the survival of macrophages and monocytes.	The different CSFs influence varied processes. M-CSF is implicated in atherosclerosis, as well as kidney failure. G-CSF stimulates bone marrow so that stem cells and granulocyte production is elevated. It fosters the proliferation of neutrophils and promotes neurogenesis (growth of neurons) and neuroplasticity. GM-CSF released by diverse immune cells acts as a cytokine that increases macrophages and dendritic cells. It has been associated with joint damage in rheumatoid arthritis and may be a way of enhancing immune activity when immune functioning is compromised.
Erythropoietin (EPO)	Produced in kidney and liver. Promotes the production of red blood cells in bone marrow, and hence increases oxygen supply.	Used to increase blood oxygen carrying capacity in patients with anemia stemming from chronic kidney disease and in patients treated with chemotherapy. However, it may also promote tumor growth. It is probably best known as an illicit means of increasing performance in endurance sports (e.g., long-distance cycling).

Modified from Anisman, H., 2021. Health Psychology: A Biopsychosocial Approach. Sage, London.

Growth factors linked to cancer

Following the initial discoveries showing that nerve growth stimulating factors could affect tumor processes, other growth factors were implicated in various phases of the cancer process (see Fig. 5.5). These growth factors influence cell dissemination, angiogenesis, tumor progression, tumor intravasation (leakage of cancer cells into blood or the lymphatic system), extravasation (movement of tumor cells from circulation into host tissues), tumor cell migration, and metastasis (Witsch et al., 2010). Some growth factors in the brain (e.g., BDNF and NGF) can also affect immune functioning, and could thereby affect tumor growth. As well, growth factors may affect RNA gene transcription and hence cancer processes, and through actions related to growth factors, miRNAs may promote similar outcomes.

Several members of the large fibroblast growth factor (FGF) family play an important role in the regulation of cell proliferation and survival and may drive cancer progression (Dianat-Moghadam and Teimoori-Toolabi, 2019). In addition to influencing immune functioning by interacting with VEGF and inflammatory cytokines, FGF may affect angiogenesis and inhibit tumor suppressor mechanisms, thereby influencing breast, lung, gastric, and several other forms of cancer. As overactivation of FGF receptors (FGFR) can directly or indirectly affect tumor growth, efforts were made to target these receptors to attenuate cancer cell proliferation. Much like other growth factors, EGF is involved in cell signaling

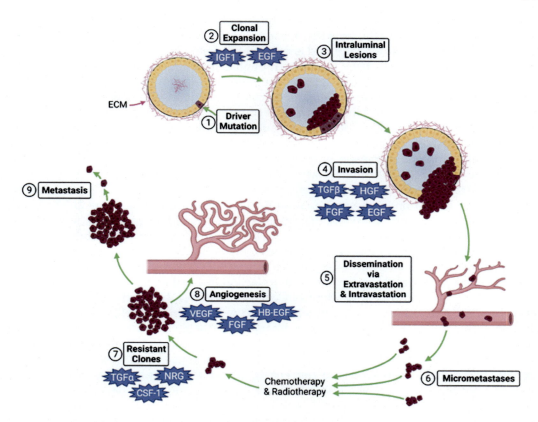

FIG. 5.5 Numerous growth factors play a significant role in the progression of cancer. Somatic mutations initially instigate cancer occurrence (1) that are fostered and supported by growth factors, such as EGF and IGF1, so clones with the mutation are increased (2), and intraluminal lesions occur (i.e., within hollow tubular structures) (3). Thereafter, cancer cells migrate and penetrate neighboring tissue (invasion) that entails the actions of oncogenes, tumor suppressors, and the actions of several growth factors (4). The dissemination of cancer cells then occurs when cancer cells enter tissues (via extravasation from blood) and exit tissues (via intravasation) by returning to lymphatic and blood vessels, allowing them to metastasize to distant sites (5). With the support of macrophages, platelets, and endothelial cells, micrometastases occur (6) that are affected by radiotherapy and chemotherapy. With cancer progression and the occurrence of metastases, new mutations develop and through the influence of several growth factors (and other processes) these clones become more resistant (7). Still other growth factors, such as VEGF, FGF, and EGF, promote angiogenesis that provides tumor cells with a source of energy (8), which after a considerable time will give rise to significant metastases to distal organs (9). *CSF-1*, colony stimulating factor 1; *EGF*, epidermal growth factor; *FGF*, fibroblasts growth factor; *HB-EGF*, heparin-binding EGF; *NRG*, neuregulin; *TGF*, transforming growth factor; *VEGF*, vascular endothelial growth factor. *Figure based on Witsch, E., Sela, M., Yarden, Y., 2010. Roles for growth factors in cancer progression. Physiology 25, 85–101 and created using BioRender.com.*

pathways that influence cell division and survival and may contribute to the progression and metastasis of numerous types of cancer.

Owing to mutations of the gene that codes for the epidermal growth factor receptor (EGFR) gene, receptor proteins are produced in great amounts on certain types of cancer cells, resulting in their more rapid division and migration. The progression of cancers related to EFGR can also come about through autocrine and paracrine signaling (in which hormones or growth factors stimulate the cells that release them or nearby cells, respectively), and can promote cell trafficking, autophagy, and energy metabolism (Sigismund et al., 2018). As well, EGFRs can be transported to distal locations through extracellular vesicles thereby promoting metastasis (Frawley and Piskareva, 2020). Given the actions of mutations of EFGR in nonsmall cell lung cancer, colorectal cancer, and carcinoma of the head and neck, these receptors have been targeted to enhance the benefits that could be provided by immunotherapy.

Being fundamental for new blood vessel formation (angiogenesis) that feed tumors, VEGF has received considerable attention, which was reinforced by its contribution to vessel permeability and consequent transfer of malignant cells outside of the bloodstream. By blocking receptors on cancer cells that are activated by VEGF, their ability to metastasize can be diminished (Kong et al., 2021). However, other growth factors, such as FGF, may compensate for VEGF inhibition so the actions of anti-VEGF treatments may be relatively limited. Nonetheless, treatments to diminish the actions of VEGF could be a valuable tool when administered in connection with other treatments, such as immune checkpoint inhibitors.

Because of the links that exist between obesity, type 2 diabetes, and cancer, it might not be surprising to learn that insulin-like growth factor (IGF-1) would be involved in these processes and tied to the development of cancer. Indeed, IGF-1 and its binding proteins may contribute to a variety of different types of cancer, such as breast, lung, colorectal, and prostate cancer, and they are related to mortality associated with cancer. As well, IGF-1 has been implicated in the development of a second primary cancer (e.g., head and neck cancer) among individuals who had survived an initial cancer experience and was thought to be a significant marker in the prognosis of nonsmall cell lung cancer (Kotsantis et al., 2019). Considerable excitement had initially been expressed concerning the possibility of targeting IGF-1 in cancer therapies, but these hopes haven't been fulfilled, largely owing to the development of treatment resistance. Better outcomes could perhaps be attained through combination therapies.

Transforming growth factor-β (TGF-β) comprises a superfamily of proteins ordinarily involved in multiple cellular processes and might play a dual role in cancer. While TGF-β can have tumor-suppressive actions, in advanced tumors it may promote cancer progression (e.g., ovarian cancer) by influencing evasion of immune protection, cancer cell proliferation, the promotion of angiogenesis, invasion, and metastatic spread (e.g., Goulet and Pouliot, 2021). In addition to playing a role in ovarian cancer, it is thought to contribute to several other forms of cancer progression (e.g., breast, colon, and hepatocellular cancer) and metastasis. Preclinical studies have provided promising effects on tumor progression by manipulating TGF-β but this hasn't yet translated to effective cancer management in clinical settings (Ciardiello et al., 2020).

Being the master regulator of macrophages, colony-stimulating factors (CSFs) can enhance immune responses to combat cancer and promote the regeneration of protective lymphocytes that had been damaged by chemotherapy. As well, CSFs have frequently been used to increase white blood cells that had declined in patients who had received chemotherapy. The progression of some types of solid tumors can be influenced by granulocyte colony-stimulating factor (G-CSF) and granulocyte-macrophage colony-stimulating factor (GM-CSF). However, because they can also influence cytokine-mediated immune suppression and angiogenesis, they may contribute to cancer relapse. The use of GM-CSF is complicated by the existence of intertumor heterogeneity, which can lead to diverse outcomes (Moshe et al., 2020). While some studies indicated that GM-CSF can act against tumor growth and metastasis, many reports have indicated that GM-CSF can promote tumor progression. It is clearly necessary to develop better approaches to predict those cancers that would be amenable to CSF-based strategies.

Influence of stressors on growth factors

The impact of stressors on growth factors related to cancer has been relatively limited, although as we saw regarding brain neurotrophin changes, circulating growth factor changes (VEGF, EGF, and BDNF) are affected by stressors, which can influence cancer-related processes. Sympathetic nerve fibers and the release of norepinephrine that are readily affected by stressors can affect several growth factors that act on angiogenesis and DNA repair within the tumor microenvironment (Zahalka et al., 2017). For instance, lung tumor development was accelerated by a chronic stressor in mice that expressed IGF-1R in the lung, which could be attenuated by β-norepinephrine antagonists (Jang et al., 2016). Consistent with studies in animals, incidental use of a β-blocker was associated with a superior response to an EGFR inhibitor among patients being treated for metastatic lung adenocarcinoma. It was similarly noted that stimulation of $β_2$-norepinephrine receptors can engender resistance to EGFR inhibitors in lung cancer treatment. Indeed, several mutations in the EGFR gene that had been linked to nonsmall cell lung cancer could be attenuated through the administration of β-blockers or IL-6 inhibition (Nilsson et al., 2017). Few studies, however, have been reported concerning the influence of psychological stressors on EGF or EGFR. Yet, it did appear that stress-related exhaustion disorder, more commonly known as burnout, which stems from chronic stressful experiences, was associated with diminished levels of circulating EGF and BDNF.

Stressor-elicited norepinephrine release can cause the release of VEGF that favors cell growth and can activate oncogenic viruses and stimulate oncogenic signaling pathways, such as HER2 and Src (e.g., Armaiz-Pena et al., 2013). In fact, β-norepinephrine antagonists attenuated the growth of ovarian cancer, presumably by inhibiting the actions of Src. Hormones other than norepinephrine may likewise interact with growth factors in determining cancer progression. In ovarian tumor cell lines, oxytocin successfully inhibited cancer metastasis through its inhibitory action on VEGF and matrix metalloproteinase-2 (MMP-2), which are involved in tissue remodeling as well as the development of metastasis (Ji et al., 2018). Chronic stressors in rodents were accompanied by altered levels of VEGF and BDNF that were tied to depressive-like behaviors (Nowacka and Obuchowicz, 2013). As growing tumors are associated with increased vascularization and elevated levels of VEGF, it was proposed that stressor-elicited norepinephrine may influence tumor progression through its action on VEGF. Limited information is, however, available concerning the impact of psychological stressors on this growth factor and its implications for cancer in humans.

As we've already seen, stressors can affect FGF-2, and it seems that such experiences can influence FGF21 in numerous organs. Stressors in the form of nutrient excess or starvation can have such effects, and exercise can affect FGF21. Indeed, FGF21 may act like a stress hormone that influences metabolic diseases, such as diabetes, and may thus have downstream consequences on cancer. However, scant information is available concerning the influence of traditional psychogenic or neurogenic stressors on the activity of circulating FGFs.

Stressors can influence colony-stimulating factor and neuronal remodeling owing to effects on microglia and could thereby influence depressive-like behaviors in mice, which can be attenuated by glucocorticoid receptor antagonism (Horchar and Wohleb, 2019). Aside from such effects, GM-CSF administered before a surgical challenge diminished postoperative colon cancer cell proliferation. Once again, scant data are available concerning the contribution of CSFs to stressor-elicited effects on cancer progression.

Overall, it seems that stressful events can affect various growth factors, which could potentially influence cancer progression and responses to therapeutic agents. It is certain that stressful events affect several growth factors within the brain and play a significant role in determining the development of psychopathology, and treatments that influence neurotrophin functioning can have marked therapeutic benefits. Curiously, however, the attention was devoted to stressor effects on peripheral growth factors and the consequent actions on cancer progression and treatment resistance has been limited. Of course, it is possible that studies assessing these links were made but hadn't shown meaningful effects and thus were never published. Alternatively, many researchers in neurosciences studying the effects of stressors focused on brain processes and didn't evaluate peripheral growth factor changes relevant to cancer, whereas among researchers who focused on the role of growth factors on cancer processes, the influence of psychogenic stressors simply wasn't on the radar.

Cellular stress responses

During normal cellular functioning, individual cells form reactive oxygen species (ROS) and reactive oxygen as part of the metabolism of oxygen. These serve essential functions, such as cell cycling, signal transduction, and host defense against pathogens. Ordinarily, a balance exists between oxidant and antioxidant formation, but if this balance is excessively disturbed, several negative outcomes can occur, such as DNA damage, and cells may undergo apoptosis or necrosis. Hydrogen peroxide, hydroxyl radicals, and nitric oxide are among the ROS members derived from mitochondria that can promote damage, but these actions can be offset by superoxide dismutases that act as antioxidants, thereby limiting negative outcomes. Considerable evidence has shown that imbalances related to mitochondrial ROS contribute to inflammatory processes, cancer cell proliferation, and tumor cell survival. This may come about because elevated ROS can interfere with tumor suppressors, facilitate immune evasion, and drive signals that encourage cell migration.

Cellular stress can be elicited by environmental challenges that range from toxicants (e.g., solvents, inorganic salts, acids, or bases), physical and mechanical damage, ultraviolet radiation, genotoxins (e.g., carcinogens), and temperature extremes. As well, ROS-related cellular stress can be engendered by psychological stressors, physical inactivity, poor diet, nutrient deficiencies, or nutrient overload, which can lead to disease states, including those associated with immune functioning (Yang and Lian, 2020). Conversely, diets that limit oxidant-antioxidant imbalances were related to lower occurrence of diseases, including cancer (Saha et al., 2017), although this does not imply that taking antioxidant supplements (beyond that obtained from proper diets) is beneficial. It is possible that methods could be developed through ROS manipulations to enhance the effectiveness of cancer therapies. This is a laudable goal, but its realization may be some time off. In the interim, diminishing the effects of stressors on mitochondrial ROS or the adoption of lifestyles that affect these processes may be reasonable ways of dealing with the negative effects stemming from oxidative stress.

The production of ROS occurs within the cells' cytosol and within several organelles, including mitochondria and the endoplasmic reticulum (ER). The ER, which comprises a continuous membrane system situated within the cytoplasm of cells is fundamental for cell functioning, including the movement of proteins and other molecules, and is essential for polypeptide folding that contributes to the shape of proteins and their functioning. If ER malfunctioning occurs, then unfolded or misfolded proteins can result, which is referred to as ER stress. These cellular stress responses can serve in an adaptive capacity so that cell integrity is maintained. Specifically, when the presence of unfolded or misfolded proteins becomes excessive, the resulting unfolded protein response (UPR) serves to diminish new protein translation, activate molecular chaperones that can either facilitate or diminish protein folding (or unfolding), and reduce the presence of misfolded proteins. As well, rather than having misshapen proteins created, heat shock proteins (HSPs) may contribute to protein production being shut down, which leads to changes in the ER and hence the functioning of mitochondria that produce needed energy for cells. Members of the HSP family also act as immunomodulators, and depending on their concentrations they can have either proinflammatory or antiinflammatory functions that influence healing processes. In the absence of mechanisms to keep the UPR functioning, persistent ER stress may favor the emergence of diseases, such as diabetes, heart disease, neurodegenerative disorders, and various forms of cancer (e.g., Hetz and Papa, 2018).

In addition to being affected by physical challenges, members of the HSP family can be induced by psychological challenges. Stressful events could affect glucocorticoids and HSP70 in specific organs, such as the liver, and regulation of glucocorticoid receptors can also occur through HSP90, and thus are fundamental in stressor-provoked HPA functioning. Likewise, stressors can affect HSP72 together with specific miRNA variations that may affect immune functioning. It also appears that HSP70 can have important prenatal actions, protecting mouse embryo development from the effects of chronic stressors (Li et al., 2015).

Prenatal and early-life stressor effects

Like the profound teratogenic effects that are produced by certain drugs taken during pregnancy, hazardous effects can be instigated by various environmental pollutants, such as polychlorinated biphenyls (PCBs), second-hand smoke, and estrogen-related products that have increasingly been appearing in sources of drinking water. It has similarly become apparent that stressful experiences among pregnant women can have lasting repercussions on their offspring, even being evident in adulthood. These behavioral and biological disturbances may develop in response to severe events such as the loss of a loved one and traumatic war-related experiences as well as severe natural disasters and may also develop because of commonly experienced stressors, such as those encountered in the workplace and social/home situations.

Promotion of premature birth

Of the many consequences of prenatal challenges, one of the most common is premature birth and low birth weight, which occurs in more than 10% of pregnancies. These outcomes have been attributed to exposure to pollutants and tend to be more frequent among women who smoke, drink alcohol, or use cocaine while pregnant. Less often is it mentioned that premature birth and low birth weight may be connected to domestic violence and is more common among women experiencing high levels of anxiety or depression. Ordinarily, progesterone serves to diminish uterine contractions to prevent premature birth. However, between weeks 37 and 42 of pregnancy, maternal stress may result in elevated expression of FKBP51, which regulates glucocorticoid functioning, so the binding to progesterone receptors is increased, thereby diminishing its action in limiting preterm birth (Guzeloglu-Kayisli et al., 2021).

The influence of prenatal challenges and the resulting premature delivery has been associated with delayed fetal neuromuscular maturation and fetal brain growth, including diminished gray matter volume, cortical maturation, and hippocampal neuronal connectivity, which may be accompanied by impaired cognitive development. The downstream psychological consequences of prenatal challenges may be apparent in adolescence or early adulthood in the form of depression, schizophrenia, drug addiction, and eating-related disturbances. Beyond cognitive and emotional disturbances, children with low birth weight frequently experience later physical health disturbances, such as obesity, metabolic syndrome, and type 2 diabetes, which can influence the development of some types of cancer.

As stressful experiences and the rumination associated with such events reflect dynamic and persistent challenges, negative outcomes in humans are not readily ascribed to events that occurred during precise (critical) prenatal periods. Indeed, stressors experienced just before pregnancy can markedly influence fetal health, possibly owing to actions secondary to the initial challenges, such as the distress that often accompanies rumination. The occurrence of low birth weight can, however, be diminished by adequate maternal social support.

Hormonal consequences of prenatal stress

Numerous hormonal and immune changes can be elicited by prenatal stressful experiences, some of which might be responsible for the subsequent behavioral disturbances or physical illnesses that emerge in offspring. The children of women who had been chronically stressed during pregnancy displayed elevated circulating CRH (possibly of placental origin) and cortisol levels (Ghaemmaghami et al., 2014). Infants of about 1 year of age who were born to mothers that experienced prenatal intimate partner violence similarly exhibited a marked increase of salivary cortisol in response to a mild stressor comprising their arm being briefly restrained. This outcome was not linked to postnatal intimate partner violence or maternal mental health, suggesting that these effects were unique to prenatal adverse experiences. The hormonal variations associated with prenatal challenges appear to be long-lasting in that the subsequent adult cortisol and ACTH profile was reminiscent of that associated with PTSD. Although these findings don't imply a causal connection to prenatal stressors, these experiences may influence the response to adult stressors so the propensity for this disorder was elevated just as this occurs in association with early-life stressor challenges.

Epigenetic changes related to prenatal stress

Along with other processes, prenatal epigenetic changes occur in utero based on mom's experiences, which might promote "gestational programming" to prepare the fetus for challenges that it will ultimately experience in the postnatal environment. This programming may be either beneficial, harmful, or insignificant, depending on the external environment ultimately experienced by the newborn (Goyal et al., 2019). For instance, poor nutritional intake by mom might produce metabolic changes that prepare the fetus for postnatal food scarcity. However, if food is actually plentiful, the very same metabolic changes stemming from the prenatal epigenetic alterations may dispose offspring to obesity and hence metabolic syndrome and type 2 diabetes. Moreover, these epigenetic actions may be carried across generations.

Turning to a very different scenario, intimate partner violence experienced by pregnant women, led both mothers and offspring to subsequently experience high rates of psychological disturbances that were accompanied by frequent epigenetic changes related to glucocorticoid functioning (Serpeloni et al., 2019). It is pertinent that violence experienced by women during pregnancy was accompanied by epigenetic changes that were seen in their grandchildren, once again pointing to transgenerational epigenetic alterations (Senaldi and Smith-Raska, 2020). Just as negative events can have persistent effects that could be disadvantageous, positive prenatal experiences can have the opposite effect reflected by elevated brain neuroplasticity that can enhance emotional and cognitive functioning.

Several epigenetic variations associated with prenatal stressors appeared on genes that influenced HPA functioning. These outcomes were detected in the placental tissue and umbilical cord blood of offspring who had been born to mothers who experienced chronic stressors and war-related conflict within the Democratic Republic of Congo (Kertes et al., 2017). General distress experienced by women during pregnancy was also associated with an increase in the methylation of the glucocorticoid pathway gene FKBP5 in placental tissue, which is in line with the epigenetic effect stemming from negative experiences in adults (Monk et al., 2016). Two other glucocorticoid-related genes, NR3C1 and HSD11, which have been implicated in depressive-like states, were similarly influenced by epigenetic alterations associated with maternal adversity that comprised nutrition problems, preeclampsia, smoking, and diabetes. A meta-analysis revealed the prominence of epigenetic changes linked to prenatal stressors, particularly within the NR3C1 promoter. It was cautioned, however, that the findings could potentially be confounded by pharmacological treatments that moms had received during pregnancy and were also dependent on the sex of the offspring.

A systematic review of the data supported the conclusion that prenatal stressors influenced DNA methylation of genes that code for corticoid functioning but greater attention to other biological processes was called for as these could affect infant and later adult well-being (Sosnowski et al., 2018). In fact, in human newborns that had prenatally experienced war trauma, epigenetic changes of BDNF genes were found in umbilical cord blood, placental tissue, and maternal venous blood (Kertes et al., 2017). Maternal stress during pregnancy was similarly associated with epigenetic changes within genes that code for insulin-like growth factor-2 that may contribute to slowed development and low birth weight. Children born after a natural disaster also displayed epigenetic changes that were associated with altered T cells and peripheral blood mononuclear cells. It has been maintained that some long-lasting effects of prenatal stress on psychological and physical illnesses were linked to epigenetic changes related to circulating cytokines (Cao-Lei et al., 2020).

While the actions of prenatal stressors have usually been attributed to biological changes that occur in utero, prenatal stressors could independently affect the behavior of mothers wherein postnatal maternal care was disturbed. Based on studies that involved in vitro fertilization in humans, it was concluded that some pathologies were primarily linked to prenatal events, whereas others were more closely tied to postnatal influences or genetic factors (Rice et al., 2010). Considering the wide-ranging effects of prenatal stressors on behavioral and biological systems, it would be imprudent to suggest one-to-one links between specific biological alterations and the appearance of pathological conditions. The broad actions of prenatal stressors may set the stage for the emergence of any number of illnesses, and in this sense, these events might create a "general susceptibility" to pathology that would be more likely to develop in response to postnatal hits that are encountered.

Impact of prenatal infection

Systemic stressors, such as those that activate inflammatory processes, may promote neurobiological changes reminiscent of those provoked by psychogenic and neurogenic stressors. From this perspective, infectious illnesses can be considered as a challenge that favors later illness occurrence, including those stemming from the infection-related immune changes. Indeed, various infections during pregnancy in both rodents and humans were shown to produce profound and lasting repercussions on the offspring. Considerable data have amassed indicating that viral infection during pregnancy increases the occurrence of autism and schizophrenia, at least to the extent that these can be simulated in a rodent model, and may be associated with reduced cortical volume in the offspring. Importantly, epidemiological studies were consistent with the suggestion that contracting a viral illness during pregnancy was associated with the later development of schizophrenia (Davies et al., 2020). The relationship was not specific to any particular infection (although it was notable in relation to influenza) and instead might have been related to fever and cytokine elevations that were common to all of them, perhaps interacting with genetic factors (Allswede et al., 2020). Resilience to such disturbances was ascribed to the effectiveness of antiinflammatory and antioxidant processes, together with appropriate levels of mom's iron, zinc, choline, vitamin D, and omega-3 fatty acids (Meyer, 2019). Moreover, even though maternal immune activation could elicit cytokine and glucocorticoid functioning and behavioral disturbances among offspring during their later life, these outcomes could be attenuated by continued environmental enrichment (Núñez Estevez et al., 2020).

There is still limited information concerning the influence of maternal infection on the development of cancer in offspring. Still, influenza and cytomegalovirus infection during pregnancy was associated with elevated risk of acute lymphoblastic leukemia in offspring (He et al., 2020). Analyses of data of individuals born in Sweden during the 1919 Spanish flu epidemic, revealed greater morbidity during the 1968–2012 period, and cancer occurrence and heart disease in males occurred at higher-than-expected rates, although these actions were relatively modest (Helgertz and Bengtsson, 2019).

Recent pandemics, such as MERS and SARS-CoV-1, have been associated with varied obstetrical complications, and this was similarly apparent among pregnant women who had contracted COVID-19 (SARS-CoV-2) infection. Of course, it is still too soon to know what long-term consequences may occur among the children of women infected while they had been pregnant, but it will no doubt be monitored carefully.

Influence of early-life stressors

The early postnatal period is a highly malleable one, being a time during which considerable biological plasticity is taking place, including neural development, connectivity (synaptic growth), and remodeling. Accordingly, stressors encountered early in life can have profound and lasting morphological, metabolic, immune, and inflammatory consequences that create risks for subsequent physical disorders. Numerous studies have indeed indicated that abuse or neglect encountered during early life are associated with marked and lasting behavioral (e.g., depression and anxiety) and physical repercussions, as well as the response to subsequently encountered stressors. Repeated evaluation of children revealed that victimization (exposure to domestic violence, physical maltreatment, sexual abuse, emotional abuse and neglect, physical neglect, and frequent bullying by peers) was accompanied by an increased presence of circulating inflammatory markers when youth were 18 years of age. The effect was stronger with repeated victimization but appeared primarily in females, whereas a smaller relationship was evident in males (Baldwin et al., 2018).

A study in about one million Danish individuals indicated that early-life adversity was associated with an increased incidence of mortality in early adulthood (i.e., between 16 and 36 years of age). The increased mortality was often reflected by elevated occurrence of accidents and suicide as well as through elevated cancer risk. Among children who experienced isolated episodes of adversity (43% of children) that largely comprised poverty or illness in the family, early mortality was 1.3–1.8 times greater than that of individuals who had not experienced these adverse events. In those who had experienced appreciable adversity throughout early life, which occurred in about 3% of children, early mortality was 4.5 times higher than that of children who had not encountered adverse events (Rod et al., 2020).

One of the powerful stressors experienced by people of any age is that of being bullied, and these experiences may be especially marked in children and adolescents. Being bullied between the ages of 7–11 was associated with increased levels of CRP and clinically relevant inflammation during subsequent mid-life, as well as increased obesity (Takizawa et al., 2015), which could affect disease occurrence. A prospective study that followed almost 1.5 million children and adolescents (5–19 years of age) for about 12 years indicated that those who had been depressed subsequently experienced an increase of 66 of 69 somatic medical conditions as well as a 600% increase in the occurrence of premature death (Leone et al., 2021). Clearly, negative events that are encountered during the adolescent period, like those encountered during early life, can cause lasting health risks that shouldn't be dismissed as "just one of those things that kids experience while growing up."

Some of the lasting effects of early-life stressors are likely due to the direct and indirect actions of mental illnesses, the endorsement of poor lifestyles, or other behavioral changes that favor the development of illnesses. The persistent effects of early-life stressors may also be due to the sensitization (priming) of neurobiological processes or to epigenetic changes that had been introduced as well as the maturation of brain networks being undermined, which might contribute to cognitive and emotional disturbances. It is unlikely that any single factor is responsible for the development of disorders stemming from early-life adversities. Instead, genetic factors, early-life stressors, and stressors encountered subsequently, may come together to influence the individual's ability to deal with later challenges, which might then favor the development of illnesses. Fig. 5.6 shows a few of the many downstream consequences

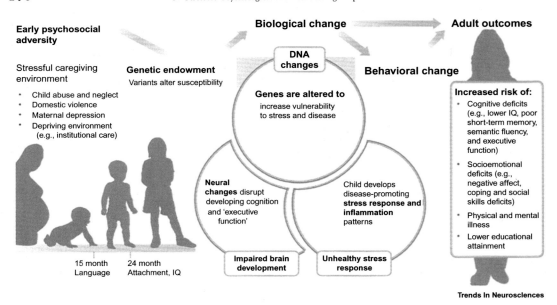

FIG. 5.6 Prenatal and different forms of early-life adversity can interact with genetic influences, thereby influencing diverse biological processes, including epigenetic changes, which can influence a constellation of biological changes across multiple levels. Together with behavioral changes provoked, the biological alterations can engender multiple adult pathological conditions. *From Nelson III, C.A., Gabard-Durnam, L.J., 2020. Early adversity and critical periods: neurodevelopmental consequences of violating the expectable environment. Trends Neurosci. 43, 133–143.*

of childhood adversity and how these might come to engender adult pathology (Nelson and Gabard-Durnam, 2020).

Negative early-life experiences may affect brain cytokines and neurotrophins and the presence of gene mutations or epigenetic marks related to glucocorticoid functioning were linked to the effects of negative experiences on later psychological disturbances. For instance, adolescents who carried genetic variations related to the glucocorticoid receptor protein FKBP5 and who encountered early-life trauma, later reported experiencing more pronounced rumination and catastrophizing, which can be a precursor to depression and anxiety (Halldorsdottir et al., 2017). In the same fashion, alterations of cortisol activity may affect microglial functioning within the brain, which can then affect synaptogenesis, synaptic pruning, axonal growth, and myelination, culminating in behavioral disturbances (Johnson and Kaffman, 2018). Of particular significance, early-life stressors may influence regulatory immune functioning and the propensity for autoimmune disorders, processes related to pain perception, and interactions between the brain and gut microbiota. It is pertinent that these early experiences may foster the development of obesity (discussed in a later chapter), which may contribute to multiple health disturbances, particularly type 2 diabetes, heart disease, immune functioning, and cancer.

While prenatal and early-life stressors have received the greatest attention, adolescence in rodents is also a stress-sensitive period. Stressors encountered at this time can engender pronounced and lasting neurobiological actions, including disturbance of hippocampal nerve cell growth factor, altered CRH receptor functioning, variations of GABA and its receptors, and persistent variations of glutamate receptors (Yohn and Blendy, 2017). Among rats that had experienced chronic social instability during adolescence later cognitive changes and diminished BDNF were apparent. However, a diet supplemented with omega-3 and vitamin A at that time, altered microbiota and prevented the cognitive impairments that might otherwise appear in adulthood (Provensi et al., 2019).

In humans, the adolescent period is accompanied by continued growth and reorganization of neuronal and hormonal systems, and importantly, it is a transitional period that is critical for adequate socialization. Youth encounter multiple challenges during this phase of life including the need to "fit in." Unfortunately, they may experience powerful stressors in the form of social rejection or bullying that can have long-term ramifications on mental health. Given the comorbidities that exist between these conditions and physical illnesses related to immune and inflammatory processes, such findings speak to the need to deal with social problems early.

Among the most powerful social stressors encountered are those that comprise unsupportive behaviors (i.e., not obtaining support from friends when it was reasonably expected) as well as social rejection, both of which may be encountered during late adolescence. Even in a laboratory situation in which social rejection could be produced through a computer game, self-reported distress was accompanied by elevated neuronal activity within the dorsal anterior cingulate cortex, which has been implicated as being important for feelings of social pain (Eisenberger, 2012). Along with the elevated neuronal activity within the cingulate cortex, rejection was also associated with reduced neuronal activity in the ventral striatum (a brain region closely tied to reward perception), perhaps reflecting a decline in the rewarding value that might be obtained from otherwise positive stimuli or events. These brain changes were most prominent in those individuals who were especially sensitive to rejection and who might also be most vigilant about peer acceptance. However, having a strong social network and spending time with friends could prevent the adverse effects of later peer rejection (Masten et al., 2012). Beyond the psychological effects of social exclusion, increased inflammation has been associated with this social stressor, whereas adequate social contacts were accompanied by enhanced antiviral immunity (Leschak and Eisenberger, 2019). Consistent with the powerful effects of social rejection, among adolescent women at risk for depression who were followed for a 2.5-year period, targeted social exclusion was accompanied by elevated expression of genes that code for inflammation (Murphy et al., 2013). Social rejection and bullying have serious mental health repercussions and increased suicide risk and may undermine physical health. Numerous neurobiological processes likely contribute to these outcomes as might elevated inflammatory processes.

Summary and conclusions

The behavioral and physiological responses to stressors are influenced by the nature of the challenge encountered and how individuals perceived and cope with these events. Negative outcomes of these experiences have been associated with previous encounters with adverse events, their chronicity, as well as a constellation of personality and psychosocial factors. Many of these same variables may contribute to chronic illnesses, likely being moderated by genetic and epigenetic factors that influence multiple endogenous processes. Because of the many biological changes that are elicited by chronic stressors, it can be difficult to determine which of these contributes to the development of specific illnesses, and in most instances, multiple biological disturbances generated by stressors are involved.

Chronic stressors and the accumulation of independent life stressors, perhaps involving more frequent epigenetic variations, appear to be tied to biological aging and may lend themselves to age-related diseases. Human studies provided relevant findings supporting these ties, but they have typically not permitted causal connections to be drawn. Animal studies pointed to the causal role of prenatal and early-life challenges in the instigation of adult pathology, even though the stressors delivered in these studies were nowhere near the trauma humans often endure.

Considerable movement has been realized concerning the stressor conditions that can lead to pathology, as well as the processes responsible for such outcomes. The involvement of various hormones, immune alterations, and inflammatory factors is foremost in accounting for stressor-related mental and physical illnesses. Predictably, variations of mitochondrial ROS and growth factors also contribute to diverse illnesses, including cancer development and progression. Most of the stressor-related biological processes that can affect inflammatory-related conditions are modifiable by lifestyles endorsed, and behavioral change strategies could be incorporated to prevent adverse outcomes and may be adopted in therapeutic efforts to deal with some of these illnesses.

References

Abizaid, A., 2019. Stress and obesity: the ghrelin connection. J. Neuroendocrinol. 31, e12693.

Allswede, D.M., Yolken, R.H., Buka, S.L., Cannon, T.D., 2020. Cytokine concentrations throughout pregnancy and risk for psychosis in adult offspring: a longitudinal case-control study. Lancet Psychiatry 7, 254–261.

Anisman, H., Hayley, S., Kusnecov, A., 2018. The Immune System and Mental Health. Academic Press, London.

Armaiz-Pena, G.N., Allen, J.K., Cruz, A., Stone, R.L., Nick, A.M., et al., 2013. Src activation by β-adrenoreceptors is a key switch for tumour metastasis. Nat. Commun. 4, 1–12.

Baldwin, J.R., Arseneault, L., Caspi, A., Fisher, H.L., Moffitt, T.E., et al., 2018. Childhood victimization and inflammation in young adulthood: a genetically sensitive cohort study. Brain Behav. 67, 211–217.

Berrocal Montiel, C., Rivas Moya, T., Venditti, F., Bernini, O., 2016. On the contribution of psychological flexibility to predict adjustment to breast cancer. Psicothema 28, 266–271.

Black, N., Stockings, E., Campbell, G., Tran, L.T., Zagic, D., et al., 2019. Cannabinoids for the treatment of mental disorders and symptoms of mental disorders: a systematic review and meta-analysis. Lancet Psychiatry 6, 995–1010.

Borba, V.V., Shoenfeld, Y., 2019. Prolactin, autoimmunity, and motherhood: when should women avoid breastfeeding? Clin. Rheumatol. 38, 1263–1270.

Borgi, M., Collacchi, B., Ortona, E., Cirulli, F., 2020. Stress and coping in women with breast cancer: unravelling the mechanisms to improve resilience. Neurosci. Biobehav. Rev. 119, 406–421.

Braile, M., Marcella, S., Marone, G., Galdiero, M.R., Varricchi, G., et al., 2021. The interplay between the immune and the endocannabinoid systems in cancer. Cells 10, 1282.

Brivio, E., Lopez, J.P., Chen, A., 2020. Sex differences: transcriptional signatures of stress exposure in male and female brains. Genes Brain Behav. 19, e12643.

Cao-Lei, L., De Rooij, S., King, S., Matthews, S., Metz, G., et al., 2020. Prenatal stress and epigenetics. Neurosci. Biobehav. Rev. 117, 198–210.

Cathomas, F., Murrough, J.W., Nestler, E.J., Han, M.-H., Russo, S.J., 2019. Neurobiology of resilience: interface between mind and body. Biol. Psychiatry 86, 410–420.

Chevalier, G., Siopi, E., Guenin-Macé, L., Pascal, M., Laval, T., et al., 2020. Effect of gut microbiota on depressive-like behaviors in mice is mediated by the endocannabinoid system. Nat. Commun. 11, 1–15.

Ciardiello, D., Elez, E., Tabernero, J., Seoane, J., 2020. Clinical development of therapies targeting TGFβ: current knowledge and future perspectives. Ann. Oncol. 31, 1336–1349.

Cohen, S., Evans, G.W., Stokols, D., Krantz, D.S., 1986. Behavior, Health, and Environmental Stress. Plenum Publishing Corporation, New York.

Davies, C., Segre, G., Estradé, A., Radua, J., De Micheli, A., et al., 2020. Prenatal and perinatal risk and protective factors for psychosis: a systematic review and meta-analysis. Lancet Psychiatry 7, 399–410.

Daviu, N., Füzesi, T., Rosenegger, D.G., Rasiah, N.P., Sterley, T.L., et al., 2020. Paraventricular nucleus CRH neurons encode stress controllability and regulate defensive behavior selection. Nat. Neurosci. 23, 398–410.

De Dreu, C.K., 2012. Oxytocin modulates cooperation within and competition between groups: an integrative review and research agenda. Horm. Behav. 61, 419–428.

Dhabhar, F.S., Malarkey, W.B., Neri, E., McEwen, B.S., 2012. Stress-induced redistribution of immune cells—from barracks to boulevards to battlefields: a tale of three hormones—Curt Richter Award winner. Psychoneuroendocrinology 37, 1345–1368.

Dianat-Moghadam, H., Teimoori-Toolabi, L., 2019. Implications of fibroblast growth factors (FGFs) in cancer: from prognostic to therapeutic applications. Curr. Drug Targets 20, 852–870.

Dickerson, S.S., Kemeny, M.E., 2004. Acute stressors and cortisol responses: a theoretical integration and synthesis of laboratory research. Psychol. Bull. 130, 355.

DiPatrizio, N.V., 2016. Endocannabinoids in the gut. Cannabis Cannabinoid Res. 1, 67–77.

Duman, R.S., Monteggia, L.M., 2006. A neurotrophic model for stress-related mood disorders. Biol. Psychiatry 59, 1116–1127.

Duman, R.S., Sanacora, G., Krystal, J.H., 2019. Altered connectivity in depression: GABA and glutamate neurotransmitter deficits and reversal by novel treatments. Neuron 102, 75–90.

Eisenberger, N.I., 2012. The pain of social disconnection: examining the shared neural underpinnings of physical and social pain. Nat. Rev. Neurosci. 13, 421–434.

Finnell, J.E., Lombard, C.M., Padi, A.R., Moffitt, C.M., Wilson, L.B., et al., 2017. Physical versus psychological social stress in male rats reveals distinct cardiovascular, inflammatory and behavioral consequences. PLoS One 12, e0172868.

Frawley, T., Piskareva, O., 2020. Extracellular vesicle dissemination of epidermal growth factor receptor and ligands and its role in cancer progression. Cancers 12, 3200.

Gabrys, R.L., Tabri, N., Anisman, H., Matheson, K., 2018. Cognitive control and flexibility in the context of stress and depressive symptoms: the cognitive control and flexibility questionnaire. Front. Psychol. 9, 2219.

Ghaemmaghami, P., Dainese, S.M., La Marca, R., Zimmermann, R., Ehlert, U., 2014. The association between the acute psychobiological stress response in second trimester pregnant women, amniotic fluid glucocorticoids, and neonatal birth outcome. Dev. Psychobiol. 56, 734–747.

Gigerenzer, G., 2008. Why heuristics work. Perspect. Psychol. Sci. 3, 20–29.

Gone, J.P., Kirmayer, L.J., 2020. Advancing indigenous mental health research: ethical, conceptual and methodological challenges. Transcult. Psychiatry 57, 235–249.

Goulet, C.R., Pouliot, F., 2021. TGFβ signaling in the tumor microenvironment. Adv. Exp. Med. Biol. 1270, 89–105.

Goyal, D., Limesand, S.W., Goyal, R., 2019. Epigenetic responses and the developmental origins of health and disease. J. Endocrinol. 242, T105–T119.

Guzeloglu-Kayisli, O., Semerci, N., Guo, X., Larsen, K., Ozmen, A., et al., 2021. Decidual cell FKBP51–progesterone receptor binding mediates maternal stress–induced preterm birth. Proc. Natl. Acad. Sci. 118. e2010282118.

Halldorsdottir, T., de Matos, A.P.S., Awaloff, Y., Arnarson, E.Ö., Craighead, W.E., et al., 2017. FKBP5 moderation of the relationship between childhood trauma and maladaptive emotion regulation strategies in adolescents. Psychoneuroendocrinology 84, 61–65.

Haslam, C., Jetten, J., Cruwys, T., Dingle, G.A., Haslam, S.A., 2018. The New Psychology of Health: Unlocking the Social Cure. Routledge, Oxfordshire.

Haslam, C., Haslam, S.A., Jetten, J., Cruwys, T., Steffens, N.K., 2021. Life change, social identity, and health. Annu. Rev. Psychol. 72, 635–661.

He, J.R., Ramakrishnan, R., Hirst, J.E., Bonaventure, A., Francis, S.S., et al., 2020. Maternal infection in pregnancy and childhood leukemia: a systematic review and meta-analysis. J. Pediatr. 217, 98–109.e108.

Heim, C., Newport, D.J., Mletzko, T., Miller, A.H., Nemeroff, C.B., 2008. The link between childhood trauma and depression: insights from HPA axis studies in humans. Psychoneuroendocrinology 33, 693–710.

Helgertz, J., Bengtsson, T., 2019. The long-lasting influenza: the impact of fetal stress during the 1918 influenza pandemic on socioeconomic attainment and health in Sweden, 1968–2012. Demography 56, 1389–1425.

Hetz, C., Papa, F.R., 2018. The unfolded protein response and cell fate control. Mol. Cell 69, 169–181.
Hill, M.N., Campolongo, P., Yehuda, R., Patel, S., 2018. Integrating endocannabinoid signaling and cannabinoids into the biology and treatment of posttraumatic stress disorder. Neuropsychopharmacology 43, 80–102.
Hinton, D.E., Kirmayer, L.J., 2017. The flexibility hypothesis of healing. Cult. Med. Psychiatry 41, 3–34.
Horchar, M.J., Wohleb, E.S., 2019. Glucocorticoid receptor antagonism prevents microglia-mediated neuronal remodeling and behavioral despair following chronic unpredictable stress. Brain Behav. Immun. 81, 329–340.
Huang, J., Xie, Y., Sun, X., Zeh III, H.J., Kang, R., et al., 2015. DAMPs, ageing, and cancer: the 'DAMP Hypothesis'. Ageing Res. Rev. 24, 3–16.
Ip, C.K., Zhang, L., Farzi, A., Qi, Y., Clarke, I., et al., 2019. Amygdala NPY circuits promote the development of accelerated obesity under chronic stress conditions. Cell Metab. 30 (111–128), e116.
Jang, H.J., Boo, H.J., Lee, H.J., Min, H.Y., Lee, H.Y., 2016. Chronic stress facilitates lung tumorigenesis by promoting exocytosis of IGF2 in lung epithelial cells. Cancer Res. 76, 6607–6619.
Jeppsson, S., Srinivasan, S., Chandrasekharan, B., 2017. Neuropeptide Y (NPY) promotes inflammation-induced tumorigenesis by enhancing epithelial cell proliferation. Am. J. Physiol. Gastrointest. Liver Physiol. 312, G103–G111.
Ji, H., Liu, N., Yin, Y., Wang, X., Chen, X., et al., 2018. Oxytocin inhibits ovarian cancer metastasis by repressing the expression of MMP-2 and VEGF. J. Cancer 9, 1379–1384.
Johnson, F.K., Kaffman, A., 2018. Early life stress perturbs the function of microglia in the developing rodent brain: new insights and future challenges. Brain Behav. Immun. 69, 18–27.
Kahneman, D., 2011. Thinking, Fast and Slow. Farrar, Straus & Giroux, New York.
Kelly, O., Matheson, K., Ravindran, A., Merali, Z., Anisman, H., 2007. Ruminative coping among patients with dysthymia before and after pharmacotherapy. Depress. Anxiety 24, 233–243.
Kertes, D.A., Bhatt, S.S., Kamin, H.S., Hughes, D.A., Rodney, N.C., et al., 2017. BNDF methylation in mothers and newborns is associated with maternal exposure to war trauma. Clin. Epigenetics 9, 1–12.
Klumpers, F., Kroes, M.C.W., Baas, J.M.P., Fernández, G., 2017. How human amygdala and bed nucleus of the stria terminalis may drive distinct defensive responses. J. Neurosci. 37, 9645–9656.
Kong, D., Zhou, H., Neelakantan, D., Hughes, C.J., Hsu, J.Y., et al., 2021. VEGF-C mediates tumor growth and metastasis through promoting EMT-epithelial breast cancer cell crosstalk. Oncogene 40, 964–979.
Kotsantis, I., Economopoulou, P., Psyrri, A., Maratou, E., Pectasides, D., et al., 2019. Prognostic significance of IGF-1 signalling pathway in patients with advanced non-small cell lung cancer. Anticancer Res. 39, 4185–4190.
Kruger, J., Dunning, D., 1999. Unskilled and unaware of it: how difficulties in recognizing one's own incompetence lead to inflated self-assessments. J. Pers. Soc. Psychol. 77, 1121.
Kruk, J., Aboul-Enein, B.H., Bernstein, J., Gronostaj, M., 2019. Psychological stress and cellular aging in cancer: a meta-analysis. Oxidative Med. Cell. Longev. 2019, 1270397.
Kudielka, B.M., Kirschbaum, C., 2005. Sex differences in HPA axis responses to stress: a review. Biol. Psychol. 69, 113–132.
Kummer, S., Walter, F.M., Chilcot, J., Emery, J., Sutton, S., et al., 2019. Do cognitive heuristics underpin symptom appraisal for symptoms of cancer?: a secondary qualitative analysis across seven cancers. Psychooncology 28, 1041–1047.
Kushner, H.S., 1981. When Bad Things Happen to Good People. Anchor Books, New York.
Kvillemo, P., Bränström, R., 2014. Coping with breast cancer: a meta-analysis. PLoS One 9, e112733.
Lazarus, R.S., Folkman, S., 1984. Stress, Appraisal, and Coping. Springer, New York.
LeDoux, J.E., 2014. Coming to terms with fear. Proc. Natl. Acad. Sci. 111, 2871–2878.
Leone, M., Kuja-Halkola, R., Leval, A., D'Onofrio, B.M., Larsson, H., et al., 2021. Association of youth depression with subsequent somatic diseases and premature death. JAMA Psychiatry 78, 302–310.
Leschak, C.J., Eisenberger, N.I., 2019. Two distinct immune pathways linking social relationships with health: inflammatory and antiviral processes. Psychosom. Med. 81, 711.
Li, X.H., Pang, H.Q., Qin, L., Jin, S., Zeng, X., et al., 2015. HSP70 overexpression may play a protective role in the mouse embryos stimulated by CUMS. Reprod. Biol. Endocrinol. 13, 1–7.
Maier, S.F., Seligman, M.E.P., 2016. Learned helplessness at fifty: insights from neuroscience. Psychol. Rev. 123, 349–367.
Masten, C.L., Telzer, E.H., Fuligni, A.J., Lieberman, M.D., Eisenberger, N.I., 2012. Time spent with friends in adolescence relates to less neural sensitivity to later peer rejection. Soc. Cogn. Affect. Neurosci. 7, 106–114.
Matheson, K., Anisman, H., 2012. Biological and psychosocial responses to discrimination. In: Jetten, J., Haslam, C., Haslam, S.A. (Eds.), The Social Cure. Psychology Press, New York, NY, pp. 133–154.

References

Matheson, K., Asokumar, A., Anisman, H., 2020. Resilience: safety in the aftermath of traumatic stressor experiences. Front. Behav. Neurosci. 14, 596919.

Matheson, K., Pierre, A., Foster, M.D., Kent, M., Anisman, H., 2021. Untangling racism: stress reactions in response to variations of racism against Black Canadians. Humanit. Soc. Sci. Commun. 8, 1–12.

McEwen, B.S., 2020. The untapped power of allostasis promoted by healthy lifestyles. World Psychiatry 19, 57–58.

McEwen, B.S., Akil, H., 2020. Revisiting the stress concept: implications for affective disorders. J. Neurosci. 40, 12–21.

McQuaid, R.J., McInnis, O.A., Abizaid, A., Anisman, H., 2014. Making room for oxytocin in understanding depression. Neurosci. Biobehav. Rev. 45, 305–322.

Meyer, U., 2019. Neurodevelopmental resilience and susceptibility to maternal immune activation. Trends Neurosci. 42, 793–806.

Monk, C., Feng, T., Lee, S., Krupska, I., Champagne, F.A., et al., 2016. Distress during pregnancy: epigenetic regulation of placenta glucocorticoid-related genes and fetal neurobehavior. Am. J. Psychiatry 173, 705–713.

Moshe, A., Izraely, S., Sagi-Assif, O., Malka, S., Ben-Menachem, S., et al., 2020. Inter-tumor heterogeneity—melanomas respond differently to GM-CSF-mediated activation. Cells 9, 1683.

Muller, A., Sirianni, L.A., Addante, R.J., 2021. Neural correlates of the Dunning–Kruger effect. Eur. J. Neurosci. 53, 460–484.

Murphy, M.L., Slavich, G.M., Rohleder, N., Miller, G.E., 2013. Targeted rejection triggers differential pro-and anti-inflammatory gene expression in adolescents as a function of social status. Clin. Psychol. Sci. 1, 30–40.

Nelson III, C.A., Gabard-Durnam, L.J., 2020. Early adversity and critical periods: neurodevelopmental consequences of violating the expectable environment. Trends Neurosci. 43, 133–143.

Nigro, E., Formato, M., Crescente, G., Daniele, A., 2021. Cancer initiation, progression and resistance: are phytocannabinoids from *Cannabis sativa* L. promising compounds? Molecules 26, 2668.

Nilsson, M.B., Sun, H., Diao, L., Tong, P., Liu, D., et al., 2017. Stress hormones promote EGFR inhibitor resistance in NSCLC: implications for combinations with β-blockers. Sci. Transl. Med. 9.

Nolen-Hoeksema, S., 1998. Ruminative coping with depression. In: Motivation and Self-Regulation Across the Life Span. Cambridge University Press, Cambridge, pp. 237–256.

Nowacka, M., Obuchowicz, E., 2013. BDNF and VEGF in the pathogenesis of stress-induced affective diseases: an insight from experimental studies. Pharmacol. Rep. 65, 535–546.

Núñez Estevez, K.J., Rondón-Ortiz, A.N., Nguyen, J.Q.T., Kentner, A.C., 2020. Environmental influences on placental programming and offspring outcomes following maternal immune activation. Brain Behav. Immun. 83, 44–55.

Palma-Gudiel, H., Fañanás, L., Horvath, S., Zannas, A.S., 2020. Psychosocial stress and epigenetic aging. Int. Rev. Neurobiol. 150, 107–128.

Parker, L.A., 2017. Cannabinoids and the Brain. MIT Press, Cambridge, MA.

Provensi, G., Schmidt, S.D., Boehme, M., Bastiaanssen, T.F., Rani, B., et al., 2019. Preventing adolescent stress-induced cognitive and microbiome changes by diet. Proc. Natl. Acad. Sci. 116, 9644–9651.

Rainville, J.R., Tsyglakova, M., Hodes, G.E., 2018. Deciphering sex differences in the immune system and depression. Front. Neuroendocrinol. 50, 67–90.

Rice, F., Harold, G.T., Boivin, J., Van den Bree, M., Hay, D.F., et al., 2010. The links between prenatal stress and offspring development and psychopathology: disentangling environmental and inherited influences. Psychol. Med. 40, 335–345.

Rod, N.H., Bengtsson, J., Budtz-Jørgensen, E., Clipet-Jensen, C., Taylor-Robinson, D., et al., 2020. Trajectories of childhood adversity and mortality in early adulthood: a population-based cohort study. Lancet 396, 489–497.

Sah, R., Ekhator, N.N., Jefferson-Wilson, L., Horn, P.S., Geracioti Jr., T.D., 2014. Cerebrospinal fluid neuropeptide Y in combat veterans with and without posttraumatic stress disorder. Psychoneuroendocrinology 40, 277–283.

Saha, S.K., Lee, S.B., Won, J., Choi, H.Y., Kim, K., et al., 2017. Correlation between oxidative stress, nutrition, and cancer initiation. Int. J. Mol. Sci. 18, 1544.

Sapolsky, R.M., Romero, L.M., Munck, A.U., 2000. How do glucocorticoids influence stress responses? Integrating permissive, suppressive, stimulatory, and preparative actions. Endocr. Rev. 21, 55–89.

Sayed, S., Van Dam, N.T., Horn, S.R., Kautz, M.M., Parides, M., et al., 2018. A randomized dose-ranging study of neuropeptide Y in patients with posttraumatic stress disorder. Int. J. Neuropsychopharmacol. 21, 3–11.

Senaldi, L., Smith-Raska, M., 2020. Evidence for germline non-genetic inheritance of human phenotypes and diseases. Clin. Epigenetics 12, 1–12.

Serpeloni, F., Radtke, K.M., Hecker, T., Sill, J., Vukojevic, V., et al., 2019. Does prenatal stress shape postnatal resilience?—an epigenome-wide study on violence and mental health in humans. Front. Genet. 10, 269.

Sigismund, S., Avanzato, D., Lanzetti, L., 2018. Emerging functions of the EGFR in cancer. Mol. Oncol. 12, 3–20.

Simard, S., Shail, P., MacGregor, J., El Sayed, M., Duman, R.S., et al., 2018. Fibroblast growth factor 2 is necessary for the antidepressant effects of fluoxetine. PLoS One 13, e0204980.

Soon, C.S., He, A.H., Bode, S., Haynes, J.D., 2013. Predicting free choices for abstract intentions. Proc. Natl. Acad. Sci. 110, 6217–6222.

Sosnowski, D.W., Booth, C., York, T.P., Amstadter, A.B., Kliewer, W., 2018. Maternal prenatal stress and infant DNA methylation: a systematic review. Dev. Psychobiol. 60, 127–139.

Southwick, S.M., Charney, D.S., 2018. Resilience: The Science of Mastering Life's Greatest Challenges. Cambridge University Press, Cambridge.

Tabak, B.A., Vrshek-Schallhorn, S., Zinbarg, R.E., Prenoveau, J.M., Mineka, S., et al., 2016. Interaction of CD38 variant and chronic interpersonal stress prospectively predicts social anxiety and depression symptoms over 6 years. Clin. Psychol. Sci. 4, 17–27.

Takizawa, R., Danese, A., Maughan, B., Arseneault, L., 2015. Bullying victimization in childhood predicts inflammation and obesity at mid-life: a five-decade birth cohort study. Psychol. Med. 45, 2705–2715.

Taylor, S.E., Klein, L.C., Lewis, B.P., Gruenewald, T.L., Gurung, R.A., et al., 2000. Biobehavioral responses to stress in females: tend-and-befriend, not fight-or-flight. Psychol. Rev. 107, 411.

Tibensky, M., Mravec, B., 2021. Role of the parasympathetic nervous system in cancer initiation and progression. Clin. Transl. Oncol. 23, 669–681.

Tversky, A., Kahneman, D., 1974. Judgment under uncertainty: heuristics and biases. Science 185, 1124–1131.

Wang, M., Wu, X., Chai, F., Zhang, Y., Jiang, J., 2016. Plasma prolactin and breast cancer risk: a meta-analysis. Sci. Rep. 6, 1–7.

Witsch, E., Sela, M., Yarden, Y., 2010. Roles for growth factors in cancer progression. Physiology 25, 85–101.

Yang, S., Lian, G., 2020. ROS and diseases: role in metabolism and energy supply. Mol. Cell. Biochem. 467, 1–12.

Yehuda, R., 2002. Current status of cortisol findings in post-traumatic stress disorder. Psychiatr. Clin. North Am. 25, 341–368. vii.

Yohn, N.L., Blendy, J.A., 2017. Adolescent chronic unpredictable stress exposure is a sensitive window for long-term changes in adult behavior in mice. Neuropsychopharmacology 42, 1670–1678.

Ysseldyk, R., Matheson, K., Anisman, H., 2010. Religiosity as identity: toward an understanding of religion from a social identity perspective. Personal. Soc. Psychol. Rev. 14, 60–71.

Zahalka, A.H., Arnal-Estapé, A., Maryanovich, M., Nakahara, F., Cruz, C.D., et al., 2017. Adrenergic nerves activate an angio-metabolic switch in prostate cancer. Science 358, 321–326.

CHAPTER 6

Stress, immunity, and cancer

OUTLINE

Blame it on stressor experiences	178
Brain and immune system interactions	178
Communication between immune processes and the brain	179
The brain's immunity	180
Stressor influences on immunity	182
Leukocyte changes elicited by stressors	182
Acute versus chronic challenges	183
Stressor-provoked responses to infection in rodents	184
Stressor-elicited immune changes in humans	185
Stressor-provoked responses to infection in humans	186
Cytokine variations associated with stressors	187
Concurrent actions of pro- and antiinflammatory cytokines	188
Stressor-elicited cytokine variations in humans	188
Impact of early-life experiences	189
Stress and microbiota	191
Stress, psychological alterations, and microbiota	192
Early-life experiences influence adolescent and adult microbiota	193
Stressful events and cancer	195
Immune processes linking stressful events to cancer	195
Stress, hormones, and cancer progression	196
Influence of psychosocial stressors on immunity and cancer	202
Studies in mice	202
Studies in humans: Retrospective analyses	202
Studies in humans: Prospective analyses	203
Prenatal and early-life stressors in relation to cancer	204
Influence of early-life stressors	205
Influence of prenatal stressors	205
Dealing with cancer in children	206
Influence of diverse life stressors	207
Job-related stress	207
The distress of loneliness	208
The distress of chronic illnesses	209
Surgery as a stressor	210
Psychological factors associated with cancer treatment	212
Depression as a predictor of cancer occurrence	212
The distress of cancer treatment	213
Treating cancer-related depression	214
Linking neuropsychiatric and neurodegenerative disorders to cancer	217
Summary and conclusions	218
References	219

Blame it on stressor experiences

Long before it was understood how stressful events might come to affect illnesses, there seemed to be a broad public consensus—a conventional wisdom, if you will—that this was a credible perspective. If the source of a malady wasn't immediately apparent, it was often attributed to stressful experiences (unless the malady involved the stomach, in which case "it must have been something you ate"). Even if these relationships had simply been based on folk wisdom as opposed to science-based evidence, it isn't necessarily a good enough reason to dismiss them out of hand. After all, the absence of evidence might not always reflect evidence of absence. Indeed, with time it was understood that stressors affect immune functioning and inflammatory processes and might thus have contributed to varied illnesses. This included several autoimmune disorders, allergies, and related respiratory conditions, and affected conditions that have been linked to inflammatory processes, such as cardiovascular disease, stroke, and stroke recovery, as well as neurodegenerative disorders.

The notion that stressors could be linked to cancer occurrence or progression had at one time captured appreciable interest (primarily during the 1970s and 1980s), but without being able to identify the processes by which this occurred, scientific interest in this topic waned. At that time, the notion that exercise and sleep might influence cancer development and progression was similarly viewed as an interesting epiphenomenon but without much clinical significance. Psychosocial factors certainly weren't ignored, but they were primarily considered in relation to the general well-being of cancer patients rather than being cogent in affecting disease progression. Since then, interest in the potential influence of stressors on cancer processes has reemerged but it is usually thought that these experiences do not cause cancer occurrence, although there are reasons to think otherwise. There is far less debate, in contrast, as to whether stressful encounters in animal models cause the progression of existent cancers. The larger questions that have emerged are whether the studies in animals translate to humans, what conditions are necessary to promote such stressor effects, which biological mechanisms generate them, and can this understanding be harnessed to avert cancer progression?

Brain and immune system interactions

It had long been accepted that stressful experiences could affect psychological well-being, but in retrospect, it is somewhat surprising that the involvement of stressful events in promoting immune-related disorders was ignored. In part, this may have stemmed from the narrow perspective that immune system functioning was independent of the central nervous system. Although hormones, such as cortisol could affect immunity, this was thought to occur in response to pharmacological doses and not those that were naturally introduced by stressors. It seems that the view that had been taken was that the contribution of the brain and related processes to immunity was relatively insignificant and thus could be disregarded without consequence. Besides, even if immune-related processes could affect brain functioning didn't necessarily mean that the brain could affect immunity. The notion that brain functioning could affect cancer occurrence or progression was met with far more skepticism. It was, of course, accepted that cancer patients experienced psychological problems, such as chronic anxiety and depression, but this was seen as having little bearing on cancer processes.

It's probably fair to say that at that time a connection between the mind (thoughts, feelings, beliefs, and attitudes) and cancer-related processes was viewed as "soft science" with little merit.

Over the decades, however, progressively more data amassed showing that multidirectional communication existed between brain functioning, immune activity, immune signaling molecules (cytokines and chemokines), various hormones and growth factors, and microbiota. Thus, it was appreciated that these processes could directly or indirectly affect cancer progression and the notion emerged that psychological factors, through their triggering of systemic biological actions, might affect mechanisms that promoted cancer progression and that benefits could be reaped through stress reduction procedures.

Communication between immune processes and the brain

The immune system is sensitive to signals that come from several biological factors. Various hormones (e.g., glucocorticoids, estrogen), and neurotransmitters from the sympathetic nervous system (norepinephrine), affect immune functioning, as do circulating neuropeptides, which include vasoactive intestinal peptide (VIP), substance P, and endorphins. As we've seen, cytokines released by immune cells can influence brain functioning and considerable interplay also exists between microbiota and immune functioning. Communication between these processes is essential for maintaining broad homeostatic functioning, in a sense assuring that these components are all working from the same playbook. Based on the intra- and intersystem communication that occurs, it was suggested that the immune system could be viewed as providing a "sixth sense" that informs the brain of pathogenic threats (Blalock, 2005).

Many routes exist by which psychological factors, particularly stressors, might come to affect immune functioning, and how immune activity could affect brain neuronal activity. Brain endothelial cells that line the interior of blood vessels are connected through tight junctions, and together with astrocytes and pericytes form the blood-brain-barrier (BBB) that prevents certain substances from gaining entry to capillary walls and brain tissue. In essence, the BBB protects the brain from toxins or pathogens, while still permitting entry by certain nutrients. Small molecules, fat-soluble molecules, and some gasses readily enter the brain, and several larger molecules (e.g., glucose) can access the brain through proteins that transport them across the BBB.

It was thought that the brain was immunologically privileged in that immune cells were absent, and cytokine molecules were deemed to be too large to gain ready access to the brain. Thus, it was surmised that stimulation of the vagus nerve by peripheral factors, including cytokines and neurotransmitters, served as a route by which the brain was informed of pathogen presence. This perspective seemed to go by the wayside for a while but reemerged in light of the vagus being an important pathway between gut microbiota and the brain (Fülling et al., 2019).

Despite their large size, cytokine access to the brain was possible at sites where the BBB is relatively porous, particularly at circumventricular sites comprising the subfornical organ, organum vasculosum of the lamina terminalis, and the area postrema. Moreover, cytokines could affect the pituitary, median eminence, and pineal gland that are relatively more accessible. Active transport mechanisms were also identified that could ferry cytokines into and through the brain (Banks, 2019) where cells express receptors for these cytokines. Besides these mechanisms, the effectiveness of the BBB can be influenced by hormones such as leptin and insulin, as well as by endocrine-related diseases (e.g., diabetes) and by the presence of inflammation, and could be affected by stressful events (Banks, 2019). This can occur through the loss of tight junction protein claudin-5 (cldn5), so that entry of circulating proinflammatory cytokines into the brain becomes possible, affecting neuronal functioning at limbic sites, such as the nucleus accumbens (a major reward center), hence favoring the occurrence of depressive-like outcomes in stress-susceptible mice (Menard et al., 2017). Furthermore, epigenetic variations of *cldn5* expression coupled with limited endothelium expression of cldn5-related transcription factor *foxo1* could diminish the actions of stressors, thereby producing increased resilience (Dudek et al., 2020).

It appeared that the membrane that surrounds the brain parenchyma (meninges) isn't as protective as once thought. Immune cells and cytokines, as well as waste products and antigens, can travel into and out of the brain through distinct drainage paths within the meninges (Rustenhoven and Kipnis, 2019). This "backdoor" can also be used for large molecules to access and attack otherwise difficult to reach glioblastoma cells. Such an effect could be accomplished by increasing the capacity of this drainage system by administration of the growth factor VEGF into the cerebrospinal fluid, which facilitates the entry of immunotherapeutic compounds to the brain and thus for T cells to attack a tumor (Song et al., 2020).

The brain's immunity

To a considerable extent, the brain has its own immune features, although it would be incorrect to paint a picture of this being precisely like that of the peripheral immune system. The brain essentially houses an innate immune apparatus, and this comprises astrocytes and microglia, both of which are part of the category of *glial* brain cells. Microglia, in particular, are the resident immune cells and can phagocytose plaques, apoptotic neurons that have been degenerating, and may be fundamental for the pruning of synapses that are not being used regularly. As depicted in Fig. 6.1, microglia serve multiple other functions within the brain. Among other things, and along with astrocytes, they are a critical reservoir of cytokines that can be central to neuroinflammatory processes, which could have downstream consequences for inflammatory-related diseases.

Stressors can provoke microglial activation and consequent cytokine release. This varied between males and females, and proinflammatory cytokines released by microglia were particularly marked within stress-susceptible mice and rats. At low concentrations, proinflammatory cytokines within the brain may have beneficial consequences by enhancing neuroplasticity, but at higher concentrations associated with stressors, they can be detrimental, provoking inflammatory-related diseases, and increasing pain sensitivity.

Like hormones and neurotransmitter processes, stressors could result in the sensitization of microglial processes so that they can be activated more readily by later stressor challenges, being especially notable among older animals (Niraula et al., 2017). With regard to cancer relevance, it is noteworthy that brain injury has been related to the later development of glioblastoma. It was suggested that some glioblastomas may begin to form with normal tissue repair that may be engendered by the injury. If healing is disturbed, for example by specific mutations, the glioblastoma cells continue to multiply since normal stop mechanisms are inoperative (Richards et al., 2021).

Brain and immune system interactions 181

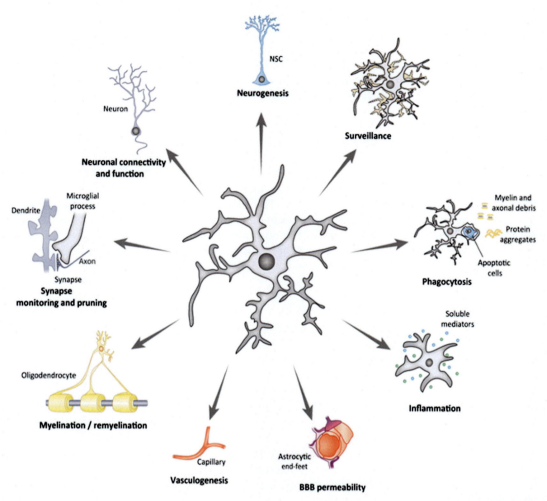

FIG. 6.1 In pathological conditions, microglia perform classical immune functions, such as the release of inflammatory mediators and phagocytosis of cellular debris (apoptotic cells, axonal, and myelin waste). In addition, microglia perform several physiological chores that are to some extent altered in pathological conditions, such as synapse monitoring and pruning. Microglia are capable of interacting with other brain cells, impacting their function: neurons and their connectivity; neural stem cells (NSCs) and neurogenesis; oligodendrocytes, and myelination/remyelination; endothelial cells and vasculogenesis/revascularization; and astrocytes and blood-brain barrier (BBB) permeability. *Figure and caption from Sierra, A., Paolicelli, R.C., Kettenmann, H., 2019. Cien anos de microglia: milestones in a century of microglial research. Trends Neurosci. 42, 778–792.*

Commensurate with the brain changes observed, sensitization effects can be instigated in response to either proinflammatory cytokine or stressor treatments so that an exaggerated response can be reinduced long after the initial challenge. These effects were evident in response to a subsequent systemic cytokine challenge so that a single exposure to either IL-1β or TNF-α sensitized stress circuitry such that augmented cortisol and sickness-like behavioral responses were apparent upon reexposure to these cytokines a month later. Conversely, a single stressor experience enhanced the later response to a cytokine challenge (i.e., cross-sensitization). It has been surmised that the sensitization of inflammatory processes might contribute to the development of depressive disorders, as well as to illnesses that emanate from depressive disorders (Anisman et al., 2018). Similar sensitization processes might similarly affect brain cytokine activity, thereby influencing neuronal functioning.

Stressor influences on immunity

Several components of the immune system are influenced by stressors and these actions are moderated by individual difference factors (e.g., sex, age), previous stressor experiences, as well as the characteristic of the stressors. The immune system is highly compartmentalized, involving distinct regulatory components that operate in circulating immune cells, local lymph nodes, and those present within the common mucosal system of the gut, lung, and urogenital tract, although the processes responsible for the impact of stressors at these different sites are not necessarily identical to one another.

Leukocyte changes elicited by stressors

Because of the importance of the first line of immune defenders, notably macrophages and dendritic cells, attention initially focused on the effects of stressors on these processes. These immune cells are drawn to sites of inflammation to act as an immediate neutralizing force before further adaptive responses are engaged. Thus, it is significant that macrophages and dendritic cell functioning were disturbed by stressors, as was the functioning of NK cells and the trafficking of immune cells to various tissues (e.g., spleen, lung, and brain). As monocytes are capable of trafficking to the brain parenchyma and perivascular spaces, they also contribute to altered psychological functioning.

These actions can vary with the features of the stressor, including its severity and chronicity, as well as the presence of anxiety. Acute stressors of moderate severity increased dendritic cell functioning, which contrasts with the effects of more intense stressors. Moreover, the stressor-provoked immune changes vary with individual difference factors, such as between stress-sensitive and stress-resilient mice. Following social defeat in mice, splenic dendritic cell reductions were more pronounced among submissive animals than in dominant mice. The submissive mice that were deemed more susceptible to illness, also exhibited elevated inflammatory neutrophils, monocytes, and spleen cells that released proinflammatory cytokines (Ambree et al., 2018). In addition to affecting macrophages and dendritic cells, acute social stressors influenced T cell presence, once again varying between submissive animals and those that were dominant in these situations. In response to chronic social defeat, susceptible mice also produced greater and more persistent proinflammatory changes (Audet et al., 2010) and a more pronounced decline of T cell populations than did resilient mice, which appeared to be mediated by microRNA variations (Pfau et al., 2019). Individual differences in response to stressors, including those related to social hierarchy, could have implications for stress-related pathological outcomes, including the progression of transplanted tumors.

Even subtle features of the stressor context, including the timing of its appearance, can have pronounced actions on immune responses triggered by infection. In particular, the humoral immune response may be diminished by a stressor applied 1 day after infection when the immune response was being mounted, whereas the effects on immunity were less likely to occur if the stressor was experienced a day earlier. In fact, an immunosuppressive response might not develop if the stressor was applied after memory cells had been formed (Kusnecov and Rabin, 1993). When we discuss the impact of stressors on cancer progression in animal models, it will be apparent that the timing of stressors can markedly affect these outcomes, and the effectiveness of cancer treatments can vary with timing and circadian rhythms.

Acute versus chronic challenges

Just as moderate and severe stressors differentially influence immune functioning, as described in Fig. 6.2, the enhanced immune activity stemming from acute, moderately intense stressors, is replaced by immunosuppression in response to chronic stressors (Dhabhar, 2014). The impacts of diverse stressors in this figure are generally accurate, but somewhat simplistic since the effects observed are moderated by a constellation of organismic factors (sex, age, strain of animal) and experiential factors (e.g., earlier stressful experiences). This said, it generally appeared that while mild and moderate acute stressors typically do not undermine immune functioning, chronic stressors could promote apoptotic death of macrophages, and in a mouse breast cancer model, a chronic social stressor suppressed CD8+ T cell proliferation and macrophage-derived IFN-γ (Muthuswamy et al., 2017) and typically disturbed NK cell functioning. A chronic stressor regimen likewise increased T_{reg} and B_{reg} cells, as well as tumor-associated macrophages, which could diminish the actions of CD8+ T (Antoni and Dhabhar, 2019). Beyond these peripheral actions, repeated social stressor encounters can promote increased monocyte trafficking to the brain, which can affect microglia, perhaps owing to the release of IL-6 or actions on GABA neurotransmission. It is obviously essential to distinguish mild or moderate (tolerable) stressors that can enhance resilience, from intense stressors that are typically immunosuppressive. At the same time,

	Acute moderate stressor	Acute strong stressor	Chronic stressor
Immune functioning NK & CD8 cells	↑	↓	↓↓
Inflammatory response	↑	↑↑	↑↑↑
Infection Wound healing	↑	↓	↓↓
Infection susceptibility	—	↑↑	↑↑↑
Cancer progression & metastasis	—	↔	↑↑↑

FIG. 6.2 General influence of diverse stressors. The actions of stressors vary with multiple other features of the stressor, as well as a constellation of genetic and epigenetic, age, sex, and earlier experience.

the impact of stressors might also vary with the immune compartment being examined (e.g., splenic versus circulating lymphocytes) and the specific immune factors evaluated (e.g., NK cells, T or B cell number) as well as their proliferation in response to a mitogen (i.e., agents that promote cell division).

The immune changes stemming from stressors are moderated by several biological actions that are elicited. Glucocorticoids released by stressors can limit immune functioning thereby preventing excessive immune activity. With repeated stressor experiences the immunosuppressive effects of cortisol are abated, possibly owing to epigenetic actions being instigated, which may be advantageous, as we saw concerning the potentially damaging effects of excessive activation of hippocampal glucocorticoid receptors leading to allostatic overload. But this can also influence the capacity to deal with immune-related disorders. Aside from the changes of cortisol, stressor-promoted epinephrine, dopamine, and prostaglandin changes could promote NK cell activity that may be beneficial in attacking viruses and cancer cells. Once again, with strong or chronic stressors, the influence of these hormones can diminish so that NK cells are less readily activated, thereby limiting the clearance of viruses and NK-sensitive neoplastic cells. In effect, if a stressor persists long enough, the actions of hormones on components of the immune system may diminish, thereby favoring potential adverse outcomes.

The immunosuppression that comes with chronic stressors might reflect the failure of an adaptive response being sustained so that susceptibility to illness is elevated and the progression of diseases, such as cancer, are exacerbated (Dhabhar, 2014). Yet, the position could be taken that the downregulated immune response associated with a chronic stressor reflects an adaptation to prevent excessive energy expenditure that might be needed by other defensive systems or to prevent the development of allostatic overload. The primary function (or consideration) of biological systems in the stress context is to assure immediate survival, even if there are downstream negative consequences, including the eventual development of diminished immune effectiveness that favors later cancer occurrence.

Stressor-provoked responses to infection in rodents

In considering the effects of stressors on immune functioning, a fundamental question that arises is to what extent immune system functioning needs to be reduced or elevated before this is clinically meaningful. The immune system may be exceptionally well adapted so that within a certain range, variations of immune activity may be of little consequence. As such, the salient question might not be whether stressors affect immune responses, as much as whether stressors can actually influence emerging illnesses, causing malaise to be longer lasting than ordinarily expected, or limiting the effectiveness of therapies.

In addition to the effects observed in response to a bacterial endotoxin, extended stressor exposure, even of moderate intensity, can disturb antiviral immunity. As we've already seen, however, the direction of the stressor's effect on immunity is complicated and not always easy to predict. While acute social defeat augmented immune functioning, and chronic stressors had the opposite action, chronic social defeat before viral infection also promoted elevated immunological memory to influenza virus. Further to this same paradoxical outcome, intense stressors may initially suppress lymphocyte proliferation but can have the opposite effect on immune cells that had previously been sensitized (e.g., to cholera toxin) (Kusnecov and Rabin, 1993).

Sex differences in stressor-provoked immune responses

Like other biological differences between males and females in response to stressors, immune functioning in rodents differed between the sexes. Generally, immune functioning was greater in females than in males, which could contribute to susceptibility, prevalence, and severity of illnesses (Klein and Flanagan, 2016). This was apparent in response to several pathogens, including viral and bacterial illnesses as well as parasitic and fungal infections. These differences may be related to sex hormones that can influence receptors situated on immune cells. As well, sex hormones may also be differentially effective in their interactions with a viruses' genome, as in the case of hepatitis B, and may also differ in their capacity to influence inflammatory damage, apoptosis, and oxidative stress.

The effects of stressors in rodents can operate through neuroinflammatory processes that differ between males and females both within the brain and the periphery, which may contribute to the female bias that appears with both physical and psychological disturbances (Martinez-Muniz and Wood, 2020). To a considerable degree, several hormones, including estrogens and androgens, moderate immune responses and the sex dimorphism that exists in response to glucocorticoids. Additionally, sex differences related to the microbiota-immune-brain axis may be integral to the impact of stressors in response to infection.

Stressor-elicited immune changes in humans

Understanding the impact of stressors in natural settings has been challenging given the diverse stressors that may be encountered, often differing in severity and chronicity, as well as the timing of these experiences relative to immune changes being assessed. Thus, it has been necessary to resort to the evaluation of stressors in laboratory contexts where challenges can be applied under strictly controlled conditions. Not unexpectedly, laboratory stressors were tied to neuroendocrine and immune responses that could affect health outcomes, varying with coping strategies endorsed as well as with the emotional and cognitive responses that were elicited. Generally, brief laboratory stressors, such as public speaking were associated with an increase of NK cell numbers and cytotoxicity but could have the opposite effect on other immune measures, such as T cell proliferation and antibody responses to influenza vaccines. However, at times the findings from these analyses differed from those seen in naturalistic contexts, which may speak to the limitations of some laboratory manipulations, especially as these don't allow for consideration of important elements of stressors, such as anticipatory distress and rumination, lifestyles that accompany many naturalistic stressors (e.g., altered diet, exercise, and disturbed sleep), and the fact that most serious challenges experienced by humans are chronic and frequently vary over time.[a]

As chronic diseases are more likely to occur in older individuals, it is meaningful that with aging, perceived stressors can promote especially marked immune disturbances. A prospective analysis of individuals 64–92 years of age had, in fact, indicated that stressors exacerbated immunological aging, thereby increasing disease susceptibility (Reed et al., 2019). In

---

[a] The "relevance" of research findings may be inversely related to the "rigor" of the study. This applies to the relevance of stressors in a laboratory context given that the procedures frequently don't map on well to stressors encountered in real-world situations.

older individuals, social isolation was accompanied by the increased presence of inflammatory markers, whereas engaging in positive social behaviors was associated with lower white blood cell counts and lower levels of the inflammatory marker C-reactive protein (Walker et al., 2019).

The profound stressor of caregiving, which often is left to older individuals (e.g., caring for a sick partner) was accompanied by many aspects of immune functioning being disturbed. This was reflected by poor responses to vaccines, accelerated cellular aging, reduced lymphocyte sensitivity to glucocorticoids, presence of inflammatory cytokines, disturbed NK cell functioning, slower recovery of immune dysfunction, and delays of restorative processes (Kiecolt-Glaser et al., 2003). The distress of caregiving has been tied to epigenetic changes of HPA axis regulation that affect immune-related genes that affect inflammation (Palma-Gudiel et al., 2021), which could affect disease occurrence. As observed with other stressor-provoked biological alterations, the impact of caregiving on immune and inflammatory processes can be exacerbated by other life experiences, such as the caregiver's early-life encounters with adversity.

As much as acting in a caregiving capacity can be exceptionally challenging, this is not the case among all individuals. It is important to consider the caregiver's suitability for this difficult role, and whether they feel they are gaining from this experience. Some individuals find comfort and meaning from caregiving, whereas others are entirely unsuited for this role and see it as a chore foisted upon them. These differences might have profound implications for the health of the caregiver and could also account for the diverse responses apparent across studies that have evaluated the impact of caregiving.

Stressor-provoked responses to infection in humans

The influence of chronic life stressors is apparent in diverse immune-related outcomes, ranging from wound healing through to the responses to viral and bacterial challenges, and the development of infectious diseases. Negative life experiences were reliably associated with the appearance of the common cold or influenza and in the responses to vaccination that ordinarily produces an immune response (Cohen et al., 2001). A prospective study conducted over 1 year indicated that stressful life events and perceived stress together with negative affect were related to the incidence of the common cold (Takkouche et al., 2001). The administration of low doses of cold virus that would ordinarily cause illness in about half of individuals produced more frequent colds among individuals with greater stressor experiences and these actions were linked to higher intranasal IL-6 and TNF-α levels, together with elevated glucocorticoid resistance (Cohen et al., 2012). Once more, such outcomes were more pronounced among individuals with greater negative affect and the absence of social support.

The immune response to vaccination using dead or inactivated viruses has served as a proxy for infectious agents and has been instructive in linking adverse life experiences and potential susceptibility to illness. Elevated daily life stressors were associated with lower antibody titers in response to influenza vaccination among individuals who had encountered stressors soon after vaccination when an immune response was being mounted. Diminished antibody titers also accompanied chronic caregiving and similar changes were associated with the distress of loneliness and having a limited social network. Furthermore, while Herpes Simplex and Epstein Barr Virus, which reside in most people, are ordinarily kept in check by cell-mediated immune responses, following intense life stressors, this restraint is undermined so that antibody titers for such viruses are altered (Glaser and Kiecolt-Glaser, 2005).

To an extent, conclusions derived from such studies may be constrained by the need for a very large number of participants to appreciate the influence of the intensity, chronicity, and the timing of individuals' stressor experiences as well as characteristics of the individuals being assessed (e.g., age, sex/gender, lifetime stressor encounters). Nonetheless, the data from animal and human studies have made it clear that immune changes are provoked by life stressors and might thus be causally linked to the emergence of viral illnesses.

Cytokine variations associated with stressors

Stressors promote elevations of circulating proinflammatory cytokines such as TNF-α, IL-6, and IFN-γ, which have been implicated in the promotion of inflammatory diseases as well as psychological disturbances, such as depression. The IL-6 alterations seemed to be tied to the behavioral vulnerability elicited by an uncontrollable stressor and like the effects on other aspects of immune and hormone functioning, the effects on IL-6 were especially pronounced in response to a chronic series of different stressors (Gibb et al., 2013). Moreover, social disruption in mice was accompanied by the elevated entry of peripheral IL-6 into the brain (Menard et al., 2017). In addition to the immediate effects of stressors, both pro- and antiinflammatory cytokine variations were subject to sensitization so that more rapid cytokine variations were provoked upon later reexposure to the stressor.

Like the effects on other cytokines, stressors can increase levels of IL-18, which is produced by both immune cells and brain microglia and could thereby influence illness occurrence. The actions of stressors promoted by IL-18 may come about through other cytokines, such as IL-6, and by activation of the NLRP3 inflammasome that is instigated by a chronic stressor, giving rise to hippocampal neuroinflammation and the appearance of depressive-like behaviors. Effects such as these have also been observed in the offspring of rats that had been stressed during pregnancy and were linked to the later development of anxiety- and depression-like behaviors.

The influence of stressors on members of the IL-17 family of cytokines has only recently garnered attention. Uncontrollable stressors and a chronic mild stressor regimen increased this cytokine in several depression-related brain regions of mice and were elevated in the serum of depressed patients. These actions may be especially pertinent as IL-17 influences the actions of myeloid cells and T cells and was associated with cancer progression (Chang, 2019). Early in cancer development, IL-17 may favor tumor growth, whereas later in tumor development, IL-17 may have a suppressive effect. Despite the biphasic actions of IL-17, targeting this cytokine might be useful in cancer treatment.

To summarize, the mood changes associated with stressors have typically been attributed to brain neurotransmitter and peptide functioning and could potentially have downstream peripheral consequences. These actions can also affect the immune activity and inflammatory cytokine release that can affect brain processes, thereby promoting emotional responses. Concurrently, by affecting inflammatory processes, peripheral cytokine variations can also affect diverse organs, increasing vulnerability to physical illnesses, thus accounting for the frequent comorbidities that occur between psychological and physical disorders. This perspective, sometimes referred to as the common soil hypothesis, assumes that the same or similar processes underly several illnesses, but the role of inflammation has taken on increasing allure in view of the broad actions observed in relation to multiple chronic diseases.

Concurrent actions of pro- and antiinflammatory cytokines

The temporal trajectories of proinflammatory and antiinflammatory may differ over time so that a proinflammatory profile may predominate at one phase following a stressor, but an antiinflammatory bias may exist at a later phase. These balances may differ between stress-reactive and nonreactive strains of mice, and with chronicity of a stressor. It also seems that the cytokine profiles apparent in blood weren't necessarily recapitulated in the brain, and could differ yet again in other tissues, which might have implications for diverse pathological conditions.

In the main, balances are maintained between the pro- and antiinflammatory cytokines. However, the equilibrium is not only affected by chronic stressors but also by the animal's previous stressor encounters. The priming of the neuroinflammatory process may come about secondary to the sensitization of neurochemical systems or a signal cascade comprising DAMPs and variations of the NLR inflammasome (Anisman et al., 2018). Since NLRs can detect a broad range of biological and physical stimuli, they may serve as the interface between stressors and the development of inflammation.

Stressor-elicited cytokine variations in humans

Negative experiences encountered by humans are accompanied by multiple proinflammatory cytokine changes in blood (e.g., IL-6 and TNF-α) and by elevations of the antiinflammatory cytokine IL-10, thereby maintaining a balance between the pro- and antiinflammatory influences. Whether or not cytokine changes that occurred was related to cognitive and emotional responses elicited by a laboratory stressor as well as the significance of this stressor to participants' lives. Specifically, emotion-inducing manipulations provoked an increase of plasma IL-1β, IL-6, and IL-8 among individuals who lacked adequate cognitive control. The link between stressor-elicited emotions and IL-6 levels similarly occurred in response to the distress of public speaking. While anger related to the stressor was associated with increased cortisol levels, this was not apparent for IL-6, whereas fear and anxiety were related to IL-6 levels (Moons and Shields, 2015). Following exposure to a scenario depicting abuse, plasma IL-6 was likewise elevated among women, particularly if they reported high levels of anger or sadness, but not that of shame and anxiety. However, if these women had previously been in an abusive relationship, then IL-6 levels also increased in association with self-reported shame and anxiety (Danielson et al., 2011).

In addition to laboratory-based experiments, cytokine changes have been evaluated under naturally occurring stressful experiences. The anticipation of academic examinations was associated with elevated IL-6 and IL-10, while IFN-γ was reduced and the production of the antiinflammatory cytokine IL-4 (in response to a mitogen) was diminished by both academic examinations and by psychosocial stressors. Once again, as observed in a laboratory context, the distress of academic stressors was tied to the emotional responses elicited.

Elevated IL-1β and IL-18 were most prominent among individuals who expressed anger, whereas IL-6 was most closely related to anxiety (La Fratta et al., 2018). These cytokine changes stand in contrast to the absence of cortisol variations associated with academic examinations. Why the differential corticoid and cytokine responses occur is uncertain, but it may have to do with the very basic nature of the immune system's capacity to act defensively in anticipation of a threatening event, whereas glucocorticoid variations are more readily promoted once a stressor is encountered.

The very intense and protracted distress that occurs among parents of young cancer patients was accompanied by elevated IL-6 production owing to cortisol losing its capacity to suppress inflammatory processes. Likewise, the distress that can be created by caregiving for ill or elderly individuals, as we discussed earlier, may also be especially taxing on immune processes, especially among older individuals. Caregiving for a patient with cancer was predictably associated with altered cytokine levels among those caregivers showing the greatest anxiety. Once again, the ties between stressors and cytokine levels were moderated by personality factors (e.g., optimism or self-esteem) and the individual's mood state. In this respect, the distress associated with caregiving was most closely associated with elevated IL-6 among individuals with low self-esteem or low self-efficacy.

Spousal bereavement was associated with elevated circulating proinflammatory cytokine levels, being directly related to the extent of the grief displayed. Ordinarily, the influence of powerful stressors can be moderated by how individuals cope, which can influence grief responses and the actions of inflammatory variations that could potentially contribute to the development of grief-related diseases. Indeed, elevated proinflammatory cytokine levels in bereaved individuals were most pronounced among individuals who adopted emotional regulation as a way of coping (Lopez et al., 2020) and the cytokine profile exhibited 3 months following spousal loss predicted the continuation of depressive symptoms 6 months after the loss (Wu et al., 2021).

The data from human studies were not entirely congruent with the data from animal-based research, possibly because human studies were limited to immune analyses in peripheral blood. Moreover, human studies typically did not control for many variables (e.g., genetic factors, and stressor history) that influence stressor-promoted immune responses. If nothing else, however, the accumulated findings point to the complexity of understanding the actions of stressors on cytokine expression and underscore the importance of considering the contribution of stressor-provoked emotional responses in determining cytokine variations. As described in Fig. 6.3, numerous other variables come into play in accounting for the immune and cytokine variations associated with stressful experiences, which need to be considered to fully appreciate how stressors may come to affect disease processes.

Impact of early-life experiences

As we've seen, psychological and systemic insults encountered during early-life predicted lasting vulnerability to a variety of illnesses, likely operating through several distinct processes, including actions on immune and inflammatory functioning. The persistent consequences of stressors may come about due to changes in the organism's developmental trajectory comprising diverse biological alterations that evolve progressively. These outcomes could also occur owing to the sensitization (priming) of neuronal or immune-related

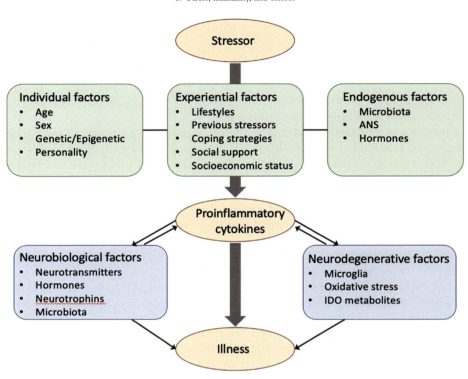

FIG. 6.3 A schematic representation of the relations between stressors, proinflammatory cytokines, and illness, with a particular focus on the individual, experiential, and endogenous factors that moderate these effects. The capacity of social stressors to promote inflammatory variations that might lead to depression (or other illnesses) may be influenced by the presence of genetic and personality factors, sex, and age. Individuals carrying specific gene combinations or polymorphisms (e.g., variants of IL-6, IL-1β, TNF-α) may be more vulnerable to the depressive effects of inflammatory activation associated with social stressors. Moreover, earlier stressor encounters both prenatally and during early life, and coping processes endorsed, can influence inflammatory processes and sensitize immune responses to subsequent stressors. The activation of proinflammatory processes may directly or indirectly influence depressive states. Elevations of cytokines may influence monoamine (e.g., 5-HT, NE), hormone (e.g., CRH), and growth factor (e.g., BDNF) activity which might favor the evolution of depression. As well, stressors can affect microglia and promote oxidative stress, as well as promoting the accumulation of neurotoxic metabolites of indoleamine 2,3-dioxygenase (IDO), culminating in neuropsychiatric illnesses.

processes, and thus would primarily be apparent when stressors were again encountered at some later time. Irrespective of the mechanisms involved, it seemed that the impact of early stressors could be attenuated by the adoption of effective coping methods, including the availability of social support.

Paralleling the immune cell changes, variations of cytokine gene expression or proteins in rodents are brought about by early-life challenges, such as separating pups from their mom. This procedure also promoted later depressive-like features along with microglial activation and elevated proinflammatory cytokine expression within the hippocampus, which could be attenuated or prevented by the antibiotic, minocycline. Much like the effects of psychological stressors, rodents treated with an immune challenge during early life, later displayed more pronounced suppression of lymphocyte proliferation and NK cell functioning, together with increased central cytokine responses upon stressor exposure. Consistent with a role for cytokines in eliciting the anxiety and depressive profile, antidepressant treatment during adolescence reduced the behavioral disturbances brought about by early-life stressors, while concurrently diminishing IL-1β levels and increasing that of the antiinflammatory cytokine IL-10 (Wang et al., 2017).

As in animals, adverse childhood experiences comprising family disturbances were accompanied by disturbed immune functioning and cytokine levels, along with indications that autoimmune response may be increased. Likewise, stressors encountered during early-life (e.g., abuse) favored inflammation, characterized by elevated circulating proinflammatory cytokines and acute-phase proteins (C-reactive protein, fibrinogen), although the effect sizes were relatively small, varying somewhat with the nature of the abuse experienced (Baumeister et al., 2016). For instance, elevated in vitro production of IL-6 from stimulated monocytes was seen in adolescent girls who had experienced early-life adverse events, possibly owing to reduced glucocorticoid sensitivity to challenges (Ehrlich et al., 2016). Older adults who reported childhood adverse events were also likely to present elevated levels of C-reactive protein and IL-6, especially if the early-life stressor comprised sexual abuse. Childhood adversity in the form of abuse, neglect, and a chaotic home environment, was similarly linked to elevated IL-6 among breast cancer survivors, potentially placing them at elevated risk for further inflammatory-related disorders (Crosswell et al., 2014). Commensurate with the proactive actions of distressing early-life experiences, the influence of early-life trauma on IL-6 among older adults was most notable among individuals who had experienced a further stressor in the days prior to testing. In effect, on a day-to-day basis, the effects of early-life trauma might only be marginal but would emerge more fully in response to further challenges.

Stress and microbiota

The well-documented relation between microbiota and other biological operations, particularly immune functioning, coupled with reports that a wide range of environmental challenges can introduce microbial changes, has invigorated efforts to determine their role in mediating numerous diseases and the contribution of stressors in this regard. The impact of psychological and physical insults on microbiota have included a wide range of challenges that comprised environmental extremes (high altitude, heat, and cold), disruption of sleep or circadian cycles, toxicants, pollutants, noise, varied pharmacological agents, and psychological stressors. These challenges can affect the gut microbiota through their hormonal or inflammatory actions, as well as by the oxidative stress produced, culminating in the development and progression of a variety of stress-related disorders. As we'll see, stress hormones, such as epinephrine, norepinephrine, and dopamine, directly influence some forms of bacteria and can affect bacterial growth and virulence-related factors that can affect the outcomes of infections.

Stressors of a purely psychological nature and strong naturalistic stressors, such as social defeat, produced varied changes in the diversity of microbiota and the genes regulating them (Galley and Bailey, 2014). This included a diminished abundance of genes involved in the biosynthesis and metabolism of short chain fatty acids (SCFAs), as well as the serotonin and norepinephrine precursors, tryptophan and tyrosine, respectively. Stressful events likewise influenced the translocation of microbiota from cutaneous and mucosal surfaces into regional lymph nodes, thereby producing neuroendocrine and immune alterations. Social stressors were also capable of promoting the translocation of Gram-positive bacteria from the gut into the circulation, where they could augment the production of IL-1 by spleen cells.

Stress, psychological alterations, and microbiota

The microbial diversity that ordinarily exists in healthy individuals can be disturbed by stressor encounters so that an unstable and potentially harmful microbial community may emerge. Gut bacteria influence HPA activity and might thus contribute to stress-related conditions. In germ-free mice, basal corticosterone levels were elevated, which could be further exaggerated by stressors, and could be prevented by microbial reconstitution with *Bifidobacteria* species. Likewise, the greater neuroendocrine response in relatively young germ-free mice was partially reversed by reconstitution with feces from pathogen-free mice (i.e., mice free of specific pathogens, but that were not germ-free). As this treatment did not produce these changes when the germ-free mice were somewhat older suggests that microbe exposure at an early developmental stage was essential for the regulation of HPA functioning.

Attesting to the significance of microbiota to health disturbances, the transfer of microbiota from stress-sensitive rats (those that were most affected in a social defeat paradigm) to otherwise naïve rats, resulted in the appearance of depressive symptoms accompanied by elevated microglial density as well as IL-1β presence in the ventral hippocampus (Pearson-Leary et al., 2020). Conversely, a prebiotic treatment that increased gut SCFAs diminished signs of anxiety and depression provoked by a chronic stressor regimen. There has also been evidence, albeit limited, that probiotics can reverse the emotion-related brain changes stemming from the prolonged separation of rat pups from their mom.

It is generally thought that balances in the abundance of a diverse constellation of microbiota are linked to beneficial or harmful actions related to health. However, specific bacteria have also been implicated in promoting these outcomes, together with increased inflammation and hormonal changes. Several reports pointed to a reduction of *Lactobacillus* being provoked by social defeat in mice, which could elicit pronounced effects on immune functioning (Gur and Bailey, 2016). These ordinarily beneficial bacteria are exquisitely sensitive to stressors, and as little as 2h of social disruption altered their abundance, and more pronounced disturbances were provoked when this procedure was repeated over 6 days. Predictably, stressors other than social disruption (e.g., physical restraint) also elicited gut microbial alterations that were particularly marked in response to chronic unpredictable and diverse stressors applied over 3–5 weeks.

Aligning with the causal role of microbiota in determining emotional responses, treatment with *Lactobacillus* over several weeks attenuated stressor-provoked anxiety and social withdrawal, coupled with the activation of dendritic cells and elevated regulatory T cells. Similarly, behavioral disturbances provoked by the stressor could be attenuated by the restoration of *Lactobacillus* while concurrently altering stressor-provoked metabolic alteration. In addition, a moderately severe stressor (physical restraint) diminished the presence of gut *Bifidobacterium*, which is also considered to be a beneficial bacterium. When germ-free mice were colonized with the microbiota of *Bifidobacterium*-depleted mice, a proinflammatory immune response was elicited and colonic pathology was exacerbated in response to a pathogen (Galley et al., 2017).

Stressor effects on microbial functioning may come about through effects on sympathetic nervous system functioning. As well, signaling associated with microbial changes may occur through variations within the enteric nervous system and activation of the vagus nerve (Fülling et al., 2019). Whatever the case, stressor-provoked microbiota changes were associated with increased T_h cells within mesenteric lymph nodes, which could be attenuated by the inhibition of bacterial gene regulation, which is known to be sensitive to norepinephrine (Werbner et al., 2019).

Consistent with the effects apparent in rodents, stressors can influence microbiota in humans. Specifically, gut microbiota diversity and richness were diminished in association with marital strife as were indices of leaky gut syndrome and inflammation (Kiecolt-Glaser et al., 2021). Microbiota alterations were likewise apparent in university students undergoing academic exams, varying with participant's perceived distress. Strong or chronic stressors that produce PTSD were similarly accompanied by reduced abundance of several microbiota phyla (Hemmings et al., 2017). It had been considered that microbial processes may promote the emergence and maintenance of PTSD and, conversely, probiotic or prebiotic intake, alone or in combination with more usual treatments, might attenuate PTSD symptoms, but data supporting this position are lacking.

Studies in humans indicated that stressor and gut microbiota manipulations can affect neuroendocrine functioning and brain activity and may thereby contribute to anxiety and depression (Cryan et al., 2019). For instance, in response to a laboratory stressor in which participants were made to feel socially excluded, neuronal activity changes occurred within the prefrontal cortex and left anterior cingulate, which could be diminished by a diet supplemented with *Bifidobacterium longum* and exacerbated by an antibiotic. As these brain regions contribute to executive functioning and mood states, these findings support the contention that gut microbiota may contribute to the actions of stressors on processes related to psychological disorders. In line with this view, a prebiotic diet maintained for several weeks could attenuate the morning cortisol response that is ordinarily elevated among individuals experiencing stressful life events. However, efforts to diminish anxiety through microbiota manipulations generally yielded inconsistent outcomes. A metaanalysis of 21 relevant studies revealed that only 11 reports indicated positive effects of microbial modifications (Yang et al., 2019b). Despite such variable results, it is conceivable that a probiotic diet based on individual microbiota features, together with standard antianxiety treatments would yield better outcomes.

Early-life experiences influence adolescent and adult microbiota

During the early dynamic developmental period, the microbial community may be unstable and susceptible to numerous experiential and environmental challenges. Stressors encountered during this fragile period markedly affect the composition of gut bacteria and can affect later responses to stressors (De Palma et al., 2015). A strong stressor comprising separation of pups from their mom, which has been aligned with variations of stress hormones, elicited marked gut bacterial alterations, and even seemingly modest challenges, such as disturbing a rodent's nest, may provoke erratic maternal behaviors together with a reduction of microbial diversity in her pups.

The repercussions on adult microbiota were especially marked in animals that had experienced multiple stressors early in life (Rincel et al., 2019), which was associated with behavioral perturbations and altered brain serotonin activity that varied with sex. Significantly, the protracted effect of early-life stressors on the behavioral alterations and exaggerated hormonal response to later challenges were precluded if young animals were treated with a dietary supplement that comprised milk fat globule membrane and a particular prebiotic blend (O'Mahony et al., 2020). Supplementation with *Bifidobacterium pseudocatenulatum* CECT7765 similarly reversed the neuroendocrine and immune disturbances and the elevated anxiety that stemmed from early-life maternal separation.

Along with altered HPA functioning, separating germ-free mice from their mom for 3 h disturbed colonic cholinergic neural regulation and engendered marked bacterial dysbiosis. Upon subsequent bacterial recolonization, the microbial profile was altered, and anxiety that was otherwise present was diminished (De Palma et al., 2015). Aside from these actions, prebiotic supplementation with galacto-oligosaccharides (plant sugars that are naturally found in milk and beans) during the neonatal period influenced levels of the neurotrophin BDNF in young adulthood. Micronutrient supplements provided early in life similarly limited later stressor-provoked corticosterone changes and cognitive disturbances. The varied endocrine and mood changes associated with early-life stressors and the impact of microbiota manipulations, as we'll see shortly, have pronounced implications for the later development of somatic diseases, including cancer.

Microbiota in centenarians

There has long been interest in determining the factors that allow some individuals to have very long, healthy lives. Philosophers had opined on the possible existence of a fountain of youth, artists imagined and painted the scene, and a few explorers searched for its physical location. The media regularly report on why centenarians believe that they've lived so long. Several attributed their long life to their daily consumption of a glass of red wine, whereas others indicated that their healthy lifestyles and maintaining a positive perspective were responsible.

Scientifically based studies have attributed longevity to a wide variety of biological factors. As females across species live longer than males it was frequently maintained that longevity was in some fashion related to having two X chromosomes. Longevity was similarly attributed to the presence of adequate levels of sirtuins (a family of proteins that contributes to cellular aging), having high levels of human growth hormone, or some other constituent present in blood, perhaps the protein GDF11 (which encouraged "vampire therapy" that comprised transfusion of young blood to older people), and the sequelae of exercise and diet. Among supercentenarians (individuals living to at least 110), the levels of circulating cytotoxic T and CD4+ T cells were elevated, raising the possibility of immune factors being fundamental for long life. In fact, an attempt was made based on the presence of inflammatory indices (e.g., the levels of the chemokine CXCL9) to decipher biological aging that could potentially be used to predict the occurrence of inflammatory illnesses that contribute to earlier death. A structural protein of nerve cells, Nfl, has similarly been implicated in longevity. Nfl, which is elevated in neurological disorders, was found to be reduced in centenarians and those approaching this milestone. It was similarly observed that Nfl increases with aging in mice, which can be reduced through dietary restriction that favors longevity. Whether Nfl is responsible for longevity is uncertain, but it does appear to be a good biomarker in predicting healthy aging (Kaeser et al., 2021).

Because of the health benefits of specific microbiota, it wouldn't be unreasonable to suspect that these bacterial species might also directly or indirectly contribute to longevity. Throughout normal aging, the abundance of some bacterial species diminishes, whereas others become more common; those bacteria that had been subordinate replaced the initially dominant species. With aging, it is not uncommon for gut dysbiosis to occur, especially in response to stressors, leading to intestines becoming leaky, so that released bacterial products promote inflammation. Broadly speaking, aging may be accompanied by a microbial shift that favors a proinflammatory profile that is characteristic of that seen in inflammatory diseases, which may contribute to affective and neurocognitive disturbances. Longitudinal analyses revealed that the abundance of core bacterial genera declined with age (e.g., Bacteroides) and the profile of bacteria became increasingly unique across individuals. These microbial changes gave rise to metabolomic signatures that were linked to lower inflammation that predicted longevity (Wilmanski et al., 2021). It may be telling that in centenarians who were in good health, their microbiota profile was reminiscent of that characteristic of healthy young people.

Like the vulnerability of early life, the effects of stressors on microbiota can be particularly profound among aging rodents and humans. A chronic stressor provoked microbiota changes, and these effects were especially pronounced among aged animals, as was inflammatory dysregulation. The greater vulnerability to inflammation of aged animals was also apparent in young rodents that received gut bacteria transplants from old mice, as was elevated leakage of inflammatory bacterial factors into circulation. These findings are congruent with the position that gut microorganisms play a fundamental role in stress-related disturbances and point to microbiota as being a feature of vulnerabilities related to aging.

Stressful events and cancer

The notion has been tossed around for decades that cancer might develop because of stressful experiences, but this view had not been strongly supported. In contrast, considerable evidence has supported the position that stressful events could exacerbate the progression of an already existing cancer, increase the likelihood of metastasis, and negatively influence the effectiveness of cancer treatments. In rodents, intense acute or chronic stressful experiences could worsen the growth of induced, transplanted, and carcinogen-elicited tumors, potentially operating through effects on sympathetic nervous system activity and exacerbation of inflammatory processes. The types of cancer exacerbated by stressors involved a wide variety of different organs, including breast, ovaries, prostate, pancreas, and could promote brain and lymphatic metastasis.

Immune processes linking stressful events to cancer

The gamut of stress-related processes that could potentially influence tumor progression is exceptionally large, including changes in hormones or growth factors, heat-shock proteins, microbiota changes, altered angiogenesis, and disturbed DNA repair. Understandably, altered immune functioning and elevated inflammation induced by chronic stressors have often taken center stage concerning tumor progression. Chronic stressors disturbed T cell functioning while increasing the growth of transplanted syngeneic tumor cells (i.e., those identical to the host) and fostered metastasis (Zhang et al., 2020). Among mice that had been subjected to a chronic social stressor, the growth of transplanted fibrosarcoma cells was linked to increased actions of myeloid-derived suppressor cells and T_{reg} cells that otherwise limit cancer growth.

Not unexpectedly, the tumorigenic actions of stressors, like the effects observed in immune and cytokine variations, may depend on the features of the individual. Relative to low anxious people, those who displayed high trait anxiety exhibited greater corticosterone levels, a higher tumor burden, greater tumor-recruited immunosuppression, and elevated levels of vascular endothelial growth factor (VEGF), which may be related to angiogenesis, invasion, and metastases (Dhabhar et al., 2012).

As much as immune changes may contribute to the effects of chronic stressors on cancer progression, many other factors have been implicated in these actions, some of which are depicted in Fig. 6.4. Increasing evidence implicated alterations of the tumor suppressor p53 in the tumorigenic actions of stressors. Moreover, immune modulation of Toll-like receptor 4 (TLR-4) mediation of PI3K/Akt signaling that is involved in regulating the cell cycle can also be modified by stressor experiences and might thereby affect tumor progression. Relatedly, stressor-activated enzymes, such as Nqo1 that diminishes free radicals ought to have beneficial actions, but they could also have harmful effects. For instance, in the presence of liver cancer, this enzyme could activate PI3K/Akt and MAPK/ERK pathways, thereby influencing metabolic processes so that cancer cells can make optimal use of glucose to support their growth. There is also evidence supporting a role for autophagy in immunosuppression promoted by chronic stressors. Beyond these actions, the expression of an appreciable number of cancer-related genes may be altered by chronic stressors, including those that act as transcription factors, growth factor receptors, and extracellular matrix components. In essence, although various hormones and immune processes may be pivotal in accounting for the effects on tumor processes, there is no shortage of other factors that can act in this capacity.

Stress, hormones, and cancer progression

Involvement of norepinephrine

A view has evolved that tumor growth and metastasis could be influenced by circulating norepinephrine owing to an exceptionally broad number of processes affected by this hormone. As indicated in Chapter 5 and described in Fig. 6.5, this included altered cell trafficking, and markedly reduced NK cell activity and macrophage functioning. As well, circulating norepinephrine can influence growth factors, promote apoptosis, and may serve in the provocation of angiogenic processes that are instrumental in providing an energy source for cancer cells (Servick, 2019).

As expected, β-norepinephrine antagonists (that block the β-adrenergic receptor stimulated by norepinephrine, and which are also referred to as β-blockers) could act against tumor growth and could attenuate the effects of stressors. This was demonstrated with stressor-provoked exacerbation of acute lymphoblastic leukemia, pancreatic cancer, and colorectal carcinoma, as well as breast cancer metastasis (e.g., Kim-Fuchs et al., 2014). These positive actions could be further enhanced by cyclooxygenase 2 inhibition, particularly during or shortly after surgery (Ricon et al., 2019). As well, a chronic stressor promoted liver metastasis, which could be attenuated by a β-norepinephrine antagonist (Zhao et al., 2015).

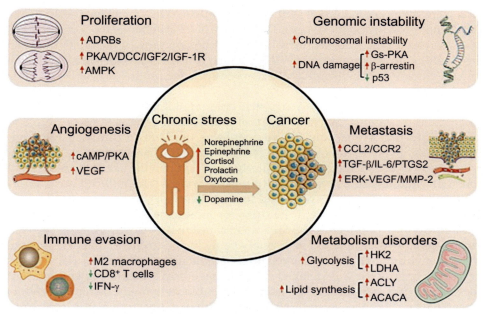

FIG. 6.4 Biological mechanisms underlying chronic stress-induced cancer progression. Chronic stress activates ADRBs, the PKA/VDCC/IGF axis and AMPK signaling pathways to promote cancer cell proliferation. Chronic stress leads to cancer chromosomal instability and DNA damage through activation of Gs-PKA/β-arrestin and suppression of p53. Furthermore, chronic stress promotes angiogenesis through activation of the cAMP/PKA signaling pathway, elevating VEGF expression. Chronic stress promotes cancer metastasis by increasing expression of TGF-β, IL-6, and PTGS2, activating the CCL2/CCR2 axis and triggering the ERK-VEGF/MMP2 pathway. In addition, chronic stress causes immune evasion by increasing the number of M2 macrophages and decreasing CD8+ T cells and IFN-γ release. Chronic stress also promotes tumor progression by inducing metabolic disorders such as ACACA- and ACLY-mediated lipid synthesis and HK2- and LDHA-regulated glycolysis. Abbreviation. *ACACA*, acetyl-CoA carboxylase alpha; *ACLY*, ATP citrate lyase; *ADRBs*, β-adrenergic receptors; *AMPK*, adenosine monophosphate-activated protein kinase; *cAMP*, cyclic adenosine monophosphate; *CCL2*, C-C motif chemokine ligand 2; *CCLR2*, C-C motif chemokine receptor 2; *ERK*, extracellular regulated MAP kinase; *Gs*, guanyl nucleotide regulatory proteins; *HK2*, hexokinase 2; *IFN-γ*, interferon-γ; *IGF-1R*, insulin-like growth factor 1 receptor; *IGF2*, insulin-like growth factor 2; *IL-6*, interleukin 6; *LDHA*, lactate dehydrogenase A; *MMP2*, matrix metalloproteinase 2; *PKA*, protein kinase A; *PTGS2*, prostaglandin-endoperoxide synthase 2; *TGF-β*, transforming growth factor-β; *VDCC*, voltage-dependent calcium channels; *VEGF*, vascular endothelial growth factor. *Figure and caption from Cui, B., Peng, F., Lu, J., He, B., Su, Q., et al., 2021. Cancer and stress: NextGen strategies. Brain Behav. Immun. 93, 368–383.*

Stressors and the ensuing norepinephrine variations not only affect tumor progression but also affect the response to treatments that diminish cancer growth. For instance, chronic stressor exposure attenuated the antiangiogenic actions of drugs that have been used in treating colorectal cancer (Liu et al., 2015). The effectiveness of the antiandrogen compound bicalutamide in reducing prostate tumors was likewise diminished in mice that had been exposed to a chronic stressor regimen.

Less attention has been devoted to the actions of α2-adrenergic blockers, although this receptor subtype, along with β2 receptor stimulation has been predictive of increased tumor size and metastatic relapse. Through a negative feedback mechanism, α2 presynaptic receptor blockade increases norepinephrine synthesis and release. Thus, blocking these receptors may promote tumor growth and metastasis. The actions of α2-adrenergic blockers might also arise owing to increased angiogenesis provoked by actions on growth factors (VEGF and FGF2) and the resulting downregulation of PPARγ, which is involved in the regulation of fatty acid storage and glucose metabolism that ultimately affects the progression of breast and pancreatic cancer (Zhou et al., 2020).

Consistent with the findings in rodents, epidemiological studies had suggested that the progression of various cancers was promoted through the actions of norepinephrine. Individuals who had been using β-norepinephrine blockers to diminish hypertension or anxiety were less likely to develop some forms of cancer (Krizanova et al., 2016). A 12-year follow-up study of 24,238 individuals similarly indicated that the use of the β-norepinephrine antagonist propranolol for at least 6 months was associated with reduced occurrence of head and neck, esophageal, stomach, colon, and prostate cancers, and this was particularly prominent among individuals who used the drug for more than 1000 days (Chang et al., 2015). Progression-free and long-term survival among cancer patients was likewise correlated with the use of β-norepinephrine blockers, possibly being linked to effects that occurred within the tumor microenvironment. This treatment inhibited the progression of ovarian cancer and the time to cancer recurrence was delayed in early-stage breast cancer patients. Accumulating evidence had, in fact, supported the position that β-blockers might be repurposed for use in triple-negative breast cancer, even though the quality of many studies was deemed to be low (Spini et al., 2019).

In addition to affecting cancer occurrence, β-norepinephrine antagonists could influence the effects of different cancer therapies. Survival among patients with nonsmall-cell lung cancer treated with radiation was improved in those who were taking β-blockers and could influence the effectiveness of immunotherapy. The data from human studies were largely observational hence precluding conclusions regarding potential causal connections to cancer occurrence and progression. Nonetheless, collectively, the available data were consistent with the view that through their actions on immune system functioning, as well as changes in tumor biology itself, norepinephrine could affect several stages in the growth of existent tumors. At the same time, it might be considered that the beneficial effects β-norepinephrine antagonists might, in part, occur through its anxiety-reducing actions, which may also diminish poor lifestyle choices and enhance treatment compliance.

Cortisol

Acute stressors ordinarily cause brief activation of inflammatory immune processes, presumably acting in a protective capacity. At the same time, elevated levels of cortisol ordinarily elicited by acute stressors can regulate inflammation by blunting signal transduction that occurs downstream of cytokine receptors and pattern recognition receptors (PRRs) and through inhibition of lymphocyte and macrophage functioning (Cain and Cidlowski, 2017). However, with a chronic stressor, cortisol receptor sensitivity is diminished, causing dysregulation of inflammatory responses so that persistent activation of inflammatory cytokines and downstream signaling pathways (e.g., MAPK, STAT, and NFκB) may provoke increased tumor progression. As well, during tumor growth, glucocorticoids can stimulate receptors at distal sites, which favor tumor cell colonization and metastasis (Obradović et al., 2019).

How glucocorticoids come to affect tumor progression may differ across different types of tumors. For instance, the elevated glucocorticoid levels stemming from social defeat may

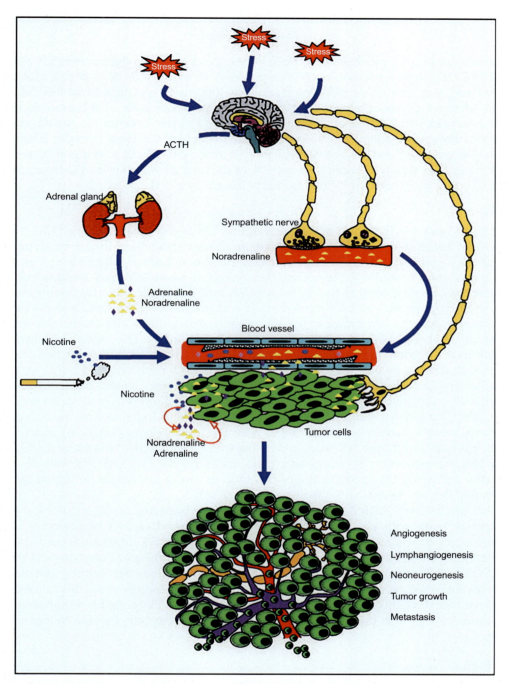

FIG. 6.5 Stress hormones in cancer development and progression. A variety of physical and lifestyle stress induces the elevation of stress hormones in the body which are mainly released from the adrenal gland and sympathetic nervous system. The adrenal gland is stimulated by adrenocorticotrophic hormone (ACTH) resulting from the activation of the pituitary byhypothalamic activation by stressors. Nicotine from cigarette smoking is also able to stimulate tumor cells to directly synthesize and release stress hormones to form an autocrine loop. Additionally, tumor cells might be innervated by nerve fibers which could also release relevant neurotransmitters, notably norepinephrine. Finally, stress hormones originating from different systems contribute to tumor development and progression such as angiogenesis, lymphangiogenesis, neurogenesis, tumor growth, and metastasis. *Figure and caption from Tang, J., Li, Z., Lu, L., Cho, C.H., 2013. Beta-adrenergic system, a backstage manipulator regulating tumour progression and drug target in cancer therapy. Semin. Cancer Biol. 23, 533–542.*

foster increased tumor progression by augmenting the expression of glucocorticoid-inducible factor TSC22D3, which antagonizes IFN responses of dendritic cells and activation of IFN-γ+ T cells (Yang et al., 2019c). Cortisol could similarly diminish the expression of BRCA1, which would otherwise act to suppress breast and ovarian cancer. As with most hormones, cortisol does not act independently of other actions elicited by aversive experiences, including effects on other hormones. As depicted in Fig. 6.6, stressors cause activation of hypothalamic processes that increase the release of cortisol and promote sympathetic nervous system norepinephrine release, which then causes a cascade of changes, as described in the figure caption, many of which have been implicated in cancer progression and metastasis.

Estrogen

Estrogen receptors that come in three forms, Erα, Erβ, and G-protein coupled ER1 (GPER1) are widely distributed within the brain. Aside from regulating reproductive behaviors, they also affect cognitive processes, mood states, and can influence inflammatory processes. Estrogen receptors are also expressed on the connective tissue of organs (stromal cells) and immune cells, thereby contributing to the regulation of the innate and adaptive immune systems. Women generally exhibit stronger innate and adaptive immune responses to infection than do men, but this advantage in fighting some diseases might also render women more vulnerable to the greater occurrence of autoimmune disorders. The elevated occurrence of autoimmune disorders in women may evolve owing to estrogen effects on T cell activation and proliferation as well as by the loss of T_{reg} cell regulatory functioning.

While estrogen-related processes may contribute to diverse cancers, particular attention has focused on its involvement in breast and ovarian cancer, especially the actions of estrogen on immune activity within the tumor microenvironment. Together with the activation of epidermal growth factor receptors, estrogen may also affect nonsmall-cell lung cancer occurrence by affecting multiple cell types within the tumor cell microenvironment (Smida et al., 2020). The overexpression of estrogen receptors, including the ERα and ERβ receptors, might be markers to predict the progression of this form of cancer and might be important in the selection of therapy to deal with the disease.

Other mechanisms related to variations of ERα have been described, including the DNA damage incurred with prolonged estrogen elevations produced by stressors as well as through estrogen actions on inflammation (Fan and Jordan, 2019). Interactions may also occur between glucocorticoids and sex hormones that could affect disease occurrence and progression. While antihormone therapy that targets estrogen receptors significantly enhanced the survival of affected women, resistance to the treatment develops in about 25% of cases. The resistance may develop owing to epigenetic changes of genes that code for estrogen receptors or because new growth pathways develop so that the cancer cells become independent of estrogen.

Oxytocin

Aside from its prosocial actions or those related to enhancing the salience of social stimuli, oxytocin may provide numerous other health benefits by influencing immune functioning, limiting inflammation, acting in an antioxidant capacity, affecting gut bacteria, and facilitating stress-coping processes, which led to it being referred to as "nature's medicine" (Carter et al., 2020). In considering the links between hormones and cancer, however, the involvement

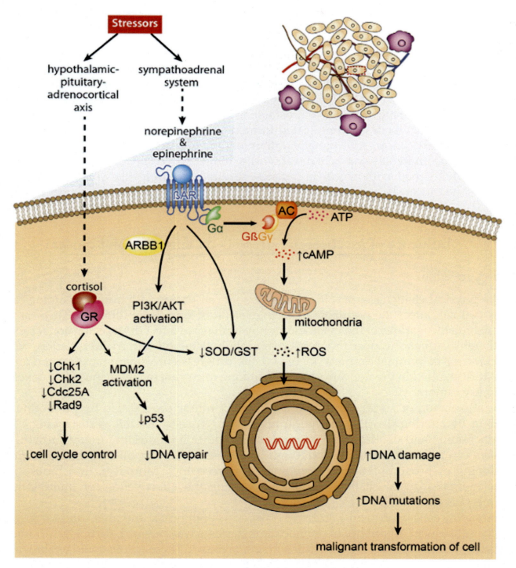

FIG. 6.6 Schematic depiction of molecular mechanisms interconnecting stressors and carcinogenesis. Psychosocial stress increases the release of norepinephrine, epinephrine, and cortisol by the sympathoadrenal system and the hypothalamic-pituitary-adrenocortical axis. Activation of β-adrenergic receptors (β-AR) by norepinephrine or epinephrine increases the synthesis of cAMP that exaggerates the generation of reactive oxygen species (ROS) by mitochondria. Activation of β-adrenergic and glucocorticoid receptors (GR) also reduces levels of ROS eliminating enzymes, superoxide dismutase (SOD), and glutathione-S-transferase (GST). Increased ROS levels induce damage to various cell molecules, including DNA. Simultaneously, activation of β-adrenergic and glucocorticoid receptors leads to activation of phosphatidylinositol 3-kinase (PI3K), Akt/Protein Kinase B (AKT), and murine double minute 2 protein (MDM2) decreases levels of p53, reducing the cell's capacity to repair DNA damage. These processes may lead to increased DNA mutation and a consequent malignant transformation of cells. Moreover, glucocorticoids also reduce gene expression of the cell cycle controlling proteins Chk1, Chk2, Cdc25A, and Rad9, which further promotes carcinogenesis. These abovementioned processes promote the transformation of cells as well as sensitize the cell to the effects of other carcinogens (e.g., ionizing radiation, chemicals). *AC*, adenylate cyclase; *ARBB1*, arrestin beta 1; *Gα, Gβ, Gγ*, subunits of G-protein couplet to β-adrenergic receptor. *Figure and caption from Mravec, B., Tibensky, M., Horvathova, L., 2020. Stress and cancer. Part I: mechanisms mediating the effect of stressors on cancer. J. Neuroimmunol. 346, 577311.*

of oxytocin probably is not "top of mind." Nonetheless, receptors for oxytocin are highly expressed in breast and ovarian cells, and the presence of oxytocin within a tumor's microenvironment was associated with diminished systemic and tumor-associated IL-6 as well as longer survival (Cuneo et al., 2019). There have also been indications that oxytocin's action in elevating the inflammatory mediator prostaglandin E2 could influence endometrial epithelial cancer cells and oxytocin was implicated in mediating the beneficial effects of intermittent exercise on breast cancer. Additionally, inflammation within the ovarian tumor microenvironment can be limited by oxytocin thus favoring longer survival (Cuneo et al., 2019). Finally, while closely aligned with cancer in females, serum oxytocin levels were elevated in men with prostate cancer, as were the number of oxytocin receptors in prostate cancer tissue.

Influence of psychosocial stressors on immunity and cancer

Studies in mice

Like physical challenges, psychosocial stressors can profoundly influence biological processes that can affect the occurrence or progression of numerous illnesses. These stressors may include the features of the macroenvironment that can affect diverse hormonal processes and thereby influence cancer progression. Simply housing mice in social isolation, which is a stressor for communal animals, increased the progression of mammary cancer, and this procedure resulted in greater cancer growth among mice genetically disposed to developing mammary tumors.

Social isolation or social disruption could affect tumor progression by promoting autophagy or by altering the phenotypic expression of NK cells, as well as disturbing CD8+ T cell functioning and dendritic cell maturation. In addition to affecting tumor growth, social stressors can influence metastases through hormones (e.g., cortisol) that diminish immune functioning or by stimulation of norepinephrine receptors. Moreover, in a mouse model of melanoma, a social challenge could undermine the effectiveness of immunotherapy (Sommershof et al., 2017).

Among mice exposed to an aggressive conspecific both before and after tumor transplantation, the subsequent occurrence of pulmonary metastases was elevated, particularly if mice had adopted a passive coping strategy to deal with the aggressor. These actions could be abrogated by pretreating mice with a β-norepinephrine blocker, whereas the β-norepinephrine agonist (which stimulates β-adrenergic receptors) isoproterenol augmented infiltration of macrophages to the lung (Chen et al., 2018). In addition to altering the effects of a previously administered stressor, the frequency of metastases was diminished by pretreating mice with a β-norepinephrine antagonist or CRH receptor antagonist several days before an aggressive social encounter, supporting the view that both sympathetic and HPA axis functioning mediate the stressor actions.

Studies in humans: Retrospective analyses

Several types of cancer had been linked to self-reported stressful events that patients had experienced. Early-stage breast cancer, for instance, was associated with previous distress, anxiety, and depressive feelings, accompanied by diminished NK cell activity, reduced IFN production, and immune-related epigenetic changes within peripheral blood mononuclear cells (Mathews et al., 2011). As interesting as such retrospective studies might be, attempting to link pathology to previous stressor experiences is notorious for being contaminated with subject bias. Aside from common recall difficulties, an individual's view of the past may be colored following diagnosis of a severe illness. Some individuals may seek a reason for their misfortune, and in doing so might conclude that earlier stressful experiences were responsible for their current lot. A serious but infrequently considered concern relevant to many retrospective studies is that they typically focused on recent events encompassing the preceding 6–24 months, although several studies examined a potential link between stress and cancer over longer periods. The cancer process, it will be recalled, is typically lengthy, and malignant cells can remain latent for many years before entering a phase of rapid multiplication. Thus, it may not be productive to search for causal links between cancer occurrence and recent stressors. Perhaps, the question that ought to be addressed is whether stressors influenced cancer multiplication following the first cancer cells appearing or during the phase of equilibrium when the battle between the cancer cells and immune functioning was at a stalemate. Determining the link between cancer occurrence and antecedent stressors requires analyses of cellular events that had occurred much earlier, coupled with a second hit encountered some time later.

Studies in humans: Prospective analyses

Relatively few prospective studies assessed the stress-cancer relationship over extended periods, and the findings from these studies have not convincingly pointed to a connection between the two. Studies assessing the links between breast cancer and the war in Croatia (1991–95) indicated that cancer occurrence and spread to axillary lymph nodes was aligned with war experiences (Fajdic et al., 2009). The Nurses' Health Study that followed individuals over many years similarly indicated that the presence of PTSD symptoms predicted a twofold increase of ovarian cancer almost a decade later (Roberts et al., 2019). However, this stands in marked contrast to a population-based Danish study indicating that PTSD was not linked to the occurrence of cancer (Gradus et al., 2015).

Based on a review that included 38 prospective studies, it was maintained that no single stressor or psychosocial factor was uniquely predictive of cancer occurrence, although having limited social support and a personality that favored feelings of "helplessness" was accompanied by somewhat faster cancer progression (Garssen, 2004). A subsequent systematic review that included both retrospective and prospective analyses yielded a confusing profile of outcomes. Specifically, 26 independent reports indicated that personality traits and stressful events were linked to breast cancer, where 18 reports provided contrary results and 8 were inconclusive (Chiriac et al., 2018). In other studies, modest increases in cancer development occurred among individuals who experienced chronic high stressor levels, but this relationship was most prominent if other poor health behaviors had been adopted. In essence, stressful experiences did not directly influence cancer development or progression but instead promoted lifestyle changes that encouraged the cancer process. The inconsistent findings across studies emphasize the need for more incisive long-term prospective studies, including evaluation of lifestyle changes secondary to the stressful experiences as well as assessment of macrolevel stressors (economic hardships, societal turmoil).

Unlike the inconsistent findings concerning the relation between stressor experiences and subsequent cancer development, a strong case was made that stressful encounters are marked by increased cancer progression and cancer-related mortality. A meta-analysis confirmed abbreviated survival across different types of cancer among individuals with elevated self-reported distress, poor coping styles, low quality of life, or having a stress-prone personality (Chida et al., 2008). Conversely, optimism, effective coping abilities, benefit finding, and social support, were directly related to NK cell activity and lymphocyte proliferative responses among individuals being treated for a variety of malignancies. A report of site-specific cancer mortality comprising more than 163,000 participants based on the pooling of 16 prospective cohort studies indicated that the most distressed individuals experienced greater cancer-related mortality (Batty et al., 2017).

A systematic review indicated that stressful experiences (based on retrospective evaluations using life-events questionnaires) probably do not promote cancer recurrence (Todd et al., 2014). However, basing such conclusions primarily on major life events experienced by cancer survivors might not only be plagued by retrospective biases but might also miss the general distress among patients that could encourage cancer recurrence. The persistent distress, fear, and anxiety that some individuals may experience about "it coming back again" may be sufficiently potent and pervasive to call for psychosocial and psychotherapeutic interventions, which have been associated with greater survival rates. Indeed, the presence of depression and anxiety was associated with later cancer recurrence as well as cancer-specific and all-cause mortality (Wang et al., 2020).

Given the problems inherent in retrospective studies coupled with the difficulties of conducting prospective studies, an alternative approach was adopted to determine the stress-cancer linkage. This entailed recording the patient's stressor history when a tumor was suspected and a biopsy performed, but before the results of the biopsy were available. Irrespective of the subsequent diagnosis, all participants ought to have been subject to similar fears and worries, and hence retrospective analyses related to stressor experiences would be less suspect. As it turned out, malignant (but not benign) tumors occurred more frequently in patients who reported greater distressing experiences. Similarly, vegetative (eating, sleep) and affective depressive symptoms and nocturnal plasma cortisol and IL-6 levels were increased to a greater extent among women who were diagnosed with advanced ovarian cancer relative to that evident among women with tumors of low malignant potential (Lutgendorf et al., 2008). In fact, individuals with higher depressive scores before biopsy were more likely to develop cancer recurrence and contralateral breast cancer over the ensuing 25 years (Eskelinen et al., 2017).

Prenatal and early-life stressors in relation to cancer

As we've seen repeatedly, early development is an exceptionally malleable period so that stressors encountered during this time can have profound effects on the infant. The biological changes elicited by such experiences can have multiple negative effects that cause health disturbances, even disturbing well-being years later. Aside from the actions of early-life adverse events, stressors experienced by a pregnant woman can affect her fetus so that the health of the offspring can be jeopardized and like the influence of early postnatal stressors the impact of prenatal challenges can emerge in adulthood.

Influence of early-life stressors

Perinatal stressors and infection in rodents may have profound and lasting effects on multiple biological systems. Among other changes elicited by these experiences, elevated levels of glucocorticoids can be promoted, which can disturb immune activity, particularly T cell functioning, thus causing stressed mice to be more susceptible to bacterial infections and the progression of some tumors (Hong et al., 2020). The early-life stressor of separating pups from their mom led to reduced NK cytotoxicity and increased tumor colonization in rats inoculated with tumor cells. These early-life experiences were similarly associated with increased tumors provoked by the carcinogen [7,12-dimethylbenz(*a*)anthracene] administration in later life. Paralleling the effects of a psychogenic stressor, neonatal exposure to bacterial endotoxin, such as LPS, diminished NK cell activity in adulthood and increased the growth of transplanted tumor cells in adult male rats (Hodgson and Knott, 2002).

As reported in rodents, early-life stressful experiences in humans were related to cancer development years later, possibly operating through the adoption of pro-tumorigenic behaviors, such as diet and the development of obesity (Hofseth et al., 2020). In the same way, microbial factors and adverse psychosocial experiences in infancy and childhood were linked to inflammation related to epigenetic changes, which could potentially promote later illnesses (McDade et al., 2017). Early childhood adversity may serve as a first hit that encourages physiological changes that can be exacerbated by a second hit comprising subsequent stressor experiences that culminate in tumorigenesis. Even though early-life negative events were frequently associated with cancer development in adulthood, this doesn't necessarily imply that the occurrence of cancer among individuals who experienced early-life adverse events stemmed from biological processes directly linked to these experiences. In fact, a review of 155 studies indicated that early adverse experiences were accompanied by increased risk factors for cancer development, such as obesity as well as alcohol and tobacco use, which could lead to elevated cancer development (Ports et al., 2019). Likewise, early-life adverse events were frequently related to later sleep disorders, which could be tied to cancer occurrence. And the appearance of depression has also frequently been found to predict later cancer development. Since early-life adversities have been tied to later development of depression, the occurrence of altered mood states could be responsible for the increased incidence of metabolic and inflammatory-related disorders, including cancer.

Influence of prenatal stressors

Like the actions of early-life negative experiences, maternal stressors that occurred during pregnancy could affect illness onset in offspring once they reach adulthood. These outcomes have been related to disturbed immune cell proliferation and NK cell functioning together with elevated circulating levels of both pro- and antiinflammatory cytokines, which could persist for years. The immune and cytokine changes might be responsible for the increased vulnerability to allergies, asthma, and infectious diseases seen in offspring, and the occurrence of cancer was similarly elevated.

The data concerning the influence of prenatal stressors on the development of some forms of cancer in humans are understandably limited. Nevertheless, low socioeconomic status during childhood as well as adversity encountered by a pregnant woman was predictive of later systemic inflammation in offspring. Furthermore, epidemiological studies frequently indicated that diverse prenatal insults (e.g., gestational diabetes, tobacco, alcohol, poor diet, endocrine-disrupting chemicals, environmental pollutants, and stressors) were linked to the development of several later illnesses. This included cancer occurrence, possibly due to epigenetic changes, particularly those related to immune and microbial processes (Zhang et al., 2018). Similarly, a mother's social resources, which presumably affect her infant, were associated with the methylation of MEG3, a gene that ordinarily acts to suppress cancer cell growth and increase the tumor suppressor p53, which was linked to later occurrence of malignant melanoma (King et al., 2016).

Beyond these actions, maternal prenatal stressful experiences were associated with elevated levels of pathogenic bacteria in the infants together with a reduced abundance of beneficial bacteria, which together favored the occurrence of inflammation as well as the elevated occurrence of allergies and gastrointestinal issues. Interestingly, when infants of prenatally stressed mothers were assessed at 2.5 months of age, they displayed an elevated abundance of *Proteobacteria* that are associated with increased inflammation and subsequent health risks (Aatsinki et al., 2020). Although the microbial changes could have emanated directly from the prenatal experiences, postnatal maternal factors might have contributed to these outcomes.

Early-onset colorectal cancer (occurring prior to 50 years of age) has been increasing over the past decades, raising understandable concerns. The progressively more frequent occurrence of early-onset colorectal cancer has not only been attributed to increased frequency of prenatal stressors, but also to low levels of maternal prenatal physical inactivity, as well as to antibiotic use, Western-style diets, and obesity (Akimoto et al., 2021). Conversely, diets that contain isothiocyanates (from broccoli), genistein (from soybean), resveratrol (in grape), epigallocatechin-3-gallate (in green tea), and ascorbic acid (in fruits), can promote changes within the epigenome that may diminish the odds of cancer occurring owing to prenatal insults, such as exposure to pollutants (Li et al., 2019). It was also ventured that in the offspring of maternally stressed women the early-onset form of the disease might be dictated by epigenetic changes together with more frequent chromosomal and microsatellite instability that influence gut health and immune functioning (Zhang et al., 2018). There probably isn't a single cause for the upsurge of colorectal cancer. Instead, it may reflect a concatenation of environmental and psychosocial influences and unhealthy dietary changes that have become more prominent.

Dealing with cancer in children

As devastating as cancer occurrence is in older adults, cancer occurrence in children is in so many ways much more horrifying and may have lasting physical and psychological repercussions. By the age of 50, survivors of childhood cancer were almost twice as likely to have experienced other chronic health conditions in comparison with individuals who had not been afflicted with cancer. Approximately 60% of children treated for cancer were at elevated risk for adult cardiac, pulmonary, endocrine, neurocognitive, and auditory system problems, although the occurrence of coronary artery disease has been declining over the past few decades owing to improved treatments. Too often, a second type of cancer develops in adulthood, possibly reflecting a genetically determined link to cancer occurrence or long-term effects of the earlier treatment. Illnesses may also come about owing to the early-life distress creating neurobiological and immunological actions that appear much later, or because the treatments that had been administered years earlier affected specific organs, such as the heart. Analyses of circulating immune cells led to the suggestion that childhood cancer or its treatments accelerated the aging process, perhaps by as much as 25 years, often being accompanied by elevated frailty and cellular senescence (Smitherman et al., 2020). Whatever the source, the increased cellular aging could potentially account for the multiple health risks experienced by childhood cancer survivors.

In addition to physical ailments, cancer patients may experience PTSD, more commonly occurring among patients with prior trauma experiences, preexisting psychiatric conditions, and poor social support. The harsh treatments for cancer among children were similarly accompanied by PTSD features in 20% of cases, and as adults, the risk for PTSD was elevated. Ordinarily, one of the best predictors of PTSD is that the individual had encountered earlier trauma irrespective of its source. The traumatic nature of childhood cancer can thus be expected to have especially pronounced repercussions in the face of further trauma. Considering these connections, it might be beneficial to have patients receive stress reduction or PTSD-focused therapies in conjunction with cancer treatment, but this might be particularly pertinent for children who haven't yet developed effective coping strategies. A systematic review of pediatric oncology studies indicated that stress intervention strategies reduced anxiety and depressive symptoms and improved quality of life. In some instances, these treatments diminished physical symptoms, including a reduction in procedural pain and symptom distress.

Influence of diverse life stressors

Many stressors encountered by humans that involve social, family, and work-related strains are especially disturbing and might favor pathological outcomes, but all stressors aren't equally distressing and could promote differences in biological responses that contribute to illness and might not have comparable effects across cancer types. Accordingly, in this section, we turn to the impact of several common stressors that are experienced among large segments of society.

Job-related stress

An individual's job can be satisfying in numerous positive ways, apart from putting food on the table. It can be a source of comfort that provides intellectual and social stimulation, even helping to handle many of life's stressors. For others, however, their job may be an unrelenting and unbearable challenge. The Whitehall studies that began in 1967 within the UK civil service had revealed that mortality related to heart disease was linked to rank within the organization, being lower among those in more senior ranks than among those at the bottom. Of the numerous features of the job that predicted poor well-being, *job strain*, defined as high job demands coupled with low decision latitude (not being in a position to make decisions) was closely related to heart disease. Such outcomes were exacerbated if individuals also perceived a lack of justice (unfairness), especially if these feelings were not expressed ("covert coping").

Such findings were observed in many subsequent studies, and it appeared that lower rank was also related to certain forms of cancer, gastrointestinal illnesses, diabetes, chronic lung disease, chronic back pain, as well as depression and suicide. Elevated job strain was associated with reduced NK cell activity, but this depended on the nature of the individual's job and the presence of job insecurity. A systematic review that included several sources of distress in addition to job strain (job dissatisfaction, overcommitment, burnout, and job insecurity) revealed that these experiences were associated with disturbed immune functioning, such as lower NK and T cell subsets, reduced NK cell activity, altered CD4+/CD8+ ratio, and increased inflammatory markers (Nakata, 2012).

Simply knowing whether individuals had experienced some form of job-related stress may be insufficient to appraise the stress-cancer linkage adequately, particularly as this strain may interact with other job features and lifestyle factors in predicting tumorigenesis. For reasons that have yet to be determined, the tie between work stress and some forms of cancer, particularly colorectal, lung, and esophageal cancers, varied between individuals in Europe versus North America (Yang et al., 2019a), although job demands in different countries and lifestyle differences may have contributed to this disparity. A cogent analysis of the ties between cancer and stressful job experiences requires greater attention to the nature and duration of the stressors experienced. For instance, work stress has been associated with an increased risk of several forms of cancer (colorectal, lung, esophageal) if the distress had been experienced for at least 15 years (Blanc-Lapierre et al., 2017). Consideration of the links between work stress and cancer ought to be accompanied by concurrent analyses of nonjob stressors encountered, and a constellation of other psychosocial factors. These might include coping styles used (benefit finding and social support), the individual's general quality of life, or having a stress-prone personality, as well as numerous nonmodifiable factors (e.g., age). As well, more attention needs to be devoted to sex differences since women may experience the double burden stemming from their job and concurrently dealing with work at home. To be sure, a person's job can be a source of achievement and may facilitate self-actualization, but when the job is a source of strain, especially when coupled with disturbed home life, the stress burden may be sufficient to cause allostatic overload and disease occurrence and progression.

The distress of loneliness

Loneliness is a condition that develops or is harbored over an extended period, posing an exceptional challenge for most affected individuals, and has been related to a variety of diseases. This is particularly notable among lonely people who are chronically ill and thus in a psychologically and physically fragile state, and psychosocial interventions are needed to enhance their well-being. Social isolation and feelings of loneliness, not surprisingly, have been associated with HPA, sympathetic nervous system, and cytokine alterations, all of which can undermine health (Cacioppo et al., 2015). Conversely, with greater social engagement, inflammatory markers were diminished and levels of IGF-1 were elevated in older individuals (Walker et al., 2019).

Numerous genes and gene polymorphisms have been related to loneliness, such as those coding for serotonin, dopamine, oxytocin, and BDNF. More to the point, altered expression of a set of genes associated with loneliness was associated with the occurrence of bone, lung, and ovarian cancer, as well as leukemia, and was tied to newly diagnosed colorectal cancer, possibly acting through the effects of VEGF. A long-term follow-up study conducted over 30 years indicated that loneliness was accompanied by an overall increase of cancer by about

10% in men, primarily being linked to lung cancer. While these effects were apparent across ages and numerous lifestyle factors, when controlling for depression scores, the link between loneliness and lung cancer was diminished, pointing to the critical role of depression in accounting for the observed relations (Kraav et al., 2021). Yet, loneliness can produce negative consequences above and beyond those provoked by depression itself.

Diminishing feelings of loneliness is no easy matter, and simply offering social support might not be effective in this regard. As we have described earlier, social support can come with its own hazards, particularly the occurrence of unsupportive behaviors coming from close others. When the patient's symptoms are minimized or when they are criticized for the way they are managing symptoms, patients may become reluctant to discuss these issues, which may create barriers to well-being. In effect, the development of these social constraints (e.g., avoidant behaviors) may be related to feelings of loneliness and symptoms expressed. Among patients with lung cancer, the feelings of loneliness and depression may be particularly pronounced, likely operating through social constraints together with the experiences of stigma that are exacerbated by related negative social expectations (Hyland et al., 2019).[b]

The distress of chronic illnesses

Experiencing chronic illness is frequently a powerful psychological stressor particularly as individuals realize that the disease may go on for an extended period and might require marked changes in their lifestyles. Pain, fatigue, depressive features, and disturbed sleep may accompany these chronic illnesses, and the features of these conditions, such as their unpredictable course, uncertainties regarding the future, and the individual's lack of control over events, are major ingredients that can further undermine well-being and increase the likelihood of comorbid conditions emerging.

Cancer risk has been linked to several chronic conditions, including diabetes, cardiovascular disease, as well as kidney disease and arthritis. These relations have been attributed to the presence of inflammation in each of these conditions, and they could also be linked to hormonal or growth factor changes that accompany chronic stressors. Even the knowledge of having some form of cancer and concerns regarding its treatment may be especially distressing and was associated with impaired NK cell functioning. These outcomes might also be due to other biological changes secondary to chronic strain and could be exacerbated among individuals without effective coping supports. Having flexible coping strategies and adequate social supports, as well as adopting healthy lifestyles (sleep, diet, exercise) can limit

---

[b] As lung cancer is often related to smoking, the attitude of "you only have yourself to blame" has marked negative repercussions on those afflicted with the disease, which understandably increases their distress. However, lung cancer also appears in a substantial number of individuals who never smoked. The incidence of lung cancer in never-smokers has been increasing, currently standing at about 12% in the United States, and is much higher in women than in men (Siegel et al., 2021). Why never smokers develop cancer isn't certain, although the genetic signature in patients suggested several distinct processes that may be involved, which could might point to targeted therapies (Zhang et al., 2021a). The stigma related to lung cancer is bad enough irrespective of how the disease came about, but never-smokers may suffer doubly since it is unlikely that they will be screened to catch the disease early.

the distress of chronic illnesses, and psychological therapies in the form of either individual counseling or support groups may also provide benefits.

Surgery as a stressor

Surgical procedures have unquestionably been lifesaving and have extended disability-free years for many people in developed countries. Yet, in large parts of the world, particularly in low- and middle-income countries, the availability of essential surgeries has been lacking, and even relatively simple procedures have been accompanied by high mortality rates. The need for adequate surgical care is essential in these countries and it is hoped that by 2030, with adequate resourcing, such care will be available to the large number of people who require these procedures (estimated to be 5% of the population of these countries). However, progress in global health care involving surgical procedures has not been advancing at an adequate pace.

Within developed countries, this has rarely been a problem, although increasing delays of some surgical procedures have been encountered, which can be distressing and in some cases, the delays may favor deleterious outcomes. This said, as much as surgery is most often essential, it is a source of considerable distress for most individuals, stemming from uncertainties about outcomes, the pain expected during recovery, concerns about future functioning, or the loss of control associated with being "put under." Whatever the case, patients have good cause to dread surgery as 7.7% of deaths globally occur within 30 days of surgical procedures. In addition, surgery was accompanied by a modest cognitive decline in some older individuals, and about 7% of patients aged 65 or older who had undergone noncardiac surgery experienced silent stroke over the following year, which may contribute to cognitive decline. Among some individuals that experienced tissue damage, a systemic inflammatory response syndrome may develop that can culminate in multiple organ dysfunction syndromes, possibly developing through proinflammatory cytokine elevations that can be predicted by increased levels of heat shock proteins.

Surgery is an essential component in the successful treatment of many forms of cancer, but there is the question as to whether the biological stress reactions, such as elevated inflammation, contribute to cancer recurrence and metastasis. Much like other strong stressors, surgery can promote immunosuppression and can diminish tumor cell clearance by disrupting NK cell functioning secondary to the formation of fibrin and platelet clots around tumor cell emboli, thereby increasing the occurrence of tumor metastasis. These effects are abetted by multiple hormones (cortisol, prostaglandins, norepinephrine), growth factors, and variations in the functioning of myeloid-derived suppressor cells (Angka et al., 2017). Aside from the impact on NK cell changes, cancer progression and recurrence associated with a surgical stressor in rodents was also related to a reduction of cytokine-producing CD8+ T cells. As well, a surgical stressor in the form of laparotomy increased T_{reg} cells and downregulated the chemokine CXCL4 in colon tumor tissue (Xu et al., 2018). These outcomes could be attenuated by the knockdown of CCL18, a chemokine produced by tumor-associated macrophages (Sun et al., 2019).

Among mice with transplanted tumor cells, modest dermal wounding resulted in tumor inflammatory gene expression and the development of a greater tumor mass and could cause aggravation of a primary tumor located at a distal site (Pyter et al., 2018). Thus, the enhanced tumor growth was not a matter of tumor cells being dislodged during surgery, but instead reflected the biological repercussions of the surgery. Indeed, the systemic inflammatory response that occurs following surgery could favor the appearance of tumors that were ordinarily restricted by a tumor-specific T cell response, such that tumor growth was diminished in mice that had received perioperative antiinflammatory treatment (Krall et al., 2018).

The data in humans have generally mapped on well to the effects evident in animals. Distress associated with surgery in lung cancer patients was accompanied by reduced T and NK cells, together with increased IFN-α and other inflammatory factors and by elevated immune checkpoints (Xu et al., 2015). To an extent, a preoperative diet that had immune-modulating effects was associated with fewer postoperative infection-related complications and reduced stay in hospital (Adiamah et al., 2019). In line with these findings, altered Th_1/Th_2 balance and T_{reg} expression provoked by surgery were reversible by diets containing supplemental arginine and omega-3 fatty acids (Marik and Flemmer, 2012).

Despite the improvements in surgical procedures (e.g., keyhole surgery) that allow for more rapid healing and better outcomes, surgery can still be a major stressor leading to immune alterations, including redistribution of monocytes and lymphocytes, which predicted the rapidity of recovery. Surgery or the stress associated with it also favors angiogenesis, thereby enhancing tumor growth, raising the possibility that antiangiogenesis therapy may be a means of countering the negative impact of surgery.

In addition to the systemic inflammatory response created by surgery, the adverse effects observed may be related to the nature of the anesthetic used. General anesthetics could promote cancer recurrence and metastasis more readily than in response to regional anesthetics (Byrne et al., 2016). Ketamine induces T-lymphocyte apoptosis and volatile (inhalation) anesthetics have similar actions in suppressing NK cell activity and enhancing angiogenesis. Local anesthetics (e.g., lidocaine), in contrast, can increase NK cell activity, and agents such as propofol decrease surgery-related sympathetic and neuroendocrine responses, thereby promoting diminished immunosuppression. As a result, the recurrence of cancer may be less prominent relative to that associated with other anesthetic and opioid treatments (Kim, 2018).[c]

The perioperative period can be especially stressful and may encourage metastases owing to immune and neuroendocrine alterations, hypothermia, tissue damage, altered nutritional status, as well as the actions of anesthetics, analgesics, and other drugs that could affect immune functioning. By understanding the processes that lead to the surgery-related protumorigenic actions it might be possible to thwart these outcomes through pharmacological manipulations, such as administration of β-norepinephrine blockers (Horowitz et al., 2015). It is germane that in rodents, perioperative treatment with a synthetic TLR-4 agonist, which activates innate and adaptive immunity, limited metastasis of mammary adenocarcinoma and colon carcinoma. Because the perioperative is one in which psychological distress is high, it was suggested that treatments, including immunotherapy, should be undertaken soon after surgery to limit the cumulative effects of feelings of distress (Matzner et al., 2020).

---

[c] Among elderly patients, surgery has frequently been associated with marked cognitive problems. It isn't unusual for this to be attributed to the anesthesia allowing the unmasking of cognitive decline that was already present. There has been increasing evidence indicating that this might occur owing to markedly elevated levels of circulating inflammatory factors and cytokine release by brain microglia.

Perioperatively blocking norepinephrine and/or cyclooxygenase 2 (COX-2) functioning could similarly limit metastasis. The latter actions could come about because diminishing β-adrenergic and COX-2 functioning reduced the activity of prometastatic, proinflammatory transcription factors and the levels of proinflammatory cytokines, while concurrently increasing tumor-infiltrating B cells (Ricon et al., 2019). Additionally, within the excised tumor of patients who had received treatment with a β-norepinephrine antagonist and a COX-2 inhibitor, the expression of the proliferation marker Ki-67 was reduced, and its transcription factors SP1 and AhR were enhanced. These treatments similarly produced positive effects related to transcription factors that have been associated with poor prognosis of primary tumors, angiogenesis, cell proliferation, and altered activity of the oncogenes c-MYB and N-MYC (Haldar et al., 2018). Together, these findings indicate that drug manipulations that affect inflammatory processes could diminish the adverse effects of surgery on cancer-related processes.

Psychological factors associated with cancer treatment

Depression as a predictor of cancer occurrence

There had long been the view (for centuries, actually) that the presence of depression may act as a risk factor for later cancer occurrence. Rather than assuming that this connection was due to the presence of black bile as ancient philosophers had maintained, recent perspectives focused on biological features that are common to these illnesses. Aside from neurotransmitter and neurotrophin alterations, depressive disorders have been attributed to stressor provoked variations of immune functioning, inflammatory cytokines, hormonal changes, and activation of oxidative and reactive nitrogen species. These same processes may similarly contribute to compromised cellular antioxidant defenses, which may lead to DNA damage, mitochondrial impairments, as well as alterations of gene expression that favor cancer occurrence, angiogenesis, and metastases.

Although there is good reason to suspect that depression may be a cancer-promoting condition, a meta-analysis concluded that the available information was not sufficiently strong to support this conclusion. In part, this stemmed from methodological problems in the studies conducted as well as a paucity of data coming from prospective studies of chronic depression that controlled for various lifestyle factors. Once again, this doesn't necessarily imply that depression is not a precursor to cancer, but only that better studies are needed to assess this relationship. Further, the tie between depression and the development of cancer could be indirect, occurring by effects on lifestyles that are aligned with affective disorders. Aside from the contribution of biological processes, behavioral changes secondary to depression, such as altered eating, exercise, and sleep, and the adoption of unhealthy behaviors, such as smoking and alcohol intake (possibly as efforts to self-medicate), might favor increased cancer occurrence.

As mentioned in Chapter 3, depression frequently precedes physical illness and may portend poorer disease outcomes. While it is conceivable that the distress of depression promoted physical illnesses, the antecedents of depression, which often stem from intense stressful experiences, could independently provoke physical disorders. For instance, oxidative stress, chronic inflammation, altered hormonal, and immune functioning associated with chronic stressors could independently favor the occurrence of mood disorders, cancer, and heart disease. But these biological disturbances likely produce depression more readily and sooner than that of somatic illnesses, making it appear as if depression is a precursor for these conditions.

Depression isn't a unitary illness

In discussing the depression-cancer link, an important issue needs to be brought up. The term "depression" may be a misnomer as this is not a unitary illness, but instead comes in multiple flavors (e.g., typical depression, atypical depression, melancholic depression, chronic low-grade depression (dysthymia), seasonal affective disorder, acute depression) with diverse antecedent conditions, and somewhat different symptoms that vary in severity. Typical depression is accompanied by symptoms such as reduced sleep and reduced eating and weight loss, whereas atypical depression is diagnosed with increased sleep and increased eating. There is reason to believe that these subtypes also involve diverse genetic and other neurochemical factors, and the efficacy of treatments may differ between the subtypes. Given the diverse symptoms of these forms of depression and the different biological mechanisms that might contribute to these conditions, analyses linking depression to cancer ought to consider these differences. Rarely, however, has this been done. For that matter, it is often ignored in other aspects of depression research, such as in the evaluation of the efficacy of antidepressant drugs and the relationship between depression and comorbid illnesses.

The distress of cancer treatment

Soon after cancer is suspected, individuals are subjected to a constellation of challenges comprising repeated medical tests that can be arduous and exceptionally taxing. Uncertainties and lack of control over events may haunt the individual, as might the many unforeseen events that may be encountered. More than half of cancer patients experience pain owing to pressure on nerves, or cancer cell infiltration into sensitive regions (e.g., bone) causing inflammation.[d] Delays in treatment initiation may occur, which can diminish the effectiveness of therapies, stumbles in treatments are not uncommon, surgical procedures are distressing, and the side effects of therapies can make a good number of people regret that they ever agreed to be treated. It takes little time for the person's individuality and identity to be undermined so that they are transformed from that of an independent and self-sufficient person to one of a "care recipient" who is dependent on the goodwill of others.

Having been "treated successfully" doesn't necessarily mean that patients will feel entirely cured but it certainly beats the alternative. As "cancer survivors," individuals often discover that this doesn't imply a return to life as it had been, and they describe their situation as comprising a "new normal." The social supports that had been in place may disappear as friends have their own lives that need attention. Patients must often change their usual routines to spend time in physical or occupational rehabilitation, and their leisure time and work-life

---

[d] Patients in large portions of the world do not receive the advanced treatments that are provided to patients in developed countries. It is unconscionable that patients in many regions (e.g., parts of the Middle East, Africa, Asia, and Latin America) do not receive adequate pain medications, despite the modest cost of these treatments. This occurs, in part, because of overregulation concerning their legitimate use. Regulatory policies regarding the use of opioid acting agents in these countries clearly need to be reconsidered.

may need to be managed differently. Periodic tests are needed to determine whether adverse effects of the earlier treatments have materialized (e.g., damage to other organs), as well as to ascertain whether cancer has returned or metastasized. Understandably, individuals may remain in a state of heightened vigilance and uncertainty, waiting for the other shoe to drop. The accumulated distress experienced can have marked repercussions on well-being, and indeed, normal diurnal cortisol variations may be replaced by a flattened cortisol curve, much as it is in association with PTSD. The very same profile was observed among women assessed 5 years after treatment for ovarian cancer, with the flattening of the cortisol curve being more pronounced with greater stressor severity or lifetime stressor experiences.

There was a time when the treatment of cancer patients focused primarily or exclusively on reducing or eliminating the disease and limiting physical discomfort as much as possible. Although it was acknowledged that greater attention ought to be devoted to the patient's psychological well-being, in some circles the view was that while psychological health was essential to well-being, it had little bearing on the primary illness. Fortunately, the position was increasingly accepted that in addition to focusing on the biological changes associated with cancer, the behavioral signs expressed by patients ought to be evaluated and acted upon.

Depression stemming from cancer diagnosis and treatment

It isn't surprising that being diagnosed with a severe chronic condition would predict the development of depressive illnesses, either because of the malaise experienced, or perhaps because individuals have been confronted by their limitations and potential mortality. The distress of a cancer diagnosis, as well as the treatment itself, can be especially disturbing—psychologically and physically—and may promote depression and negatively influence disease progression in some patients. Exacerbating an already dire situation, mood disturbances may have lasting cancer-related effects by undermining treatment compliance and efficacy and predicted shortened survival in some patients.

Predictably, symptoms of depression following treatment among adults were related to how threatening they perceived their situation to be, their ability to cope, and the presence of personality factors, such as trait optimism. Effective coping (and self-efficacy in coping with cancer) and the use of relaxation skills were accompanied by less distress during treatment, enhanced quality of life, and superior adaptation once cancer therapy had ended. Cognitive changes likewise influenced appraisals of the illness, hence affecting mood states and quality of life for some time after treatment, and social support profoundly influenced patients' ability to adapt to their illness condition.

Treating cancer-related depression

Even though depressive illness among cancer patients was frequently self-limiting, often declining within 6 or 7 months, disturbed NK cell functioning in some individuals was more persistent. A prospective analysis indicated that depression among early-stage lung cancer patients was accompanied by a poorer prognosis, and a 25-year cohort study likewise indicated that depression severity predicted relapse and overall survival among breast cancer patients (Eskelinen et al., 2017). Postsurgical depressive symptoms similarly predicted survival as long as 11 years later (Antoni et al., 2017), and distress related to anxiety and depression was predictive of poorer responses to neoadjuvant chemotherapy. There is a definite need for psychological treatments to be available for many cancer patients and survivors. Yet, the majority of patients who experienced cancer-related depression went untreated for their psychological well-being, and for a time collaborative care was infrequently provided to reduce depression.

The development of depression in cancer patients may come about because of neuronal and glial disturbances introduced by the therapy. Chemotherapeutic agents can affect neurogenesis within depression-related brain regions, such as the hippocampus, and treatments that affect sex hormones may have marked psychological consequences. The distress related to cancer and its treatment has also been tied to inflammatory and nitrosative mechanisms (involving nitrous oxide and oxygen radicals), autonomic alterations, and variations of HPA axis activity. These processes may contribute to specific features of depression, particularly neurovegetative symptoms, such as sleep disturbances, circadian problems, altered eating, and fatigue, which together could shorten the life span of cancer patients.

Some pharmacological treatments of depression were accompanied by longer survival, although such findings have not been uniformly apparent across studies. An overview of meta-analyses and systematic reviews indicated that in some cases antidepressants might be useful in treating severe depression associated with cancer. However, it was cautioned that further data were needed to determine whether these actions were unique to only some forms of cancer and in particular settings, and special attention was needed given potential interactions with cancer therapies (Grassi et al., 2018). The diverse outcomes notwithstanding, the administration of SSRIs used to treat depression could increase survival of mice with advanced pancreatic or colon tumors, possibly because of the capacity of elevated peripheral serotonin to diminish PD-L1 that would otherwise protect cancer cells from destruction by T cells (Schneider et al., 2021). Although it may be premature to fully endorse the usefulness of antidepressant drugs in the treatment of affective illness in cancer patients or with respect to survival benefits. In fact, the variable actions attributable to the actions of these agents in diminishing disordered mood are not at all surprising given that antidepressants have only modest effects on depression among noncancer patients. The diverse responses to antidepressants may stem from the illness being biochemically heterogeneous, and consequently, the effectiveness of treatments may be specific to a subset of patients. There have been calls for an individualized approach in the treatment of psychiatric disorders, and this might similarly be needed to produce optimal results among depressed cancer patients.

It has generally been assumed that antidepressants are safe for use in cancer patients and could be beneficial as an additional treatment to diminish cancer-related symptoms. In fact, as observed in mice, there have been indications that selective serotonin reuptake inhibitors (SSRIs) were accompanied by reduced time to disease progression in patients with ovarian cancer, possibly owing to serotonin changes within the tumor's microenvironment (Christensen et al., 2016). Moreover, some types of cancer can be affected by the neurobiological and immune alterations instigated by antidepressant medications.

Given the limited effectiveness of pharmacological treatments to treat depression, coupled with potential adverse effects that may come about, it might be of greater benefit to incorporate behavioral, cognitive, or psychosocial approaches to treat depressive illnesses among cancer patients, particularly those individuals who are at high risk for depression (e.g., among individuals with a previous depressive episode, or a family history of depression). Considerable progress has been realized in the development of psychosocial treatments to enhance well-being among cancer patients, and impressive health enhancements have come from the formation of treatment teams that are attuned to patient needs. The improved psychological well-being associated with such remedies may be accompanied by altered neuroendocrine functioning and increased lymphocyte proliferation (McGregor and Antoni, 2009). These biological changes might act against cancer and serve to extend life, although they might simply be markers of diminished distress. Indeed, while varied psychotherapeutic approaches to diminish depression have been successful in enhancing mood among cancer patients, such outcomes could vary with numerous factors, such as the nature of cancer, the stage of disease, patient sex and age, as well as treatment compliance. Even though cognitive and behavioral treatments may be effective in reducing depression, this didn't necessarily translate as enhanced survival (Mulick et al., 2019). It has frequently been observed that cognitive behavioral therapy in combination with SSRI treatments is more effective in diminishing depression than either treatment alone. It is possible that comparable outcomes would occur in relation to cancer progression or in enhancing the effects of cancer therapies.

In theory, it should be possible to identify the genes that are common in subtypes of depression and those that appear in specific forms of cancer. For instance, do depressed patients with certain forms of cancer, and those who are experiencing only one condition share common genes or epigenetic alterations? However, even if these genes could be identified, this might not be particularly meaningful as the influence of these genes could be moderated by different second hits. This is probably a moot point as finding genes responsible for depression has proven to be exceptionally difficult. To be sure, the heritability of depression is relatively high, and candidate genes have been reported that might be fundamental for the occurrence of these illnesses, such as those related to serotonin reuptake (SLC6A4) and the serotonin 2A receptor (HTR2A), as well as many others (Howard et al., 2019). The difficulty in determining core genes (or gene combinations) related to depression is made difficult since subtypes of the illness may involve different gene interactions, lifestyle factors, early adverse experiences, and several other cogent variables. So, as much as depressive illnesses may be linked to forms of cancer, determining genes common to these comorbid illnesses remain to be identified.

Treating cancer-related depression through (magic) mushrooms

Increasingly more data have accumulated showing that psychedelic compounds may be effective in alleviating treatment-resistant depression and may diminish the fear of dying. Psilocybin (magic mushrooms) can promote pronounced, rapid, and lasting action for treatment-resistant depressive illness and may similarly be effective in ameliorating depressive symptoms among individuals with life-threatening diseases (Vargas et al., 2020). Other compounds, notably lysergic acid diethylamide (LSD) and MDMA (ecstasy) have also been used to treat depressive disorders, typically in conjunction with the guidance of a psychiatrist. These agents could potentially be effective in diminishing distress among cancer patients (Ross, 2018). Although the safety of these compounds among cancer patients remains to be fully determined, their rapid actions might ultimately make them an important component of therapy for individuals with cancer who are struggling with anxiety and depression.

Linking neuropsychiatric and neurodegenerative disorders to cancer

Several neuropsychiatric conditions other than depression were related to cancer occurrence. Early studies had indicated that schizophrenia was associated with a lower incidence of several types of cancer relative to the remainder of the population except for breast and lung cancer which were elevated. Interestingly, the incidence of cancer was reduced among siblings and parents of schizophrenic patients, which led to the suggestion that genes related to schizophrenia might have a protective effect. Yet, lifestyles of affected individuals (e.g., schizophrenia is accompanied by increased smoking, reduced exercise, poorer food choices, obesity) ought to have favored elevated cancer occurrence as observed in the case of lung cancer. Moreover, the incidence of some forms of cancer (e.g., breast cancer) was elevated among individuals with schizophrenia, possibly owing to genetic factors such as mutations of tumor suppressor genes (Zhuo et al., 2019).

Despite the reduced occurrence of some forms of cancer, overall, patients affected with schizophrenia have an abbreviated lifespan, with numerous lifestyle factors playing into this. But it can't go without saying that these individuals may not receive adequate healthcare for physical illnesses (e.g., the occurrence of type 2 diabetes and heart disease), and this might also be relevant to cancer (Seeman, 2019). Owing to diminished care, cancer may not be detected early, which may account for their poorer survival rates. Moreover, complaints from patients relevant to cancer may not be taken as seriously as they should be (diagnostic overshadowing), although patients might be less likely to complain about their symptoms and might infrequently visit their family physician.

Much as it may be cogent to consider instances in which cancer is comorbid with other illnesses, it may be instructive to determine why cancer in particular groups of individuals occurs at rates lower than in the general population. The fact that diseases cluster as they do may point to mechanistic factors that could have both diagnostic and therapeutic implications. The occurrence of cancer among people with Alzheimer's disease is about half that expected based on simple probabilities (Musicco et al., 2013). This curious relationship wasn't attributable to people dying of one disease before the other had an opportunity to emerge, although people with cancer who also suffer dementia, died sooner than cancer patients without dementia. There is the question of whether a focus on Alzheimer's disease obscured or precluded the symptoms of the other condition from being detected. Dementia patients receive fewer cancer screening tests, making it less likely that both illnesses would be detected, which may account for the diminished incidence of cancer reported in those with dementia. Physicians may also be less likely to recommend aggressive cancer therapy for patients with dementia (given the hardships this would create), hence their shorter survival time. Yet, there may be more to the inverse relationship than a lack of screening or treatment.

Several genes have been identified that may contribute to Alzheimer's disease and cancer, operating in different ways in these conditions. The actions of the p53 tumor suppressor illustrate the relations between cancer and Alzheimer's disease. Although p53 is well known to suppress tumor formation, it also contributes to cellular aging through its action on the cell cycle, and disturbance of apoptotic cascades, and might thereby influence the development of Alzheimer's disease. The p53 protein and poly ADP ribose polymerase (PARP) genes were also implicated in differentially contributing to the programming of immune cells in cancer and Alzheimer's disease (Salech et al., 2018).

As in the case of Alzheimer's disease, patients who develop some forms of cancer are less likely to develop Parkinson's disease, even after correcting for lifestyles related to cancer. A gene, Parkin, that is altered in Parkinson's disease, ordinarily removes damaged mitochondria thereby protecting the brain, but when a mutation exists in this gene, this protection is eliminated. It seems that Parkin also serves to regulate cancer cell metabolism and may be important in glucose metabolism as well as in the removal of damaged mitochondria that are fundamental in generating energy for cells. The overlapping roles of Parkin in relation to Parkinson's disease and some form of cancer may account for the relationship between these diseases and supports the proposition that regulating energy metabolism is important in cancer progression (Agarwal et al., 2021).

The curious relationship between the two disorders is still more complicated given that a direct relationship exists between Parkinson's disease and some forms of cancer, such as melanoma. Furthermore, common mediators of cancer and neurological pathology may depend on interactions with environmental toxicants, such as pesticides. Specifically, α-synuclein, which is the primary component of the pathological Lewy body aggregates that characterize the Parkinsonian brain, were implicated in melanoma and mammary cancers. It was similarly reported that the herbicide, paraquat, which has been linked to the occurrence of Parkinson's disease, was associated with increased incidence of non-Hodgkin's lymphoma in agricultural workers regularly exposed to this agent. How and why these relationships exist are difficult to discern given the numerous factors that play into their development.

Some of the observed relationships between Parkinson's disease and cancer have been puzzling but may indicate the inverse relationship between these diseases. Smoking that is known to promote lung and other forms of cancer was associated with diminished risk of Parkinson's disease, although individuals with Parkinson's disease generally smoke less than other people. This certainly doesn't imply a causal connection, but the link to Parkinson's disease was related to the amount of smoking (i.e., an inverse dose relationship). Moreover, while former smokers were somewhat protected from Parkinson's disease (20% lower), this relation was stronger in current smokers, which is consistent with a causal relationship (Gallo et al., 2019). However, the diminished occurrence of cancer in Parkinsonian patients was apparent with respect to both smoking-related and unrelated forms of cancer (Zhang et al., 2021a,b). It is tempting to ascribe the inverse relation between Parkinson's disease and cancer to the treatment of Parkinson's disease with l-DOPA, which could act against cancer occurrence. As it happens, however, some forms of cancer were associated with a diminished occurrence of later Parkinson's disease, raising the possibility that cancer processes can limit the development of the neurological disorder. Yet, both these diseases may take years to develop, thus it remains possible that gene mutations that are linked to increased occurrence of Parkinson's disease may be responsible for the diminished occurrence of some forms of cancer.

Summary and conclusions

Unlike the potentially beneficial effects of the immune changes introduced by acute stressors of moderate severity, chronic stressors diminish immune functioning and may consequently favor the occurrence of various illnesses that otherwise are suppressed by adequate immune functioning. It is conceivable that chronic stressors may influence the development of some forms of cancer, but the available data have not allowed for this to be adequately assessed, especially as initial cancer cells or their break from immune control may have occurred years before frank cancer appearance. In contrast, the progression of cancers has frequently been linked to stressful encounters, possibly owing to immune disturbances, but many other processes may also contribute to this. The mood disturbances that occur in conjunction with cancer, notably anxiety and depression might also foster biological changes that encourage cancer progression, and depression itself may be a forerunner of cancer development by virtue of the inflammatory changes that occur. Unfortunately, the distress of cancer and some of the treatments that ought to be beneficial (e.g., surgery and chemotherapy) can, through their stressor actions, act against the treatment's effectiveness. Ultimately, best practices might include broader treatment approaches to deal with the multiple needs of patients during and after treatment. Even if this doesn't enhance the efficacy of treatments, by their stress-reducing effects they ought to enhance well-being and quality of life.

References

Aatsinki, A.K., Keskitalo, A., Laitinen, V., Munukka, E., Uusitupa, H.M., et al., 2020. Maternal prenatal psychological distress and hair cortisol levels associate with infant fecal microbiota composition at 2.5 months of age. Psychoneuroendocrinology 119, 104754.

Adiamah, A., Skorepa, P., Weimann, A., Lobo, D.N., 2019. The impact of preoperative immune modulating nutrition on outcomes in patients undergoing surgery for gastrointestinal cancer: a systematic review and meta-analysis. Ann. Surg. 270, 247–256.

Agarwal, E., Goldman, A.R., Tang, H.Y., Kossenkov, A.V., Ghosh, J.C., 2021. A cancer ubiquitome landscape identifies metabolic reprogramming as target of Parkin tumor suppression. Sci. Adv. 7, eabg7287.

Akimoto, N., Ugai, T., Zhong, R., Hamada, T., Fujiyoshi, K., et al., 2021. Rising incidence of early-onset colorectal cancer—a call to action. Nat. Rev. Clin. Oncol. 18, 230–243.

Ambree, O., Ruland, C., Scheu, S., Arolt, V., Alferink, J., 2018. Alterations of the innate immune system in susceptibility and resilience after social defeat stress. Front. Behav. Neurosci. 12, 141.

Angka, L., Khan, S.T., Kilgour, M.K., Xu, R., Kennedy, M.A., et al., 2017. Dysfunctional natural killer cells in the aftermath of cancer surgery. Int. J. Mol. Sci. 18, 1787.

Anisman, H., Hayley, S., Kusnecov, A., 2018. The Immune System and Mental Health. Academic Press, San Diego.

Antoni, M.H., Dhabhar, F.S., 2019. The impact of psychosocial stress and stress management on immune responses in patients with cancer. Cancer 125, 1417–1431.

Antoni, M.H., Jacobs, J.M., Bouchard, L.C., Lechner, S.C., Jutagir, D.R., et al., 2017. Post-surgical depressive symptoms and long-term survival in non-metastatic breast cancer patients at 11-year follow-up. Gen. Hosp. Psychiatry 44, 16–21.

Audet, M.C., Mangano, E.N., Anisman, H., 2010. Behavior and pro-inflammatory cytokine variations among submissive and dominant mice engaged in aggressive encounters: moderation by corticosterone reactivity. Front. Behav. Neurosci. 4, 156.

Banks, W.A., 2019. The blood-brain barrier as an endocrine tissue. Nat. Rev. Endocrinol. 15, 444–455.

Batty, G.D., Russ, T.C., Stamatakis, E., Kivimaki, M., 2017. Psychological distress in relation to site specific cancer mortality: pooling of unpublished data from 16 prospective cohort studies. BMJ 356, j108.

Baumeister, D., Akhtar, R., Ciufolini, S., Pariante, C.M., Mondelli, V., 2016. Childhood trauma and adulthood inflammation: a meta-analysis of peripheral C-reactive protein, interleukin-6 and tumour necrosis factor-alpha. Mol. Psychiatry 21, 642–649.

Blalock, J.E., 2005. The immune system as the sixth sense. J. Intern. Med. 257, 126–138.

Blanc-Lapierre, A., Rousseau, M.C., Weiss, D., El-Zein, M., Siemiatycki, J., et al., 2017. Lifetime report of perceived stress at work and cancer among men: a case-control study in Montreal, Canada. Prev. Med. 96, 28–35.

Byrne, K., Levins, K.J., Buggy, D.J., 2016. Can anesthetic-analgesic technique during primary cancer surgery affect recurrence or metastasis? Can. J. Anaesth. 63, 184–192.

Cacioppo, J.T., Cacioppo, S., Capitanio, J.P., Cole, S.W., 2015. The neuroendocrinology of social isolation. Annu. Rev. Psychol. 66, 733–767.

Cain, D.W., Cidlowski, J.A., 2017. Immune regulation by glucocorticoids. Nat. Rev. Immunol. 17, 233–247.

Carter, C.S., Kenkel, W.M., MacLean, E.L., Wilson, S.R., Perkeybile, A.M., et al., 2020. Is oxytocin "nature's medicine"? Pharmacol. Rev. 72, 829–861.

Chang, S.H., 2019. T helper 17 (Th17) cells and interleukin-17 (IL-17) in cancer. Arch. Pharm. Res. 42, 549–559.

Chang, P.Y., Huang, W.Y., Lin, C.L., Huang, T.C., Wu, Y.Y., et al., 2015. Propranolol reduces cancer risk: a population-based cohort study. Medicine 94, e1097.

Chen, H., Liu, D., Guo, L., Cheng, X., Guo, N., et al., 2018. Chronic psychological stress promotes lung metastatic colonization of circulating breast cancer cells by decorating a pre-metastatic niche through activating beta-adrenergic signaling. J. Pathol. 244, 49–60.

Chida, Y., Hamer, M., Wardle, J., Steptoe, A., 2008. Do stress-related psychosocial factors contribute to cancer incidence and survival? Nat. Clin. Pract. Oncol. 5, 466–475.

Chiriac, V.F., Baban, A., Dumitrascu, D.L., 2018. Psychological stress and breast cancer incidence: a systematic review. Clujul Med. 91, 18–26.

Christensen, D.K., Armaiz-Pena, G.N., Ramirez, E., Matsuo, K., Zimmerman, B., et al., 2016. SSRI use and clinical outcomes in epithelial ovarian cancer. Oncotarget 7, 33179.

Cohen, S., Miller, G.E., Rabin, B.S., 2001. Psychological stress and antibody response to immunization: a critical review of the human literature. Psychosom. Med. 63, 7–18.

Cohen, S., Janicki-Deverts, D., Doyle, W.J., Miller, G.E., Frank, E., et al., 2012. Chronic stress, glucocorticoid receptor resistance, inflammation, and disease risk. Proc. Natl. Acad. Sci. 109, 5995–5999.

Crosswell, A.D., Bower, J.E., Ganz, P.A., 2014. Childhood adversity and inflammation in breast cancer survivors. Psychosom. Med. 76, 208–214.

Cryan, J.F., O'Riordan, K.J., Cowan, C.S.M., Sandhu, K.V., Bastiaanssen, T.F.S., et al., 2019. The microbiota-gut-brain axis. Physiol. Rev. 99, 1877–2013.

Cuneo, M.G., Szeto, A., Schrepf, A., Kinner, E.M., Schachner, B.I., et al., 2019. Oxytocin in the tumor microenvironment is associated with lower inflammation and longer survival in advanced epithelial ovarian cancer patients. Psychoneuroendocrinology 106, 244–251.

Danielson, A.M., Matheson, K., Anisman, H., 2011. Cytokine levels at a single time point following a reminder stimulus among women in abusive dating relationships: relationship to emotional states. Psychoneuroendocrinology 36, 40–50.

De Palma, G., Blennerhassett, P., Lu, J., Deng, Y., Park, A.J., et al., 2015. Microbiota and host determinants of behavioural phenotype in maternally separated mice. Nat. Commun. 6, 7735.

Dhabhar, F.S., 2014. Effects of stress on immune function: the good, the bad, and the beautiful. Immunol. Res. 58, 193–210.

Dhabhar, F.S., Saul, A.N., Holmes, T.H., Daugherty, C., Neri, E., et al., 2012. High-anxious individuals show increased chronic stress burden, decreased protective immunity, and increased cancer progression in a mouse model of squamous cell carcinoma. PLoS One 7, e33069.

Dudek, K.A., Dion-Albert, L., Lebel, M., LeClair, K., Labrecque, S., et al., 2020. Molecular adaptations of the blood-brain barrier promote stress resilience vs. depression. Proc. Natl. Acad. Sci. 117, 3326–3336.

Ehrlich, K.B., Ross, K.M., Chen, E., Miller, G.E., 2016. Testing the biological embedding hypothesis: is early life adversity associated with a later proinflammatory phenotype? Dev. Psychopathol. 28, 1273–1283.

Eskelinen, M., Korhonen, R., Selander, T., Ollonen, P., 2017. Beck Depression Inventory as a predictor of long-term outcome among patients admitted to the breast cancer diagnosis unit: a 25-year cohort study in Finland. Anticancer Res. 37, 819–824.

Fajdic, J., Gotovac, N., Hrgovic, Z., Fassbender, W.J., 2009. Influence of stress related to war on biological and morphological characteristics of breast cancer in a defined population. Adv. Med. Sci. 54, 283–288.

Fan, P., Jordan, V.C., 2019. New insights into acquired endocrine resistance of breast cancer. Cancer Drug Resist. 2, 198.

Fülling, C., Dinan, T.G., Cryan, J.F., 2019. Gut microbe to brain signaling: what happens in vagus. Neuron 101, 998–1002.

Galley, J.D., Bailey, M.T., 2014. Impact of stressor exposure on the interplay between commensal microbiota and host inflammation. Gut Microbes 5, 390–396.

Galley, J.D., Mackos, A.R., Varaljay, V.A., Bailey, M.T., 2017. Stressor exposure has prolonged effects on colonic microbial community structure in *Citrobacter rodentium*-challenged mice. Sci. Rep. 7, 45012.

Gallo, V., Vineis, P., Cancellieri, M., Chiodini, P., Barker, R.A., et al., 2019. Exploring causality of the association between smoking and Parkinson's disease. Int. J. Epidemiol. 48, 912–925.

References

Garssen, B., 2004. Psychological factors and cancer development: evidence after 30 years of research. Clin. Psychol. Rev. 24, 315–338.

Gibb, J., Al-Yawer, F., Anisman, H., 2013. Synergistic and antagonistic actions of acute or chronic social stressors and an endotoxin challenge vary over time following the challenge. Brain Behav. Immun. 28, 149–158.

Glaser, R., Kiecolt-Glaser, J.K., 2005. Stress-induced immune dysfunction: implications for health. Nat. Rev. Immunol. 5, 243–251.

Gradus, J.L., Farkas, D.K., Svensson, E., Ehrenstein, V., Lash, T.L., et al., 2015. Posttraumatic stress disorder and cancer risk: a nationwide cohort study. Eur. J. Epidemiol. 30, 563–568.

Grassi, L., Nanni, M.G., Rodin, G., Li, M., Caruso, R., 2018. The use of antidepressants in oncology: a review and practical tips for oncologists. Ann. Oncol. 29, 101–111.

Gur, T.L., Bailey, M.T., 2016. Effects of stress on commensal microbes and immune system activity. Adv. Exp. Med. Biol. 874, 289–300.

Haldar, R., Shaashua, L., Lavon, H., Lyons, Y.A., Zmora, O., et al., 2018. Perioperative inhibition of beta-adrenergic and COX2 signaling in a clinical trial in breast cancer patients improves tumor Ki-67 expression, serum cytokine levels, and PBMCs transcriptome. Brain Behav. Immun. 73, 294–309.

Hemmings, S.M.J., Malan-Muller, S., van den Heuvel, L.L., Demmitt, B.A., Stanislawski, M.A., et al., 2017. The microbiome in posttraumatic stress disorder and trauma-exposed controls: an exploratory study. Psychosom. Med. 79, 936–946.

Hodgson, D.M., Knott, B., 2002. Potentiation of tumor metastasis in adulthood by neonatal endotoxin exposure: sex differences. Psychoneuroendocrinology 27, 791–804.

Hofseth, L.J., Hebert, J.R., Chanda, A., Chen, H., Love, B.L., et al., 2020. Early-onset colorectal cancer: initial clues and current views. Nat. Rev. Gastroenterol. Hepatol. 17, 352–364.

Hong, J.Y., Lim, J., Carvalho, F., Cho, J.Y., Vaidyanathan, B., et al., 2020. Long-term programming of CD8 T cell immunity by perinatal exposure to glucocorticoids. Cell 180, 847–861 e815.

Horowitz, M., Neeman, E., Sharon, E., Ben-Eliyahu, S., 2015. Exploiting the critical perioperative period to improve long-term cancer outcomes. Nat. Rev. Clin. Oncol. 12, 213–226.

Howard, D.M., Adams, M.J., Clarke, T.K., Hafferty, J.D., Gibson, J., et al., 2019. Genome-wide meta-analysis of depression identifies 102 independent variants and highlights the importance of the prefrontal brain regions. Nat. Neurosci. 22, 343–352.

Hyland, K.A., Small, B.J., Gray, J.E., Chiappori, A., Creelan, B.C., et al., 2019. Loneliness as a mediator of the relationship of social cognitive variables with depressive symptoms and quality of life in lung cancer patients beginning treatment. Psychooncology 28, 1234–1242.

Kaeser, S.A., Lehallier, B., Thinggaard, M., Häsler, L.M., Apel, A., et al., 2021. A neuronal blood marker is associated with mortality in old age. Nat. Aging 1, 218–225.

Kiecolt-Glaser, J.K., Preacher, K.J., MacCallum, R.C., Atkinson, C., Malarkey, W.B., et al., 2003. Chronic stress and age-related increases in the proinflammatory cytokine IL-6. Proc. Natl. Acad. Sci. 100, 9090–9095.

Kiecolt-Glaser, J.K., Wilson, S.J., Shrout, M.R., Madison, A.A., Andridge, R., et al., 2021. The gut reaction to couples' relationship troubles: a route to gut dysbiosis through changes in depressive symptoms. Psychoneuroendocrinology 125, 105132.

Kim, R., 2018. Effects of surgery and anesthetic choice on immunosuppression and cancer recurrence. J. Transl. Med. 16, 8.

Kim-Fuchs, C., Le, C.P., Pimentel, M.A., Shackleford, D., Ferrari, D., et al., 2014. Chronic stress accelerates pancreatic cancer growth and invasion: a critical role for beta-adrenergic signaling in the pancreatic microenvironment. Brain Behav. Immun. 40, 40–47.

King, K.E., Kane, J.B., Scarbrough, P., Hoyo, C., Murphy, S.K., 2016. Neighborhood and family environment of expectant mothers may influence prenatal programming of adult cancer risk: discussion and an illustrative DNA methylation example. Biodemography Soc. Biol. 62, 87–104.

Klein, S.L., Flanagan, K.L., 2016. Sex differences in immune responses. Nat. Rev. Immunol. 16, 626–638.

Kraav, S.L., Lehto, S.M., Kauhanen, J., Hantunen, S., Tolmunen, T., 2021. Loneliness and social isolation increase cancer incidence in a cohort of Finnish middle-aged men. A longitudinal study. Psychiatry Res. 299, 113868.

Krall, J.A., Reinhardt, F., Mercury, O.A., Pattabiraman, D.R., Brooks, M.W., et al., 2018. The systemic response to surgery triggers the outgrowth of distant immune-controlled tumors in mouse models of dormancy. Sci. Transl. Med. 10, eean3464.

Krizanova, O., Babula, P., Pacak, K., 2016. Stress, catecholaminergic system and cancer. Stress 19, 419–428.

Kusnecov, A.W., Rabin, B.S., 1993. Inescapable footshock exposure differentially alters antigen- and mitogen-stimulated spleen cell proliferation in rats. J. Neuroimmunol. 44, 33–42.

La Fratta, I., Tatangelo, R., Campagna, G., Rizzuto, A., Franceschelli, S., et al., 2018. The plasmatic and salivary levels of IL-1beta, IL-18 and IL-6 are associated to emotional difference during stress in young male. Sci. Rep. 8, 3031.

Li, S., Chen, M., Li, Y., Tollefsbol, T.O., 2019. Prenatal epigenetics diets play protective roles against environmental pollution. Clin. Epigenetics 11, 82.

Liu, J., Deng, G.H., Zhang, J., Wang, Y., Xia, X.Y., et al., 2015. The effect of chronic stress on anti-angiogenesis of sunitinib in colorectal cancer models. Psychoneuroendocrinology 52, 130–142.

Lopez, R.B., Brown, R.L., Wu, E.L., Murdock, K.W., Denny, B.T., et al., 2020. Emotion regulation and immune functioning during grief: testing the role of expressive suppression and cognitive reappraisal in inflammation among recently bereaved spouses. Psychosom. Med. 82, 2–9.

Lutgendorf, S.K., Weinrib, A.Z., Penedo, F., Russell, D., DeGeest, K., et al., 2008. Interleukin-6, cortisol, and depressive symptoms in ovarian cancer patients. J. Clin. Oncol. 26, 4820–4827.

Marik, P.E., Flemmer, M., 2012. The immune response to surgery and trauma: implications for treatment. J. Trauma Acute Care Surg. 73, 801–808.

Martinez-Muniz, G.A., Wood, S.K., 2020. Sex differences in the inflammatory consequences of stress: implications for pharmacotherapy. J. Pharmacol. Exp. Ther. 375, 161–174.

Mathews, H.L., Konley, T., Kosik, K.L., Krukowski, K., Eddy, J., et al., 2011. Epigenetic patterns associated with the immune dysregulation that accompanies psychosocial distress. Brain Behav. Immun. 25, 830–839.

Matzner, P., Sandbank, E., Neeman, E., Zmora, O., Gottumukkala, V., et al., 2020. Harnessing cancer immunotherapy during the unexploited immediate perioperative period. Nat. Rev. Clin. Oncol. 17, 313–326.

McDade, T.W., Ryan, C., Jones, M.J., MacIsaac, J.L., Morin, A.M., et al., 2017. Social and physical environments early in development predict DNA methylation of inflammatory genes in young adulthood. Proc. Natl. Acad. Sci. 114, 7611–7616.

McGregor, B.A., Antoni, M.H., 2009. Psychological intervention and health outcomes among women treated for breast cancer: a review of stress pathways and biological mediators. Brain Behav. Immun. 23, 159–166.

Menard, C., Pfau, M.L., Hodes, G.E., Kana, V., Wang, V.X., et al., 2017. Social stress induces neurovascular pathology promoting depression. Nat. Neurosci. 20, 1752–1760.

Moons, W.G., Shields, G.S., 2015. Anxiety, not anger, induces inflammatory activity: an avoidance/approach model of immune system activation. Emotion 15, 463–476.

Mulick, A., Walker, J., Puntis, S., Symeonides, S., Gourley, C., et al., 2019. Is improvement in comorbid major depression associated with longer survival in people with cancer? A long-term follow-up of participants in the SMaRT Oncology-2 and 3 trials. J. Psychosom. Res. 116, 106–112.

Musicco, M., Adorni, F., Di Santo, S., Prinelli, F., Pettenati, C., et al., 2013. Inverse occurrence of cancer and Alzheimer disease: a population-based incidence study. Neurology 81, 322–328.

Muthuswamy, R., Okada, N.J., Jenkins, F.J., McGuire, K., McAuliffe, P.F., et al., 2017. Epinephrine promotes COX-2-dependent immune suppression in myeloid cells and cancer tissues. Brain Behav. Immun. 62, 78–86.

Nakata, A., 2012. Psychosocial job stress and immunity: a systematic review. Methods Mol. Biol. 934, 39–75.

Niraula, A., Sheridan, J.F., Godbout, J.P., 2017. Microglia priming with aging and stress. Neuropsychopharmacology 42, 318–333.

Obradović, M.M.S., Hamelin, B., Manevski, N., Couto, J.P., Sethi, A., et al., 2019. Glucocorticoids promote breast cancer metastasis. Nature 567, 540–544.

O'Mahony, S.M., McVey Neufeld, K.A., Waworuntu, R.V., Pusceddu, M.M., Manurung, S., et al., 2020. The enduring effects of early-life stress on the microbiota-gut-brain axis are buffered by dietary supplementation with milk fat globule membrane and a prebiotic blend. Eur. J. Neurosci. 51, 1042–1058.

Palma-Gudiel, H., Prather, A.A., Lin, J., Oxendine, J.D., Guintivano, J., et al., 2021. HPA axis regulation and epigenetic programming of immune-related genes in chronically stressed and non-stressed mid-life women. Brain Behav. Immun. 92, 49–56.

Pearson-Leary, J., Zhao, C., Bittinger, K., Eacret, D., Luz, S., et al., 2020. The gut microbiome regulates the increases in depressive-type behaviors and in inflammatory processes in the ventral hippocampus of stress vulnerable rats. Mol. Psychiatry 25, 1068–1079.

Pfau, M.L., Menard, C., Cathomas, F., Desland, F., Kana, V., et al., 2019. Role of monocyte-derived microRNA106b approximately 25 in resilience to social stress. Biol. Psychiatry 86, 474–482.

Ports, K.A., Holman, D.M., Guinn, A.S., Pampati, S., Dyer, K.E., et al., 2019. Adverse childhood experiences and the presence of cancer risk factors in adulthood: a scoping review of the literature from 2005 to 2015. J. Pediatr. Nurs. 44, 81–96.

Pyter, L.M., McKim, D.B., Husain, Y., Calero, H., Godbout, J.P., et al., 2018. Effects of dermal wounding on distal primary tumor immunobiology in mice. J. Surg. Res. 221, 328–335.

Reed, R.G., Presnell, S.R., Al-Attar, A., Lutz, C.T., Segerstrom, S.C., 2019. Perceived stress, cytomegalovirus titers, and late-differentiated T and NK cells: between-, within-person associations in a longitudinal study of older adults. Brain Behav. Immun. 80, 266–274.

Richards, L.M., Whitley, O.K.N., MacLeod, G., Cavalli, F.M.G., Coutinho, F.J., et al., 2021. Gradient of developmental and injury response transcriptional states defines functional vulnerabilities underpinning glioblastoma heterogeneity. Nat. Cancer 2, 157–173.

Ricon, I., Hanalis-Miller, T., Haldar, R., Jacoby, R., Ben-Eliyahu, S., 2019. Perioperative biobehavioral interventions to prevent cancer recurrence through combined inhibition of beta-adrenergic and cyclooxygenase 2 signaling. Cancer 125, 45–56.

Rincel, M., Aubert, P., Chevalier, J., Grohard, P.A., Basso, L., et al., 2019. Multi-hit early life adversity affects gut microbiota, brain and behavior in a sex-dependent manner. Brain Behav. Immun. 80, 179–192.

Roberts, A.L., Huang, T., Koenen, K.C., Kim, Y., Kubzansky, L.D., et al., 2019. Posttraumatic stress disorder is associated with increased risk of ovarian cancer: a prospective and retrospective longitudinal cohort study. Cancer Res. 79, 5113–5120.

Ross, S., 2018. Therapeutic use of classic psychedelics to treat cancer-related psychiatric distress. Int. Rev. Psychiatry 30, 317–330.

Rustenhoven, J., Kipnis, J., 2019. Bypassing the blood-brain barrier. Science 366, 1448–1449.

Salech, F., Ponce, D.P., SanMartin, C.D., Rogers, N.K., Henriquez, M., et al., 2018. Cancer imprints an increased PARP-1 and p53-dependent resistance to oxidative stress on lymphocytes of patients that later develop Alzheimer's disease. Front. Neurosci. 12, 58.

Schneider, M.A., Heeb, L., Beffinger, M.M., Pantelyushin, S., Linecker, M., et al., 2021. Attenuation of peripheral serotonin inhibits tumor growth and enhances immune checkpoint blockade therapy in murine tumor models. Sci. Transl. Med. 13, eabc818.

Seeman, M.V., 2019. Schizophrenia mortality: barriers to progress. Psychiatry Q. 90, 553–563.

Servick, K., 2019. War of nerves. Science 365, 1071–1073.

Siegel, D.A., Fedewa, S.A., Henley, S.J., Pollack, L.A., Jemal, A., 2021. Proportion of never smokers among men and women with lung cancer in 7 US states. JAMA Oncol. 7, 302–304.

Smida, T., Bruno, T.C., Stabile, L.P., 2020. Influence of estrogen on the NSCLC microenvironment: a comprehensive picture and clinical implications. Front. Oncol. 10, 137.

Smitherman, A.B., Wood, W.A., Mitin, N., Ayer Miller, V.L., Deal, A.M., et al., 2020. Accelerated aging among childhood, adolescent, and young adult cancer survivors is evidenced by increased expression of p16(INK4a) and frailty. Cancer 126, 4975–4983.

Sommershof, A., Scheuermann, L., Koerner, J., Groettrup, M., 2017. Chronic stress suppresses anti-tumor TCD8+ responses and tumor regression following cancer immunotherapy in a mouse model of melanoma. Brain Behav. Immun. 65, 140–149.

Song, E., Mao, T., Dong, H., Boisserand, L.S.B., Antila, S., et al., 2020. VEGF-C-driven lymphatic drainage enables immunosurveillance of brain tumours. Nature 577, 689–694.

Spini, A., Roberto, G., Gini, R., Bartolini, C., Bazzani, L., et al., 2019. Evidence of beta-blockers drug repurposing for the treatment of triple negative breast cancer: a systematic review. Neoplasma 66, 963–970.

Sun, Z., Du, C., Xu, P., Miao, C., 2019. Surgical trauma-induced CCL18 promotes recruitment of regulatory T cells and colon cancer progression. J. Cell. Physiol. 234, 4608–4616.

Takkouche, B., Regueira, C., Gestal-Otero, J.J., 2001. A cohort study of stress and the common cold. Epidemiology 12, 345–349.

Todd, B.L., Moskowitz, M.C., Ottati, A., Feuerstein, M., 2014. Stressors, stress response, and cancer recurrence: a systematic review. Cancer Nurs. 37, 114–125.

Vargas, A.S., Luis, A., Barroso, M., Gallardo, E., Pereira, L., 2020. Psilocybin as a new approach to treat depression and anxiety in the context of life-threatening diseases: a systematic review and meta-analysis of clinical trials. Biomedicines 8, 331.

Walker, E., Ploubidis, G., Fancourt, D., 2019. Social engagement and loneliness are differentially associated with neuro-immune markers in older age: time-varying associations from the English Longitudinal Study of Ageing. Brain Behav. Immun. 82, 224–229.

Wang, Q., Dong, X., Wang, Y., Liu, M., Sun, A., et al., 2017. Adolescent escitalopram prevents the effects of maternal separation on depression- and anxiety-like behaviours and regulates the levels of inflammatory cytokines in adult male mice. Int. J. Dev. Neurosci. 62, 37–45.

Wang, X., Wang, N., Zhong, L., Wang, S., Zheng, Y., et al., 2020. Prognostic value of depression and anxiety on breast cancer recurrence and mortality: a systematic review and meta-analysis of 282,203 patients. Mol. Psychiatry 25, 3186–3197.

Werbner, M., Barsheshet, Y., Werbner, N., Zigdon, M., Averbuch, I., et al., 2019. Social-stress-responsive microbiota induces stimulation of self-reactive effector T helper cells. mSystems 4.

Wilmanski, T., Diener, C., Rappaport, N., Patwardhan, S., Wiedrick, J., et al., 2021. Gut microbiome pattern reflects healthy ageing and predicts survival in humans. Nat. Metab. 3, 274–286.

Wu, E.L., LeRoy, A.S., Heijnen, C.J., Fagundes, C.P., 2021. Inflammation and future depressive symptoms among recently bereaved spouses. Psychoneuroendocrinology 128, 105206.

Xu, P., Zhang, P., Sun, Z., Wang, Y., Chen, J., et al., 2015. Surgical trauma induces postoperative T-cell dysfunction in lung cancer patients through the programmed death-1 pathway. Cancer Immunol. Immunother. 64, 1383–1392.

Xu, P., He, H., Gu, Y., Wang, Y., Sun, Z., et al., 2018. Surgical trauma contributes to progression of colon cancer by downregulating CXCL4 and recruiting MDSCs. Exp. Cell Res. 370, 692–698.

Yang, T., Qiao, Y., Xiang, S., Li, W., Gan, Y., et al., 2019a. Work stress and the risk of cancer: a meta-analysis of observational studies. Int. J. Cancer 144, 2390–2400.

Yang, B., Wei, J., Ju, P., Chen, J., 2019b. Effects of regulating intestinal microbiota on anxiety symptoms: a systematic review. Gen. Psychiatry 32, e100056.

Yang, H., Xia, L., Chen, J., Zhang, S., Martin, V., et al., 2019c. Stress-glucocorticoid-TSC22D3 axis compromises therapy-induced antitumor immunity. Nat. Med. 25, 1428–1441.

Zhang, Q., Berger, F.G., Love, B., Banister, C.E., Murphy, E.A., et al., 2018. Maternal stress and early-onset colorectal cancer. Med. Hypotheses 121, 152–159.

Zhang, L., Pan, J., Chen, W., Jiang, J., Huang, J., 2020. Chronic stress-induced immune dysregulation in cancer: implications for initiation, progression, metastasis, and treatment. Am. J. Cancer Res. 10, 1294–1307.

Zhang, X., Guarin, D., Mohammadzadehhonarvar, N., Chen, X., Gao, X., 2021a. Parkinson's disease and cancer: a systematic review and meta-analysis of over 17 million participants. BMJ Open 11, e046329.

Zhang, T., Joubert, P., Ansari-Pour, N., Zhao, W., Hoang, P.H., et al., 2021b. Genomic and evolutionary classification of lung cancer in never smokers. Nat. Genet. 53, 1348–1359.

Zhao, L., Xu, J., Liang, F., Li, A., Zhang, Y., et al., 2015. Effect of chronic psychological stress on liver metastasis of colon cancer in mice. PLoS One 10, e0139978.

Zhou, J., Liu, Z., Zhang, L., Hu, X., Wang, Z., et al., 2020. Activation of beta2-adrenergic receptor promotes growth and angiogenesis in breast cancer by down-regulating PPARgamma. Cancer Res. Treat. 52, 830–847.

Zhuo, C., Wang, D., Zhou, C., Chen, C., Li, J., et al., 2019. Double-edged sword of tumour suppressor genes in schizophrenia. Front. Mol. Neurosci. 12, 1.

CHAPTER 7

Eating and nutrition links to cancer

OUTLINE

Then and now	225	Diet and nutrition in relation to cancer	239
The digestive process	227	Food constituents and cancer	241
		Fiber	242
Hormonal and brain processes underlying eating	229	Nonmeat proteins and the Mediterranean diet	244
Leptin and ghrelin	229	Fats and meat proteins	245
Insulin and related peptides	230	Balancing nutritional needs and starving cancer cells: The case of cancer cachexia	247
Incretins	231		
Diets and weight loss	234	Summary and conclusions	248
Microbiota, short-chain fatty acids, fibers, and polyphenols	234	References	249
Setpoint, diet, and fasting	235		
Impact of periodic fasting	238		

Then and now

Back in the day, hominids spent a good deal of time and effort searching for food, with survival probably tied to their capacity to store fat. Since seeking food was replete with risks, it would have been advantageous for stress hormone output to be elevated during these periods. The levels of such hormones (akin to cortisol) would have affected eating processes and body fat distribution, although it is doubtful that obesity would have been a problem. Natural selection ought to have assured that digestion, including the presence of microbiota, would have been compatible with ingested foods, thereby maintaining health. The advent of the agricultural revolution triggered changes in social structures along with genetic adaptations. Selection pressures that had existed for our ancient ancestors were different from those present today, where the hunt for food amounts to searching through supermarket aisles and reading nutrition labels on food packaging. As a result, some of the benefits gained from the hunt have been lost, and for the nondiscriminating consumer ultra-processed foods may abet the accumulation of unhealthy abdominal fat, which is unlikely to get burned off with the next hunt or excursion to gather food. The balance between the exertion of the hunt and consumption of the foods obtained has been lost to a convenience of excess that has threatened the health of millions of people in the developed world.

Examination of supermarket shelves, which likely reflect consumer preferences, confirms that processed and ultra-processed foods are highly desired despite the many warnings concerning the health risks they pose. Consistent with this casual observation, over the past two decades calories obtained from ultra-processed foods have increased, whereas those obtained from whole foods have declined. Many people have fixed opinions concerning dietary factors and ignore the advice of experts. As well, the choice of poor diets may stem from socioeconomic challenges or the lack of knowledge concerning healthy and unhealthy foods. Food marketing has also been enormously successful at steering individuals to counterproductive food choices. No matter how they came about, poor diets are health risks that occur owing to their contribution to the development of obesity and excessive levels of inflammatory mediators, numerous microbial variations, and marked metabolic perturbations.

Being overweight and having obesity has increased dramatically in developed countries and has been linked to diseases such as metabolic syndrome, type 2 diabetes, cardiovascular illnesses, cancer, and premature death. Women appear to be more prone to developing obesity-related type 2 diabetes, whereas men are at elevated risk of developing obstructive pulmonary disease and chronic kidney disease. In both sexes, obesity is among the leading causes of death stemming from noninfectious illnesses, making it certain that broader and more effective public health strategies are needed to derail this ongoing hazard.[a]

Within many Western countries over 50%–60% of people are currently overweight or have obesity. Four decades ago, obesity was infrequent in children and adolescents (4%), whereas about 20% are now experiencing these conditions, increasing by 1%–2% annually. The weight increase is likely due to children often exceeding the recommended levels of sugar consumption, abetted by added sugars in many products marketed for infants and toddlers along with a lack of exercise. There is also good reason to believe that the development of obesity might also be encouraged by prenatal factors related to the gut microbiota, metabolic changes, and immune alterations that were tied to factors, such as the method of delivery (Caesarean vs natural birth), being bottle or breast fed, and obesity present in the mom. Treating obesity is certainly possible. However, it is made difficult for several reasons, not least the fact that 40% of overweight individuals and 10% of those with obesity fail to perceive themselves as suffering from a weight issue. This misperception may stem from a general increase of weight within the population so when individuals make social comparisons, they believe that their weight is in line with that of others.

Innate and learned food preferences

Certain taste preferences may be prewired, such as the aversion toward sour and bitter tastes and the preference for sweet and salty tastes. These preferences may be biologically significant, as many sweet-tasting foods provide a rapid source of energy. In addition, experience and learning

---

[a] From a biomedical perspective, obesity is frequently considered a disease, much like any other condition, such as heart disease, type 2 diabetes, and cancer. Depending on the context, it is probably more accurate to describe "an obese individual" as an "individual who *has* obesity."

have a considerable influence on dietary choices. We like what we're used to, and if a food deviates too far from what we're accustomed to, it's rejected. Those who are a bit older will remember the days before milk was homogenized when a layer of fat was skimmed from the surface. When milk pasteurization (to remove some bacteria) and homogenization were introduced whole milk consumption was more appealing and quickly became the norm. The introduction of 2% milk was deemed to be a healthier option, but for those who had been whole milk drinkers, this fat-reduced concoction was met with revulsion. With experience, 2% became acceptable, and eventually so was 1% milk (although skim milk may have been a step too far). The result was that whole milk actually became distasteful—too creamy, too fatty. We now have multiple alternatives to cow's milk, including "plant milks" derived from soy, almonds, coconut, hemp, and several others, that have been gaining popularity. No doubt, if healthy eating begins early in life, the brain and palate can be trained. There is also the issue of food habits or constituents influencing reward processes. Swirling sugary foods in their mouth causes certain reward circuits in the brain of children to be activated, more prominently in obese than in nonobese children. Fortunately, these neuronal circuits can be altered by intensive experiences in a structured weight loss program, to the extent that previously shunned foods become the choice of snack.

Several brain regions contribute to hunger processes, food selection, and satiety. It has long been known that strains of mice differ markedly in their preference for certain nutrients (e.g., fat intake), which may be related to both central and peripheral processes. It was similarly observed that in humans, genetic influences contribute to individual preferences for fat, protein, or carbohydrates that may be related to brain neuronal functioning (Merino et al., 2021). Predictably, there's more to eating, digestion, and energy formation than brain neurochemical and hormonal influences. To understand how diet- and obesity-related illnesses come about, we need to briefly examine the digestive system.

The digestive process

Some of the primary components of food digestion are provided in Fig. 7.1. Digestion is something we take for granted, hardly giving it a thought. It is, however, a highly complex process that involves many factors necessary for cells to obtain the requisite energy to maintain survival and physiological integrity. Each organ in this process has its own function, and often they have multiple influences beyond that of digestion itself.

The digestion of food begins in the mouth through mechanical disruption (chewing or *mastication*) and moistening with saliva, which contains the enzyme amylase for the digestion of starch. As a soft, moist mass food then travels down the esophagus and into the stomach by successive contraction and relaxation of muscles (*peristalsis*). In the lower part of the stomach gastric juice containing both hydrochloric acid and the enzyme pepsin break down ingested food, while stomach muscles mix it in preparation for absorption in the intestine. The resulting thick liquid (chyme) enters the duodenum, the first part of the small intestine, where a hormone, cholecystokinin signals the gallbladder to release bile and promotes the release of digestive enzymes from the pancreas. These pancreatic enzymes break down proteins, fats, and carbohydrates, and specialized cells within the gastrointestinal tract, stomach, and

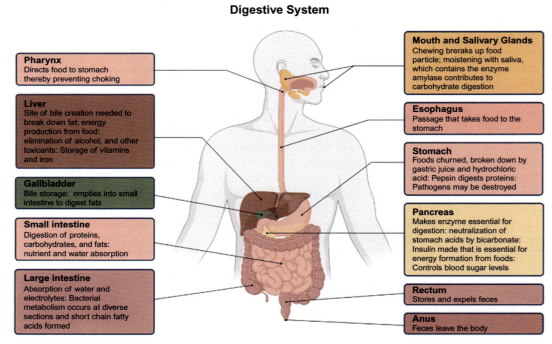

FIG. 7.1 Organs involved in the digestive process and their specific functions. *Created with BioRender.com (2021).*

pancreas (enteroendocrine cells) produce and secrete hormones that signal brain processes, thereby recruiting neural regulation of appetite and energy-related processes. Bile acids made in the liver and stored in the gallbladder also facilitate digestion by digestion of fatty foods and oils and by affecting microbiota that foster a hospitable environment for different microbes. As the chyme travels through the duodenum, several vitamins are absorbed, and in the middle portion of the small intestine (jejunum) various nutrients, such as fatty acids, amino acids, and sugars are absorbed. One of the amino acids present, taurine, may act against pathogen-provoked infection, and indeed it seems that taurine supplementation may be particularly useful in this regard.

By the time food and digestive fluids reach the large intestine a significant portion of absorption has taken place, but several important functions are still undertaken. Water and salts are absorbed, and complex carbohydrates (polysaccharides), such as the fiber and starch present in grains, potatoes, and legumes that are difficult to break down serve as a source of energy for microbiota. Through anaerobic fermentation, these microorganisms form metabolites, short-chain fatty acids (SCFAs), namely butyrate, propionate, and acetate. These SCFAs together with succinate have multiple beneficial effects; they can act against nonalcoholic fatty liver disease and type 2 diabetes, and as stated later, they can act against the development and progression of cancer, and can influence the response to therapies.

The gut is elegantly designed and constructed to maximize the extraction of energy. Certain portions of the digestive system primarily deal with nutrient absorption, while other regions contain nodes filled with immune cells that eliminate harmful pathogens. Microbiota communities also vary across the different portions of the digestive system, thereby allowing for diverse adaptive immune changes within the host.

Hormonal and brain processes underlying eating

The enteric (gut) nervous system governs the functioning of the gastrointestinal tract, sensing the presence of significant stimuli. This system involves neuronal processes that in some ways are reminiscent of those of the CNS and is sometimes referred to as our "second brain." Communication between the enteric nervous system and the brain occurs through the vagal nerve and sympathetic nervous system activity and is influenced by microbiota that reside within the gut. Once the brain receives messages from the periphery, further actions are taken by central processes that affect eating and satiety.

It had at one time been assumed that distension of the gastric wall was the main reason for feeling sated. With the discovery that manipulations of the hypothalamus could affect eating and satiety, attention shifted to focus primarily on biological processes involving hypothalamic neuronal mechanisms. It is now understood that eating and energy utilization are governed to some extent by gastric distension, as well as by numerous peripheral gut hormones and brain neurochemical processes. Macronutrients (fat and proteins) influence receptors within the small intestine leading to appetite being reduced. Furthermore, information transmitted to the brain regarding food intake occurs through activation of specific hormone receptors located on vagal nerves as well as hormone transport through the bloodstream and may be affected by the presence of microbiota (Ameku et al., 2020). Unfortunately, in individuals with obesity, critically ill patients, those with bulimia nervosa, and in some older individuals, appetite suppressing processes may not operate effectively, resulting in overeating or repeated bouts of eating, which may ultimately lead to obesity.

Leptin and ghrelin

Two hormones, leptin and ghrelin, produced by adipocytes (fat cells) and by cells in the gastrointestinal tract, respectively, are especially significant to eating and energy regulation. These hormones act on peripheral organs and through brain processes they modulate food intake, energy expenditure, and together with other hormones (e.g., cortisol), they contribute to adiposity (accumulation of more body fat), including abdominal fat storage (Abizaid et al., 2013). Over the day, ghrelin levels rise, peaking before mealtimes, thereby promoting eating, after which levels of this hormone decline. Conversely, elevations of leptin increase during and following food intake, thereby signaling satiety. For instance, among mice consuming a high-fat diet, the gut hormone, gastric inhibitory polypeptide (GIP), increases so some of it enters the blood and eventually reaches the brain, where it inhibits the actions of leptin so eating persists, thereby contributing to obesity (Kaneko et al., 2019). Disturbances of these hormones may contribute to eating-related pathologies and may thereby contribute to other diseases.

Eating-related hormones are intricately linked with hypothalamic processes, influencing HPA axis activity and affect neurochemical activity in other brain regions. Through their influence on dopamine and serotonin functioning within brain regions relevant to reward and cognitive processes (e.g., the ventral tegmentum, nucleus accumbens, and frontal cortical regions), they may affect food consumption and the choice of foods consumed and may influence mood states (Abizaid et al., 2013). The action of ghrelin on reward circuitry may also promote food craving and anticipatory responses to food. Indeed, the hedonic and nutritional value of foods may be affected by dopamine release from the ventral and dorsal striatum in ways that can be distinguished from one another (Tellez et al., 2016). Brain imaging studies also revealed that activation of dopamine receptors and the expressed preference for sweet foods in lean people, were not apparent among obese individuals, possibly reflecting disturbed dopamine neuronal connections that affect food choices and/or signals to stop eating (Pepino et al., 2016). Beyond actions on reward-related processes, ghrelin is markedly affected by stressors and through its central actions, it affects sympathetic nervous system activity, thereby influencing appetite, energy expenditure, and the use of carbohydrates as a source of fuel (Abizaid, 2019).

A comprehensive model was advanced in which it was maintained that a coalition of several factors led to obesity. These comprised (a) excessive functioning of a neuronal circuit (an amygdala-striatum circuit) that was linked to reward anticipation and processing in response to food-related stimuli, (b) diminished functioning of a prefrontal cortex inhibitory control system that impaired cognitive and executive functioning, thereby permitting unrestrained (or impulsive) food-related responses, and (c) a disturbed interoceptive awareness system (i.e., in response to stimuli within the organism) involving the insular cortex, so reward processes become unduly activated, promoting craving through the functioning of yet another hormone, agouti-related protein (Chen et al., 2018). At the same time, ghrelin stimulation of orexin (hypocretin) neurons may influence craving and stressor-related selection of comfort foods. In essence, the strong inclination favoring food consumption hijacks inhibitory systems that might otherwise limit consumption. When animals were maintained on a high-fat diet, lateral hypothalamic processes that ordinarily serve as a brake to diminish eating, seemed to lose their functioning, thereby permitting excessive food consumption leading to obesity (Rossi et al., 2019).

Various brain regions and hormones in addition to those involving the hypothalamus may be part of a complex loop in which regulatory processes are essential to maintain appropriate food intake based on energy expenditure. Specific brain regions, such as the orbital frontal cortex may play a regulatory role in food intake by integrating food-related sensory modalities, such as taste, smell, and vision. The proposition was offered that this brain region may be fundamental in orchestrating food-motivated behaviors, which is in keeping with its role in decision-making and guiding goal-oriented behaviors (Seabrook and Borgland, 2020), although dysfunctions of other brain regions, such as those related to reward processes, could foster the development of obesity.

Insulin and related peptides

Insulin produced by pancreatic beta cells is essential for the cells to take up glucose from the blood so that it can be converted to glycogen, which is then stored in cells. When insulin is not produced by pancreatic beta cells, as in the case of type 1 diabetes, glucose is not readily removed from the blood, and the resulting glucose excess can be toxic to normal cells.

Type 2 diabetes, in contrast, does not result from an absolute reduction of beta cells. Rather, it typically occurs when the cells in muscle, liver, and fat fail to respond to insulin (called "insulin resistance") and thus do not effectively regulate the glucose in the blood. Numerous factors contribute to the emergence of type 2 diabetes, notably genetic influences, being overweight, beta-cell disturbances, inappropriate cellular responses to messages related to eating and satiety, lack of exercise, and chronic stressor experiences. Insulin also acts in conjunction with many other hormones—including glucocorticoids, leptin, neuropeptide Y, orexin, agouti-related peptide, glucagon-like peptide-1, α-melanocyte-stimulating hormone—to promote obesity and metabolic disturbances. Importantly, interactions between leptin and insulin influence energy regulation and dopamine-related reward processes, which may explain why altered mood states occur when insulin is low.

Glucose is a major fuel for all cells. Hypoglycemic conditions can produce proinflammatory states that favor multiple diseases, including myocardial infarction. And when extended for long periods, it can impair neurocognitive functioning, promote acute cerebrovascular disease, and induce retinopathy and vision loss. The availability of energy sources is fundamental to cancer growth (Warburg effect), so not surprisingly insulin-related processes have been linked to cancer progression. Accordingly, some auxiliary treatments that discourage elevated glucose levels could limit cancer growth.

Incretins

Gut peptides, such as incretin hormones are secreted after nutrient intake, serving to promote insulin secretion. The best-known incretins are glucagon-like peptide-1 (GLP-1) and glucose-dependent insulinotropic polypeptide (GIP). GLP-1 agonists serve as powerful antidiabetic agents and have gained recent attention because of their ability to promote weight loss (Nauck and Meier, 2018).

In addition to being stimulated by glucose, gut microbes may produce proteins that signal the release of the GLP-1 from cells in the lower gastrointestinal region. This acts on the pancreas to stimulate insulin release and modify pancreatic glucagon secretion. Treatment with GLP-1 receptor agonists has therefore been used to reduce high glucose levels associated with type 2 diabetes and as a bonus, these agents produce food-related anhedonia leading to weight loss. When combined with behavior therapy, GLP-1 agonists produced weight reductions that were maintained for 24 weeks, although appetite was back to pretreatment levels by 52 weeks (Tronieri et al., 2020). Although there had been concerns that GLP-1 agonists might favor the occurrence of thyroid and pancreatic cancer, detailed analyses indicated that these agents did not pose risks for these or other forms of cancer.

The various gut peptides interact with one another to produce eating and satiety, and as we will discuss shortly, they may also influence brain processes that affect the reinforcing value of food and may thus contribute to obesity. In essence, while eating may be essential for survival, cues associated with food can elicit "wanting" or "liking" that may involve different neural circuits. As well, hormones, such as ghrelin, are sensitive to the actions of stressors and through interactions between energy signals and immune processes may affect the regulation of metabolic processes (Abizaid, 2019). Fig. 7.2 shows the involvement of several peptides in regulating appetite and hunger.

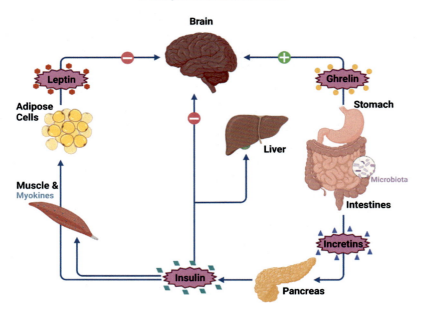

FIG. 7.2 Leptin and ghrelin are key hormones that influence appetite, hunger, and satiety. Elevated levels of ghrelin (right side of the figure) produced by specialized cells that line the stomach and the pancreas stimulate eating, and then decline after food has been consumed. As this hormone affects brain reward processes, certain foods will eventually be favored. Leptin secreted by white adipocytes [adipose (fat) cells] *acts to suppress appetite and regulate energy balances* (shown on left side). When leptin levels decline, appetite and food cravings increase. Moreover, microbiota within the gut that are influenced by diet and exercise (and other psychosocial influences) can affect energy regulation and may influence weight changes. In addition to these hormones, incretin peptides, such as glucagon-like peptide-1 (GLP-1) and glucose-dependent insulinotropic polypeptide (GIP) secreted from enteroendocrine cells in the intestinal epithelium, facilitate the coordination of metabolic responses associated with food ingestion, and thus can be used to treat diabetes and can be used to promote weight loss. Insulin produced and secreted from the pancreas allows glucose to be taken up into cells and affects brain regulatory processes associated with reward processes. Insulin also influences the liver that is necessary to process the nutrients absorbed from the small intestine. The bile secreted from the liver into the small intestine also serves in the digestion of fat and certain vitamins, as well as brain mechanisms related to reward processes. Through its action on protein synthesis and the availability of glucose, insulin affects the availability of energy to muscles. The muscle cells influence adipocytes and in response to exercise skeletal muscles release myokines (e.g., irisin and IL-6) that contribute to energy regulation and the presence of inflammation. The adipose cells release leptin that affects feelings of satiety. Several other hormones (not shown in the figure), such as cortisol, insulin, neuropeptide Y, and Neuromedin B, affect eating processes, as do serotonin and dopamine. Aside from these processes, hormones (e.g., estrogen and testosterone) and growth factors also contribute to energy-related processes and fat accumulation and distribution. *Created using BioRender.com.*

Forms of fat, fibers, and processed and unprocessed foods

To keep people healthy, legislation across many countries requires that food labels appear on packaged food products. This provides consumers with nutritional content for a more informed dietary decision. However, it is not unusual for people to either ignore these labels or misunderstand the information provided. These labels also do not contain complete information about the food in question. Specifically, food labels show the fat content of food, but different types of fat have different implications for health. Fat is an essential ingredient for well-being, serving as a source of energy, a key structural element of cell membranes, and a requirement for muscle movement, blood clotting, and optimal inflammatory responses. But not all forms of fat are equally beneficial, and some can be harmful when their levels are too high. The main distinction is between what are known as saturated and unsaturated fats.

Unsaturated fats often appear in a liquid form, such as vegetable oils (e.g., olive and sunflower oil, but not palm or coconut oil) and are obtained from a variety of seeds and nuts, particularly almonds, hazelnuts, and pecans. These fats may have health benefits, such as reducing low-density lipoproteins (LDL) cholesterol, thereby reducing the risk of some illnesses. Polyunsaturated fats (PUFAs), a form of unsaturated fat, come from foods such as fatty fish (e.g., wild salmon, tuna, sardines, herring, trout, and mackerel), some nuts and seeds (sunflower, hemp, chia, and flax seeds), and soybeans. PUFAs come in several forms, which can be distinguished from one another by their chemical signatures. Omega-3 PUFAs, which are high in fish, may reduce disease-causing inflammation, whereas omega-6 is suspected to have a proinflammatory effect, although it has been maintained that it can facilitate heart health.

Saturated fats typically appear in a solid form. These come from fat cuts of meat (beef, pork, lamb, and processed meats), dairy products (whole milk, cream, butter, and some cheeses), and numerous baked and processed foods. Saturated fat has long been viewed as creating health risks, increasing "bad" low-density lipoprotein (LDL) cholesterol levels which contributes to heart disease—although there is debate on this issue. In contrast, there is no debate at all concerning the health risks associated with trans fat consumption. Trans fat is created through an industrial process that adds hydrogen to liquid vegetable oils to make them more solid. Its use increased appreciably as a substitute for lard and butter in bread, cookies, and pies and is found in common fast foods (e.g., french fries, fried chicken, battered fish, and fried noodles). By increasing LDL and reducing "good" high-density lipoproteins (HDL), trans fat increases the risk of type 2 diabetes, heart disease, and stroke. Moreover, trans fats can increase reactive oxygen species (ROS) production by mitochondria, which may cause DNA damage.

Food labels also indicate fiber content, but the amount of fiber indicated hardly gives the consumer the information that is needed to determine whether the fiber is or is not beneficial. Fiber is broadly classified as being soluble in water (e.g., dried beans, oats, barley, bananas, potatoes, and soft parts of apples and pears) or insoluble (whole bran, whole grain products, nuts, corn, carrots, grapes, berries, and the peels of apples and pears). In general, soluble fibers diminish cholesterol levels, and by slowing down the speed of digestion, blood sugars will not rise quickly. Insoluble fibers, which serve as roughage, helps regulate bowel movement by drawing water into the colon and may reduce food intake by promoting a sense of fullness. By affecting the movement of food through the intestine, this type of fiber may also limit the development of intestinal disorders. Of particular importance (as described earlier) is that the enzymes needed to break down fiber are absent in humans. Therefore, these foods reach the gut undigested, where bacterial enzymes break them down. As such they serve as food for beneficial bacteria, promoting the presence of SCFAs that can enhance health. Consumers, however, typically do not know the function of different fibers or the health benefits that these provide. They are likely unaware that owing to regulatory changes, manufacturers can use synthetic carbohydrates as a component of the fiber count for foods. Moreover, reference to the use of terms such as "containing whole grain" or "multigrain" says nothing about the amount or source of the grain. The net result has been that people frequently consume a diet comprising low-quality carbohydrates.

Even examining the calories present on food labels can be misleading. Unprocessed foods that reach the consumer relatively unmodified (e.g., plants, fruits, and edible parts of animals), require more energy to digest and absorb them. Processed foods are modified with sugars, salts, pasteurization (which can limit bacterial infection), and other interventions that lower the energetic cost to the body which contributes to weight gain, even if the calorie content is the same as for an equivalent serving of unprocessed food. Common processed foods include salted, cured, or smoked meats, canned fish, cheeses, and "preserved" foods, such as vegetables, fruits, and legumes, and salted or sugared nuts and seeds. These foods are generally considered unhealthy, and if there are excessive levels of added sugars and salt their consumption may well create health risks. Adding to the confusion for consumers is that the term "processed foods" is deceptive and not necessarily indicative of reflecting a poor dietary choice. Just about anything done to foods, labels them "processed." Technically, frozen fish, as well as frozen and canned vegetables, are processed foods, but they are not unhealthy.

Ultra-processed foods and beverages that are spiked with multiple additives, including more sugar and salt, are unquestionably health hazards. Yet, such foods constitute about 70% of the products in the typical supermarket: these tantalizing "comfort" foods comprise pastries, cookies, chicken nuggets, frozen pizza, hot dogs, and carbonated or energy beverages. Some additives are meant to make foods look better or allow them to sit on the shelves longer without losing their moisture. The consumption of ultra-processed foods among children and adolescents has increased appreciably over the past two decades, while that of unprocessed foods has declined. The presence of the preservative tert-butylhydroquinone (TBHQ), may disturb immune functioning, and the safety of titanium dioxide nanoparticles, which are used in many foods, is poorly documented, despite evidence that it contributes to the formation of biofilms (in which bacteria stick together) that potentially have adverse consequences.

In view of the multiple issues related to ultra-processed foods, trans fat, added sugars and salt, and low-quality carbohydrates, it isn't at all surprising that they have contributed to a rise in obesity, metabolic diseases, inflammatory bowel diseases, and several forms of cancer (Gourd, 2018).

Diets and weight loss

Microbiota, short-chain fatty acids, fibers, and polyphenols

Food digestion and weight change are known to be affected by gut microorganisms. These can also drive the development or prevention of a variety of diseases, affect immune functioning and inflammatory processes, supply essential enzymes, and enhance gut barrier functioning. The composition of probiotics includes beneficial bacteria abundant in foods such as yogurt, sauerkraut, pickles, sourdough bread, miso, some cheeses, as well as kefir, kombucha, and kimchi. Prebiotics are also present in a variety of foods, but they serve to promote the growth or activity of gut bacteria and fungi that are already present. Through immunoenhancing effects, they can ward off pathogens, as well as aid in mineral absorption, metabolism, and bowel functioning (Sanders et al., 2019). Different prebiotic fibers serve as substrates for different bacteria and consequently influence the nature of bacterial communities in the gut. Numerous prebiotic foods or food components (e.g., glucose, human milk, polyphenols, oligomers of mannose, pectin, starches, and xylose) might also turn out to have positive actions. Prebiotic fibers such as fructans and galactooligosaccharides increase the abundance of *Bifidobacterium* and *Lactobacillus*, bacteria thought to promote positive health effects. Conversely, depriving mice of a fiber diet alters microbiota diversity throughout the colon (Riva et al., 2019) that may impair health. Before going further, it is important to add that food and microbes in relation to health has become a veritable marketplace of ideas as would be expected given the variety of other natural products, such as plants and mushrooms, that are thought to diminish obesity and type 2 diabetes. But, in evaluating the benefits of pro- and prebiotics it is essential to distinguish between those obtained from foods versus those that are obtained as supplements, typically in pill form.

A diet high in polyphenols—common in citrus fruits, cocoa, red wine, tea, and coffee—has pronounced microbial actions, with multiple potential health benefits via antioxidant and antiinflammatory actions. Most dietary polyphenols accumulate in the large intestine, where they are converted to less complex metabolites that can have similar actions to prebiotics, affecting the growth of specific forms of gut microbiota. It also seems that polyphenols can protect against stressor-provoked neuronal damage associated with neuroinflammation and oxidative challenges, possibly through influences on corticoid processes related to the adrenal hormone, cortisol. In addition to affecting brain processes, polyphenols may also work against the development of heart disease and some forms of cancer. Polyphenols may also interact with other food substances, such as polysaccharides, so their nutrient properties are enhanced. Regrettably, polyphenols can also promote the growth of "bad" bacteria—*Helicobacter pylori*, *Staphylococcus aureus*, *Salmonella typhimurium*, and *Listeria monocytogenes*, emphasizing again that the balance of different gut bacteria is fundamental to well-being. With so many polyphenols found in individual diets, it can seem insurmountable to determine which provide the greatest health benefits, and under what conditions this can be expected.

The actions of different carbohydrates, fats, and proteins on various microbes can affect health through immune, inflammatory, and metabolic processes. As we have mentioned on several occasions, there frequently isn't a one-to-one correspondence between specific microbes and illness (or wellness) although there are core microbes that play a significant role in this respect. Most often, food consumption is not restricted to a single food category, producing diets that reflect a blend of diverse microbes. It may be more important to think less of specific bacteria and more about their diversity, particularly as the variability or plasticity of the gut microbiota throughout a diet regimen may be fundamental in determining long-term weight loss and health benefits.

Setpoint, diet, and fasting

Dieting or food restriction is the most common approach to weight loss. Too often, however, this is ineffective, and in some instances may have effects opposite to those intended. It is thought that a "set point" exists, varying across individuals, such that when the weight falls to critical levels, compensatory metabolic adjustments occur to limit further weight changes. In response to dieting, body cells may not understand that this is intentional and thus go into a "famine preparedness mode." As an example, peptides like agouti-related peptides can affect neurons in the hypothalamus that integrate information obtained from exteroceptive and interoceptive food-related cues to determine whether calories should be burned or conserved (Burke et al., 2017). To the dismay of dieters, once they choose to eat normally again, body cells seem not to be in on this decision and thus continue in the famine mode so calories continue to be conserved. As a result, weight gain may occur even in the absence of excessive eating. The response to this might be that of further dieting, which only perpetuates the cycle of weight loss and gain (a pattern referred to as "yo-yo" dieting).

Speaking to this, it has been suggested that the nature of foods consumed may operate on similar principles in driving weight gain. The notion generally adopted is that more energy consumed relative to that expended results in weight gain. However, from what has already been described, this notion is overly simplistic. Among other things, to account for weight gain it is necessary to consider the metabolic and hormonal actions of the foods consumed. It has been maintained that the progressive rise of obesity that has been apparent over decades is more appropriately attributed to the increase in the consumption of processed foods that comprise readily digestible carbohydrates that influence metabolic processes and fat storage. These foods markedly increase insulin secretion and concurrently limit the release of glucagon, which then causes fat cells to increase their storage of calories, which diminishes energy available to muscles and metabolic functioning. Under these conditions, the brain misinterprets these signals as a need for food, but even as foods are consumed hunger will persist and at the same time the imperative to conserve foods continues unabated (Ludwig et al., 2021).

Choosing a diet: No diet works for everyone, and every diet works for someone

With so many diets potentially having positive effects, the obvious question arises as to which diet is best. This will be dictated by the individual's goals and motivation. People might wish to lose weight to look better, whereas others might diet to maintain or restore good health. Individuals often make successive efforts to reduce weight, and even when they are successful in fitting into new clothing and take pride in their achievement, often they readily regain the weight that had been lost. Thus, in selecting an appropriate diet, it isn't enough to choose one that facilitates weight loss, but also to consider an approach that helps individuals maintain weight loss. Temptation is a terrible enemy, and cravings are a potent driver in undermining diet adherence. Individuals shouldn't randomly select a diet, but instead, calibrate their desires to what they can do and what is appealing so that temptation can be averted. There is also ample reason to believe that diets should be recommended based on individual differences, including psychological factors, as well as the nature of the individual's gut milieu. Recommendations to promote behavioral change to sustain healthy living are provided in Chapter 13, so we will leave this aspect aside for the moment.

It's certainly difficult to make the right diet choices given that diet gurus often have vested interests as to which diet is best. Some common diet programs have focused on reducing fat consumption, whereas others recommended reduced carb intake. Among individuals with obesity, each of these diets did some good, even after considering the influence of social support obtained and exercise regimens engaged (Johnston et al., 2014). The essential ingredient to weight loss was to adhere to the diet consistently. Predictably, if a particular diet is too stringent or if it is too unappealing, individuals will fall off the wagon, often never returning.

To make matters more problematic for those wanting to engage in a healthy diet, recommendations offered by health agencies seem to change every few years. The notion that coffee was a carcinogen had been around for ages, but the idea had been percolating that through its antiinflammatory and antioxidant actions, as well as by acting on sex hormone levels, it may be beneficial in limiting the occurrence of liver, endometrial, oral/pharyngeal, colorectal, and advanced prostate cancer.

At one time, eggs had been viewed as having health benefits, but then the tide turned, and eggs were implicated as a culprit for bad cholesterol accumulation. Not long afterward, eggs again went to the good side of the ledger, but now they're back on the negative side. Salt, which had been warned against for decades to prevent heart disease is now considered less culpable unless very high levels are consumed, and admonitions against high fat intake, including that obtained through milk products, have met serious challenges. Although milk had been suggested as a way of enhancing bone health, this may not be accurate and there have been indications that consumption of whole milk may be related to prostate and endometrial cancer owing to the hormones present in milk. Given the diversity of opinions and the ever-changing recommendations, is it any wonder that confusion often exists among many people regarding which foods are healthy.

Not surprisingly, numerous herbal treatments have been used to lose weight. These have ranged broadly from turmeric, cinnamon, cayenne pepper, through to ginger, rosemary, saffron, and garlic. Most of these have not been found effective, although there is some evidence favoring the benefits of curcumin that is part of turmeric preparations, and formulations containing multiple herbs and spices were also said to be effective. A meta-analysis that assessed 54 randomized, placebo-controlled trials indicated, not unexpectedly, that most herbal medicines that entail whole plants or combinations of such plants, such as *Camellia sinensis* (green tea), *Garcinia cambogia* (Malabar tamarind), *Phaseolus vulgaris* (white kidney bean), and *Ephedra sinica* were largely ineffective in reducing weight (Maunder et al., 2020).

For decades short cuts to weight loss have been attempted through the development of a variety of drugs. These treatments often produced serious adverse effects, as in the case of fenfluramine/phentermine (often called fen-phen), which provoked mitral valve dysfunction and pulmonary hypertension. Other attempts have evolved to fool the body so that individuals feel full. This entails consuming a cellulose compound together with half a liter of water, resulting in the cellulose swelling and creating a feeling of "fullness" so that less food is consumed. Another approach to producing a state of fullness comprised the use of the drug fexaramine that causes bile acids to be released, thereby causing eating cessation. This reduced blood sugar and cholesterol levels and increased brown fat that burns energy quickly (Fang et al., 2015). However, the effects in humans are uncertain and clinical trials are not presently planned.

Brain serotonin processes were long thought to be important in determining food intake, and manipulations that influenced serotonin functioning were offered as possible ways of altering food intake. Peripheral serotonin manipulations could also be a player in acting against obesity. Among mice deficient in the enzyme tryptophan hydroxylase 1, which is necessary for serotonin formation, a high-fat diet did not lead to obesity or insulin resistance because of the promotion of brown adipose tissue thermogenesis (Crane et al., 2015). Agents such as lorcaserin (Belviq), which acts on certain serotonergic receptors, reduced weight by about 5%–10% and had received approval as an antiobesity agent (Higgins et al., 2020), but the FDA has since reconsidered given the possibility of this agent promoting several forms of cancer. As indicated earlier, GLP-1 receptor agonists that are used to diminish glucose levels in type 2 diabetic patients are effective in producing weight loss in patients. It also appeared that when used as an adjunct to exercise and diet, some GLP-1 agonists can promote weight loss among overweight individuals who were not diabetic (Wilding et al., 2021). The fat reduction promoted by a GLP-1 agonist in overweight people showed that this occurred primarily in abdominal tissue and within the liver, which has implications for the development of illnesses stemming from inflammatory factors released from visceral fat (Neeland et al., 2021).

Impact of periodic fasting

Despite the well-known harms created by yo-yo dieting, periodic fasting has become ever more popular as an effective way of reducing weight and may have an array of health benefits. Relative to simple diet restriction, some forms of intermittent fasting were accompanied by improvements in glucose levels, LDL cholesterol levels, triglycerides, and free fatty acids. As described in their detailed account, de Cabo and Mattson (2019) indicated that intermittent fasting can diminish age-related diseases, such as cardiovascular disease, type 2 diabetes, neurodegenerative disorders, tissue injury stemming from traumatic or ischemic injury, autoimmune conditions, and cancer.

The benefits of intermittent fasting, which has been seen across species, may come about through the effects on a variety of biological processes. In the small worm *C. elegans*, diet restriction can influence mitochondrial homeostasis, leading to increased longevity. In phylogenetically higher species, such as mice, a restricted diet enhanced the functioning of T memory cells, which exhibited stronger responses after reexposure to previously experienced bacteria and protected them from illness (Collins et al., 2019). Finally, enhanced longevity and delayed illness onset were apparent in the relatively long-lived rhesus monkey that had been maintained on a restricted diet (Mattison et al., 2017).

Fasting can also act against the development of illnesses elicited by bacterial infection. For example, fasting mice on a repeated, periodic water-only diet exhibited diminished chemokine functioning related to B cells but augmented memory T cell responses together with altered aspects of gut immune function. More to the point, mice that had been fasted for 48 h prior to the oral administration of *Salmonella enterica* subsequently presented with diminished signs of infection, inflammation, and intestinal tissue damage (Graef et al., 2021). The diminished food intake that often appears in conjunction with an infectious illness is often thought to reflect a symptom of illness; however, it may also comprise an adaptive response to diminishing the actions of the pathogen.

A variant of intermittent fasting has been introduced in which eating is restricted to a set period, often about 10 h a day (8:00 a.m. to 6:00 p.m.). This diet may align with circadian rhythms, producing weight loss in obese humans and ultimately favoring reduced visceral fat, enhanced glucose regulation, increased stress resistance, and suppressed inflammation. These fasting protocols not only limited physical health and disease processes but also affected neuroplasticity and brain health (Mattson et al., 2017).

Not surprisingly, fasting or calorie restriction can produce marked enteric changes. This includes an increase in butyrate presence as well as lipid biosynthesis, fatty acid catabolism, glycogen production, and hence gluconeogenesis. This also coaxed some microbial factors to assume a healthier state. In line with the view that microbiota are responsible for weight loss associated with calorie restriction, when mice received fecal transplants from other mice that had been calorie restricted, they were resistant to the weight gain associated with a high-fat diet and metabolic improvements were evident in the fecal recipients (Wang et al., 2018). A 4-day diet that mimicked fasting conditions similarly promoted the presence of protective gut microbiota and diminished intestinal inflammation. These diet regimens increased the abundance of beneficial bacteria *Lactobacillus* and *Bifidobacterium* and reduced *Helicobacter*, which has been aligned with negative health outcomes, as well as increasing microbial pathways that promote antioxidant actions and reduced IL-17 producing T cells (Cignarella et al., 2018).

The health benefits of fasting diets might come about because a moderate (25%) reduction in daily calorie intake elicits antiinflammatory actions, limiting risks for inflammatory-related diseases (Youm et al., 2015). In Chapter 14, we'll discuss the influence of diets, including ketogenic diets, on cancer progression in some detail. It will suffice for the moment that caloric restriction may have its beneficial effects on cancer through effects on hormonal processes as well as by limiting inflammatory (cytokine) conditions owing to reduction of the regulatory protein p38. Intermittent fasting also diminished oxidative stress, stimulated reparatory processes, and was accompanied by gene expression changes associated with longevity. Significantly, a 50% reduction of calorie intake was sufficient to augment protection against cancer (Collins et al., 2019) and provoked T cell-dependent killing of cancer cells. Moreover, a fast-mimicking diet enhanced the effects of chemotherapy in breast cancer patients (Caffa et al., 2020). The mechanisms for these actions have not been discerned with certainty, but they may involve changes in metabolic factors such as leptin, insulin, and insulin-like growth factor 1 (IGF1), immune variations, and reduced inflammation.

It will be recalled that autophagy is a general process that regulates innate immunity and cancer survival. This process can be triggered by relatively prolonged fasting (usually exceeding 24 h) or calorie restriction. The induction of autophagy by pharmacologic or dietary interventions may be useful as an adjunctive treatment to diminish cancer progression, although the viability of this approach is still uncertain. However, while autophagy instigated by fasting may act against cancer and could potentially be used as an adjunct to standard treatments, it could also have pro-tumorigenic actions, depending on the stage of tumorigenesis and the mutational load present (Levy et al., 2017). Therefore, the induction of autophagy that works against cancer remains a challenge.

Diet and nutrition in relation to cancer

Numerous studies have assessed nutritional contributions to obesity and other diseases, but from the outset, it needs to be reemphasized that the reliability and validity of some of these studies were considered suspect. Without disparaging all the relevant research, Schoenfeld and Ioannidis (2013) argued that cancer risk or benefits attributed to specific foods and diets have been claimed far too often, and many of the single studies provided effects that were simply too impressive to be credible. Indeed, self-reports of caloric intake were in some cases so extreme as to be "incompatible with life" (Ioannidis, 2013). Aside from the difficulty in keeping track of actual foods consumed, studies frequently failed to consider the different ingredients that are used in food preparations and didn't adequately distinguish between different alcoholic drinks. Findings from diet studies were also challenged based on the statistical methods used, the frequent studies that comprised too few participants, and logical fallacies that permeated the conclusions drawn from the empirical findings.

In their critique of the literature regarding diet and diseases, Schoenfeld and Ioannidis (2013) indicated that virtually every food product has been evaluated and each was frequently linked with illness development or reduction. In this regard, many common foods (tea, milk, coffee, eggs, corn, butter, and beef) have been associated with either increased or decreased cancer incidence. To be sure, most studies indicated that olives, onions, and wine have most often been associated with reduced cancer risk, whereas pork and beef were linked to greater cancer occurrence, as was too much alcohol consumption. In fact, the single most significant epidemiological link to some forms of cancer (e.g., breast cancer) was alcohol consumption, irrespective of whether this occurred relatively early or late in life. Moreover, alcohol drinking was associated with the initial occurrence and subsequent recurrence of breast cancer in postmenopausal women, which may stem from actions on estrogen activity or alcohol breaking down into acetaldehyde that can act as a carcinogen. Additionally, alcohol causes oxidative stress which can damage cells and may disturb the ability to convert nutrients from food as well as changes of energy metabolism and the consequent emergence of obesity. The obvious question is how much alcohol consumption is too much. Predictably, the risk for many cancers increased in direct relation to the amount of alcohol consumed, varying across different types of cancer (Bagnardi et al., 2015), but even light alcohol consumption can promote cancer occurrence among women.

Needed evaluation of the anticancer actions of diet ingredients

Criticisms of specific diet ingredients shouldn't be (mis)interpreted as suggesting that particular foods never have positive actions, nor that "biotherapeutics" cannot be developed to enhance the effects of cancer therapies. Taxol obtained from the bark of the Pacific yew tree is among the best-known natural agents used to treat breast cancer. As a result, the roots, leaves, or needles of other trees and bushes were advanced for similar purposes. Graviola and its relative Pawpaw were enlisted in this effort, which provided fodder for conspiracy theorists who informed readers that these compounds were a breast cancer breakthrough remedy that was banned by government agencies. Unfortunately, flakey ideas permeate internet sites that offer untested natural remedies to fight a range of cancers. Every food imaginable seems to be on the list of cancer-beating drugs. Blueberries, artichokes, green juices, carrots, and a carrot-apple mix, and garlic seems to be a favorite (of course, garlic may be especially useful, doubling as a way to ward off vampires…). Many foods may well have health benefits (e.g., perhaps through their antioxidant properties), but naturally occurring agents have frequently been advanced to treat cancer based on little or no experimental data.

Most of these "natural therapies" haven't gone through adequate testing, although in some instances, evidence-based information has been provided, as in the case of curcumin, a primary component of turmeric. This said, it becomes a bit suspicious when curcumin, like several other natural products, has been purported to have beneficial effects that range from anticancer actions, a way of diminishing signs of Alzheimer's disease, and a treatment for erectile dysfunction. But, unlike so many other natural remedies, curcumin has frequently been assessed for its remedial actions.

It was maintained that curcumin can have anticancer effects by inhibiting inflammation-related transcription factors NFκB and STAT3 or by inhibiting the gene Sp-1, which through actions on other genes can favor cancer occurrence (Vallianou et al., 2015). The effects of curcumin on inflammatory diseases, including cancer, may also come about through inhibition of extracellular TLR2 and TLR4, as well as intracellular TLR9 (Boozari et al., 2019). Curcumin was also alleged to act against breast and lung cancer by affecting a variety of signaling pathways, including those mediated by p53, Ras, PI3K, protein kinase B, Wnt-β catenin, and mTOR, although it is uncertain which if any of these might be uniquely responsible for the actions of curcumin.

Even though findings pertaining to curcumin have been encouraging, in many instances in which single types of foods have been proposed to have anticancer properties, the supporting evidence in humans has been limited, and much more data are necessary before these positions are given appreciable credibility. This has also been apparent with curcumin. A meta-analysis of more than 120 trials questioned the health benefits of curcumin, indicating that not a single double-blind placebo-controlled trial supported any benefits in relation to illnesses (Nelson et al., 2017). Moreover, curcumin makes up only a small fraction of turmeric, and other ingredients may act along with curcumin in providing beneficial actions. Finally, even if curcumin were to have positive effects this would more likely be as an adjuvant with other therapies rather than a stand-alone treatment. In this regard, curcumin has multiple effects on processes related to cancer and could thus offer benefits in cancer prevention as well as being an effective adjunct along with standard therapies. This said, the correctness of these assertions needs to be fully evaluated in well-controlled trials.

Food constituents and cancer

A very large number of retrospective and prospective studies have been conducted assessing the relationship between consumption of specific foods (and broad diets) in the development and progression of diverse forms of cancer, the response to different cancer therapies, as well as during the post-therapy period. While unanimity has not been reached (it rarely ever is) as to which diets are healthiest, many systematic reviews and meta-analyses have been conducted to summarize the existent data, and we'll rely on some of these to make sense of the information available.

Prospective cohort studies have generally indicated that increased intake of whole grains, vegetables, fruits, nuts, and fish was accompanied by reduced risk of early death, whereas increased consumption of red meat and processed meat was accompanied by earlier all-cause mortality (Schwingshackl et al., 2017). In calculating the impact of particular foods on lifespan, it is not only important to determine the direct effects of food on life but also to consider the environmental impact of producing these foods that can indirectly affect lifespan. An analysis that ranked 5853 foods based on their nutritional disease burden to humans and the environmental effects produced, suggested that substituting just 10% of daily caloric intake from beef and processed meats to a combination of fruits, vegetables, nuts, legumes, and certain seafood could add 48 min of healthy minutes a day and concurrently reduce the carbon footprint by one-third (Stylianou et al., 2021).

Studies that tracked individuals over many years have indicated that total diet rather than any single food was tied to all-cause and cancer-related mortality. Diets with high inflammatory potential were most closely linked to increased cancer risk, especially when accompanied by diabetes and smoking.[b] Sustained maintenance of some diets can engender persistent variations of microbial factors that could lend themselves to broad disease susceptibility

---

[b] Some food products or their modifications may represent lurking dangers that frequently escape detection. The loads of oil present in fried food are well known to be bad for our health, but some oils are worse than others. "Thermally abused frying oil," referring to oil that has been repeatedly reused (as occurs in most chip stands) resulted in cancer cells becoming more invasive and aggressive in mice. This might have occurred because the thermally abused oil promoted altered expression of genes, such as those associated with oxidative stress. It also seems that the thermally abused oils awaken dormant tumor cells or alter the microenvironment so it is more hospitable to cancer cells (Cam et al., 2019). Besides, when cooking oil is reused, a carcinogenic substance, acrolein, is released, and toxic polymer molecules accumulate.

including disorders involving inflammatory processes. A 22-year prospective study among nurses indicated that diets that promoted inflammation during adolescence and early adulthood favored the occurrence of premenopausal breast cancer (Harris et al., 2017). Congruent with these findings, data from a prospective analysis as part of the Women's Health Initiative indicated that the "diet inflammatory index," which was based on a food questionnaire, was aligned with estrogen and progesterone negative and HER2 positive subtypes of breast cancer. Significantly, breast cancer incidence was correlated with baseline inflammatory diet, even after controlling for several other variables (Tabung et al., 2016). Proinflammatory diets were similarly associated with elevated colorectal cancer among men and women who had been followed over 26 years.

The notion that food consumption could be tied to cancer is eminently reasonable but attempting to make these linkages to specific food items has been difficult given how many different foods most individuals consume, which may also be influenced by other variables that could directly or indirectly affect the cancer process. Added to this is that the ties between specific foods and cancer occurrence may be confounded by other factors tied to food choices. Stressors could promote carbohydrate craving and consumption, which could affect weight gain and obesity, and hence cancer development. Living in poverty can influence the consumption of unhealthy foods and may come with other risk factors that may favor cancer occurrence. The source of foods might also affect the development of illnesses. Fruits and vegetables that are sprayed with pesticides, such as the widely used chlorpyrifos, can diminish the burning of calories within brown fat, causing diet-induced thermogenesis so calories are stored, thereby favoring the development of obesity (Wang et al., 2021). Thus, while fruits and vegetables may be an important component of a healthy diet, the presence of pesticides may have adverse effects. A large-scale study similarly indicated that consumption of organic foods (which do not contain manufactured pesticide residues) was associated with a lower overall occurrence of cancer relative to consumption of the same foods that were not organically grown (Baudry et al., 2018).

There has been a concerted movement to identify genes and other markers that predict the impact of foods on disease occurrence and progression. The simple fact is that there is likely no optimal health-promoting diet or specific food that is appropriate for *all* individuals. The case hasn't been closed as to which specific foods increase the risk for cancer occurrence, although it is believed that nutritional factors and the development of obesity can affect multiple biological processes that lend themselves to disease occurrence. Genetic and epigenetic factors, microbiota, eating-related hormones, and metabolic processes may all play into diets that are favorable or unfavorable, once again speaking to the possible need for a precision medicine approach for health and disease (e.g., Kolodziejczyk et al., 2019). To do so it will first be necessary to determine the doses (amounts) of specific (or combined) dietary components that favor positive health changes without eliciting adverse effects. Some of the many factors implicated in cancer occurrence stemming from diet, nutrition, sedentary behaviors, and obesity are summarized in Fig. 7.3.

Fiber

Systematic reviews and meta-analyses of prospective studies and clinical trials repeatedly indicated that high dietary fiber or whole-grain consumption was associated with reduced

FIG. 7.3 Lifestyles (poor nutrition, lack of exercise, disturbed sleep, chronic stressors) influence a constellation of biological processes that favor the promotion and exacerbation of several types of cancer. Some of these entail actions that favor the development of cancer cells, whereas others contribute to the behavior of these cells.

occurrence of colorectal, breast, and liver cancer, type 2 diabetes, heart disease, and stroke (e.g., Reynolds et al., 2019). In a large Japanese cohort that had been followed for approximately 18 years, the consumption of plant proteins was accompanied by reduced cancer- and heart-related mortality relative to that of individuals who consumed greater amounts of meat products (Budhathoki et al., 2019). Reduced cancer risk and daily consumption of foods high in fiber (particularly nuts) as well as fruits and vegetables—and avoiding red meat, processed meats, and salt—was associated with a diminished occurrence of mouth, esophageal, stomach, and bowel cancer. Fibers and sulforaphane (a sulfur-rich compound) obtained through cruciferous vegetables (broccoli, kale, cabbage, collard greens, brussels sprouts, and cauliflower) could similarly help in the prevention of prostate cancer.

Based on 43 cohort or randomized controlled trials it was estimated that a modest increase in fiber consumption could reduce cancer risk by 10% (Aune et al., 2011). More recent analyses that had been based on numerous clinical trials had similarly indicated that a diet containing high dietary fiber may be accompanied by a 15%–30% reduction of mortality related to colorectal and breast cancer as well as other chronic health conditions, such as heart disease and type 2 diabetes (Reynolds et al., 2019). Aside from affecting cancer occurrence and progression, a review of 29 prospective studies indicated that diets that comprised low fat, but high in fiber (vegetables and high-quality protein) were associated with reduced risk of mortality following treatment of some forms of cancer (Rinninella et al., 2020). This study, however, indicated that food categories, such as meat and dairy, needn't be eliminated following therapy to obtain the advantages of high fibers.

The benefits obtained from foods high in fiber have been attributed to the actions of SCFAs, but other factors can contribute to health. For instance, cruciferous vegetables contain a compound that influences *WWP1*, a gene that increases the presence of the tumor suppressor phosphatase and tensin homolog (PTEN), which is ordinarily diminished in many cancers. Indole-3-carbinol, another compound present in cruciferous vegetables, similarly acted on PTEN by affecting *WWP1*. This might turn out to be an important discovery for future drug development, but for the moment the positive effects have only been shown in mice. It would be ideal if eating cruciferous vegetables could produce such an outcome, but this isn't a viable option as humans would have to eat more than they actually could to obtain any benefits. Thus, these foods need to simply be part of a healthy diet, including avoidance of those foods that foster cancer development.

We have repeatedly seen that there can be too much of a good thing. This is true of diets, exercise, and sleep, which supports the adage "everything in moderation." Biological processes are often subject to hormesis in which biphasic changes occur in response to various agents so progressively more positive effects are evident with increasing doses, but beyond a certain dose, adverse actions develop. Essentially, a goldilocks zone exists in which beneficial effects of an agent are present but straying outside of this zone can have negative consequences. It may also be the case that some treatments, such as antioxidants can have beneficial effects concerning some health conditions, but might concurrently undermine other health conditions (e.g., metabolic syndrome, bone loss). Thus, the use of supplements to reduce ROS to preclude illnesses may also have unintended negative consequences (Salehi et al., 2018).

Are antioxidants beneficial in cancer prevention and treatment?

In addition to the value of fibers present in cruciferous vegetables, their anticancer actions have also been attributed to their pronounced antioxidant properties, although recent studies have questioned this long-held notion, even suggesting that several common antioxidants can promote melanoma metastasis (Le Gal et al., 2015). Irrespective of whether they are naturally produced in the body or obtained through dietary supplements, antioxidants can promote the spread of lung cancer to other organs. Moreover, antioxidants can increase the presence of Bach 1, which augments glucose uptake, hence feeding cancer cells. Evidently, despite the potential advantages of antioxidants and vitamins on most processes that would be expected to fight cancer, excessive levels of these compounds can have pro-tumor consequences.

Nonmeat proteins and the Mediterranean diet

With the realization that people living around the Mediterranean Sea often lived longer than did those in other countries, increased attention focused on what contributed to their longevity. Ultimately, it was suggested that the low incidence of heart disease and cancer in these countries might be attributable to the diet that was so often maintained. Of course, the countries surrounding the Mediterranean differ in the foods consumed, but many common features figured into the "Mediterranean diet," typically comprising low consumption of red meats, sugar, and saturated fat, and high intake of whole grains, beans, nuts, fruits, vegetables, herbs, spices, and healthy fats, particularly in the form of extra virgin olive oil.

Mediterranean diets have been associated with a diminished occurrence of some forms of cancer possibly owing to effects on immune system functioning and in limiting inflammatory responses. Most of the research has focused on colorectal cancer, but distinct microbial signatures are also apparent in breast tissue and could thereby act on the development of breast cancer (Newman et al., 2019). Indeed, in nonhuman primates, a Mediterranean diet increased the abundance of beneficial bacteria, such as *Lactobacillus* abundance as well as levels of bile acid metabolites within the mammary gland to a greater extent than did a Western diet (Shively et al., 2018). These findings raise the possibility that the diet directly affects microbiota outside of the gut so cancer could be affected at distal sites.

Adherence to a Mediterranean diet was accompanied by greater bacterial diversity, diminished levels of proinflammatory cytokines otherwise engendered by stressors or by sedentary behaviors among adolescents, and reduced oxidative stress that could otherwise promote tumorigenesis. With age and the increased occurrence of exogenous and endogenous stressors, the appearance of beneficial bacteria may decline, but with the adoption of a Mediterranean diet, these bacteria became more abundant, accompanied by elevations of SCFAs (Ghosh et al., 2020). If an appropriate diet is maintained the imbalance between proinflammatory and anti-inflammatory factors may be diminished so that adverse health effects are less likely to arise.

A Mediterranean diet can have indirect actions on physical illnesses by reducing the occurrence of depressive illnesses. A meta-analysis of 21 studies across 10 countries concluded that a Mediterranean diet was accompanied by reduced risk of depression, whereas a diet high in red meat, processed meats, high-fat dairy products, butter, potatoes, and refined grains, together with limited consumption of fruits and vegetables was associated with an elevated risk of depression (Fond et al., 2020). Conversely, Mediterranean diets can diminish depressive symptoms. Based on findings such as these, it was recommended that psychological disorders can be prevented (and perhaps partially treated) by a plant-based diet high in grains, fiber, and fish (Dinan et al., 2019). As indicated earlier (and will be repeated frequently), since depressive disorders are frequently comorbid with cancer, the impact of diet may indirectly come to affect cancer occurrence.

Fats and meat proteins

It is probably not a coincidence that the incidence of colorectal cancer varies appreciably across developed countries, typically being greater in men than women, and exceeding the incidence in most parts of Africa, Asia, and India. While numerous factors could account for the occurrence of colorectal cancer and its regional distribution, specific diets consumed may have a say in the development of this form of cancer. Western diets that largely comprise a high intake of calories, simple sugars, animal proteins, and saturated fat, have been associated with multiple health risks, whereas plant proteins may have health benefits that are not typically obtained from meat proteins. Some dietary fats had been appropriately demonized, especially trans fats, although debate has continued about the benefits and risks engendered by saturated fats relative to carbohydrates. Ordinarily, adipocytes (fat cells) remove fat from circulation and then release it slowly when it is needed. With high-fat diets (and lack of exercise) adipocytes can be overburdened, resulting in fat accumulating in the liver and elsewhere, thereby promoting toxic reactions so the cells die and insulin efficiency is undermined. Fat intake and obesity are believed to affect the development of type 2 diabetes and have been linked to some forms of cancer and increased risk of all-cause mortality.

Elevated saturated fat consumption was accompanied by the reduction of gut bacteria levels and diversity together with the elevated abundance of *Clostridium bolteae* and *Blautia*, which favored greater body mass and insulin resistance. A diet rich in polyunsaturated fats, in contrast, was accompanied by microbial variations that were deemed to be beneficial. Once more, it is necessary to consider the source of saturated fat, as these can operate differently in determining the microbial consequences and can differentially influence different types of cancer. In rodents, fatty acids or their residue derived from lard (pork fat) was accompanied by somewhat different microbiota communities relative to that associated with a fish oil-based diet (Li et al., 2017). In mice, a high-fat diet containing lard increased the progression of prostate cancer accompanied by elevated proinflammatory cytokine levels. Likewise, when a high-fat diet was provided for an extended period that began the following weaning, the occurrence of triple-negative breast cancer was increased in a mouse model (Mustafi et al., 2017).

Red meat consumption has often been castigated owing to its links to various diseases, such as heart functioning and cancer. An analysis conducted among more than 81,000 people indicated that elevated red meat consumption over 8 years was associated with a 10% increase in mortality over the ensuing 8 years (Zheng et al., 2019). Prospective studies had similarly indicated that red meat consumption was associated with overall cancer occurrence, especially colorectal and breast cancer (Diallo et al., 2018). A meta-analysis that included 11 cohorts and 32 case-control studies confirmed these assertions, showing that consumption of red meats and processed meats were dose-dependently related to gastric cancer, whereas white meat was not linked to increased gastric cancer occurrence (Kim et al., 2019).

A challenge to this common belief that red meat is unhealthy was provided by a review of the relevant literature which concluded that moderate consumption of lean red meat was actually not accompanied by an elevated risk of cardiovascular disease or colon cancer (McAfee et al., 2010). Moreover, mortality was neither related to red meat nor processed meat consumption. As expected, these views initially gained little traction, but follow-up reviews and large-scale meta-analyses supported the position that the purported health risks of red meat consumption were likely overestimated. A series of reports indicated that dietary patterns and meat consumption were only weakly related to cancer incidence and mortality (Vernooij et al., 2019), and reducing red meat intake did not have appreciable positive effects (Han et al., 2019). An analysis from this same group of researchers indicated that randomized controlled studies reinforced the position that red meat was of minimal risk for heart disease or cancer incidence and mortality (Zeraatkar et al., 2019).

What can we make of these inconsistent and often contradictory findings, especially as these reports appear to be convincing?[c] Many of the earlier studies were often not compelling as they often failed to distinguish between the influence of diets that had been in effect over the lifetime and those that reflected more recent food choices. Furthermore, few efforts were made to assess individual difference factors that might have made some people more (or less) vulnerable to the adverse influence of specific dietary factors, with less consideration of

---

[c] Aside from the health implications of these reports, they could have obvious repercussions for food producers. Enormous pushback was instigated, and strong efforts were made to prevent publication of these reports. Like so many other controversies that embroiled the food industry, this debacle could only serve to undermine trust in the science promoted by the food industry.

genetic factors or microbiota analyses—although the latter is a relatively recent addition to the study of disease determinants. Importantly, analyses that focus on red meat consumption without considering other dietary elements are too narrow to allow for definitive conclusions to be made. At the same time, genetic influences and lifestyle factors may be fundamental for the links between food intake and health, and both sex and age ought to be considered as well. In this regard, among individuals 50–65 years of age, high protein intake was associated with a marked increase of cancer death and a 75% increase of earlier mortality over an 18-year evaluation period. Conversely, individuals older than 65 showed a decline in cancer-related death (Levine et al., 2014). Why this occurred is a mystery. Perhaps because microbial communities change with aging the links between diet and illness might vary accordingly. Of course, this is just conjecture but emphasizes the complexities of studying dietary influences on health and cancer across the lifespan.

Balancing nutritional needs and starving cancer cells: The case of cancer cachexia

Diet and energy needs among cancer patients is a complicated issue. There is a fine line between the effectiveness of treatments to deprive cancer cells of nutrients versus the fundamental needs of patients to maintain their strength. Cancer itself, as well as some of the treatments (e.g., radiation and chemotherapy) often promote malnutrition, metabolic disturbances, and disturbed repair of tissues damaged by treatments. These outcomes vary over the course of the disease and thus nutrition needs vary accordingly.

Of the numerous disturbing effects associated with many forms of cancer, cachexia has been near the top of the list, occurring in about 50% of patients with advanced cancer, more often affecting males than females. Cachexia is characterized by intense fatigue and weakness related to severe weight loss, atrophy of adipose tissue, and deterioration of skeletal muscles. Altered lipolytic processes associated with triglyceride formation were related to diminished adipose tissue, whereas reduced protein synthesis, elevated protein degradation, and increased resting energy expenditure stemming from elevated thermogenesis contributed to skeletal muscle deterioration. Moreover, chronically altered visceral adiposity, together with cytokines released from adipose cells promote metabolic disturbances that result in insulin resistance and the ensuing weight loss. Making things still worse, cancer cachexia interferes with the efficacy of therapies and increases patient vulnerability to the toxic effects of chemotherapy and radiation therapy.

With so many potential factors being involved in cachexia, it has been difficult to develop effective prevention or treatment strategies. Altering diets or using specific nutrients has largely been ineffective in attenuating cachexia. However, detailed reviews have suggested that omega-3 may promote some positive effects in this regard (Gorjao et al., 2019). In fact, omega-3 supplementation was effective in producing weight and appetite stabilization in pancreatic cancer patients and even enhanced physical activity and quality of life. As variations of several proinflammatory cytokines were implicated in aspects of cachexia, attempts had been made to tackle this condition using anticytokine treatments with promising outcomes being apparent. Given the potential influence of insulin resistance in promoting cachexia, treatments with GLP-1 receptor agonists—used to treat type 2 diabetes—could potentially limit muscle atrophy associated with cancer cachexia (Hong et al., 2019).

It had been maintained that rather than adopting narrow procedures, managing cachexia would best be served through multimodal methods that incorporate nutritional support, exercise, agents that diminish inflammation, and ought to be based on individualized features of the patient, including the presence of biomarkers that might be relevant to cachexia. As much as this might be an ideal approach, there have been few well-controlled studies in humans that have evaluated the most effective multimodal strategies (McKeaveney et al., 2021). As a result, cancer cachexia has been and continues to be a problem in the treatment of cancer patients.

An extensive report concerning nutrition recommendations has been provided (Arends et al., 2017), which ought to be exceptionally useful for patients, health care workers managing the needs of cancer patients, and family members uncertain of diets that are best for patients. Among many other recommendations offered was that screening for malnutrition ought to be a component of cancer care so that their nutritional needs can be met. These diets should be assigned on an individual basis, varying with disease progression. Nutrients rich in proteins may be especially beneficial, particularly when supplemented by energy-enriched fluids. Other foods, in contrast, ought to be avoided, such as those containing added sugars, saturated fats, and high levels of sodium—and alcohol use should be eschewed. Vitamin and mineral intake should only be provided to reverse specific deficiencies that may have developed. These suggestions obviously run counter to the position regarding the adoption of food-restricted diets to thwart cancer occurrence and progression. Thus, the recommendations offered ought to be considered based on the stage of the disease, the individual's general health, the person's sex and age, and other individual characteristics relevant to metabolic and digestive processes.

Summary and conclusions

Eating processes are regulated by multiple hormones and brain processes that signal when to initiate food intake and when to stop eating. These regulatory processes in themselves aren't sufficient to maintain proper eating styles, which are subject to psychological influences and the brain processes that govern them. It is tempting to blame the person with obesity for their ill health, but their behaviors might have been determined by prenatal experiences or those encountered early in life over which they had no control. Similarly, genetic and epigenetic factors might have contributed to the development of obesity, and related processes might have contributed to diseases stemming from being overweight. It also should be considered that after years of maintaining poor health habits and the neurobiological changes that accompany them, regulatory brain processes might have been hijacked so these counterproductive behaviors are maintained, such as occurs with substance use disorders. Specifically, neural circuits related to the expectation of positive feelings obtained by certain foods might overwhelm control circuits that ought to promote better food choices and limit excessive food intake.

The consumption of certain foods, such as those containing trans fats or those that foster the development of obesity, can be challenging to biological functioning and not unlike allostatic overload that may occur in response to chronic stressors. Some systems (e.g., pancreatic beta cells within the pancreas) may become overly taxed in response to poor diets leading to systemic inflammation and the development of various diseases. It is often recommended that so-called antiinflammatory diets (e.g., Mediterranean diets) be endorsed to prevent the occurrence of diseases, and foods high in fiber may be instrumental in disease prevention through their actions on gut microbiota and the formation of SCFAs.

Still, it is both surprising and unfortunate that after so many years of research, debate continues concerning which foods are healthy and which are not. Diet experts seem to alter their opinions on a fairly regular basis. And it is similarly unclear which type of diets are best to produce weight loss and promote good health, the public being overwhelmed by various methods to reduce weight, with each offering the fastest route to becoming thin. In the main, however, the effectiveness of these diets may be limited since compensatory weight gain may soon follow. Accordingly, programs that focus on changing behavioral patterns rather than simply promoting weight loss are likely to be more beneficial.

For cancer survivors, there is good reason to adopt or select a dietary approach that provides some confidence that it promotes health. Nonetheless, whatever their chosen diet, it seems that adherence to recommended diets among cancer survivors is often poor, varying with sociodemographic factors, educational level, and body mass index. Multiple factors can limit or prevent full compliance regarding healthy food choices. Following therapy, breast cancer patients may encounter fatigue, distress, pain, and discomfort, as well as altered eating habits secondary to loss of appetite, change in taste, and craving unhealthy foods. This serves to emphasize that while individuals may be motivated to consume recommended food choices, they may encounter fundamental barriers to adherence that fortunately are not insurmountable.

References

Abizaid, A., 2019. Stress and obesity: the ghrelin connection. J. Neuroendocrinol. 31, e12693.
Abizaid, A., Luheshi, G., Woodside, B.C., 2013. Interaction between immune and energy-balance signals in the regulation of feeding and metabolism. In: Kusnecov, A., Anisman, H. (Eds.), The Wiley-Blackwell Handbook of Psychoneuroimmunology. Wiley-Blackwell, Hoboken, NJ, pp. 488–503.
Ameku, T., Beckwith, H., Blackie, L., Miguel-Aliaga, I., 2020. Food, microbes, sex and old age: on the plasticity of gastrointestinal innervation. Curr. Opin. Neurobiol. 62, 83–91.
Arends, J., Bachmann, P., Baracos, V., Barthelemy, N., Bertz, H., et al., 2017. ESPEN guidelines on nutrition in cancer patients. Clin. Nutr. 36, 11–48.
Aune, D., Chan, D.S., Lau, R., Vieira, R., Greenwood, D.C., et al., 2011. Dietary fibre, whole grains, and risk of colorectal cancer: systematic review and dose-response meta-analysis of prospective studies. BMJ 343, d6617.
Bagnardi, V., Rota, M., Botteri, E., Tramacere, I., Islami, F., et al., 2015. Alcohol consumption and site-specific cancer risk: a comprehensive dose-response meta-analysis. Br. J. Cancer 112, 580–593.
Baudry, J., Assmann, K.E., Touvier, M., Alles, B., Seconda, L., et al., 2018. Association of frequency of organic food consumption with cancer risk: findings from the NutriNet-Sante Prospective Cohort Study. JAMA Intern. Med. 178, 1597–1606.
Boozari, M., Butler, A.E., Sahebkar, A., 2019. Impact of curcumin on toll-like receptors. J. Cell. Physiol. 234, 12471–12482.
Budhathoki, S., Sawada, N., Iwasaki, M., Yamaji, T., Goto, A., et al., 2019. Association of animal and plant protein intake with all-cause and cause-specific mortality in a Japanese cohort. JAMA Intern. Med. 179, 1509–1518.
Burke, L.K., Darwish, T., Cavanaugh, A.R., Virtue, S., Roth, E., et al., 2017. mTORC1 in AGRP neurons integrates exteroceptive and interoceptive food-related cues in the modulation of adaptive energy expenditure in mice. eLife 6, e22848.
Caffa, I., Spagnolo, V., Vernieri, C., Valdemarin, F., Becherini, P., et al., 2020. Fasting-mimicking diet and hormone therapy induce breast cancer regression. Nature 583, 620–624.

Cam, A., Oyirifi, A.B., Liu, Y., Haschek, W.M., Iwaniec, U.T., et al., 2019. Thermally abused frying oil potentiates metastasis to lung in a murine model of late-stage breast cancer. Cancer Prev. Res. 12, 201–210.

Chen, R., Li, D.P., Turel, O., Sorensen, T.A., Bechara, A., et al., 2018. Decision making deficits in relation to food cues influence obesity: a triadic neural model of problematic eating. Front. Psychiatry 9, 264.

Cignarella, F., Cantoni, C., Ghezzi, L., Salter, A., Dorsett, Y., et al., 2018. Intermittent fasting confers protection in CNS autoimmunity by altering the gut microbiota. Cell Metab. 27, 1222–1235 e1226.

Collins, N., Han, S.J., Enamorado, M., Link, V.M., Huang, B., et al., 2019. The bone marrow protects and optimizes immunological memory during dietary restriction. Cell 178, 1088–1101 e1015.

Crane, J.D., Palanivel, R., Mottillo, E.P., Bujak, A.L., Wang, H., et al., 2015. Inhibiting peripheral serotonin synthesis reduces obesity and metabolic dysfunction by promoting brown adipose tissue thermogenesis. Nat. Med. 21, 166–172.

de Cabo, R., Mattson, M.P., 2019. Effects of intermittent fasting on health, aging, and disease. N. Engl. J. Med. 381, 2541–2551.

Diallo, A., Deschasaux, M., Latino-Martel, P., Hercberg, S., Galan, P., et al., 2018. Red and processed meat intake and cancer risk: results from the prospective NutriNet-Sante cohort study. Int. J. Cancer 142, 230–237.

Dinan, T.G., Stanton, C., Long-Smith, C., Kennedy, P., Cryan, J.F., et al., 2019. Feeding melancholic microbes: MyNewGut recommendations on diet and mood. Clin. Nutr. 38, 1995–2001.

Fang, S., Suh, J.M., Reilly, S.M., Yu, E., Osborn, O., et al., 2015. Intestinal FXR agonism promotes adipose tissue browning and reduces obesity and insulin resistance. Nat. Med. 21, 159–165.

Fond, G., Young, A.H., Godin, O., Messiaen, M., Lancon, C., et al., 2020. Improving diet for psychiatric patients: high potential benefits and evidence for safety. J. Affect. Disord. 265, 567–569.

Ghosh, T.S., Rampelli, S., Jeffery, I.B., Santoro, A., Neto, M., et al., 2020. Mediterranean diet intervention alters the gut microbiome in older people reducing frailty and improving health status: the NU-AGE 1-year dietary intervention across five European countries. Gut 69, 1218–1228.

Gorjao, R., Dos Santos, C.M.M., Serdan, T.D.A., Diniz, V.L.S., Alba-Loureiro, T.C., et al., 2019. New insights on the regulation of cancer cachexia by N-3 polyunsaturated fatty acids. Pharmacol. Ther. 196, 117–134.

Gourd, E., 2018. Ultra-processed foods might increase cancer risk. Lancet Oncol. 19, e186.

Graef, F.A., Celiberto, L.S., Allaire, J.M., Kuan, M.T.Y., Bosman, E.S., et al., 2021. Fasting increases microbiome-based colonization resistance and reduces host inflammatory responses during an enteric bacterial infection. PLoS Pathog. 17, e1009719.

Han, M.A., Zeraatkar, D., Guyatt, G.H., Vernooij, R.W.M., El Dib, R., et al., 2019. Reduction of red and processed meat intake and cancer mortality and incidence: a systematic review and meta-analysis of cohort studies. Ann. Intern. Med. 171, 711–720.

Harris, H.R., Willett, W.C., Vaidya, R.L., Michels, K.B., 2017. An adolescent and early adulthood dietary pattern associated with inflammation and the incidence of breast cancer. Cancer Res. 77, 1179–1187.

Higgins, G.A., Fletcher, P.J., Shanahan, W.R., 2020. Lorcaserin: a review of its preclinical and clinical pharmacology and therapeutic potential. Pharmacol. Ther. 205, 107417.

Hong, Y., Lee, J.H., Jeong, K.W., Choi, C.S., Jun, H.S., 2019. Amelioration of muscle wasting by glucagon-like peptide-1 receptor agonist in muscle atrophy. J. Cachexia Sarcopenia Muscle 10, 903–918.

Ioannidis, J.P., 2013. Implausible results in human nutrition research. BMJ 347, f6698.

Johnston, B.C., Kanters, S., Bandayrel, K., Wu, P., Naji, F., et al., 2014. Comparison of weight loss among named diet programs in overweight and obese adults: a meta-analysis. JAMA 312, 923–933.

Kaneko, K., Fu, Y., Lin, H.Y., Cordonier, E.L., Mo, Q., et al., 2019. Gut-derived GIP activates central Rap1 to impair neural leptin sensitivity during overnutrition. J. Clin. Invest. 129, 3786–3791.

Kim, S.R., Kim, K., Lee, S.A., Kwon, S.O., Lee, J.K., et al., 2019. Effect of red, processed, and white meat consumption on the risk of gastric cancer: an overall and dose(-)response meta-analysis. Nutrients 11, 826.

Kolodziejczyk, A.A., Zheng, D., Elinav, E., 2019. Diet-microbiota interactions and personalized nutrition. Nat. Rev. Microbiol. 17, 742–753.

Le Gal, K., Ibrahim, M.X., Wiel, C., Sayin, V.I., Akula, M.K., et al., 2015. Antioxidants can increase melanoma metastasis in mice. Sci. Transl. Med. 7, 308re308.

Levine, M.E., Suarez, J.A., Brandhorst, S., Balasubramanian, P., Cheng, C.W., et al., 2014. Low protein intake is associated with a major reduction in IGF-1, cancer, and overall mortality in the 65 and younger but not older population. Cell Metab. 19, 407–417.

Levy, J.M.M., Towers, C.G., Thorburn, A., 2017. Targeting autophagy in cancer. Nat. Rev. Cancer 17, 528–542.

Li, H., Zhu, Y., Zhao, F., Song, S., Li, Y., et al., 2017. Fish oil, lard and soybean oil differentially shape gut microbiota of middle-aged rats. Sci. Rep. 7, 826.

Ludwig, D.S., Aronne, L.J., Astrup, A., de Cabo, R., Cantley, L.C., et al., 2021. The carbohydrate-insulin model: a physiological perspective on the obesity pandemic. Am. J. Clin. Nutr. 114, 1873–1885.

Mattison, J.A., Colman, R.J., Beasley, T.M., Allison, D.B., Kemnitz, J.W., et al., 2017. Caloric restriction improves health and survival of rhesus monkeys. Nat. Commun. 8, 14063.

Mattson, M.P., Longo, V.D., Harvie, M., 2017. Impact of intermittent fasting on health and disease processes. Ageing Res. Rev. 39, 46–58.

Maunder, A., Bessell, E., Lauche, R., Adams, J., Sainsbury, A., et al., 2020. Effectiveness of herbal medicines for weight loss: a systematic review and meta-analysis of randomized controlled trials. Diabetes Obes. Metab. 22, 891–903.

McAfee, A.J., McSorley, E.M., Cuskelly, G.J., Moss, B.W., Wallace, J.M., et al., 2010. Red meat consumption: an overview of the risks and benefits. Meat Sci. 84, 1–13.

McKeaveney, C., Maxwell, P., Noble, H., Reid, J., 2021. A critical review of multimodal interventions for cachexia. Adv. Nutr. 12, 523–532.

Merino, J., Dashti, H.S., Sarnowski, C., Lane, J.M., Todorov, P.V., et al., 2021. Genetic analysis of dietary intake identifies new loci and functional links with metabolic traits. Nat. Hum. Behav. https://doi.org/10.1038/s41562-021-01182-w. In press.

Mustafi, D., Fernandez, S., Markiewicz, E., Fan, X., Zamora, M., et al., 2017. MRI reveals increased tumorigenesis following high fat feeding in a mouse model of triple-negative breast cancer. NMR Biomed. 30. https://doi.org/10.1002/nbm.3758.

Nauck, M.A., Meier, J.J., 2018. Incretin hormones: their role in health and disease. Diabetes Obes. Metab. 20 (suppl 1), 5–21.

Neeland, I.J., Marso, S.P., Ayers, C.R., Lewis, B., Oslica, R., et al., 2021. Effects of liraglutide on visceral and ectopic fat in adults with overweight and obesity at high cardiovascular risk: a randomised, double-blind, placebo-controlled, clinical trial. Lancet Diabetes Endocrinol. 9, 595–605.

Nelson, K.M., Dahlin, J.L., Bisson, J., Graham, J., Pauli, G.F., et al., 2017. The essential medicinal chemistry of curcumin: miniperspective. J. Med. Chem. 60, 1620–1637.

Newman, T.M., Vitolins, M.Z., Cook, K.L., 2019. From the table to the tumor: the role of Mediterranean and western dietary patterns in shifting microbial-mediated signaling to impact breast cancer risk. Nutrients 11, 2565.

Pepino, M.Y., Eisenstein, S.A., Bischoff, A.N., Klein, S., Moerlein, S.M., et al., 2016. Sweet dopamine: sucrose preferences relate differentially to striatal D2 receptor binding and age in obesity. Diabetes 65, 2618–2623.

Reynolds, A., Mann, J., Cummings, J., Winter, N., Mete, E., et al., 2019. Carbohydrate quality and human health: a series of systematic reviews and meta-analyses. Lancet 393, 434–445.

Rinninella, E., Mele, M.C., Cintoni, M., Raoul, P., Ianiro, G., et al., 2020. The facts about food after cancer diagnosis: a systematic review of prospective cohort studies. Nutrients 12, 2345.

Riva, A., Kuzyk, O., Forsberg, E., Siuzdak, G., Pfann, C., et al., 2019. A fiber-deprived diet disturbs the fine-scale spatial architecture of the murine colon microbiome. Nat. Commun. 10, 4366.

Rossi, M.A., Basiri, M.L., McHenry, J.A., Kosyk, O., Otis, J.M., et al., 2019. Obesity remodels activity and transcriptional state of a lateral hypothalamic brake on feeding. Science 364, 1271–1274.

Salehi, B., Martorell, M., Arbiser, J.L., Sureda, A., Martins, N., et al., 2018. Antioxidants: positive or negative actors? Biomolecules 8, 124.

Sanders, M.E., Merenstein, D.J., Reid, G., Gibson, G.R., Rastall, R., 2019. Probiotics and prebiotics in intestinal health and disease: from biology to the clinic. Nat. Rev. Gastroenterol. Hepatol. 16, 605–616.

Schoenfeld, J.D., Ioannidis, J.P., 2013. Is everything we eat associated with cancer? A systematic cookbook review. Am. J. Clin. Nutr. 97, 127–134.

Schwingshackl, L., Schwedhelm, C., Hoffmann, G., Lampousi, A.M., Knuppel, S., et al., 2017. Food groups and risk of all-cause mortality: a systematic review and meta-analysis of prospective studies. Am. J. Clin. Nutr. 105, 1462–1473.

Seabrook, L.T., Borgland, S.L., 2020. The orbitofrontal cortex, food intake and obesity. J. Psychiatry Neurosci. 45, 304–312.

Shively, C.A., Register, T.C., Appt, S.E., Clarkson, T.B., Uberseder, B., et al., 2018. Consumption of mediterranean versus western diet leads to distinct mammary gland microbiome populations. Cell Rep. 25, 47–56 e43.

Stylianou, K.S., Fulgoni, V.L., Jolliet, O., 2021. Small targeted dietary changes can yield substantial gains for human health and the environment. Nat. Food 2, 616–627.

Tabung, F.K., Steck, S.E., Liese, A.D., Zhang, J., Ma, Y., et al., 2016. Association between dietary inflammatory potential and breast cancer incidence and death: results from the Women's Health Initiative. Br. J. Cancer 114, 1277–1285.

Tellez, L.A., Han, W., Zhang, X., Ferreira, T.L., Perez, I.O., et al., 2016. Separate circuitries encode the hedonic and nutritional values of sugar. Nat. Neurosci. 19, 465–470.

Tronieri, J.S., Wadden, T.A., Walsh, O., Berkowitz, R.I., Alamuddin, N., et al., 2020. Effects of liraglutide on appetite, food preoccupation, and food liking: results of a randomized controlled trial. Int. J. Obes. 44, 353–361.

Vallianou, N.G., Evangelopoulos, A., Schizas, N., Kazazis, C., 2015. Potential anticancer properties and mechanisms of action of curcumin. Anticancer Res. 35, 645–651.

Vernooij, R.W.M., Zeraatkar, D., Han, M.A., El Dib, R., Zworth, M., et al., 2019. Patterns of red and processed meat consumption and risk for cardiometabolic and cancer outcomes: a systematic review and meta-analysis of cohort studies. Ann. Intern. Med. 171, 732–741.

Wang, S., Huang, M., You, X., Zhao, J., Chen, L., et al., 2018. Gut microbiota mediates the anti-obesity effect of calorie restriction in mice. Sci. Rep. 8, 13037.

Wang, B., Tsakiridis, E.E., Zhang, S., Llanos, A., Desjardins, E.M., et al., 2021. The pesticide chlorpyrifos promotes obesity by inhibiting diet-induced thermogenesis in brown adipose tissue. Nat. Commun. 12, 5163.

Wilding, J.P.H., Batterham, R.L., Calanna, S., Davies, M., Van Gaal, L.F., et al., 2021. Once-weekly semaglutide in adults with overweight or obesity. N. Engl. J. Med. 384, 989.

Youm, Y.H., Nguyen, K.Y., Grant, R.W., Goldberg, E.L., Bodogai, M., et al., 2015. The ketone metabolite beta-hydroxybutyrate blocks NLRP3 inflammasome-mediated inflammatory disease. Nat. Med. 21, 263–269.

Zeraatkar, D., Johnston, B.C., Bartoszko, J., Cheung, K., Bala, M.M., et al., 2019. Effect of lower versus higher red meat intake on cardiometabolic and cancer outcomes: a systematic review of randomized trials. Ann. Intern. Med. 171, 721–731.

Zheng, Y., Li, Y., Satija, A., Pan, A., Sotos-Prieto, M., et al., 2019. Association of changes in red meat consumption with total and cause specific mortality among US women and men: two prospective cohort studies. BMJ 365, l2110.

C H A P T E R

8

Dietary components associated with being overweight, having obesity, and cancer

OUTLINE

Being overweight or having obesity as a health risk	253	Dietary components in relation to cancer and its treatment	262
Linking the development of obesity to cancer	254	Glucose	262
		Fat and lipid metabolism	266
		Fatty acids	268
Relations between obesity, immunity, inflammation, and cancer	257	Polyphenols	270
Ties between leptin, ghrelin, and cancer	258	Dietary amino acids	271
		Caveats concerning diet manipulations and cancer	271
Cholesterol involvement in cancer	259	Summary and conclusions	274
Genetic influences on obesity	260	References	275

Being overweight or having obesity as a health risk

Many years of investigation led to the belief that being overweight carries multiple health risks, such as metabolic syndrome, type 2 diabetes, and chronic heart disease. Having obesity at middle age was associated with lifespan being reduced by 5.8 and 7.1 years in males and females, respectively, and simply being overweight was accompanied by a 3.2-year reduction of life. Predictably, shorter life span and reduced health span (disease-free years) related to obesity are said to vary with socioeconomic status, physical activity, and smoking habits (Nyberg et al., 2018). A metaanalysis that included 2.88 million people enrolled in 97 independent studies concluded that mortality varied with the extent of the obesity. While being seriously obese (defined as having a BMI above 35) predicted earlier mortality, this was not apparent among overweight individuals and those who had moderate obesity (having a BMI of 25–30 and 30–35, respectively). Even in men with suspected or confirmed heart disease, obesity was most closely aligned with mortality if individuals were also deemed not to be physically fit (McAuley et al., 2012). Findings like these have been instrumental in fueling the perspective that "I may be overweight, but I'm perfectly healthy," which has helped sustain the poor lifestyles that led to such individuals becoming obese.

Not every overweight person or even those having obesity will succumb to cardiovascular disease or other weight-related illnesses. This resulted in the creation of a category known as "metabolically healthy obesity," which is characterized by relatively low levels of inflammatory markers, enhanced cardiorespiratory fitness, lower liver and visceral fat, insulin sensitivity, normal adipose tissue function, and greater subcutaneous leg fat. While reassuring, such features may not endure, since "metabolically fit" people may still be at elevated risk for later problems (Blüher, 2020). In a 20-year prospective study one-third of people with obesity initially seemed to be healthy based on their blood pressure, fasting blood sugar, cholesterol, and insulin resistance. But after 10 years, 40% of these ostensibly "healthy obese" people were deemed to be "unhealthy obese," and after another 10 years 50% were at risk for disease (Bell et al., 2015). These findings were supported by a 43-year prospective analysis which indicated that those most at risk for health problems were individuals who were overweight when young and continued to gain weight. In contrast, individuals who had normal weight and then transitioned to being overweight later in life did not show comparable elevated health problems and early mortality, and in some individuals this trajectory was accompanied by improved survival (Zheng et al., 2021).

The relationship between weight (or body mass index) and cancer occurrence needs to consider weight changes that had occurred over time and how long individuals had been overweight. Being overweight before the age of 40 and sustaining the elevated weight for more than 3 years was associated with a 29% increase of colon cancer, a 70% elevation of endometrial cancer, and a 58% increase of renal cancer in males (Bjørge et al., 2019). In fact, it was estimated that for each 5 kg gain, the risk of colorectal cancer increased by 4%. Of course, this is to be expected as the influence of biological changes, particularly inflammation, may have cumulative actions, making it essential to consider when and for how long obesity had been present. Evidently, if obesity persists over a lengthy period, it can potentially increase disease risk, and the trajectory of weight changes over the lifetime may be fundamental in predicting illness occurrence.

The case has been made that obesity in many ways recapitulates the health risks created by aging. This includes epigenetic changes and gene mutations that can undermine immune functioning together with altered apoptosis and multiorgan autophagy. The persistence of obesity over years can work to accelerate aging, shortening both life and health span.

Linking the development of obesity to cancer

The initial studies concerning the consequences of obesity largely focused on type 2 diabetes and heart disease, but it wasn't long before it became abundantly clear that obesity was also predictive of cancer. Elevated body-mass index (BMI) was associated with the increased presence of diverse types of cancer, and high BMI and diabetes were linked to more than 800,000 new cancer cases worldwide (Pearson-Stuttard et al., 2018).

It is difficult to fully dissociate the contribution of diet from obesity in predicting cancer occurrence, although it was estimated that 16% of diet-related cancers could be attributed to obesity. These relations were more prominent in some groups and some countries than in others. Cancer occurrence stemming from diet was particularly marked in low- and middle-income countries and was elevated among individuals who moved from urban areas to cities, possibly reflecting the altered food choices and microbial changes that accompany these transitions.

Many illnesses have their origins in diets consumed during early life. The offspring of obese mothers were at greater risk of developing childhood leukemia, although the absolute risk of this occurring was small. At the same time, the children of obese mothers were more likely to become obese themselves, and hence might subsequently be at increased risk for cancers in adulthood. But more than this, as obesity has increased in younger people, cancer occurrence has been seen at progressively younger ages. Obesity-related cancers, such as colon cancer, that had typically occurred in individuals older than 65, have been appearing in the 50–64 years age group, as well as within those 20–49 years of age (Koroukian et al., 2019). Several forms of cancer that have been linked to obesity, such as kidney, uterine, and colorectal cancer, have also been increasing in frequency in adolescents and young adults. For that matter, the frequency of several obesity-related cancers among young adults (millennials) was twice that observed within an earlier cohort (baby boomers), which could be linked to obesity occurring more often in younger people over the past few decades. These effects could also be due to increased consumption of certain foods or food additives (e.g., fructose in soft drinks).

What people know and don't know

Are most people aware of the risk factors associated with cancer, such as the influence of genetic and epigenetic influences or even what symptoms might be indicative of surreptitious onset of cancer? Are people aware that unexplained fatigue, bleeding, or weight loss are possible hints of cancer being present? And are they aware of the possible significance of a chronic cough or trouble swallowing? The American Institute for cancer research (AICR) conducts surveys concerning individual's awareness of factors that favor cancer occurrence. Most participants (78%) in such surveys knew that smoking and the use of tobacco were major risk factors for cancer, and most people were aware (66%) that sun exposure was a serious risk (but only 48% used sun blockers as a preventative measure). Far fewer were aware that poor food choices and alcohol consumption were linked to cancer (38% and 30%, respectively), still fewer were aware of the benefits that could be obtained through exercise (25%), and knowledge that certain viruses could cause cancer often seemed to be off the radar as fewer than 20% of individuals were aware of this risk factor.

The fact that most people understand that there are links between lung cancer and smoking as well as sun exposure and melanoma can likely be attributed to the wide dissemination of this information through the media, as well as school curricula. Perhaps the same approach ought to be adopted in policies aimed at diminishing the health risks associated with other factors, including the risks of being overweight and having obesity.

Getting at the processes that might be linked to cancer has been difficult given that most studies in humans have been correlational, and typically involved retrospective analyses. Asking cancer patients about their earlier food intake, as we saw concerning the recall of stressful experiences, may result in exceptionally biased responses, and the reliability of retrospective analyses is highly questionable under any conditions. The alternative, which comprises prospective analyses, is often impractical and difficult to conduct (e.g., involving a diary of food intake). Still, a prospective analysis of about 900,000 individuals indicated that among those who were overweight at the start of the study, death rates over 10 years from all cancers were elevated by 52% and 62% in men and women, respectively (Calle et al., 2003). In fact, colon cancer among overweight individuals was elevated by more than 50% (Pearson-Stuttard et al., 2018), and in postmenopausal women elevated body fat at baseline predicted later breast cancer occurrence.

Just as obesity was associated with cancer occurrence, weight loss has been linked to diminished risk of cancer development. Findings from the Women's Health Initiative revealed that a 5% weight reduction was associated with a lower incidence of breast cancer relative to women who maintained stable weight (Pan et al., 2019). An analysis based on 10 prospective studies of women who were more than 50 years of age similarly indicated that sustained weight reduction during middle or later adulthood was accompanied by reduced incidence of breast cancer over 10 years, and even a modest weight loss (2.0–4.5 kg) was associated with a positive effect (Teras et al., 2020).

Congruent with these findings, a randomized controlled trial revealed that women who commenced a low-fat diet exhibited lower breast cancer occurrence by 8% relative to women who continued with their usual diets. Sticking to a low-fat diet comprised consumption of vegetables, fruits, legumes, and whole grains, and avoiding milk and high-fat meats seemingly provided an advantage to warding off cancer. Still, it is difficult to distinguish between the relative contribution of weight loss and the consumption of specific foods in limiting cancer occurrence, especially as individuals may also have adopted other lifestyle changes in their weight reduction efforts, which might also have contributed to the benefits observed. An extensive metaanalysis indicated that weight loss and physical activity were the important elements that accompanied reduced breast cancer risk, irrespective of whether women were premenopausal or postmenopausal (Hardefeldt et al., 2018). Parenthetically, while bariatric surgery might not be the ideal way to produce weight loss, there have been reports indicating that this procedure was accompanied by the reduced occurrence of breast and endometrial cancer in postmenopausal women, and might influence colorectal cancer, although there has been some debate in this regard (Bruno and Berger, 2020).

Obesity, sex, and disease

Men are more frequently diagnosed as obese than women, and they also differ in the distribution of adipose tissue. Males generally develop more visceral fat that serves as an energy storage site, but this is also a source of inflammatory factors that can promote illness. In females, fat accumulation prior to menopause is greater in subcutaneous tissue, but with menopause fat deposition shift to the viscera, which is accompanied by elevated risk for metabolic disorders, just as in men. Of the many forms of cancer that have been tied to excessive weight, sex differences were reported for esophageal, liver, and colorectal cancers. The mechanisms for these sex differences are now better understood, including the role of sex hormones, but our understanding of how these factors operate across ethnic and cultural groups is still weak.

Genome-wide analyses showed independent associations for 98 genes and fat distribution within the body. Genetic differences between males and females were apparent within adipose and musculoskeletal tissues as well as in tissues related to female reproduction. Among women, the distribution of fat to the legs and trunk was largely driven by hormonal processes that were tied to sex-specific genetic influences, which could directly contribute to obesity and could also stimulate receptors on adipocytes within fat tissues (Palmer and Clegg, 2015).

Aside from these factors, sex differences related to obesity may also be tied to differential responses to stressful experiences as cortisol release associated with stressors promotes visceral adiposity. Stressful encounters at varied times over the course of development can also affect adipose tissue metabolism through epigenetic changes or appetite regulation by hormones, thereby affecting energy balances, which can affect cancer occurrence and progression.

Relations between obesity, immunity, inflammation, and cancer

Numerous mechanisms have been linked to the ties between obesity and occurrence of the wide range of cancers that have been observed. Fig. 8.1 depicts some of these mechanisms, including several hormones and metabolic process discussed in Chapter 7, gut microbiota and SCFAs, features of the extracellular matrix (ECM), together with adipokines released from fat as well as various immune and cytokine variations. These factors are selective in the sense

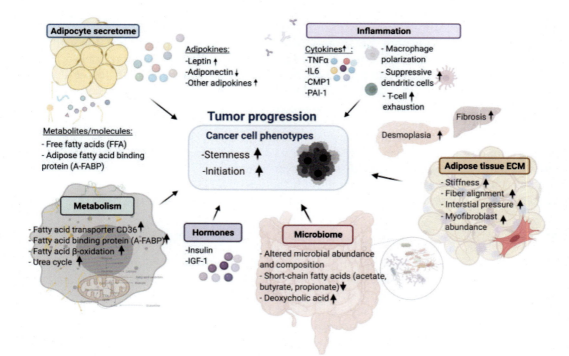

FIG. 8.1 Tumor progression is influenced by a constellation of biological changes associated with high-fat diets and with obesity. *Figure and caption from Liu, X.Z., Pedersen, L., Halberg, N., 2021. Cellular mechanisms linking cancers to obesity. Cell Stress 5, 55–72.*

that obesity is not associated with all forms of cancer, and some of these biological processes are likely related to some forms of cancer but not others.

Several observations linking obesity to immune-related changes have been made. Promotion of cancer arising from obesity can involve diverse immune alterations comprising altered dendritic cell functioning, variations of macrophage and regulatory CD8+ T cell accumulation, reduced T cell proliferation and effector efficacy, diminished NK cells, and elevated inflammation. And metabolic alterations due to obesity may affect inflammatory processes related to white fat, as well as the promotion of a microenvironment that favors cancer cell growth. Conversely, diet restrictions can augment immune functioning through modification of metabolic regulation.

The link between obesity and several forms of cancer may involve many of these immune-related factors, but is widely believed to stem, in part, from the release of inflammatory cytokines from fat cells (adipocytes) and interactions with specific hormones. Prostate cancer, for instance, is more frequent in obese than in lean men, with fat-derived inflammatory factors playing a mediating role in this regard (Fujita et al., 2019). Obesity and type 2 diabetes were similarly implicated in the development of pancreatic cancer, probably owing to interactions between inflammatory factors, metabolic processes, and microbial activity. Consistent with this supposition, increased cancer growth associated with obesity was tied to stimulation of inflammatory processes through estrogen receptor-α, and breast cancer recurrence among women with obesity could be attenuated by antiinflammatory agents (Bowers et al., 2014).

Alterations of various growth factors, noted earlier, can affect inflammation and might thereby influence cancer occurrence and progression. Obesity has been associated with changes of the growth factors, VEGF, and insulin-like growth factor-1 (IGF-1), as well as the metabolic hormones, insulin, leptin, and adiponectin. Furthermore, among obese patients with endometrial cancer, weight loss was accompanied by altered VEGF, lower IL-6, adiponectin, and insulin-like growth factor binding protein 3 (IGFBP), which could serve as relevant biomarkers for cancer risk reduction.

Increasing evidence has pointed to cross talk between adipose tissue and tumor cells that promotes their progression and metastasis. Furthermore, fat depots adjacent to the prostate, pancreas, and breast—by virtue of the presence of inflammatory factors—might render them more likely to promote metastatic dissemination (Annett et al., 2020). Beyond influencing the course of illness, obesity may interfere with the effectiveness of chemotherapy in limiting cancer growth, depending on the amount and features of visceral adipose tissue.

Ties between leptin, ghrelin, and cancer

The connection between cancer and obesity has been explored through hormonal mechanisms that regulate food intake. Appreciable evidence has implicated the eating-related peptides leptin and ghrelin in the occurrence and progression of several types of cancer. Leptin affects inflammatory processes and contributes to immune system disorders (Abella et al., 2017), and may act as a signaling and growth factor for breast cancers, including the progression of triple-negative breast cancer (Bowers et al., 2018). This hormone was also implicated in glucose uptake and cell proliferation in pancreatic cancer, which could be attenuated by silencing leptin receptor functioning. Leptin and leptin receptors are highly expressed in colon and adrenocortical cancers, raising the possibility of these receptors playing a regulatory role in these forms of cancer.

Like leptin, ghrelin may influence cancer cell proliferation and metastasis, once more being attributed to its action on energy processes and on growth hormone receptors situated on cancer cells. However, the data are not all that clear given that the influence of ghrelin may differ across cancer cell lines, and studies conducted in vivo and in vitro haven't uniformly agreed with one another. Still, low serum ghrelin levels were associated with increased gastric cancers and esophageal squamous cell carcinoma and could potentially serve as a biomarker to predict the later occurrence of such cancers (Pritchett et al., 2020).

Cholesterol involvement in cancer

Cholesterol made by the liver is ordinarily present in all cells. It serves multiple functions related to the production of hormones, such as estrogen and testosterone, and contributes to digestion by creating fat dissolving bile acids. The integrity and fluidity of cell membranes is also dependent on cholesterol, allowing lipids and proteins to diffuse from sites in which they are synthesized.

Cholesterol comes in two forms, low-density lipoproteins (LDL) and high-density lipoproteins (HDL). The LDL, sometimes referred to as "bad cholesterol," makes up most of the cholesterol in the body. Excessive levels of LDL may occur because of the foods being consumed, such as saturated fat and trans fat. This form of cholesterol may combine with triglycerides that come from unburned calories stored in fat cells. This contributes to fatty deposits on the arterial wall, thereby contributing to heart disease and ischemic stroke. The value of HDL is that it carts off about 25%–30% of LDL back to the liver, where it is degraded and then eliminated from the body. In this sense, HDL acts in a protective capacity and is referred to as "good" cholesterol, but since it only eliminates a fraction of LDL, its benefits are limited.

Discussions of cholesterol and triglycerides most often center around heart disease, but several epidemiological studies pointed to elevated LDL levels being related to increased cancer occurrence, and the use of statins ("cholesterol busters") was accompanied by reduced cancer occurrence (Chimento et al., 2019). Epidemiological studies indicated that elevated LDL cholesterol was linked to melanoma, colorectal, pancreatic, lung, kidney, bladder, breast, and ovarian cancers. The use of statins has been associated with diminished occurrence of several forms of cancer, and among women who were incidental statin users improved survival was observed in response to triple negative breast cancer (Nowakowska et al., 2021). Derivatives of cholesterol, oxysterols, that contribute to lipid metabolism, apoptosis, and autophagy have also been linked to diverse forms of cancer (Chimento et al., 2019). And high triglycerides, and the ratio between triglycerides and HDL cholesterol, were also related to clinical outcomes in patients with gastric cancer.

The relationship to cancers could be due to multiple factors that affect cholesterol. Cancer may evolve owing to mutations within genes associated with cholesterol metabolism, as well as disturbed cholesterol homeostasis, potentially caused by epigenetic changes. Cholesterol is not only important for metabolic processes and immune functioning but may also play a fundamental role in regulating angiogenesis and it could also have effects on cancer processes through effects on estrogens, androgens, and glucocorticoids that affect cell proliferation, migration, invasion, and apoptosis. As such, by reducing cholesterol synthesis and altering LDL through statins, the risk for several forms of cancer was reduced and statins were proposed as an adjuvant treatment for some forms of cancer (Borgquist et al., 2018).

As it happens, tumor cells release cholesterol and oxysterols that undermine the effectiveness of NK cells (Assmann et al., 2017) so that high levels of cholesterol were accompanied by increased tumor growth and metastasis, which could be reduced by statins. A similar effect was obtained by inhibiting the formation of oxysterols, the derivatives of cholesterol, or by-products of cholesterol biosynthesis. The cholesterol derivative, 27-hydroxycholesterol (27-HC), might be key to the negative actions of cholesterol (Kloudova et al., 2017). It appeared that 27-HC can be taken up readily by certain cells, which are more likely to be tumorigenic owing to actions on ferroptosis (cell death involving presence of iron). When the influence of ferroptosis was diminished the tumorigenic effects otherwise linked to 27-HC were diminished (Liu et al., 2021). If nothing else, these findings raise the possibility that cholesterol or 27-HC synthesis might serve as targets to limit cancer development. Furthermore, the occurrence of heart failure secondary to treatment with anthracyclines (particularly doxorubicin) in the treatment of breast cancer was markedly reduced among women taking statins, and there were indications that similar outcomes were apparent with trastuzumab (Herceptin) treatment of breast cancer. Together, these findings are consistent with reports concerning the involvement of cholesterol in some forms of cancer as well as in moderating side effects of cancer therapies.

Although red meats had been implicated in the development of some forms of cancer, possibly through actions on cholesterol levels, the idea that red meats acted in this way became a highly contentious issue as we saw in Chapter 7. While not dismissing the merits of reports suggesting that red meat is not as bad as first maintained, it ought to be acknowledged that there are ways by which red meat could affect cancer occurrence. Ordinarily, gut microbiota convert protein from red meat into trimethylamine (TMA), which is then converted to trimethylamine oxide (TMAO). Circulating TMAO has been tied to the development of obesity and elevated inflammation and might therefore contribute to heart disease and cancer, particularly with increasing age. Yet, TMAO is also produced by fish consumption, which has not typically been associated with cancer occurrence. It seems that factors other than or in addition to TMAO might be pertinent in the development of cancer. In this respect, long-term consumption of red meat was tied to the presence of high levels of the sialic acid, *N*-glycolylneuraminic acid (Neu5Gc), which isn't readily synthesized in humans. When derived from the diet, Neu5Gc could promote inflammation and consequently tumorigenesis (Zaramela et al., 2019).

We have indicated repeatedly that it is difficult to ascribe any single food to disease occurrence and consideration of the links between red meats and cancer ought to be conducted in the context of other foods individuals consume. Steak with a side of fries together with a fructose-laced soft drink obviously has very different implications than does a steak together with a green salad washed down with plain soda water. Furthermore, was this main course followed by a dessert replete with trans fats (e.g., some cakes or pastries) or did the dessert simply comprise a cup of tea or coffee? The moral here is that the company some foods keep on the evening dinner plate may make a considerable difference to a person's health.

Genetic influences on obesity

It is not uncommon for discrimination and stigma to be attached to obesity. In some instances, this may stem from the belief that overweight people are simply uninterested in controlling their impulses, although there are certainly other factors that promote such biases. It is absolutely clear, however, that multiple factors contribute to obesity over which individuals have little say, such as prenatal and early life negative experiences and their family environment. Furthermore, genetic and/or epigenetic factors may contribute to obesity and related disorders.

Studies in twins have indicated that heritability of BMI was high, although shared environmental factors (lifestyle) could influence obesity. In fact, numerous genes have been implicated in obesity, as have specific polymorphisms and epigenetic changes, including those linked to leptin and its receptor, BDNF, and proopiomelanocortin (POMC). Variants of the melanocortin-4-receptor (MC4R) gene have received particular attention in this regard, as have mutations in the coding region of MC3R. And a polymorphism for the fat mass and obesity-associated (FTO) gene was also implicated in obesity (Zhao et al., 2019). In each instance these genetic alterations have downstream effects on hormones regulated by neurons in specific hypothalamic regions (e.g., paraventricular and arcuate nuclei) that affect eating processes.

Obesity may arise owing to overeating or eating certain foods, but it has also been considered to be a multisystem metabolic illness that involves the interplay between numerous processes, including genetic factors. It is probably not productive to single out any specific gene in promoting obesity. Indeed, polygenic factors were highly predictive of weight differences and risk for obesity, and over the course of development, polygenic factors were increasingly more aligned with weight elevations, peaking when individuals were about 18 years of age (Khera et al., 2019). The influence of gene effects and the presence of certain polymorphisms tied to eating processes also interact with lifestyles (e.g., balanced diet and regular exercise) in affecting weight control and hence cancer development. As much as genetic factors might be tied to cancer occurrence, the adoption of healthy lifestyles can diminish the risk of cancer occurrence changes, irrespective of genetic background (Cho et al., 2019).

Epigenetic changes have been identified in a variety of human tissues associated with metabolic functioning (adipose tissue, skeletal muscle, pancreatic islets, liver), which may be relevant to the promotion of obesity and the development of type 2 diabetes (Ling and Rönn, 2019). Such variations were similarly noted in a gene linked to lipid metabolism among individuals who had obesity together with a metabolic disease. The disturbed lipid functioning may have promoted inflammation, which favored illness development. Epigenetic changes that influence microbiota may also contribute to obesity as might epigenetic modifications caused by microbial metabolites. Similarly, epigenetic changes associated with energy and metabolic processes, such as insulin and leptin, were reported among obese individuals and those with type 2 diabetes (Ling and Rönn, 2019).

Just like other experiences and biological changes, disturbed metabolic functioning during pregnancy can promote epigenetic changes within the fetus, thereby contributing to obesity that develops in the years following birth (Stolzenbach et al., 2020). Nutrients consumed by a pregnant woman can influence fetal programming that can affect whether cancer would occur in offspring at a later time (Kaur et al., 2013). Speaking to this, a high-fat diet among pregnant mice was associated with epigenetic changes that led to mammary cancer in offspring. However, these actions could be attenuated by a combination of drugs that attenuated the epigenetic alterations and could thus limit cancer occurrence (Andrade et al., 2019). Further to this, among obese pregnant women who experienced gestational diabetes, multiple epigenetic changes were detected that could potentially affect the well-being of the offspring following birth. But when women exercised and controlled their diet, the frequency of these epigenetic modifications was reduced (Antoun et al., 2020).

Dietary components in relation to cancer and its treatment

There are multiple ways by which nutrients can affect the prevention or provocation of various illnesses, as well as the response to therapies. In this section we will attend to common nutritional components fundamental to energy processes. This includes glucose, fats, fatty acids, polyphenols, dietary amino acids, and a variety of supplements. In discussing these, we will address their role in cancer occurrence and the possibility of manipulating energy-related processes during cancer treatment.

Glucose

Multiplying as quickly as they do, cancer cells require appreciable energy in the form of glucose, which we have discussed earlier in relation to the Warburg effect. Glucose may contribute to cancer initiation, foster cancer progression, and may contribute to the development of treatment resistance. As glucose is so highly desired by cancer cells, depriving them of this energy source, such as by diet or pharmacological intervention, might allow their proliferation to be stalled. But cancer cells are tenacious and capable of thwarting the very best efforts to destroy them. They are adept at obtaining energy through multiple pathways and can also engage in actions to undermine the functioning of healthy cells, as well as cannibalize their own neighbors to obtain nutrients. At the same time, when immune cells are engaged in battling a threat, they also require considerable energy that can be supplied by glucose. Should glucose not be available, then immune cells are more likely to be dysregulated and have a more difficult time fighting cancer cells (Lawless et al., 2017).

In some instances, the ability of healthy cells to consume glucose is undermined, thus allowing for more glucose to be available for cancer cells. This occurs because cancer cells push fat cells into excessive production of IGFBP-3, a molecule which makes ordinary cells less sensitive to insulin and hence less likely to take up glucose. Even if it proved ineffective in limiting cancer progression, starving cancer cells might augment their sensitivity to cancer therapies. Consequently, benefits could potentially be derived by reducing glucose intake both before and during chemotherapy to deprive cancer cells of their energy needs.

If only it were that simple! The position has been adopted that in the presence of obesity a combination of factors may conspire to foster cancer progression. High levels of insulin-like growth factor (IGF), insulin resistance, hypercholesterolemia, and excessive oxidative stress may be potent drivers of obesity-related breast cancer among postmenopausal women (Engin, 2017). These actions may be further exacerbated by obesity-related estrogen together with variations of leptin and adiponectin concentrations. Other processes may also come into this combination. Cancer cells can inhibit gut serotonin production, which would otherwise be needed for pancreatic insulin production, thereby diminishing glucose absorption by normal cells. This may be abetted by the capacity of cancer cells to reduce gut microbiota abundance that can downregulate insulin production, which, in turn, may lower metabolic hormones, such as incretins that promote a decline of blood glucose levels. The net result of these hormonal actions is that glucose availability to cells would be diminished, and even when glucose levels were increased the limited insulin presence would undermine the glucose being taken up by normal cells, making it more available to feed the cancer cells. Concurrently, owing to the starvation of healthy cells, fatigue and weight loss may be promoted, and perhaps cancer-related cachexia engendered.

Bait and Switch

Sugar, perhaps more than any other dietary factor, has been implicated in obesity and a variety of diseases, including some forms of cancer. Accordingly, there has been a push to better regulate soft drink consumption, including increasing taxation rates on these products. The counter argument has been that while sugar intake has been declining over the last two decades, obesity has been increasing, seemingly reflecting their independence. However, the development of obesity may have its roots in sugar-spiked foods consumed decades earlier when individuals were young (Alexander Bentley et al., 2020).

The sugar industry engaged in relentless attacks on the notion that their products contribute to obesity. They have sponsored research assessing the impact of sugar on health-related processes, but this research has been argued to have a history of bad science behind it and poor recommendations related to health followed. It was maintained that results from industry-sponsored food trials favored their own positions relative to those trials that were not financially supported by industry. Indeed, the food industry could have done a whole lot of good, but they chose not to engage in these efforts despite their claims to the contrary (Ludwig and Nestle, 2008).

The alternative to sugar, which comprised nonnutritive sweeteners, initially seemed like a good option; artificial sweeteners captured a substantial slice of the market, and the sale of diet drinks spiked. Moderate consumption of these drinks precluded blood sugar spikes and probably had beneficial effects in preventing tooth decay, but we shouldn't be misled about the benefits derived from "diet" drinks. Among older individuals, diet soda consumption was linked to increased waist size, accompanied by more frequent incidence of type 2 diabetes and elevated cardiometabolic risk. Artificial sweeteners may have some negative effects by altering gut microbes, thereby encouraging metabolic syndrome, and the development of obesity and diabetes (Suez et al., 2014). Like sugary drinks, artificial sweeteners were also linked to diminished hippocampal volume, poorer cognitive functioning, dementia, and increased risk of stroke. Understandably, concerns have been raised regarding the influence of nonnutritive sweeteners in children, and maternal consumption of these artificial sweeteners influenced the microbiota and metabolism in offspring (Olivier-Van Stichelen et al., 2019). Admittedly, it had been difficult to make sense of some of the reported findings concerning the benefits versus drawbacks of nonnutritive sweeteners. Parsing out studies that are unbiased, and those funded by one food industry or another would be a good start.

It is of practical significance that type 2 diabetes has been linked to several forms of cancer—particularly cancer of the pancreas, liver, breast, endometrium, bladder, colorectal, and kidney. This is possibly due to greater accrual of DNA damage, which is less likely to be repaired when blood sugar concentrations are too high. Hyperglycemia also increases intestinal barrier permeability, leading to an increase of microbial factors and their products, which could foster enteric infection (Thaiss et al., 2018) and favor colorectal cancer development. Reducing circulating glucose levels may therefore have beneficial effects on gut barrier permeability and reduced incidence of cancer.

A metaanalysis of 10 studies supported a role for reduced blood glucose in cancer processes and pointed to several other endogenous substrates that were linked to breast cancer survival, including altered lipid profile, insulin and insulin resistance, sex hormones and sex hormone-binding globulin Ki67 (which might be needed for cell proliferation), caspase-3, p-Akt, and the inflammatory marker C-reactive protein (Zhang et al., 2019). In line with the perspective that energy substrates are linked to cancer, the type 2 diabetes medication metformin, which reduces glucose made by the liver and sugar absorption from foods, positively influenced several forms of cancer (Zhang et al., 2019). For example, treatment of triple-negative breast cancer could be augmented by a combination of metformin and heme (an iron-containing compound present in hemoglobin). It seems that heme affects the transcription factor BACH1 so cancer cells sensitive to metformin will suppress mitochondrial respiration, thereby undermining their growth and multiplication.[a]

Given the diverse ways by which cancer cells may affect cancer progression, it was considered that metformin may have its beneficial actions through other routes. Specifically, it can diminish the odds of developing pancreatic cancer by reducing IGF-1 signaling, thereby inhibiting the mammalian target of rapamycin (mTOR) pathway, which is fundamental for numerous anabolic and catabolic processes that contribute to tumor cell replication (Gong et al., 2014). The positive effects of metformin may also be tied to the activation of AMPK (5′ AMP-activated protein kinase), leading to the inhibition of mTOR. Essentially, through the actions on AMPK pathways, altered metabolic processes may limit cancer cell proliferation and this could be aided by the actions of metformin in reducing growth factor signaling. Beyond AMPK variations, metformin can also affect other kinases that facilitate the restoration of metabolic processes that could potentially influence cancer development. It also appears that metformin can attenuate epigenetic changes that have been associated with some brain cancers in children. Metformin also engenders gut microbial changes, possibly accounting for a portion of the beneficial effects of the treatment. In fact, when fecal samples were transferred from metformin-treated donors to germ-free mice, glucose tolerance was augmented (Wu et al., 2017).

The involvement of insulin in cancer may come about through processes other than those involving glucose availability. The presence of insulin is sensed by cancer cells and upon being stimulated the signaling pathway, PI3K, is activated and can affect the cell cycle, autophagy, and anabolic processes (Saxton and Sabatini, 2017). Some cancer cells that are resistant to the actions of dietary restriction carry mutations that promote PI3K activation so these cells can multiply in the absence of insulin or IGF-1. As the PI3K pathway contributes to many forms of human cancer, it was maintained that targeting this pathway may be an effective strategy to diminish cancer growth (Yang et al., 2019). Unfortunately, numerous clinical trials using PI3K inhibitors failed to yield positive outcomes owing to their poor tolerability or because resistance developed to the treatments. Yet, selective forms of PI3K inhibitors may yield better outcomes with fewer side effects. There is already some evidence indicating that these agents may have better effects in patients with PI3K mutations, but this factor alone was

---

[a] Metformin, which has been viewed as a miracle drug, has been around for almost a century. In fact, the beneficial effects coming from leaves of the French lilac (that were used as long as three centuries ago) have been attributed to compounds similar to metformin. Aside from its use in treating type 2 diabetes, metformin is also used off label in the treatment of obesity, polycystic ovarian syndrome, infertility, nonalcoholic fatty liver disease, and even acne. In addition, it regulates metabolic and nonmetabolic processes and could potentially be used to increase life span and health span. Parenthetically, it was also observed that among individuals with type 2 diabetes metformin reduced death related to COVID-19 infection.

insufficient to predict whether PI3K inhibitors would be effective. Ultimately, more precise biomarkers will need to be developed, which might come about with greater understanding of how PI3K and its inhibitors operate (Yang et al., 2019).

There is one further comment that needs to be made concerning the link between insulin and cancer. Some cancer therapies, such as hormonal manipulations and chemotherapy, can promote insulin resistance, and the subsequent hyperglycemia favors disease progression and disturbed efficacy of cancer therapy (Ariaans et al., 2015). This raises the uncomfortable possibility that some of the treatments that ought to act against cancer might engender untoward effects on cancer progression.

The tyranny of fructose

Sugars, such as sucrose, are often frowned upon owing to their challenges to physical well-being. But it is not uncommon for individuals to rationalize that "fructose isn't too bad, since it comes from fruits and vegetables and so it must be good for health—right?" It's true that at one time fructose in the human diet was obtained primarily from certain fruits. However, with the development of high fructose corn syrup, which is used in soft drinks and some prepackaged foods, these have become the primary source of "fructose overload," with children and adolescents likely being most affected.

Fructose is thought to affect health through diverse processes. Unlike glucose that is generally used by all cells in the body, fructose is primarily metabolized by the liver, and hence high amounts of fructose can overload this organ (Jegatheesan and De Bandt, 2017). Additionally, fructose can also be metabolized within the small intestine and may cause deterioration of the intestinal epithelial barrier so the interaction between the liver and gut may contribute to pathology (Febbraio and Karin, 2021). When fructose intake is excessive the production of triglycerides (lipogenesis) is elevated, as is the production of reactive oxygen species. These actions may contribute to insulin resistance and hepatic inflammation leading to metabolic disorders and nonalcoholic fatty liver disease (Jegatheesan and De Bandt, 2017). Individuals who consumed a fructose-sweetened drink daily over a 7-week period displayed fat production within the liver that was twice that of individuals who consumed a glucose sweetened drink over this period (Geidl-Flueck et al., 2021). As well these alterations, fructose may promote leptin resistance, thereby further contributing to obesity and the health risks that this creates.

The possibility that beverages sweetened with high-fructose corn syrup might contribute to cancer has been considered due to the increasing occurrence of colorectal cancer in young and middle-aged individuals since the 1970s, when high-fructose corn syrup was broadly introduced. This is not to say that other factors are not responsible for the increased colorectal cancer in young people. For it has been linked to high-fat diets, possibly reflecting a double hit comprising a mutation in a tumor suppressor gene (APC) and fat-elicited bile acid that affects intestinal stem cells (Fu et al., 2019). It is significant that in vulnerable mice, chronic daily consumption of the equivalent of a single can of soda sweetened with fructose was sufficient to promote colorectal cancer (Goncalves et al., 2019). Moreover, data from the Nurses Study indicated that women who consumed sugary drinks containing fructose during adolescence were at elevated risk of developing early-onset colorectal cancer. Although the occurrence of early onset colorectal cancer is infrequent, it has been increasing at an alarming rate, with about half of these cases being unrelated to familial or genetic factors. No doubt dietary influences and the increased occurrence of obesity may to some extent account for this rise; each daily serving of a fructose-laced drink between 13 and 18 years of age was associated with a 32% rise of colorectal cancer prior to the age of 50 (Hur et al., 2021). Excessive fructose consumption was also linked to other tissue cancers, including, breast, pancreatic, prostate, liver, breast, and lung cancer (e.g., Jin et al., 2019) as well as breast cancer metastasis (Kim et al., 2020).

An important feature related to fructose is that it can affect glycolysis and its metabolism can be driven in the presence of low levels of oxygen. Through downregulated mitochondrial respiration and increased anaerobic glycolysis (Warburg effect) fructose may fuel cancer development and metastasis (Nakagawa et al., 2020). The actions of fructose on cancer development may also be exacerbated by affecting the pentose phosphate pathway, which among other things is a metabolic pathway that parallels glycolysis. It can also promote negative consequences by the production of uric acid and lactate that may promote mitochondrial reactive oxygen species, blocking fatty acid oxidation, and by inhibiting the tricarboxylic acid cycle (TCA cycle; also referred to as the Krebs cycle) through which normal cells ordinarily obtain energy in the presence of oxygen (Nakagawa et al., 2020). The enzyme ketohexokinase, which normally breaks down fructose into fructose-1-phosphate, is also present within tumors where it can alter cancer cell metabolism and promote increased tumor growth (Goncalves et al., 2019). This suggested that dietary manipulations that influence the ketohexokinase-A signaling pathway might limit the occurrence of breast cancer metastasis (Kim et al., 2020).

Fructose absorption by cells is determined by GLUT2 and GLUT5 that serve as fructose transporters within the small intestine as well as in fat tissue, kidney, and brain. The presence of GLUT5 was exceptionally elevated in ovarian cancers and was predictive of poor survival (Jin et al., 2019), which is in keeping with the presumed role of GLUT5 in fructose provoked tumorigenesis. Elevated GLUT5 was likewise elevated in glioma tissues, which was tied to poorer survival. While fructose intake in animal models increased glioma growth, knockdown of GLUT5-inhibited tumor progression and similar effects were observed in animal models of lung cancer (Chen et al., 2020). In view of the links between fructose and varied cancers in animal models and in humans, much greater attention ought to be devoted to the consumption of products with high fructose content and should be reduced in cancer patients.

Fat and lipid metabolism

Among their varied capabilities, cancer cells can adapt to changes in their microenvironment by recalibrating their metabolism to favor rapid division, which contrasts with that of immune cells that are less capable in making this adjustment. Aside from being enamored by sugars, cancer cells make exceptional use of fat that contains great amounts of energy. In mice placed on a high-fat diet, cancer cells could reprogram their metabolism so they more readily consumed fat. Concurrently, the presence of cytotoxic CD8+ T cells declined within the tumor microenvironment and those that were present were less capable of multiplying, thereby favoring tumor growth (Ringel et al., 2020). Because of the links between fat and cancer, efforts have been made to control cancer growth by disrupting lipid metabolism in these cells, including the use of calorie restriction as an adjunctive treatment. However, such approaches are met with uncanny adaptive skills by cancer cells. As part of their tactic to survive and grow, when faced with treatments that limit fat synthesis cancer cells can acquire the ability to obtain *exogenous* fat particles from the surrounding environment (Lupien et al., 2020). It's a battle of wits that exemplifies the tenacious drive for survival among cancer cells, and the challenges faced by researchers to find their Achille's heel.

But back to fat. While fat is often thought of in negative ways, not all forms of fat are equally pernicious and related to disease occurrence. White fat serves as an energy storage tissue that is relatively resistant to being altered by diet or exercise, whereas brown fat, which is limited in adults, is a readily burned energy source. As such, brown fat is seen as a "good" form of fat. Abdominal obesity is particularly harmful owing to the prevalence of white fat, which as we've seen, releases inflammatory factors. Some individuals (more so women than men) have a greater ability to activate brown fat, and it seems that several genetic variants contribute to the distribution of fat within the body (Justice et al., 2019). Unfortunately, inflammatory factors released from white fat limit the conversion of white fat to brown among obese people so their weight issues persist as do the detrimental effects of inflammation. In the presence of chronic inflammation stemming from the sustained secretion of proinflammatory cytokines from white fat, elevated insulin stimulates receptors located on T_{reg} cells, resulting in suppression of the inhibitory cytokine IL-10. The net result of this proinflammatory state is the furthering of disease progression. It also seems that excessive white fat is associated with increased aggressiveness of some cancers, likely owing to the recruitment of adipose stromal cells, which vary with the presence of the chemokine CXCL1 (Zhang et al., 2016).

Given the abundance of inflammatory factors in white fat that might promote the development and progression of various types of cancer, it was thought that coaxing white fat to become brown or beige (an intermingling of white and brown fat) could act against illnesses linked to obesity. Indeed, certain agents delivered via nanoparticles could encourage white fat cells to change to brown, engendering appreciable weight loss, a decline of cholesterol and triglycerides, and enhanced sensitivity to insulin (Xue et al., 2016). A targeted CRISPR interference delivery system was similarly effective in limiting white fat, thereby attenuating obesity in mice by 20% without affecting food intake, and predictably limited insulin resistance and inflammation (Chung et al., 2019). In this regard, knockdown of two genes, IRX3 and IRX5, related to the FTO gene believed to be involved in obesity-related food intake and impulsivity, affected fat storage or burning (Claussnitzer et al., 2015) and might provide important clues to the development of antiobesity treatments.

Brain processes, notably hypothalamic neurons that contribute to hunger, may also be involved in the conversion of white to brown or beige fat. During the hours following a meal, circulating insulin stimulates hypothalamic processes that provoke the browning of fat, whereas related processes operate in reverse after a fasting period. For whatever reason, the necessary switch responsible for converting white fat to brown may not be operating efficiently among some obese people, so energy expenditure processes are less likely to be enabled. If this switching process could be manipulated, benefits could be obtained to diminish abdominal obesity.

Besides effects on energy processes, brown fat may affect well-being through other routes. Ordinarily, branch-chain amino acids, such as leucine, isoleucine, and valine, are obtained from a variety of foods (e.g., meat, fish, chicken, eggs, and milk). These amino acids are ordinarily beneficial to health, but when excessive they have been linked to diabetes and heart disease. In addition to its other functions, brown fat may serve as a filter to reduce the concentrations of branch-chain amino acids. Should methods be found to influence branch-chain amino acid uptake by brown fat, it might serve as a boon for various diseases that have been linked to obesity, although the limited availability of brown fat in adults might be a problem for this intervention strategy.

Fatty acids

Polyunsaturated fatty acids, such as omega-3 fatty acids, have received an unusual amount of attention, having reached a huge audience for their purported health benefits. Two omega-3 fatty acids, eicosapentaenoic acid (EPA) and docosahexaenoic acid (DHA), have been evaluated extensively and have been *alleged* to influence just about every known psychiatric illness and a great number of physical illnesses. Although these fatty acids are often considered together, they may have somewhat different actions. Specifically, DHA appeared to engage more factors that lower proinflammatory actions than did EPA, whereas the balance between pro- and antiinflammatory factors was more reliant on EPA. In contrast to omega-3 fatty acids, its cousin omega-6 fatty acids do not have such benefits and may have adverse effects owing to their proinflammatory actions. The involvement of microbial changes introduced by omega-3 in the provocation of illnesses hasn't been widely examined. Nonetheless, the data available have indicated that gut microbiota dysbiosis can instigate endotoxemia and inflammatory processes that favor the development of several conditions that can promote illnesses as described in Fig. 8.2 (Costantini et al., 2017).

It is tempting to dismiss the attention to omega-3 as a fad, yet omega-3 fatty acids in fish oil, as well as polyphenolics in fruits and vegetables, influence NK and T_{reg} cells, and the balance between the two. Moreover, in mice maintained on a high-fat diet that produced obesity, omega-3 supplementation through diet enhanced immune functioning that was otherwise disturbed, including phagocytosis, NK cell activity, and lymphoproliferation (Hunsche et al., 2020). In humans, a diet rich in omega-3 fatty acids provoked an increase of beneficial bacteria that produced SCFAs and was linked to reduced inflammation that could be enhanced by the concurrent administration of an antiinflammatory agent. In line with these actions, omega-3 fatty acids in fish could affect cancer occurrence, such as hematological malignancies and fewer instances of breast cancer (e.g., Nindrea et al., 2019) and was recommended as an adjuvant therapy for colorectal cancer. Elevated risk of breast cancer and mortality was likewise related to inflammation that was linked to poor dietary habits comprising low intake of omega-3 fatty acids, natural antioxidants, and fiber, coupled with relatively high intake of refined starches, sugar, and both saturated and trans fats (Seiler et al., 2018).

Research in this field is extensive, and this brief review can't possibly do it justice. Nonetheless, a detailed discussion concerning mechanisms by which omega-3 may have beneficial effects, in which types of cancer, and under what conditions has been provided in several excellent reviews of this topic. Studies across species indicated that the benefits of omega-3 fatty acids may come about by affecting lipid uptake and metabolism, enhancing insulin signaling and sensitivity, reducing glucose and serum lactate levels, diminishing inflammation, promoting tumor cell apoptosis, and by limiting angiogenesis and metastasis. Moreover, in the presence of excessive amounts of the omega-3 fatty acid, DHA, tumor cells may be unable to store the DHA, and the resulting oxidization of the free omega-3 promotes the death of the tumor cells (Dierge et al., 2021).

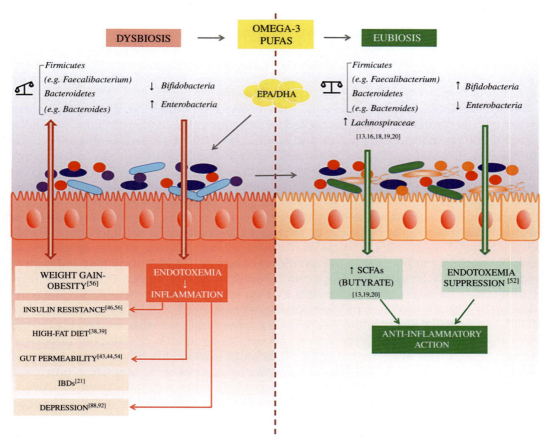

FIG. 8.2 Potential action of omega-3 polyunsaturated fatty acid (PUFA) in restoring eubiosis in gut microbiota. Dysbiosis of the Firmicutes/Bacteroidetes ratio is associated with several conditions, such as weight gain and obesity, insulin resistance, high fat, gut permeability, IBDs, and depression. Similarly, a Bifidobacteria decrease combined with an Enterobacteria increase leads to the establishment of endotoxemia that causes a chronic low-grade inflammation associated with some pathological conditions like insulin resistance, gut permeability, and depression. Initial evidence shows that omega-3 PUFAs are able to reverse this condition by restoring the Firmicutes/Bacteroidetes ratio, and increasing Lachnospiraceae taxa, both associated with an increased production of the antiinflammatory short-chain fatty acid (SCFA). Moreover, animal studies showed the ability of omega-3 PUFAs to increase lipopolysaccharide (LPS)-suppressing bacteria, Bifidobacteria, and to decrease LPS-producing bacteria, Enterobacteria, negating the endotoxemia phenomenon. For all these actions, omega-3 PUFAs can be considered as prebiotics, able to restore gut eubiosis in some pathological conditions. *Figure and caption from Costantini, L., Molinari, R., Farinon, B., Merendino, N., 2017. Impact of omega-3 fatty acids on the gut microbiota. Int. J. Mol. Sci. 18, 2645.*

Despite the apparent benefits of omega-3 fatty acids, support for the health benefits of omega-3 intake is not universal. A randomized placebo-controlled trial with 25,871 participants followed for a median of 5.3 years indicated that supplementation with marine omega-3 did not influence the occurrence of cancers or heart disease (Manson et al., 2019). A systematic review concerning prostate cancer similarly concluded that it may be premature to attribute beneficial actions to omega-3 (Aucoin et al., 2017). Indeed, a metaanalysis revealed that omega-3 intake did not improve glucose metabolism and at high doses could even have negative effects in relation to diabetes (Brown et al., 2019), which has been linked to cancer occurrence. There have even been indications that omega-3 intake could disturb immune functioning (e.g., suppression of CD8+ T cells) that would otherwise attack tumor cells and could therefore favor the progression of high-grade prostate cancer. Questions also emerged as to whether omega-3 augments or disrupts the effectiveness of cancer therapies (Murphy et al., 2013).

Despite the increasing information that has evolved regarding the actions of polyunsaturated fatty acids, further data are needed to determine their actions in conjunction with standard and novel cancer treatments and in relation to cancer-related complications. There are many factors that need to be considered before maximal effectiveness can be obtained, such as whether there is an ideal mix or concentration of omega-3 that is effective. Moreover, are the responses more impressive when omega-3 comes from specific foods (e.g., fatty fish) rather than supplements?

Therefore, as in the case of so many other cancer medications, different responses associated with omega-3 might not simply be related to their efficacy so much as to how they are being employed. Once more, a precision medicine approach, using genetic, epigenetic, and other biomarkers, might enhance the usefulness of omega-3 supplementation. Furthermore, rather than simply focusing on the amount or form of dietary omega-3 consumed, it may be profitable to consider related processes that are fundamental for fat and fatty acids to be taken up by cells. For instance, polymorphonuclear myeloid-derived suppressor cells (PMN-MDSCs) are essential in regulating the response to cancer cells, but when excessively activated they can contribute to cancer progression and have been associated with poor responses to therapeutic agents. It appears that PMN-MDSCs can increase the functioning of fatty acid transport protein 2 (FATP2) that influences cancer cell uptake of the long-chain polyunsaturated fatty acid, arachidonic acid, and can undermine immune responses so the effectiveness of cancer therapy is impaired. As expected, knocking out a gene linked to FATP2, curtailed the progression of several types of cancers (lymphoma, lung carcinoma, colon carcinoma, and pancreatic cancer), and the FATP2-inhibiting agent, lipofermata, diminished cancer growth, supporting the possibility that it might be a useful adjunctive treatment (Veglia et al., 2019).

Polyphenols

A diet high in polyphenols was offered as a possible way of acting against a wide range of cancers (Focaccetti et al., 2019). Considerable research pointed to the benefits of major classes of polyphenols (flavonoids, phenolic acids, lignans, and stilbenes) in cancer prevention. The actions of polyphenols are affected by gut microbial factors, and their effectiveness in relation to cancer and other diseases has also been ascribed to their antiinflammatory actions. It was likewise indicated that the benefits obtained from polyphenols could occur through effects on estrogen receptors (Cipolletti et al., 2018) and androgen receptor actions related to apoptosis and angiogenesis (Costea et al., 2019). The bulk of the evidence seems to be congruent with the view that polyphenols generally have positive actions in maintaining good health and can perhaps be useful as an adjunctive treatment for cancer patients. It remains to be determined which polyphenols are most effective in this regard, at which doses they should be consumed, and for how long. This is particularly pertinent as it was cautioned that they can also disturb immune functioning and thus could favor tumor growth (Focaccetti et al., 2019).

Dietary amino acids

Cancer cells need amino acids in plentiful supply in order to survive and thrive. Consequently, therapies that diminish amino acids through diets or specific enzymes may enhance cancer therapies. Attempting to deprive cancer cells of these resources for extended durations, however, might be impractical given that deficiencies of amino acids will likely occur in the rest of the body before the tumor is affected, thus potentially having negative health consequences (Kanarek et al., 2020). Still, if extreme diets can produce positive effects, then it might be reasonable to expect that when used judiciously, diets and pharmacological treatments that target amino acids can have beneficial actions. In Chapter 14 a discussion is provided concerning the effectiveness of starving cancer cells of *essential* amino acids (those that must come from the diet), notably methionine, serine, glutamine, asparagine, arginine, aspartate, histidine, mannose, and cystine, in enhancing the response to cancer treatments. These positive outcomes may occur by affecting specific microbiota as well as through actions on KRAS and by diminishing p53.

The available data concerning the influence of amino acid deprivation on tumor growth have been interesting but it is improbable that manipulation of any single amino acid will affect different types of cancer. Among other things, varied forms of energy that involve different amino are required by assorted tumors. Similarly, dietary alterations may not similarly affect nutrients and amino acids in all tissues or tumor microenvironments. Accordingly, while amino acid manipulations could facilitate the actions of chemotherapy and immunotherapy, multiple individual difference factors and those related to specific tumors will affect outcomes.

Caveats concerning diet manipulations and cancer

Numerous diet manipulations have been offered that might be useful in the treatment of cancers. To be sure, in some quarters some of these approaches are still considered to be on the fringe of being scientifically valid or useful. Yet, dietary alterations can modify mechanisms believed to be directly or indirectly related to cancer occurrence and progression. This offers justification for manipulating specific dietary components to deprive cancer of nutrients and may be a reasonable approach that could enhance the efficacy of cancer treatments. Additionally, aside from modifying components of an individual's diet, drug treatments that influence energy-related processes represent an additional line of assault on cancer growth.

Overall, there are formidable challenges in recommending unique dietary changes in cancer patients. As described earlier, food restriction diets often differ from one another with respect to the intake of carbs, fats, and proteins, making comparisons across studies difficult. Indeed, these diets may have different effects on the availability of glucose, insulin, triglycerides, and cholesterol, all of which have been related to cancer progression. As we have indicated so often, not every dietary treatment will be effective for all individuals. Different outcome may depend on genetic factors and constitutional characteristics of the individual and may vary with the type of cancer.

Vitamins and other supplements

Vitamins are essential micronutrients needed for normal cell functioning, growth, and development. In the main, vitamins are not synthesized in the body (or appear only in very small amounts), largely being obtained from the foods we consume. However, the quantity we derive from food can vary for some vitamins. Vitamin D, for example, needs to be obtained through UVB rays emitted by the sun, since insufficient amounts come from food.

Vitamins are segregated into two main classes: fat-soluble and water-soluble groups. The fat-soluble vitamins comprise vitamins A, D, E, and K, which can be stored in fatty tissue and liver where they remain bioactive or useful for many days. Water-soluble vitamins, comprising the vitamin B family and vitamin C, are less stable and are not stored. They are rapidly eliminated (e.g., through urine) and must be replaced more often than fat-soluble vitamins.

Each vitamin has unique physiological contributions that ensure good health. The regulation of mineral metabolism needed by bones and diverse organs is facilitated by vitamin D, whereas vitamin A regulates cell and tissue growth and differentiation. To combat excessive oxidation and undue oxidative stress that arises from cellular metabolism, help can be obtained through the antioxidant properties of vitamin C and E. Members of the vitamin B family—such as B_1, B_2, B_3, and B_9, better known as thiamin, riboflavin, niacin, and folate, respectively—help enzymes do what they are meant to do, such as promote the release of energy from carbohydrates and fat, as well as transporting energy-containing nutrients and oxygen throughout the body. Finally, the importance of vitamins is underscored by the consequences of specific vitamin deficiencies. We have long known that scurvy results from vitamin C deficiency, rickets from low vitamin D, pellagra from a lack of niacin, and neural tube defects in newborns may occur due to insufficient folate consumption during pregnancy.

Given the importance of vitamins for so many aspects of human health, an industry evolved that provided people with vitamin supplements. However, unless individuals are deficient of certain particular vitamins, these supplements will likely not have any benefits beyond vitamins obtained through foods. Contrary to industry efforts that promoted vitamin supplements to diminish age-related disturbances, longitudinal studies confirmed that multivitamins neither enhanced cognitive abilities among older individuals nor did they increase longevity (Macpherson et al., 2013). The possibility was also considered that vitamin supplements could be useful in the prevention and treatment of cancer, but the evidence supporting this position was largely unconvincing (Vernieri et al., 2018). Of considerable significance, an analysis of data obtained from more than 30,000 participants through multiple cycles across 10 years indicated that dietary supplements from food were associated with lower mortality, but this was not apparent with vitamin supplements in pill form (Chen et al., 2019).

One of the perennial favorite supplements, that of vitamin D, did not provide benefits for either heart disease or cancer occurrence, and based on randomized controlled trials vitamin D intake had no effect on breast cancer development in postmenopausal women (Sperati et al., 2013). However, a metaanalysis suggested that a course of vitamin D lasting 3 years or more was associated with increased longevity among cancer patients (Samji et al., 2019). Yet, overall, the case for vitamin D as a preventative or auxiliary agent in cancer treatment has been weak, just as it was for many other illnesses. In a harsh rebuke of the studies regarding the health benefits of vitamin D, it was defamed as "a promiscuous biological candidate, likely to be attracted to any passing half-baked hypothesis (it is a risk factor in search of an adverse outcome)" (McGrath, 2017).

And what about vitamin C? In high doses this has long been known for its capacity to break down into hydrogen peroxide, which can kill cells, including those that are cancerous and lack enzymes to efficiently remove hydrogen peroxide (Doskey et al., 2016). Consequently, it had been thought that it might be effective as an adjunct cancer therapy, but this didn't materialize. The effects of vitamin C may be absent because it is usually taken orally, and better effects might be achieved if the gut were by-passed, but this remains to be determined. It was reported that cancer stem cells were more readily destroyed by vitamin C if it was preceded by the antibiotic doxycycline, possibly because the combination treatment alters the cell's ability to obtain necessary energy supplies (De Francesco et al., 2015; Di Francesco et al., 2018).

Finally, vitamin A has also been implicated in diminishing cancer, such as breast and colon cancer, although the findings have been highly variable. Low doses of vitamin A (as well as vitamins C and E) were associated with lower occurrence of gastric cancer (Kong et al., 2014). And intake of the vitamin A derivative β-carotene was linked to lower breast cancer, while vitamin A and β-carotene intake were associated with diminished occurrence of lung cancer (Yu et al., 2015). An analysis based on 10 studies suggested that α-carotene intake was inversely related to the occurrence of non-Hodgkin lymphoma (Chen et al., 2017), and α-carotene, β-carotene, and β-cryptoxanthin were inversely related to colorectal cancer occurrence. In many other studies, however, β-carotene and vitamin A intake were not related to decreased cancer occurrence, and vitamin A was associated with a small, but significant increase of cancer risk (Schwingshackl et al., 2017). The overall conclusion drawn was that insufficient consistent data were available to conclude that vitamin supplements were useful in cancer prevention. Observational studies are certainly useful in suggesting that certain treatments might be efficacious in cancer prevention. But in the end, these are poor substitutes for randomized controlled trials to determe the effectiveness of these supplements.

Further concerns regarding supplements

Debate on the effects of supplements most often focused on whether they have positive effects but less often has there been attention as to whether they can have adverse actions. There is an unfortunate perspective among some individuals that "these vitamins couldn't hurt, so I'll keep taking them." Yet excessive use of certain supplements has been correlated with negative health events, including cancer occurrence. For instance, high doses of daily calcium use or high-dose vitamin D taken as supplements were associated with increased cancer death (Chen et al., 2019). And daily vitamin E intake was associated with increased risk of prostate cancer. Immunity is also disturbed, as seen after excessive vitamin A intake, which may have induced detrimental epigenetic changes.

It's not just vitamins that can have negative consequences. Older women taking daily supplements comprising iron, copper, magnesium, and zinc were at moderately elevated risk of dying. Excessive use of omega-3 fatty acids was linked to endometrial cancer, excess folate was predictive of colon cancer, β-carotene supplementation was tied to lung cancer, high doses of selenium was aligned with later prostate and skin cancer, and calcium was also linked to aggressive prostate cancer. The succession of negative findings regarding the use of supplements makes it much more unfortunate that many patients using complementary medicine, including supplements, refused at least one component of their standard treatment, and hence were more likely to fare poorly.

Some individuals will vehemently oppose the suggestion that supplements can have adverse effects on well-being. Such naysayers ought to consider that even usual or regularly consumed foods can react negatively with meds. Grapefruit affects the response to some cholesterol busters, licorice containing glycyrrhizin can undermine the effects of the immunosuppressive drug, cyclosporine, and when consumed shortly after antibiotic intake, milk products can diminish their microbicidal effects. The reality is that supplements typically aren't marketed as medicines, and thus they haven't been subject to the strict federal drug testing requirements or the protracted monitoring that follows drug approval. With only a few exceptions, the needed randomized controlled trials have not been conducted, and long-term prospective studies to assess health benefits and risks have been rare. Moreover, regulatory policies have little bearing on the supplement market, and frequently it's entirely uncertain whether the contents of the supplements actually contain what the packaging indicates that they do. Based on such considerations, it was argued that most complementary therapies certainly shouldn't be encouraged, and in some instances, they ought to be explicitly discouraged (Vernieri et al., 2018). As the expression goes, "extraordinary claims require extraordinary evidence," but in the case of vitamin supplements being useful in cancer treatments, the data have not even come close to justifying their use as a therapeutic tool against cancer.

Summary and conclusions

The long-held view that obesity may give rise to numerous health risks has met challenges. Reports suggesting that obesity was not a major problem gave rise to the counterfactual notion that "I'm fat, but fit," feeding the misconception that being obese was not a health risk. It is now fairly well accepted that obesity may be a precursor to several diseases including cancer, depending on the length of time that individuals had been overweight.

Obesity has been associated with altered expression of numerous genes that were linked to diabetes and heart disease, and considerable evidence implicated genetic and epigenetic links between obesity and several forms of cancer. The specific diets consumed can also be important in predicting outcomes. That obesity is linked to the occurrence of multiple forms of cancer is established, but the specific processes—or mediating causal factors—by which this occurs are less certain. Among other mechanisms, the relationships may be caused by processes associated with hormonal and growth factor variations, peptides related to eating processes, increased availability of energy sources for cancer cells, altered immune functioning, and elevated circulating inflammatory factors. Of course, these aren't mutually exclusive from one another. This makes it difficult to identify which of these factors is a particularly cogent target for efforts to offset the negative consequences of obesity on cancer development, although in many people (not all) the occurrence of obesity is itself a modifiable condition.

Numerous epidemiological studies indicated that several nutrients were linked to diminished cancer risk, whereas other foods favored cancer occurrence. A strong association to adult cancer was also observed with poor diets consumed relatively early in life. Such incidence of cancer might be due to epigenetic changes, such as those related to glucose metabolism or the promotion of obesity and processes linked to obesity, such as hyperglycemia, inflammation, and oxidative stress. Conducting randomized controlled trials over extended periods in humans to determine the causal connection between diet and cancer might not be on the cards simply owing to these trials not being feasible. Nonetheless, understanding whether and how dietary factors may affect cancer and its treatment will require analyses of food choices over extended periods. Beyond affecting cancer occurrence, dietary factors could also influence cancer recurrence. Cancer survivors frequently fail to obtain high-quality diets, which varies with sociodemographic features and points to the importance of educational efforts or related programs to diminish further health risks.

It is disconcerting that some forms of cancer have been occurring progressively earlier in life. Some of the outcomes, such as early colorectal cancer, have been evident globally, but particularly in developed countries. Multiple factors likely contribute to the early cancer occurrence (e.g., diet, obesity, lack of exercise, and antibiotic use) and there is ample reason to believe that microbiota perturbations contribute to this (Akimoto et al., 2021). While genetic and epigenetic changes are presumed to influence early cancer development, many of the contributing factors are modifiable. Better strategies are needed to convince young people to change their harmful behaviors before a crisis level of cancer development occurs (or have we already reached that point in the case of some cancers?)

Finally, considerable evidence has amassed concerning genetic factors related to cancer and the epigenetic modifications associated with obesity. Add to this chronic stressors, excessive alcohol consumption, dietary factors, microbial and viral influences, sleep disturbances, and environmental toxicants, and we have a veritable powder keg of factors that can offset regulatory set points in normally "healthy" physiological systems we take for granted. To some extent, increased understanding has developed as to the steps that can be taken to preclude or reverse the effects of these negative influences. These have ranged from specific diets and other lifestyle factors, which together with genetic influences, could point the way to individualized approaches to prevent or facilitate cancer therapy.

References

Abella, V., Scotece, M., Conde, J., Pino, J., Gonzalez-Gay, M.A., et al., 2017. Leptin in the interplay of inflammation, metabolism and immune system disorders. Nat. Rev. Rheumatol. 13, 100–109.
Akimoto, N., Ugai, T., Zhong, R., Hamada, T., Fujiyoshi, K., et al., 2021. Rising incidence of early-onset colorectal cancer—a call to action. Nat. Rev. Clin. Oncol. 18, 230–243.
Alexander Bentley, R., Ruck, D.J., Fouts, H.N., 2020. U.S. obesity as delayed effect of excess sugar. Econ. Hum. Biol. 36, 100818.
Andrade, F.O., Nguyen, N.M., Warri, A., Hilakivi-Clarke, L., 2019. Reversal of increased mammary tumorigenesis by valproic acid and hydralazine in offspring of dams fed high fat diet during pregnancy. Sci. Rep. 9, 20271.
Annett, S., Moore, G., Robson, T., 2020. Obesity and cancer metastasis: molecular and translational perspectives. Cancers 12, 3798.
Antoun, E., Kitaba, N.T., Titcombe, P., Dalrymple, K.V., Garratt, E.S., et al., 2020. Maternal dysglycaemia, changes in the infant's epigenome modified with a diet and physical activity intervention in pregnancy: secondary analysis of a randomised control trial. PLoS Med. 17, e1003229.
Ariaans, G., de Jong, S., Gietema, J.A., Lefrandt, J.D., de Vries, E.G., et al., 2015. Cancer-drug induced insulin resistance: innocent bystander or unusual suspect. Cancer Treat. Rev. 41, 376–384.
Assmann, N., O'Brien, K.L., Donnelly, R.P., Dyck, L., Zaiatz-Bittencourt, V., et al., 2017. Srebp-controlled glucose metabolism is essential for NK cell functional responses. Nat. Immunol. 18, 1197–1206.
Aucoin, M., Cooley, K., Knee, C., Fritz, H., Balneaves, L.G., et al., 2017. Fish-derived omega-3 fatty acids and prostate cancer: a systematic review. Integr. Cancer Ther. 16, 32–62.
Bell, J.A., Hamer, M., Sabia, S., Singh-Manoux, A., Batty, G.D., et al., 2015. The natural course of healthy obesity over 20 years. J. Am. Coll. Cardiol. 65, 101–102.
Bjørge, T., Haggstrom, C., Ghaderi, S., Nagel, G., Manjer, J., et al., 2019. BMI and weight changes and risk of obesity-related cancers: a pooled European cohort study. Int. J. Epidemiol. 48, 1872–1885.
Blüher, M., 2020. Metabolically healthy obesity. Endocr. Rev. 41, 405–420.

Borgquist, S., Bjarnadottir, O., Kimbung, S., Ahern, T.P., 2018. Statins: a role in breast cancer therapy? J. Intern. Med. 284, 346–357.

Bowers, L.W., Maximo, I.X., Brenner, A.J., Beeram, M., Hursting, S.D., et al., 2014. NSAID use reduces breast cancer recurrence in overweight and obese women: role of prostaglandin-aromatase interactions. Cancer Res. 74, 4446–4457.

Bowers, L.W., Rossi, E.L., McDonell, S.B., Doerstling, S.S., Khatib, S.A., et al., 2018. Leptin signaling mediates obesity-associated CSC enrichment and EMT in preclinical TNBC models. Mol. Cancer Res. 16, 869–879.

Brown, T.J., Brainard, J., Song, F., Wang, X., Abdelhamid, A., et al., 2019. Omega-3, omega-6, and total dietary polyunsaturated fat for prevention and treatment of type 2 diabetes mellitus: systematic review and meta-analysis of randomised controlled trials. BMJ 366, l4697.

Bruno, D.S., Berger, N.A., 2020. Impact of bariatric surgery on cancer risk reduction. Ann. Transl. Med. 8, S13.

Calle, E.E., Rodriguez, C., Walker-Thurmond, K., Thun, M.J., 2003. Overweight, obesity, and mortality from cancer in a prospectively studied cohort of U.S. adults. N. Engl. J. Med. 348, 1625–1638.

Chen, F., Du, M., Blumberg, J.B., Chui, K.K.H., Ruan, M., et al., 2019. Association among dietary supplement use, nutrient intake, and mortality among US adults: a cohort study. Ann. Intern. Med. 170, 604–613.

Chen, F., Hu, J., Liu, P., Li, J., Wei, Z., et al., 2017. Carotenoid intake and risk of non-Hodgkin lymphoma: a systematic review and dose-response meta-analysis of observational studies. Ann. Hematol. 96, 957–965.

Chen, W.L., Jin, X., Wang, M., Liu, D., Luo, Q., et al., 2020. GLUT5-mediated fructose utilization drives lung cancer growth by stimulating fatty acid synthesis and AMPK/mTORC1 signaling. JCI Insight 5, e131596.

Chimento, A., Casaburi, I., Avena, P., Trotta, F., De Luca, A., et al., 2019. Cholesterol and its metabolites in tumor growth: therapeutic potential of statins in cancer treatment. Front. Endocrinol. 9, 807.

Cho, Y.A., Lee, J., Oh, J.H., Chang, H.J., Sohn, D.K., et al., 2019. Genetic risk score, combined lifestyle factors and risk of colorectal cancer. Cancer Res. Treat. 51, 1033–1040.

Chung, J.Y., Ain, Q.U., Song, Y., Yong, S.B., Kim, Y.H., 2019. Targeted delivery of CRISPR interference system against Fabp4 to white adipocytes ameliorates obesity, inflammation, hepatic steatosis, and insulin resistance. Genome Res. 29, 1442–1452.

Cipolletti, M., Solar Fernandez, V., Montalesi, E., Marino, M., Fiocchetti, M., 2018. Beyond the antioxidant activity of dietary polyphenols in cancer: the modulation of estrogen receptors (ERs) signaling. Int. J. Mol. Sci. 19, 2624.

Claussnitzer, M., Dankel, S.N., Kim, K.H., Quon, G., Meuleman, W., et al., 2015. FTO obesity variant circuitry and adipocyte browning in humans. N. Engl. J. Med. 373, 895–907.

Costantini, L., Molinari, R., Farinon, B., Merendino, N., 2017. Impact of omega-3 fatty acids on the gut microbiota. Int. J. Mol. Sci. 18, 2645.

Costea, T., Nagy, P., Ganea, C., Szollosi, J., Mocanu, M.M., 2019. Molecular mechanisms and bioavailability of polyphenols in prostate cancer. Int. J. Mol. Sci. 20, 1062.

De Francesco, E.M., Bonuccelli, G., Maggiolini, M., Sotgia, F., Lisanti, M.P., 2015. Vitamin C and doxycycline: a synthetic lethal combination therapy targeting metabolic flexibility in cancer stem cells. Oncotarget 8, 67269–67286.

Di Francesco, A., Di Germanio, C., Bernier, M., de Cabo, R., 2018. A time to fast. Science 362, 770–775.

Dierge, E., Debock, E., Guilbaud, C., Corbet, C., Mignolet, E., et al., 2021. Peroxidation of n-3 and n-6 polyunsaturated fatty acids in the acidic tumor environment leads to ferroptosis-mediated anticancer effects. Cell Metab. 33, 1701–1715.

Doskey, C.M., Buranasudja, V., Wagner, B.A., Wilkes, J.G., Du, J., et al., 2016. Tumor cells have decreased ability to metabolize H2O2: implications for pharmacological ascorbate in cancer therapy. Redox Biol. 10, 274–284.

Engin, A., 2017. Obesity-associated breast cancer: analysis of risk factors. Adv. Exp. Med. Biol. 960, 571–606.

Febbraio, M.A., Karin, M., 2021. "Sweet death": fructose as a metabolic toxin that targets the gut-liver axis. Cell Metab. 33, 2316–2328.

Focaccetti, C., Izzi, V., Benvenuto, M., Fazi, S., Ciuffa, S., et al., 2019. Polyphenols as immunomodulatory compounds in the tumor microenvironment: friends or foes? Int. J. Mol. Sci. 20, 1714.

Fu, T., Coulter, S., Yoshihara, E., Oh, T.G., Fang, S., et al., 2019. FXR regulates intestinal cancer stem cell proliferation. Cell 176, 1098–1112.

Fujita, K., Hayashi, T., Matsushita, M., Uemura, M., Nonomura, N., 2019. Obesity, inflammation, and prostate cancer. J. Clin. Med 8, E201.

Geidl-Flueck, B., Hochuli, M., Nemeth, A., Eberl, A., Derron, N., et al., 2021. Fructose- and sucrose- but not glucose-sweetened beverages promote hepatic de novo lipogenesis: a randomized controlled trial. J. Hepatol. 75, 46–54.

References

Goncalves, M.D., Lu, C., Tutnauer, J., Hartman, T.E., Hwang, S.K., et al., 2019. High-fructose corn syrup enhances intestinal tumor growth in mice. Science 363, 1345–1349.

Gong, J., Robbins, L.A., Lugea, A., Waldron, R.T., Jeon, C.Y., et al., 2014. Diabetes, pancreatic cancer, and metformin therapy. Front. Physiol. 5, 426.

Hardefeldt, P.J., Penninkilampi, R., Edirimanne, S., Eslick, G.D., 2018. Physical activity and weight loss reduce the risk of breast cancer: a meta-analysis of 139 prospective and retrospective studies. Clin. Breast Cancer 18, e601–e612.

Hunsche, C., Martinez de Toda, I., Hernandez, O., Jimenez, B., Diaz, L.E., et al., 2020. The supplementations with 2-hydroxyoleic acid and n-3 polyunsaturated fatty acids revert oxidative stress in various organs of diet-induced obese mice. Free Radic. Res. 54, 455–466.

Hur, J., Otegbeye, E., Joh, H.K., Nimptsch, K., Ng, K., et al., 2021. Sugar-sweetened beverage intake in adulthood and adolescence and risk of early-onset colorectal cancer among women. Gut, 323450.

Jegatheesan, P., De Bandt, J.P., 2017. Fructose and NAFLD: the multifaceted aspects of fructose metabolism. Nutrients 9, 230.

Jin, C., Gong, X., Shang, Y., 2019. GLUT5 increases fructose utilization in ovarian cancer. OncoTargets Ther. 12, 5425–5436.

Justice, A.E., Karaderi, T., Highland, H.M., Young, K.L., Graff, M., et al., 2019. Protein-coding variants implicate novel genes related to lipid homeostasis contributing to body-fat distribution. Nat. Genet. 51, 452–469.

Kanarek, N., Petrova, B., Sabatini, D.M., 2020. Dietary modifications for enhanced cancer therapy. Nature 579, 507–517.

Kaur, P., Shorey, L.E., Ho, E., Dashwood, R.H., Williams, D.E., 2013. The epigenome as a potential mediator of cancer and disease prevention in prenatal development. Nutr. Rev. 71, 441–457.

Khera, A.V., Chaffin, M., Wade, K.H., Zahid, S., Brancale, J., et al., 2019. Polygenic prediction of weight and obesity trajectories from birth to adulthood. Cell 177, 587–596 e589.

Kim, J., Kang, J., Kang, Y.L., Woo, J., Kim, Y., et al., 2020. Ketohexokinase-A acts as a nuclear protein kinase that mediates fructose-induced metastasis in breast cancer. Nat. Commun. 11, 5436.

Kloudova, A., Guengerich, F.P., Soucek, P., 2017. The role of oxysterols in human cancer. Trends Endocrinol. Metab. 28, 485–496.

Kong, P., Cai, Q., Geng, Q., Wang, J., Lan, Y., et al., 2014. Vitamin intake reduce the risk of gastric cancer: meta-analysis and systematic review of randomized and observational studies. PLoS One 9, e116060.

Koroukian, S.M., Dong, W., Berger, N.A., 2019. Changes in age distribution of obesity-associated cancers. JAMA Netw. Open 2, e199261.

Lawless, S.J., Kedia-Mehta, N., Walls, J.F., McGarrigle, R., Convery, O., et al., 2017. Glucose represses dendritic cell-induced T cell responses. Nat. Commun. 8, 15620.

Ling, C., Rönn, T., 2019. Epigenetics in human obesity and type 2 diabetes. Cell Metab. 29, 1028–1044.

Liu, W., Chakraborty, B., Safi, R., Kazmin, D., Chang, C.Y., et al., 2021. Dysregulated cholesterol homeostasis results in resistance to ferroptosis increasing tumorigenicity and metastasis in cancer. Nat. Commun. 12, 5103.

Ludwig, D.S., Nestle, M., 2008. Can the food industry play a constructive role in the obesity epidemic? JAMA 300, 1808–1811.

Lupien, L.E., Bloch, K., Dehairs, J., Traphagen, N.A., Feng, W.W., et al., 2020. Endocytosis of very low-density lipoproteins: an unexpected mechanism for lipid acquisition by breast cancer cells. J. Lipid Res. 61, 205–218.

Macpherson, H., Pipingas, A., Pase, M.P., 2013. Multivitamin-multimineral supplementation and mortality: a meta-analysis of randomized controlled trials. Am. J. Clin. Nutr. 97, 437–444.

Manson, J.E., Cook, N.R., Lee, I.M., Christen, W., Bassuk, S.S., et al., 2019. Marine n-3 fatty acids and prevention of cardiovascular disease and cancer. N. Engl. J. Med. 380, 23–32.

McAuley, P.A., Artero, E.G., Sui, X., Lee, D.C., Church, T.S., et al., 2012. The obesity paradox, cardiorespiratory fitness, and coronary heart disease. Mayo Clin. Proc. 87, 443–451.

McGrath, J.J., 2017. Vitamin D and mental health—the scrutiny of science delivers a sober message. Acta Psychiatr. Scand. 135, 183–184.

Murphy, R.A., Clandinin, M.T., Chu, Q.S., Arends, J., Mazurak, V.C., 2013. A fishy conclusion regarding n-3 fatty acid supplementation in cancer patients. Clin. Nutr. 32, 466–467.

Nakagawa, T., Lanaspa, M.A., Millan, I.S., Fini, M., Rivard, C.J., et al., 2020. Fructose contributes to the Warburg effect for cancer growth. Cancer Metab. 8, 16.

Nindrea, R.D., Aryandono, T., Lazuardi, L., Dwiprahasto, I., 2019. Protective effect of omega-3 fatty acids in fish consumption against breast cancer in Asian patients: a meta-analysis. Asian Pac. J. Cancer Prev. 20, 327–332.

Nowakowska, M.K., Lei, X., Thompson, M.T., Shaitelman, S.F., Wehner, M.R., et al., 2021. Association of statin use with clinical outcomes in patients with triple-negative breast cancer. Cancer 127, 4142–4150.

Nyberg, S.T., Batty, G.D., Pentti, J., Virtanen, M., Alfredsson, L., et al., 2018. Obesity and loss of disease-free years owing to major non-communicable diseases: a multicohort study. Lancet Public Health 3, e490–e497.

Olivier-Van Stichelen, S., Rother, K.I., Hanover, J.A., 2019. Maternal exposure to non-nutritive sweeteners impacts progeny's metabolism and microbiome. Front. Microbiol. 10, 1360.

Palmer, B.F., Clegg, D.J., 2015. The sexual dimorphism of obesity. Mol. Cell. Endocrinol. 402, 113–119.

Pan, K., Luo, J., Aragaki, A.K., Chlebowski, R.T., 2019. Weight loss, diet composition and breast cancer incidence and outcome in postmenopausal women. Oncotarget 10, 3088–3092.

Pearson-Stuttard, J., Zhou, B., Kontis, V., Bentham, J., Gunter, M.J., et al., 2018. Worldwide burden of cancer attributable to diabetes and high body-mass index: a comparative risk assessment. Lancet Diabetes Endocrinol. 6, e6–e15.

Pritchett, N.R., Maziarz, M., Shu, X.O., Kamangar, F., Dawsey, S.M., et al., 2020. Serum ghrelin and esophageal and gastric cancer in two cohorts in China. Int. J. Cancer 146, 2728–2735.

Ringel, A.E., Drijvers, J.M., Baker, G.J., Catozzi, A., Garcia-Canaveras, J.C., et al., 2020. Obesity shapes metabolism in the tumor microenvironment to suppress anti-tumor immunity. Cell 183, 1848–1866 e1826.

Samji, V., Haykal, T., Zayed, Y., Gakhal, I., Veerapaneni, V., et al., 2019. Role of vitamin D supplementation for primary prevention of cancer: meta-analysis of randomized controlled trials. J. Clin. Oncol. 37 (15_suppl), 1534.

Saxton, R.A., Sabatini, D.M., 2017. mTOR signaling in growth, metabolism, and disease. Cell 168, 960–976.

Schwingshackl, L., Boeing, H., Stelmach-Mardas, M., Gottschald, M., Dietrich, S., et al., 2017. Dietary supplements and risk of cause-specific death, cardiovascular disease, and cancer: a systematic review and meta-analysis of primary prevention trials. Adv. Nutr. 8, 27–39.

Seiler, A., Chen, M.A., Brown, R.L., Fagundes, C.P., 2018. Obesity, dietary factors, nutrition, and breast cancer risk. Curr. Breast Cancer Rep. 10, 14–27.

Sperati, F., Vici, P., Maugeri-Sacca, M., Stranges, S., Santesso, N., et al., 2013. Vitamin D supplementation and breast cancer prevention: a systematic review and meta-analysis of randomized clinical trials. PLoS One 8, e69269.

Stolzenbach, F., Valdivia, S., Ojeda-Provoste, P., Toledo, F., Sobrevia, L., et al., 2020. DNA methylation changes in genes coding for leptin and insulin receptors during metabolic-altered pregnancies. Biochim. Biophys. Acta Mol. Basis Dis. 1866, 165465.

Suez, J., Korem, T., Zeevi, D., Zilberman-Schapira, G., Thaiss, C.A., et al., 2014. Artificial sweeteners induce glucose intolerance by altering the gut microbiota. Nature 514, 181–186.

Teras, L.R., Patel, A.V., Wang, M., Yaun, S.S., Anderson, K., et al., 2020. Sustained weight loss and risk of breast cancer in women 50 years and older: a pooled analysis of prospective data. J. Natl. Cancer Inst. 112, 929–937.

Thaiss, C.A., Levy, M., Grosheva, I., Zheng, D., Soffer, E., et al., 2018. Hyperglycemia drives intestinal barrier dysfunction and risk for enteric infection. Science 359, 1376–1383.

Veglia, F., Tyurin, V.A., Blasi, M., De Leo, A., Kossenkov, A.V., et al., 2019. Fatty acid transport protein 2 reprograms neutrophils in cancer. Nature 569, 73–78.

Vernieri, C., Nichetti, F., Raimondi, A., Pusceddu, S., Platania, M., et al., 2018. Diet and supplements in cancer prevention and treatment: clinical evidences and future perspectives. Crit. Rev. Oncol. Hematol. 123, 57–73.

Wu, H., Esteve, E., Tremaroli, V., Khan, M.T., Caesar, R., et al., 2017. Metformin alters the gut microbiome of individuals with treatment-naive type 2 diabetes, contributing to the therapeutic effects of the drug. Nat. Med. 23, 850–858.

Xue, Y., Xu, X., Zhang, X.Q., Farokhzad, O.C., Langer, R., 2016. Preventing diet-induced obesity in mice by adipose tissue transformation and angiogenesis using targeted nanoparticles. Proc. Natl. Acad. Sci. U. S. A. 113, 5552–5557.

Yang, J., Nie, J., Ma, X., Wei, Y., Peng, Y., et al., 2019. Targeting PI3K in cancer: mechanisms and advances in clinical trials. Mol. Cancer 18, 26.

Yu, N., Su, X., Wang, Z., Dai, B., Kang, J., 2015. Association of dietary vitamin A and beta-carotene intake with the risk of lung cancer: a meta-analysis of 19 publications. Nutrients 7, 9309–9324.

Zaramela, L.S., Martino, C., Alisson-Silva, F., Rees, S.D., Diaz, S.L., et al., 2019. Gut bacteria responding to dietary change encode sialidases that exhibit preference for red meat-associated carbohydrates. Nat. Microbiol. 4, 2082–2089.

Zhang, T., Tseng, C., Zhang, Y., Sirin, O., Corn, P.G., et al., 2016. CXCL1 mediates obesity-associated adipose stromal cell trafficking and function in the tumour microenvironment. Nat. Commun. 7, 11674.

Zhang, Z.J., Yuan, J., Bi, Y., Wang, C., Liu, Y., 2019. The effect of metformin on biomarkers and survivals for breast cancer- a systematic review and meta-analysis of randomized clinical trials. Pharmacol. Res. 141, 551–555.

Zhao, N.N., Dong, G.P., Wu, W., Wang, J.L., Ullah, R., et al., 2019. FTO gene polymorphisms and obesity risk in Chinese population: a meta-analysis. World J. Pediatr. 15, 382–389.

Zheng, H., Echave, P., Mehta, N., Myrskyla, M., 2021. Life-long body mass index trajectories and mortality in two generations. Ann. Epidemiol. 56, 18–25.

CHAPTER

9

Microbiota in relation to cancer

OUTLINE

Finding the right microbial mix	279	Breast and female reproductive system cancers	296
Foods associated with microbial changes	280	Lung cancer	297
Beneficial and nonbeneficial bacteria	282	Parasites and cancer	298
Microbiota in relation to cancer	282	The inflammatory link between microbiota and cancer	298
Epigenetic involvement in the effects of microbiota	283	Impact of the prenatal and perinatal diet	301
Influence of antibiotics and germ-free conditions	284	Intergenerational and transgenerational actions	303
Impact of fecal transplants	286	The other side of microbiota: Is there value to prebiotic and probiotic supplements?	304
Short-chain fatty acids	288	Summary and conclusions	304
Immune factors, cytokines, and free fatty acids	288	References	305
Butyrate	291		
Propionate	291		
Acetate	292		
Influence of specific microbiota in diverse forms of cancer	292		
Colorectal cancer	293		

Finding the right microbial mix

Despite the enormous interest in the health benefits and harms that can be produced by microbiota, considerable uncertainty exists concerning what constitutes a *eubiotic* (positive) or a *dysbiotic* (negative) state. To some extent, this uncertainty stems from the sheer number of microbes that inhabit the body, many of which remain to be identified. In addition to bacteria, the gut is replete with viruses, fungi, protozoa, and archaea that can also influence health and ought to be considered in defining what constitutes a healthy gut environment.

With so many microbes being present, identifying the "right" microbial milieu is difficult. At the same time, it does seem that certain bacterial genera (*Lactobacillus*, *Bifidobacterium*, and *Saccharomyces*) have health benefits, as are particular bacterial species. Other bacteria (e.g., genera including *Staphylococcus*, *Streptococcus*, *Enterococcus*, *Klebsiella*, *Enterob*acter, and *Neisseria*), in contrast, may disturb the gut ecosystem, thereby fostering ill health. Knowing which individual microbes (or sets of microbes) are beneficial or harmful may facilitate the development of strategies to enhance health and may play a role in developing illness prevention strategies and might be incorporated into treatments for various diseases. At the same time, substantial microbial differences exist among healthy people, again making it difficult to define what constitutes a beneficial microbiome. Nonetheless, foods that act in a prebiotic or probiotic capacity can normalize disturbances of microbial communities and act against pathogenic organisms. We have learned that this can occur through the production of SCFAs, vitamin synthesis, neutralization of toxins, and regulation of cytokines and several hormones (e.g., estrogen).

Commercialization of probiotics has grown exponentially in recent years and off-the-shelf supplements containing live bacteria could have some benefits, but very often these are sold without experimental evidence supporting the claims that have been made. For that matter, the use of supplements in the treatment of conditions such as obesity, irritable bowel syndrome, atopic dermatitis, and diarrhea secondary to the use of antibiotics, typically have transient actions (if any at all) and the experimental evidence in humans is still insufficient to recommend their use. Among individuals in whom the gut is adequately colonized, it is questionable whether any further health benefits are derived from probiotic supplements, and it is likely that the additional bacteria obtained do not colonize the gut (Zmora et al., 2018). Neither the uncertainty concerning the value of prebiotic and probiotic supplements in promoting wellness nor the possible negative interactions of supplements with prescribed drugs have deterred various firms from offering their products to people seeking a healthier life. Given the progressively increasing sales of supplements in capsule form, it seems that consumers are largely unaware of what may or may not be beneficial.

Foods associated with microbial changes

Considerable data have pointed to the development and progression of several illnesses being linked to ingested foodstuffs and pharmaceutical agents and the resultant microbiota changes. The ordinary renewal of epithelial cells serves as a barrier against pathogen infiltration, but gut insults and injuries, ranging from the presence of foreign substances, infections, trauma, and dietary factors can have negative consequences on the body's mucosal barriers. Breaches within epithelial cells that line the intestinal wall are typically repaired readily, but should the effective repair not occur, the microbiota present could affect immune functioning, specific cytokines, and metabolic processes, which could favor the emergence of illnesses. Gut immune functioning serves as a general defense and gut microbes can trigger B cell populations to produce antibodies that remember numerous microbial species, thereby providing protracted protection against infection (Li et al., 2020). Conversely, gut immune functioning can facilitate the emergence of beneficial microbiota. For instance, T follicular helper cells affect B cell functioning that can affect microbiota components, such as *Clostridia*, which act against obesity and metabolic syndrome (Petersen et al., 2019).

Although there has been debate as to whether meat products are linked to the development of some diseases, less debate exists concerning the value of some foods that influence microbes and hence favor good health. Adherence to a Mediterranean diet was generally accompanied by greater bacterial diversity and higher SCFA levels relative to that apparent in association with low adherence to this diet, varying somewhat based on whether individuals were overweight. Diets that included fibers were related to the gut microbiome constituency and to metabolomic variations, which could have downstream health benefits. Likewise, diets heavier in seafood were associated with limited weight gain, often being attributed to the provision of omega-3 fatty acids. It also appears that some kinds of seafood influence the amino acid taurine that is involved in fat metabolism and could limit diet-induced obesity and altered microbiota and SCFAs, thereby favoring good health. Moreover, proteins obtained from milk, notably casein and whey, limited obesity otherwise introduced by high-fat diets and altered the abundance of beneficial gut bacteria. Figure 9.1 provides an overview of food components and their actions on microbiota and their metabolites, gut barrier functioning, and ultimately on well-being.

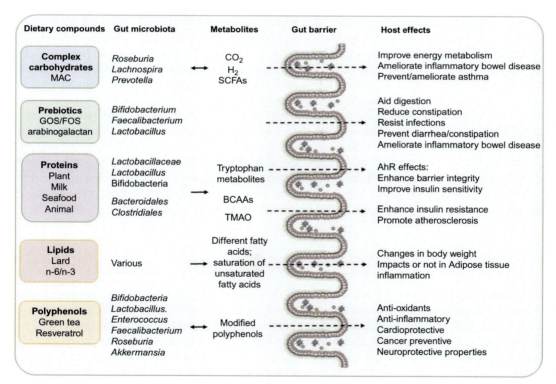

FIG. 9.1 Overview of the interplay between food components, gut microbiota, metabolites, and host health. Dietary compounds may elicit changes in the composition and the activity of the gut microbiota resulting in the generation of secondary metabolites that modulate host responses. *Arrows* indicate the interaction between gut microbiota and metabolites. *AhR*, aryl hydrocarbon receptor; *BCAAs*, branched-chain amino acids; *MAC*, microbiota accessible carbohydrates; *SCFAs*, short-chain fatty acids; *TMAO*, trimethylamine oxide. *Figure and caption from Danneskiold-Samsøe, N.B., Dias de Freitas Queiroz Barros, H., Santos, R., Bicas, J.L., Cazarin, C.B.B., et al., 2019. Interplay between food and gut microbiota in health and disease. Food Res. Int. 115, 23–31.*

Beneficial and nonbeneficial bacteria

Just as ingested food may affect the presence of microbiota, being lean or heavy was accompanied by variations of microorganisms. Indeed, lean and obese individuals could be distinguished from one another based on the relative abundance of *Bacteroidetes* and *Firmicutes*. The microbiota varied further as individuals gained or lost weight, possibly because of differences in microbiota being able to extract energy from certain foods. The fecal abundance of *Akkermansia muciniphila* in both humans and rodents was correlated with being lean and the effectiveness of bariatric surgery was tied to the presence of this type of bacteria. It was similarly possible to predict weight loss in a structured diet program based on the presence of genes that facilitated bacterial growth, the ability of gut microbiota to break down starches, and degradation of complex polysaccharides and proteins. Thus, while individual motivations to lose weight no doubt contribute to the ability to lose weight, features of the microbiome contribute markedly to these efforts (Diener et al., 2021).

Observational studies in themselves do not necessarily implicate these bacteria as causal agents for obesity, but manipulations of gut bacteria could produce weight changes, thus supporting a causal connection between microbiota and obesity. When the fecal microbiota from twins discordant for obesity was transferred to mice, these animals subsequently showed a body profile commensurate with that of the fecal donors. The differences in body composition were linked to variations of SCFAs and the metabolism of branched-chain amino acids (leucine, isoleucine, and valine). It was likewise demonstrated that altering specific microbiota can affect weight. Providing *Bifidobacterium longum* to mice maintained on a high-fat diet or overweight/obese humans, resulted in a decline of body fat (Schellekens et al., 2021). Moreover, treating mice with pasteurized *A. muciniphila* reduced fat mass and concurrently diminished insulin resistance. And providing this treatment to overweight individuals or those with obesity over 3 months reduced insulin resistance (Depommier et al., 2019). Nonetheless, there are limits to what certain bacteria can achieve. For instance, levels of *A. muciniphila* increased following bariatric surgery among women with severe obesity but this wasn't necessarily accompanied by enhanced metabolic health (Dao et al., 2019).

Microbiota in relation to cancer

Microbial factors were linked to cancer occurrence, operating by disturbing immunosurveillance and promoting inflammation, and could be relevant to cancer therapies (Garrett, 2015). It has indeed been maintained that diet might affect well-being to a greater extent than the influence of host genetics, although this perspective is certainly debatable. At the same time, it should be underscored that while dietary factors influence microbiota that provoke diverse biochemical processes linked to health, these actions may be moderated by characteristics of the host, varying considerably across individuals.

Health disturbances have often been viewed as being related to gut dysbiosis or the confluence of several "bad" bacteria. Of the pernicious bacteria, one of the best known is *Escherichia coli*, which produces colibactin that can damage DNA. Being present in 20% of humans, this bacterium may have broad consequences. Using a human intestine grown in a lab (an organoid) it was demonstrated that 5 months of exposure to *E. coli* bacteria that affected colibactin, provoked mutational patterns (signatures) that might be related to the development of colon cancer. Relatedly, analyses of about 5000 tumors of various types indicated that the same signature was detected among 5% of colorectal cancer patients, whereas in other forms of cancer this profile was rare, although signs of this signature were evident in oral and bladder cancers (Pleguezuelos-Manzano et al., 2020).

Diet and microbial factors have been tied to a variety of cancer types beyond those of the gut. High-fat diets favor the development of liver cancer, whereas reducing certain microbiota could act against these effects of the diet. Interactions between bile acid and microbiota were likewise associated with both colorectal and hepatocellular cancers, possibly acting through actions on immune factors and the promotion of inflammation (Ma et al., 2018). A pooled analysis of 10 prospective studies that comprised more than 1,445,000 individuals across several countries indicated that the combination of high fiber, a major source of prebiotics, as well as a probiotic, was associated with diminished lung cancer risk, supporting the importance of the combination of pre- and probiotic components in cancer prevention (Yang et al., 2020).

Microbiota present in a variety of different tumors (e.g., breast, pancreas, colon, lung, prostate, ovarian) could be differentiated from those seen in surrounding healthy tissues (Urbaniak et al., 2016). Moreover, in mice and humans with pancreatic cancer, the microbiome was different from that otherwise apparent, accompanied by suppression of aspects of immune functioning that ought to act against cancer occurrence (Pushalkar et al., 2018). Among the relatively long-term survivors of pancreatic adenocarcinoma, the diversity of microbiota, particularly *Pseudoxanthomonas*, *Streptomyces*, *Saccharopolyspora*, and *Bacillus clausii*, was elevated within the intratumoral environment. Further to this, among patients with resected pancreatic adenocarcinoma who survived longer than 5 years, which is relatively uncommon, diversity in the tumor microbiome was elevated and the intratumoral microbiome signature could be distinguished from that apparent in short-term survivors.

Epigenetic involvement in the effects of microbiota

Epigenetic processes may contribute to the occurrence of cancer, and it appears that microbial metabolites can instigate epigenetic changes that affect cancer processes. Numerous foods could promote epigenetic variations that lead to cancer but these have yet to be fully and systematically assessed. Even so, dietary factors can promote epigenetic effects that influence microbiota thereby favoring the development of colorectal cancer (Ideraabdullah and Zeisel, 2018). It also appears that gut microbes that promote the development of obesity may reprogram the intestinal epigenome so colon cancer is more likely to occur.

Several nutrient components that are found in plants are thought to act against cancer occurrence through their epigenetic actions. These include genistein, curcumin, resveratrol, sulforaphane, and other isothiocyanates (present in cruciferous vegetables), and epigallocatechin-3-gallate (present in green and black tea) that diminish free oxygen radicals (Gao and Tollefsbol, 2015). Some of these foods, particularly plant-derived phytochemicals, can influence epigenetic processes that were related to cancer and might thus be useful in protective capacity. As well, some gut bacteria can enhance responses to cancer therapy and diminish the side effects of therapy by metabolizing chemotherapeutic agents that would otherwise harm beneficial bacteria.

Influence of antibiotics and germ-free conditions

The general view has frequently been expressed that cancer was more likely to develop in the presence of certain microorganisms or when the abundance of harmful bacteria exceeded that of beneficial bacteria. To this point, little was said concerning the consequences of bacteria being eliminated, including those suspected of being harmful as well as those that might be beneficial. One approach to examining this question is to assess illness vulnerability that might be present among germ-free mice (i.e., bred and maintained to lack any microorganisms). Immune functioning is altered in these animals, including diminished dendritic cells, macrophages, and B cells. The appearance of CD4+ T lymphocytes was also reduced in lymphoid organs and the ratio of T_h1 and T_h2 cells was altered. Likewise, T_h17 and T_{reg} cells were diminished in germ-free mice, which could be restored by administering fecal material from ordinary mice (Kennedy et al., 2018).

Given the diminished immune response in germ-free mice, it might have been expected that tumor development would appear more readily than in conventional mice. Instead, several studies had indicated that solid tumors occurred less often in germ-free mice, and similar outcomes were subsequently observed in the analysis of intestinal tumors induced by carcinogenic compounds. Perhaps germ-free mice were better prepared to respond to evolving cancer cells since they had not been exposed to antigens earlier and hence were less likely to develop immune tolerance. In fact, among mice that were born germ-free and maintained under these conditions until weaning, subsequent transfer to a clean (but not germ-free) environment, colorectal cancers developed more frequently and were larger than in conventionally raised mice. The increased tumor development was accompanied by elevated proinflammatory gene expression and more myeloid-derived suppressor cells that inhibit immune cells that would ordinarily attack the tumor (Harusato et al., 2019). Speaking to the importance of controlling inflammatory factors in cancer inhibition, when germ-free mice were engineered so that inflammatory responses were increased, colon cancer was elevated.

While the use of germ-free mice has unquestionably been a good way to assess the links between microbiota and various diseases, it needs to be kept in mind that other processes related to illnesses may be preserved in these mice, such as the enhanced HPA response to stressors. Furthermore, some of the variations of gastrointestinal physiology seen in germ-free mice may reflect an adaptation that develops owing to the absence of microorganisms, which could influence the impact of later challenges. It should be added that raising mice in a germ-free environment, essentially having them "grow up in a bubble," may deprive them of necessary constituents, so permanent developmental deficits may be engendered that have repercussions for later well-being. Accordingly, despite the frequent use of germ-free mice to evaluate the contribution of microbiota to various disease conditions, it has drawbacks that are not readily overcome.

An alternative to using germ-free mice is the evaluation of the effects of antibiotics that cause the elimination of microbiota, irrespective of whether they are good or bad. Antibiotics have been essential in controlling bacterial infections and have been among the most important and commonly prescribed medications available to eliminate otherwise fatal illnesses. A range of antibiotics have been developed (e.g., penicillin, tetracyclines, cephalosporins, quinolones), which have similar chemical and pharmacological properties that are differentially effective in treating varied illnesses. Because of the effectiveness of these compounds, their use increased exponentially over decades and most people have been prescribed an antibiotic at some time.

As vital as antibiotics have been, like many medications, they can produce negative outcomes, especially if taken by patients with kidney or liver diseases, as well as in the elderly and infants. The elimination of bacteria by antibiotics can result in altered biliary acids and the metabolism of foods. Since microbiota affect immune processes and may be an important element in the regulation of antiinflammatory cytokines, eliminating these bacteria by antibiotics can elicit an exaggerated inflammatory response and could also render mice more susceptible to intestinal pathogens. Moreover, children who were treated with antibiotics prior to 2 years of age were at increased risk for later development of respiratory allergies, eczema, celiac disease, and obesity. An extensive and informative review (Becattini et al., 2016) indicated that several illnesses involving different organs and immune processes can be instigated by antibiotics (see Fig. 9.2).

Several reports have supported the view that microbiota depletion can promote cancer occurrence. Repeated antibiotic use was accompanied by increased incidents of breast cancer, and an analysis of 10 studies that included more than 4,850,000 individuals indicated that relatively extensive antibiotic treatment (more than 5 prescriptions and 60 days of antibiotic use) predicted elevated colon cancer occurrence (Qu et al., 2020). Curiously, while antibiotic use was associated with colon cancer, this was not mirrored by rectal cancer occurrence or cancer in the most distal part of the colon. In fact, rectal cancer was reduced in association with antibiotic use (Zhang et al., 2019). It is uncertain why these differential actions occurred, but they might reflect microbiota populations varying across different aspects of the intestine.

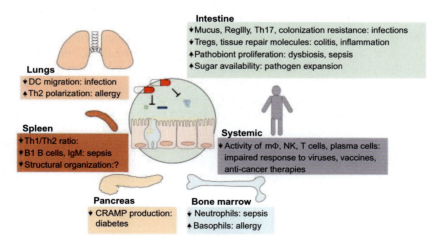

FIG. 9.2 Antibiotic-mediated microbiota depletion causes disease in multiple organs. Antibiotics act on the gut microbiota by decreasing its density and modifying its composition in a long-lasting fashion. This causes reduced signaling to the intestinal mucosa and peripheral organs, which results in impaired functioning of the immune system. Depicted are examples of organs that can be negatively affected as a consequence of antibiotic treatment in mouse models. *Figure and caption from Becattini, S., Taur, Y., Pamer, E.G., 2016. Antibiotic-induced changes in the intestinal microbiota and disease. Trends Mol. Med. 22, 458–478.*

It is especially significant that in some instances, antibiotic use preceded cancer detection by more than 10 years. A metaanalysis that included 7,947,200 participants within 25 studies revealed that cancer occurrence was more common among individuals who had used antibiotics at some time relative to those who had never used an antibiotic. The increased risk was most prominent with lymphomas as well as lung cancer, pancreatic cancer, renal cell carcinoma, and multiple myeloma (Petrelli et al., 2019). As the incidence of colon cancer associated with antibiotics was marked among individuals who were less than 50 years of age, raises the possibility that the increasing incidence of colon cancer in recent decades may have stemmed from the use of certain antibiotics (Perrott et al., 2021).

Cancer patients are often at elevated risk of bacterial infection either because of the malignancy or owing to the immune suppression provoked by therapies, thus they are frequently treated with antibiotics. Despite the pro-tumorigenic effects of earlier antibiotic use, depleting gram-positive bacteria in the gut through an antibiotic (vancomycin) enhanced the effectiveness of adoptive T cell therapy, possibly through the indirect actions of IL-12 (Uribe-Herranz et al., 2018). However, the sword cuts both ways and there have been indications that antibiotic treatments could undermine the effectiveness of cancer therapies, including the actions of immune checkpoint inhibitors. This was the case among patients who received antibiotics during cancer therapy, as well as individuals who had received antibiotics sometime earlier (Martins et al., 2021). There are still only a few studies, usually with a small number of patients, that have pointed to this. However, as we'll see in Chapters 14 and 15, the gut microbial population can markedly influence the efficacy of several cancer therapies, including immunotherapies.

The excessive use of antibiotics, sometimes inappropriately (e.g., in treating viral infections), has unfortunately led to antibiotic resistance. Therefore, some bacterial infections cannot be effectively controlled, including *Clostridium difficile*, methicillin-resistant *Staphylococcus aureus* (MRSA), penicillin-resistant Enterococcus, and multidrug-resistant *Mycobacterium tuberculosis*. The development of antibiotic resistance has also been attributed to their use in farm animals, which eventually reach humans. Less often have there been discussions as to whether the indirect accumulation of antibiotics in humans may contribute to the subsequent development of cancer.

Owing to the increasing resistance to antibiotics as well as the negative effects on gut microbiota, methods have been sought as alternatives to antibiotics. Increased attention has focused on using bacteriophages (viruses that infect bacteria) to deal with infectious illnesses and the treatment of some forms of cancer (Krut and Bekeredjian-Ding, 2018). Because of their small size, their ability to carry therapeutic agents, and the possibility of genetically engineering them to target specific tumors, their use in cancer therapies has been encouraging.

Impact of fecal transplants

As microbiota play an important role in health, it was reasoned that positive effects could be obtained by transferring fecal microbiota from a healthy individual to one who was experiencing a particular illness. The benefits of fecal transfers are best documented in relation to *C. difficile* that is unaffected by usual antibiotic treatment and may be effective in dealing with other resistant bacteria, such as *Enterococcus faecium* and *Klebsiella pneumoniae* (Suez et al., 2018). Fecal transplants from healthy donors also diminish signs of inflammatory bowel disease and cardiometabolic diseases, and in mice, the enhanced metabolic effects of diet and exercise could be transferred to diet-induced obese animals.

Evidence from preclinical studies showed that fecal transplants could affect cancer processes and might also be useful in cancer treatment. When fecal samples from colorectal cancer patients were transferred to conventional or germ-free mice that had been treated with a carcinogen, they developed intestinal polyps, accompanied by elevated expression of inflammatory cytokines and upregulation of genes associated with cell proliferation, apoptosis, angiogenesis, invasiveness, and metastasis (Wong et al., 2017). Conversely, in mice with colitis-related colorectal cancer, fecal microbiota transplantation from a healthy donor reduced tumor size accompanied by diminished proinflammatory and elevated antiinflammatory factors (Wang et al., 2019). In preclinical studies, this procedure was effective for cancer management and complications related to cancer treatment and could augment the actions of cancer immunotherapy (Chen et al., 2019) as well as serve in a supportive capacity to limit some side effects among patients being treated for cancer.

Despite reports that fecal transfer was generally safe, this procedure has not been recommended for conditions other than *C. difficile* and inflammatory bowel disease pending additional data becoming available supporting its usefulness and safety. Indeed, along with good bacteria, fecal transplants could potentially contain multidrug-resistant bacteria, thereby harming recipients. Fecal microbiota obtained from a healthy donor can also be transplanted in a purified form by colonoscopy or through the nasogastric route, while a less distasteful procedure for those individuals who tend to be squeamish has entailed transfer through acid-resistant capsules (dubbed "crapsules"). Ultimately, a complete fecal transfer might not be necessary as a small cocktail of four bacteria that can affect bile acids may be sufficient to produce beneficial outcomes, at least concerning *C. difficile* (Buffie et al., 2015).

Together, the available data are consistent with the position that gut microbiota may contribute to cancer development and fecal transplants could potentially produce positive therapeutic actions. The approach of enhancing health through fecal transplants has been used increasingly more often, but it might be still more useful if it were possible to identify the specific microbiota balances that were present in each individual (Zmora et al., 2018).[a]

Deterrents to using microbiota manipulations in a therapeutic capacity exist but some are being overcome. Aside from the appreciable individual variability, the microbiota constituency within an individual can vary somewhat over time, although with repeated sampling it ought to be possible to define an individual's microbiota signature. This, however, is complicated by the fact that mutations may occur within the bacterial genome, thereby influencing gut microbial colonization, supporting the position that treatment of illnesses through diet or microbial manipulations requires that these be tailored to the individual (Bashiardes et al., 2018).

---

[a] Not all feces are equally endowed with beneficial actions. Some individuals seem to be "super-donors" whose feces seems to be appreciably more effective than that of other donors. Their stool may contain keystone bacteria as well as viruses that can enhance the effectiveness of the bacteria present, possibly being related to their diet. It is possible that immune responses to transplanted fecal matter may contribute to whether the procedure will be effective, raising the possible importance of genetic differences between the donor and recipient.

Short-chain fatty acids

Diet composition affects the types of nutrients available to gut microbes and contributes to the diverse changes that are induced in host immunity. For effective immune functioning, essential vitamins and minerals (zinc, selenium, iron, copper, and folic acid) that are ordinarily obtained from foods need to be present. As well, lack of fiber-containing foods affects gut microbial diversity, particularly as these might not readily be reconstituted on their own. Operating through several routes, short-chain fatty acids (SCFAs) can facilitate or promote the health and effective functioning of diverse body organs, whereas gut dysbiosis can lead to several diseases that involve hormonal and inflammatory processes. Concurrently, as shown in Fig. 9.3, and operating through several routes, SCFAs may influence processes that affect the functioning of brain cells, including neurons, microglia, and astrocytes, which can facilitate the maintenance of a healthy brain. Conversely, the adoption of counterproductive lifestyles (improper diet, sedentary behaviors, sleep disturbances, ineffective means of coping with stressors) can result in poor cognitive abilities, emotional reactivity, psychological disorders, and neurodegenerative conditions. The connections between microbiota and the brain are multidirectional so that brain changes can impact diverse peripheral hormonal and immune processes that favor multiple diseases.

The data supporting a microbial link to different types of cancer have been impressive and progressively more information has emerged pointing to the possible use of microbes in dealing with such cancers. The entire microbial milieu and the balances that exist between bacteria probably determine well-being, but as previously noted in other contexts, certain commensal bacteria (e.g., *Lactobacillus* and *Bifidobacterium*) produce SCFAs that have several beneficial actions, including working against tumorigenesis. The abundance of other bacteria, such as *Fusobacterium nucleatum* was elevated in colorectal cancer, breast, and liver cancer as well as liver metastases (e.g., Bullman et al., 2017).

Immune factors, cytokines, and free fatty acids

The influence of microbiota and SCFAs on cancer occurrence and progression likely occurs by enhancing gut barrier functioning as well as promoting processes that suppress the development and growth of pathogens, regulate gut enzymatic activity, and enhance immune functioning. These actions, as depicted in Fig. 9.4, can be promoted by SCFAs, particularly butyrate, propionate, and acetate. Several immune changes, beyond those shown in the figure, can be facilitated by microbiota. For example, *Lactobacillus* has been associated with increased NK cells, MHC class II antigen-presenting cells, and CD8+ T cells. Metabolites of *Lactobacillus plantarum* likewise produced antiproliferative effects and promoted apoptosis of some types of cancer cells, while sparing normal cells. Additionally, SCFAs may also promote the production of the inhibitory cytokine IL-10, thereby attenuating other gut-related disorders, including the development of colitis that has been associated with cancer development.

In addition to alterations of immune and inflammatory processes, SCFAs can influence the development of cancer through several other processes. Specifically, SCFAs can moderate the occurrence of DNA damage and oxidative stress and may affect cell apoptosis, diminish the invasiveness of tumor cells, and inhibit cell proliferation. Probiotic diets that promote the production of butyrate could also inhibit cancer cells' preference for glucose (Warburg effect), hence affecting cancer progression (Eslami et al., 2020).

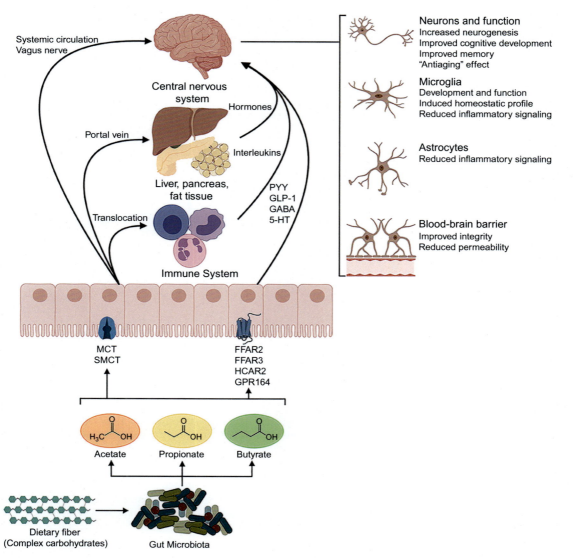

FIG. 9.3 Potential pathways through which SCFAs influence gut-brain communication. Short-chain fatty acids (SCFAs) are the main metabolites produced by the microbiota in the large intestine through the anaerobic fermentation of indigestible polysaccharides such as dietary fiber and resistant starch. SCFAs might influence gut-brain communication and brain function directly or indirectly. Following their production, SCFAs are absorbed by colonocytes, mainly via H+-dependent monocarboxylate transporters (MCTs) or sodium-dependent monocarboxylate transporters (SMCTs). Through binding to G protein-coupled receptors (GPCRs) such as free fatty acid receptor 2 and 3 (FFAR2 and FFAR3), as well as GPR109a/HCAR2 (hydrocarboxylic acid receptor) and GPR164 or by inhibiting histone deacetylases, SCFAs influence intestinal mucosal immunity, and barrier integrity and function. SCFA interaction with their receptors on enteroendocrine cells promotes indirect signaling to the brain via the systemic circulation or vagal pathways by inducing the secretion of gut hormones such as glucagon-like peptide 1 (GLP1) and peptide YY (PYY), as well as γ-aminobutyric acid (GABA), and serotonin (5-HT). Colon-derived SCFAs reach the systemic circulation and other tissues, leading to brown adipose tissue activation, regulation of liver mitochondrial function, increased insulin secretion by β-pancreatic cells, and whole-body energy homeostasis. Peripherally, SCFAs influence systemic inflammation mainly by inducing T regulatory cell (T$_{reg}$) differentiation and by regulating the secretion of interleukins. SCFAs can cross the blood-brain barrier (BBB) via monocarboxylate transporters located on endothelial cells and influence BBB integrity by upregulating the expression of tight junction proteins. Finally, in the central nervous system (CNS) SCFAs also influence neuroinflammation by affecting glial cell morphology and function, as well as by modulating the levels of neurotrophic factors, increasing neurogenesis, contributing to the biosynthesis of serotonin, and improving neuronal homeostasis and function. Together, the interaction of SCFAs with these gut-brain pathways can directly or indirectly affect emotion, cognition, and pathophysiology of brain disorders. *Figure and caption from Silva, Y.P., Bernardi, A., Frozza, R.L., 2020. The role of short-chain fatty acids from gut microbiota in gut-brain communication. Front. Endocrinol. 11, 25.*

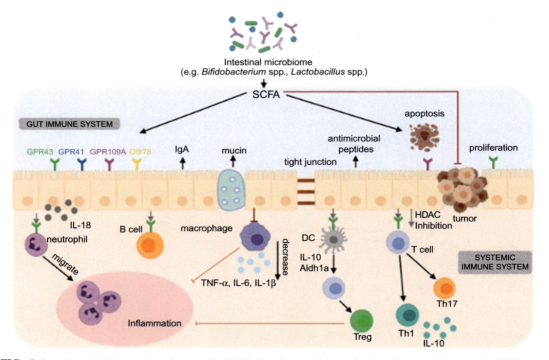

FIG. 9.4 The role of short-chain fatty acids (SCFAs) in the regulation of gut and systemic immunity. SCFAs are produced by the intestinal microbiome in the fermentation of nondigestible carbohydrates (NDCs), undigested food fiber, or resistant starch (RS). SCFAs regulate intestinal barrier function by inducing intestinal epithelial cell secretion of interleukin-18 (IL-18), mucin (MUC2), antimicrobial peptides, and upregulating the expression of the tight junction. Moreover, SCFAs regulate the T cell function through G protein-coupled receptors (GPR41, GPR43, GPR109A), Olfr78 receptor signaling, and through inhibition of histone deacetylase (HDAC) which affects inhibition of transcription factors (nuclear factor-κB; NFκB). SCFAs play an important role in inducing neutrophil migration to inflammatory sites and enhancing their phagocytosis. Moreover, SCFAs inhibit intestinal macrophage production of proinflammatory cytokines (e.g., IL-6, IL-8, IL-1β, and TNFα), and possibly induce intestinal IgA production of B cells. SCFAs affect the differentiation of regulatory T (T_{reg}) cells and the production of interleukin-10 (IL-10). The direct impact of SCFAs, as well as SCFA regulation of dendritic cells (DCs), are mediated in the differentiation of T cells. SCFAs regulate the generation of T_{reg}, T_h1, and T_h17 in different cytokine environments. Finally, SCFAs promote apoptosis and suppress the proliferation of tumor cells resulting in inhibiting carcinogenesis. Abbreviation: Aldh1A, aldehyde dehydrogenase 1A2. From Śliżewska, K., Markowiak-Kopeć, P., Śliżewska, W., 2020. The role of probiotics in cancer prevention. Cancers 13, 20.

The influence of SCFAs may also come about through effects on G protein-coupled receptors (GPCRs) that comprise membrane receptors that are involved in transducing extracellular stimuli into intracellular signals. Free fatty acids (FFA), the primary fuel within the body, act like signaling molecules by binding to various types of GPCRs, primarily exciting FFA2 (GPR43) and FFA3 (GPR41) receptors. Through actions on FFA2 receptors, SCFAs can diminish inflammatory processes, thereby limiting the occurrence of colon cancer. Such an outcome may result from actions on an inflammatory suppressor, possibly through epigenetic alterations. As well, FFA2 and FFA3 receptors (FFARs) may be reduced in triple-negative breast cancer and metastasis, leading to the suggestion that SCFAs ordinarily drive breast cancer cells to a noninvasive phenotype, thereby limiting cancer metastasis (Thirunavukkarasan et al., 2017). It seems likely that in women with obesity, FFARs may interact with estrogens in determining cancer occurrence. For that matter, circulating FFARs might promote the rewiring of metabolic processes that favor the development and aggressiveness of breast cancer cells.

Unlike SCFAs, medium and long-chain FFAs stimulate FFA1 and FFA4 receptors, which behave very differently from short-chain FFARs, and in some instances may even have opposing actions. Activation of receptors associated with medium- and long-chain FFAs contribute to cell regulation associated with the promotion of some forms of colon cancer, so the inhibition of FFA1 can limit the growth of these cancer cells. Targeting FFA4 might likewise be effective in regulating tumorigenesis as well as cancer cell migration and invasion, just as this free fatty acid has been targeted to diminish metabolic disorder.

Butyrate

A diet rich in fiber was accompanied by elevated levels of gut bacteria that promoted greater butyrate levels, particularly if the diet contained fructans and galactooligosaccharides, leading to the suggestion that the fermentability of fibers (in which glucose is broken down anaerobically) is essential in determining microbiota variations. A high-fat diet, in contrast, was accompanied by a reduction of bacteria that ordinarily increase butyrate, together with the elevated abundance of bacteria that promote disturbed glucose metabolism. This diet was also associated with an increase of several markers of inflammation, notably C-reactive protein, thromboxane B_2, leukotriene B_4, and prostaglandin E_2 (Wan et al., 2019), which have been associated with the appearance of multiple adverse outcomes, including cancer. The link between butyrate and responses to high-fat diets is likely causal, given that mice maintained on a high-fat diet that was supplemented with butyrate exhibited improved microbiota constitution together with a better intestinal barrier and lower circulating inflammatory factors.

The possibility was repeatedly mentioned that ways of increasing butyrate concentrations might be a viable approach to promote anticancer outcomes, at least with respect to colon cancer. This said, the concentrations of butyrate produced by specific diets may be sufficient to aid in preventing the development of cancers but may be insufficient as a therapeutic strategy. Accordingly, alternative delivery systems of butyrate will be necessary (e.g., through nanoparticles aimed at the tumor) or this SCFA can be combined with other anticancer agents, including phytochemicals.

Propionate

Like butyrate, variations of propionate may be instrumental in controlling metabolic disorders and excessive weight by triggering the secretion of gut peptides, such as GLP-1 and peptide YY, which contribute to appetite regulation and glucose metabolism. Propionate also influences dendritic cell functioning and modulates CD8+ T cell activity (Nastasi et al., 2017) as well as augmenting T_{reg} cell development, thereby limiting the actions of proinflammatory T_h17 cells (Arpaia et al., 2013). Aside from effects on immune functioning, the availability of propionate might arrest the growth of colon cancer cells by promoting apoptosis and autophagy.

Of the different foods that affect the gut microbiota, inulin (a soluble fiber present in many plants) is especially significant. This fiber is consumed by bacteria, which then release propionate. Thus, diet supplementation with inulin can suppress appetite, diminish glycemia, reduce the accumulation of abdominal fat, and inhibit the secretion of the proinflammatory cytokine IL-8. In mice given inulin-type fructans in their water supply so that propionate was increased, malignant cell proliferation in the liver was inhibited (Bindels et al., 2012). Supplementation of the diet of mice with inulin likewise had antitumor effects and could delay the emergence of cancer drug resistance. A randomized, double-blind study indicated that inulin-type prebiotic supplementation was accompanied by alterations of specific microbiota that differed among individuals who were habitual low or high fiber consumers (Healey et al., 2018). These findings speak to the variability that exists regarding microbiota across individuals and point to the essential nature of specific foods in producing these differences.

Acetate

Like other SCFAs, acetate may affect immune functioning and can have distinct benefits on metabolic processes and energy production. Much like butyrate, acetate binds to FFA (GPCR) receptors, thereby promoting several downstream effects, including glucose-stimulated insulin secretion, and like propionate, it can affect eating-related processes and obesity through its effects on GLP-1 and peptide YY. Further, owing to its effects on parasympathetic nervous system activity and the consequent insulin and ghrelin secretion, hyperphagia, and obesity, acetate can affect metabolic syndrome, type 2 diabetes as well as heart disease, and along with other SCFAs, acetate inhibits inflammatory processes, thus diminishing cancer occurrence (Sivaprakasam et al., 2016). It was similarly suggested that pyruvate, the end product of glycolysis, may promote acetate production thereby limiting disease occurrence (Bose et al., 2019). In essence, acetate together with several other factors can affect the amount of energy cancer cells require for multiplication.

Influence of specific microbiota in diverse forms of cancer

Studies in humans linking microbial factors and cancer occurrence were largely correlational, precluding causal conclusions in this regard. Even though elevated levels of harmful bacteria at boundary surfaces (e.g., respiratory, digestive, and urogenital tract, as well as the skin and oropharynx) were associated with cancer, microorganisms may be attracted to these sites simply owing to these being a well-resourced niche (e.g., oxygen or carbon availability). Nevertheless, it would be reasonable to postulate that by affecting immune functioning, these gut microbiota are linked to the development of certain cancers, especially those related to gastrointestinal processes (Levy et al., 2017; Zitvogel et al., 2016). This could similarly occur because microbiota can alter inflammatory processes and various pattern recognition receptors (PRRs)—such as NOD-2, NLRP3, NLRP6, and NLRP12—or by directly damaging DNA (Garrett, 2015). This outcome could also develop through actions on β-catenin, which regulates several fundamental cellular functions, including cell-cell adhesion, gene transcription, cell proliferation and differentiation, migration, apoptosis, and stem cell renewal (Rubinstein et al., 2019).

Specific bacteria have been implicated as being involved in several forms of cancer. Yet, given the diversity of bacteria that have been implicated in various types of cancer, and the different ways in which they can have such consequences, the view was frequently adopted that rather than searching for unique microbial contributions, it might be more profitable to consider the cumulative actions of multiple microorganisms in attempting to predict and prevent cancer development and progression. The imbalance between beneficial and harmful bacteria (dysbiosis) through actions on immune functioning may be key in promoting cancer or allowing cancer cells to flourish, as shown in Fig. 9.5 (Wong and Yu, 2019). In the sections that follow we provide a brief overview of the microbial involvement in several of the most common forms of cancer.

Colorectal cancer

The involvement of microbiota in the production or prevention of colon cancer has received more attention than have other forms of cancer, although cancer of other organs involved in digestive processes has been linked to specific microbiota. In addition to symbiosis or dysbiosis, certain bacteria may be especially effective in fostering colon cancer, doing so by interfering with immune responses that ought to act against cancer, promoting an inflammatory environment, and producing toxins or carcinogenic metabolites that elicit DNA damage. Epidemiological studies indicated that high fiber diets were accompanied by an increase in the abundance of *Firmicutes* and a reduction of *Bacteroides* leading to elevated concentrations of SCFAs that could inhibit the development of colon cancer. Likewise, a constellation of 11 bacterial strains can drive the production of CD8+ T cells, thereby engendering enhanced anticancer outcomes (Tanoue et al., 2019). It was similarly reported that gut bacteria of colon cancer patients were distinguishable from those of healthy individuals. Moreover, microbiota within colorectal adenomas could be differentiated from those present in surrounding healthy tissue, varying still further over the course of colorectal cancer progression (Nakatsu et al., 2015).

The consumption of high-fat, high-protein diets, as well as excessive red meat and processed meat, was accompanied by increased *F. nucleatum*, *E. coli*, or *Bacteroides fragilis*, which were related to increased intestinal cancer occurrence (Huang and Liu, 2019). It was similarly reported that relative to healthy controls, in colorectal cancer patients several bacterial species related to bacteroids were enriched and that of butyrate-producing bacteria, notably *Faecalibacterium* and *Roseburia*, were less abundant (Wu et al., 2013). It may be of considerable significance that the abundance of *B. fragilis* and the accompanying inflammation differed among individuals without or with colon polyps, which could evolve into colon cancers. Thus, the early detection of *B. fragilis* could potentially offer clues as to who might develop polyps and be at risk for colorectal cancer (Kordahi et al., 2021).

While numerous microbiota have been linked to colon cancer, *F. nucleatum* has received particular attention. These bacteria, which were exceptionally elevated in colorectal tumor tissues, reduce T cell infiltration of colorectal carcinoma and may predict the occurrence of metastasis as well as shorter survival (Bullman et al., 2017). It may be that in the presence of somatic mutations, *F. nucleatum* could be a second hit that promoted intestinal tumors through β-catenin signaling leading to inflammation. While not dismissing a two-hit perspective concerning the influence of *F. nucleatum* on cancer, it can have multiple actions that favor

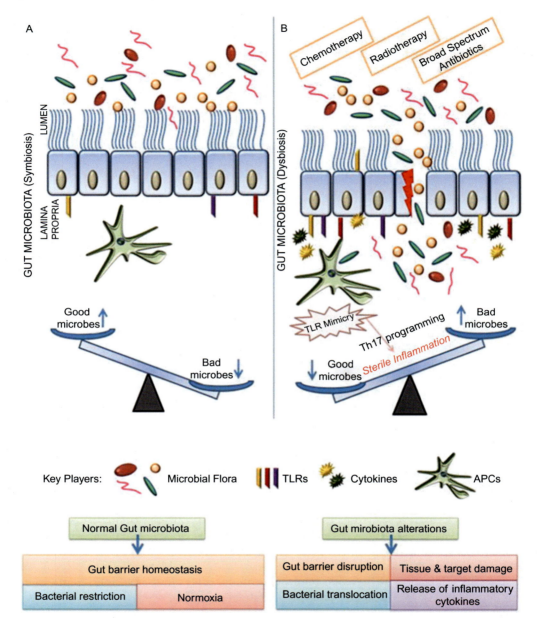

FIG. 9.5 Disruption of normal gut tissue micromilieu including intestinal villi and gut epithelium (A) during dysbiosis (B) perturb gut immune homeostasis and GI colony index, which is manifested by TLR mimicry, T_h17 programming, and hypoxia, which sensitizes normalized gut tissue micromilieu (A) for the progression of gastric inflammatory and tumor diseases. *Figure and caption from Toor, D., Wsson, M.K., Kumar, P., Karthikeyan, G., Kaushik, N.K., et al., 2019. Dysbiosis disrupts gut immune homeostasis and promotes gastric diseases. Int. J. Mol. Sci. 20, 2432.*

cancer development. Specifically, *F. nucleatum* can influence myeloid cell tumor infiltration, and changes of NK and T cell cytotoxicity might also influence cancer development (Routy et al., 2018). This bacterium could similarly orchestrate the influence of multiple other factors (e.g., toll-like receptor, microRNAs, and autophagy) in the promotion of colorectal cancer and recurrence of the illness. As we have indicated earlier, bacterial species may congregate, so *F. nucleatum* may interact with other microbes, such as *Campylobacter* and *Leptotrichia*, in promoting the development of cancer.

It may be especially significant that *F. nucleatum* appears early in colorectal cancer (Yachida et al., 2019), perhaps pointing to its potential in cancer initiation and possibly serving as an early marker of this type of cancer. As it has been recommended that individuals over the age of 50 years undergo colonoscopy every few years, identifying very early microbial changes could serve as practical biomarkers that predict later cancer, even before polyps are present. Among individuals undergoing colonoscopy, an elevated abundance of *F. nucleatum* was not alone in being associated with subsequent intramucosal carcinoma. Elevated, *Atopobium parvulum* and *Actinomyces odontolyticus* were associated with multiple polypoid adenomas as well as intramucosal carcinomas, as were branched-chain amino acids and phenylalanine (Yachida et al., 2019). As such, their elevated abundance might also serve as markers of increased colon cancer risk.

Gut microbiota likely do not act independently of genetic factors in promoting cancer. While the combination of *B. fragilis* and *E. coli* may be associated with colon cancer, these bacteria may be most prominent among individuals with a familial history of colon cancer (Dejea et al., 2018). Based on a metaanalysis it was indicated that colorectal cancer was accompanied by a set of 29 genes that were common among people living in various geographical locations, pointing to the global significance of these microbial signatures. This analysis linked colorectal cancer to the enrichment of proteins and genes involved in the catabolism of mucin (a glycoprotein secreted by salivary glands and epithelial cells) together with a reduction of genes associated with carbohydrate degradation (Wirbel et al., 2019).

Not all portions of the gut are affected by microbiota in the same way, and as we saw earlier, antibiotics also differentially influence colon vs rectal cancer. Only a very small percentage of gut cancers appear in the small intestine (2%) and the remaining 98% appear in the colon. It seems that mutations of p53 which could promote cancer occurrence had different actions across aspects of the gut. Within the large intestine, mutated p53 had oncogenic effects, but this was not apparent within the small intestine. It seems that colonic gut microbes that produce antioxidants do so through the microbiota metabolite gallic acid that can influence cancer promotion related to p53 (Kadosh et al., 2020).

Lateral transmission of microbiota and cancer

Since members of a given inbred mouse strain are genetically identical to one another, they ought to share phenotypes, but epigenetic changes can occur so they differ from one another in several ways. Despite being genetically identical, marked phenotypic differences occur within inbred mouse strains depending on the supplier of these mice. Mice obtained from a variety of vendors were differentially susceptible to Salmonella infection, which could be modified by cohousing mice (leading to transfer of commensal bacteria to one another), thereby altering vulnerability to infection (Velazquez et al., 2019).

The very same differences in colon cancer development have been observed in two colonies of genetically identical mice. In response to a carcinogen or an agent that prompts gastrointestinal inflammation mice of one colony developed three times as many colon tumors than in the second colony accompanied by distinct microbial communities. When gut bacteria from these two colonies were transplanted to germ-free mice, colon tumor development was again observed pointing to microbiota accounting for the differences that were observed in the two colonies. The different microbiota observed were associated with variations in the appearance of CD8+ T cells, being elevated in the tumor susceptible mice, possibly suggesting that these immune cells may have become overly activated by specific bacteria so they eventually suffered exhaustion and were thus less able to contend with cancer cells. Given the differential effects observed in genetically identical mice, this begs the question as to what environmental factors accounted for the differences between these microbial factors, immune cells, and cancer? Furthermore, are these influences relevant to cancer occurrence in humans?

Breast and female reproductive system cancers

Diets that are high in fat and low in fibers have been associated with the development of breast cancer, possibly owing to variations of microbial factors together with changes in estrogen; and it seems that estrogen and microbiota may act synergistically in affecting obesity and cancer occurrence. The bacteria present in breast tumors, notably *Enterobacteriaceae*, *Staphylococcus*, *Escherichia*, and *Staphylococcus epidermidis*, differed from that present in adjacent tissue as well as in breast tissue of women without breast cancer (Urbaniak et al., 2016). Breast tissue of women who had undergone mastectomy also contained greater amounts of the bacterial species *Methylobacterium* when compared with women who had cosmetic breast surgery. These data may be indicative of this bacterium being a biomarker of cancer development and might turn out to be a potential target for treatment. In a mouse model, a toxin secreted by enterotoxigenic *B. fragilis* was implanted in mouse mammary ducts and shown to promote breast cancer growth and metastasis—however, a similar outcome was not produced by nontoxigenic *B. fragilis* (Parida et al., 2021). This provides support for the hypothesis that certain microbiota, and the toxins they release, may cause breast cancer.

Hormonal interactions may also be important. Of relevance are gut microbial changes that interact with estrogen-related factors, and that can influence the development of breast cancer (Kwa et al., 2016). Such interactions could also affect cancer development in other estrogen-sensitive tissues. The female reproductive tract, for instance, is home to distinct microbial species that may be beneficial, but when dysbiosis occurs, chronic inflammation and several other biological changes may ensue. Among the potential changes are altered genetic stability, breach of the epithelial barrier, increased cell proliferation, and altered apoptosis and angiogenesis. These variations within the reproductive tract could favor the development and occurrence of various cancers, including cervical and endometrial cancer (Łaniewski et al., 2020). And although the involvement of microbiota has not been as extensively examined in ovarian cancer, specific bacterial phyla elevations are known to be present in ovarian cancer. In fact, beyond bacterial changes, there are alterations in viral, fungal, and parasitic communities. Overall, the particular signature of the microorganism community might help develop therapeutic strategies, although admittedly, identifying the role of these diverse organisms in cancer occurrence and determining which can be harnessed in a therapeutic capacity will likely be a challenge.

Lung cancer

The development of lung cancers is influenced by multiple factors, including those that stem from environmental toxicants over which individuals have little influence (e.g., inhaled particulate matter, age) and preventable factors over which we have considerable control (smoking). The lungs are continuously exposed to the presence of environmental toxicants and meet numerous microbes that could affect lung health. When microbiota dysbiosis occurs, lung cancer could develop possibly because of effects on immune activity and several inflammatory cytokines. Microbiota could similarly influence miRNA expression thereby affecting the regulation of lung immunity so lung-related diseases could be promoted, including COPD and cancer (Casciaro et al., 2020). This tripartite view of lung cancer does not imply that these components are independent of one another. Age and genetic susceptibility can affect microbiota as can toxicants, and the biochemical actions of these factors can interact to favor cancer occurrence. From this perspective, microbiota represent one of the primary players in health outcomes rather than simply being one of many ancillary factors.

Lung cancer patients and control participants differed with respect to the presence of respiratory microbiota, which might have contributed to the disease by affecting metabolic, immune, and inflammatory pathways. In fact, specific bacteria might serve as biomarkers for this form of cancer. Smoking increases numerous harmful microbes while concurrently reducing protective microbiota within the lung, hence promoting inflammation and DNA changes that culminate in lung cancer. The microbiota disturbances associated with lung cancer are not limited to those present within the lungs but have also been linked to those within the gut (Botticelli et al., 2020). Of importance, and covered in Chapter 15, the presence of microbiota and their SCFA metabolites were related to the response to therapies (e.g., checkpoint inhibitors) for nonsmall cell lung cancer. As even the most advanced lung cancer therapies have achieved limited success, identifying specific microbes linked to lung cancer might be instrumental in selecting therapeutic procedures that would promote optimal outcomes.

Periodontal diseases

The mouth is replete with numerous bacteria and gingival inflammation may be determined by the interaction between microbiota present and specific IL-17 producing cells that reside in the epithelium of the mouth. The presence of oral bacterial flora, perhaps secondary to factors such as tobacco and actions of alcohol, favor the development of oral cancers. Among oral cancer patients who were smokers or who chewed tobacco, more epigenetic changes were also detected in the cancer tissue than in adjacent tissues (Roy et al., 2019), although periodontal disease and oral microbiota have also been associated with oral carcinomas that seemed to be independent of tobacco and alcohol use.

The varied bacteria within the mouth affect those present throughout the complete gastrointestinal system and could have implications for the emergence of diseases beyond the oral cavity. Microbiota within the mouth can interact with bacteria in the gut, and changes in nitric oxide synthesis and antioxidant processes can affect colonic health. Oral bacteria associated with periodontal disease have also been linked to esophageal cancer as well as pancreatic cysts that may subsequently form pancreatic cancer (Gaiser et al., 2019). Periodontal disease has also been accompanied by an increased risk of breast cancer and among nonsmokers mouth microbiota, particularly Firmicutes, were linked to the later occurrence of lung cancer (Hosgood et al., 2021). Clearly, what happens in the mouth doesn't just stay in the mouth.

Parenthetically, in addition to various cancers, the presence of gum disease was also linked to numerous other inflammatory-related disorders, such as type 2 diabetes, heart disease, the severity of heart attack, ischemic stroke, autoimmune disorders, and Alzheimer's disease. As specific microbes vary with multiple systemic diseases, it might be considered that salivary microbiota could potentially serve in the development of personalized treatment strategies.

Parasites and cancer

The focus of gut-related processes related to cancer has primarily dealt with microbiota, but helminths (parasitic worms) and protozoa that can be present within the gut were also related to cancer occurrence. Many of these factors are especially pertinent within parts of Africa as well as in the Caribbean, East Mediterranean, and some South American countries where schistosomiasis is common, as well as in countries in which clonorchiasis has become endemic (e.g., Vietnam, China, Taiwan, Japan, as well as North and South Korea).

Parasitological examination revealed that several protozoa (especially Blastocystis) were elevated in some cancer patients. Likewise, the presence of helminths (e.g., blood and liver flukes) as well as helminth-based diseases, such as schistosomiasis, opisthorchiasis, and clonorchiasis, may be associated with the development of bile duct cancer, hepatocellular carcinoma, squamous cell carcinoma, and urinary bladder cancer (Scholte et al., 2018). This might occur because helminths promote genomic instability, disturb suppressor tumor proteins, limit apoptosis, and stimulate angiogenesis. Moreover, a prime action of helminths in fostering diseases occurred through their action on regulatory T cells, thereby affecting inflammatory processes.

Unlike helminths, as described in Fig. 9.6, some parasites seem to have anticancer and antimetastatic actions, provoking immune activation and apoptosis, regulating inflammatory processes, as well as limiting angiogenesis (Callejas et al., 2018). Certain parasites may also promote changes in gut microbial balances causing a shift away from *Bacilli* to *Clostridia* species. Still, other protozoa, such as *Trypanosoma cruzi*, which are responsible for Chagas disease, can act as either a pro- or anticancer inducing agent, as can other helminths.

The inflammatory link between microbiota and cancer

Microbiota can affect a diversity of cells of the innate and adaptive immune systems, and it was frequently indicated that the microbiota-cancer link might be mediated by inflammatory factors. Conversely, SCFAs related to microbiota can diminish cancer occurrence by promotion of an inflammatory suppressor or by affecting specific cytokines and immune cells, such as CD8+ T and T_{reg} cells (Nastasi et al., 2017). Aspects of the immune system located within the gut mucosa, notably PRRs comprising Toll-like and Nod-like receptors, are fundamental in affecting gut-related inflammation and play a key role in illness development. While some factors related to inflammation may primarily have local effects, other actions may affect disease processes at distal sites, thereby promoting a wide range of illnesses.

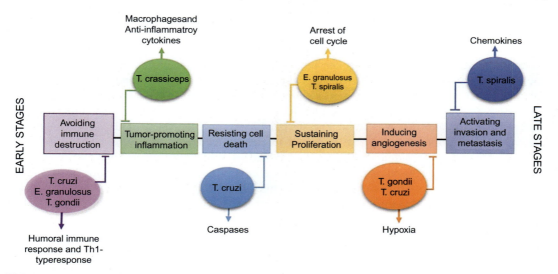

FIG. 9.6 Protozoa such as *Toxoplasma gondii* and *Trypanosoma cruzi* have an antitumor effect on some types of cancer through their antiangiogenic capacity, reactivation of the immune response, and induction of apoptosis. In contrast, *Taenia crassiceps* is able to regulate the cancer-promoting inflammatory response. *Echinococcus granulosus* have different antitumor mechanisms, such as reactivation of the immune response and antiproliferative effect on transformed cells, and *Trichinella spiralis* has regulating effects that affect invasion and metastasis. *Figure and caption from Callejas, B.E., Martinez-Saucedo, D., Terrazas, L.I., 2018. Parasites as negative regulators of cancer. Biosci. Rep. 38, BSR20180935.*

Through interactions with the immune system, microorganisms situated within the mucosal microenvironment can either foster or inhibit cancer growth and spread (Schwabe and Jobin, 2013). Certain bacteria may infiltrate the mucus layer of the colon where they obtain necessary nutrients, which can promote chronic inflammation and perhaps engender DNA damage that favors colon cancer. By promoting inflammation, microbiota could affect epithelial cells, thus allowing for an environment that supports tumor growth as well as facilitating the development of resistance to drugs that otherwise kill cancer cells. Inflammation could also drive carcinogenesis by producing oxidative stress and the generation of reactive oxygen species that produce mutagenesis and could also provoke epigenetic changes that favor neoplastic transformation.

Not surprisingly, genetic influences on components of the intestinal immune system can affect gut dysbiosis, which stimulates inflammatory processes, thereby increasing the risk for cell transformation tied to colorectal cancer (e.g., Brennan and Garrett, 2016). In this regard, while genetic features of microbiota have a substantial impact on health and well-being, genetic differences within the host also contribute to the actions of microbiota. A chronic inflammatory state may itself promote the excessive accumulation of bacteria, such as *E. coli*, or the presence of metabolites that are carcinogenic (Arthur et al., 2014). Predictably, treatment with microbiota that act against inflammation (e.g., *Lactobacillus reuteri*) can diminish cancer occurrence in highly vulnerable mice. A diet comprising a probiotic mixture could likewise diminish intestinal cancer, potentially operating through actions on inflammatory processes and by limiting angiogenesis.

Inflammatory bowel disease and cancer

Inflammatory bowel diseases, primarily ulcerative colitis and Crohn's disease are not very common illnesses, but their impact can be devastating. The occurrence of these illnesses have been increasing steadily and in 2017 the global incidence was about 6.8 million people. The prevalence of IBD varied substantially across countries. Whereas IBD in wealthy countries, such as in the US and UK, exceeded 400 cases per 100,000 people, the prevalence within less wealthy Latin-American countries, was about 30 cases per 100,000 people (Alatab et al., 2020). However, in second-generation offspring that migrated to wealthy countries, the incidence of IBD was comparable to that of residents within the host countries. While these disorders typically appear when individuals are 20–40 years their occurrence in children has been increasing. The factors and processes that cause IBD are not fully understood, but it is thought that in genetically at-risk individuals a second hit in the form of excessive immune activity and inflammation may instigate disease occurrence. There is a reason to suppose that the presence of certain microbial factors serve in the provocation of these illnesses perhaps secondary to smoking, certain medications, and stressful experiences (Ye et al., 2015).

Individuals living with inflammatory bowel diseases are at a considerably elevated risk of developing colon cancer. Inflammation associated with these disorders as well as the occurrence of bacterial and viral infection may contribute to IBD. These actions may likewise come about through the presence of specific bacteria, such as a select subset of *E. coli* that produce a toxin that engenders DNA damage and hence the occurrence of colon cancer. Not unexpectedly, the individual's diet and the presence of obesity may contribute to colon cancer secondary to IBD, as are age and family history of IBD and cancer. Through precision editing of the gut microbiome, specific effects can be exacted that are not obtained through the more common approaches to treat this condition.

Up to this point, we have primarily discussed the links between microbiota and inflammatory processes that can influence cancers at peripheral sites, but gut microbiota may also affect brain processes that influence inflammation. Indeed, aside from the multiple peripheral actions of specific foods in promoting health, diets that influence short-chain fatty acids can affect brain functioning (Dalile et al., 2019). Dietary factors, as we saw earlier, can influence mood states owing to their influence on several microbial species and their metabolic products that affect brain neuronal functioning and inflammatory processes. Some of the gut bacteria and inflammatory processes linked to depression have also been associated with varied cancers and it is conceivable that these factors underlie the frequent comorbidity that exists between depressive disorders and cancer. In line with this, gut microbiota may affect IFN-γ which regulates TRAIL and LAMP1, proteins that can act together with brain astrocytes to diminish inflammation, thereby potentially influencing inflammatory-related disorders, including brain cancers (Sanmarco et al., 2021). If these antiinflammatory actions could be harnessed through microbiota manipulations or in other ways, then benefits could be gained in dealing with brain cancers that largely remain untreatable.

Impact of the prenatal and perinatal diet

Like the effects of prenatal stressors, maternal diet can influence fetal maturation of both central and peripheral immune cells, the formation of neural circuits, as well as metabolic and developmental transcription programs, ultimately affecting childhood behaviors and the emergence of illnesses. The foods mom eats and the weight she gains during pregnancy could affect her microbiota and immune functioning, which might affect the fetus (Macpherson et al., 2017). Through actions on SCFAs, consumption of a high fiber diet during pregnancy and the lactation period can influence the abundance of T_{reg} cells in the offspring, which might influence their later disposition toward various disorders (Nakajima et al., 2017).

Experimentally manipulating the diet of pregnant women hasn't been an option, thus investigators have had to rely on animal studies or analysis of the offspring of pregnant women who chose to consume one sort of diet or another. Studies in humans are compromised in several ways, including the fact that living in impoverished environments may be responsible for the foods consumed, thus the outcome among offspring may reflect the broad effects of living conditions. It also appeared that the individual's coping strategies may be moderated by socioeconomic factors so those people who were socioeconomically disadvantaged may have resorted to emotional eating to deal with distress. As we've seen, eating processes and those involving stress reactivity are tightly intertwined, and food intake, especially comfort foods, might reflect an effort at self-medication to deal with stressful experiences.

It's not just what mom eats while pregnant that affects the fetus, as her diet before pregnancy and having obesity could also have marked consequences. Considerable data have indicated that obesity prior to pregnancy influences offspring and specific diets may have similar effects, such as being linked to pregnancy complications (e.g., gestational hypertension) that can affect the offspring. Diet at around the time of conception could also promote epigenetic changes of a gene linked to cancer suppression thereby increasing the lifelong risk of cancer in offspring.

Studies in rodents provided a more precise portrayal concerning how maternal diet could affect offspring. Mice fed a high-fat diet while pregnant displayed variations of their microbiota, including a greater abundance of *Akkermansia* and *Clostridium*, together with a concomitant reduction in the abundance of *Bifidobacterium* and several genera of the butyrate-producing families. Likewise, the fetal gut barrier was affected if pregnant dams were maintained on the high-fat diet, and there were indications of elevated gut inflammation that might contribute to gut bacterial and metabolic disturbances (Gohir et al., 2019).

Although there is considerable debate about whether microbial communities naturally exist in utero, bacterial challenges that influence hormonal processes in a pregnant woman could indirectly affect the fetus. It was also maintained that microbes present in the intrauterine environment may be transferred to the fetus, which could affect immune functioning (Chen and Gur, 2019). Given the links between gut microbiota and immune functioning, it is predictable that antibiotic-induced disturbances of gut microbiota during gestation would influence commensal bacterial colonization within the offspring and affect the presence of CD4+ T cells and B cells (Nyangahu and Jaspan, 2019). This treatment also led to altered bacterial colonization in the pups, accompanied by diminished weight gain that persisted into adulthood. It likewise appeared that antibiotic use during pregnancy was associated with numerous inflammatory illnesses in offspring, and when used during the first trimester of pregnancy the overall cancer risk was elevated among offspring (Ye et al., 2019).

The offspring of rats that had been maintained on a high-fat/high-sugar diet during pregnancy were likely to gain excessive weight during the suckling period and were more prone to developing a diabetes-like condition. A diet of this sort during gestation and lactation also resulted in mouse pups expressing an epigenetic profile comparable to that associated with diabetes, which could be reversed by a low-fat postweaning diet (Moody et al., 2017). It is also noteworthy that maintaining rodents on a Western diet during pregnancy affected the brain reward circuitry of their offspring, making them more attuned to rewards so elevated eating and weight gain ensued (Paradis et al., 2017). Evidently, prenatal diet and metabolic processes, as well as obesity, can profoundly influence fetal and postnatal health, likely interacting with host genetics, mode of delivery (e.g., Caesarean birth), infections encountered, drugs consumed, such as antibiotics, and later breastfeeding practices. In view of the strong tie between obesity and cancer occurrence, the obesogenic action of prenatal diet could indirectly affect adult cancer occurrence.

Paralleling the influence of prenatal diets, early life feeding may have lasting adverse consequences. Mice that were overnourished during the suckling period frequently became overweight throughout their lives and type 2 diabetes often developed, possibly owing to epigenetic alterations that favor pancreatic islet cell dysfunction. Maintaining rodent dams on a high-fat diet during lactation also influenced dopamine functioning within the nucleus accumbens (a key brain reward center) and altered HPA activity in the offspring in response to a chronic stressor and could thus influence the risk for obesity. Commensurate with these studies in rodents, a large study of mother-infant pairs indicated that among moms who were overweight, microbiota variations and a marked increase in the likelihood of obesity was present in their offspring as early as 1 year of age (Tun et al., 2018). Given the links between the duration of obesity and cancer occurrence, these early life experiences have obvious implications for the later development of varied types of cancer.

The impact of pre- and postnatal diets is often evaluated independently, but prenatal and early postnatal development and health may be affected through overlapping processes. Maternal gut microbiota during pregnancy provides varied substrates fundamental for fetal growth, including expansion and maturation of neural circuits as well as both central and peripheral immune cell functioning. Thereafter, during the early postnatal period, maternal microbiota can provide immunostimulatory actions in offspring that influence metabolic and developmental programs, provide the necessary substrates for effective brain and peripheral metabolic functioning, and could be essential in priming the immune system to react to later encounters with microorganisms. Consistent with this, bacterial translocation from the maternal gut can influence breast milk, thereby affecting offspring's immune functioning, and may thus contribute to risk for illnesses (Nyangahu and Jaspan, 2019). As such, experiential and environmental factors (e.g., malnutrition, stressors, and infection that affect microbiota) encountered prenatally or during the early postnatal period can profoundly influence the well-being of offspring that can carry through to adulthood.

Immune and hormonal factors are obtained by infants through breast milk, which is fundamental for physical and psychological/cognitive development and the view was entertained that in addition to genetic factors, maternal pathologies can be passed down to offspring by virtue of the hormonal and inflammatory factors obtained from breast milk. Breast milk contains various hormones, such as leptin and insulin, polyunsaturated fatty acids, including omega-3, as well as proinflammatory cytokines, notably IL-6 and TNF-α, that can affect infant health and development. For whatever reason, among mothers who had obesity during pregnancy, the signals coming from these milk-borne factors were not operating in the usual way, hence slowing development in offspring.

Microbial and immune processes are not only functionally related to one another but also mature in parallel. Accordingly, perturbations of one system may affect the other, so prenatal and early postnatal microbiota disturbances can instigate profound and lasting effects on vulnerability to immune-related adult illnesses. Diminished maternal omega-3 fatty acids could produce persistent effects on several components of the microbiota profile that can influence metabolic processes and weight gain in offspring. As it happens, omega-3 fatty acids consumed at this time of development can limit later stressor-induced anxiety and depression, as well as the occurrence of subsequent physical illness. Moreover, in mice prone to develop breast cancer, the development of this form of cancer was delayed in the offspring of mothers that had been fed a diet rich in omega-3 fatty acids (Abbas et al., 2021).

It is well known that drugs taken during pregnancy can profoundly influence the fetus, potentially producing teratogenic actions. Thus, women tend to avoid toxicants, such as alcohol and cigarettes, which can affect the fetus. Too often, however, the influence of dietary factors has not been as judiciously considered. Obstetricians and nurse practitioners typically provide advice about proper diets and the risks of obesity for mom and offspring, yet more than 20% of pregnant women are obese (this exceeds 30% in some countries, such as India). The incidence of obesity and poor dietary choices may extend to the postnatal period, which may adversely affect offspring. With the increasing incidence of being overweight or having obesity, the health risks to children have been increasing proportionately.

Intergenerational and transgenerational actions

Early life nutrition and the maternal diet during fetal development of rodents were accompanied by obesity-related epigenetic changes, which could promote obesity and diseases later in life. Dietary fat increases DNA methylation in adipocytes (Perfilyev et al., 2017), and metabolic syndrome and related disorders were accompanied by epigenetic changes measured in blood DNA. Metabolic disturbances that stemmed from epigenetic changes related to the diet could be transmitted intergenerationally and transgenerationally (Kaspar et al., 2020). For instance, microbiota diversity in mice was diminished with low consumption of microbiota accessible carbohydrates, which persisted even after mice had been transferred to a diet high in this carbohydrate. The disappearance of certain gut microbiota may occur incrementally over generations, which may be linked to health-related cultural differences (Sonnenburg et al., 2016). As described in Chapter 3, with industrialization the microbial community has been changing, so it became less compatible with normal human physiological processes. It is possible that adaptive processes might emerge in the form of epigenetic changes provoked by the changing human diet, so adverse effects will be mitigated, but this remains to be established.

The other side of microbiota: Is there value to prebiotic and probiotic supplements?

Even though they can readily be obtained from foods, probiotics in the form of supplements have increasingly found their way into the marketplace. Unproven probiotics have been offered to cure every imaginable illness even though there is ample reason to question whether these commercially available prebiotic and probiotic supplements have any beneficial effects. In fact, as indicated earlier, in the presence of a healthy microbiome, the microbiota ordinarily present in the gut may create resistance to the actions of these supplements so they are eliminated with bowel movements (Zmora et al., 2018).

Despite there being no guarantee that probiotic supplements will have beneficial effects, a case has been made for their use in diminishing cancer risk and might influence the efficacy of cancer therapies, particularly as they have antiinflammatory actions. Given the uncertainties concerning the actions of probiotic supplements, it's somewhat surprising that supplements have found their way into hospital treatment regimens (Cohen, 2018). In some instances, they are used to rejuvenate microbiota after antibiotic treatment, but contrary to what is frequently believed, the gut mucosal microbiome is not readily reconstituted by probiotic supplements, and may actually be slowed, although antibiotic-associated diarrhea may be diminished (Suez et al., 2018). At the same time, when examined in mice, transplanted human microbiota were relatively resilient and recovered quickly following antibiotic treatment, unless mice had been maintained on a low fiber diet or when mice did not have other critters around who could provide lateral microbial transmission (Ng et al., 2019). But mice aren't necessarily a perfect proxy for humans, and in severely ill patients, probiotics could in some manner allow (or cause) gut bacteria translocation to the blood where mutations may promote bacteremia if immune system functioning is compromised (Yelin et al., 2019).

Some supplements could potentially have positive effects, but as in the case of vitamins, the value of probiotics for generally healthy people is uncertain. As we've now said repeatedly, it isn't certain what the right microbial balance ought to be, but even if we did, individuals short on some commensal bacteria wouldn't know it and would be unaware of what supplements might remedy this. It isn't unreasonable to assume that supplementation with SCFAs may have benefits by affecting the multiple interactions that exist between diet and various gut bacteria that influence health. But, before pre- or probiotic supplements are recommended, it is essential that further randomized placebo-controlled human trials be undertaken, and that extensive follow-up analyses be conducted to determine the long-term actions of these treatments. We'll have much more to say about the use of supplements in Chapters 14 and 15. Suffice it to say that perhaps in the end, it might turn out that some of these supplements taken in pill form are beneficial, although it may be more productive—and gratifying—to rely on *diet-based* microbial improvements.

Summary and conclusions

The centrality of the gut microbiome in determining disease occurrence is underscored by the fact that dietary influences, together with genetic factors, affect host metabolism, immune functioning, and the response to infection. The specific microbial constituents that act in one way or another haven't yet been fully deduced but may ultimately be cogent in defining the ideal treatments to treat particular cancers, and in considering the interindividual treatments that would be most suitable. Several reports made it clear that the glycemic response elicited by specific foods (i.e., how slowly or quickly foods produce an increase of glucose) varies considerably across individuals, likely owing to the microbiota present. Given the importance of glucose to cancer and microbial involvement in modulating immune functioning, diverse microbiota ought to be considered in the analysis and development of personalized strategies to deal with specific forms of cancer. This said, developing a precision approach to cancer treatment won't be easy given that microbiota communities vary dramatically among healthy individuals, making it difficult to define what a healthy microbiome looks like. Still, a precision medicine approach using machine learning has shown that the presence of numerous molecules could have anticancer actions, which could be adapted to mesh with the needs of individuals. It's still early in these efforts, but with the development of a broader array of biomarkers, including concurrent analyses of genetic and microbial indices, it may be possible to determine best practices concerning eating and exercise in relation to specific illnesses (Zeevi et al., 2015).

The use of live probiotic bacteria, or enriching the microbiota through prebiotic consumption, has been advanced for many diseases, but in the main, only modest positive effects have been seen in randomized control trials. Moreover, the microbial benefits accrued with respect to immune functioning may be relatively short-lived. Still, a cocktail of diverse bacteria that act on immune processes could turn out to have beneficial effects in an adjuvant capacity. A therapeutic approach involving these microbes and their metabolites (e.g., varied SCFAs) may be a good option if these are based on individual difference factors (Suez and Elinav, 2017). A retrospective analysis of the effectiveness of medically tailored meals among ill individuals (diabetes, HIV, cancer) under the supervision of a registered dietitian or nutritionist was associated with fewer hospitalizations, speaking to the feasibility of this approach.

Ideally, individuals would be able to deal with obesity without resorting to treatments other than diet and exercise, facilitated further through effective social support. The reality is, however, that this often isn't in the cards, and for many individuals, surgical or pharmacological treatments for obesity might be needed. The increasing focus on the development of medications to treat obesity has been making headway and there may be ways of browning or beiging white fat, or increasing fat-burning in some other way, thereby increasing endurance and further weight loss. At the same time, we shouldn't fool ourselves that these treatments are suitable alternatives for exercise. In view of the multiple benefits of exercise to heart health and immune functioning, counting on drug-based shortcuts to achieve weight loss could in multiple ways prove counterproductive.

References

Abbas, A., Witte, T., Patterson, W.L., Fahrmann, J., Guo, K., et al., 2021. Epigenetic reprogramming mediated by maternal diet rich in omega-3 fatty acids protects from breast cancer development in F1 offspring. Front. Cell Dev. Biol. 9, 1517.

Alatab, S., Sepanlou, S.G., Ikuta, K., Vahedi, H., Bisignano, C., et al., 2020. The global, regional, and national burden of inflammatory bowel disease in 195 countries and territories, 1990–2017: a systematic analysis for the Global Burden of Disease Study 2017. Lancet Gastroenterol. Hepatol. 5, 17–30.

Arpaia, N., Campbell, C., Fan, X., Dikiy, S., Van Der Veeken, J., et al., 2013. Metabolites produced by commensal bacteria promote peripheral regulatory T-cell generation. Nature 504, 451–455.

Arthur, J.C., Gharaibeh, R.Z., Mühlbauer, M., Perez-Chanona, E., Uronis, J.M., et al., 2014. Microbial genomic analysis reveals the essential role of inflammation in bacteria-induced colorectal cancer. Nat. Commun. 5, 1–11.

Bashiardes, S., Godneva, A., Elinav, E., Segal, E., 2018. Towards utilization of the human genome and microbiome for personalized nutrition. Curr. Opin. Biotechnol. 51, 57–63.

Becattini, S., Taur, Y., Pamer, E.G., 2016. Antibiotic-induced changes in the intestinal microbiota and disease. Trends Mol. Med. 22, 458–478.

Bindels, L.B., Porporato, P., Dewulf, E., Verrax, J., Neyrinck, A.M., et al., 2012. Gut microbiota-derived propionate reduces cancer cell proliferation in the liver. Br. J. Cancer 107, 1337–1344.

Bose, S., Ramesh, V., Locasale, J.W., 2019. Acetate metabolism in physiology, cancer, and beyond. Trends Cell Biol. 29, 695–703.

Botticelli, A., Vernocchi, P., Marini, F., Quagliariello, A., Cerbelli, B., et al., 2020. Gut metabolomics profiling of non-small cell lung cancer (NSCLC) patients under immunotherapy treatment. J. Transl. Med. 18, 1–10.

Brennan, C.A., Garrett, W.S., 2016. Gut microbiota, inflammation, and colorectal cancer. Annu. Rev. Microbiol. 70, 395–411.

Buffie, C.G., Bucci, V., Stein, R.R., McKenney, P.T., Ling, L., et al., 2015. Precision microbiome reconstitution restores bile acid mediated resistance to *Clostridium difficile*. Nature 517, 205–208.

Bullman, S., Pedamallu, C.S., Sicinska, E., Clancy, T.E., Zhang, X., et al., 2017. Analysis of Fusobacterium persistence and antibiotic response in colorectal cancer. Science 358, 1443–1448.

Callejas, B.E., Martinez-Saucedo, D., Terrazas, L.I., 2018. Parasites as negative regulators of cancer. Biosci. Rep. 38. BSR20180935.

Casciaro, M., Di Salvo, E., Pioggia, G., Gangemi, S., 2020. Microbiota and microRNAs in lung diseases: mutual influence and role insights. Eur. Rev. Med. Pharm. Sci. 24, 13000–13008.

Chen, H.J., Gur, T.L., 2019. Intrauterine microbiota: missing, or the missing link? Trends Neurosci. 42, 402–413.

Chen, J., Zhao, K.N., Vitetta, L., 2019. Effects of intestinal microbial–elaborated butyrate on oncogenic signaling pathways. Nutrients 11, 1026.

Cohen, P.A., 2018. Probiotic safety—no guarantees. JAMA Intern. Med. 178, 1577–1578.

Dalile, B., Van Oudenhove, L., Vervliet, B., Verbeke, K., 2019. The role of short-chain fatty acids in microbiota–gut–brain communication. Nat. Rev. Gastroenterol. Hepatol. 16, 461–478.

Dao, M.C., Belda, E., Prifti, E., Everard, A., Kayser, B.D., et al., 2019. *Akkermansia muciniphila* abundance is lower in severe obesity, but its increased level after bariatric surgery is not associated with metabolic health improvement. Am. J. Physiol. Endocrinol. Metab. 317, E446–E459.

Dejea, C.M., Fathi, P., Craig, J.M., Boleij, A., Taddese, R., et al., 2018. Patients with familial adenomatous polyposis harbor colonic biofilms containing tumorigenic bacteria. Science 359, 592–597.

Depommier, C., Everard, A., Druart, C., Plovier, H., Van Hul, M., et al., 2019. Supplementation with *Akkermansia muciniphila* in overweight and obese human volunteers: a proof-of-concept exploratory study. Nat. Med. 25, 1096–1103.

Diener, C., Qin, S., Zhou, Y., Patwardhan, S., Tang, L., et al., 2021. Baseline gut metagenomic functional gene signature associated with variable weight loss responses following a healthy lifestyle intervention in humans. mSystems 6, e00964-21.

Eslami, M., Sadrifar, S., Karbalaei, M., Keikha, M., Kobyliak, N.M., et al., 2020. Importance of the microbiota inhibitory mechanism on the Warburg effect in colorectal cancer cells. J. Gastrointest. Cancer 51, 738–747.

Gaiser, R.A., Halimi, A., Alkharaan, H., Lu, L., Davanian, H., et al., 2019. Enrichment of oral microbiota in early cystic precursors to invasive pancreatic cancer. Gut 68, 2186–2194.

Gao, Y., Tollefsbol, O.T., 2015. Impact of epigenetic dietary components on cancer through histone modifications. Curr. Med. Chem. 22, 2051–2064.

Garrett, W.S., 2015. Cancer and the microbiota. Science 348, 80–86.

Gohir, W., Kennedy, K.M., Wallace, J.G., Saoi, M., Bellissimo, C., et al., 2019. High-fat diet intake modulates maternal intestinal adaptations to pregnancy and results in placental hypoxia, as well as altered fetal gut barrier proteins and immune markers. J. Physiol. 597, 3029–3051.

Harusato, A., Viennois, E., Etienne-Mesmin, L., Matsuyama, S., Abo, H., et al., 2019. Early-life microbiota exposure restricts myeloid-derived suppressor cell–driven colonic tumorigenesis. Cancer Immunol. Res. 7, 544–551.

Healey, G., Murphy, R., Butts, C., Brough, L., Whelan, K., et al., 2018. Habitual dietary fibre intake influences gut microbiota response to an inulin-type fructan prebiotic: a randomised, double-blind, placebo-controlled, cross-over, human intervention study. Br. J. Nutr. 119, 176–189.

Hosgood, H.D., Cai, Q., Hua, X., Long, J., Shi, J., et al., 2021. Variation in oral microbiome is associated with future risk of lung cancer among never-smokers. Thorax 76, 256–263.

Huang, P., Liu, Y., 2019. A reasonable diet promotes balance of intestinal microbiota: prevention of precolorectal cancer. Biomed. Res. Int. 2019.

Ideraabdullah, F.Y., Zeisel, S.H., 2018. Dietary modulation of the epigenome. Physiol. Rev. 98, 667–695.

Kadosh, E., Snir-Alkalay, I., Venkatachalam, A., May, S., Lasry, A., et al., 2020. The gut microbiome switches mutant p53 from tumour-suppressive to oncogenic. Nature 586, 133–138.

Kaspar, D., Hastreiter, S., Irmler, M., de Angelis, M.H., Beckers, J., 2020. Nutrition and its role in epigenetic inheritance of obesity and diabetes across generations. Mamm. Genome 31, 119–133.

Kennedy, E.A., King, K.Y., Baldridge, M.T., 2018. Mouse microbiota models: comparing germ-free mice and antibiotics treatment as tools for modifying gut bacteria. Front. Physiol. 9, 1534.

Kordahi, M.C., Stanaway, I.B., Avril, M., Chac, D., Blanc, M.P., et al., 2021. Genomic and functional characterization of a mucosal symbiont involved in early-stage colorectal cancer. Cell Host Microbe 29, 1589–1598.

Krut, O., Bekeredjian-Ding, I., 2018. Contribution of the immune response to phage therapy. J. Immunol. 200, 3037–3044.

Kwa, M., Plottel, C.S., Blaser, M.J., Adams, S., 2016. The intestinal microbiome and estrogen receptor–positive female breast cancer. J. Natl. Cancer Int. 108.

Łaniewski, P., Ilhan, Z.E., Herbst-Kralovetz, M.M., 2020. The microbiome and gynaecological cancer development, prevention and therapy. Nat. Rev. Urol. 17, 232–250.

Levy, M., Kolodziejczyk, A.A., Thaiss, C.A., Elinav, E., 2017. Dysbiosis and the immune system. Nat. Rev. Immunol. 17, 219–232.

Li, H., Limenitakis, J.P., Greiff, V., Yilmaz, B., Schären, O., et al., 2020. Mucosal or systemic microbiota exposures shape the B cell repertoire. Nature 584, 274–278.

Ma, C., Han, M., Heinrich, B., Fu, Q., Zhang, Q., et al., 2018. Gut microbiome–mediated bile acid metabolism regulates liver cancer via NKT cells. Science 360.

Macpherson, A.J., de Agüero, M.G., Ganal-Vonarburg, S.C., 2017. How nutrition and the maternal microbiota shape the neonatal immune system. Nat. Rev. Immunol. 17, 508–517.

Martins, D., Mendes, F., Schmitt, F., 2021. Microbiome: a supportive or a leading actor in lung cancer? Pathobiology 88, 198–207.

Moody, L., Chen, H., Pan, Y.X., 2017. Postnatal diet remodels hepatic DNA methylation in metabolic pathways established by a maternal high-fat diet. Epigenomics 9, 1387–1402.

Nakajima, A., Kaga, N., Nakanishi, Y., Ohno, H., Miyamoto, J., et al., 2017. Maternal high fiber diet during pregnancy and lactation influences regulatory T cell differentiation in offspring in mice. J. Immunol. 199, 3516–3524.

Nakatsu, G., Li, X., Zhou, H., Sheng, J., Wong, S.H., et al., 2015. Gut mucosal microbiome across stages of colorectal carcinogenesis. Nat. Commun. 6, 1–9.

Nastasi, C., Fredholm, S., Willerslev-Olsen, A., Hansen, M., Bonefeld, C.M., et al., 2017. Butyrate and propionate inhibit antigen-specific CD8+ T cell activation by suppressing IL-12 production by antigen-presenting cells. Sci. Rep. 7, 1–10.

Ng, K.M., Aranda-Díaz, A., Tropini, C., Frankel, M.R., Van Treuren, W., et al., 2019. Recovery of the gut microbiota after antibiotics depends on host diet, community context, and environmental reservoirs. Cell Host Microbe 26, 650–665.e654.

Nyangahu, D., Jaspan, H., 2019. Influence of maternal microbiota during pregnancy on infant immunity. Clin. Exp. Immunol. 198, 47–56.

Paradis, J., Boureau, P., Moyon, T., Nicklaus, S., Parnet, P., et al., 2017. Perinatal western diet consumption leads to profound plasticity and GABAergic phenotype changes within hypothalamus and reward pathway from birth to sexual maturity in rat. Front. Endocrinol. 8, 216.

Parida, S., Wu, S., Siddharth, S., Wang, G., Muniraj, N., et al., 2021. A procarcinogenic colon microbe promotes breast tumorigenesis and metastatic progression and concomitantly activates notch and β-catenin axes. Cancer Discov. 11, 1138–1157.

Perfilyev, A., Dahlman, I., Gillberg, L., Rosqvist, F., Iggman, D., et al., 2017. Impact of polyunsaturated and saturated fat overfeeding on the DNA-methylation pattern in human adipose tissue: a randomized controlled trial. Am. J. Clin. Nutr. 105, 991–1000.

Perrott, S., McDowell, R., Murchie, P., Cardwell, C., Samuel, L., 2021. SO-25 global rise in early-onset colorectal cancer: an association with antibiotic consumption? Ann. Oncol. 32, S213.

Petersen, C., Bell, R., Klag, K.A., Lee, S.H., Soto, R., et al., 2019. T cell–mediated regulation of the microbiota protects against obesity. Science 365, eaat935.

Petrelli, F., Ghidini, M., Ghidini, A., Perego, G., Cabiddu, M., et al., 2019. Use of antibiotics and risk of cancer: a systematic review and meta-analysis of observational studies. Cancers 11, 1174.

Pleguezuelos-Manzano, C., Puschhof, J., Huber, A.R., van Hoeck, A., Wood, H.M., et al., 2020. Mutational signature in colorectal cancer caused by genotoxic pks+ E. coli. Nature 580, 269–273.

Pushalkar, S., Hundeyin, M., Daley, D., Zambirinis, C.P., Kurz, E., et al., 2018. The pancreatic cancer microbiome promotes oncogenesis by induction of innate and adaptive immune suppression. Cancer Discov. 8, 403–416.

Qu, G., Sun, C., Sharma, M., Uy, J.P., Song, E.J., et al., 2020. Is antibiotics use really associated with increased risk of colorectal cancer? An updated systematic review and meta-analysis of observational studies. Int. J. Color. Dis. 35, 1397–1412.

Routy, B., Gopalakrishnan, V., Daillère, R., Zitvogel, L., Wargo, J.A., et al., 2018. The gut microbiota influences anticancer immunosurveillance and general health. Nat. Rev. Clin. Oncol. 15, 382–396.

Roy, R., Chatterjee, A., Das, D., Ray, A., Singh, R., et al., 2019. Genome-wide miRNA methylome analysis in oral cancer: possible biomarkers associated with patient survival. Epigenomics 11, 473–487.

Rubinstein, M.R., Baik, J.E., Lagana, S.M., Han, R.P., Raab, W.J., et al., 2019. Fusobacterium nucleatum promotes colorectal cancer by inducing Wnt/β-catenin modulator Annexin A1. EMBO Rep. 20, e47638.

Sanmarco, L.M., Wheeler, M.A., Gutiérrez-Vázquez, C., Polonio, C.M., Linnerbauer, M., et al., 2021. Gut-licensed IFNγ+ NK cells drive LAMP1+ TRAIL+ anti-inflammatory astrocytes. Nature 590, 473–479.

Schellekens, H., Torres-Fuentes, C., van de Wouw, M., Long-Smith, C.M., Mitchell, A., et al., 2021. *Bifidobacterium longum* counters the effects of obesity: partial successful translation from rodent to human. EBioMedicine 63, 103176.

Scholte, L.L.S., Pascoal-Xavier, M.A., Nahum, L.A., 2018. Helminths and cancers from the evolutionary perspective. Front. Med. 5, 90.

Schwabe, R.F., Jobin, C., 2013. The microbiome and cancer. Nat. Rev. Cancer 13, 800–812.

Sivaprakasam, S., Prasad, P.D., Singh, N., 2016. Benefits of short-chain fatty acids and their receptors in inflammation and carcinogenesis. Pharmacol. Ther. 164, 144–151.

Sonnenburg, E.D., Smits, S.A., Tikhonov, M., Higginbottom, S.K., Wingreen, N.S., et al., 2016. Diet-induced extinctions in the gut microbiota compound over generations. Nature 529, 212–215.

Suez, J., Elinav, E., 2017. The path towards microbiome-based metabolite treatment. Nat. Microbiol. 2, 1–5.

Suez, J., Zmora, N., Zilberman-Schapira, G., Mor, U., Dori-Bachash, M., et al., 2018. Post-antibiotic gut mucosal microbiome reconstitution is impaired by probiotics and improved by autologous FMT. Cell 174, 1406–1423.e1416.

Tanoue, T., Morita, S., Plichta, D.R., Skelly, A.N., Suda, W., et al., 2019. A defined commensal consortium elicits CD8 T cells and anti-cancer immunity. Nature 565, 600–605.

Thirunavukkarasan, M., Wang, C., Rao, A., Hind, T., Teo, Y.R., et al., 2017. Short-chain fatty acid receptors inhibit invasive phenotypes in breast cancer cells. PLoS One 12, e0186334.

Tun, H.M., Bridgman, S.L., Chari, R., Field, C., Guttman, D.S., et al., 2018. Roles of birth mode and infant gut microbiota in intergenerational transmission of overweight and obesity from mother to offspring. JAMA Pediatr. 172, 368–377.

Urbaniak, C., Gloor, G.B., Brackstone, M., Scott, L., Tangney, M., et al., 2016. The microbiota of breast tissue and its association with breast cancer. Appl. Environ. Microbiol. 82, 5039–5048.

Uribe-Herranz, M., Bittinger, K., Rafail, S., Guedan, S., Pierini, S., et al., 2018. Gut microbiota modulates adoptive cell therapy via CD8α dendritic cells and IL-12. JCI Insight 3.

Velazquez, E.M., Nguyen, H., Heasley, K.T., Saechao, C.H., Gil, L.M., et al., 2019. Endogenous Enterobacteriaceae underlie variation in susceptibility to Salmonella infection. Nat. Microbiol. 4, 1057–1064.

Wan, Y., Wang, F., Yuan, J., Li, J., Jiang, D., et al., 2019. Effects of dietary fat on gut microbiota and faecal metabolites, and their relationship with cardiometabolic risk factors: a 6-month randomised controlled-feeding trial. Gut 68, 1417–1429.

Wang, Z., Hua, W., Li, C., Chang, H., Liu, R., et al., 2019. Protective role of fecal microbiota transplantation on colitis and colitis-associated colon cancer in mice is associated with Treg cells. Front. Microbiol. 10, 2498.

Wirbel, J., Pyl, P.T., Kartal, E., Zych, K., Kashani, A., et al., 2019. Meta-analysis of fecal metagenomes reveals global microbial signatures that are specific for colorectal cancer. Nat. Med. 25, 679–689.

Wong, S.H., Yu, J., 2019. Gut microbiota in colorectal cancer: mechanisms of action and clinical applications. Nat. Rev. Gastroenterol. Hepatol. 16, 690–704.

Wong, S.H., Zhao, L., Zhang, X., Nakatsu, G., Han, J., et al., 2017. Gavage of fecal samples from patients with colorectal cancer promotes intestinal carcinogenesis in germ-free and conventional mice. Gastroenterology 153, 1621–1633.e1626.

Wu, N., Yang, X., Zhang, R., Li, J., Xiao, X., et al., 2013. Dysbiosis signature of fecal microbiota in colorectal cancer patients. Microb. Ecol. 66, 462–470.

Yachida, S., Mizutani, S., Shiroma, H., Shiba, S., Nakajima, T., et al., 2019. Metagenomic and metabolomic analyses reveal distinct stage-specific phenotypes of the gut microbiota in colorectal cancer. Nat. Med. 25, 968–976.

Yang, J.J., Yu, D., Xiang, Y.B., Blot, W., White, E., et al., 2020. Association of dietary fiber and yogurt consumption with lung cancer risk: a pooled analysis. JAMA Oncol. 6, e194107.

Ye, Y., Pang, Z., Chen, W., Ju, S., Zhou, C., 2015. The epidemiology and risk factors of inflammatory bowel disease. Int. J. Clin. Exp. Med. 8, 22529–22542.

Ye, X., Monchka, B.A., Righolt, C.H., Mahmud, S.M., 2019. Maternal use of antibiotics and cancer incidence risk in offspring: a population-based cohort study in Manitoba, Canada. Cancer Med. 8, 5367–5372.

Yelin, I., Flett, K.B., Merakou, C., Mehrotra, P., Stam, J., et al., 2019. Genomic and epidemiological evidence of bacterial transmission from probiotic capsule to blood in ICU patients. Nat. Med. 25, 1728–1732.

Zeevi, D., Korem, T., Zmora, N., Israeli, D., Rothschild, D., et al., 2015. Personalized nutrition by prediction of glycemic responses. Cell 163, 1079–1094.

Zhang, J., Haines, C., Watson, A.J., Hart, A.R., Platt, M.J., et al., 2019. Oral antibiotic use and risk of colorectal cancer in the United Kingdom, 1989–2012: a matched case–control study. Gut 68, 1971–1978.

Zitvogel, L., Ayyoub, M., Routy, B., Kroemer, G., 2016. Microbiome and anticancer immunosurveillance. Cell 165, 276–287.

Zmora, N., Zilberman-Schapira, G., Suez, J., Mor, U., Dori-Bachash, M., et al., 2018. Personalized gut mucosal colonization resistance to empiric probiotics is associated with unique host and microbiome features. Cell 174, 1388–1405.e1321.

CHAPTER 10

Exercise

OUTLINE

The broad effects of exercise	311	Impact of exercise on cancer features and side effects of treatments	322
Immune and cytokine changes associated with exercise	312	Cancer-related fatigue and quality of life	323
		Cancer cachexia	325
Influence of moderate vs strenuous exercise	313	Sleep disturbances	325
		Neuropathy	326
Cytokine changes associated with exercise	314	Depression	326
Myokines in relation to cancer	315	Fueling cancer progression	327
Exercise and cancer prevention	316	Exercise and microbiota	329
Impact of exercise on existent cancers	318	Microbiota changes and amount of exercise	329
Safety of exercise among cancer patients	318	Sedentary behaviors	332
Variations of cancer progression produced by exercise	318	The social element	334
Intensity of exercise in relation to cancer	319	Roadblocks to exercise and how to get around them	334
Mechanisms related to exercise benefits	319	Summary and conclusions	336
Influence of epigenetic changes stemming from exercise	321	References	337

The broad effects of exercise

Some readers may recall the words of Saturday Night Live's Fernando (played by Billy Crystal) "It's more important to look good than to feel good." As it happens, exercise can do both: make you look good *and* feel great. Moderate exercise can diminish the influence of stressors and has been associated with reduced anxiety and depressed mood as well as enhanced sleep quality. Additionally, engaging in regular exercise training was accompanied by cognitive benefits, such as modestly improved attention, executive functioning, and memory, and can limit age-related cognitive problems. Indeed, through changes in brain glial functioning, exercise combined with diet and stress reduction methods may have a significant impact on neurodegenerative diseases.

Besides the benefits to mental and cognitive functioning, aerobic exercises, such as bicycling, jogging, and distance running, can reduce the risk of chronic diseases, such as metabolic syndrome, type 2 diabetes, cardiovascular illness, and COPD, as well as facilitating the ability to overcome existing illnesses. Paralleling these health benefits, aerobic exercise could potentially reduce the incidence of some types of cancer, improve surgical outcomes, enhance physical functioning, and limit the side effects of cancer therapies. Additionally, an exercise regimen can improve psychological health that might otherwise be disturbed by cancer occurrence and cancer therapy. The adoption of anaerobic exercises (e.g., resistance training that comprises weight lifting, high-intensity interval training) may similarly confer multiple health benefits. These exercises can break down glucose rapidly without the use of large amounts of oxygen, burn fat efficiently, and improve muscle strength, including the heart muscle.

It is hardly surprising that maintaining a moderate exercise schedule throughout life, from adolescence through to later adulthood, enhances well-being and diminishes all-cause mortality (29%–36%) relative to individuals who maintained inactive lifestyles. It is good news, however, that individuals who had been inactive in their youth and then took up regular exercise at midlife benefited appreciably so that the likelihood of early mortality was diminished by 32%–35%. In contrast, the benefits of exercise obtained by exercising during later adolescence were lost if individuals subsequently stopped exercising (Saint-Maurice et al., 2019). In this particular study as well as in many others, the focus was on all-cause mortality and heart disease, but exercise can affect a remarkably wide assortment of diseases (26 at last count), so that "Exercise as Medicine" has become a common refrain among exercise physiologists. Routine exercise regimens most certainly enhance various attributes of wellness and fitness, although marked individual differences are apparent in the extent to which these benefits are attained. A considerable portion of the variance between individuals could be accounted for based on specific genes activated, and it may be possible to predict the effectiveness of exercise in producing phenotypic differences based on genetic profiles. As discussed in the context of dietary factors, it was suggested that tailoring exercise for specific individuals based on genetic signatures could yield optimal effects of diverse exercise routines (Chung et al., 2021).

Immune and cytokine changes associated with exercise

While exercise is recognized as having multiple health benefits, we need to dig a bit deeper to appreciate what forms of exercise—and how much—need to be undertaken to realize fully the positive actions of exercise. Maintaining a regular exercise schedule has most often been seen as important in limiting the occurrence of chronic illnesses by influencing metabolic and immune functioning as well as by reducing inflammation associated with exercise. Preclinical studies indicated that exercise in the form of wheel running, which is voluntarily adopted by rodents, resulted in the slowing of cancer growth. Furthermore, transferring T cells from exercised mice to those that had not received this training, enhanced their survival (Rundqvist et al., 2020). These actions may come about because of epinephrine activation of NK cells and their intratumoral infiltration (Pedersen et al., 2016), as well as by epinephrine signaling of CD8+ T cells and monocyte subtypes. Studies in rodents also indicated that exercise undertaken over several weeks in the form of swim training reduced carcinogen-induced mammary tumor development and engendered an immune profile that was aligned with an antitumor T_h1 pattern that included the increased presence of immune cells that produce levels of cytokines, such as IFN-γ.

Influence of moderate vs strenuous exercise

Despite the obvious advantages of exercise, there can be too much of a good thing. The relationship between exercise and immune changes varies as an inverted U-shaped (or J-shaped) function, wherein immune activity is progressively enhanced with increasingly greater exercise but then declines if it is excessive. In humans, as little as 20–30 min of exercise can limit circulating inflammatory factors and was sufficient to increase the levels of neutrophil, monocyte, NK, and T cells and promote their redistribution from storage sites into circulation. The translocation of immune cells brings them to sites, such as the gut, lung, and lymph nodes, where the likelihood of encountering antigens is relatively high. A moderate exercise regimen was also accompanied by elevated phagocytic activity, NK cytotoxicity, together with T-cell proliferation (Simpson et al., 2015), potentially diminishing susceptibility to infectious illnesses and perhaps limiting the development and progression of cancers. The value of moderate exercise may be especially notable among older individuals in whom immune functioning ordinarily declines, and once more, moderate exercise daily was sufficient to reduce early mortality risk by as much as a third (Matthews et al., 2020).

Among individuals who had undertaken intense exercise, several negative immune changes were observed, although they were generally short lived. These included disturbed T-cell, NK-cell, and neutrophil functioning, as well as blunted immune responses to specific antigens, and imbalances between pro- and antiinflammatory cytokines (Simpson et al., 2015). With excessive exercise, mitochondrial functioning can also be disturbed so that glucose control is undermined (Flockhart et al., 2021). Although high-performance athletes are generally viewed as being fit, it had been demonstrated that their intense exercise regimens can disturb immune functioning, as reflected by lymphocyte loss. Findings such as these gave rise to the position that immune disturbances wrought by extreme exercise rendered individuals more susceptible to immune-related disorders, particularly viral illnesses (the "Open Window Hypothesis").

This view had been widely accepted for some time, but it encountered vigorous challenges (Campbell and Turner, 2018). Aside from limited data being available supporting this position, it was maintained that immune disturbances that occur are typically transient, making it unlikely for individuals to be infected by opportunistic viruses. Moreover, some of the immune cell variations that had been reported during the brief period following intense exercise likely reflected the redistribution of immune cells to peripheral tissues so that they could take on a more prolific role in hunting infection. Yet, the immune changes promoted by exercise vary over time so there may be periods during which vulnerability to infection is elevated. Following a 2-h intense exercise session, high-level cyclists displayed elevated T cell counts for several hours before falling below baseline, and NK cell activity and neutrophil phagocytic functioning were diminished for up to 8h (Kakanis et al., 2010). To be sure, immune suppression for these durations is not excessive, and it is unclear to what extent they translated to increased illness vulnerability. More than that, these findings might not be particularly relevant to most people who are certainly not as fit as elite athletes and do not engage in high-intensity exercise.

The influence of exercise on immune and inflammatory functioning may depend on contextual and psychosocial factors. Whereas high-intensity exercise in noncompetitive settings was accompanied by diminished immune cell proliferation, immunoenhancement was reported in a competitive situation. Much like the effects reported in other situations, psychosocial factors can have multidimensional influences on biological processes provoked by exercise that can ultimately affect well-being. Accordingly, it may not be productive to consider the effects of exercise in athletes in the absence of other factors that affect immunity and risk of infection, such as altered mood states, sleep quality and sleep disturbances, nutritional influences, and exposure to environmental extremes (Simpson et al., 2020).

Cytokine changes associated with exercise

The influence of exercise on inflammatory processes has been viewed as being fundamental to the effects on tumor progression. The inflammatory effects of exercise vary as a function of the intensity and chronicity of the exercise experienced. In mice that had the opportunity to engage in moderate exercise in the form of freewheel running, the activity of several proinflammatory cytokines was reduced, including IL-6 and TNF-α, particularly in older mice (Packer and Hoffman-Goetz, 2012). Relatedly, among mice that had been maintained on a high-fat diet, exercise inhibited macrophage infiltration into fatty tissue and promoted a macrophage switch from the ostensibly inflammatory M1 to the antiinflammatory M2 phenotype within adipose tissue, so that ultimately inflammation was reduced (Kawanishi et al., 2010).

In contrast to the diminished inflammation associated with moderate levels of exercise, extreme exercise in mice gave rise to elevations of intestinal proinflammatory cytokines (Pervaiz and Hoffman-Goetz, 2012), whereas antiinflammatory cytokines were reduced. Paralleling these effects, a systematic review that comprised the analysis of 18 studies indicated that intense, lengthy periods of exercise in humans were associated with elevated levels of inflammatory markers (Cerqueira et al., 2020). Effects such as these may vary at different times after an exercise session, depending on the intensity of the exercise. As temporal cytokine changes could have implications for susceptibility to pathology, it is not only important to determine the extent of biological changes that are provoked, but also how long it takes for normalization of immune and cytokine functioning to occur and what factors, such as diet and sleep, facilitate recovery processes.

The influence of acute exercise sessions on immune and cytokine functioning is instructive, but to understand the ties to illnesses it is more important to assess such changes in response to sustained exercise regimens. Diminished inflammation, as reflected by lower C-reactive protein, has been observed in young, healthy individuals following a lengthy (e.g., 12–15 months) combined endurance and resistance training regimen, and after a 1-year exercise program, this outcome was similarly observed among obese postmenopausal women (Campbell et al., 2009). Predictably, the antiinflammatory actions were more pronounced when exercise was accompanied by a calorie-restricted diet. This was also apparent among individuals with obesity in that a 15-week lifestyle intervention that comprised a hypocaloric diet and daily exercise reduced low-grade inflammation reflected by diminished macrophage-specific markers and IL-6, IL-8, and TNF-α levels in adipose tissue. As the reduced inflammation associated with exercise was most notable with a decline of BMI and reduced percentage of fat, raises the possibility that these features, rather than the exercise alone, contributed to the positive actions associated with exercise. Speaking to this, exercise-related weight loss coupled with healthy dietary practices was accompanied by altered expression of genes coding for adipokines (cytokines released by fat tissue) together with a decline of inflammatory markers within subcutaneous adipose tissue, which could predict cancer risk (e.g., Campbell et al., 2017). Indeed, the diminished C-reactive protein variations associated with exercise were largely absent when statistically controlling for changes in dietary fiber intake, and it seemed the ties between exercise and reduced C-reactive protein were mediated by fat loss. In essence, aside from the direct effects of exercise, the benefits obtained may be tied to other lifestyles individuals had adopted.

Myokines in relation to cancer

Among the many actions of exercise, one of the most significant is the release of myokines, a type of cytokine produced by skeletal muscles. More than 3000 myokines have been identified, with irisin and IL-6 having received particular attention in relation to well-being. Most often, IL-6 is discussed in terms of its role in inflammatory processes elicited by immune cells in response to tissue injury or infection. However, it also functions as a myokine released by skeletal muscles and is associated with beneficial antiinflammatory actions and metabolic effects (Villar-Fincheira et al., 2021).

Myokines are thought to mediate the beneficial effects of exercise through their capacity to regulate energy homeostasis within several organs as shown in Fig. 10.1. Myokines are

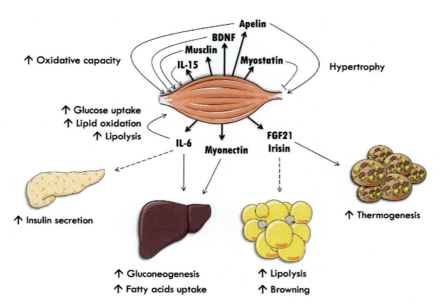

FIG. 10.1 Skeletal muscle as an endocrine organ. Interleukin-15 (IL-15), brain-derived neurotrophic factor (BDNF), and interleukin-6 (IL-6) stimulate lipid oxidation and oxidative metabolism in an autocrine fashion. IL-6 also stimulates intramyocellular triacylglycerol lipolysis and glucose uptake. IL-6 stimulates pancreatic insulin secretion and exercise-induced hepatic neoglucogenesis through an endocrine communication. Myostatin inhibits muscle hypertrophy. Finally, fibroblast growth factor-21 (FGF-21) stimulates white adipose tissue lipolysis and brown adipose tissue thermogenesis. *From Laurens, C., Bergouignan, A., Moro, C., 2020. Exercise-released myokines in the control of energy metabolism. Front. Physiol. 11, 9.*

involved in complex interactions with multiple organs and tissues through actions on cytokines and immune processes, microbiota (and SCFAs), and various neurotrophins (Gubert and Hannan, 2021). By virtue of these effects, exercise may act against the development of chronic diseases, such as type 2 diabetes, heart disease, and neurodegenerative disorders (e.g., Alzheimer's disease) and may also influence the development and progression of cancer. To this end, myokines can operate like hormones that affect visceral fat and promote antiinflammatory actions, and can influence myokine-cancer cell interactions, including within the tumor microenvironment (Kim et al., 2021). Myokines could also have anticancer actions by promoting apoptotic cell death, and by altering migration and viability of cancer cells.

Exercise and cancer prevention

The prophylactic actions of exercise on diverse types of cancer have become increasingly more apparent, typically varying with the nature and intensity of the exercise regimen undertaken. An extensive metaanalysis that included 71 prospective cohort studies revealed that moderate exercise could limit cancer occurrence (Li et al., 2016), and an analysis of pooled data from 12 prospective studies similarly revealed that of 26 cancers evaluated, leisure-time physical activity was associated with a lower frequency of 13 types of cancer, in some cases by as much as 20% (Moore et al., 2016). A subsequent review of prospective cohort studies confirmed these findings, indicating that modest amounts of exercise were accompanied by the diminished occurrence of 7 of 15 types of cancer that had been included in this analysis (Matthews et al., 2020). Supporting the view that extreme exercise was not needed to achieve these outcomes, so long as comparable MET values were achieved (i.e., the ratio of metabolic rate while exercising relative to the resting metabolic rate) the outcomes were the same irrespective of the intensity of the exercise.

While not diminishing the importance of keeping exercise within tolerable limits, vigorous physical exercise was accompanied by reduced mortality related to heart disease and cancer. Physical activity that appreciably exceeded the minimum recommended levels was accompanied by reduced risk of both breast (21%) and colon cancer (28%) (Kyu et al., 2016) and a prospective analysis conducted over 20 years revealed that brisk walking was associated with reduced breast cancer occurrence among postmenopausal women. Likewise, among 43,479 cancer-free men assessed over 26 years, higher physical activity was accompanied by the reduced occurrence of cancer of the digestive tract, being most apparent with aerobic exercise. It has been estimated that 46,000 yearly cancer cases could be prevented if individuals exercised just 5 h each week. These outcomes varied across regions of the United States and were more frequent in marginalized groups that had less access to safe exercise facilities.

In evaluating the prophylactic actions of exercise on illnesses in general, and cancer specifically, it should be considered that those who religiously engage in exercise might also adopt other positive lifestyles (eating properly, obtaining sufficient sleep) and avoid high-risk behaviors that can affect cancer occurrence. Fig. 10.2 provides a brief description of some of the immune, inflammatory, and endocrine changes that are produced by acute exercise and those that occur among well-trained individuals who exercise on a routine basis (Hojman et al., 2018).

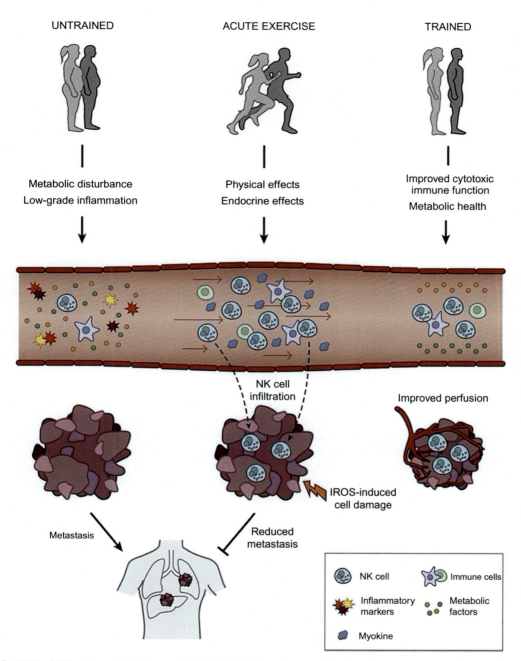

FIG. 10.2 Molecular and cellular mechanisms linking exercise to cancer protection. Exercise consisting of acute sessions (middle section) leads to physical changes (increased blood flow, shear stress on the vascular bed, temperature increases, sympathetic activation) and endocrine responses (release of catecholamines and exercise hormones, myokine secretion) that results in increased tumor perfusion, oxygen delivery, intratumoral metabolic stress, cellular damage, and ROS production. Although these actions can promote tumor growth, they can paradoxically stimulate signaling pathways that prevent metastasis. Chronic training (far right section) can promote adaptations comprising systemic alterations with improved immune function, reduced systemic inflammation, and improved metabolic health, as well as intratumoral changes in the form of enhanced blood perfusion, immunogenic profile, and immune cell infiltration. *From Hojman, P., Gehl, J., Christensen, J.F., Pedersen, B.K., 2018. Molecular mechanisms linking exercise to cancer prevention and treatment. Cell Metab. 27, 10–21.*

Impact of exercise on existent cancers

Safety of exercise among cancer patients

Aerobic and resistance exercise undertaken during and after cancer treatment was safe even among individuals with advanced cancer, often being accompanied by improved quality of life and enhanced physical and social functioning. Likewise, exercise regimens typically did not disturb chemotherapy completion rates, and among early-stage breast cancer patients exercise reportedly enhanced completion rates. Supervised exercise (e.g., coaching by a physiotherapist) comprising moderate to vigorous exercise enhanced clinical and functional outcomes as well as survival, irrespective of the type of cancer examined. The improvements of functioning associated with supervised exercise not only comprised enhanced physical fitness, but also augmented social, mental, and cognitive functioning. Of course, the feasibility of most exercise programs varies with the nature and stage of cancer, characteristics of the individual, and the presence of comorbid conditions. Thus, exercise protocols need to be tailored to the specific features of each individual and modification of exercise regimens may be necessary over the course of cancer therapy.

Variations of cancer progression produced by exercise

Exercise can potentially influence every aspect of the cancer process; it not only prevents its occurrence but also enhances the response to therapies and improves well-being after therapy had been completed. Exercise could also limit some of the negative sequelae of cancer and cancer therapy, including fatigue, sleep disturbances, cognitive impairments, and could act against the development of cardiovascular problems related to the therapy.

An analysis of 26 studies indicated that while sedentary behaviors were associated with increased mortality among cancer patients, moderate levels of activity were associated with lower mortality related to breast, prostate, and colorectal cancer, in some instances by almost 40% (e.g., Friedenreich et al., 2016). Paralleling the exercise-provoked immune changes, as little as 3–5 h of walking a week was sufficient to diminish mortality risk among women with breast cancer, and such outcomes were noted after controlling for a wide array of psychosocial and physical variables. The positive effects of exercise in men with colorectal cancer were likewise observed across ages, body mass index, and prediagnostic physical activity.

The benefits of exercise have been observed in numerous studies that involved different research approaches. A systematic review that included 12 prospective studies that followed individuals for 4.3–12.7 years revealed that breast cancer progression as well as all-cause and breast cancer-related mortality was lower among women who had engaged in regular exercise before or after diagnosis than in women who had not done so (Lahart et al., 2018). Not unexpectedly, the impact of exercise varied with the features of the individual (age, sex, and their fitness), and the effects of exercise on breast cancer patients differed as a function of the hormone responsiveness of the tumor, with greater effects attained if the tumor was relatively small. This doesn't imply that exercises undertaken during later stages of cancer development are without any benefit. Patients with metastatic colorectal cancer experienced a slowing of cancer progression and diminished adverse side effects with as little as 4 h of exercise a week. Exercise can even have positive effects among glioblastoma patients, possibly through effects on myokines (Huang et al., 2020), although the short survival time associated with this form of brain cancer, together with the side effects of the treatments, might make a sustained exercise program untenable.

Intensity of exercise in relation to cancer

Commensurate with the effects on immune functioning, it was maintained that intense exercise was not needed to produce benefits on cancer survival and that virtually any amount of aerobic exercise was accompanied by a 30% reduction of cardiovascular problems secondary to cancer and diminished all-cause and cancer mortality. Understandably, compliance with a high-intensity exercise regimen among cancer patients can be poor, particularly among individuals who were markedly overweight, and in patients who experienced high levels of physical fatigue. Sustained exercise regimens, in contrast, were maintained when sessions were not too intensive, possibly because individuals may have felt a sense of agency, as well as being able to engage in activities without being distressed, which might have led to the activities being perceived as rewarding. Moreover, randomized controlled trials indicated that both moderate and intense exercise regimens during adjuvant therapy among women with breast cancer diminished psychological disturbances and fatigue while enhancing physical strength. In some individuals, several difficulties stemming from chemotherapy can be lessened by high-intensity aerobic interval training, such as improving muscle strength and attenuating the decline in cardiorespiratory fitness, so patients could return to work sooner than patients who received usual care. Once more, combined efforts could yield better outcomes than single intervention modes, and supervised exercise to avoid patients "overdoing it" was generally superior to unsupervised exercise in promoting physical functioning and quality of life.

As surgery is frequently used in dealing with cancer and may itself have adverse consequences, it is significant that exercise can influence the response to this procedure. Randomized controlled trials generally indicated that several forms of preoperative exercise among non-small cell lung cancer patients were associated with enhanced walking endurance and peak exercise capacity, while concurrently reducing postoperative pulmonary problems and risk of hospitalization (Rosero et al., 2019). Furthermore, treatment complications were reduced by 50%, and hospitalization among lung cancer patients was less frequent. This said, it is somewhat surprising that a better handle hasn't been obtained concerning the many moderators that influence the impact of exercise interventions. Despite several trials being undertaken to evaluate the effects of perioperative and postoperative exercise among patients with non-small cell lung cancer, a consensus has not been reached concerning optimal strategies to limit the side effects of treatments. In line with this conclusion, a systematic review indicated that the influence of exercise among individuals facing dual hits (neoadjuvant treatment and surgery) was unclear given the limited number of controlled trials that had been conducted (Loughney et al., 2016).

Mechanisms related to exercise benefits

The development of general health enhancements or moderation of host- and tumor-related factors are presumed to underlie the beneficial effects attributable to exercise, some of which are provided in Fig. 10.3. These actions include enhanced immune functioning and diminished inflammation, together with altered metabolic processes, beiging of white fat, and elevation of antioxidant activity. Exercise similarly influenced features of the tumor microenvironment in that vascularity and tumor perfusion were altered, hypoxia was induced, and immune responsivity enhanced (Buss and Dachs, 2020).

In women with breast cancer who attained 80% of their exercise intervention goals, including weight loss, there were diminished C-reactive protein and IL-6 levels, perhaps pointing to the importance of combined exercise and weight loss in promoting antiinflammatory benefits. In preclinical models in which other factors could be controlled, exercise produced an influx of immune cells into tumors, and a corresponding reduction of tumor growth (e.g., Idorn and Thor Straten, 2017). It seems that the positive outcomes of exercise may be due to the creation of an environment that is generally more hospitable to immune cells as well as altered oxygenation. In the latter regard, moderate exercise could occur by reducing oxidative stress, whereas exhaustive exercise could potentially have the opposite effect on malignant cells (Arena et al., 2019).

Many other factors may figure into the favorable effects of exercise, including epigenetic actions and DNA repair processes, variations of insulin-like growth factor and vasoactive intestinal peptide, heat shock proteins, and energy metabolism and insulin resistance (Thomas et al., 2017). For instance, exercise can act against triple-negative breast cancer through actions on mammalian target of rapamycin (mTOR) signaling, which is a primary nutrient-sensitive regulator that is fundamental in metabolic and aging processes (Agostini et al., 2018).

In addition to moderating the actions of cancer therapy, exercise undertaken following treatments may have multiple positive actions. Elevated levels of inflammatory biomarkers predicted lower survival among breast cancer survivors. It is therefore pertinent that a meta-analysis of 26 studies confirmed that aerobic and resistance exercise among breast and prostate cancer survivors was associated with diminished markers of inflammation (Khosravi et al., 2019).

Is there a best time to exercise?

All cells in the body (including cancer cells) have circadian cycles during which they are more active, which will be considered in greater detail in Chapter 11. Owing to these circadian variations as well as the time-dependent changes of immune and hormonal functioning, the effectiveness of cancer treatments could vary with the time of day that these are delivered. Metabolic changes related to exercise similarly varied with time of day, which was related to the activation of pathways associated with fatty acid oxidation and glycolysis, which contribute to sugar metabolism and energy production. Based on muscle glycolysis and lipid oxidation (fat burning) functioning, the optimal time for exercise corresponds to late morning in humans. In line with these reports, morning exercise appeared to be best to prevent the occurrence of breast and prostate cancer.

But the effects of exercise could vary depending on whether the individual is "a morning or a night person" as well as whether the person is a night shift worker. Additionally, it could also vary with other lifestyle factors (e.g., eating, sleep). In the final analysis, it would be prudent to obtain information about individual rhythms in order to maximize the benefits of exercise therapies.

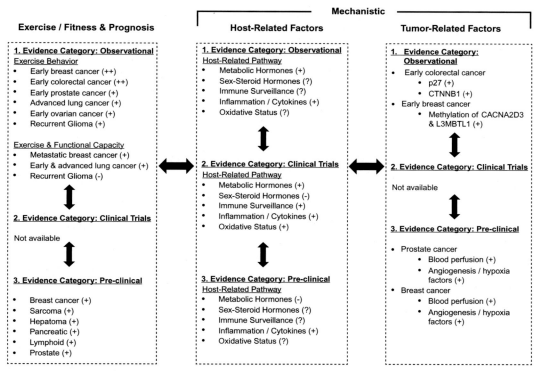

FIG. 10.3 Evidence-based representation of the known effects and mechanisms of exercise on tumor progression adopting a bidirectional translational research or scientific discovery paradigm. **Exercise/fitness and prognosis**: evidence supporting an association between self-reported exercise behavior, objective measures of exercise capacity or functional capacity, and cancer prognosis; **host-related factors**: postulated systemic (host-related) pathways mediating the association between exercise behavior and exercise/functional capacity and cancer prognosis; **tumor-related factors**: intratumoral factors shown to mediate the association between exercise and prognosis or factors shown to be modulated in response to exercise. +++, strong evidence; ++, moderate evidence; +, weak evidence; —, null; ?, unknown at present. *From Betof, A.S., Dewhirst, M.W., Jones, L.W., 2013. Effects and potential mechanisms of exercise training on cancer progression: a translational perspective. Brain Behav. Immun. 30 Suppl, S75–S87.*

Influence of epigenetic changes stemming from exercise

Exercise can engender epigenetic variations related to NK cell functioning as well as inflammatory and metabolic processes and could thus contribute to cancer occurrence and the response to cancer therapies. In this regard, acute aerobic exercise was accompanied by decreased methylation of 19 genes and increased methylation of 14 genes involved in NK cell regulation, although the functional relevance of all these genes was uncertain (Schenk et al., 2019). Nonetheless, it may be of interest that in a small sample of healthy men, a resistance exercise training program was associated with diverse genome-wide epigenetic marks, some of which might be transmitted across generations and could thus contribute to inherited health benefits (Denham, 2018).

The positive effects of exercise on cancer survival aren't apparent in everyone and can be moderated by numerous other factors. A prospective analysis conducted over 15 years indicated that among women who engaged in recreational physical activity those who survived longest following breast cancer treatment were more likely to express epigenetic changes on promoters of several genes relevant to cancer. Evidently, exercise may be associated with beneficial effects, but this occurred primarily if the exercise regime was linked to particular epigenetic alterations. It is uncertain, however, what led to these actions occurring only in a subset of women but might be tied to changes of inflammatory factors or signaling pathways related to tumor suppression.

The impact of exercise on processes that could affect cancer occurs, to some extent, even after a single exercise session. Brief exercise was accompanied by altered miRNA expression patterns within the population of circulating NK cells that have been tied to cancer processes. This was accompanied by changes in the p53 signaling pathway, which is involved in cell division, apoptosis, and genomic stability, and can stimulate DNA repair processes (Radom-Aizik et al., 2013). Further to this, among non-Hodgkin lymphoma patients, a single exercise session was complemented by elevated IL-6 levels and CD8+ T-lymphocyte histone acetylation, which were taken to reflect epigenetic modifications. It was similarly suggested that a single exercise session might increase hematopoietic stem cells that could, in theory, enhance the effectiveness of cell transplantation among patients with a blood cancer that leaves them deficient in circulating cell numbers. Of course, this by no means implies that a single exercise session will have positive consequences but should instead be seen as exercise sessions potentially having cumulative benefits so "every session might count" (Dethlefsen et al., 2017).

The influence of exercise on cancer processes may be enhanced by changes in food intake, and it seems that the combination of exercise and a hypocaloric diet provoked changes in the ASC gene, which encodes a protein that is related to the NLRP3 inflammasome (Barrón-Cabrera et al., 2020). Given that exercise may be accompanied by dietary changes, it is difficult to discern the relative contributions of diet and exercise (and weight loss) in producing beneficial actions related to cancer. Further to this, exercise programs were accompanied by better adherence to a diet and diminished overeating, perhaps owing to enhanced appetite regulation. Thus, while epigenetic changes associated with exercise are correlated with reduced cancer occurrence and progression, the actions of exercise are confounded with several other factors that can act in this capacity.

Impact of exercise on cancer features and side effects of treatments

Exercise may be an important component of well-being during aftercare and can act against multiple adverse effects that are otherwise attributable to the cancer or its treatment. This comprises diminished fatigue, cardiotoxicity, neurotoxicity, loss of muscle mass and bone density, and improved cardiorespiratory fitness. The available evidence has also supported the contention that exercise could have positive effects on cancer-related emotional and social functioning, symptoms of anxiety, and quality of life. There were indications that children with cancer gained from exercise, being reflected by improved cardiorespiratory fitness, muscle strength, and health-related quality of life. However, the data were generally not sufficiently extensive to form a definitive conclusion regarding the therapeutic benefits of exercise.

The negative effects provoked by cancer that can be diminished by exercise in adults are fairly broad, but not all the side effects of cancer treatments can be abrogated. Fatigue, insomnia, and dyspnea were generally diminished, whereas other features, such as nausea/vomiting, loss of appetite, constipation, and diarrhea were not affected. Pain following radiation therapy in breast cancer patients was tempered by exercise, possibly being related to cytokine variations. Several months of aerobic and resistance training (or their combination) following chemotherapy or radiation therapy to treat breast cancer also produced small-to-moderate beneficial effects on perceived physical, emotional, and social functioning, cardiorespiratory fitness, and quality of life. Such effects were also seen when exercise was adopted sometime after therapy was completed. Among women who had been treated for early-stage breast cancer 3–18 months earlier, a combined exercise and individualized hypocaloric eating program was associated with diminished depressive symptoms, normalized morning cortisol levels, and elevated leukocyte, neutrophil, and lymphocyte counts.

There are limits to the value of exercise undertaken during therapy. While light and moderate exercise during chemotherapy was associated with diminished toxicity related to the treatment and improved physical functioning, it was neither accompanied by reduced myelosuppression (reduced bone marrow functioning and hence diminished red and white blood cells, as well as platelets) nor was an augmented response to treatment or survival necessarily realized. In several instances, as in acute myeloid leukemia, exercise was a feasible option to enhance general well-being, but mixed results were apparent in preventing more prominent side effects of therapy. Further, cancer and cancer therapies may promote a constellation of features that are not equally amenable to the benefits that could be introduced by exercise regimens. Having highlighted the difficulty of applying exercise therapy during treatment, we will now turn to the impact of exercise on some of the more common behavioral features associated with cancer.

Cancer-related fatigue and quality of life

Debilitating fatigue is a frequent feature of the cancer experience that may come about because of several biological changes. These include, but are not limited to, alterations of peripheral adenosine triphosphate and muscle contractile functioning, cytokine dysregulation, HPA axis disturbances, variations of serotonin, disruptions of circadian rhythm, and sleep disturbances. Moreover, inflammation associated with cancer may trigger a reduction in the availability of cellular energy owing to a switch of metabolic functioning from energy-efficient oxidative phosphorylation to rapidly acting aerobic glycolytic energy production, resulting in diminished glucose availability and correspondingly diminished cellular energy.

Irrespective of how it comes about, it was concluded based on systematic reviews and metaanalyses that moderate exercise diminishes fatigue related to cancer and its treatment. A systematic review that included 32 studies of patients receiving adjuvant therapy for breast cancer indicated that both aerobic and resistance exercise enhanced fitness and reduced fatigue but did not uniformly improve cancer-related quality of life (Furmaniak et al., 2016). When quality of life was enhanced in association with exercise among cancer survivors, this relationship was mediated by improved cardiorespiratory functioning and the related diminution of fatigue.

Cancer-related fatigue was especially pronounced among obese women with breast cancer. While numerous factors may account for the fatigue, the elevated presence of proinflammatory cytokines can play a prominent role in this outcome. Thus, it is pertinent that combined aerobic and resistance exercise in a 16-month program in breast cancer survivors with obesity promoted an appreciable decline of inflammatory markers, coupled with a reduction of proinflammatory macrophages in adipose tissue (Dieli-Conwright et al., 2018). However, contrary to the idea that IL-1β and IL-6 were responsible for the fatigue associated with cancer, the reduction of fatigue stemming from exercise within several animal models was independent of variations of IL-1. Likewise, although variations of IL-1 and IL-6 in humans were moderately linked with fatigue in cancer patients, the cytokine changes that occurred with supervised exercise were not uniformly aligned with altered fatigue.

Even though the data have been extensive and impressive, some caveats ought to be considered concerning the influence of exercise on fatigue related to cancer and its treatment. Reiterating earlier comments, the influence of exercise may vary with the form of cancer being considered as well as the stage at which exercise was undertaken. Furthermore, the possibility was raised that subtle biases might have occurred within some studies that could have affected the outcomes. For instance, patients in the exercise conditions may have received more attention than control participants, and simply being enrolled in the exercise group could have affected outcomes because patients adjusted their behaviors when they were being assessed (Hawthorne effect). Finally, not all patients offered the opportunity to engage in the exercise studies chose to do so. Did systematic differences exist between patients who opted to be included in the studies and those who didn't, and likewise were unique characteristics present in those patients who dropped out midway through the exercise regimen?

These limitations regarding the impact of exercise are all certainly important, but overall, it did appear that exercise had positive effects concerning fatigue and mood states, and the findings have clear clinical implications. Ultimately, to diminish fatigue and other side effects, a key focus might need to be that of enhancing the patients' propensity to engage in exercise programs while concurrently focusing on mood and self-efficacy. To this end, it might be advantageous to broaden the nature of the exercise programs available to patients. Among patients with various cancers who were receiving adjuvant chemotherapy, multiple positive outcomes were observed through multimodal exercise programs that included cardiovascular and resistance training, relaxation, and body awareness training, as well as massage therapy. Other strategies can be included in a multimodal exercise program, such as psychotherapy that might be effective in fostering improved self-efficacy and quality of life. Further to this, it might be useful to adopt physical activity programs that are more interesting than those most often adopted. Most studies examining side effects of cancer therapy have focused on aerobic and resistance exercise, but improved health-related quality of life, reduced fatigue, and diminished sleep problems could also be obtained by engaging in yoga.

Ideally, the benefits of exercise would be lasting, but frequently the positive actions were somewhat limited. Enhancements of several aspects of quality of life were apparent 12 weeks and 6 months following an exercise regimen but declined thereafter. Other components of quality of life, such as emotional well-being, social functioning, perceived pain, and sleep improvements were also short lived, no longer being apparent after 6 months. These findings not only point to the efficacy of diverse forms of exercise but also indicate that analyses of the impact of these programs ought to evaluate the long-term benefits obtained as well as the need to maintain exercise regimens, preferably having these incorporated into individual lifestyles.

Cancer cachexia

Cachexia is among the most common and debilitating effects of cancer, but as we described in Chapter 8, treatments to diminish this condition have largely been ineffective. Although moderate exercise in rodents diminished muscle wasting and mitochondrial alterations provoked by chemotherapy, this finding was not the rule. A review of animal studies indicated that exercise generally did not diminish features of cachexia, although an exercise regimen undertaken before a tumor being induced could diminish later cachexia severity (Niels et al., 2020). It had been maintained that exercise could attenuate cachexia in humans, but the evidence supporting this was weak. However, when aerobic exercise was maintained on a sustained basis for weeks or months, the production and release of free fatty acids and inflammatory cytokines by white adipose tissue were moderately diminished, which could reduce features of cancer cachexia. Moreover, combination treatments that included exercise and diet during therapy in conjunction with other therapeutic agents (e.g., progesterone, ghrelin, omega-3-fatty acids), produced superior effects in patients with cachexia (Aoyagi et al., 2015).

Cachexia may be brought about by numerous inflammatory cytokines through their actions on mitochondrial functioning, muscle mass and strength, as well as adipose tissues. Particular attention has focused on the involvement of IL-6 in the provocation of cachexia given that this cytokine serves to regulate white adipose tissue lipolysis during early-stage cachexia as well as the browning of fat in late-stage cachexia. It will be recalled that IL-6 is among the myokines released from muscles during exercise, in this case serving as an antiinflammatory agent, and could thereby affect cachexia (Daou, 2020). Thus, while targeting IL-6 diminished a few aspects of cachexia, these actions could be augmented by exercise. Taken together, it seems exercise, diet, and several other variables can have modest effects on cachexia but this is a difficult beast to tackle so that a multimodal approach probably offers the greatest likelihood of achieving optimal actions.

Sleep disturbances

Sleep problems are common among cancer patients, often persisting for years after therapy had been completed. These disturbances may comprise alterations of sleep patterns and sleep-wake cycles arising because of anxiety and depression, pain, and side effects of treatment, such as nausea and vomiting, frequent urination, night sweats, and general malaise. As sleep is necessary for the effective operation of biological systems and is essential for the restoration of biological resources utilized through the day, disturbed sleep can have marked health repercussions (see Chapter 11).

Improved sleep and cardiorespiratory fitness were observed among breast cancer survivors who engaged in a sustained intervention comprising physical activity and behavior change therapies. In other studies, exercise interventions only modestly alleviated sleep disturbances, but sleep quality (e.g., sleeping with few awakenings and feeling refreshed following sleep) was not reliably improved. The inconsistent findings that have been reported regarding the effects of exercise undertaken during cancer therapy may be related to the form and intensity of exercise undertaken. Whereas modest exercise regimens among breast cancer patients, even if supervised, had little effect, a more intense (50–60 min three times a week) combined regimen of aerobic and resistance exercise had lasting effects (6–12 months) on health-related fitness and sleep quality (An et al., 2020). Aside from the effects of exercise undertaken before and during cancer therapy, a supervised multimodal exercise regimen following completion of chemotherapy enhanced sleep quality in breast cancer patients. However, whether the long-term well-being of cancer survivors, including stress reduction, was attributable to improved sleep, has not been adequately addressed. In fact, when the effects of exercise on sleep disturbances were observed, this was most often accompanied by improvements in other aspects of well-being. Thus, even if exercise improved sleep quantity and exercise, this may have been secondary to diminished depression rather than the direct actions of exercise on biological functioning.

Neuropathy

Several drugs used to treat cancer can damage peripheral nerves, causing the emergence of neuropathy characterized by pain or discomfort, altered peripheral sensations (e.g., diminished ability to feel), and impaired gross motor movements, such as walking and writing, as well as fine motor control (e.g., picking up a coin). Regulated exercise that entailed a 6-week program of escalating walking and resistance training reduced self-reported peripheral neuropathy. Measures of balance and mobility were likewise improved after an 8-week exercise program (three times a week), although neither sensory nor motor neurophysiological changes were noted. A small metaanalysis of five available reports suggested that exercise that focused on muscle strengthening and balance could potentially be useful for peripheral neuropathy produced by chemotherapy (Duregon et al., 2018). Even in stage IV colorectal cancer patients who had undergone extensive chemotherapy, an exercise regimen comprising endurance, exercise, and balance training prevented the worsening of neuropathic signs that were apparent in a nonexercise group (Zimmer et al., 2018). As there are no treatment strategies that are widely accepted as being useful for peripheral neuropathy, even limiting the worsening of this condition should be seen as meaningful.

Depression

As stated earlier, depression not only occurs because of the distress associated with a cancer diagnosis and the treatments patients have to undergo, but in cancer-free individuals the occurrence of depression may also predict the later development of cancer. Exercise may diminish depressive symptoms and could act prophylactically in limiting mood disorders. It has been estimated that among noncancer patients the occurrence of clinical depression could be reduced by 12% through regular exercise comprising as little as 1 or 2 h a week. These positive outcomes could occur through diverse actions of exercise, some of which have been associated with cancer occurrence and progression. In fact, exercise influences circulating cytokines, and as described earlier, inflammatory factors may contribute to the comorbidity that exists between depression and cancer. Exercise was also reported to increase brain neurotrophins, such as BDNF and FGF, thereby providing a hypothesized mechanism for reducing depression. Similarly, NGF and IGF-1 elevations may mediate antidepressant effects in breast cancer survivors (Meneses-Echávez et al., 2016). Exercise in rodents can affect brain neuroplasticity and influence circulating growth factors (e.g., VEGF), and there have been reports of such effects among older people who are especially vulnerable to both depression and cancer. The position was also advanced that exercise could engender antidepressant-like effects by increasing the activity of erythropoietin (EPO), a cytokine that stimulates the production of oxygen-carrying red blood cells.[a] In rodents, EPO promotes antiinflammatory processes, increases neuronal functioning, and influences the neurotrophin BDNF, which then elicits antidepressant actions. As it happens, EPO may attenuate cancer-related fatigue, and a combination of mild exercise plus EPO can act against cancer muscle disturbances associated with cachexia. Unfortunately, EPO might contribute to angiogenesis in lung cancer models and consequently might increase the risk of tumor growth, invasion, and metastasis (Liu et al., 2020a).

As several growth factors have been linked to depression and the progression of multiple cancers, they may also account for the effects of exercise on both illnesses. Although, this doesn't necessarily imply that they are causally related, the mere fact that exercise is able to diminish inflammation and depression speaks to their importance in relation to cancer and leaves open the possibility that some of the positive effects of exercise on cancer occurrence and progression come from its actions in attenuating depression. Furthermore, depression has been associated with some features of cancer, including cachexia, neuropathy, and sleep disturbances that were discussed in the preceding sections, which raises the possibility that at least some of these may be interrelated.

Fueling cancer progression

As elevated glucose uptake and the release of lactate is apparent in most forms of cancer (Warburg effect), exercise might have some of its helpful action by influencing energy-related processes. In essence, the glycolytic pathway and the formation of lactic acid are adopted for energy production [through adenosine triphosphate (ATP) production] by cancer cells. Exercise could affect tumor growth by influencing glycolytic metabolism. Despite numerous studies pointing to light and moderate exercise being effective in diminishing cancer progression, relatively intense exercise might be more effective in affecting glycolytic processes. As well, a combination of exercise and diet that influences energy-related processes may produce effects superior to that of exercise alone (Koelwyn et al., 2017). But several other features regarding exercise regimens, as described in Fig. 10.4, are relevant to obtaining optimal outcomes and precluding the emergence of adverse effects. In developing exercise-based strategies to deal with cancer, several important considerations ought to be addressed that take into account: (a) individual differences in the biological actions of exercise, (b) the nature (specificity) of the exercise based on its physiological effects, (c)

---

[a] The 2019 Nobel laureates, William G. Kaelin Jr., Sir Peter J. Ratcliffe, and Gregg L. Semenza, played principal roles in showing how EPO regulated oxygen levels in virtually all cells within the human body. The finding that hypoxia-inducible factor (HIF) was key in this process as was the VHL gene, led to discoveries concerning their importance in how oxygen sensing is fundamental in various diseases, including the survival and proliferation of cancer cells.

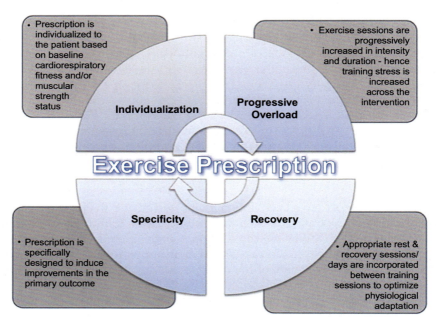

FIG. 10.4 The principles of exercise training for cancer patients. *From Sasso, J.P., Eves, N.D., Christensen, J.F., Koelwyn, G.J., Scott, J., et al., 2015. A framework for prescription in exercise-oncology research. J. Cachexia Sarcopenia Muscle 6, 115–124.*

that the load on physiological systems can be increased gradually (progressive load) so that physiological adaptation develops without the risk of damaging overload, and (d) that a rest and recovery period are necessary so that biological systems have the opportunity to be replenished (Sasso et al., 2015).

Given the "sweet-tooth" characteristic of cancer cells, it was surmised that some forms of cancer may be influenced by altered levels of insulin-like growth factor-1 (IGF-1) and elevated insulin-like growth factor-binding protein-3 (IGFBP-3), which could be influenced by exercise. Although IGF functioning was tied to the positive effects of exercise, a consistent pattern was not uniformly detected in all cancer survivors. Again, this is not unexpected as multiple moderating variables can determine the influence of exercise on insulin-related processes and links to cancer. Moreover, the benefits of exercise might not stem from the same mechanisms across all individuals. For instance, exercise could affect other factors that have been associated with several forms of cancer, such as high-density lipoprotein and total cholesterol, and the presence of leptin. Thus, individual changes of IGF might be more relevant for a subset of people, whereas different factors may be more relevant for others.

Among breast cancer survivors, exercise programs in which participants had lost weight were tied to changes in fasting levels of insulin and inflammatory factors (Kang et al., 2017); and aerobic exercise coupled with a hypocaloric diet provoked greater effects on cancer-related biomarkers than did exercise alone. Since the treatment combination also elicited greater weight loss, it was considered that this, rather than exercise alone, was fundamental for biomarker changes. The effects of exercise on immune functioning also differed between men and women, which might be tied to alterations of circulating sex hormones. For that matter, among postmenopausal women, a reduction in total body fat owing to exercise or diet was accompanied by variations of estradiol, free testosterone, sex hormone-binding globulin, and enhanced leptin regulation, all of which have been linked to breast cancer.

Exercise and microbiota

Being only a relatively recent focus of attention, the data concerning the action of exercise on microbial processes are understandably less extensive than those related to diet. Nevertheless, studies in rodents and a modest number of reports in humans revealed that exercise elicits multiple microbial changes that lend themselves to wellness and act against mental health and cognitive disturbances, metabolic syndrome, obesity, several gut-related diseases (e.g., inflammatory bowel disease), and colorectal cancer (Mailing et al., 2019). Physical exercise in rodents enhanced gut integrity and gave rise to an increase in the abundance of beneficial bacteria that could influence immune functioning. Exercise might also be effective in protecting the intestinal barrier, which limits the occurrence of nonalcoholic fatty liver disease that might otherwise be linked to hepatocellular carcinoma and colorectal cancer. Conversely, as in the case of immune changes and inflammation, microbial processes that have been linked to cancer were aligned with sedentary behaviors, poor diet, disturbed sleep processes, and stressful experiences. Fig. 10.5 describes several microbial pathways by which obesity, physical activity, and diet can affect inflammation and hence cancer development (Song et al., 2020).

Microbiota changes and amount of exercise

Congruent with the other effects associated with varying amounts of exercise, an inverted U-shaped relationship was reported between the amount and intensity of exercise and gut microbiota alterations. Not unexpectedly, the influence of exercise interacted with diet and as a function of being overweight, although inconsistent findings have been reported. Exercise in rodents limited weight gain due to a high-fat diet and produced a shift of microbial species otherwise observed in association with this diet (Evans et al., 2014). It also appeared that exercise in obese rats was superior to calorie restriction in sedentary rats (that led to weight loss comparable to that of the exercised rats) in modifying microbiota and in reducing inflammation, improving insulin resistance, fat oxidation, and altered brown adipose tissue (Welly et al., 2016). Of course, attempting to reduce weight and diminish diseases related to obesity isn't an either/or choice as both regimens can obviously be adopted concurrently.

In humans, the amount and intensity of exercise associated with microbial changes may vary with different exercise routines. Moderate exercise (3h a week) among women was associated with elevated levels of several butyrate-producing microbes. A 12-week moderate exercise regimen (brisk walking) among older participants similarly increased intestinal *Bacteroides*, whereas a more limited (5-week) endurance exercise program had limited effects on microbiota diversity. The impact of endurance exercise may also vary as a function of whether participants were lean or overweight. Prior to undertaking an exercise regimen that

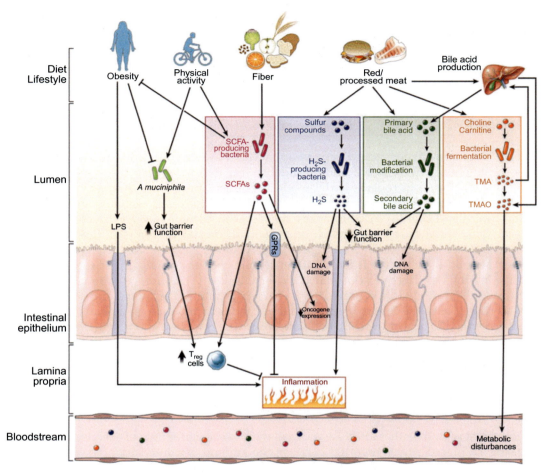

FIG. 10.5 Pathways by which dietary and environmental factors affect the intestinal microbiome and their roles in colorectal carcinogenesis. Obesity may promote CRC through LPS-mediated systemic inflammation and depletion of *Akkermansia muciniphila* and SCFA-producing bacteria, whereas physical activity might protect against CRC by increasing the abundance of *A. muciniphila* and SCFA-producing bacteria. The benefit of dietary fiber might be mediated by an enrichment of SCFA-producing bacteria and increased production of SCFAs that inhibit CRC development and modulation of the immune and metabolic response. Red and processed meat may increase CRC risk by increased bacterial production of secondary bile acids, H_2S, and TMAO. T_{reg}, T regulatory. *From Song, M., Chan, A.T, Sun, J., 2020. Influence of the gut microbiome, diet, and environment on risk of colorectal cancer. Gastroenterology 158, 322–340.*

became progressively more intense over 6 weeks, lean and obese individuals differed in their microbiota and SCFA composition, but following the endurance exercise, the changes from baseline were more prominent among lean individuals and then returned to their earlier state when they again adopted their sedentary lifestyles (Allen et al., 2018).

Consistent with the advantages provided by exercise, the diversity of microbial species among athletes was appreciably greater than in a healthy control group and such differences were accompanied by elevated SCFAs, amino acid synthesis, and carbohydrate metabolism (Barton et al., 2018). The greater gut microbe diversity among athletes relative to controls was also accompanied by differences in metabolic and inflammatory markers. The effects were further enhanced by specific diets. For example, among elite race walkers engaged in an intensified training program, the inclusion of a ketogenic diet increased the abundance of specific gut and oral microbiota, including SCFA-producing bacteria.

In evaluating the effects of exercise and diet on physiological processes and health, the outcomes observed using fecal transplants suggested that the benefits of exercise could be attributable to metabolic and endocrine changes that might have been induced. Fecal microbiota transplantation from mice that had been exercised or those kept on a high-fat diet, influenced metabolic profiles in recipient mice (Lai et al., 2018). The actions of exercise on microbiota were also gleaned from the analysis of the effects of microbiota transfer from prediabetic men, who had exercised, to naïve mice. Exercise in some of these men (dubbed "exercise responders") promoted gut microbiota alterations that were aligned with enhanced glucose regulation and insulin sensitivity. Of particular interest was that fecal transfers from exercise responders to obese mice enhanced their insulin sensitivity, which did not occur if the transplants came from men who were nonresponders (Liu et al., 2020b).

The influence of exercise on tumorigenesis likely involves a constellation of microbial changes. The exercise-provoked changes of SCFAs and bile acids could promote a range of immune and inflammatory variations together with the functioning of NK cells. Several of the presumed factors involved in the series of changes provoked by exercise and their action on

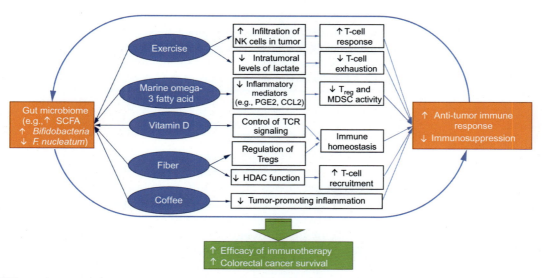

FIG. 10.6 Potential immune and microbial mechanisms underlying the benefit of exercise and nutritional factors for colorectal cancer survival. In addition to the local effects, these lifestyle factors also reduce systemic inflammation induced by cancer and treatment. *CCL2*, C-C motif chemokine ligand 2; *HDAC*, histone deacetylase; *MDSC*, myeloid-derived suppressor cells; *NK*, natural killer; *PGE2*, prostaglandin E_2; *SCFA*, short-chain fatty acid; *TCR*, T-cell receptor; *Treg*, T regulatory cells. *From Song, M., Chan, A.T., 2018. The potential role of exercise and nutrition in harnessing the immune system to improve colorectal cancer survival. Gastroenterology 155, 596–600.*

cancer development and progression are described in Fig. 10.6. These interactions have been extensively assessed in association with colorectal cancer (e.g., Song and Chan, 2018), but some of these processes likely contribute to cancer of other organs as well as other features of the disease, such as increased tumor cell apoptosis, reduced angiogenesis, augmented insulin signaling, and sensitivity. What Fig. 10.6 also makes clear is that exercise and diet may affect cancer occurrence through independent processes and thus a combination of lifestyles ought to lead to better outcomes than focusing on a single approach.

As we saw with stressful experiences, exercise that was initiated during early life promoted especially enhanced microbial changes (Mika and Fleshner, 2016), and exercise was more effective in altering microbiota in juveniles than in adult rats. Moreover, the mom's exercise experiences while she was pregnant affected the fetus and could have long-term actions on the offspring. Wheel running exercise in rats before and during pregnancy altered gut microbial diversity and the abundance of numerous microbial taxa in her offspring, but this was far less apparent if mom had been maintained on an obesogenic diet (Bhagavata Srinivasan et al., 2018). Evidently, while exercise in pregnant rats could enhance microbiota in offspring, these effects were moderated by the mother's physical condition.

Microbiota in cancer cachexia

As microbiota have been tied to diets, digestive processes, and cancer occurrence and progression, it would be reasonable to hypothesize that microorganisms might also be related to cancer cachexia. In rodents, cancer cachexia was accompanied by reduced levels of Lactobacillales, whereas levels of Enterobacteriaceae and Parabacteroides were elevated (Herremans et al., 2019). Moreover, the presence (or the absence) of certain microorganisms was associated with increased gut permeability and translocation of proinflammatory microbial factors and it appeared that beneficial bacteria could inhibit cachexia in animal models (Ziemons et al., 2021). It is still premature to assume that similar processes occur in humans, but there has been increasing interest in the possibility that cancer cachexia might be modifiable by prebiotics, probiotics, symbiotics, and fecal microbiota transfer.

Sedentary behaviors

If exercise is medicine, then sedentary behaviors that entail excessive sitting, reclining, or lying about can be viewed as dangerous as toxicants and were implicated in cancer production and progression. The damaging effects of inactivity have been linked to 35 health conditions, including the 10 primary causes of death in the developed countries (Booth et al., 2017). Predictably, the relationships between sedentary time and illness occurrence were dose dependent and were exacerbated in the presence of other lifestyle factors that ordinarily increased the risk of illness as well as the presence of other preexisting health conditions.

In a high-risk population that comprised individuals with type 2 diabetes, greater sedentary time was accompanied by elevated levels of IL-6, C-reactive protein, and the satiety hormone leptin (Henson et al., 2013). Interventions to reduce sedentary time, as expected, reduced C-reactive protein levels in women who had been newly diagnosed with type 2 diabetes. Among individuals with obesity, replacing 30 min of sedentary behaviors with moderate to vigorous exercise was able to reduce several indices of inflammation, and bouts of exercise comprising leisurely cycling were sufficient for mitochondria to burn ~ 12% more fuel that came from fat and 14%–17% of fuel that came from sugars (Newsom et al., 2021). Even increasing the amount of time spent standing, say by about an hour a day, can influence insulin sensitivity and thus can influence the development of metabolic syndrome that affects other chronic illnesses.

A metaanalysis that included 17 prospective studies revealed that sedentary behaviors were linked to the later occurrence of endometrial, colorectal, breast, and lung cancer (Shen et al., 2014). The elevated risk of cancer was remarkably pronounced, being increased from 28%–44% for colon cancer, while endometrial cancer risk was elevated by 28%–36%, and breast cancer occurrence was elevated by 8%–17% (Jochem et al., 2019). Not unexpectedly, the greater incidence of aggressive prostate cancer and lung cancer among sedentary individuals was most prominent among those who were overweight (e.g., Berger et al., 2019) and the relationship between sedentary behavior and cancer occurrence was further moderated by dietary factors. It was provisionally suggested that the elevated cancer occurrence was also tied to metabolic processes, variations of sex hormones, and the presence of chronic low-grade inflammation (Jochem et al., 2019).

Just as exercise can enhance immune activity, adopting a sedentary lifestyle can limit immune functioning. In sedentary older humans, several aspects of immune functioning were generally inferior to that of physically active individuals. Unfortunately, longitudinal studies suggested that in previously sedentary individuals, an exercise training program did not fully restore immune functioning. Although more extensive analyses are needed to determine the reliability of these findings and whether there are conditions in which the effects of sedentary behavior can be reversed, for the moment it seems likely that the damage might not readily be undone. That said, in previously sedentary older men, multiple inflammatory-related genes were differentially expressed after 24 weeks of exercise, perhaps reflecting greater illness resilience (Chen et al., 2020). Furthermore, in a randomized controlled trial among previously sedentary women who had survived breast cancer, resistance training increased NK cells and diminished the presence of inflammation related to TNF-α (Hagstrom et al., 2016).

Wanting vs needing to exercise

In rodents that voluntarily engage in taking opioids (self-administration) the rewarding effects observed are very different from those associated with forced intake. Indeed, based on behavioral responses emitted, it seemed that forced drug intake was perceived as being aversive. In the same way, in a mouse model of colitis, voluntary exercise (wheel running) had positive effects, whereas forced exercise can instigate inflammation, and comparable effects were also reported concerning gut microbiota presence (Allen et al., 2015). In effect, in evaluating the impact of exercise in rodents, it is important to distinguish between forced exercise that may be an uncontrollable stressor from voluntary exercise that rodents seem to enjoy and adopt readily.

Diverse attitudes toward exercise similarly exist in humans; some individuals perceive exercise as being rewarding, but others see it as a chore or even a stressor. Thus, it might be expected that these individual differences might also be recapitulated in the physiological changes that follow exercise training. Studies that assessed the effects of different forms of exercise in humans frequently did so through randomized controlled trials. This, of course, is the standard in clinical trials, but at the same time, this ignores patient's preferences and abilities, which may act against the positive effects of this treatment. Allowing patients the option of exercise methods may enhance compliance and diminish attrition, particularly as the effectiveness of the form of exercise used may depend on multiple individual difference factors, such as the individual's age, frailty, and the presence of other chronic conditions, as well as general fitness. While individuals need to exercise and avoid sedentary behaviors, this might not happen if individuals lack the desire and motivation to do so.

The social element

Maintaining exercise regimens can be enhanced when individuals perceive that benefits will be obtained when they have social support for this endeavor, and when they identify with an exercise culture (Haslam et al., 2020). As we saw in our discussion of coping processes, adopting exercise as a component of group behaviors may be of particular benefit by enhancing social identity, which could translate into enhanced social functioning and improved engagement in the prescribed activities (Stevens et al., 2017). These positive effects may have evolved owing to group-based activities cementing a shared social identity that can enhance health and create a social environment in which exercise would be rewarding, and thereby would diminish attrition.

The importance of social identity in various human endeavors can't be overstated, and an especially convincing case has been made concerning the benefits derived by group exercise (and social connectedness) in maintaining good psychological and physical health (Haslam et al., 2020). The health benefits obtained from group exercise may be greater than that gained from individual efforts since individuals may be more likely to exercise regularly, to the extent that it becomes a habit. No doubt, the benefits of group exercise can be enhanced further through a leader who can encourage the group so that they form a shared identity (Haslam et al., 2020). The case for social support in promoting exercise among cancer patients, however, isn't always attained. An extensive systematic review had indicated that among cancer survivors mixed results were reported as to whether social support had positive actions in the promotion of physical exercise (McDonough et al., 2019). Such varied results are not entirely unexpected given the diverse experiences of cancer survivors, together with their health following treatments. This said, social integration may be a better predictor of diminished mortality than are any other health-related behaviors (Holt-Lunstad et al., 2017), and investment in social connectedness through sporting activities may pay marked dividends.

Roadblocks to exercise and how to get around them

Many people would probably like to engage in exercise, but sometimes "things get in the way." Common mantras might include "I just don't have the time" or "after a tough workday, I simply don't have the energy and would rather just sit back and relax." Even when people are sufficiently inspired to begin an exercise program, the motivation may be fleeting, and they soon drop out. It might simply be that their chosen exercise activity wasn't right for them, or the routines became boring. Not infrequently, individuals may have set unattainable goals and perhaps they expected too much, too soon, which could lead to a self-defeating attitude and reduced motivation to exercise.

A likely factor that limits the adoption of exercise as part of a healthy lifestyle is that doing nothing is simply easier than doing something. Still, if the will is initially present, then those individuals with high self-efficacy or who wish to take control and determine their fates (self-determination) will likely be relatively successful. Accordingly, exercise programs that include behavior change methods might diminish attrition and may turn out to be relatively effective (see Chapter 12).

Depressed individuals and those who had experienced intense stressors may be less motivated to engage in exercise and may adopt sedentary behaviors that undermine health. As such, methods to diminish depression and enhance effective coping may facilitate the adoption of exercise and the physical health benefits that come with this. However, as exercising is frequently not engaged by healthy nonstressed individuals, it can reasonably be expected that a good number of cancer patients experiencing distress and depression would still be less inclined to do so. Many other factors act as roadblocks that deter cancer patients from engaging in routine exercise, so much so that across studies to evaluate the impact of exercise, 25%–50% declined to participate. Factors that hindered the adoption of regular exercise comprised lower levels of education together with self-reported fatigue and low health-related quality of life. These features were accompanied by negative attitudes toward exercise, low social support, a poor sense of self-efficacy, fewer perceived benefits, and greater perceived barriers (van Waart et al., 2016).

Too often behavioral change doesn't come about simply based on the available scientific data, no matter how convincing these might be—witness the current (and previous) resistance to vaccination against covid-19, and the reliance on unproven supplements and oddball procedures to alleviate a broad variety of diseases. Still, when individuals accept that exercise and other lifestyles are essential to well-being and are motivated to alter their lifestyles, the chasm between intentions and actions is more easily bridged.

Several approaches have been developed to encourage the adoption of healthy behaviors and avoid those that are counterproductive. It is understood that great individual differences exist so that no single approach is suitable to produce a healthy behavioral change in everyone nor is there a magic bullet that will be effective for all individuals. Still, with the progressively increasing prevalence of unhealthy lifestyles that increase disease risk, the mounting presence of obesity being one example, there is a critical need to develop widely applicable strategies. Some of these might focus on changing behaviors among individuals with a high risk of illness. Others, such as the Groups 4 Health approach advocated by Haslam et al. (2016) are relevant to enhance self-efficacy and group-based social identifications, which ultimately favors good health. Thus, it is of practical importance that community-based exercise programs that promote group behaviors may influence well-being and enhance quality of life in adult cancer survivors. In the end, simple exercises need to become a component of lifestyles rather than a chore to be endured.

Summary and conclusions

Considerable data have amassed supporting the position that moderate exercise can act to prevent the occurrence of some forms of cancer, limit some of the common symptoms associated with cancer, and reduce the side effects of therapies. Among other things, exercise can enhance muscle strength, including heart functioning, and may thereby augment an individual's ability to get through a cancer treatment regimen successfully. Most of the available data concerning the influence of exercise is based on observational studies; however, enough prospective studies and randomized controlled trials have revealed multiple beneficial effects for cancer patients. The exercise protocol that is optimal to provide benefits likely depends upon the nature and stage of cancer and may vary across individuals; it may also be moderated by factors such as the presence of obesity, the adoption of other healthy behaviors, and individual attitudes concerning exercise.

The influence of exercise could be moderated by genetic and epigenetic factors, and the influence of exercise may be linked to changes in foods consumed or other lifestyles that may be adopted in conjunction with exercise regimens. To address these and other questions regarding the efficacy of exercise, further high-quality randomized control trials will be required. Importantly, with an increased understanding of the processes by which exercise produces beneficial outcomes, including the variations of microbial, immune, and inflammatory functioning, as well as variations of hormones or growth factors, it may be possible to develop tailored strategies that yield optimal outcomes for individual patients. An extensive and detailed review of structured exercise programs led to the suggestion that exercise ought to be incorporated as part of standard cancer care, but given the many variables that could affect outcomes, this needs to be done on an individual basis (Christensen et al., 2018).

With the breadth of problems that stem from poor health styles, broader approaches will also be needed to achieve population-wide benefits. Social media is an obvious method that could be used to this end, despite the flood of misinformation that inundates us. That said, impressive websites have been created (e.g., yourdiseaserisk.wustl.edu) to translate scientific data related to cancer prevention into accurate data-driven, engaging, and practical messages and advice for the public (Colditz and Dart, 2021). To be sure, adopting lifestyle changes that could affect cancer progression—such as altering diet and food preferences—can be difficult; getting enough sleep might not always be an activity that can easily be modified, and nor is keeping away from environmental toxicants entirely under an individual's control (although communities are always trying). But to some extent, if some cancers are potentially preventable through avoidance of risk factors (and we know that telling a smoker to give up can certainly do that by eliminating known carcinogens in cigarette smoke), then what could be simpler than a prescription that urges avoiding sedentary behaviors and engaging in brief daily exercise amounting to a brisk daily 30-min walk? Given the overwhelming mental and physical benefits of exercise, it would appear that to do nothing is commensurate with poor health behavior. And exercising poor health behaviors only invites trouble.

References

Agostini, D., Natalucci, V., Baldelli, G., De Santi, M., Donati Zeppa, S., et al., 2018. New insights into the role of exercise in inhibiting mTOR signaling in triple-negative breast cancer. Oxidative Med. Cell. Longev. 2018, 1–19.

Allen, J.M., Berg Miller, M.E., Pence, B.D., Whitlock, K., Nehra, V., et al., 2015. Voluntary and forced exercise differentially alters the gut microbiome in C57BL/6J mice. J. Appl. Physiol. 118, 1059–1066.

Allen, J.M., Mailing, L.J., Niemiro, G.M., Moore, R., Cook, M.D., et al., 2018. Exercise alters gut microbiota composition and function in lean and obese humans. Med. Sci. Sports Exerc. 50, 747–757.

An, K.Y., Morielli, A.R., Kang, D.W., Friedenreich, C.M., McKenzie, D.C., et al., 2020. Effects of exercise dose and type during breast cancer chemotherapy on longer-term patient-reported outcomes and health-related fitness: a randomized controlled trial. Int. J. Cancer 146, 150–160.

Aoyagi, T., Terracina, K.P., Raza, A., Matsubara, H., Takabe, K., 2015. Cancer cachexia, mechanism and treatment. World J. Gastrointest. Oncol. 7, 17–29.

Arena, S.K., Doherty, D.J., Bellford, A., Hayman, G., 2019. Effects of aerobic exercise on oxidative stress in patients diagnosed with cancer: a narrative review. Cureus 11, e5382.

Barrón-Cabrera, E., González-Becerra, K., Rosales-Chávez, G., Mora-Jiménez, A., Hernández-Cañaveral, I., et al., 2020. Low-grade chronic inflammation is attenuated by exercise training in obese adults through down-regulation of ASC gene in peripheral blood: a pilot study. Genes Nutr. 15, 1–11.

Barton, W., Penney, N.C., Cronin, O., Garcia-Perez, I., Molloy, M.G., et al., 2018. The microbiome of professional athletes differs from that of more sedentary subjects in composition and particularly at the functional metabolic level. Gut 67, 625–633.

Berger, F.F., Leitzmann, M.F., Hillreiner, A., Sedlmeier, A.M., Prokopidi-Danisch, M.E., et al., 2019. Sedentary behavior and prostate cancer: a systematic review and meta-analysis of prospective cohort studies. Cancer Prev. Res. 12, 675–688.

Bhagavata Srinivasan, S.P., Raipuria, M., Bahari, H., Kaakoush, N.O., Morris, M.J., 2018. Impacts of diet and exercise on maternal gut microbiota are transferred to offspring. Front. Endocrinol. 9, 716.

Booth, F.W., Roberts, C.K., Thyfault, J.P., Ruegsegger, G.N., Toedebusch, R.G., 2017. Role of inactivity in chronic diseases: evolutionary insight and pathophysiological mechanisms. Physiol. Rev. 97, 1351–1402.

Buss, L.A., Dachs, G.U., 2020. Effects of exercise on the tumour microenvironment. Adv. Exp. Med. Biol. 1225, 31–51.

Campbell, J.P., Turner, J.E., 2018. Debunking the myth of exercise-induced immune suppression: redefining the impact of exercise on immunological health across the lifespan. Front. Immunol. 9, 648.

Campbell, P.T., Campbell, K.L., Wener, M.H., Wood, B., Potter, J.D., et al., 2009. A yearlong exercise intervention decreases CRP among obese postmenopausal women. Med. Sci. Sports Exerc. 41.

Campbell, K.L., Landells, C.E., Fan, J., Brenner, D.R., 2017. A systematic review of the effect of lifestyle interventions on adipose tissue gene expression: implications for carcinogenesis. Obesity (Silver Spring) 25 (Suppl. 2), S40–S51.

Cerqueira, E., Marinho, D.A., Neiva, H.P., Lourenco, O., 2020. Inflammatory effects of high and moderate intensity exercise—a systematic review. Front. Physiol. 10, 1550.

Chen, L., Bai, J., Li, Y., 2020. The change of interleukin-6 level-related genes and pathways induced by exercise in sedentary individuals. J. Interf. Cytokine Res. 40, 236–244.

Christensen, J.F., Simonsen, C., Hojman, P., 2018. Exercise training in cancer control and treatment. Compr. Physiol. 9, 165–205.

Chung, H.C., Keiller, D.R., Roberts, J.D., Gordon, D.A., 2021. Do exercise-associated genes explain phenotypic variance in the three components of fitness? A systematic review & meta-analysis. PLoS One 16, e0249501.

Colditz, G.A., Dart, H., 2021. Commentary: 20 years online with "Your Disease Risk". Cancer Causes Control 32, 5–11.

Daou, H.N., 2020. Exercise as an anti-inflammatory therapy for cancer cachexia: a focus on interleukin-6 regulation. Am. J. Physiol. Regul. Integr. Comp. Physiol. 318, R296–R310.

Denham, J., 2018. Exercise and epigenetic inheritance of disease risk. Acta Physiol. 222, e12881.

Dethlefsen, C., Pedersen, K.S., Hojman, P., 2017. Every exercise bout matters: linking systemic exercise responses to breast cancer control. Breast Cancer Res. Treat. 162, 399–408.

Dieli-Conwright, C.M., Parmentier, J.H., Sami, N., Lee, K., Spicer, D., et al., 2018. Adipose tissue inflammation in breast cancer survivors: effects of a 16-week combined aerobic and resistance exercise training intervention. Breast Cancer Res. Treat. 168, 147–157.

Duregon, F., Vendramin, B., Bullo, V., Gobbo, S., Cugusi, L., et al., 2018. Effects of exercise on cancer patients suffering chemotherapy-induced peripheral neuropathy undergoing treatment: a systematic review. Crit. Rev. Oncol. Hematol. 121, 90–100.

Evans, C.C., LePard, K.J., Kwak, J.W., Stancukas, M.C., Laskowski, S., et al., 2014. Exercise prevents weight gain and alters the gut microbiota in a mouse model of high fat diet-induced obesity. PLoS One 9, e92193.

Flockhart, M., Nilsson, L.C., Tais, S., Ekblom, B., Apro, W., et al., 2021. Excessive exercise training causes mitochondrial functional impairment and decreases glucose tolerance in healthy volunteers. Cell Metab. 33, 957–970 e956.

Friedenreich, C.M., Neilson, H.K., Farris, M.S., Courneya, K.S., 2016. Physical activity and cancer outcomes: a precision medicine approach. Clin. Cancer Res. 22, 4766–4775.

Furmaniak, A.C., Menig, M., Markes, M.H., 2016. Exercise for women receiving adjuvant therapy for breast cancer. Cochrane Database Syst. Rev. 9, CD005001.

Gubert, C., Hannan, A.J., 2021. Exercise mimetics: harnessing the therapeutic effects of physical activity. Nat. Rev. Drug Discov. 20, 862–879.

Hagstrom, A.D., Marshall, P.W., Lonsdale, C., Papalia, S., Cheema, B.S., et al., 2016. The effect of resistance training on markers of immune function and inflammation in previously sedentary women recovering from breast cancer: a randomized controlled trial. Breast Cancer Res. Treat. 155, 471–482.

Haslam, C., Cruwys, T., Haslam, S.A., Dingle, G., Chang, M.X., 2016. Groups 4 Health: evidence that a social-identity intervention that builds and strengthens social group membership improves mental health. J. Affect. Disord. 194, 188–195.

Haslam, S.A., Fransen, K., Boen, F., 2020. The Social Identity Approach to Sport and Exercise. Sage, London.

Henson, J., Yates, T., Edwardson, C.L., Khunti, K., Talbot, D., et al., 2013. Sedentary time and markers of chronic low-grade inflammation in a high risk population. PLoS One 8, e78350.

Herremans, K.M., Riner, A.N., Cameron, M.E., Trevino, J.G., 2019. The microbiota and cancer cachexia. Int. J. Mol. Sci. 20, 6267.

Hojman, P., Gehl, J., Christensen, J.F., Pedersen, B.K., 2018. Molecular mechanisms linking exercise to cancer prevention and treatment. Cell Metab. 27, 10–21.

Holt-Lunstad, J., Robles, T.F., Sbarra, D.A., 2017. Advancing social connection as a public health priority in the United States. Am. Psychol. 72, 517.

Huang, C.W., Chang, Y.H., Lee, H.H., Wu, J.Y., Huang, J.X., et al., 2020. Irisin, an exercise myokine, potently suppresses tumor proliferation, invasion, and growth in glioma. FASEB J. 34, 9678–9693.

Idorn, M., Thor Straten, P., 2017. Exercise and cancer: from "healthy" to "therapeutic"? Cancer Immunol. Immunother. 66, 667–671.

Jochem, C., Wallmann-Sperlich, B., Leitzmann, M.F., 2019. The influence of sedentary behavior on cancer risk: epidemiologic evidence and potential molecular mechanisms. Curr. Nutr. Rep. 8, 167–174.

Kakanis, M.W., Peake, J., Brenu, E.W., Simmonds, M., Gray, B., et al., 2010. The open window of susceptibility to infection after acute exercise in healthy young male elite athletes. Exerc. Immunol. Rev. 16, 119–137.

Kang, D.W., Lee, J., Suh, S.H., Ligibel, J., Courneya, K.S., et al., 2017. Effects of exercise on insulin, IGF axis, adipocytokines, and inflammatory markers in breast cancer survivors: a systematic review and meta-analysis. Cancer Epidemiol. Biomark. Prev. 26, 355–365.

Kawanishi, N., Yano, H., Yokogawa, Y., Suzuki, K., 2010. Exercise training inhibits inflammation in adipose tissue via both suppression of macrophage infiltration and acceleration of phenotypic switching from M1 to M2 macrophages in high-fat-diet-induced obese mice. Exerc. Immunol. Rev. 16, 105–118.

Khosravi, N., Stoner, L., Farajivafa, V., Hanson, E.D., 2019. Exercise training, circulating cytokine levels and immune function in cancer survivors: a meta-analysis. Brain Behav. Immun. 81, 92–104.

Kim, J.S., Galvao, D.A., Newton, R.U., Gray, E., Taaffe, D.R., 2021. Exercise-induced myokines and their effect on prostate cancer. Nat. Rev. Urol. 18, 519–542.

Koelwyn, G.J., Quail, D.F., Zhang, X., White, R.M., Jones, L.W., 2017. Exercise-dependent regulation of the tumour microenvironment. Nat. Rev. Cancer 17, 620–632.

Kyu, H.H., Bachman, V.F., Alexander, L.T., Mumford, J.E., Afshin, A., et al., 2016. Physical activity and risk of breast cancer, colon cancer, diabetes, ischemic heart disease, and ischemic stroke events: systematic review and dose-response meta-analysis for the Global Burden of Disease Study 2013. BMJ 354, i3857.

Lahart, I.M., Metsios, G.S., Nevill, A.M., Carmichael, A.R., 2018. Physical activity for women with breast cancer after adjuvant therapy. Cochrane Database Syst. Rev. 1, CD011292.

Lai, Z.L., Tseng, C.H., Ho, H.J., Cheung, C.K.Y., Lin, J.Y., et al., 2018. Fecal microbiota transplantation confers beneficial metabolic effects of diet and exercise on diet-induced obese mice. Sci. Rep. 8, 15625.

Li, T., Wei, S., Shi, Y., Pang, S., Qin, Q., et al., 2016. The dose-response effect of physical activity on cancer mortality: findings from 71 prospective cohort studies. Br. J. Sports Med. 50, 339–345.

Liu, X., Tufman, A., Behr, J., Kiefl, R., Goldmann, T., et al., 2020a. Role of the erythropoietin receptor in lung cancer cells: erythropoietin exhibits angiogenic potential. J. Cancer 11, 6090–6100.

Liu, Y., Wang, Y., Ni, Y., Cheung, C.K.Y., Lam, K.S.L., et al., 2020b. Gut microbiome fermentation determines the efficacy of exercise for diabetes prevention. Cell Metab. 31, 77–91 e75.

Loughney, L., West, M.A., Kemp, G.J., Grocott, M.P., Jack, S., 2016. Exercise intervention in people with cancer undergoing neoadjuvant cancer treatment and surgery: a systematic review. Eur. J. Surg. Oncol. 42, 28–38.

Mailing, L.J., Allen, J.M., Buford, T.W., Fields, C.J., Woods, J.A., 2019. Exercise and the gut microbiome: a review of the evidence, potential mechanisms, and implications for human health. Exerc. Sport Sci. Rev. 47, 75–85.

Matthews, C.E., Moore, S.C., Arem, H., Cook, M.B., Trabert, B., et al., 2020. Amount and intensity of leisure-time physical activity and lower cancer risk. J. Clin. Oncol. 38, 686–697.

McDonough, M.H., Beselt, L.J., Daun, J.T., Shank, J., Culos-Reed, S.N., et al., 2019. The role of social support in physical activity for cancer survivors: a systematic review. Psychooncology 28, 1945–1958.

Meneses-Echávez, J.F., Jimenez, E.G., Rio-Valle, J.S., Correa-Bautista, J.E., Izquierdo, M., et al., 2016. The insulin-like growth factor system is modulated by exercise in breast cancer survivors: a systematic review and meta-analysis. BMC Cancer 16, 682.

Mika, A., Fleshner, M., 2016. Early-life exercise may promote lasting brain and metabolic health through gut bacterial metabolites. Immunol. Cell Biol. 94, 151–157.

Moore, S.C., Lee, I.M., Weiderpass, E., Campbell, P.T., Sampson, J.N., et al., 2016. Association of leisure-time physical activity with risk of 26 types of cancer in 1.44 million adults. JAMA Intern. Med. 176, 816–825.

Newsom, S.A., Stierwalt, H.D., Ehrlicher, S.E., Robinson, M.M., 2021. Substrate-specific respiration of isolated skeletal muscle mitochondria after 1 h of moderate cycling in sedentary adults. Med. Sci. Sports Exerc. 53, 1375–1384.

Niels, T., Tomanek, A., Freitag, N., Schumann, M., 2020. Can exercise counteract cancer cachexia? A systematic literature review and meta-analysis. Integr. Cancer Ther. 19. 1534735420940414.

Packer, N., Hoffman-Goetz, L., 2012. Exercise training reduces inflammatory mediators in the intestinal tract of healthy older adult mice. Can. J. Aging 31, 161–171.

Pedersen, L., Idorn, M., Olofsson, G.H., Lauenborg, B., Nookaew, I., et al., 2016. Voluntary running suppresses tumor growth through epinephrine- and IL-6-dependent NK cell mobilization and redistribution. Cell Metab. 23, 554–562.

Pervaiz, N., Hoffman-Goetz, L., 2012. Immune cell inflammatory cytokine responses differ between central and systemic compartments in response to acute exercise in mice. Exerc. Immunol. Rev. 18, 142–157.

Radom-Aizik, S., Zaldivar, F., Haddad, F., Cooper, D.M., 2013. Impact of brief exercise on peripheral blood NK cell gene and microRNA expression in young adults. J. Appl. Physiol. 114, 628–636.

Rosero, I.D., Ramirez-Velez, R., Lucia, A., Martinez-Velilla, N., Santos-Lozano, A., et al., 2019. Systematic review and meta-analysis of randomized, controlled trials on preoperative physical exercise interventions in patients with non-small-cell lung cancer. Cancers 11, 944.

Rundqvist, H., Velica, P., Barbieri, L., Gameiro, P.A., Bargiela, D., et al., 2020. Cytotoxic T-cells mediate exercise-induced reductions in tumor growth. eLife 9, e59996.

Saint-Maurice, P.F., Coughlan, D., Kelly, S.P., Keadle, S.K., Cook, M.B., et al., 2019. Association of leisure-time physical activity across the adult life course with all-cause and cause-specific mortality. JAMA Netw. Open 2, e190355.

Sasso, J.P., Eves, N.D., Christensen, J.F., Koelwyn, G.J., Scott, J., et al., 2015. A framework for prescription in exercise-oncology research. J. Cachexia Sarcopenia Muscle 6, 115–124.

Schenk, A., Koliamitra, C., Bauer, C.J., Schier, R., Schweiger, M.R., et al., 2019. Impact of acute aerobic exercise on genome-wide DNA-methylation in natural killer cells—a pilot study. Genes 10, 380.

Shen, D., Mao, W., Liu, T., Lin, Q., Lu, X., et al., 2014. Sedentary behavior and incident cancer: a meta-analysis of prospective studies. PLoS One 9, e105709.

Simpson, R.J., Kunz, H., Agha, N., Graff, R., 2015. Exercise and the regulation of immune functions. Prog. Mol. Biol. Transl. Sci. 135, 355–380.

Simpson, R.J., Campbell, J.P., Gleeson, M., Kruger, K., Nieman, D.C., et al., 2020. Can exercise affect immune function to increase susceptibility to infection? Exerc. Immunol. Rev. 26, 8–22.

Song, M., Chan, A.T., 2018. The potential role of exercise and nutrition in harnessing the immune system to improve colorectal cancer survival. Gastroenterology 155, 596–600.

Song, M., Chan, A.T., Sun, J., 2020. Influence of the gut microbiome, diet, and environment on risk of colorectal cancer. Gastroenterology 158, 322–340.

Stevens, M., Rees, T., Coffee, P., Steffens, N.K., Haslam, S.A., et al., 2017. A social identity approach to understanding and promoting physical activity. Sports Med. 47, 1911–1918.

Thomas, R.J., Kenfield, S.A., Jimenez, A., 2017. Exercise-induced biochemical changes and their potential influence on cancer: a scientific review. Br. J. Sports Med. 51, 640–644.

van Waart, H., van Harten, W.H., Buffart, L.M., Sonke, G.S., Stuiver, M.M., et al., 2016. Why do patients choose (not) to participate in an exercise trial during adjuvant chemotherapy for breast cancer? Psychooncology 25, 964–970.

Villar-Fincheira, P., Sanhueza-Olivares, F., Norambuena-Soto, I., Cancino-Arenas, N., Hernandez-Vargas, F., et al., 2021. Role of Interleukin-6 in vascular health and disease. Front. Mol. Biosci. 8, 7.

Welly, R.J., Liu, T.W., Zidon, T.M., Rowles 3rd, J.L., Park, Y.M., et al., 2016. Comparison of diet versus exercise on metabolic function and gut microbiota in obese rats. Med. Sci. Sports Exerc. 48, 1688–1698.

Ziemons, J., Smidt, M.L., Damink, S.O., Rensen, S.S., 2021. Gut microbiota and metabolic aspects of cancer cachexia. Best Pract. Res. Clin. Endocrinol. Metab. 35, 101508.

Zimmer, P., Trebing, S., Timmers-Trebing, U., Schenk, A., Paust, R., et al., 2018. Eight-week, multimodal exercise counteracts a progress of chemotherapy-induced peripheral neuropathy and improves balance and strength in metastasized colorectal cancer patients: a randomized controlled trial. Support. Care Cancer 26, 615–624.

CHAPTER 11

Sleep and circadian rhythms

OUTLINE

A brief history of thoughts and research related to sleep	341	Sleep duration in predicting later cancer occurrence	352
Functions of sleep	343	The curse of sleep disorders	353
Consequences of sleep loss	344	Mechanisms linking sleep and circadian disturbances to cancer	355
Links between sleep and other lifestyle factors	345	Focusing on clock genes in cancer	355
Sleep and exercise	345	Occupations associated with altered sleep cycles and cancer occurrence	358
Sleep and eating	346		
Neurobiological aspects related to sleep and circadian rhythms	347	Sleep disruption and diurnal variations in relation to microbiota	361
Circadian clocks	347	Implications of circadian rhythmicity to cancer treatments	363
Circadian rhythms: Immune and cytokine changes	349	Summary and conclusions	364
Sleep, circadian rhythms, and cancer progression	352	References	365

A brief history of thoughts and research related to sleep

Early philosophers had considered the processes that might produce sleep and its recuperative functions, and they were apparently fascinated by dreaming. This might have been encouraged by dreams frequently appearing in the bible seemingly as a way by which God chose to communicate with common people, providing them with a glimpse of the future or a means of providing warnings of imminent dangers. The fascination with dreams persisted for many centuries afterward, becoming a component of Freudian and Jungian theory.

Physicians and philosophers of the time had opined upon the value of sleep based on their understanding of the body's functions coupled with some creative thinking. The Greek physician Alcmaeon of Croton (~ 500 BCE) was particularly interested in internal processes associated with diseases, asserting that good health came when internal forces and humors were in a state of balance. Illnesses, he claimed, emerged when disequilibrium occurred owing to negative environmental influences, such as poor nutrition or lifestyles. He also had a say on how sleep came about, suggesting that it occurred when blood on or near the body's surface was withdrawn into the interior through large blood vessels, and awakening occurred when blood was restored to the periphery. Hippocrates and Aristotle may have adopted some of Alcmaeon's views concerning body humors and the importance of brain functioning, focusing on the operation of the senses. Aristotle accepted the importance of blood flow through the body and postulated that following the digestion of food particles, vapors created in the stomach rose to the brain, which promoted sleep.

The healing and recuperative power of sleep continued to be assessed by 17th-century physicians who believed that sleep was fundamental for the health of the mind and body. Sleep was thought to be important to purge the body of "foul humors" leading to the suggestion that individuals ought to sleep with their mouths open to allow easy escape for these humors. Sleeping barefoot was recommended since shoes could block the escape of harmful vapors through their downward path. And the fact that individuals typically needed to urinate upon waking was taken to reflect that sleep had been effective in purging the body of harmful substances. Indeed, these early sleep theorists were resolute in suggesting the importance of sleep being aligned with natural day-night cycles, even maintaining that light at night could have detrimental actions. To be sure, some of these assertions might seem odd from today's perspectives, but others were remarkably astute and were important in cementing the view that sleep was fundamental for good health.

The somewhat commonsense belief that sleep was needed because of its recuperative powers had begun to be widely adopted but like so many other common human activities, societal and economic pressures modified and shaped sleep patterns. In earlier times, it wasn't unusual for people to sleep in two shifts, each of about 4 h, interrupted by a waking period of about an hour (segmented sleep), during which odd jobs could be done, such as stoking the fireplace. With the industrial revolution that focused on productivity, and hence the requirement that people begin their work early, segmented sleep patterns were abandoned. Owing to the need for labor, long sleep was dismissed as being unproductive and a sign of laziness. And to make matters worse, with the eventual introduction of electric lighting, altered sleep patterns and habits were adopted, and shift work was instituted that was incompatible with normal rhythms.

Formal sleep research began in earnest during the late 1800s. Aside from analyses of the benefits of sleep, researchers began to systematically address the consequences of sleep loss. It was during this time that the concept of insomnia as an illness and a source for other medical conditions first appeared, frequently being attributed to industrial pressures and those that accompanied modernity. At that time, other sleep disorders were identified (e.g., obstructive sleep apnea, narcolepsy, restless leg syndrome), and their consequences were ultimately assessed.

Sleep research came into its own during the early-mid portion of the 20th century with Nathaniel Kleitman opening the first sleep laboratory in the United States. The work of one of his students Eugene Aserinsky gave rise to early efforts to use electroencephalogram (EEG) recordings of sleep patterns and the discovery of rapid eye movement (REM) sleep during which dreaming occurs. At about this time, concerted efforts were made to identify the processes leading to sleep and its capacity to promote well-being. The many physical values of sleep were thus acknowledged, its healing powers were examined broadly, and its ability to influence memory processes, creativity, and emotional regulation was widely recognized. This was likely encouraged by findings that a third of the population experienced some sort of sleep disorder and that average sleep durations had declined by 20% over the preceding century. Sleep medicine eventually evolved as a subfield of medicine that not only assessed sleep processes but also invited other subfields to study the impact of sleep disturbances on diverse physical illnesses.

With the basic understanding of neurotransmission, serotonin was seen as the key driver related to sleep, but sleep is a complex process that involves numerous other neurotransmitters and hormones. In a detailed analysis of sleep processes, Irwin (2015) indicated that sleep and sleep disturbances are linked to alterations of numerous neurotransmitters (GABA, glutamate, norepinephrine, serotonin, nitric oxide), hormones (e.g., GHRH, CRH, cortisol, insulin, prostaglandin), neurotrophins, immune factors, and cytokines. Owing to such neurobiological changes, sleep disturbances and not getting enough sleep, not obtaining good sleep quality, and even inconsistencies of sleep habits can have negative health repercussions, including the instigation of metabolic and immune-related disorders, heart disease, and cancer processes.

The links between sleep and illness have come from epidemiological studies, prospective analyses, and randomized controlled trials. These studies have largely confirmed the view that sleep disturbances stemming from biological alterations (or changes in circadian rhythms) could favor cancer development and progression, and limit the effectiveness of therapies. The actions of sleep disturbances on cancer may be an indirect one, as sleep loss and circadian alterations may have stressor-like actions and may contribute to allostatic overload and the illnesses that follow (McEwen and Karatsoreos, 2015). Furthermore, sleep loss can affect eating processes and may affect sedentary behaviors that promote various illnesses.

Not unexpectedly, a cancer diagnosis can disrupt sleep owing to distress, rumination, anxiety, and depression, which can aggravate tumor growth. As cancer progresses and therapies are initiated, further sleep disturbances may emerge that can persist well after treatments had ended, which may foster the appearance of other illnesses.

Functions of sleep

Sleep is a fundamental need, much as eating and drinking are essential for survival. Among its other actions, sleep acts in a recuperative capacity allowing for the replenishment of resources that had been used during the day. Adenosine, which serves as a principal contributor to energy regulation, affects sleep promotion during which glycogen energy stores can recover so that skeletal and muscular systems can rejuvenate, and hormones and neurotransmitters can be restored. Sleep is important for the repair of damaged tissues and to prevent adverse physiological changes that would otherwise occur, such as diminishing accumulating DNA damage within neurons (Zada et al., 2019). In addition, clearance of waste that had accumulated over the day can be flushed out of the brain during deep sleep. Specifically, toxins within the brain are retrieved and drained into lymph nodes through a complex series of channels that make up the *glymphatic* system (which consists of glia transporting molecules to the lymphatic vessels in the meninges, a membrane that covers the brain), allowing the brain to remain in a relatively pristine state. This seems to be especially important among individuals who had experienced traumatic brain injuries in which the accumulation of waste increases appreciably, making sleep that much more essential. It isn't just sleep that determines the removal of waste through the glymphatic system since its functioning varies with circadian rhythms. It seems that among individuals working on night shifts or who rely on cat naps during the day, the removal of toxins occurs less readily, consequently increasing the risk for developing neurological disorders (Hablitz et al., 2020).

Consequences of sleep loss

The frequency of sleep problems varies across countries, but globally about 20% of individuals experience some type of sleep disorder. Individuals with sleep disturbances experience numerous comorbid conditions particularly cognitive difficulties and depressive illness; conversely, depressive illness may promote sleep problems. Not obtaining enough sleep is apt to promote classic features of depressive illness in which some people become more reactive to negative events (arguments, social tensions, as well as work and family stress), and may promote diminished positive responses to events or stimuli that would ordinarily be viewed as being pleasant. The impaired psychological well-being associated with sleep loss, particularly anxiety can be diminished after individuals have obtained sufficient sleep; and among individuals experiencing a chronic health problem, enhanced sleep was accompanied by improved well-being.

Through actions on microglial processes, sleep contributes to synaptic plasticity and serves in synaptic pruning wherein synapses that have not been used and are no longer needed get eliminated. With sleep loss these actions may be diminished, resulting in impaired neuronal connectivity and hippocampal atrophy, which may be accompanied by memory and cognitive disturbances. The neuronal changes within the hippocampus that are associated with sleep loss are accompanied by reduced BDNF functioning, which also occurs within the frontal cortex, potentially contributing to depressive symptoms. Sleep loss similarly disturbed glutamatergic transmission between the medial prefrontal cortex and nucleus accumbens, thereby undermining reward processes that may be integral to the development of depression. Moreover, sleep deprivation disrupted neuronal functioning within the ventromedial prefrontal cortex, which could affect decision-making so individuals relied more on habits that are not productive rather than on planned goal-directed behaviors.

Beyond these more extensively examined consequences of sleep loss, elevated reactivity of neurons within primary sensing regions may increase pain perception while concurrently blunting neuronal activity in brain regions important for the modulation of pain processing (Krause et al., 2019). Thus, reduced quality and efficiency of sleep, as well as unrefreshing sleep, may have serious biological sequelae that promote pain perception. Because of pain experienced by cancer patients even long after therapy has been completed, it may be particularly important to determine whether impaired sleep contributes to this chronic condition. In line with this, patients with poor sleep quality before breast surgery experienced greater postoperative pain, making it important to evaluate sleep efficiency before treatments are administered.

Sleep disturbances and variations of sleep patterns may undermine immune, neuroendocrine, and autonomic processes so vulnerability to communicable and noncommunicable diseases can be affected (Irwin et al., 2015). Through HPA and sympathetic nervous system variations, chronic sleep loss can disturb immune functioning, similar to the effects of chronic stressors (Irwin, 2019) and may impair immunological memory. With chronic sleep disturbances, HPA axis functioning may be persistently activated, leading to allostatic overload as well as immune cells becoming glucocorticoid resistant so immune activation and the corresponding inflammation may be continuously elevated. Indeed, an extensive meta-analysis indicated that sleep disturbances and sleep duration are generally associated with elevated levels of C-reactive protein and IL-6 (Irwin et al., 2016). These sleep alterations have also been associated with insulin resistance, impaired glucose tolerance, and risk of type 2 diabetes, and may also increase the frequency of DNA breaks and diminished DNA repair, which may favor cancer appearance.

NREM and REM sleep

Sleep is characterized by distinct brain activity changes that can be determined through EEG recordings measured using electrodes gently fixed on the skull surface of a sleeping individual. As individuals transition from the waking state through to light sleep and then to deep sleep (i.e., moving from Stage 1 to Stage 4 sleep), fewer brain electrical oscillations occur and the amplitude of brain waves becomes larger. During sleep periods, the brain's electrical pattern is occasionally interrupted by transient high-frequency (sleep spindles) and high-amplitude waves (K-complex) that are linked to brief awakenings. During these periods, there is typically little response to external stimuli, but awakening can be provoked by particularly significant stimuli, such as the sound of a baby crying or the person's name being called. It is during deep sleep (Stage 4), which makes up 15%–20% of the sleep period, that body repair primarily occurs. During this sleep stage brain electrical activity is particularly slow, brain temperature declines, breathing and heart rate are slowed, and blood pressure is reduced. Periods of rapid eye movement (REM) occur accompanied by a brain electrical activity pattern resembling wakefulness (even though the person is still asleep). These REM periods occur approximately every 90 min, and typically are associated with dreaming. For this reason, REM sleep has been referred to as "dream" sleep, and it appears to be a fundamental need for the brain to undergo this stage of sleep. The REM period involves an interplay between several brain regions, including neuronal groups (called "nuclei") in the brainstem, hypothalamus, limbic brain regions, and several cortical areas. But when REM is disrupted, this can produce disturbed emotional processes, impaired memory consolidation, and a subjective sense of not feeling rested. Moreover, disturbed REM sleep has been linked to numerous neurological disorders, including Huntington's disease, amyotrophic lateral sclerosis (ALS), Parkinson's disease, and Alzheimer's disease, and may be an early marker of these disorders. While considerable data are available showing REM and non-REM sleep disorders in cancer survivors, there is a paucity of information concerning REM sleep disturbances in predicting later cancer occurrence.

Links between sleep and other lifestyle factors

Sleep and exercise

In evaluating the influence of sleep on health, it needs to be kept in mind that many lifestyle factors operate in tandem to affect health. As described in Chapter 10, modest exercise (e.g., walking, yoga) was accompanied by improved sleep quantity and quality among women with breast cancer. Moreover, among patients undergoing treatment for either breast or prostate cancer, augmented sleep duration and efficiency (proportion of time asleep relative to the time in bed) associated with exercise were accompanied by variations of IL-6. These and similar findings led to the suggestion that the sleep improvements accounted for the reduced cancer-related fatigue produced by exercise and that cytokine regulation may be responsible for improvements in sleep quality. However, exercise was not uniformly tied to sleep improvements irrespective of whether subjective or objective sleep measures were used.

Sleep and eating

The ties between sleep and eating are not intuitively obvious. Like the actions of moderate stressors, circadian disturbances or sleep loss may promote increased food intake and obesity, perhaps because food odors become more enticing, or because judgment is disturbed, impulsivity is increased, and consequently the ability to deal with temptation is impaired. These actions may stem from disturbed serotonin functioning within the frontal and insular cortex and endocannabinoid variations within these brain regions may likewise affect decision-making regarding food choices.

Sleep can be affected by stressful experiences that, in turn, affect eating processes. Alternatively, sleep loss may cause feelings of distress, which can promote increased food consumption and poor food choices. For example, distressing feelings (e.g., loneliness, depression) associated with sleep loss can engender increased consumption of calorie-dense comfort foods, possibly to self-medicate. But more than this, stressors can disturb the functioning of genes related to circadian rhythms that may be aligned with alterations of the eating-related hormones leptin and ghrelin, which may be accompanied by obesity and type 2 diabetes.

Weight gain ordinarily occurs among adolescents, but in some young people, the gain is excessive. Rapid weight increases have been ascribed to hormonal changes that occur as well as the adoption of poor food choices (ultra-processed and fast foods). In some instances, the weight gain may be tied to altered sleep patterns. Adolescents who tended to be night owls and who suffered "social jet lag" owing to late-night weekend activities were relatively prone to adiposity, and to illnesses associated with being overweight. Contrary to the common myth that sleep loss during workdays could be compensated by subsequently sleeping longer (e.g., on weekends), this tactic does not uniformly allow for proper functioning to be fully regained.

Besides the psychological influences on eating that may come from sleep loss, metabolic changes that occur during sleep can influence weight loss. Specifically, although sleep is a fasting period and would thus be expected to contribute to increased metabolism in the form of fat burning, this was the case for the first few hours of sleep and then rebounded. Individual differences were apparent in this regard, such that variability in metabolism was accompanied by less carbohydrates than fat being burned relative to that seen among individuals with less variable metabolism (Zhang et al., 2021). Whether this actually translates into improved ability to lose weight or maintain appropriate weight levels and thus favor well-being remains to be established.

Neurobiological aspects related to sleep and circadian rhythms

Ordinarily, behaviors are synchronized with neurobiological processes that vary over the circadian cycle, but can be misaligned by chronic challenges, such as shift work and psychological factors (e.g., worry and rumination), thereby promoting hormonal changes (e.g., cortisol, glucose, insulin) that can engender or worsen the course of diseases. For that matter, illness can itself affect sleep and circadian processes, which can exacerbate these conditions or even cause the appearance of comorbid conditions.

Circadian rhythmicity and sleep are obviously linked, but they could independently contribute to the adverse effects of sleep loss on well-being (Irwin, 2015) and it seems that those brain regions supporting sleep and circadian processes can be distinguished from one another. Whereas subcortical regions might primarily be tied to the circadian cycle, cortical regions are more closely aligned with sleep processes and sleep loss (Muto et al., 2016). Sleep is associated with altered T cell subsets, changes of numerous hormones, including an initial decline of epinephrine and cortisol, and an increase of growth hormone, prolactin, and melatonin (Besedovsky et al., 2016). Like virtually every other hormone, these are subject to circadian fluctuations. Hormones related to eating processes, such as insulin and ghrelin are also subject to circadian clock variations and their dysregulation could influence the emergence of metabolic diseases (Stenvers et al., 2019). Circadian variations of gonadal steroids may also account for sex differences that occur in disease occurrence.

In view of the multiple regulatory processes influenced by sleep and circadian factors, including those related to immune and inflammatory processes, their dysregulation could affect well-being. In this regard, while many hormones have been implicated in sleep processes, melatonin has received particular attention, having been found to affect circadian processes and sleep promotion, as well as being useful in diminishing circadian rhythm-related sleep disorders and diminishing sleep-related features of jet lag and shift work (Zisapel, 2018). The significance of melatonin to illnesses may also be tied to its antiinflammatory, antioxidant, and protective mitochondrial actions, as well as its actions on immune cell viability and functioning, including within the tumor microenvironment (Mirza-Aghazadeh-Attari et al., 2020).

Circadian clocks

The timing and regulation of sleep are determined by internal clocks within several brain regions and within the periphery, often being primarily controlled by neurons situated within the suprachiasmatic nucleus of the hypothalamus. These clocks have typically been linked to neuronal circuits, but astrocytes also contribute to clock regulation (Brancaccio et al., 2019) through the actions on cytokines, such as TNF-α and transforming growth factor-β (TGF-β).

Numerous clock genes, such as periodic genes *Per1* and *Per2* and cryptochrome genes *Cry1* and *Cry2* are integral to circadian rhythmicity, as are cell cycle genes *c-Myc*, *Wee1*, *cyclin D*, and *p21*. Clock gene activation varies throughout sleep so gene transcripts related to metabolic processes were noted in the period preceding dawn, whereas those that affected synaptic functioning were most active before dusk. The time-dependent variations of gene translation could be disturbed by sleep deprivation, pointing to the essential role of sleep (or demands created by sleep loss) in regulating circadian gene fluctuations (Noya et al., 2019).

Clock genes can operate within large networks that regulate stress and metabolic pathways, and the great variability between rhythms might provide flexibility to deal with diverse challenges. These clock genes may be influenced by epigenetic changes thereby promoting metabolic diseases as well as aging processes and age-related conditions, such as neurodegenerative disorders and cancer. Likewise, numerous miRNAs appear to influence clock genes, which may be relevant to the development of diseases associated with circadian rhythm disturbances. Dysregulation of clock genes has indeed been associated with several forms of cancer, particularly those involving endocrine tissues. As depicted in Fig. 11.1, circadian timing is aligned with multiple clock genes that comprise loops that feed on one another

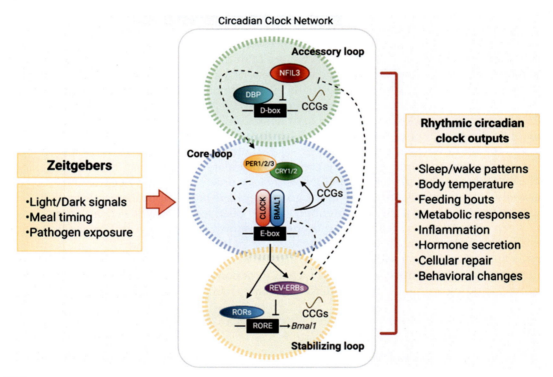

FIG. 11.1 Circadian networks set the metronome of life. The highly conserved molecular circadian gene network is made up of a core negative feedback loop of transcription factors that respond to environmental Zeitgebers over a ~24h period. The feedback loop consists of two primary activators (CLOCK and BMAL1) that dimerize and bind to E-Box promoter regions which induce transcription of clock-controlled genes (CCGs), including repressors Period 1–3 (PER) and Cryptochrome 1/2 (CRY). These proteins heterodimerize to prevent further transcription induction by CLOCK:BMAL1. Stabilizing loop components contribute to molecular clock regulation, including retinoic acid-related orphan nuclear receptors REV-ERBα and RORα, which induce or repress Bmal1 transcription via RORE promoter binding. The accessory loop is made up of the activator albumin-D-box-binding protein (DBP) and repressor nuclear factor interleukin 3 (Nfil3), which influence Per gene expression via D-box promoter binding. These loops drive and maintain rhythmic expression of the underlying circadian clock network and each contributes to overlapping and mutually exclusive sets of CCGs that drive rhythmic physiological responses. *Figure and caption from Frazier, K., Frith, M., Harris, D., Leone, V.A., 2020. Mediators of host–microbe circadian rhythms in immunity and metabolism. Biology 9, 41.*

to maintain rhythms related to sleep, feeding, metabolic, immune, and cellular repair processes. Disruptions of the circadian loops can ultimately influence the occurrence of multiple illnesses (Frazier et al., 2020).

Circadian rhythms: Immune and cytokine changes

As in the case of other endogenous processes, immune functioning varies over the circadian cycle, operating in conjunction with hormones that can affect immune activity and inflammatory immune signaling (Besedovsky et al., 2016). The number and functioning of circulating lymphocyte subsets, granulocytes, monocytes, and NK cell activity vary over the course of a night's sleep (Irwin, 2019), and circadian factors are likewise associated with the response of immune cells (e.g., CD8+ T cells) to antigenic challenges, possibly being controlled by the clock gene *Bmal1* (Nobis et al., 2019). In line with the immune cell alterations, several cytokine changes were associated with nocturnal sleep, notably the production of IL-2, IFN-γ, and augmented IL-12 production by dendritic cells and monocytes. The elevated cytokine levels that occur toward the end of the sleep period might be important to prepare the organism for the challenges (e.g., pathogens) that could be experienced in the day ahead (Irwin, 2019).

The relations between sleep and immune/cytokine functioning are bidirectional. Just as cytokine alterations may affect sleep, both sleep and circadian disruptions were accompanied by changes of proinflammatory cytokines, such as IL-6 and TNF-α (Besedovsky et al., 2019; Irwin et al., 2015). In general, elevated levels of proinflammatory cytokines IL-1β, IL-6, and IL-17 occurred with several nights of partial sleep deprivation (Besedovsky et al., 2019), which normalized once proper sleep was obtained. These cytokines are differentially sensitive to sleep loss as several cytokine changes are elicited with a day or two of sleep loss, whereas others only appear with more extensive sleep loss (Irwin and Opp, 2017). Through effects on immune and inflammatory processes, sleep can diminish the risk of infection and can lead to better outcomes if infection is present. In fact, infection may promote sleep, presumably as an adaptive response to enhance the capacity of the immune system to deal with pathogens, although infection under some conditions can also disrupt sleep.

Cytokine variations might not simply be a consequence of sleep being aligned with circadian rhythms, although IL-6 variations that were related to circadian processes were also tied to REM sleep. It is also relevant that variations of IL-6 may contribute to the fatigue associated with illness, possibly serving as an adaptive response to promote sleep. In rodents, administration of proinflammatory cytokines increased non-REM sleep, whereas inflammatory cytokine antagonists and antiinflammatory cytokines, notably IL-4 and IL-10, elicited the opposite effect (Irwin and Opp, 2017). Fig. 11.2 shows the effects of various types of challenges (stressors, microbial, and pathogens) on immune-related processes and their effects on REM and NREM sleep.

Elevated levels of inflammatory cytokines and peripheral antimicrobial peptides in the periphery can stimulate inflammatory cytokine activity within the brain, thereby altering sleep patterns, duration, and efficiency, and a concurrent increase of slow-wave sleep together with diminished REM sleep duration. As described in Fig. 11.3, which was adapted from Irwin (2019), chronic stressor experiences or infection influence sleep continuity and sleep characteristics, which can promote chronic, mild inflammation, and the accompanying cytokine

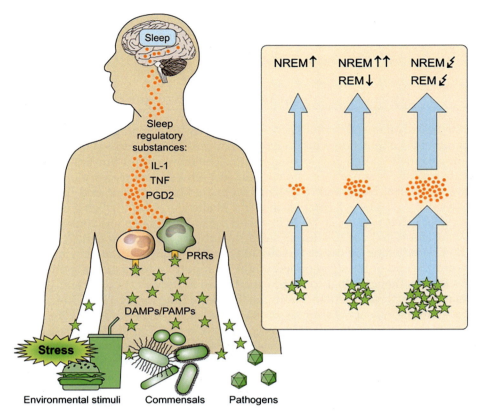

FIG. 11.2 Conceptual model of sleep changes in response to immune activation and underlying mechanisms. Environmental stimuli (e.g., food intake, stress), commensal bacteria, and infectious pathogens (here illustrated as viruses) are recognized by the immune system as damage- and pathogen-associated molecular patterns (DAMPs and PAMPs, *green stars*), which activate pattern recognition receptors (PRRs, *orange polygon*) on innate leukocytes. This PRR activation induces an inflammatory response with the production of sleep regulatory substances, such as interleukin (IL)-1 and tumor necrosis factor (TNF) (both represented by *orange dots*), which reach the brain and promote nonrapid eye movement (NREM) sleep *(left arrow)*. In higher doses (e.g., during an infection; *middle arrow*), these sleep regulatory substances may also suppress rapid eye movement (REM) sleep. Prostaglandin (PG) D_2 is shown as a potential further mediator of sleep changes in response to immune activation. These sleep responses to immune activation are assumed to be adaptive. Subtle immune activation may be involved in homeostatic NREM sleep regulation that in turn could serve to restore immune homeostasis. More pronounced immune activation during an infection can induce a sleep response that in turn may support host defense and immunological memory formation. However, an extreme immune activation (e.g., during severe infection; *right arrow*) seems to disrupt both NREM and REM sleep, often accompanied by sleep fragmentation, feelings of nonrestorative sleep, and daytime fatigue. Notably, most of our knowledge is based on animal research, and confirmation in humans is still needed. *Figure and caption from Besedovsky, L., Lange, T., Haack, M., 2019. The sleep-immune crosstalk in health and disease. Physiol. Rev. 99, 1325–1380.*

variations. These challenges, in turn, may affect brain processes, which could favor the production of diseases, such as depression and neurodegenerative diseases. Concurrently, the varied immune disturbances can favor greater susceptibility to viral infection, provocation of biological aging, and the repair of DNA damage may be undermined. The diverse actions of chronic stressors or chronic sleep loss may also contribute to cancer progression.

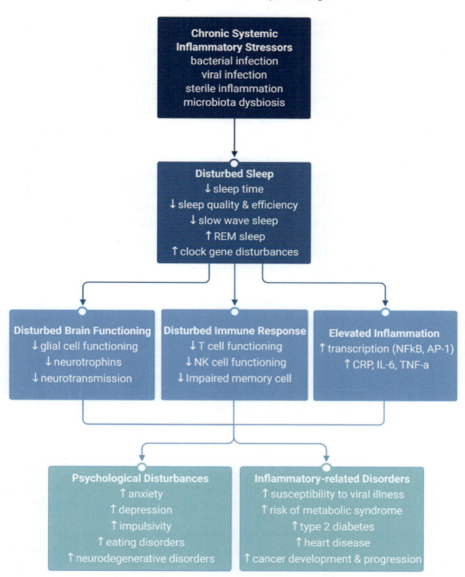

FIG. 11.3 Chronic psychological and systemic challenges promote varied disturbances related to sleep. The sleep disturbances directly increase circulation inflammatory factors and disturb immune functioning and can exacerbate the actions induced by chronic stressor experiences. These actions may disrupt brain functioning, thereby promoting psychological disorders. The multiple immune- and inflammatory variations stemming from disturbed sleep related disorders likewise favor the development of several physical illnesses. *Based on Irwin, M.R., 2019. Sleep and inflammation: partners in sickness and in health. Nat. Rev. Immunol. 19, 702–715.*

Sleep, circadian rhythms, and cancer progression

In considering the relationship between sleep disturbances and cancer, it is again important to distinguish between specific types of sleep disorders. Sleep disturbances may entail specific conditions (e.g., sleep apnea, narcolepsy), interrupted or shortened sleep, and sleeping too long or excessively. Some sleep problems may presage cancer occurrence, whereas others are a result of the cancer being present. Sleep disturbances in cancer patients might drive inflammation, thereby worsening cancer progression and might promote depressive features among cancer survivors as well as the fatigue common among patients. The processes responsible for the cancer-related sleep disturbances or those that persist following therapy have not been fully identified, but at least some that stem from cancer may have been fueled by changes of IL-6 levels that developed. Among women assessed 6 months after treatment for ovarian cancer, variations were apparent related to plasma IL-6 together with reduced nocturnal cortisol secretion and altered diurnal cortisol rhythm. These physiological measures were associated with fatigue, vegetative symptoms of depression, and general disability. In fact, genome-wide association studies identified hundreds of genes that predicted varying degrees of insomnia, many of which were the same as those that had been linked to depression, metabolic disturbances, and heart disease.

Sleep duration in predicting later cancer occurrence

Sleep duration and sleep disorders were frequently found to be predictive of subsequent illness development. Both severely shortened (< 5 h) and extended sleep (> 8.5 h) were associated with elevated levels of several proinflammatory cytokines, and a detailed analysis based on the findings of 153 studies indicated that abbreviated sleep was accompanied by increased risk of obesity, type 2 diabetes, varied heart diseases, and earlier mortality (Itani et al., 2017). Features of sleep and circadian disturbances similarly operate in tandem in promoting breast cancer occurrence (Samuelsson et al., 2018). Prospective studies indicated that frequent sleeping difficulty (on more than four nights a week) was associated with elevated risk of overall and postmenopausal breast cancer, while light exposure at night (having light or television on during sleep) was accompanied by increased estrogen-related breast cancer. For men, the risk of prostate cancer was appreciably elevated if there were problems both falling and staying asleep.

In line with the immune changes that had been reported, a J-shaped relationship existed between sleep duration and cancer occurrence and cancer-related mortality. Both short (< 7 h) and long sleep durations (> 9 h) predicted the occurrence of some forms of cancer, as well as greater risk of all-cause mortality, but a detailed analysis of prospective studies indicated that this relationship was far more prominent with long sleep durations (Ma et al., 2016). It also appeared that sleep duration was more closely aligned with some forms of cancer than others, although there have been inconsistent reports in this regard. A systematic review revealed that both short and long sleep durations were associated with mortality related to breast and lung cancer but not colorectal, ovarian, or prostate cancer (Stone et al., 2019). Yet a review of prospective studies that followed individuals over 7.5–22 years indicated that long sleep duration was associated with a greater risk of colorectal cancer while being inversely related to hormone-related cancers (Zhao et al., 2013). It was similarly concluded based on the

Health Professionals Follow-Up Study that long sleep duration (> 9 h) was accompanied by an elevated risk of colorectal cancer, occurring more robustly in men than in women. Among postmenopausal women, long sleep duration was accompanied by a moderate increase of liver cancer, but this was only evident among obese women (Royse et al., 2017).

A large-scale meta-analysis that included data from 10 studies supported a J-shaped relation between sleep duration and breast cancer among women who developed estrogen receptor-positive breast cancer but not among women who developed estrogen-negative breast cancer (Lu et al., 2017). In addition to predicting cancer occurrence, a 30-year follow-up within the Nurses' Health Study indicated that both regular occurrence of sleep problems and long sleep duration following diagnosis were associated with an elevated risk of earlier all-cause and breast cancer-specific mortality (Trudel-Fitzgerald et al., 2017).

Although the data have generally pointed to the risks of cancer being more closely aligned with long than short sleep durations, this has not been the rule as several studies indicated that abbreviated sleep (< 6 h) represented a risk factor for chronic diseases, including cancer, and cancer-related mortality was especially notable with severe sleep loss (e.g., sleeping 4–5 h a night). As so often occurs in other contexts, this relationship appeared to be moderated by several features of the individual. For instance, a 19-year follow-up study revealed that short sleep duration associated with increased incidence of breast cancer was most notable among postmenopausal women and in those with low parity (low number of children). Ethnicity may also be tied to the sleep duration and cancer relationship. Incident breast cancer was associated with sleep disturbances in the extensive Women's Health Initiative, and a subset of data from this study indicated that abbreviated sleep (< 6 h) was associated with more aggressive breast cancer in African Americans than in white people of European descent (Soucise et al., 2017).

During scheduled colonoscopies, the detection of colon cancer was 50% more common among individuals who slept less than 6 h a night (Thompson et al., 2011) and in an 11-year follow-up of the Women's Health Initiative observational study that included more than 75,000 postmenopausal women, both short (< 5 h) and long sleep (> 9 h) were accompanied by increased colorectal cancer risk (Jiao et al., 2013). Similarly, short sleep duration and elevated cancer risk were particularly prominent among Asian people, and long sleep duration was specifically associated with increased risk of colorectal cancer (Chen et al., 2018).

Why specific forms of cancer were more likely to be associated with sleep problems in particular ethnic or cultural groups is uncertain despite the diversity of factors that have been assessed. Among other things, the incidence of particular gene polymorphisms that exist across cultural groups could interact with the actions of sleep disturbances. Moreover, differences in lifestyles (e.g., smoking, alcohol consumption, specific diets), weight gain, physical inactivity, may contribute to ethnic or cultural differences.

The curse of sleep disorders

Sleep apnea in which breathing repeatedly stops and starts can be a serious and sometimes deadly disorder. The most common form of this disorder, obstructive sleep apnea, has been linked to excessive weight, a narrowed airway, being an older male, having a family history of this condition, and it has also been tied to smoking and the use of alcohol and sedatives. The risk of obstructive sleep apnea was also predicted based on chronic conditions, such as asthma, high blood pressure, type 2 diabetes, and other metabolic disorders. Another form of apnea, termed as central sleep apnea, is less frequent and appears to occur owing to periodic failures of brain signals to properly modulate muscles associated with breathing.

Sleep apnea was associated with the subsequent development of hypertension and cardiovascular problems, as well as complications secondary to surgery, and was also predictive of the greater incidence of cancer and cancer-related mortality, possibly because this condition was associated with increased inflammation. The forms of cancer related to severe apnea included greater incidence of kidney, breast, uterine cancer, and malignant melanoma. Sleep apnea was also linked to lung cancer becoming more deadly, potentially because intermittent hypoxia results in the release of exosomes that may augment energy supplies to cancer cells (Almendros et al., 2016). The actions of intermittent hypoxia might also stem from the activation of processes (e.g., HIF-1 and VEGF pathways) that promote increased blood supply that favors tumor growth. Like the outcomes apparent in adults, sleep-related disturbances were reported in children or adolescents with leukemia or brain cancer, and obstructive sleep apnea among younger patients predicted earlier cancer mortality.

Another sleep disorder, restless leg syndrome, a condition characterized by an uncontrollable urge of individuals to move their legs owing to uncomfortable sensations, occurs among cancer patients undergoing chemotherapy at about twice the rate seen in the general population. Because of the distress created by restless leg syndrome, mood disorders and poorer quality of life were relatively common. The consequences of restless leg syndrome on cancer progression have not been systematically assessed but given the distress created by this syndrome and the occurrence of depression, it warrants further attention.

Narcolepsy is a chronic neurological condition in which brain processes seem incapable of maintaining sleep-wake cycle so individuals experience exceptionally high levels of daytime drowsiness and sudden episodes of sleep. As narcolepsy occurs infrequently, affecting about 1 in 2000 individuals (although it likely is often undiagnosed so its occurrence may be considerably higher), it hasn't received as much attention—in relation to cancer—as other sleep disorders. Nonetheless, narcolepsy has been associated with increased occurrence of cancer, being more prominent in females than males, especially in gastric and head and neck cancers (Tseng et al., 2015).

Distinguishing between specific sleep disorders is particularly important given that they might not be equally distressing and may involve different mechanisms and might thus not be similarly related to the occurrence of different types of cancer. A large epidemiological analysis across 13 countries indicated that in many studies valid conclusions might have been hindered by misclassifications of sleep problems (Erren et al., 2016). Considering the pernicious effects of sleep disturbances, incorporating sleep measures in assessments of cancer patients may have important ramifications for cancer care.[a]

---

[a] Given the importance of circadian rhythms and sleep for health and recovery, it's more than a bit perturbing that hospitals are among the most uncomfortable places to obtain sleep. Aside from being in a strange bed in new surroundings, visitors come by even when they're not supposed to, sounds emanate from hallways, and once sleep finally comes, patients might be awoken to have tests done. More than this, little attention is devoted to the timing at which meds might have their optimal effects (based on circadian factors) and instead are often administered on preordained schedules (Ruben et al., 2019).

Mechanisms linking sleep and circadian disturbances to cancer

While sleep and circadian factors are intertwined, to some extent their influence on health can be distinguished from one another. Apart from effects on sleep quality and quantity, disturbances of circadian rhythms are related to metabolic changes that can produce multiple health disturbances, including cancer occurrence and progression (Masri and Sassone-Corsi, 2018). Disturbed circadian functioning can prompt processes that affect cell senescence, metabolic processes, DNA damage repair, apoptosis, angiogenesis, and ultimately cancer occurrence (Sancar et al., 2015). Furthermore, several types of cancer can promote a hyperactive cell cycle so tumor cell multiplication is facilitated, presumably owing to the modification of inflammatory factors that influence cell growth, differentiation, and survival. Some of the linkages between breast cancer and both sleep and circadian disturbances are outlined in Table 11.1 (Samuelsson et al., 2018). This listing is by no means exhaustive, and the profile related to other cancers can be very different. A reciprocal relationship exists between the circadian clock and epigenetic processes that can affect the transcription and translation of core circadian genes, which can then affect further epigenetic variations. The epigenetic and clock gene variations are believed to contribute to breast, colorectal, and blood cancers (Hernández-Rosas et al., 2020).

Circadian changes have been linked to metabolic enzymes as well as oncogenes that could in some way increase the occurrence of pathology. Similarly, orexin and adenosine, which vary with circadian cycles, as well as cortisol and estrogen, which affect immune functioning, could be linked to cancer (Samuelsson et al., 2018). It is difficult to say which circadian-related processes are responsible for the links to specific types of cancer.

Focusing on clock genes in cancer

Virtually every hormone is regulated by circadian clocks, such that some peak during different portions of the night—this includes vasopressin, growth hormone, melatonin, leptin, thyroid-stimulating hormone, prolactin, ghrelin, and FGF-21. During the day, however, others reach their zenith or peak either early in the day (e.g., cortisol) or somewhat later (adiponectin, insulin). This means that the tissues or cells targeted by hormones are more (or less) affected at different times of the 24-hour cycle. In fact, most physiological processes are rhythmic in nature, and metabolic and immune processes are no exception, given their sensitivity to neurohormonal factors. As we've already seen, immune and metabolic function is tied to circadian cycles, and this could potentially be related to the occurrence of some types of cancer. Numerous hormones and circadian variations of these hormones (e.g., melatonin, cortisol, epinephrine, estrogen) have also been linked to cancer occurrence and might influence the effectiveness of cancer therapies. Positive effects of these hormones can, however, emerge owing to actions other than those associated with sleep and circadian processes.

Genes that act as either positive or negative regulators of circadian functioning have been linked to various forms of cancer. A favorable prognosis for patients with a form of nonsmall-cell lung carcinoma (adenocarcinoma) was tied to overexpression of several genes, such as *Cry2*, *BMAL1*, and *RORA* together with underexpression of *Timeless* and *NPAS2* (Qiu et al., 2019). Disturbances of other core clock genes (*Per1*, *Per2*, and *Per3*) were similarly found in oral cancers, head and neck, kidney, renal cell, pancreatic, and breast

TABLE 11.1 Pathways between circadian and sleep disruption and mammary oncogenesis.

Mechanistic Pathway	Evidence for Effects Circadian Disruption	Evidence for Effects of Sleep Disruption
DNA DAMAGE & OXIDATIVE STRESS		
Increased oxidative stress, oxidative DNA damage	Limited	✓✓
Decreased antioxidant effects	✓	Theoretical
MELATONIN & ESTROGEN		
Decreased melatonin release	✓✓	Limited
Increased mammotropic hormone production and/or release	✓✓	✓
INFLAMMATION & IMMUNE FUNCTION		
Decreased NK cells & cytotoxicity		✓✓
Shift to Th-2 cytokine production & tumor cell survival		✓
Decreased Th-1 cytokine production	✓✓	✓✓
Chronic inflammation via pro-inflammatory activation	✓	✓✓
Increased oxidative stress, oxidative DNA damage	Limited	✓✓
Decreased antioxidant effects	✓	Theoretical
Increased repeated SNS activation		✓✓ (especially via sleep-disordered breathing)
Changes in glucocorticoid production & feedback	Limited	✓
METABOLIC FUNCTION		
Metabolic disruption (e.g. insulin, glucose, leptin, ghrelin)	✓✓	✓✓ (especially via sleep-disordered breathing)
Increased adiposity & obesity	✓✓	✓✓

Notes: ✓✓ = strong research support; ✓ = some/mixed research support; "limited" = support from 1 to 2 studies; "theoretical" = no known support, compelling theoretical association.

From Samuelsson, L.B., Bovbjerg, D.H., Roecklein, K.A., Hall, M.H., 2018. Sleep and circadian disruption and incident breast cancer risk: an evidence-based and theoretical review. Neurosci. Biobehav. Rev. 84, 35–48.

cancer as well as squamous cell carcinoma. For instance, reduced expression of *Per2* and *Per3* was associated with increased aggressiveness and a poorer prognosis of colorectal cancer, whereas increased expression of *BMAL1* was linked to reduced survival (Karantanos et al., 2014). It had been suggested that *Per2*, which is present within just about every cell within the body, acts as a tumor suppressor, and disturbances of this gene's functioning allows for uncontrolled cell proliferation, genomic instability, and inflammation that promotes tumor progression.

Consistent with the fundamental role of clock genes in the promotion of cancer, polymorphisms in the *CCRN4L* and *Per3* genes were associated with increased risk of nonsmall-cell lung carcinoma, and polymorphisms of several clock genes were linked to breast, prostate, and ovarian carcinogenesis. Paralleling the profiles associated with gene mutations, epigenetic processes related to clock genes may contribute to cancer development and progression (Masri et al., 2015). Among women with breast cancer, hypermethylation was observed in *PER 1, 2*, and *3* within cancer tissue, whereas hypomethylation was detected in several other genes, namely that of *CLOCK, BMAL1*, and *Cry2*, varying with the features of the cancer (Lesicka et al., 2019). Another circadian factor, CRY-1, which is induced by androgen receptors, was similarly elevated in prostate cancer. It appeared that CRY-1 regulates processes involved in DNA repair and might thus serve as a target to diminish prostate cancer, and such procedures might best be undertaken at times that correspond to the functioning of this circadian cryptochrome (Shafi et al., 2021).

The notion that manipulations of biological clocks could be used to slow cancer progression has been enticing, particularly since circadian clock disturbances have been tied to changes within genes that code for tumor suppressor cells. Individual cancer cells, like any other cells, have their own clocks; alterations of these clock genes can favor tumor development and so manipulating these genes or their actions could affect tumor progression. Mutations of clock genes were accompanied by accelerated growth of tumors in mice, and in genetically engineered mice that develop lung adenocarcinoma, simulated jet lag, and mutations of the central circadian clocks, notably core circadian genes *Per2* and *Bmal1*, favored lung tumor progression and reduced survival. These clock genes could act in a tumor-suppressive capacity in that their loss increased c-Myc expression that has been associated with many types of cancer (Papagiannakopoulos et al., 2016). Similarly, 1 month following injection of melanoma cells, mice that had been maintained under conditions in which their light and dark periods were shifted every 6h over 2days, thereby simulating travel across time zones, subsequent tumors were three times larger than in mice that had not experienced the light-dark changes. This was accompanied by the rhythms of immune cells being disrupted and macrophages within the tumor microenvironment were more accommodating to cancer cells (Aiello et al., 2020).

Of the many hallmarks of cancer that have been identified, a considerable portion is subject to circadian variations (see Fig. 11.4). Manipulations that influence clock genes or the proteins related to them could potentially serve to influence the effectiveness of cancer therapies. It is uncertain why individuals with specific clock gene variations may exhibit different types of cancers, especially as so many factors could be interacting with circadian rhythms. Whatever relations exist between these variables, they are not straightforward, potentially being moderated by other essential factors, such as whether individuals had a morning or evening preference, and any number of other lifestyle and experiential influences. Assuming that circadian variations are linked to chronic illness, it is significant that it is possible to

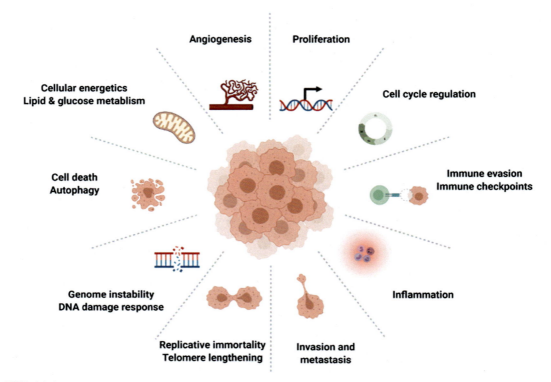

FIG. 11.4 Circadian clock genes and circadian functioning contribute to many of the factors that are frequently considered to comprise the hallmarks of cancer. *Based on Hadadi, E., Acloque, H., 2021. Role of circadian rhythm disorders on EMT and tumour-immune interactions in endocrine-related cancers. Endocr. Relat. Cancer 28, R67–R80.*

normalize circadian rhythms through manipulation of mealtimes, behavioral therapy, and drug treatments. Furthermore, circadian disturbances may have prognostic value concerning some forms of cancer, such as triple-negative breast cancer, and could potentially inform the most effective treatment strategies. As we'll see, it is possible to take advantage of circadian variations so that therapeutic agents can be administered at times when cancer cells are most vulnerable to being destroyed.

Occupations associated with altered sleep cycles and cancer occurrence

It is possible to distinguish night owls and early risers based on their genetic profiles. When an individual's genetic profile related to their sleep pattern and their actual activities were misaligned (e.g., being a genetic night owl, but having to arise early), the occurrence of anxiety and depression was more likely to be evident. Jobs that entail shift work can similarly be a serious challenge to biological rhythms and have been associated with increased vulnerability to a variety of illnesses, such as gastrointestinal disorders, obesity, metabolic syndrome, diabetes, cardiovascular diseases, stroke, and cancer (Kecklund and Axelsson, 2016).

At least some of these conditions may arise owing to sleep or circadian rhythm disturbances that provoke elevated cortisol, diminished insulin sensitivity, as well as altered immunity, cytokine elevations, and oxidative stress. Even 3 days of simulated night shift work among healthy participants were sufficient to disturb the daily rhythms of cancer-related genes so DNA damage would be more likely to occur, and DNA repair mechanisms would be mistimed (Koritala et al., 2021).

Based on the relationships between sleep loss and cancer, it could reasonably be hypothesized that occupations that entail shift work, including occupations that produce jet lag, would be related to the elevated occurrence of some forms of cancer. While this was observed in some studies, this relation was not consistently apparent. A population-based case-control study in several thousand participants revealed that night shift work was unrelated to prostate cancer occurrence, and prospective studies reported that shift work was not predictive of breast or prostate cancer (Travis et al., 2016). An extensive systematic review and meta-analysis that included 53 studies and about 8.47 million respondents suggested that a history of having been engaged in shift work was not related to the increased occurrence of several different forms of cancer (Dun et al., 2020). Given the discordant results reported, it is important to dig deeper into the links between shift work and cancer occurrence, although it should be said that the report of Dun et al. (2020) was remarkably detailed.

Two prospective studies (The Nurses' Health Studies) that tracked about 78,000 and about 114,000 women might shed light on why contradictory findings regarding night shift work were observed in the incidence of breast cancer. In the first of these studies, breast cancer occurrence was not elevated (Poole et al., 2011), whereas, in the second study, which assessed women who began night shift work at a younger age, cancer risk was elevated (Wegrzyn et al., 2017). It is possible that the different outcomes were due to the combination of age of participants and the duration of the night shift work experience. Aside from breast cancer, slightly or moderately increased risk for prostate, colon, endometrial malignancies, and non-Hodgkin's lymphoma was also observed among longtime night workers and shift workers (Lin et al., 2015). Shift work was likewise accompanied by elevated risk of melanoma, increasing by 2% for every year of shift work (Yousef et al., 2020), and cancer incidence among men who had engaged in night shift work for 20 years or more and who slept for more than 10 h a day was elevated by 27% (Bai et al., 2016). Once again, however, the data concerning the links between cancer and duration of shift work have been inconsistent. A 10-year follow-up study as part of the Nurses' Health Study II indicated that long-term night shift work was not accompanied by elevated melanoma or squamous cell carcinoma and among nurses who also slept less than 6 h a night the incidence of melanoma was diminished, particularly among naturally dark-haired women, a physical attribute that is associated with a lower risk for skin malignancies (Heckman et al., 2017). These effects were not related to the duration of shift work, menopausal status, or UV radiation exposure during adulthood. Thus, it is uncertain why the link between shift work and melanoma and basal cell carcinoma was so different from other forms of cancer.

In evaluating the relation between shift work and the occurrence of diseases, yet another related feature ought to be brought into the mix. Among some shift workers, their routine is maintained unaltered for years or even decades, whereas in other instances shift work routines may vary every few weeks. Individuals might spend 2 weeks on the night shift and then be transferred to the day shift. This might appear as an equitable strategy so no individual is consistently burdened by an undesirable schedule, but there may be a downside to this. Shift workers may experience circadian misalignment comprising the incongruity between the brain and peripheral rhythms, which could be brought about by changes of shift work schedules as well as the mismatch between sleep/wake cycles and feeding rhythms. While cells in the brain might adjust to changes of shifts in work periods relatively quickly, cells in the body might not keep pace, profoundly affecting diverse biological processes that could promote illnesses (Khalyfa et al., 2020). Consistent with this possibility, a 28-year follow-up study indicated that circadian disturbances stemming from a rotating day-night work schedule were linked to fatal ovarian cancer (Carter et al., 2014). Data obtained through the Nurses' Health Study also revealed that colorectal cancer was associated with rotating night shift work, particularly among nurses who had been engaged in this for more than 15 years. A modest increase of lung cancer was also apparent under these conditions, although this outcome was evident only in those who smoked, suggesting that variable shift work was a second hit that promoted lung cancer in smokers (Schernhammer et al., 2013).

Several studies attempted to link night shift work to inflammatory processes relevant to cancer. Circulating IL-6 was elevated in night shift workers and inflammatory cytokine levels in nurses were greater after night shifts than after day shifts (Bjorvatn et al., 2020). Levels of IL-1β and IL-6 were most elevated among nurses who experienced greater variability in sleep cycles and duration (Slavish et al., 2020). The processes that promote these outcomes are uncertain, although variable sleep patterns were associated with elevated self-reported perceived stress that could engender multiple pro-tumorigenic actions. Whatever the source, these findings point to the importance of unstable sleep patterns in determining inflammatory changes that could favor disease occurrence.

With so many inconsistent reports regarding the link between shift work and cancer occurrence, it is difficult to be confident as to whether there is a real connection. Which can make the search difficult when trying to identify the key factors that might be responsible for the negative consequences that have been reported. Nonetheless, one can be fairly confident that if there is a relationship between shift work and cancer, it is likely moderated by a constellation of lifestyles (increased calorie intake and a preference for high carbohydrate foods) and socioeconomic factors, as well as comorbid conditions (e.g., obesity). Despite the uncertainties concerning the links between shift work and cancer, the International Agency for Research on Cancer (IARC) classified night shift work as a *probable* carcinogen to humans (Zhang and Papantoniou, 2019).

Light at night

Understandably, the focus of research in humans has been on factors that act to promote sleep disturbances, sleep loss, and circadian rhythm disruptions related to shift work. Less attention has been devoted to the impact of light at night (produced by artificial light), which is present in and out of the house, and has come to be known as "light pollution," a mainstay of urban living. Although an obvious practical invention ever since the burning candle stick or lantern graced people's dwellings, there is reason to believe that the excesses of light exposure that allows dusk to dawn night life to persist can have marked physiological consequences. In many animal species light at night, even at low levels, can affect sleep and various hormonal processes (e.g., melatonin) and can influence the behavior of nocturnal mammals, alter the foraging behaviors of birds, and pollination by insects. Humans in cities are flooded by light long into the night, and lighting in our homes is out of whack with natural light. Studies in humans have pointed to adverse effects of light at night, including the development of cancer possibly operating through interactions with core clock genes (Walker et al., 2020). To be sure, the data concerning the health risks created by light at night have not been entirely consistent, in part owing to methodological issues. Well-controlled prospective studies are needed to determine the breadth of these effects, and whether steps need to be taken to limit the adverse effects of artificial light.

Sleep disruption and diurnal variations in relation to microbiota

A strong case was made that sleep and circadian disruptions and illnesses that follow may be mediated by microbial changes (Nobs et al., 2019). For instance, chronic sleep fragmentation (recurrent brief arousal during sleep) was accompanied by the increased presence of microbial families that were tied to greater inflammation produced by visceral white adipose tissue, and hence altered insulin sensitivity. Sleep deprivation has comparable effects, and partial sleep deprivation stemming from obstructive sleep apnea was linked to microbial changes.

Gut microbiota have their own daily rhythms, and through their metabolites (e.g., butyrate, polyphenolic derivatives, vitamins) influence host metabolic homeostasis and circadian clock functioning within various tissues. These microbial rhythms are likely entrained by the pineal gland and by melatonin, as well as by *PER1* and *PER2* genes. Variations of microbiota may also be important in triggering sleep. Administration of tributyrin, which essentially is a form of butyrate, elicited a 50%–70% increase of NREM sleep (Szentirmai et al., 2019). Bacterial cell wall factors can similarly trigger sleep, possibly by provoking PAMP recognition receptors, which instigate cytokine release (Krueger and Opp, 2016).

The abundance of *Lactobacillus* ordinarily declines during the active phase of the diurnal cycle and increases during the resting phase of the cycle. However, a shift of the light–dark cycle can promote gut microbiota changes that cause disturbed gut barrier integrity (Deaver et al., 2018). Like these microbial circadian changes, certain immune cells within the gut, such as innate lymphoid cells (e.g., ILC3), have clock genes that are sensitive to sleep perturbations, so when sleep is disturbed, inflammation is provoked. In addition, disturbed migration of immune cells in and out of the gut, which is controlled by brain circadian clock processes, may affect the ability to fight gut pathogens (Godinho-Silva et al., 2019). It also appeared that microbiota diversity and richness of *Bacteroidetes* and *Firmicutes* were directly related to IL-6 as well as to sleep efficiency and sleep duration, whereas several taxa (*Lachnospiraceae*, *Corynebacterium*, and *Blautia*) were negatively related to sleep indices. As depicted in Fig. 11.5, numerous environmental factors affect diurnal variations of microbiota that can ultimately affect physical and psychological well-being (Murakami and Tognini, 2020).

Like so many other intrinsic and extrinsic factors, food consumption is tied to circadian mechanisms related to various hormones, such as ghrelin. Disturbed sleep and microbial diversity elicited by a stressor in rodents could be attenuated by an earlier prebiotic diet over 5 weeks. The actions of the diet were also accompanied by variations of secondary bile acids and neuroactive steroids that were altered by the stressor. By providing rats with a prebiotic diet that contained specific bacteria, disturbances of sleep-wake cycles that were

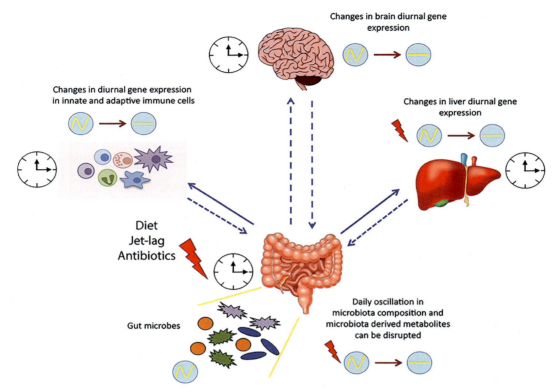

FIG. 11.5 Microbial regulation of circadian physiology. Diurnal oscillation of microbiota synchronizes the whole body system through microbiota-derived signals and metabolites. Dysbiosis induced by various factors, such as diet and jet lag, alters the expression of several circadian genes, resulting in a detrimental impact on host health. *clock shape circle*, endogenous clock of different tissues; *red lightning bolts* denote disruption in diurnal gene expression; *circle with wave* denotes schematic representation of diurnal oscillation; *circle with horizontal line* refers to diurnal variations being absent. *Figure and caption from Murakami, M., Tognini, P., 2020. The circadian clock as an essential molecular link between host physiology and microorganisms. Front. Cell. Infect. Microbiol. 9, 469.*

disturbed (e.g., by stressors or by altering light-dark cycles) could be realigned. Furthermore, as immune changes instigated by microbiota may be linked to circadian rhythms, they may occur in anticipation of food arrival (Murakami and Tognini, 2020), just as this occurs with hormones, such as ghrelin. This could, in turn, affect obesity and metabolic syndrome, which influence immune activity and may thereby affect enteric infections.

In addition to these broad relationships, the actions of some clock genes may be linked to specific microbiota that interact with dietary factors. For instance, knockout of a clock gene (*delta19*) was accompanied by altered intestinal microbiota, and disturbances of diurnal rhythmicity had similar consequences. Furthermore, disturbing microbiota circadian rhythms may promote a proinflammatory state that has downstream health consequences. Such actions might come about through epigenetic effects that occur with gut microbial disturbances (Kuang et al., 2019), perhaps by affecting receptor processes (e.g., G-coupled receptors), or through effects on pattern recognition receptors that influence inflammatory processes.

To summarize, sleep and circadian rhythms, which are influenced by several clock genes, can affect gut microbiota and hence immune functioning and inflammatory processes that may promote tumorigenesis. Shift work or jobs that produce jet lag result in microbial changes, and the development of diseases linked to inflammation, including obesity, which may affect cancer occurrence. Understanding more about circadian variations may have important ramifications regarding the ties between specific clock genes and the appearance of certain types of cancer, and may even be instrumental in determining strategies to prevent cancer occurrence.

Implications of circadian rhythmicity to cancer treatments

Chronotherapy entails the administration of therapeutics based on biological rhythms to obtain maximal benefits and the fewest undesirable effects. This approach has taken on increasing allure in the treatment of cancers, and it was considered that circadian dynamics could influence the efficacy of various agents used to treat cancer (Harper and Talbot, 2019), as well as the toxicity associated with treatments. Essentially, what this means is that cancer treatments ought to be administered at times that coincide with phases of the circadian cycle when cancer cells are most vulnerable to destruction. For instance, in cell cultures, inhibition of p38 mitogen activated protein kinase, which contributes to the aggressive and invasive features of glioblastoma, can be somewhat deterred if this is done at a time when cancer cells were particularly active. In human glioblastoma patients the administration of therapies at specific times was effective in extending life for several months relative to that evident when treatment was administered at other times. Likewise, in rodents, cancer cells multiply most prodigiously during their active phase (at night), and inhibition of EGFR to treat breast cancer might be most effective during this phase of the diurnal cycle (Lauriola et al., 2014). It has also been reported that epigenetic changes may occur that allow for macrophages to become more proficient in attacking cancer cells and it may be possible to train macrophages to be more effective in carrying out these duties by exposing them to extracellular stimuli, such as a pathogen or a vaccine. However, training macrophages to do this may vary with the nature of the training and when this occurs. It seems that the inflammatory factor NFκB may play a role in allowing access to tightly compacted DNA, doing so in response to specific threats. However, NFκB goes through different cycles so at some phases it is stable, but at other times it oscillates, essentially being unstable. It is uncertain what governs these phases, although it may be regulated by circadian factors. More to the point, when NFκB is in a steady state it is more apt to allow access to specific genes that would be subject to epigenetic changes which could permit macrophages to become more aggressive in response to threats (Cheng et al., 2021).

It has been known for at least five decades that the timing of treatments can affect their efficacy, but only during the past decade have recommendations related to this been seriously considered. In this regard, a greater understanding of individuals' circadian rhythms may benefit precision medicine approaches to treatment and to limit the side effects of therapy. For instance, the migration of dendritic cells from skin to lymph nodes occurs just prior to the organism's active phase (morning for humans) and therapies administered at this time could yield optimal effects (Holtkamp et al., 2021). It similarly appeared that in a mouse model, side effects, such as heart damage, varied with the timing of radiation therapy (Dakup et al., 2020). The appropriate timing of treatments might also make it possible to use lower drug doses, thereby reducing potential short- and long-term side effects as well as limiting the development of resistance (Ruben et al., 2019). Thus, cancer treatment delivery (e.g., using programmable-in-time infusion pumps and devices to monitor rhythmic physiology) could be adjusted to map onto an individual's circadian rhythms, thereby augmenting treatment efficacy and tolerability. And it may turn out to be especially significant if cancer-related insomnia and impaired circadian rhythms might be retrained through behavioral interventions and could then be adopted to enhance treatment responses (Sulli et al., 2019).

Summary and conclusions

Loss of sleep and particular sleep-related disorders can foster a range of psychological and physical illnesses, including some forms of cancer. Disturbances of circadian rhythms and control clock genes may likewise affect cancer-related processes owing to immune dysregulation and inflammatory actions, metabolic changes, microbial variations, effects on apoptosis, and impaired DNA repair processes. Importantly, sleep alterations can serve as a stressor, promote obesity, and favor the appearance of depression, which could promote multiple microbial, endocrine, and proinflammatory biases that could affect cancer progression and the efficacy of cancer therapies.

Sleep disturbances are certainly among the most problematic secondary features associated with some forms of cancer. The distress created by cancer treatments can have marked repercussions on immune and microbial processes, which could exacerbate cancer progression and undermine therapeutic responses, as well as promoting greater side effects of chemotherapy. Because of the significance of sleep problems associated with many forms of cancer, greater attention has been focused on ways to diminish such difficulties in the management and care of cancer patients.

The occurrence of sleep disturbances may vary with cancer type, and it may be of practical importance that sleep disturbance and fatigue are frequently evident in advance of diagnosis and might serve as an early marker of disease occurrence and progression. Among individuals with insomnia prior to chemotherapy, their propensity to exhibit later immune disturbances was elevated and they were more likely to develop infections related to the treatment. Not surprisingly, dramatically altered sleep and circadian rhythms have been observed in critically ill individuals and have been notable among cancer patients (Masri and Sassone-Corsi, 2018). Following diagnosis, marked stress-related microbiota changes may persist and the symptom cluster comprising sleep disturbances, fatigue, and depression predicted the aggravation of symptoms that subsequently emerged following treatment. In essence, sleep disturbances and the associated microbial variations could be a forerunner of cancer progression, even predicting the continuation of symptoms following therapy.

It is now commonplace that most cells within the mammalian body are subject to rhythmic variations. Immune functioning and cytokine presence also vary with the circadian cycle, and commensurately, circadian clock genes have a say in the healing process associated with surgery as well as the effectiveness of antiinflammatory agents (the use of antiinflammatories in the morning and analgesic at night seem to provide the best effects) (Scheiermann et al., 2018). As such, it might be possible to develop strategies aimed at realigning rhythmicity to diminish sleep problems and enhance well-being. Given that sleep disturbances and circadian rhythms can be altered by behavioral change strategies or by altering lifestyles, taking advantage of these factors might be effective in preventing the development of pathology. It also seems that manipulations of sleep and circadian processes or clock genes can influence the effects of standard therapeutic methods to deal with illness and it should be possible to manipulate circadian variations among individual patients to exact the most beneficial actions from therapeutic methods.

References

Aiello, I., Fedele, M.L.M., Roman, F., Marpegan, L., Caldart, C., et al., 2020. Circadian disruption promotes tumor-immune microenvironment remodeling favoring tumor cell proliferation. Sci. Adv. 6, eaaz4530.

Almendros, I., Khalyfa, A., Trzepizur, W., Gileles-Hillel, A., Huang, L., et al., 2016. Tumor cell malignant properties are enhanced by circulating exosomes in sleep apnea. Chest 150, 1030–1041.

Bai, Y., Li, X., Wang, K., Chen, S., Wang, S., et al., 2016. Association of shift-work, daytime napping, and nighttime sleep with cancer incidence and cancer-caused mortality in Dongfeng-tongji cohort study. Ann. Med. 48, 641–651.

Besedovsky, L., Dimitrov, S., Born, J., Lange, T., 2016. Nocturnal sleep uniformly reduces numbers of different T-cell subsets in the blood of healthy men. Am. J. Phys. Regul. Integr. Comp. Phys. 311, R637–R642.

Besedovsky, L., Lange, T., Haack, M., 2019. The sleep-immune crosstalk in health and disease. Physiol. Rev. 99, 1325–1380.

Bjorvatn, B., Axelsson, J., Pallesen, S., Waage, S., Vedaa, Ø., et al., 2020. The association between shift work and immunological biomarkers in nurses. Front. Public Health 8, 415.

Brancaccio, M., Edwards, M.D., Patton, A.P., Smyllie, N.J., Chesham, J.E., et al., 2019. Cell-autonomous clock of astrocytes drives circadian behavior in mammals. Science 363, 187–192.

Carter, B.D., Diver, W.R., Hildebrand, J.S., Patel, A.V., Gapstur, S.M., 2014. Circadian disruption and fatal ovarian cancer. Am. J. Prev. Med. 46, S34–S41.

Chen, Y., Tan, F., Wei, L., Li, X., Lyu, Z., et al., 2018. Sleep duration and the risk of cancer: a systematic review and meta-analysis including dose-response relationship. BMC Cancer 18, 1149.

Cheng, Q.J., Ohta, S., Sheu, K.M., Spreafico, R., Adelaja, A., et al., 2021. NF-κB dynamics determine the stimulus specificity of epigenomic reprogramming in macrophages. Science 372, 1349–1353.

Dakup, P.P., Porter, K.I., Gajula, R.P., Goel, P.N., Cheng, Z., et al., 2020. The circadian clock protects against ionizing radiation-induced cardiotoxicity. FASEB J. 34, 3347–3358.

Deaver, J.A., Eum, S.Y., Toborek, M., 2018. Circadian disruption changes gut microbiome taxa and functional gene composition. Front. Microbiol. 9, 737.

Dun, A., Zhao, X., Jin, X., Wei, T., Gao, X., et al., 2020. Association between night-shift work and cancer risk: updated systematic review and meta-analysis. Front. Oncol. 10, 1006.

Erren, T.C., Morfeld, P., Foster, R.G., Reiter, R.J., Gross, J.V., et al., 2016. Sleep and cancer: synthesis of experimental data and meta-analyses of cancer incidence among some 1,500,000 study individuals in 13 countries. Chronobiol. Int. 33, 325–350.

Frazier, K., Frith, M., Harris, D., Leone, V.A., 2020. Mediators of host–microbe circadian rhythms in immunity and metabolism. Biology 9, 41.

Godinho-Silva, C., Domingues, R.G., Rendas, M., Raposo, B., Ribeiro, H., et al., 2019. Light-entrained and brain-tuned circadian circuits regulate ILC3s and gut homeostasis. Nature 574, 254–258.

Hablitz, L.M., Pla, V., Giannetto, M., Vinitsky, H.S., Staeger, F.F., et al., 2020. Circadian control of brain glymphatic and lymphatic fluid flow. Nat. Commun. 11, 4411.

Harper, E., Talbot, C.J., 2019. Is it time to change radiotherapy: the dawning of chronoradiotherapy? Clin. Oncol. 31, 326–335.

Heckman, C.J., Kloss, J.D., Feskanich, D., Culnan, E., Schernhammer, E.S., 2017. Associations among rotating night shift work, sleep and skin cancer in Nurses' Health Study II participants. Occup. Environ. Med. 74, 169–175.

Hernández-Rosas, F., Lopez-Rosas, C.A., Saavedra-Velez, M.V., 2020. Disruption of the molecular circadian clock and cancer: an epigenetic link. Biochem. Genet. 58, 189–209.

Holtkamp, S.J., Ince, L.M., Barnoud, C., Schmitt, M.T., Sinturel, F., et al., 2021. Circadian clocks guide dendritic cells into skin lymphatics. Nat. Immunol. 22, 1375–1381.

Irwin, M.R., 2015. Why sleep is important for health: a psychoneuroimmunology perspective. Annu. Rev. Psychol. 66, 143–172.

Irwin, M.R., 2019. Sleep and inflammation: partners in sickness and in health. Nat. Rev. Immunol. 19, 702–715.

Irwin, M.R., Opp, M.R., 2017. Sleep health: reciprocal regulation of sleep and innate immunity. Neuropsychopharmacology 42, 129–155.

Irwin, M.R., Olmstead, R., Carroll, J.E., 2015. Sleep disturbance, sleep duration, and inflammation: a systematic review and meta-analysis of cohort studies and experimental sleep deprivation. Biol. Psychiatry 80, 40–52.

Irwin, M.R., Olmstead, R., Carroll, J.E., 2016. Sleep disturbance, sleep duration, and inflammation: a systematic review and meta-analysis of cohort studies and experimental sleep deprivation. Biol. Psychiatry 80, 40–52.

Itani, O., Jike, M., Watanabe, N., Kaneita, Y., 2017. Short sleep duration and health outcomes: a systematic review, meta-analysis, and meta-regression. Sleep Med. 32, 246–256.

Jiao, L., Duan, Z., Sangi-Haghpeykar, H., Hale, L., White, D.L., et al., 2013. Sleep duration and incidence of colorectal cancer in postmenopausal women. Br. J. Cancer 108, 213–221.

Karantanos, T., Theodoropoulos, G., Pektasides, D., Gazouli, M., 2014. Clock genes: their role in colorectal cancer. World J. Gastroenterol. 20, 1986–1992.

Kecklund, G., Axelsson, J., 2016. Health consequences of shift work and insufficient sleep. BMJ 355, i5210.

Khalyfa, A., Gaddameedhi, S., Crooks, E., Zhang, C., Li, Y., et al., 2020. Circulating exosomal miRNAs signal circadian misalignment to peripheral metabolic tissues. Int. J. Mol. Sci. 21, 6396.

Koritala, B.S.C., Porter, K.I., Arshad, O.A., Gajula, R.P., Mitchell, H.D., et al., 2021. Night shift schedule causes circadian dysregulation of DNA repair genes and elevated DNA damage in humans. J. Pineal Res. 70, e12726.

Krause, A.J., Prather, A.A., Wager, T.D., Lindquist, M.A., Walker, M.P., 2019. The pain of sleep loss: a brain characterization in humans. J. Neurosci. 39, 2291–2300.

Krueger, J.M., Opp, M.R., 2016. Sleep and microbes. Int. Rev. Neurobiol. 131, 207–225.

Kuang, Z., Wang, Y., Li, Y., Ye, C., Ruhn, K.A., et al., 2019. The intestinal microbiota programs diurnal rhythms in host metabolism through histone deacetylase 3. Science 365, 1428–1434.

Lauriola, M., Enuka, Y., Zeisel, A., D'Uva, G., Roth, L., et al., 2014. Diurnal suppression of EGFR signalling by glucocorticoids and implications for tumour progression and treatment. Nat. Commun. 5, 5073.

Lesicka, M., Jablonska, E., Wieczorek, E., Seroczynska, B., Kalinowski, L., et al., 2019. A different methylation profile of circadian genes promoter in breast cancer patients according to clinicopathological features. Chronobiol. Int. 36, 1103–1114.

Lin, X., Chen, W., Wei, F., Ying, M., Wei, W., et al., 2015. Night-shift work increases morbidity of breast cancer and all-cause mortality: a meta-analysis of 16 prospective cohort studies. Sleep Med. 16, 1381–1387.

Lu, C., Sun, H., Huang, J., Yin, S., Hou, W., et al., 2017. Long-term sleep duration as a risk factor for breast cancer: evidence from a systematic review and dose-response meta-analysis. Biomed. Res. Int. 2017, 4845059.

Ma, Q.Q., Yao, Q., Lin, L., Chen, G.C., Yu, J.B., 2016. Sleep duration and total cancer mortality: a meta-analysis of prospective studies. Sleep Med. 27–28, 39–44.

Masri, S., Sassone-Corsi, P., 2018. The emerging link between cancer, metabolism, and circadian rhythms. Nat. Med. 24, 1795–1803.

Masri, S., Kinouchi, K., Sassone-Corsi, P., 2015. Circadian clocks, epigenetics, and cancer. Curr. Opin. Oncol. 27, 50–56.

McEwen, B.S., Karatsoreos, I.N., 2015. Sleep deprivation and circadian disruption: stress, allostasis, and allostatic load. Sleep Med. Clin. 10, 1–10.

Mirza-Aghazadeh-Attari, M., Mohammadzadeh, A., Mostavafi, S., Mihanfar, A., Ghazizadeh, S., et al., 2020. Melatonin: an important anticancer agent in colorectal cancer. J. Cell. Physiol. 235, 804–817.

Murakami, M., Tognini, P., 2020. The circadian clock as an essential molecular link between host physiology and microorganisms. Front. Cell. Infect. Microbiol. 9, 469.

Muto, V., Jaspar, M., Meyer, C., Kusse, C., Chellappa, S.L., et al., 2016. Local modulation of human brain responses by circadian rhythmicity and sleep debt. Science 353, 687–690.

Nobis, C.C., Dubeau Laramee, G., Kervezee, L., Maurice De Sousa, D., Labrecque, N., et al., 2019. The circadian clock of CD8 T cells modulates their early response to vaccination and the rhythmicity of related signaling pathways. Proc. Natl. Acad. Sci. 116, 20077–20086.

Nobs, S.P., Tuganbaev, T., Elinav, E., 2019. Microbiome diurnal rhythmicity and its impact on host physiology and disease risk. EMBO Rep. 20, e47129.

Noya, S.B., Colameo, D., Brüning, F., Spinnler, A., Mircsof, D., et al., 2019. The forebrain synaptic transcriptome is organized by clocks but its proteome is driven by sleep. Science 366.

Papagiannakopoulos, T., Bauer, M.R., Davidson, S.M., Heimann, M., Subbaraj, L., et al., 2016. Circadian rhythm disruption promotes lung tumorigenesis. Cell Metab. 24, 324–331.

Poole, E.M., Schernhammer, E.S., Tworoger, S.S., 2011. Rotating night shift work and risk of ovarian cancer. Cancer Epidemiol. Biomark. Prev. 20, 934–938.

Qiu, M., Chen, Y.B., Jin, S., Fang, X.F., He, X.X., et al., 2019. Research on circadian clock genes in non-small-cell lung carcinoma. Chronobiol. Int. 36, 739–750.

Royse, K.E., El-Serag, H.B., Chen, L., White, D.L., Hale, L., et al., 2017. Sleep duration and risk of liver cancer in postmenopausal women: the women's health initiative study. J. Women's Health 26, 1270–1277.

Ruben, M.D., Hogenesch, J.B., Smith, D.F., 2019. Sleep and circadian medicine: time of day in the neurologic clinic. Neurol. Clin. 37, 615–629.

Samuelsson, L.B., Bovbjerg, D.H., Roecklein, K.A., Hall, M.H., 2018. Sleep and circadian disruption and incident breast cancer risk: an evidence-based and theoretical review. Neurosci. Biobehav. Rev. 84, 35–48.

Sancar, A., Lindsey-Boltz, L.A., Gaddameedhi, S., Selby, C.P., Ye, R., et al., 2015. Circadian clock, cancer, and chemotherapy. Biochemistry 54, 110–123.

Scheiermann, C., Gibbs, J., Ince, L., Loudon, A., 2018. Clocking in to immunity. Nat. Rev. Immunol. 18, 423–437.

Schernhammer, E.S., Feskanich, D., Liang, G., Han, J., 2013. Rotating night-shift work and lung cancer risk among female nurses in the United States. Am. J. Epidemiol. 178, 1434–1441.

Shafi, A.A., McNair, C.M., McCann, J.J., Alshalalfa, M., Shostak, A., et al., 2021. The circadian cryptochrome, CRY1, is a pro-tumorigenic factor that rhythmically modulates DNA repair. Nat. Commun. 12, 401.

Slavish, D.C., Taylor, D.J., Dietch, J.R., Wardle-Pinkston, S., Messman, B., et al., 2020. Intraindividual variability in sleep and levels of systemic inflammation in nurses. Psychosom. Med. 82, 678–688.

Soucise, A., Vaughn, C., Thompson, C.L., Millen, A.E., Freudenheim, J.L., et al., 2017. Sleep quality, duration, and breast cancer aggressiveness. Breast Cancer Res. Treat. 164, 169–178.

Stenvers, D.J., Scheer, F., Schrauwen, P., la Fleur, S.E., Kalsbeek, A., 2019. Circadian clocks and insulin resistance. Nat. Rev. Endocrinol. 15, 75–89.

Stone, C.R., Haig, T.R., Fiest, K.M., McNeil, J., Brenner, D.R., et al., 2019. The association between sleep duration and cancer-specific mortality: a systematic review and meta-analysis. Cancer Causes Control 30, 501–525.

Sulli, G., Lam, M.T.Y., Panda, S., 2019. Interplay between circadian clock and cancer: new frontiers for cancer treatment. Trends Cancer 5, 475–494.

Szentirmai, E., Millican, N.S., Massie, A.R., Kapas, L., 2019. Butyrate, a metabolite of intestinal bacteria, enhances sleep. Sci. Rep. 9, 7035.

Thompson, C.L., Larkin, E.K., Patel, S., Berger, N.A., Redline, S., et al., 2011. Short duration of sleep increases risk of colorectal adenoma. Cancer 117, 841–847.

Travis, R.C., Balkwill, A., Fensom, G.K., Appleby, P.N., Reeves, G.K., et al., 2016. Night shift work and breast cancer incidence: three prospective studies and meta-analysis of published studies. J. Natl. Cancer Inst. 108.

Trudel-Fitzgerald, C., Zhou, E.S., Poole, E.M., Zhang, X., Michels, K.B., et al., 2017. Sleep and survival among women with breast cancer: 30 years of follow-up within the Nurses' Health Study. Br. J. Cancer 116, 1239–1246.

Tseng, C.M., Chen, Y.T., Tao, C.W., Ou, S.M., Hsiao, Y.H., et al., 2015. Adult narcoleptic patients have increased risk of cancer: a nationwide population-based study. Cancer Epidemiol. 39, 793–797.

Walker 2nd, W.H., Bumgarner, J.R., Walton, J.C., Liu, J.A., Melendez-Fernandez, O.H., et al., 2020. Light pollution and cancer. Int. J. Mol. Sci. 21, 9360.

Wegrzyn, L.R., Tamimi, R.M., Rosner, B.A., Brown, S.B., Stevens, R.G., et al., 2017. Rotating night-shift work and the risk of breast cancer in the nurses' health studies. Am. J. Epidemiol. 186, 532–540.

Yousef, E., Mitwally, N., Noufal, N., Tahir, M.R., 2020. Shift work and risk of skin cancer: a systematic review and meta-analysis. Sci. Rep. 10, 2012.

Zada, D., Bronshtein, I., Lerer-Goldshtein, T., Garini, Y., Appelbaum, L., 2019. Sleep increases chromosome dynamics to enable reduction of accumulating DNA damage in single neurons. Nat. Commun. 10, 895.

Zhang, Y., Papantoniou, K., 2019. Night shift work and its carcinogenicity. Lancet Oncol. 20, e550.

Zhang, S., Tanaka, Y., Ishihara, A., Uchizawa, A., Park, I., et al., 2021. Metabolic flexibility during sleep. Sci. Rep. 11, 1–13.

Zhao, H., Yin, J.Y., Yang, W.S., Qin, Q., Li, T.T., et al., 2013. Sleep duration and cancer risk: a systematic review and meta-analysis of prospective studies. Asian Pac. J. Cancer Prev. 14, 7509–7515.

Zisapel, N., 2018. New perspectives on the role of melatonin in human sleep, circadian rhythms and their regulation. Br. J. Pharmacol. 175, 3190–3199.

CHAPTER 12

Adopting healthy behaviors: Toward prevention and cures

OUTLINE

Understanding counterproductive behaviors	369	Theory of Reasoned Action and the Theory of Planned Behavior	382
A brief look back	370	Stage models: The Transtheoretical Model	383
Intervention approaches	372	Behavioral and cognitive methods to alter health behaviors	385
Forms of intervention	373	Exposure therapy and systematic desensitization	385
Changing attitudes and behaviors and barriers to change	374	Applied behavior analysis	386
Education in the promotion of behavior change	374	Cognitive behavioral therapy	386
Bridging the intention—Behavior gap	376	Mindfulness	388
Nudge theory	377	Positive psychology	390
Psychosocial and cognitive approaches to enhance health behaviors	378	Critiques of positive psychology	391
Illness perceptions	378	Social support	392
Common-sense model of behavioral change	379	Change through social identity: The social cure	393
The Health Belief Model	380	Limits to the effects of social support	395
Protection Motivation Theory	381	Summary and conclusions	396
Self-efficacy Theory	382	References	396

Understanding counterproductive behaviors

It is a given that even smart people will engage in poor health behaviors (e.g., smoking, excessive alcohol consumption, to name just a couple) that promote self-harm. Indeed, many of us have a story that can back this up in some form or another. And there is no single reason why people might choose to engage in behaviors that place them at risk for diseases. Some individuals might not fully understand the risks versus benefits of their behaviors. Or their behaviors might stem from social influences, personality characteristics (e.g., intolerance of uncertainty or impulsivity), and if heredity has a say, through genetic and epigenetic factors that dispose people to behave in counterproductive ways. Whatever the reason, and even if they fully appreciate the risks, they may be unwilling to change their behaviors (e.g., eating poorly) as they simply enjoy what they're doing, even resisting change when confronted by data pointing to the self-harm being created. Substance abusers presented with neuroimaging evidence that their brains look 20 years older than they should will cry foul. And consider the struggles of health authorities during the covid-19 pandemic to convince people to take a vaccine that would keep them from being hospitalized. Among some people who are resistant to change their perceived and actual knowledge may be inversely related to one another, and they may even diminish the knowledge they attribute to others (the Kruger-Dunning effect), supporting the adage that ignorance is bliss. However, if individuals (or groups) are prepared to seek help, there are ways to produce behavior change to diminish the odds of disease occurrence as well as diminish the distress created by ongoing chronic illnesses.

A brief look back

The focus of medicine for the longest time was to exact cures for illnesses, with varying degrees of success, whereas considerably less attention had been focused on illness prevention. Yet environmental factors and human behaviors were long known to produce multiple diseases, many of which could be averted by individual and societal actions. Epidemiological analyses relevant to cancer occurrence and prevention came into their own about 60 years ago, but well before then, it was certain that illnesses could frequently be prevented through simple behavioral changes or community-based efforts. In the 18th century, the English surgeon Sir Percivall Pott noted that environmental factors, particularly exposure to soot, could lead to scrotal cancer among young boys who worked as chimney sweeps, and more than a century later, Sir Henry Butlin proposed that protective clothing could prevent such outcomes.

But even well before the observations of Percivall Pott, warnings had been issued (e.g., by Thomas Venner during the early 1600s) that tobacco could have negative consequences on the brain and heart. A century later it was noted that the use of snuff was associated with nasal cancer and that pipe smokers displayed elevated occurrence of mouth and lip cancer. Another 100 years passed before it was reported that smoking was linked to lung cancer, but preventive actions were slow in coming. By the 1940s and 1950s, considerable evidence had emerged supporting the earlier supposition that cigarette smoking was linked to the development of lung cancer, which had been a rare disease before the broad use of tobacco products. Despite epidemiological studies providing overwhelming support for this contention, effective marketing efforts by the tobacco industry countered these evidence-based conclusions, so much so that most physicians hadn't signed on to this view even by the 1960s. Ultimately, the seminal work of Doll and his associates that had begun in the 1950s, as well as that of later researchers, made it abundantly clear that smoking was associated with cancers and heart disease (Doll et al., 1994). With time and considerably more scientific evidence, the causal connection between cigarette smoking and multiple forms of cancer became incontrovertible.

Secondhand smoke was similarly linked to several forms of cancer, but it was many years before laws were enacted to protect nonsmokers from the dangers of smoke.

The use of cigarettes has declined markedly over the past few decades, thanks to better education, increased social pressures to abstain from smoking, and government-sponsored programs, together with increased taxation and outright bans of smoking within the workplace and in public spaces. Despite these efforts and the overwhelming evidence of health risks created by cigarette smoking, it remains a serious problem within North America and most of the EU countries and has been still more problematic within countries of the former Soviet Union. It would be best for smokers to cease smoking entirely, but among heavy smokers even reducing smoking by 50% diminished cancer occurrence, although more pronounced smoking reductions are needed to prevent heart disease.

e-cigarettes and disease

As an alternative to smoking cigarettes, vaping was developed as a way to provide the nicotine hit, satisfy the oral habit, and presumably eliminate the harmful components of tobacco smoke. However, it does not appear that vaping was effective in eliminating a smoking habit. In fact, smokers who switched to e-cigarettes were not only likely to return to their old habit, but relative to individuals who did not use a substitute, the occurrence of relapse was 8.5% greater (Pierce et al., 2021).

Even if e-cigarettes facilitate individual efforts to quit smoking, at what expense would this be achieved? Soon after the introduction of e-cigarettes, sufficient data were not available to make credible assertions concerning the long-term ramifications of their use. However, it is now evident that these substitutes can cause airway and lung damage and it doesn't take decades for this to occur. Nanoparticles within e-cigarettes can elicit inflammatory responses that can precipitate the development of asthma, diabetes, heart disease, and lung damage. Similarly, solvents used to create flavored e-cigarettes may contain acetaldehyde and formaldehyde as carcinogenic by-products, and the high temperature achieved through newer devices can potentially produce adverse effects related to specific components of the vape mixture, such as the conversion of vitamin E acetate to toxic ketene gas (Wu and O'Shea, 2020). There is also the possibility that the presence of chromium and nickel in second-hand vapors from e-cigarettes can render them dangerous. And it is more than a passing interest that vaping was also accompanied by oral microbiota changes, elevated levels of the proinflammatory cytokines IL-1β and IL-6, and increased occurrence of gum disease (Pushalkar et al., 2020). The latter has also been linked to chronic inflammatory diseases, such as diabetes, heart disease, and cancer, emphasizing the risks created by vaping.

Epidemiological research has documented many other instances in which environmental toxicants, dietary factors, and infectious agents were implicated in the emergence of cancer. The data obtained through basic science hasn't always translated well to public knowledge and often hasn't affected the adoption of health-related behaviors. There can hardly be a person within a developed country who is unaware of the harms that are so often created by smoking, but public knowledge concerning the links between alcohol intake and cancer seems to be less well known. As much as a good number of people would like to think otherwise, considerable evidence has amassed showing that alcohol intake is associated with cancer of the mouth, throat, larynx, esophagus, colon and rectum, liver, pancreas, and breast, being elevated by 130%–200%. Alcohol is ordinarily broken down into several components, including acetaldehyde, which causes DNA damage and prevents DNA repair. Moreover, alcohol may influence various hormones that contribute to cancer occurrence and heavy alcohol use can cause tissue damage that allows for the effects of carcinogens to be more pronounced. Through several processes stemming from alcohol metabolism, folate deficiency, and elevated reactive oxygen species, epigenetic changes could be promoted that enhance cancer occurrence (Dumitrescu, 2018).

And then there is food. Previously we emphasized that overeating, or eating the wrong foods, promotes many diseases and may be a greater risk factor for all-cause mortality than either smoking or alcohol intake. As the incidence of obesity has been increasing progressively over the years, whereas cigarette smoking has been declining, their relative impact on cancer incidence has been changing. Unfortunately, when one problem behavior (e.g., smoking) is abandoned, another (e.g., eating) may replace it, possibly reflecting the absence of eating suppression provided by nicotine, or it might simply serve as a substitute that involves oral behaviors.

Intervention approaches

Diverse strategies have been recommended to prevent health risks brought about by self-harming behaviors. Many of these programs focused on individual behaviors, such as helping individuals maintain a proper diet, obtaining regular exercise, getting sufficient sleep, and having a good work-life balance. Other strategies have been directed toward changing the behaviors of large swaths of society. These encompassed efforts to have people stop smoking, stay out of direct sunlight, be vaccinated to reduce the risk of viral infection and cancers related to viruses, and most recently to maintain social distancing and wearing a mask to limit pandemic viral spread. As successful as some of these efforts have been, many people have persisted in their risky behaviors, as observed at the time of writing with people failing to adopt behaviors to prevent the spread of COVID-19.

Most preventable chronic illnesses develop after years of exposure to environmental insults or by risky behaviors that had been maintained for extended periods. Interventions undertaken early can be effective in disease prevention, but too often the cumulative effects of toxicants experienced over years could already have produced irreversible damage. Nonetheless, even if behavioral change is adopted late, these measures may slow disease progression. Unfortunately, when risky behaviors had been maintained for years, essentially becoming habits, eliminating them can be daunting.

The processes that lead to risky behaviors being adopted and maintained often involve the confluence of multiple elements and producing behavioral change might similarly require that several factors be dealt with concurrently or sequentially. Some of these efforts ought to be directed at the individual, whereas others are concerned with societal issues that need to be addressed (i.e., health-care services, economic factors, education, social and community influences, the built environment, and issues related to policymaking). Numerous approaches have been established to provide information relevant to varied chronic illnesses. Many of the excellent recommendations are seemingly straightforward; however, to achieve success it may not be enough to offer science-based information. Instead, active intervention efforts are often necessary irrespective of whether they entail individually targeted approaches or those that involve community-wide policies.

Forms of intervention

Increasingly more programs have focused on preventing diseases among people who were deemed to be at risk, and community efforts were initiated to ensure that the opportunities for better health were more broadly available. The Institute of Medicine (IOM)/National Academy of Sciences (NAS) offered a series of recommendations regarding best practices that could be adopted to encourage behavioral changes important to limit illness (Solomon and Kington, 2002). These recommendations, which are as relevant today as they were 20 years ago, focused on the importance of social and behavioral determinants of disease, injury, and disability and it was acknowledged that social, political, and economic systems can shape the interventions that promote the adoption of healthy behaviors. The suggestion was advanced that multiple societal sectors (e.g., law, business, education, social services, and the media) ought to become more engaged in health promotion efforts and that greater focus be placed on unique needs of specific target groups (based on age, gender, race, ethnicity, social class). At the same time, since social and environmental circumstances are ever-changing, intervention approaches ought to be equally adaptable to deal with dynamic processes. Moreover, it was emphasized that the "long view" ought to be encouraged since behavior change may take years to develop.

Prevention programs that come in a variety of flavors can lessen the risks of illnesses emerging and might thereby enhance disability-free survival. *Universal prevention* approaches are geared toward strategies being instituted across segments of society irrespective of individual risk factors (e.g., vaccination programs). In other cases, *selective prevention* programs have been advanced that target individuals because they are at elevated risk of specific illnesses. This could include older people who are prone to multiple illnesses, as well as individuals who are genetically at risk for specific disorders, or those who present with early signs of a disorder yet don't meet the criteria for a clinical diagnosis (e.g., the presence of metabolic syndrome that could be a harbinger of other illnesses that might follow).

Regardless of the intervention strategy adopted, success is predicated on people following the recommendations made. Aside from behaviors that should be eliminated (smoking, alcohol consumption) people frequently fail to engage in healthy behaviors, as we've seen with diet and exercise. Yearly, fewer than 40% of individuals choose to be vaccinated against influenza, and regular screening for diseases is often ignored. Only about 50% of women obtain breast mammography even though 1 of 8 women will develop breast cancer, and still fewer men are screened for prostate cancer that will eventually affect 1 of 9 men. Even though most people (92%) agree that it is important to receive an annual medical check-up, fewer than half do so. The 1% doctrine made famous by then Vice-President Dick Cheney holds that even if there was only "a 1% chance' that a threat was real and could produce a calamitous outcome "we have to treat it as a certainty." Yearly visits to a doctor by older individuals will have more than a 1% chance of revealing something being amiss, but many people have the attitude that "I've got no reason to see doctors since there's nothing wrong with me." Even patients with chronic illnesses who do visit their physician regularly, often fail to follow prescribed treatments despite knowing their importance. Diabetic and prediabetic patients might feel that an occasional muffin "isn't that bad," and once they fall off the wagon more "cheating" may follow. Likewise, patients are known to take holidays from their antidepressant medications because these agents interfere with sexual gratification, and bipolar patients may stop using their medications in the belief that these agents stifle their creativity. The rationale in each case may be understandable, but they're most certainly not the smart thing to do.

Many broad interventions have yielded encouraging results, including those aimed at cancer prevention, but demonstrating their scalability and their effectiveness at a population level has often been challenging. Although several programs were assumed to be effective in increasing knowledge related to cancer prevention, few studies verified whether behavioral change occurred, and still fewer evaluated the long-term impact of these procedures (Shah et al., 2020).

Changing attitudes and behaviors and barriers to change

Education in the promotion of behavior change

It had at one time been assumed that producing behavior change could be achieved by providing individuals with valid and reliable information, which would result in individuals accepting and following the advice offered. However, it readily became apparent that getting people to behave in particular ways wasn't that simple. Providing effective educational programs can have positive effects, but this necessarily entails an understanding of the audience being addressed and depends on the nature of the message being delivered, as well as who was delivering the message.

Effective messages should not focus on creating negative emotions (e.g., fear), but instead should emphasize the benefits of certain behaviors (framing effect). Negative messages may be effective in demonizing political opponents but are less effective in encouraging the adoption of healthy behaviors. Rather than pointing to diseases associated with smoking or poor eating (creating fear), it might be better to indicate that quitting smoking or eating the right foods to prevent obesity may give a person extra healthy years to enjoy.

An effective message regarding a topic as important as health change ought to come from a credible source. In the context of the covid-19 pandemic, Dr. Anthony Fauci, a clinical immunologist and virologist at the National Institutes of Health in the United States, is a far more credible spokesperson for warning the public than a radiologist, ophthalmologist, economist, or wrestling coach with no background in virology or epidemiology. Moreover, the message needs to be unambiguous, simple, brief, and to the point (wear the mask, it saves lives!). At the same time, the message should not be so intense as to create pushback, especially if the intended audience is disposed not to follow the advice, which we've seen often enough in the context of advising parents who refuse to have their children vaccinated. Rather than dismissing counter opinions, these should be acknowledged but with caveats (e.g., not every vaccine ever created has been entirely safe, but on balance, the benefits of vaccinating children far outweigh the possible harms). The arguments can be strengthened further by considering the attributes and social values that are important to the other party (self-affirmation) to avoid a knee-jerk defensive response when that person encounters a threat to their sense of self. By affirming an individual's core values, greater neuronal activity occurs in brain regions associated with decision-making along with the person being able to appreciate the self-relevance and value of the message (Falk et al., 2015).

Some messages might not be at all threatening but may be confusing. The nature of the message needs to be considered in the context of what other information is being delivered. When irrelevant information is provided, individuals may not absorb the important message being delivered. Thus, the message ought to be provided in a way that allows individuals to easily filter the relevant from ancillary information and ignore what is less germane to them. The impact of messaging is still less effective if individuals have received contradictory information (witness the problems created concerning diverse messages related to which foods are healthy and which are not). This confusion was prominent in vaccination efforts during the 2009 H1N1 pandemic, as people recalled what they considered to be false alarms about earlier similar threats, particularly experiences with threats of mad cow disease, MERS, and SARS. Consequently, the new virus was not taken nearly as seriously as it should have been (Taha et al., 2013). The lack of trust and lack of confidence was exacerbated by slow vaccine development, insufficient vaccine being available, questions about vaccine safety as there hadn't been sufficient time for testing, and the ever-present media that sensationalized potential risks. We saw this reoccur with COVID-19 protection that was plagued by a "pandemic of misinformation," which made a bad situation much worse, despite the development of highly effective and safe vaccines.

It could be argued that in the face of novel threats that we know little about, there might be a good reason to mistrust the media and government organizations. The available information might initially be scant and there may be uncertainty as to what information should be disclosed to the public. On the one side, health agencies might be inclined to appear in control of a hazardous situation, but they might also be apprehensive of creating a panic about the potential risks of the disease. Transparency is probably the best call in this situation; however, since past warnings had not developed into the disasters that had been feared, new warnings might be ignored. Rather than seeing our good fortune at having dodged a bullet in the past, people may fail to take heed because government agencies had "exhausted their quota of scary utterances" (Sandman, 2009). Consequently, individuals have been less likely to trust blindly, even if this means going by their intuitions no matter how poorly informed these were. If the past is a good teacher, then we will experience pandemics in the future, and hopefully, people will be more inclined to listen to scientists who know what they are talking about rather than to politicians who don't.

Antivaccine attitudes: Not a new development

It is tempting to think of vaccine hesitancy as a new phenomenon that is restricted to a minority of individuals. However, antivaccine attitudes have a very long history (Dubé et al., 2015). Despite the horrific effects of smallpox, when a vaccine was first developed to treat this condition, its use was met with severe criticism in the United Kingdom so in the mid-1800s smallpox vaccination became compulsory, and fines were applied to those who failed to comply. In later years the very same attitudes and behaviors were seen in the United States and Europe with polio vaccination, and in recent times similar attitudes were apparent in vaccine programs meant to curb diseases, such as measles, mumps, and rubella, and this was also encountered with HPV vaccination to limit cervical and other forms of cancer. Even though most people likely to perceive themselves as part of an enlightened group, attitudes and behaviors concerning vaccination hesitancy today have often been as short-sighted as the resistance to vaccination two centuries earlier.

In an interesting piece regarding the best ways to provide an effective message, five key factors were proposed (Blastland et al., 2020), some of which go beyond those already described here. Specifically, it was suggested that the message should.

(1) simply inform the audience, rather than try to persuade them, which could be met by mistrust;
(2) offer genuine balance, especially the temptation to skirt what is important to the audience—doing otherwise could result in skepticism;
(3) disclose uncertainties concerning the message, thereby encouraging trust in the message;
(4) indicate the quality of evidence being brought to the table especially as the audience may sense when inaccurate or exaggerated information is being delivered;
(5) inoculate (forewarn) individuals against rampant misinformation, essentially debunking (or, more appropriately, prebunking) inaccurate information that might otherwise act as a barrier to the acceptance of evidence-based views.

All this said, altering opinions that are firmly entrenched and supported by an individual's social group will be very difficult to change, but if messages are heard often enough and done so persuasively, movements forward might be achievable, it just takes a lot of time and effort. In cases where time is short (e.g., in the face of an ongoing pandemic), preparedness might mean that appropriate messages were delivered well before the crisis emerged. Above all, the message and the messenger need to be trusted. The maxim needs to be kept in mind that "trust once lost can never be regained."

Bridging the intention—Behavior gap

The adage that "the road to hell is paved with good intentions" is particularly apt in reference to the adoption of healthy lifestyle choices. Few people are unaware of the risks inherent in poor lifestyle choices, but nonetheless, there are still people resistant to change. Even when they finally reach the point where they are ready and seemingly determined to change their lifestyles, this may only appear as a fleeting good intention that is not acted upon, certainly not for any meaningful duration. Healthy behaviors might not be adopted simply because doing so would be inconvenient (e.g., exercise), or it might be that resources to engage in these behaviors are not readily available, which is often a problem for individuals living in impoverished areas. It's fine for government agencies to recommend that healthy behaviors be adopted, but for these to be enacted, adequate resources need to be committed.

Various approaches have been suggested to shrink the gap between intentions and actions. A frequent suggestion is that *implementation intentions* become more salient so that goals are clear ("I will not smoke any longer" vs "I intend not to smoke eventually"), and specific plans ought to be formed on how to achieve these ends. As in so many other life endeavors, being prepared to act in a particular manner within specific situations will make it more likely that individuals will recognize these situations when they emerge so that they act on their intentions. It is common in training programs in which threatening or dangerous conditions are often met (e.g., in readying first responders) realistic practices can facilitate responses that are needed when challenges are subsequently encountered. Imagining (thinking out) intentions to act about the adoption of healthy behaviors (or rejecting those that are unhealthy) might not be as effective as actual practice, but it may nonetheless help individuals take the right steps when temptations arise—as they inevitably will.

Nudge theory

Given the difficulties of having people change their behaviors through education (or harsh legislation), Nudge theory was developed to move people gradually in a particular direction using positive reinforcement, social suggestions, and capitulation to perceived social norms (Thaler and Sunstein, 2008). Even subtle manipulations, such as making options more salient can make it more likely that a given option will be chosen over another. Although Nudge theory was initially applied by governments and businesses (e.g., in the context of green energy choices and motivating sustainable food choices), it has also found its way into health-related issues, such as encouraging hand hygiene to reduce infection. A nudge approach could be adopted to overcome the hesitancy among some people who are still on the fence about having their children receive vaccinations. Nudge could be combined with other behavioral change strategies, such as focusing on the influence of shared identities to augment the internalization of social norms, thereby promoting greater and more enduring effects.

For his work on Nudge theory, Richard Thaler was awarded the 2017 Nobel Prize in Economics, attesting to the presumed value of this behavior change method. However, this approach has had its share of detractors. There are certainly limits to what can be achieved through a nudge approach, particularly in dealing with "wicked, complex problems" (e.g., in limiting substance abuse disorders) that could potentially be dealt with through other methods. Moreover, because a nudge approach may take a considerable amount of time to have a significant societal impact, its value in time-sensitive situations, such as containing viral spread during a pandemic, seems impractical. It has also been maintained that "nudgeability" varies among individuals as some people who simply do not want to change their behaviors won't be affected by attempts at nudging them (de Ridder et al., 2021). The nudge approach has not been broadly used in clinical practice, but where it has, its efficacy was limited, possibly because situational barriers were not adequately considered before applying this strategy (Lamprell et al., 2021).

Other criticisms have been related to the ethics surrounding the approach rather than its practical or scientific merit. To begin with, who gets to decide which behaviors need or should be nudged? The nudge approach tacitly suggests that people generally don't know what is best for them, and an outside force covertly manipulating individual choices might not be well received. This Big Brother approach may cause reactance when people become aware that they have been manipulated. There have even been questions regarding the legality of using a nudge approach to produce behavioral changes. But knowing what we do now about vaccine hesitation, some people might welcome any approach that produces the essential behavioral changes.

The default option

Behaviors of individuals can be altered by taking advantage of individuals' propensity not to take active steps when given an option that entails filling out forms or making a phone call that might be met by a lengthy list of links. As we indicated earlier, doing nothing is simpler than doing something.

Consider the tactic used by newspapers and magazines that offer initial low prices for their product, which subsequently increases after some period. This practise acknowledges that many people won't exercise their "default option" to end their subscription. The same approach has been increasingly used for more serious matters that can be lifesaving. Potential organ donors in most countries are typically asked to sign their driving license if they are willing to be a donor. But changing this so that organ donation became the norm (i.e., people can opt-out by signing their license) the availability of organs increased, and perceptions regarding the value of organ donations may follow. Being able to ensure healthy lifestyles based on a default option is probably not a possibility that can readily be enforced—free will trumps all else, even if this means that some individuals will be at risk for illnesses and that they can affect others. Nevertheless, by judiciously employing ways of making the default option an easy choice, healthy behaviors might be adopted. For instance, making stairs a more desirable option than using an escalator or elevator for ascending a couple of floors, or having flu vaccines ordinarily be scheduled for eligible individuals (e.g., by family doctors or through government databases) may encourage positive behavior change.

Psychosocial and cognitive approaches to enhance health behaviors

The adoption of health behaviors is greatly affected by societal influences together with institutional policies. But when so many individuals run into health difficulties stemming from their behaviors, waiting for broad social change to occur may be impractical. Thus, while not diminishing the fundamental importance of society to promote changes, multiple programs and approaches were established to alter hazardous individual behaviors. Below we turn to several approaches that focus on factors that are fundamental in predicting and facilitating health behavior change.

Illness perceptions

Perceptions of an illness or risks concerning an illness occurring can potentially influence the actions that will be adopted. How illnesses are perceived may vary with a constellation of factors, such as a person's earlier health-related experiences, belief as to whether disease progression and outcomes can be controlled, and the context in which illnesses occur. When an individual discovers that they are at very high risk of developing a severe illness (e.g., carrying gene mutations that place them at high risk for cancer occurrence) or when a severe illness develops, a person may want to understand what caused this condition, and they make attributions regarding the cause of the illness. In some instances, appraisals and perceptions of illness are evidence-based and are likely accurate, but they can also be baseless, having been gleaned from inaccurate social media sites or from what some person had said.

Regardless of the validity of the information received, it can affect illness perceptions, beliefs regarding the dangers of an illness, and health behaviors adopted. While some individuals diagnosed with cancer attributed the illness to genetic factors, a good number believed that the disease came about owing to stressor experiences, hormonal changes, environmental factors, or simply God's will. When patients made strong causal attributions concerning the development of their cancer and believed that these factors were uncontrollable, they were at greater risk for becoming depressed and anxious and tended to be less likely to adopt adaptive health behaviors. In contrast, if they believed that recurrence of illness might involve controllable factors, then it was more likely that they would adopt healthy behaviors (Durazo and Cameron, 2019).

Regrettably, some individuals may engage in either self-blame or other blame about illnesses they developed. As chronic illnesses may develop owing to poor health behaviors there may be ample reason for self-blame, but this is impractical and could produce depression or guilt feelings, which can impair healing and lengthen the course of recovery. Illnesses may also develop because of extrinsic influences, such as pollutants that had spewed into the environment for years, thus other blame might not be an unreasonable response. In an interpersonal context, other blame may be counterproductive especially if individuals become preoccupied with this. Unlike blame directed at a specific individual, when other blame is used in the hope of finding meaning and bringing closure through actions against the party believed to be responsible for their condition, such as a large industry that contaminated the environment, there is a possibility of benefits being obtained. Realistically, however, while this may work out in movies (e.g., *Erin Brockovich*) in ordinary life this is less likely. Nevertheless, if everybody remains silent, polluters will continue in their damaging ways, so idly sitting by is counterproductive. Besides, doing nothing and suppressing anger can be more damaging than blaming others. One is reminded of the expression attributed to Buddha that "holding onto anger is like drinking poison and expecting the other person to die."

Common-sense model of behavioral change

Of the different views that had been offered to predict and produce behavioral change, one of the most influential was the Common-Sense model (Leventhal et al., 1984). Fundamental to this perspective was that appraisals of an illness and expectancies regarding treatment outcomes depend on how patients make sense of their illness, which can then affect how this condition is subsequently managed. In line with perspectives concerning the role of appraisals and coping in contending with stressors, health threats are met by cognitive and emotional changes to make sense of the situation. These may comprise attributions related to the cause of illness, expected impacts of the illness, and beliefs concerning recovery or control, often accompanied by emotional responses, such as dismay, blame, anger, and fear. When individuals initially face a health threat, they form an illness representation (schema) based on their experiences, personality, or knowledge that they had obtained. This schema then influences coping with the threat and appraisals of the efficacy of behaviors, which affects decision-making and the behaviors adopted to meet the challenge (Diefenbach and Leventhal, 1996).

Since its introduction, the model has received broad support concerning the position that illness cognitions and coping predicted subsequent psychological and social functioning (Hagger et al., 2017). Among cancer patients, subjective illness and emotional representations, expected consequences of illness, and control perceptions (problem-focused coping) were predictive of health perceptions and emotional well-being (Richardson et al., 2017). Likewise, in women who had survived cancer, their choice of treatments had been predicated, in part, on their beliefs concerning the cause of their cancer. Women who ascribed their cancer to stress and poor coping, which meant that they viewed the disease as being controllable were, for better or worse, more likely to express their agency by including complementary and alternative integrative oncology as a component of their cancer treatment (Andersen et al., 2017).

Survivors of cancer were also frequently prepared to adopt behavioral changes to enhance their physical, emotional, and spiritual well-being, and to this end, they were ready to adopt health prevention interventions that included features of the Common-sense model of health.

The common-sense model has held its ground and been modestly revamped to enhance its predictive ability. It was deemed important to distinguish between illness representations related to prior cancer experiences from those that entail concerns about the risk of recurrence, including worry and related emotions that can motivate protective behaviors in cancer survivors (Durazo and Cameron, 2019). Among survivors of endometrial cancer, for instance, negative illness perceptions were associated with greater health-care use. Yet, if individuals maintained a "patient identity" rather than that of a "cancer survivor," greater depression and poorer quality of life could ensue (Thong et al., 2018).

The Health Belief Model

One of the most widely adopted approaches to understand and modify individual behaviors, the Health Belief Model (Becker, 1974), initially addressed two fundamental questions: first, to what extent are people able to perceive health threats that are present, and second, do they believe that these threats can be attenuated through their behaviors. This necessarily considers the individual's beliefs concerning the barriers to health change (e.g., financial constraints) weighed against the potential benefits that could be gained by adopting certain behavioral strategies. With the appreciation that these beliefs may themselves not lead to the adoption of effective actions, two further components were incorporated in this model. The first of these, referred to as "cues to action," addressed the critical nature of both internal cues (e.g., illness symptoms, such as shortness of breath after exercise and increased lung congestion commonly apparent in long-term smokers) and external cues (e.g., subtle and explicit disapproval coming from others) that could act as triggers to promote behavioral changes. The second addition, which has proven fundamental to many other methods of producing behavioral change, comprised the person's belief and confidence that they could adopt and succeed in modifying their behaviors (i.e., their perceived *self-efficacy*).

Soon after the introduction of this model, support for its usefulness became evident in cross-sectional and prospective studies. Many of the key elements of the model (perceived susceptibility and severity of illness) successfully predicted behavior change and pointed to relevant barriers to action (Jones et al., 2014). The Health Belief Model has been predictive of behavioral change in a wide assortment of venues, such as enhanced adoption of vaccination during a pandemic (H1N1), the endorsement of oral health care, weight management behavior, and improved diet quality. Pertinent to our present focus is that this approach was relevant to behavior change related to cancer occurrence and management. Among other things, this involved increased health screening, reducing smoking and alcohol consumption, enhanced treatment adherence, improved weight loss measures, and facilitating self-management of chronic illnesses (e.g., Lau et al., 2020). Fig. 12.1 depicts the primary components of the Health Belief Model, including a role for self-efficacy, which was not present in the original formulation. The Health Belief

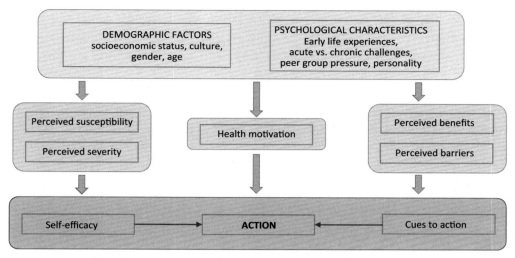

FIG. 12.1 Components of the Health Belief Model. Several demographic factors together with an individual's psychological characteristics may affect an individual's motivation for change. The motivation to adopt particular health behaviors may be dependent on the perceived harms and benefits that can be accrued as well as the barriers that exist. These factors, in turn, can influence actions being undertaken, which can be affected by cues to action as well as an individual's perceived self-efficacy. This approach has been relevant to a wide range of conditions, including health screening programs, smoking and alcohol consumption, preventive dental care, dieting, and self-management of chronic illnesses. *Figure and caption from Anisman, H., 2021. Health Psychology: A Biopsychosocial Approach. SAGE.*

Model was largely descriptive but offered little in the way of specific strategies that could be used to exact behavior change. It is certainly useful to be able to predict behavioral change but developing effective approaches that people will act on is an entirely different kettle of fish.

Protection Motivation Theory

Protection Motivation Theory (Rogers, 1983) shares several common features with the Health Belief Model but primarily focuses on behaviors that stemmed from fear-promoting health threats. According to this position, behavior change is tied to an individual's perceived vulnerability and the likelihood of threats manifesting as illnesses, as well as their belief that they can enact recommended behaviors successfully (*self-efficacy*) and whether these actions will reduce these threats (*response efficacy*).

The components of Protection Motivation Theory predicted both individual intentions and actions, and self-efficacy was deemed to be a key ingredient for this to occur, depending on the specific behaviors being considered. Because of the obvious threat that cancer poses for many individuals (e.g., those deemed to be at high risk), this theoretical framework was used to account for cancer prevention measures that were adopted. Among other things, this included the disposition of women toward genetic testing to determine the risk of breast cancer, screening for cervical cancer, prediction of skin cancer protective measures, and the adoption of cancer protection measures within the workplace (e.g., Roozbahani et al., 2020).

Self-efficacy Theory

The Self-efficacy Theory advanced by Bandura (1986) has had a marked influence on other theoretical frameworks aimed at behavior change and has offered ways by which behavior change can be encouraged. An individual's belief that they can achieve goals within a social context (self-efficacy) comprises complex reciprocal interactions between the person and the social context. This may entail the perception of personal control and experiences that affected expectations of success. However, knowing what to do, might not translate into how these behaviors are enacted. Based on traditional reinforcement theory, it was proposed that individuals can learn behavioral methods by observing and then modeling the behavior of others. Eventually, the person forms expectations regarding their behaviors and subsequent outcomes so that they will anticipate particular scenarios and would be prepared to act. Ultimately, self-confidence will be increased (self-efficacy beliefs) that will facilitate health behavior change being initiated and maintained, varying with the individual's specific goals.

The development of ways of changing attitudes, norms, and self-efficacy have been instrumental in affecting health behavior change (e.g., among individuals at high risk) and was poignant in its capacity to enhance personal agency to deal with cancer symptoms and the aftermath of cancer experiences. Specifically, self-efficacy was related to coping with cancer-related distress and quality of life (Chirico et al., 2017) and was related to both symptom management and in mediating symptom distress and quality of life in breast cancer patients and breast cancer survivors. A meta-analysis of 79 randomized controlled trials confirmed that interventions that comprised psychological methods to enhance self-efficacy facilitated personal agency and coping abilities that affected cancer-related distress, fatigue, and pain (Merluzzi et al., 2019). Additionally, self-efficacy might be useful in dealing with the side effects of cancer or its therapy. Since some patients may not be prepared to undertake the steps that could have positive effects (e.g., exercise, specific diets), devising personalized strategies for these patients could be beneficial.

Theory of Reasoned Action and the Theory of Planned Behavior

The Theory of Reasoned Action was meant to predict individual intentions to engage particular behaviors at a specific time and within diverse contexts (Ajzen and Fishbein, 1980). Like earlier formulation, behavioral intentions relevant to health change were seen as largely dictated by attitudes regarding the engagement of relevant behaviors and the expectancy that the behaviors would promote positive outcomes, which could be enhanced if these behaviors aligned with social norms. To accommodate the need for actions (rather than just intentions), as described in Fig. 12.2, this theoretical framework was modified somewhat. The Theory of Planned Behavior (Ajzen, 2011) considered that some behaviors could be initiated more readily than others. Perceived behavioral control comprising the ease (or difficulty) in engaging in these behaviors, experiences pertinent to a given situation, together with attitudes and subjective norms, were fundamental in determining behavior change. Congruent with other perspectives, behavioral intentions and actions were seen as being subject to moderation by the individual's mood state and social influences, and constraints (e.g., financial problems) that could be encountered by engaging in behavior change. Even if individuals have the best intentions and self-will, obstacles may stand in the way so that behavior changes can be difficult, making it essential to facilitate ways of maintaining motivation during the lag that may come between intention and action.

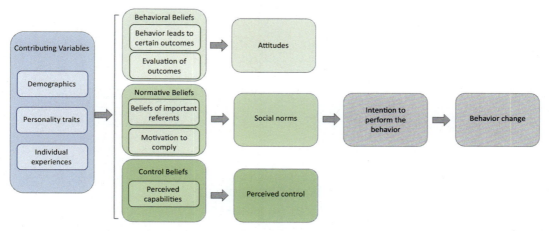

FIG. 12.2 Characteristics of individuals, such as personality factors, experiences, and attitudes can influence their beliefs related to whether the adoption of specific behaviors can be effective in producing positive health outcomes, and that individuals can adopt these behaviors successfully (control beliefs). These beliefs can affect attitudes related to the adoption of specific behaviors and can influence the intention to engage in these behaviors and ultimately in promoting behavior change.

The Theory of Planned Behavior successfully predicted various actions that were important for maintaining health and in dealing with an ongoing illness. This model anticipated cancer screening intentions and implementation of behavioral interventions among cancer patients. Moreover, it could predict treatment adherence among cancer patients, as well as intentions and actions related to follow-up care among cancer survivors (e.g., Hurtado-de-Mendoza et al., 2019).

Stage models: The Transtheoretical Model

Not only are well-ingrained behaviors difficult to modify (e.g., substance use disorders) but once the change is achieved individuals are frequently unable to sustain their new lifestyles. Accordingly, approaches were developed, such as the Transtheoretical Model (TTM) in which behavior change is promoted in small steps, incorporating different strategies to facilitate movement from one stage to the next. Should an individual fail at a particular stage (i.e., fall back into a bad habit), participants are encouraged not to drop out entirely, but instead, take a step back and continue with their efforts (Prochaska and DiClemente, 1983). Based on this model, individuals go through five stages to achieve a desired behavioral change. These stages are depicted in Fig. 12.3 along with what should be done at each stage to promote behavior change (the model shown in the figure shows six stages; the sixth is relevant to behavior relapse that can occur at any stage).

This approach has been used broadly to produce lifestyle changes. This has included efforts to quit smoking, weight management, and exercise regimens to control glucose levels in patients with diabetes, all of which have obvious implications to prevent cancer occurrence. This stage model was also useful in promoting screening for breast and cervical cancer, as well as in advanced planning to deal with cancer (Levoy et al., 2019), and could influence lifestyles among cancer survivors (Scruggs et al., 2018).

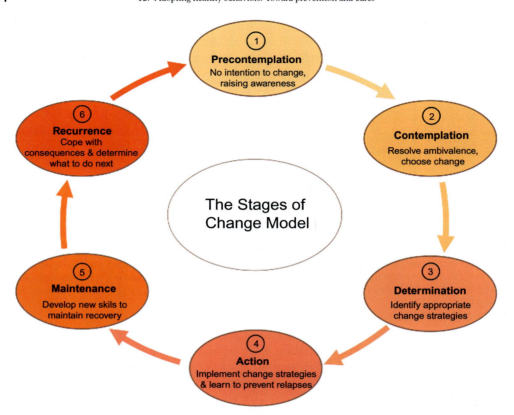

FIG. 12.3 Stages of change within the Transtheoretical Model (TTM). Specific aims are associated with each change. During the *Precontemplation* stage, individuals may be unaware that their behaviors pose significant health risks and haven't considered engaging in behavior change. During the *Contemplation* stage, individuals may have come to recognize the risks of continuing their behavior and are prepared to engage in healthy behavior in the foreseeable future (within the next 6 months). When the *Determination (Preparation)* stage is reached, individuals realize that behavioral change could enhance their well-being, and they are thus prepared to take appropriate actions imminently (within the next 30 days). During the *Action* stage attitudes are altered related to behavior change, some unhealthy behaviors are diminished, and individuals are encouraged to acquire new healthy habits, and to adopt strategies to prevent relapse. The *Maintenance* stage is defined as that in which behavior change has been sustained for an extended time (more than 6 months) and individuals intend to maintain their new outlooks and behaviors. They are still encouraged to develop new skills and strategies to sustain their motivation and new behavior. Throughout the behavioral change process, relapse may be encountered. Should *Recurrence* occur, individuals are encouraged not to abandon their efforts, but to reestablish their motivation, cope effectively with this experience, and then define strategies to move forward.

Fundamental to the stages of change is that individuals achieve *decisional balance* in which they come to understand that the benefits of change exceed the perceived disadvantages. They must also gain the confidence that they are, in fact, fully capable of initiating and then maintaining changes. To this end, it was proposed that individuals need to go through a *process of change* that comprises several cognitive and behavioral alterations. Among other things, this comprises greater awareness about the benefits of healthy behaviors, elimination of negative emotional response by adopting healthy behaviors, being committed to change through the appreciation that healthy behaviors can be successfully adopted and obtaining social support that will encourage behavioral change. In line with other models that are based on learning theory, the process of change also requires positive behaviors to be reinforced, whereas rewards are reduced in response to unwanted behaviors.

Despite its popularity in some quarters, the TTM approach was not uniformly accepted. It was seen as not adequately providing standardization regarding what defines each stage, although this may be a hollow criticism as each stage may be loosely defined to accommodate individual differences. It was also argued that TTM assumes that individuals go through logical decision-making processes that enable coherent systematic plans. Yet, decision-making is frequently not based on accurate appraisals or logical thinking, which can be disturbed further under stressful conditions. Most critically, stage-based intervention strategies may be no more effective than other approaches. As stated earlier, there is no approach that is effective for all individuals, and certainly not for all behavioral problems and the TTM model is no exception. If a given approach is ineffective, then individuals should be prepared to adopt an alternative that might be more suitable for them.

Behavioral and cognitive methods to alter health behaviors

Basic learning principles have long been the bedrock of many procedures used to promote behavioral change. Reinforcing desired behaviors results in their increased frequency, whereas failure to reinforce these behaviors causes their disappearance (extinction), although these could easily be reinstated. Aside from these instrumental (conditioning) learning methods, simply pairing a stimulus with a positive or a negative event can result in the classical conditioning of emotional and autonomic responses so the later presentation of these conditioned cues would reelicit these responses. However, if the cue was later repeatedly presented in the absence of the primary reinforcer, the response to the conditioned stimulus would disappear. These instrumental and classical conditioning principles, together with related aspects of learning theory (e.g., generalization, discrimination, secondary reinforcers) have continued to be a core component of behavior change methods within applied and clinical settings.

Exposure therapy and systematic desensitization

Getting Grover the dog to stop begging for food at the dinner table or to stop salivating whenever he sees a snack can easily be eliminated simply by not rewarding him under these conditions. However, it is much more difficult to extinguish Grover's fear and anxiety elicited by cues that had previously been associated with an adverse event. After all, Grover will initiate avoidant responses as soon as the fear-related cue is presented so that he may not learn that the primary negative event never actually occurs again. Moreover, Grover may generalize his fearful responses to relatively similar cues (e.g., fear of any stranger based on one awful experience with a human who had been abusive). Humans sometimes do precisely the same thing. Individuals who fear public speaking or who suffer a social phobia may avoid these situations and consequently may never learn that there really is little to fear. Promoting extinction of fear responses may require that individuals in some fashion experience such events and related cues without negative events occurring. In a clinical setting *"forced exposure"* to feared cues is applied in small, graded steps, through a procedure known as *"systematic desensitization."* This procedure may be beneficial in diminishing several anxiety conditions (e.g., phobias, obsessive-compulsive disorders) and could potentially be effective in alleviating anxiety related to physical illnesses and related therapies.

There are several reasons that patients with symptoms of cancer might delay in seeking a diagnosis, including the fear of facing what may transpire. The fear of cancer recurrence is likewise substantial among those who previously had been treated for cancer, and it is not infrequent for individuals to feel as if their life is running out (*future limitation perspective*). This view of the future can promote considerable psychological distress coupled with elevated neuronal activity within the ventromedial prefrontal cortex that is involved in appraisal processes (Zhou et al., 2018). A form of exposure therapy (imaginal exposure) may have some benefits, but patients may discontinue treatment finding it too distressing to even imagine cancer recurrence.

Applied behavior analysis

With greater attention dedicated to cognitive processes, many older methods to achieve behavioral change have incorporated both learning principles and cognitive approaches. Applied behavior analysis (ABA) has been widely used to alter behavior through instrumental learning methods (e.g., shaping, reinforcement, punishment, extinction). In general, ABA entails a systematic approach to promote socially acceptable behaviors in small steps (*shaping*), each of which can then be linked together (*chaining*). Conversely, undesirable behaviors can be limited by breaking the chain that holds them together. Initially, desired behaviors can be prompted by specific cues, but with therapy, they are used less and less frequently (*fading*) so that positive behaviors are adopted in the absence of specific cues. Ideally, individuals become proficient in *self-monitoring* so they will appreciate the factors that drive their negative behaviors and the necessary actions needed to reduce or eliminate them. The ABA approach has been adopted in the treatment of an exceptionally wide range of health conditions. This includes health care, mental health issues, and substance use disorders. It has also been successfully used to promote weight loss and weight management, exercise adoption, and smoking cessation, all of which have obvious implications for physical health.

Cognitive behavioral therapy

Probably more than any other approach, cognitive behavioral therapy (CBT) has gained the greatest acceptance in treating psychological illnesses, such as depressive disorders. Beck (1970) who was the originator of CBT thought of the cognitive system as comprising both primitive and mature characteristics. When the primitive system predominates, its unrealistic and idiosyncratic features favor the development of psychological disturbances. In depressed people, cognitive distortions may stem from primitive features that involve automatic thoughts or images that focus on negativity, self-debasement, and feelings of hopelessness. Individuals are thought to be enmeshed in a negative triad in which the past is viewed as negative, the present is gloomy, and prospects for the future are poor. Moreover, they may come to believe that not only are they incapable of helping themselves but also their situation is *hopeless* in that nobody else can help them.

Typically, CBT is undertaken to reduce dysfunctional behaviors, cognitions, and emotions through a goal-oriented, systematic procedure to challenge inappropriate and counterproductive beliefs, and to modify dysfunctional attitudes and belief systems ("schemas"). In this way primitive (automatic) cognitive systems are quieted, cognitive errors and distortions (e.g., magnification of negatives, minimization of positives, and catastrophizing and overgeneralizing) are eliminated and replaced with realistic and effective perspectives (Beck, 1979). During therapy, individuals are encouraged to question and test their counterproductive assumptions, evaluations and beliefs, so that they can deal directly with uncomfortable issues and find new approaches to dealing with them. While CBT has specific components and features, the way the treatment is delivered is not fixed in stone, allowing for the incorporation of other therapeutic methods (e.g., relaxation training, exposure therapy) depending on the nature and features of the condition being treated.

Cognitive behavioral therapy was initially developed to treat mood disturbances, but it has since been used in a far greater number of conditions, ranging from weight loss and eating disorders to the treatment of PTSD and anxiety. In addition to its effectiveness in these conditions, it was beneficial as an adjunct treatment to diminish the distress related to physical illnesses, such as multiple sclerosis, epilepsy and HIV. The psychological benefits and the enhanced quality of life produced by CBT were likewise apparent among patients undergoing cancer therapy as well as in cancer survivors. Among other things, CBT could reduce cancer-related depression and anxiety, as well as the fear of cancer recurrence (Rudolph et al., 2018), and was effective in attenuating cognitive dysfunctions associated with cancer, such as disturbed memory, attention, concentration, and organizational skills as well as modifying cancer features. Although CBT is most often used on a one-on-one basis, for some illnesses a group therapy approach can be effective, having the added benefit of social support.

Beyond the emotional and cognitive benefits, CBT was instrumental in ameliorating secondary features related to cancer, such as fatigue and the frequent sleep disturbances that may persist after treatments had been completed (Johnson et al., 2016). A systematic review and meta-analysis confirmed that diverse psychotherapies, including CBT, mindfulness, acceptance-based interventions, and relaxation training, modestly reduce cancer-related pain but this differed appreciably across patients. Even so, these psychotherapies diminished distress related to pain and psychosocial interventions were particularly effective in this respect (Warth et al., 2020). Once more, better pain relief could potentially be obtained by tailoring therapies to individual patients, and multimodal strategies that have been effective in alleviating pain and distress in other situations may be more effective than single strategies. Thus, it may be productive to use moderate exercise, yoga, music, and distraction procedures, along with psychotherapeutic approaches to attain optimal pain reduction.

Increasingly more studies have pointed to the efficacy of internet-delivered CBT in treating a variety of conditions. With the acute shortage of therapists, and for individuals living in remote areas, the availability and effectiveness of internet-based therapies have become increasingly more important. Obviously, during times of imposed social isolation associated with pandemic concerns, this became essential for many individuals. Fortunately, CBT delivered through internet resources could be used in the treatment of cancer-related insomnia and the resultant fatigue

(Zachariae et al., 2018), and in a subgroup of patients who adhered to treatment protocols, this approach diminished the general distress associated with the cancer experience.

Like any other treatment approach, CBT is not effective for everyone. To an extent, this might be related to patient expectancies regarding its efficacy. Much like the effects of placebos and nocebos, if patients expect that the treatment will produce positive outcomes it may do so, but if patients expect that it will be ineffective, then it is likely that beneficial effects will not be obtained. Beyond this, some patients might not be disposed to "talk therapy" whereas others may feel that CBT is too time-consuming, requiring weekly sessions over about 12 weeks, and consequently are more inclined to obtain relief through medications.

Aside from the psychological benefits promoted by behavioral and psychosocial intervention, an obviously important question is whether these procedures affect cancer-related processes. It was reported that the behavioral and cognitive changes in adults were accompanied by lower cortisol and altered production of IFN-γ and IL-2 (McGregor and Antoni, 2009). It also appeared that cognitive-behavioral stress management was associated with lower breast cancer and all-cause mortality (Stagl et al., 2015).

As much as psychotherapeutic treatments of many sorts can be used with cancer patients to enhance coping, reduce distress, and improve quality of life, some of the benefits were often short-lasting. These interventions enhanced survival assessed 1 year later but this outcome was largely absent at lengthier times (Fu et al., 2016). Of course, therapies such as CBT can be followed by other approaches so that lasting effects could potentially be obtained. For instance, CBT can be combined with supportive group therapy that itself had positive effects among women with malignant melanoma and gastrointestinal cancer, and a combination of psychotherapy and antidepressant drug treatments likewise enhanced general well-being. It would certainly be desirable for these therapies to enhance long-term survival, but diminishing distress and enhancing the quality of life over the short run is no small matter.

Mindfulness

Mindfulness meditation had been the flavor of the month (in fact, month after month for several years) in the treatment of stress-related states. This therapeutic practice stemmed from both meditation-centered practices and CBT (Brown and Ryan, 2003; Kabat-Zinn, 1990), with the goal of redirecting an individual's thoughts away from unpleasant external stressors and conflicts toward moment-to-moment internal processes. To this end, individuals are encouraged to focus on events in the present moment, and neither dwell on the past nor worry about the future. Concurrently, a person should avoid being judgmental regarding events being experienced (not an easy thing to do). Some individuals are naturally mindful and thus tend to be less vulnerable to depression and other stress-related disorders, whereas for other individuals *mindfulness-based stress reduction* (MBSR) or *mindfulness-based cognitive therapy* (MBCT) can be obtained to reduce distress.

When people focus on the present moment, it is less likely that they will negatively ruminate. Instead, they will be able to observe events unfolding without making attributions (e.g., self- or other blame) that might have negative repercussions. From the mindfulness perspective, rumination undermines effective coping, ensnaring individuals in a self-perpetuating loop that promotes depression and further rumination. By extricating themselves from this loop through a shift in perspective, including better appraisals of stressors, enhanced self-regulation, clarification of values, together with enhanced cognitive and behavioral flexibility (*reperceiving*), creative solutions to deal with problems are more likely to emerge. The benefits of mindfulness may also be tied to increased self-acceptance, mood regulation, and attention control, which may facilitate coping with daily challenges while enhancing awareness and appreciation of positive events.

It is believed that owing to these processes, mindfulness training is moderately effective in the treatment of several psychological illnesses, particularly anxiety, depression, and depressive relapse. Concurrent with mood changes, mindfulness was accompanied by altered physiological indices of distress, such as modified neuronal activity within the amygdala and anterior cingulate cortex that are presumed to mediate mood and emotional regulation, perspective-taking, and various executive functions (Tang et al., 2015).

Although the focus of mindfulness training had long been on the treatment of psychological disturbances, accumulating evidence pointed to its positive effects in the treatment of physical ailments, including the diminution of migraine headache, pain reduction, and fibromyalgia. It was also adopted in the management of distress associated with chronic illnesses, such as multiple sclerosis and cardiovascular health. A detailed annotated review of many studies that assessed stress-related behavioral and biological correlates of cancer described numerous positive actions that were produced by mindfulness training and several variants of this procedure. Among early breast cancer patients, mindfulness intervention enhanced the subjective experiences of flourishing and having a positive outlook, which was accompanied by diminished expression of inflammatory genes (Boyle et al., 2019). This approach also ameliorated distress, anxiety, and depression associated with cancer, and during cancer recovery emotional and functional quality of life was improved and posttraumatic growth was facilitated (Carlson et al., 2016). Through the adoption of a personalized approach and more targeted sessions still better outcomes could be achieved (Zimmermann et al., 2018).

As effective as mindfulness might be in alleviating the psychological distress associated with cancer, some of the positive effects, such as reduced anxiety, were not lasting, often disappearing after 12 weeks. A review of the literature regarding the effects of mindfulness-based stress reduction similarly concluded that the treatment modestly reduced anxiety and depression and enhanced quality of sleep, but while such benefits were evident 6 months after treatment, the mood-enhancing effects were absent when measured after 2 years (Schell et al., 2019).

While mindfulness interventions are effective in a subset of individuals, their value among cancer patients should not be over-inflated. A systematic review of 30 studies, most of which used modified versions of mindfulness-based stress reduction or mindfulness-based cognitive therapy, revealed considerable variability in behavioral outcomes (Shaw et al., 2018). Analyses that included randomized controlled trials similarly concluded that mindfulness training among breast cancer patients had only modest effects in altering mood disturbances (Schell et al., 2019). Comparable conclusions were drawn based on 29 randomized controlled trials, generally pointing to the procedure having modest effects in cancer patients and survivors, with the best outcome being obtained among younger patients and in those trials that adhered to preestablished mindfulness manuals (Cillessen et al., 2019). To a degree, the variable and limited effects of mindfulness procedures are hardly unexpected; mindfulness training is ordinarily effective only in some individuals, even in the absence of cancer as a triggering event. As with so many other depression treatments, there is no one treatment that is effective for all patients.

Positive psychology

If negative expectations of the future have adverse consequences, then do positive expectancies and perspectives lead to improved health? The last two decades have seen exponential growth in attention to the benefits of "positive psychology" in relation to well-being. It was maintained that positive emotions, engagement in activities that individuals enjoy, positive relationships, finding meaning in life, and appreciating individual accomplishments were related to enhanced general well-being (Seligman, 2019). Significant to the positive psychology perspective is that while happiness, character strengths, and good social relationships can buffer the effects otherwise promoted by disappointments and setbacks that may be encountered, the health benefits of positive experiences and optimistic attitudes went beyond that of just eliminating the impact of negative experiences.

The core aspects of positive psychology, some of which are reminiscent of mindfulness, were offered more than four decades ago. It had been maintained that to achieve well-being individuals ought to focus on the moment, maintain personal control or agency in diverse contexts, and experience situations as being intrinsically rewarding. In this regard, acts of kindness (e.g., charitable acts, doing things for those who are disadvantaged, and volunteering for good and meaningful causes) can go a long way in making people obtain a positive sense of self, and feelings of gratitude can act in a similar capacity (Aknin et al., 2018).

An issue addressed by positive psychology concerns the basic question of what promotes life satisfaction and what makes individuals happy; however, happiness is defined across individuals. Moreover, how do such positive states come to influence both physical and psychological well-being? It was maintained that positive experiences may broaden a person's capacity to form favorable thought-action patterns, which can then support the building of numerous enduring biological and psychosocial resources that foster good health (Fredrickson, 2001). Likewise, by enhancing positive social relationships, positivity could enhance social coping that could favor well-being and can have health benefits indirectly through the adoption of positive lifestyles.

There have been only a few reports concerning the effects of positivity on biological processes that are related to either well-being or ill health. Positive experiences were associated with cortisol changes over the day (Sin et al., 2017) and positive emotions, such as optimism, were associated with a diminished early morning cortisol increase that ordinarily occurs among distressed individuals (Jobin et al., 2014). Furthermore, finding purpose and meaning in life, and seeking and finding pleasure from life experiences, was accompanied by altered gene expression related to aspects of inflammatory immune functioning, which might affect health and well-being (Fredrickson et al., 2013).

Several reports indicated that positive dispositional characteristics, such as optimism, were associated with diminished risk for chronic illnesses (Scheier and Carver, 2018). Additionally, psychological well-being among individuals with breast cancer was related to whether an individual had a pessimistic or optimistic disposition (Carver et al., 1994). And several other personality characteristics—such as neuroticism, mastery, a sense of coherence (as well as optimism)—predicted health-related quality of life among chronically ill patients, including those with cancer. These individual characteristics could theoretically be tied to biological processes that favored good health, or they could be related to the adoption of healthy behaviors. That is, should individuals experience illness they might be more inclined to adhere to recommended treatments.

Given the presumed relationship between positivity and health, intervention approaches were designed to cultivate positivity that could limit health problems. Positive psychotherapy is the best known of these, but numerous variants of this approach have been adopted that emphasize different features of positivity, such as hope, gratitude, kindness, empathy, optimism, and humor, and could enhance psychological and physical functioning. Positive psychotherapy was also effective in promoting posttraumatic growth, and in so doing diminished emotional distress and posttraumatic stress among cancer survivors (Ochoa et al., 2017).

Positive emotions in animals: Links to immunity and cancer

It might be possible to study the biological correlates of positive feelings in dogs and cats as they express their contentment when they're scratched behind the ear or just about anywhere else (dogs) or on the head (cats). But is this the same as happiness expressed by humans? Assessing this in rodents is still more difficult as they often may seem to be less expressive in response to positive events (at least within a laboratory context), although negative emotions can readily be determined. Raising mice or rats within enriched environments has repeatedly been found to promote health benefits, but this might have little to do with their moods or positivity. When rodents receive stimulation of brain reward centers, such as the ventral tegmentum, they clearly find it rewarding and will spend considerable time and effort to obtain the brain stimulation. Again, this might not mean that the stimulation makes these critters happy as humans might define this. Nonetheless, it is intriguing that brain stimulation rewards enhanced immune functioning and diminished their bacterial load, apparently operating through changes of sympathetic nervous system functioning (Ben-Shaanan et al., 2016). Moreover, activation of brain reward circuits provoked a reduction of Lewis lung carcinoma in an animal model, again being mediated by peripheral norepinephrine modulation of immune suppressor cell production (Ben-Shaanan et al., 2018). Although highly speculative, it was tentatively suggested that the benefits of positive emotions in humans might operate by affecting brain reward processes. Evidence in humans that placebo effects are associated with activation of reward areas of the brain speaks to this as a plausible hypothesis.

Critiques of positive psychology

Few clinicians would deny that positivity can be psychologically helpful, but the extensive benefits that have been attributed to positive psychology have been seen as a bit over-the-top and was harshly criticized (Wong and Roy, 2018). Some of the published reports were challenged based on statistical problems and even conceptual flaws that may have contributed to failures to replicate research findings. Furthermore, most studies comprised a limited number of participants and the magnitude of the effects observed were small and might have been influenced by biases inherent in studies with few participants (White et al., 2019).

Even stronger criticisms have been leveled, suggesting that positive psychology is beset by circular reasoning and the assumption of causal connections where none existed. The scathing criticisms even included the suggestion that assertions made by proponents of positive psychology were incompatible with the empirical evidence, particularly those that attempted to link positive psychology to cancer processes and outcomes (Coyne and Tennen, 2010).

To be sure, positive states can favor the adoption of positive lifestyles that can favor some diseases being less likely to occur, but on some occasions, data are brought forward that seems to be at odds with what is known about illness processes.

While not dismissing these seemingly valid criticisms, in dealing with chronic and quite wicked diseases such as cancer, if individuals improve their world views, find meaning in life, or achieve *eudaimonic* well-being (happiness and contentment obtained through self-actualization and having a meaningful purpose in life), then this might foster better living (and an easier death). But is it realistic to expect that longevity will be extended? Given the many tactics that cancer cells adopt to thwart innate and acquired biological protective mechanisms, is it reasonable to believe that they will be altered by optimism or eudaimonic well-being? In the end, the case for positive psychology to favor physical health will require randomized controlled trials that prospectively examine links to illness morbidity and mortality, and define in whom the benefits are most likely to be realized. At present, however, it seems that the claims of positive psychology are in danger of providing false hope.

Social support

Human social interactions have been viewed as a basic need that may have evolved over millennia because of the survival advantages it offers. Indeed, like food cravings that occur following food deprivation, acute social isolation may instigate craving for social rewards in response to relevant environmental cues and may engender several brain neurochemical changes ordinarily associated with reward (Tomova et al., 2020). Just as feelings of hunger or pain perception cause individuals to engage in behaviors to alleviate uncomfortable sensations, feelings of loneliness may signal the need for social comfort (Cacioppo et al., 2014). In essence, maintaining "social homeostasis" can be viewed as a highly adaptive response and when social needs are unmet, health risks can ensue (Tomova et al., 2020).

Of the many coping strategies available, obtaining social support is among the most important, and may be particularly useful in dealing with the distress of a cancer diagnosis, cancer therapy, and the uncertain period after treatment cessation. As we saw earlier, social support has multiple benefits, providing a shoulder to cry on, an effective distractor, and a functional resource, such as obtaining help to get to the hospital or having food delivered. It also buffers the effects of stressors on immune functioning and enhances adherence to treatment protocols. It was less clear, however, whether social support halted cancer progression despite enhancing the quality of life in patients. Although prospective studies had indicated that effective coping and social support were associated with somewhat slower cancer progression, the findings have been inconsistent across studies. It was nevertheless generally maintained that the data were sufficiently strong to encourage further study of this issue.

In considering the influence of coping on cancer progression, it is important to consider effects beyond those attributable to the availability of social support. The support that people believe they have and their satisfaction with this support was associated with lower levels of proinflammatory factors and lower mortality risk than that observed among individuals who reported poorer levels of satisfaction with the support obtained (Boen et al., 2018). Commensurate with these findings, following breast cancer surgery, satisfaction with the available social resources (*social well-being*) was accompanied by lower levels of proinflammatory cytokines and expression of pro-metastatic genes (Jutagir et al., 2017). Similarly, among breast cancer survivors, having effective social support was linked to lower CRP levels, which were correlated with diminished left amygdala activity, pointing to possible ties between emotional responses and inflammation related to social support (Muscatell et al., 2016).

Among women living alone, the perception of low social support before colorectal cancer diagnosis was associated with greater overall and cancer-specific mortality (Kroenke et al., 2020). Being married or having a close confidante or friend, in contrast, was associated with fewer health problems and increased longevity relative to being unmarried, and this was also observed in some types of cancer, varying with sex and ethnicity. The analysis of a large cohort exceeding 734,000 patients indicated that in married people metastases were 17% lower across 10 different types of cancer, relative to that seen in unmarried patients (Aizer et al., 2013). Conversely, poorer cancer-specific survival was reported in newly divorced individuals (Dinh et al., 2018). Such outcomes could well be related to the time at which cancer was diagnosed, and indeed, single people were generally diagnosed with colon cancer later than were married people, and longer survival commensurately occurred among married colon cancer patients (Wang et al., 2011).

Findings such as these have typically been attributed to the increased social support that married people provide to one another. Yet, a persuasive case has been made that this view was far too simplistic as being married does not necessarily universally translate to receiving greater social support. For that matter, in some instances, unmarried individuals may have a greater range of supports, often having more social connections, better community ties, and particularly close bonds with their extended family. The relationship between being married and cancer survival might also be related to implicit biases among oncologists that influence the therapies provided to married vs unmarried people. Perhaps, oncologists are subject to the mistaken belief that unmarried people, lacking social support, might be less able to withstand the strains of aggressive treatments (DelFattore, 2019).

Change through social identity: The social cure

As social creatures, humans are very much influenced by their in-group. By sharing meaningful identities, individuals are more likely to adopt their group's core values, which can also facilitate individuals learning from one another, and transmitting cultural knowledge. Furthermore, while self-efficacy, self-esteem, and social status often are fundamental in decision making, in uncertain situations individuals might look to their in-group to guide their attitudes and behaviors.

The sense of belonging, together with subjective psychological bonds that people feel toward one another and to outside groups, defines social connectedness (reflected by the number and quality of groups to which an individual is linked), and which could favor the adoption of a variety of preventive health services. As expected, social connectedness was positively related to general well-being and may be important in diminishing distress (Haslam et al., 2016). In contrast, the loneliness stemming from the absence of social connections has been tied to poor health, whereas simple social group activities (e.g., arts-based groups, exercise groups) can reduce low and moderate levels of depression and could diminish illness recurrence. The position has been adopted that shared social identities can provide a "social cure" in illness prevention and in dealing with pathological conditions (Haslam et al., 2018).

Commensurate with the psychological benefits obtained, experiences that include social connections are associated with brain neurochemical changes akin to those elicited by other rewarding stimuli. Threats to connectedness, in contrast, may activate neural circuits much like other threats do, and the positive experiences related to social connections can affect immunity so that inflammation can be reduced, and antiviral responses enhanced (Leschak and Eisenberger, 2019).

Having a limited social network was accompanied by greater inflammation that was related to cancer occurrence and these relationships were notable among lower socioeconomic status groups even after adjusting for social, behavioral, and illness factors (Busch et al., 2018). The absence of a spouse/partner and limited community ties was related to earlier breast cancer-specific mortality. Conversely, among breast cancer survivors, the frequency of contacts with family and friends, attending religious services, and participation in community activities was associated with a 15%–28% decline of early all-cause mortality. Having larger social networks similarly predicted a better prognosis among women who had been diagnosed with stage I/II breast cancer, varying with the quality (or burden) of relationships, especially with family members (Kroenke et al., 2017). It similarly appeared that diminished levels of social connectedness were associated with greater allostatic overload, which was tied to the emergence of several chronic diseases, including various forms of cancer (Larrabee Sonderlund et al., 2019). As much as connectedness to multiple groups may have benefits in relation to dealing with cancer, as mentioned earlier, there is reason to suppose that simply having multiple connections may not be as important as individuals being satisfied with the support received.

Social identity may also be key in affecting cancer-related behaviors, including seeking treatment. Obtaining social support may diminish the impact of the stigma that can appear in those who engaged in behaviors (e.g., smoking) that contributed to cancer emergence as well as among individuals who experienced disfigurement owing to head and neck cancer. Importantly, many former patients who had received successful therapy may continue to identify as cancer survivors, which can undermine psychological health, although this can vary with age and the type of cancer (Thong et al., 2018). Social support and connectedness could potentially alter this "cancer" identity, thereby encouraging the adoption of positive health behaviors and the promotion of psychological growth.

The benefits of having a shared social identity and high connectedness are apparent in numerous contexts, but perhaps none as profound as in support groups in which members are dealing with especially distressing experiences, such as parents of children with cancer and in family members of those who died through suicide. Those with a shared identity in these situations not only express a cognitive sense that others "share my pain," but might also know the right things to say or do when interacting with other cancer patients. These social support conditions might also allow for a depth of support that is not readily obtained from others. There have been indications that survival time increased somewhat if patients interacted with others who survived cancer for more than 5 years, possibly by encouraging hopefulness or by diminishing stress-related processes (Lienert et al., 2017). In dealing with specific types of cancer, as in survivors of ovarian cancer, a group-based internet intervention that was targeted specifically to these patients, diminished distress and enhanced quality of life (Kinner et al., 2018).

When interventions are undertaken by a group, the common context allows for a positive social identity that can enhance compliance and successful outcomes. An intervention, dubbed Groups 4 Health (G4H), which targets both the development and maintenance of positive social group relationships, could diminish health-related disturbances stemming from disconnectedness (Haslam et al., 2016). Improved social connectedness and social identity are fundamental to well-being and could be incorporated into individual behavior change approaches described earlier. The value of social connection is so broad and profound as to have prompted the position that promoting social connections ought to be a public health priority (Holt-Lunstad et al., 2017).

Limits to the effects of social support

As much as supportive relations are remarkably important for well-being, obtaining or being offered social support can be a sword that cuts both ways. Receiving support in the context of chronic illnesses can undermine self-esteem if the individual comes to perceive themselves as being insufficiently competent in contending with their situation, or if they feel the loss of role functioning. Furthermore, although many people welcome support and closeness from others, some individuals might prefer to maintain their independence and self-efficacy.

There are occasions in which individuals might seek support and reasonably expect that it will be received, only to find that the support is not forthcoming or appears in an entirely unhelpful fashion. Such *unsupportive* responses are far worse than simply not having support and may produce negative psychological and somatic outcomes. These unsupportive responses may come in several forms, as unfair and inappropriate as these may be. Unsupportive relations may appear in the form of blame as in the case of patients with lung cancer being subject to stereotypes (e.g., smokers who are said to have "brought it on themselves"). Friends might also minimize the seriousness of the situation ("Oh, I'm sure it will all work out in the end") or they might bumble about trying to find the right thing to say, opting to say nothing. Worse still, they might make efforts to distance themselves from the person needing support or simply seem to lack empathy—there can also be the awkwardness of feeling powerless to deal with the big elephant in the room or simply not wanting to draw attention to the issue in the belief that the patient doesn't wish to talk about it. Whatever the case, meeting unsupportive responses may have profound effects on those who need and seek support.

Individuals who are dealing with chronic illnesses also need to consider that what they perceive as unsupportive might not be an accurate appraisal. When a person is initially diagnosed with cancer, their friends frequently gather to support them, and their support is typically appreciated and valued. The support will often go on until the affected person has completed therapy and seems to be on the way to recovery, not realizing that the cancer survivor needs continued support for some time afterward. The distress of the previous months had likely taken a physical and psychological toll on the cancer patient, and persistent concerns that cancer will reoccur may haunt them. The posttreatment period can be excruciatingly difficult, yet survivors also need to appreciate that the support they receive is time-limited as people need to get on with their own lives. Those who provided support to the person in need of help may also become weary of helping (the helpers "high" may have turned into a chore). They may also misinterpret the behavior of the cancer patient as being "needy," failing to realize that as the support giver, they are not being "put upon," but instead are being "relied upon." In the end, there are multiple dynamics possible in what is a highly sensitive situation, one that generally lacks answers and requires considerable courage.

Summary and conclusions

The occurrence of diseases—especially those that involve cancer—is sometimes entirely out of an individual's control, seemingly being a consequence of fate, random gene mutations, the actions of other individuals or industrial enterprises, or poor governmental policies that failed to provide protection. However, some illnesses might be preventable—or at the very least, the risks which lead to the illness—through the adoption of appropriate lifestyles or the actions of organizations whose mandate is to assure our well-being.

Numerous strategies are available to promote behavior change to prevent disease or to diminish distress associated with an illness that can also affect disease progression. Several of these psychosocial approaches were effective in enhancing positive immune changes and diminishing those that can have harmful effects (e.g., persistent proinflammatory elevations), often being evident for months after treatment. These outcomes were apparent in the context of several therapies (alone or in combination), with the actions being most pronounced with CBT (Shields et al., 2020).

Should an illness develop, its management and progression can also be affected by government policies (e.g., availability of universal health care, adequate preparation and facilities, and well-trained health professionals) and to a significant degree, individuals can either help themselves (e.g., not delaying medical help being sought and treatments being obtained) or they can obtain help in dealing with varied illnesses. There are, in fact, many approaches that could potentially help individuals deal with the impact of diseases, such as cancer, but it's difficult to know which might be the most beneficial for a given person. Each approach seems to help a subset of individuals, but none is effective in helping all people. When pharmacological treatments are evaluated and then come to market, an important aspect of testing is whether the new treatment has advantages over earlier remedies, such as whether it has greater or faster effects as well as fewer side effects. Unlike pharmacological therapies, most behavioral and cognitive therapies are not assessed on a head-to-head basis. Rather, studies are most often conducted simply to determine whether this or that treatment was effective in a certain number of people—in other words, was there a statistically significant effect? The net result is that patients may not be able to make educated choices as to which, if any, will be most appropriate for them.

Given the actions on neurobiological and inflammatory processes, it would be reasonable to hypothesize that psychological interventions that act against these processes might also influence cancer growth. Thus, for individuals diagnosed with cancer "prehabilitation" programs may be especially valuable in preparing them for the travails of treatment, potentially elevating their resilience and enhancing therapeutic outcomes. Even if these procedures don't extend the life of patients, they may improve mood and coping that adds, under the circumstances, something positive to their quality of life. As such, these programs can be an important and an exceptionally desirable option.

References

Aizer, A.A., Chen, M.H., McCarthy, E.P., Mendu, M.L., Koo, S., et al., 2013. Marital status and survival in patients with cancer. J. Clin. Oncol. 31, 3869–3876.

Ajzen, I., 2011. The theory of planned behaviour: reactions and reflections. Psychol. Health 26, 1113–1127.

References

Ajzen, A., Fishbein, M., 1980. Understanding Attitudes and Predicting Social Behaviour. Preventive-Hall. Inc., Englewood Cliffs, NJ.

Aknin, L.B., Van de Vondervoort, J.W., Hamlin, J.K., 2018. Positive feelings reward and promote prosocial behavior. Curr. Opin. Psychol. 20, 55–59.

Andersen, M.R., Afdem, K., Hager, S., Gaul, M., Sweet, E., et al., 2017. The 'cause' of my cancer, beliefs about cause among breast cancer patients and survivors who do and do not seek IO care. Psychooncology 26, 248–254.

Bandura, A., 1986. Social Foundations of Thought and Action: A Social Cognitive Theory. Prentice-Hall, Englewood Cliffs, NJ.

Beck, A.T., 1970. Cognitive therapy: nature and relation to behavior therapy. Behav. Ther. 1, 184–200.

Beck, A.T., 1979. Cognitive Therapy of Depression. Guilford Press, NY.

Becker, M.H., 1974. The health belief model and sick role behavior. Health Educ. Monogr. 2, 409–419.

Ben-Shaanan, T.L., Azulay-Debby, H., Dubovik, T., Starosvetsky, E., Korin, B., et al., 2016. Activation of the reward system boosts innate and adaptive immunity. Nat. Med. 22, 940–944.

Ben-Shaanan, T.L., Schiller, M., Azulay-Debby, H., Korin, B., Boshnak, N., et al., 2018. Modulation of anti-tumor immunity by the brain's reward system. Nat. Commun. 9, 2723.

Blastland, M., Freeman, A.L.J., van der Linden, S., Marteau, T.M., Spiegelhalter, D., 2020. Five rules for evidence communication. Nature 587, 362–364.

Boen, C.E., Barrow, D.A., Bensen, J.T., Farnan, L., Gerstel, A., et al., 2018. Social relationships, inflammation, and cancer survival. Cancer Epidemiol. Biomarkers Prev. 27, 541–549.

Boyle, C.C., Cole, S.W., Dutcher, J.M., Eisenberger, N.I., Bower, J.E., 2019. Changes in eudaimonic well-being and the conserved transcriptional response to adversity in younger breast cancer survivors. Psychoneuroendocrinology 103, 173–179.

Brown, K.W., Ryan, R.M., 2003. The benefits of being present: mindfulness and its role in psychological well-being. J. Pers. Soc. Psychol. 84, 822–848.

Busch, E.L., Whitsel, E.A., Kroenke, C.H., Yang, Y.C., 2018. Social relationships, inflammation markers, and breast cancer incidence in the Women's Health Initiative. Breast 39, 63–69.

Cacioppo, J.T., Cacioppo, S., Boomsma, D.I., 2014. Evolutionary mechanisms for loneliness., 28, 3–21. Cogn. Emot. 8, 58–72.

Carlson, L.E, Tamagawa, R., Stephen, J., Drysdale, E., Zhong, L., Speca, M., 2016. Randomized-controlled trial of mindfulness-based cancer recovery versus supportive expressive group therapy among distressed breast cancer survivors (MINDSET): long-term follow-up results. Psychooncology 25, 750–759.

Carver, C.S., Pozo-Kaderman, C., Harris, S.D., Noriega, V., Scheier, M.F., et al., 1994. Optimism versus pessimism predicts the quality of women's adjustment to early stage breast cancer. Cancer 73, 1213–1220.

Chirico, A., Lucidi, F., Merluzzi, T., Alivernini, F., Laurentiis, M., et al., 2017. A meta-analytic review of the relationship of cancer coping self-efficacy with distress and quality of life. Oncotarget 8, 36800–36811.

Cillessen, L., Johannsen, M., Speckens, A.E.M., Zachariae, R., 2019. Mindfulness-based interventions for psychological and physical health outcomes in cancer patients and survivors: a systematic review and meta-analysis of randomized controlled trials. Psychooncology 28, 2257–2269.

Coyne, J.C., Tennen, H., 2010. Positive psychology in cancer care: bad science, exaggerated claims, and unproven medicine. Ann. Behav. Med. 39, 16–26.

de Ridder, D., Kroese, F., van Gestel, L., 2021. Nudgeability: mapping conditions of susceptibility to nudge influence. Perspect. Psychol. Sci. 1745691621995183.

DelFattore, J., 2019. Death by stereotype? Cancer treatment in unmarried patients. N. Engl. J. Med. 381, 982–985.

Diefenbach, M.A., Leventhal, H., 1996. The common-sense model of illness representation: theoretical and practical considerations. J. Soc. Distress Homeless 5, 11–38.

Dinh, K.T., Aizer, A.A., Muralidhar, V., Mahal, B.A., Chen, Y.W., et al., 2018. Increased vulnerability to poorer cancer-specific outcomes following recent divorce. Am. J. Med. 131, 517–523.

Doll, R., Peto, R., Wheatley, K., Gray, R., Sutherland, I., 1994. Mortality in relation to smoking: 40 years' observations on male British doctors. BMJ 309, 901–911.

Dubé, E., Vivion, M., MacDonald, N.E., 2015. Vaccine hesitancy, vaccine refusal and the anti-vaccine movement: influence, impact and implications. Expert Rev. Vaccines 14, 99–117.

Dumitrescu, R.G., 2018. Alcohol-induced epigenetic changes in cancer. Methods Mol. Biol. 1856, 157–172.

Durazo, A., Cameron, L.D., 2019. Representations of cancer recurrence risk, recurrence worry, and health-protective behaviours: an elaborated, systematic review. Health Psychol. Rev. 13, 447–476.

Falk, E.B., O'Donnell, M.B., Cascio, C.N., Tinney, F., Kang, Y., et al., 2015. Self-affirmation alters the brain's response to health messages and subsequent behavior change. Proc. Natl. Acad. Sci. 112, 1977–1982.

Fredrickson, B.L., 2001. The role of positive emotions in positive psychology. The broaden-and-build theory of positive emotions. Am. Psychol. 56, 218–226.

Fredrickson, B.L., Grewen, K.M., Coffey, K.A., Algoe, S.B., Firestine, A.M., et al., 2013. A functional genomic perspective on human well-being. Proc. Natl. Acad. Sci. 110, 13684–13689.

Fu, W.W., Popovic, M., Agarwal, A., Milakovic, M., Fu, T.S., et al., 2016. The impact of psychosocial intervention on survival in cancer: a meta-analysis. Ann. Palliat. Med. 5, 93–106.

Hagger, M.S., Koch, S., Chatzisarantis, N.L.D., Orbell, S., 2017. The common sense model of self-regulation: meta-analysis and test of a process model. Psychol. Bull. 143, 1117–1154.

Haslam, C., Cruwys, T., Haslam, S.A., Dingle, G., Chang, M.X., 2016. Groups 4 Health: evidence that a social-identity intervention that builds and strengthens social group membership improves mental health. J. Affect. Disord. 194, 188–195.

Haslam, C., Jetten, J., Cruwys, T., Dingle, G.A., Haslam, S.A., 2018. The New Psychology of Health: Unlocking the Social Cure. Routledge, Oxfordshire.

Holt-Lunstad, J., Robles, T.F., Sbarra, D.A., 2017. Advancing social connection as a public health priority in the United States. Am. Psychol. 72, 517.

Hurtado-de-Mendoza, A., Carrera, P., Parrott, W.G., Gomez-Trillos, S., Perera, R.A., et al., 2019. Applying the theory of planned behavior to examine adjuvant endocrine therapy adherence intentions. Psychooncology 28, 187–194.

Jobin, J., Wrosch, C., Scheier, M.F., 2014. Associations between dispositional optimism and diurnal cortisol in a community sample: when stress is perceived as higher than normal. Health Psychol. 33, 382–391.

Johnson, J.A., Rash, J.A., Campbell, T.S., Savard, J., Gehrman, P.R., et al., 2016. A systematic review and meta-analysis of randomized controlled trials of cognitive behavior therapy for insomnia (CBT-I) in cancer survivors. Sleep Med. Rev. 27, 20–28.

Jones, C.J., Smith, H., Llewellyn, C., 2014. Evaluating the effectiveness of health belief model interventions in improving adherence: a systematic review. Health Psychol. Rev. 8, 253–269.

Jutagir, D.R., Blomberg, B.B., Carver, C.S., Lechner, S.C., Timpano, K.R., et al., 2017. Social well-being is associated with less pro-inflammatory and pro-metastatic leukocyte gene expression in women after surgery for breast cancer. Breast Cancer Res. Treat. 165, 169–180.

Kabat-Zinn, J., 1990. Full Catastrophe Living: Using the Wisdom of Your Body and Mind to Face Stress, Pain, and Illness. Delacourt, New York.

Kinner, E.M., Armer, J.S., McGregor, B.A., Duffecy, J., Leighton, S., et al., 2018. Internet-based group intervention for ovarian cancer survivors: feasibility and preliminary results. JMIR Cancer 4, e1.

Kroenke, C.H., Michael, Y.L., Poole, E.M., Kwan, M.L., Nechuta, S., et al., 2017. Postdiagnosis social networks and breast cancer mortality in the After Breast Cancer Pooling Project. Cancer 123, 1228–1237.

Kroenke, C.H., Paskett, E.D., Cene, C.W., Caan, B.J., Luo, J., et al., 2020. Prediagnosis social support, social integration, living status, and colorectal cancer mortality in postmenopausal women from the women's health initiative. Cancer 126, 1766–1775.

Lamprell, K., Tran, Y., Arnolda, G., Braithwaite, J., 2021. Nudging clinicians: a systematic scoping review of the literature. J. Eval. Clin. Pract. 27, 175–192.

Larrabee Sonderlund, A., Thilsing, T., Sondergaard, J., 2019. Should social disconnectedness be included in primary-care screening for cardiometabolic disease? A systematic review of the relationship between everyday stress, social connectedness, and allostatic load. PLoS One 14, e0226717.

Lau, J., Lim, T.Z., Jianlin Wong, G., Tan, K.K., 2020. The health belief model and colorectal cancer screening in the general population: a systematic review. Prev. Med. Rep. 20, 101223.

Leschak, C.J., Eisenberger, N.I., 2019. Two distinct immune pathways linking social relationships with health: inflammatory and antiviral processes. Psychosom. Med. 81, 711–719.

Leventhal, H., Nerenz, D.R., Steele, D.J., 1984. Illness representations and coping with health threats. In: Baum, A., Taylor, S.E., Singer, J.E. (Eds.), Handbook of Psychology and Health: Social psychological aspects of health. Lawrence Erlbaum, Hillsdale, NJ, pp. 219–252.

Levoy, K., Salani, D.A., Buck, H., 2019. A systematic review and gap analysis of advance care planning intervention components and outcomes among cancer patients using the transtheoretical model of health behavior change. J. Pain Symptom Manage. 57, 118–139 e116.

Lienert, J., Marcum, C.S., Finney, J., Reed-Tsochas, F., Koehly, L., 2017. Social influence on 5-year survival in a longitudinal chemotherapy ward co-presence network. Netw. Sci. 5, 308–327.

McGregor, B.A., Antoni, M.H., 2009. Psychological intervention and health outcomes among women treated for breast cancer: a review of stress pathways and biological mediators. Brain Behav. Immun. 23, 159–166.

Merluzzi, T.V., Pustejovsky, J.E., Philip, E.J., Sohl, S.J., Berendsen, M., et al., 2019. Interventions to enhance self-efficacy in cancer patients: a meta-analysis of randomized controlled trials. Psychooncology 28, 1781–1790.

Muscatell, K.A., Eisenberger, N.I., Dutcher, J.M., Cole, S.W., Bower, J.E., 2016. Links between inflammation, amygdala reactivity, and social support in breast cancer survivors. Brain Behav. Immun. 53, 34–38.

Ochoa, C., Casellas-Grau, A., Vives, J., Font, A., Borras, J.M., 2017. Positive psychotherapy for distressed cancer survivors: posttraumatic growth facilitation reduces posttraumatic stress. Int. J. Clin. Health Psychol. 17, 28–37.

Pierce, J.P., Chen, R., Kealey, S., Leas, E.C., White, M.M., et al., 2021. Incidence of cigarette smoking relapse among individuals who switched to e-cigarettes or other tobacco products. JAMA Netw. Open 4, e2128810.

Prochaska, J.O., DiClemente, C.C., 1983. Stages and processes of self-change of smoking: toward an integrative model of change. J. Consult. Clin. Psychol. 51, 390–395.

Pushalkar, S., Paul, B., Li, Q., Yang, J., Vasconcelos, R., et al., 2020. Electronic cigarette aerosol modulates the oral microbiome and increases risk of infection. iScience 23, 100884.

Richardson, E.M., Schuz, N., Sanderson, K., Scott, J.L., Schuz, B., 2017. Illness representations, coping, and illness outcomes in people with cancer: a systematic review and meta-analysis. Psychooncology 26, 724–737.

Rogers, R.W., 1983. Cognitive and physiological processes in fear appeals and attitude change: a revised theory of protection motivation. In: Cacioppo, J.T., Petty, R.E. (Eds.), Social Psychophysiology: A Sourcebook. Guilford Press, New York, pp. 153–176.

Roozbahani, N., Kaviani, A.H., Khorsandi, M., 2020. Path analysis of skin cancer preventive behavior among the rural women based on protection motivation theory. BMC Womens Health 20, 1–8.

Rudolph, B., Wunsch, A., Herschbach, P., Dinkel, A., 2018. Cognitive-behavioral group therapy addressing fear of progression in cancer out-patients. Psychother. Psychosom. Med. Psychol. 68, 38–43.

Sandman, P.M., 2009. Pandemics: good hygiene is not enough. Nature 459, 322–323.

Scheier, M.F., Carver, C.S., 2018. Dispositional optimism and physical health: a long look back, a quick look forward. Am. Psychol. 73, 1082–1094.

Schell, L.K., Monsef, I., Wockel, A., Skoetz, N., 2019. Mindfulness-based stress reduction for women diagnosed with breast cancer. Cochrane Database Syst. Rev. 3, CD011518.

Scruggs, S., Mama, S.K., Carmack, C.L., Douglas, T., Diamond, P., et al., 2018. Randomized trial of a lifestyle physical activity intervention for breast cancer survivors: effects on transtheoretical model variables. Health Promot. Pract. 19, 134–144.

Seligman, M.E.P., 2019. Positive psychology: a personal history. Annu. Rev. Clin. Psychol. 15, 1–23.

Shah, S.K., Nakagawa, M., Lieblong, B.J., 2020. Examining aspects of successful community-based programs promoting cancer screening uptake to reduce cancer health disparity: a systematic review. Prev. Med. 141, 106242.

Shaw, J.M., Sekelja, N., Frasca, D., Dhillon, H.M., Price, M.A., 2018. Being mindful of mindfulness interventions in cancer: a systematic review of intervention reporting and study methodology. Psychooncology 27, 1162–1171.

Shields, G.S., Spahr, C.M., Slavich, G.M., 2020. Psychosocial interventions and immune system function: a systematic review and meta-analysis of randomized clinical trials. JAMA Psychiatry 77, 1031–1043.

Sin, N.L., Ong, A.D., Stawski, R.S., Almeida, D.M., 2017. Daily positive events and diurnal cortisol rhythms: examination of between-person differences and within-person variation. Psychoneuroendocrinology 83, 91–100.

Solomon, S., Kington, R., 2002. National efforts to promote behavior-change research: views from the Office of Behavioral and Social Sciences Research. Health Educ. Res. 17, 495–499.

Stagl, J.M., Lechner, S.C., Carver, C.S., Bouchard, L.C., Gudenkauf, L.M., et al., 2015. A randomized controlled trial of cognitive-behavioral stress management in breast cancer: survival and recurrence at 11-year follow-up. Breast Cancer Res. Treat. 154, 319–328.

Taha, S.A., Matheson, K., Anisman, H., 2013. The 2009 H1N1 Influenza Pandemic: the role of threat, coping, and media trust on vaccination intentions in Canada. J. Health Commun. 18, 278–290.

Tang, Y.Y., Holzel, B.K., Posner, M.I., 2015. The neuroscience of mindfulness meditation. Nat. Rev. Neurosci. 16, 213–225.

Thaler, R.H., Sunstein, C.R., 2008. Nudge: improving decisions about health. Wealth Happiness 6, 14–38.

Thong, M.S., Wolschon, E.M., Koch-Gallenkamp, L., Waldmann, A., Waldeyer-Sauerland, M., et al., 2018. "Still a cancer patient"—associations of cancer identity with patient-reported outcomes and health care use among cancer survivors. JNCI Cancer Spectr. 2, pky031.

Tomova, L., Wang, K.L., Thompson, T., Matthews, G.A., Takahashi, A., et al., 2020. Acute social isolation evokes midbrain craving responses similar to hunger. Nat. Neurosci. 23, 1597–1605.

Wang, L., Wilson, S.E., Stewart, D.B., Hollenbeak, C.S., 2011. Marital status and colon cancer outcomes in US Surveillance, Epidemiology and End Results registries: does marriage affect cancer survival by gender and stage? Cancer Epidemiol. 35, 417–422.

Warth, M., Zöller, J., Köhler, F., Aguilar-Raab, C., Kessler, J., Ditzen, B., 2020. Psychosocial interventions for pain management in advanced cancer patients: a systematic review and meta-analysis. Curr. Oncol. Rep. 21, 3.

White, C.A., Uttl, B., Holder, M.D., 2019. Meta-analyses of positive psychology interventions: the effects are much smaller than previously reported. PLoS One 14, e0216588.

Wong, P.T., Roy, S., 2018. Critique of positive psychology and positive interventions. In: Brown, N.J.L., Lomas, T., Eiroa-Orosa, F.J. (Eds.), The Routledge International Handbook of Critical Positive Psychology. Routledge, pp. 142–160.

Wu, D., O'Shea, D.F., 2020. Potential for release of pulmonary toxic ketene from vaping pyrolysis of vitamin E acetate. Proc. Natl. Acad. Sci. 117, 6349–6355.

Zachariae, R., Amidi, A., Damholdt, M.F., Clausen, C.D.R., Dahlgaard, J., et al., 2018. Internet-delivered cognitive-behavioral therapy for insomnia in breast cancer survivors: a randomized controlled trial. J. Natl. Cancer Inst. 110, 880–887.

Zhou, J., Feng, P., Lu, X., Han, X., Yang, Y., et al., 2018. Do future limitation perspective in cancer patients predict fear of cancer recurrence, mental distress, and the ventromedial prefrontal cortex activity? Front. Psychol. 9, 420.

Zimmermann, F.F., Burrell, B., Jordan, J., 2018. The acceptability and potential benefits of mindfulness-based interventions in improving psychological well-being for adults with advanced cancer: a systematic review. Complement. Ther. Clin. Pract. 30, 68–78.

CHAPTER 13

Cancer therapies: Caveats, concerns, and momentum

OUTLINE

Promises, promises	401	Progress in cancer treatment development	412
Cancer screening	402	Difficulties and limitations	412
Early detection	402	*The upside of the story*	414
Sensitivity versus specificity	403	*Problems in determining treatment efficacy*	415
Cancer screening in common types of cancer	403	Complementary and alternative medicine	416
Breast cancer screening	403		
Prostate cancer screening	404	Precision treatment: Obstacles and challenges	419
Colon cancer screening	404	*Genetic and epigenetic factors in precision medicine*	420
Lung cancer screening	405	*Matching individuals to treatments*	420
Caveats concerning screening and early cancer detection	406	*Beyond the genome*	422
The ever-present problem of overdiagnosis	408	*Therapies agnostic to cancer type*	423
Delayed presentation of illness: Assessment delays and delays of treatment	408	Summary and conclusions	426
Unintended consequences and unintended benefits	411	References	427

Promises, promises

"Remarkable breakthrough," "cancer defeated," "cancer cure at hand," "new class of drugs destroys multiple cancer targets," "the future in stem cells." Every time there's a hint of an effective therapy, mainstream media *kvell* over the apparent success. Not long afterward, when it becomes apparent that the initial enthusiasm was premature, newspaper articles and various internet sites have lambasted attempts to develop cancer cures. Often, sensationalist media trumpet the words of frustrated scientists, such as the comment by one critic that "We have cured mice of cancer for decades—and it simply didn't work in humans" or paraphrasing another critic "if studies in mice were valid and reliable, we would have cured cancer a thousand times." Some science critics have gone so far as to suggest that "most medical studies are wrong" or "This is why you shouldn't believe that exciting new medical study."

To be sure, while the search for cancer cures has been beset by false alarms, it is certainly not the only illness that's fallen into this trap. This is the norm rather than the exception. The initial hype that accompanied the introduction of SSRIs to treat depression, was subsequently found to be effective in only about 50% of patients. Disappointments have repeatedly been experienced in the search for medications to treat neurodegenerative brain disorders, and it took years to discover that β-blockers designed to diminish hypertension were not nearly as effective as initially believed. Finding effective cancer therapies has been an especially gnarly problem. Nevertheless, one need only look back to see how far cancer therapies have come over the past decades. Criticism can be healthy for scientific approaches in any domain, especially if it offers better approaches. However, being overly critical might not always be productive, particularly if it undermines individual and public trust in the development of effective treatments.

It would be wonderful to have treatments to cure chronic diseases, but this will likely occur in baby steps. In this chapter, we will expand on some of the difficulties that have been encountered in finding effective cancer treatments. Unless there is a clear genetic predictor, such as with Huntington's disease, many chronic diseases, to varying degrees, leave open the possibility for their prevention. Therefore, in the absence of cures, prevention ought to be the first-line strategy in dealing with these illnesses. Actively seeking or maintaining preventative strategies is better than waiting for the dice to roll, so we will consider important aspects of prevention that we had not considered earlier.

Cancer screening

Early detection

Adopting healthy lifestyles to prevent cancer occurrence or increase survival from cancer means more than maintaining proper diets and exercising regularly. For decades it had been accepted that earlier cancer detection is accompanied by a greater likelihood of patients being treated successfully. There was no reason to challenge this common-sense perspective, and cancer screening became the exhortation that would permit better therapeutic outcomes. As a result, early detection of some forms of cancer increased dramatically, but alternative views emerged concerning the value of some screening methods, the problems that can be encountered using screening procedures, and even the significance of early cancer detection was challenged. Generally, however, screening for cancers is highly recommended especially if individuals are at elevated risk for the disease. For instance, the risk for liver cancer is linked to having chronic hepatitis B or C, liver cirrhosis, and having used anabolic steroids for extended periods. Endometrial cancer has been linked to altered menstrual cycles, early menstruation (before the age of 12), obesity and metabolic syndrome, as well as hormone dysregulation associated with estrogen use. Of course, family history and the presence of genetic factors are reasons for attention to a diverse range of cancers.

TABLE 13.1 Outcomes that could occur using a "signal detection" paradigm.

	Diagnosis: Illness present	Diagnosis: Illness absent
Disease signal: present	True positive	False positive (false alarm)
Disease signal: absent	False negative	True negative

Sensitivity versus specificity

Whether it involves the development of diagnostic tests or intervention strategies for diseases (e.g., development of vaccines), several basic criteria must be considered to ascertain their effectiveness. For instance, when screening procedures are used it is obviously desirable to accurately identify every person who has the disease (true positives), which reflects the sensitivity of the test. It is similarly important that the test has high specificity so that unaffected individuals are correctly identified as not being ill (true negatives). However, there is often a trade-off between sensitivity and specificity, where as the sensitivity of a test increases, its specificity decreases. In effect, as the ability of the test to detect the presence of illness increases, so might false positives (i.e., false alarms). Conversely, if the test is insufficiently sensitive, positive cases may be missed (false negatives), so a disease can progress untreated. Table 13.1 depicts a signal detection analysis paradigm that shows the possible combinations of having a test that is either too sensitive or too insensitive in the context of a signal (features of an illness) being present or absent and the actual presence of the illness.

Cancer screening in common types of cancer

Breast cancer screening

Breast cancer is currently the most common form of cancer in women occurring in about 13% of individuals worldwide, varying with ethnicity. Mortality rates for breast cancer have been declining owing to earlier detection and because of improved therapies. In screening for breast cancer through mammography, sensitivity may vary with several factors, such as the size of a tumor, the density of the breast tissue (since increasing tissue density may obscure the tumor), the quality of the image obtained, and the abilities of the radiologist reading and interpreting the scan. The introduction of mammography was fundamental in detecting tumors relatively early so that patients were more likely to survive, but standard mammography may miss tumors that are present (as frequently as 17%–20%). As well, false alarms were not infrequent (6%–8%), creating considerable distress among women incorrectly diagnosed. It was also not uncommon to detect small tumors that likely would never have resulted in clinically significant symptoms (Welch et al., 2016). Essentially, screening involves fine juggling to detect illness or conversely to diagnosis the absence of ill health.

The predictive power of mammograms and issues related to sensitivity and specificity have improved over the years. With the increased use of 3D mammograms comprising multiple images of breast tissue to create a three-dimensional picture of the breast, tumors could be detected somewhat more readily and earlier, and it can reasonably be expected that the benefits of the procedure will increase with still newer technologies together with the identification of biomarkers of illness.

Prostate cancer screening

Worldwide, prostate cancer is the second most common form of cancer in males and the fifth leading cause of death. Within the United States, prostate cancer affects more than 191,000 individuals yearly, leading to more than 33,300 deaths. The incidence of prostate cancer is about 12%, but increases appreciably with age, and varies with family history and ethnicity (being more common in African Americans who also develop more aggressive prostate tumors) and differs globally, being relatively prominent in developed countries (Rawla, 2019).

For men being screened for prostate cancer the difficulties encountered were in several ways at least as great as those involving mammography in women. The standard approach to determine the presence of prostate cancer had been through a digital rectal exam to assess prostate enlargement or the presence of hard, lumpy, or abnormal areas. This procedure, however, could not detect signs of cancer on the side of the prostate that was inaccessible to a gloved finger. Accordingly, the introduction of the prostate-specific antigen (PSA) test to detect this protein in blood and to signify prostate cancer was met with considerable enthusiasm. This test frequently detected cancer that was present, but false positives occurred far too frequently owing to the presence of urinary tract infection, inflammation of the prostate (prostatitis), and enlargement of the prostate that was not cancerous (benign prostatic hyperplasia). As a result, many patients went through considerable distress and further unnecessary invasive testing. Ultimately, because of the poor sensitivity and selectivity of this test, it was recommended that this procedure not be embraced. This recommendation was also based on some forms of prostate cancer in older men often being slow growing, so most men with the disease would likely die of other causes without ever being aware that they had the disease. This is not to say that some form of screening should not be undertaken since prostate cancer can begin to grow rapidly and may spread outside the prostate. Indeed, with the advantage of lead time in detecting cancer through screening, death related to prostate cancer was markedly diminished (25%–30%). With newer approaches, such as combining a blood test that uses an algorithm that considers protein markers and genetic markers, together with magnetic resonance imaging (MRI), it is possible to determine prostate cancer with fewer cases of false alarms (Nordström et al., 2021). Likewise, as prostate-specific membrane antigen (PSMA) is highly expressed in aggressive metastatic prostate cancer, targeting cells with this antigen using radioactive particles increased survival, although this amounted to only a few months (Sartor et al., 2021).

Colon cancer screening

Colorectal cancer screening for those over the age of 50 is especially important as this frequently occurring form of cancer (affecting about 4.7% of males and females) is second to lung cancer in causing mortality. Unfortunately, among individuals older than 50, colon cancer screening through colonoscopy and sigmoidoscopy (the former assess the whole large intestine, while the latter only assess the last section) was only undertaken by about two-thirds of individuals. Educational efforts to promote testing have only been moderately successful, but colonoscopy is more likely to be adopted when it is aligned with social norms or is in line with common health beliefs. Screening procedures have been associated with earlier cancer detection and across studies, cancer mortality declined by 68%–88%. With the risks of colon cancer increasing appreciably among younger individuals, there has been a push for regular colon screening to begin at 45 years of age.

For those averse to colonoscopy, yearly fecal immunochemical tests (FIT) are available in which antibodies are used to detect blood present in stool samples. Likewise, guaiac-based fecal occult blood test (gFOBT) and DNA tests can be undertaken that do not require invasive procedures. While computed tomographic colonography (CTC), a noninvasive radiographic imaging procedure can be conducted to detect intestinal cancer. Among asymptomatic patients, colonoscopy has reasonably good sensitivity and specificity (~ 93% and 73%), while that of CT colonography was around 89% and 75%, respectively. Although FIT and gFOBT are somewhat less reliable than colonoscopy in detecting certain cancers, they can be done yearly with a likelihood that early indications of cancers will be detected. Moreover, as it is less invasive, individuals are less likely to avoid testing. This said, false positives are not uncommon (upward of 15%), which would then need to be followed by colonoscopy.

Lung cancer screening

The incidence of various cancers differs across countries, but lung cancer invariably is among the most common and lethal within developed countries. The incidence of lung cancer has been declining over successive years, but still accounts for 13% of all cancer cases and 25% of cancer deaths, being only slightly more common in men than in women (Siegel et al., 2021). Generally, lung cancer is suspected and diagnosed once symptoms, such as pain or cough, have appeared, but by then it might be too late to eliminate the cancer. Should lung cancer be detected early (Stage 1), however, surgery and radiation can often be effective in removing the tumor. The fact is, however, that screening for lung cancer has been undertaken at a very low rate (fewer than 4% of individuals) even among smokers.

When lung cancer was detected relatively early, mortality over 5 years was reduced by about 16%–20% in large-scale studies, although false positives may be fairly high, ranging from 3% to 23% across studies. Given the relative risks and benefits of low-dose computerized tomography (CT), it was recommended that for individuals who smoked the equivalent of a pack a day for 20 years, CT screening for lung cancer should be undertaken among those who are older than 50 (the earlier guidelines recommended screening after the age of 55). It had been recommended previously that CT screening was unnecessary among never smokers. Yet, the incidence of lung cancer among never smokers has increased over the past two decades (Siegel et al., 2021), particularly among women in whom 20% had never smoked. Thus, there has been a push to reevaluate the criteria adopted for lung cancer screening. These cancers can be caught at a relatively early stage using low-dose computerized tomography. However, adopting screening among never smokers still holds a risk for false positives so once more the balance between risks and benefits needed to be weighed carefully.[a]

[a] Increasingly, out of the ordinary approaches are being developed to guide cancer treatments. Most of us will have heard that dogs, through their extraordinary olfactory ability might be used to sense the presence of cancer, and an eNose has been created that is thought to predict cancer presence with 85% accuracy, based on patients exhaled breath, and whether they would respond to immunotherapy (de Vries et al., 2019). More recent variations of eNose has reportedly been 95% accurate in detecting cancer.

Changing recommendations

Based on improved detection methods coupled with the age at which breast cancer was most likely to occur, the guidelines for mammograms to catch breast cancer early have been modified. The changes were largely related to the age of the individual, family history of breast cancer, or the presence of specific genetic mutations (e.g., BRCA1 or BRCA2). In the absence of these risk factors, it was recommended that mammography begin at age 45 rather than 40. The consequence, of course, might be that 8% of breast cancers that occur between women 35–44 years of age will go undetected. Guidelines for other forms of cancer have also changed. It was recommended that cervical cancer screening begin at 25 rather than 21 years of age since the frequency of cervical cancers before the age of 25 is relatively low, perhaps because of the success of HPV vaccination. But this does not alter the risks that are inherent among women 21–25 years of age who had not been vaccinated. Furthermore, given the frequency of colorectal cancer and its high rate of mortality, together with the extended time for its development, it was recommended that screening begin at the age of 45 rather than 50.

Recommendations to modify the age of screening are based on primary risk factors (age, genetics) and are economically sensible, at least in the short run. To be sure, recommendations for breast cancer screening also consider factors such as whether menstruation began before the age of 12 or who entered menopause after the age of 55. But what about the many other risk factors that predict the occurrence of some forms of cancer? The presence of obesity, harmful diets, alcohol usage, and occupations that bring people into contact with known or likely carcinogens have all been linked to cancer occurrence. Ethnicity that has been linked to various forms of cancer needs to be considered in the decision to screen for a particular cancer. Likewise, poverty, living in vicinities with high levels of fine particulate matter or the absence of green space have been correlated with cancer occurrence, and should be included in determining screening guidelines. Remarkably, it has been said that a person's postal code may be as strong a predictor of cancer occurrence as their genetic code.

Caveats concerning screening and early cancer detection

In many instances, finding and treating cancer early (e.g., cancer of the breast, prostate, testicular, thyroid, as well as melanoma and Hodgkin's lymphomas) may enhance survival. However, some types of cancer are not easily detected early (pancreatic, kidney, renal, ovarian, liver, and nonsmall-cell lung cancer) since symptoms may not appear until the disease is fairly advanced. In this regard, a 16-year longitudinal study within the general population indicated that even when ovarian cancer was detected early through annual screening, ovarian cancer-related deaths were not significantly reduced, suggesting that population-based screening may not be warranted (Menon et al., 2021).

The view has been advanced that the statistics on improved survival being associated with early detection can be misleading, as in the case of what is referred to as *lead-time bias*. Take the hypothetical case in which two individuals developed cancer at the same time (say, at the age of 60), but did not know it. Further, they also died from the disease at the same time (e.g., the age of 70). Now consider if they had been screened. In one individual, the screening was at age 62, and led to the cancer being detected, such that when their survival was calculated it was measured from the time of detection to the time of death at the age of 70 (8 years). In the second individual, their screening was not conducted until the age of 66, which meant that their survival time from detection to death was deemed to be shorter (4 years) than that of the first person even though both of them lived just as long (10 years) from the time of cancer onset (at the age of 60). In this example, early detection did not result in longer survival, but simply meant that the clock started earlier at least in so far as statistics were concerned (Welch et al., 2019). There is yet another type of situation, referred to as *length bias*, that can be misleading. If the disease progresses slowly, it is apt to be detected earlier relative to the cancer that progresses quickly. In fast-growing tumors, the asymptomatic period is shorter, and if these tumors are more likely to be fatal, then the bias related to detection and death may be badly skewed, resulting in incorrect conclusions regarding the benefit of early detection.

So, what does all this imply? Should attempts at early detection be abandoned? Certainly not, but there is the question of what "catching it early" actually means. Is detection through mammography or by scans to detect lung cancer early enough? Cancer cells can be present for many years before being detectable through most approaches. It might be possible to detect these rogue cells through sensitive biomarkers, making it possible to eliminate them more readily. Even if it were accepted that early detection did not enhance the odds of survival, the conclusion was based on yesterday's technologies to detect cancer early. With newer technologies and still earlier detection, the playing field may be altered. Improved detection methods, such as those based on cancer signatures in the blood (immunosignatures), could turn out to be very advantageous especially if these permit cancer detection before individuals become symptomatic.

Sensitive techniques have been developed which can detect as few as nine cancer cells in 200 µL of blood (Sun et al., 2020), and using a *cytophone* platform that entails the use of laser beams and sound waves, melanoma cells could be detected in circulating blood cells, thereby permitting early detection (Galanzha et al., 2019). Similarly, the presence of more than 60 mitochondrial genes may be aligned with breast cancer recurrence or metastases, which could potentially point the way for improved ways of dealing with the disease (Sotgia et al., 2017). It might similarly be possible to identify gut microbial signatures that predict colon cancer, even before polyps are present, which would likely allow for preventive approaches to be initiated. Furthermore, by evaluating methylation patterns in blood samples from colorectal cancer patients, a profile of methylation could be deduced which could potentially be used to catch cancers at a time where treatment effects would be enhanced. In fact, in a high-risk population, even single epigenetic changes seemed to have high sensitivity and specificity in detecting colorectal cancer and precancerous lesions (Luo et al., 2020). These methodologies are certainly inspiring and when combined with AI approaches still more effective detection procedures will be available to predict the efficacy of therapeutics. Even now, AI algorithms have been effective in predicting breast cancer more readily with fewer false alarms and false negatives being encountered (McKinney et al., 2020).

There is another important issue that is relevant to early detection and improved survival. Specifically, who is it that generally gets tested regularly, and is this relevant to conclusions concerning early detection and survival times? Unfortunately, people with mental health issues as well as those with physical disabilities have typically not received sufficient attention in this regard. Social inequities likewise have ramifications related to health care, which have obvious implications for early cancer detection. People who are generally better educated and wealthier are more likely to receive regular health screening and they tend to live longer than are relatively disadvantaged people. The greater longevity among wealthier, more educated people might well be due to earlier screening; but it may also reflect the opportunity to maintain healthier lifestyles and preventive care, thereby providing them with a health advantage (i.e., the "health user effect").

The ever-present problem of overdiagnosis

As much as screening has been important for the detection of illnesses, it can have negative repercussions other than those related to issues of specificity and selectivity. The vigorous quest to "catch it early" may have engendered a medical culture of excessive testing. Either because of being enamored by new technologies, fear of litigation if patients are dissatisfied with outcomes, or perhaps because of an abundance of caution, physicians have been ordering an increasingly greater number of diagnostic tests (Moynihan et al., 2014). With this comes the problem of overdiagnosis, which essentially means the diagnosis of conditions that won't actually reach the point of causing frank symptoms as we saw with some forms of breast and prostate cancer.

The term "disease epidemic" has been used more frequently, which is accurate in the case of obesity and type 2 diabetes, but in the case of some other illnesses, this may reflect more an epidemic of screening than one of diseases becoming more frequent. In fact, those skeptical of these practices suggested that the focus on early detection primarily benefitted "the business of medicine" (Welch et al., 2019). The reasonable view has been adopted that testing is based on dispassionate assessments of medical evidence, and at the same time, patients should be forewarned of the risks of overdiagnosis. This is particularly important given that clinicians and patients may tend to overestimate the benefits of treatment while underestimating the potential harms that might emerge (Hoffmann and Del Mar, 2017).

There is a disconnect between the value of early screening on the one side and the possibility of overdiagnosis on the other (see Table 13.2). A happy medium may be reached with the introduction of better methodologies to detect cancer as well as the appreciation of the factors at play that increases the risk for cancer. Yet, the benefits that can be derived from these advances are dependent on the propensity of individuals to take appropriate actions regarding their health.

Delayed presentation of illness: Assessment delays and delays of treatment

In response to threats, many individuals seem to have difficulty making proper appraisals and using effective coping strategies. Responding to bloody stool or an obvious lump in the breast by hoping that it will just go away ("*avoidant coping*") is a counterproductive strategy. Some people exhibit "*appraisal delay*," which reflects the time it takes to acknowledge and accept that symptoms are a cause for concern. And even after this is acknowledged they might still be unprepared to accept that their symptoms require medical attention ("*illness delay*"). Once medical attention is sought and a problem identified, the need for further tests to identify the source of the problem might lead to additional resistance, as some individuals might still not follow-up on testingrecommendations ("*behavioral delay*").

TABLE 13.2 Potential benefits and harms associated with cancer screening.

CANCER SCREENING
Benefits
Diminish cancer incidence. Detect precancerous lesions, thereby reducing or preventing cancer risk
Diminish cancer incidence in high-risk individuals. Early detection among high-risk individuals (e.g., those carrying specific mutations or who display biomarkers associated with cancer) may reduce cancer occurrence, or permit early treatment
Diminish risk of advanced disease. Detecting cancer early can permit therapies that prevent cancer spread
Diminish cancer-related mortality. Early detection increases the likelihood of cancer therapy being effective in reducing the death.
Harms
Adverse events. Some screening tests are invasive (e.g., colon screening) potentially leading to adverse events, although the likelihood is low
False positive results. Screening may result in tests showing cancer presence despite no cancer being evident. This requires further testing, which is sometimes more invasive and typically is associated with distress and anxiety. False positive results are not uncommon, varying with the nature of the test.
False negative results. Cancer that is present may be missed by the test so that cancer can progress. Occurrence of false negatives are relatively infrequent.
Overdiagnosis and overtreatment. Cancers that are detected may not progress and hence may not be a threat to health. Overdiagnosis may led to unnecessary therapies at the expense of patient mental and physical health.
Medical costs. Broad screening is costly, which is exacerbated with over diagnosis and overtreatment

Many reasons can play into why these delay tactics are adopted to deal with possible illness. All too often delays in seeking treatment may come about simply because of limited access to care services and is more frequent among disadvantaged people. Beyond this, however, individuals may suspect the worst and are fearful or reluctant to have this confirmed. In other instances, the signs are ambiguous, or individuals are unaware of the significance of the symptoms. Alternatively, people might be reluctant to seek treatment for symptoms that initially seem minor, especially as they might want to avoid being labeled as a catastrophizer. Procrastination might also stem from embarrassment or efforts to avoid social stigma (often seen in smokers), or when symptoms appear to be innocuous and appear stable it is easier to hold off on any action.

Delays in seeking help may also occur because the individual distrusts their doctor (or the medical field in general), or the relationship between the patient and their family physician is weak. The patient may feel that their physician is cold, aloof, or seemingly insensitive to their needs. Moreover, it is equally possible that the physician had at some time expressed frustration with the patient who had not followed the guidance that had been offered (e.g., the diabetic patient who does not adhere to the advice that had been provided). Overall, the doctor-patient relationship or bond is as much a critical feature of medicine as the prescriptions doled out to patients. When this is not well-cultivated either because of patients' attitudes or poor medical training of physicians, the repercussions can be serious. It is very much incumbent on medical practitioners to ensure that patients seeking help are comfortable in doing so. A natural tendency to delay as we have listed above is only too normal (it is the classic fear response—the person freezes and/or flees). A potential serious medical problem takes people outside the comfort zone of their everyday lives and into a world they anticipate as a threat to their well-being. Whereas the reality is that it represents a context that can aid their health and recovery.

Yet another form of delay encountered by patients, and which they have little control over, comprises the long wait between diagnosis andreceiving the needed treatment. Such delays may stem from a long and extensive diagnostic workup, as well as the time needed for the patient and their doctor to select the treatments that will be given. The delay might also be due to a shortage of treatment specialists, a medical system that is overly burdened, or unexpected circumstances (as unfortunately occurred during the COVID-19 pandemic). Regardless of why such delays arise, on average the risk of death was found to be elevated by 6%–8% for every 4-week delay in obtaining required surgery and was still greater for delays for radiotherapy (e.g., in the case of bladder, breast, colon, rectum, lung, cervix, and head and neck cancer). With delays of 8 and 12 weeks to obtain breast cancer surgery, the risk of premature death increased by 17% and 26%, respectively (Hanna et al., 2020). A delay in surgical procedures beyond 12 weeks of diagnosis of Stage I nonsmall-cell lung cancer was associated with an elevated risk of recurrence and worse survival (Heiden et al., 2021). Likewise, longer waits between surgery and subsequent adjuvant chemotherapy in women with breast cancer were associated with poorer survival.

In many countries, delays in receiving cancer therapy are not uncommon. It is reassuring that in some countries in which universal health care is the norm, patients lower in socioeconomic status are not at increased risk of delayed treatment. However, in the absence of socialized medicine people are more likely to experience such delays if they lack adequate medical insurance. Sadly, regardless of socioeconomic status, within the United States, Black women were more likely to experience delayed diagnosis and treatment, which could contribute to greater mortality (Emerson et al., 2020).

Unintended consequences and unintended benefits

We have learned from our many missteps across numerous domains that the best of intentions aren't always met with the best outcomes. People are frequently poor in making sensible decisions and have on occasion been subject to unanticipated and unintended consequences when attempts to advance well-being were instituted through broad policies. This isn't a new concept, and several causes of negative outcomes were outlined more than 80 years ago (Merton, 1936). Unintended consequences were seen as emanating from (a) ignorance that makes it unlikely that all outcomes could be predicted; (b) problems being inappropriately analyzed or approaches to problems that may have been appropriate in the past, but not for current situations; (c) the tendency to favor short-term gain without appropriate consideration of long-term consequences; (d) basic values interfering with certain behaviors even if these are likely to produce unfavorable consequences; (e) fear or panic motivating actions taken in anticipation of a problem, when this problem would never have actually emerged.

The odd behaviors of humans are particularly apparent with numerous programs that have been instituted to enhance safety, in which individuals adjust their behaviors based on the appraised risks that accompany changed policies (called "risk compensation" theory). Ordinarily, when the risk of harm is appraised as being high, cautious behavior is adopted; conversely, the perception that risks are low tends to be accompanied by individuals being less careful. When safety measures are appraised as lowering the risk of harm, this somehow seems to make some people feel that they have a pass to engage in clearly risky behaviors (the "Peltzman effect"). The introduction of safety measures to diminish vehicular accidents and injuries (e.g., use of seat belts) resulted in some individuals driving faster and closer to the car in front of them. When penicillin became available to treat gonorrhea and syphilis, riskier behaviors were more often adopted, so the rates of these conditions increased (although changes in sexual mores may also have influenced the elevated frequencies of STDs). The very same issues may be present concerning the failure of people to adopt sensible lifestyles that could limit disease occurrence (Prasad and Jena, 2014). Diabetic patients who received metformin might see their sugar levels normalize, but this can mistakenly be taken as an "all clear" signal, so they can return to the poor lifestyles that led to the illness in the first place.

There are occasions when perverse results may appear after efforts are made to deal with specific problems. Warnings that antidepressant agents may be associated with increased suicide risk among adolescents led to the understandable reduction of these drugs being prescribed for youth, but this had the unintended effect of fewer outpatient visits and hence an increase in suicide attempts (Lu et al., 2014). The markedly increased and sometimes inappropriate use of antibiotics (along with their use in farm animals) fostered resistance to these treatments. Within a somewhat different context, the introduction of electronic medical records was meant to facilitate the management of patient care (e.g., with medications being administered), although it can be argued that this had more to do with financial concerns by insurance companies or government agencies. Irrespective of the reason, the procedures have burdened physicians with excessive computer-related tasks at the expense of being able to deal with patients directly. This is valuable time lost to the quality of the doctor-patient interaction, which as we emphasized earlier is critical to the well-being of the patient.

It might also be the case that the search for novel cancer therapies produces unintended negative consequences. This was discussed in a thoughtful and thought-provoking essay.

Huge expenditures were committed to finding effective cancer therapies that had only modest effects in extending progression-free and overall survival and had unexpected costs (Fojo et al., 2014). It was maintained that disbursement of research funds tended to favor the "me-too" approach that focused on small incremental changes in treatment outcomes, which concurrently undermined innovation and creativity that might have provided better cures or ways of diminishing side effects. Moreover, the funding that went into developing these therapies could have been used more effectively in other ventures, such as cancer prevention. Likewise, the myopic views that promoted overspecialization in fields of medicine might have limited broader perspectives concerning patient care.

Not to leave the wrong impression, there is a flip side to this argument. While some approaches and treatments may have negative unintended consequences, unanticipated benefits may also be derived from new approaches. This is particularly notable in drug repurposing (drug repositioning) in which a given treatment intended for one illness has benefits for other conditions. Many drugs developed to treat depression are even more useful in diminishing anxiety, and the antimalarial hydroxychloroquine has been adopted to diminish symptom flares in lupus erythematosus patients (but, obviously, was ineffective in limiting COVID-19 symptoms). Metformin that is commonly used to treat type 2 diabetes has pronounced positive effects in preventing some forms of cancer. The horrible drug thalidomide that was developed to produce sedative effects and to diminish morning sickness among pregnant women caused pronounced teratogenic effects. It now seems that this agent might be useful in treating multiple myeloma by acting against angiogenesis. Best known, no doubt, is that sildenafil (Viagra) that was initially developed to treat hypertension and angina turned out not to be particularly effective for that purpose but was useful for males experiencing erectile dysfunction.

Repurposing older drugs, such as those used to treat heart disease and diabetes, could potentially provide benefits to cancer therapies by acting on targets that were not previously known to be linked to cancer. An analysis of 4518 drugs tested across 578 human cancer cell lines revealed a considerable number of drugs that could potentially act against cancer by operating through different mechanisms (Corsello et al., 2020). It is entirely possible that the vast search for better cancer treatments, even if this amounts to throwing stuff against the wall and seeing what sticks, may well produce unintended (serendipitous) benefits. As we've said before, there are times when "it's better to be lucky than smart."

Progress in cancer treatment development

Difficulties and limitations

The difficulties encountered in getting treatment from bench to clinical practice are significant and begin at preclinical trials. Manipulations that appeared to be effective in animal models, might not pan out within clinical trials. In fact, they usually don't. To make matters exponentially worse, even those preclinical treatments that were effective in one laboratory may be ineffective when evaluated in another. Given the complexity of the cancer process, it would reasonably be expected that some failures to replicate earlier findings would be encountered. However, when the pharmaceutical company Amgen Inc. selected studies for replication based on their apparent scientific and practical merit, only 11% (6 of 53 studies) were repeated successfully (Begley and Ellis, 2012). In a comparable analysis initiated by Bayer AG, only 25% of the studies were successfully replicated (Prinz et al., 2011).

The Reproducibility Project, which assessed the replicability of high-profile reports of successful cancer treatments, showed a mixed bag of results. Some findings were replicable, to be sure, but that of others was less clear (see Baker and Dolgin, 2017). The failures to replicate earlier findings could stem from seemingly small differences in the procedures used between experiments, or because small numbers of animals were used in the original studies, and hence the outcomes were more prone to errors. It is equally possible that the originally reported studies were tainted in some manner and might even have been the victim of confirmatory biases in which a tendency exists to search for or interpret data consistent with a hypothesized outcome.[b]

More than 250,000 cancer drug trials of some form were conducted between 2008 and 2017 of which about 16,360 reached Phase III trials. There are certainly impressive and extensive efforts being made to find effective cancer therapies. While some trials fail to produce positive effects and can be eliminated from further evaluation, a significant number appeared to provide encouraging outcomes, but were ultimately unreliable. Among industry-sponsored Phase III randomized controlled trials from 2008 to 2017, fully 87% of 362 proved to be unreliable in the treatment of lung, breast, gastrointestinal, prostate, and hematologic cancers. Of these trials, 58.4% turned out to yield false positives, which certainly created hardships for patients in these trials and exacted financial costs and time lost for the industry sponsors. It was maintained that altering the procedures in Phase III trials is not likely to produce more meaningful outcomes, and instead, these findings point to the need for better and more stringent analyses of Phase II trials before jumping into Phase III evaluations (Shen et al., 2021).

Reasonable concerns have been expressed that human trials to assess the effectiveness of therapies aren't offering what they claim (Mailankody and Prasad, 2017). When gains were made with new treatments, these were most often modest. Of the 48 anticancer treatments approved by European Medicines Agency from 2009 to 2013, the enhanced overall survival engendered by these therapies ranged from 1.0 to 5.8 months (median being 2.7 months), and the reported improvement in quality of life was limited (Davis et al., 2017). Statistically, the increased survival with new therapies has been significant, but how consequential is this from the patient's perspective? Even a few months may be precious for some patients, but it needs to be asked how meaningful this is given the ordeal individuals had to go through to obtain this brief life extension?

One can also question the extent to which all clinical trials are valid representations of treatment efficacy. Among other things, in some instances the patients in clinical trials might not be an adequate representation of cancer patients in general, tending to be younger, healthier, and more aware of being able to participate in these trials, possibly because they're better educated or wealthier. Moreover, as indicated earlier concerning genome-wide association studies, some trials might not have had adequate representation from various cultures or racial groups.

Some of the same concerns can also be applied to preclinical studies. Some mouse models simply aren't ideal to produce reliable and translatable data. Most studies are conducted in highly inbred mice in the belief that the genetic homogeneity is advantageous allowing for

---

[b] One can't help being reminded of the quote by the mathematician John Tukey "The combination of some data and an aching desire for an answer does not ensure that a reasonable answer can be extracted from a given body of data."

the analysis of specific gene manipulations (e.g., using transgenic mice). However, even in these mice random mutations and epigenetic effects can influence outcomes. More than this, inbred mice also carry numerous inherited vulnerabilities unrelated to the pathology being studied and could thereby affect the impact of the therapy of interest. Furthermore, as the characteristics of inbred mouse strains are very different from one another across multiple dimensions, including immune and microbial functioning, the generalizability of the findings may be limited. Indeed, even the experiences of an inbred strain, such as the breeding farm from which they were obtained or the conditions within testing laboratories, can affect experimental outcomes. Of particular concern is that when treatments are assessed in murine models, this is often done in relatively young, otherwise healthy mice, which is in marked contrast to human cancers that most often appear in older people who are often weakened by numerous effects of the illness (e.g., fatigue and sleep problems), and they may have other illnesses and immune disturbances present (see Klevorn and Teague, 2016). There's no question that experimental analyses of mice are and should be a key component to develop effective therapies for several forms of cancer. However, if mice are to be used in a cancer model, then greater attention ought to be directed at appropriately simulating the human condition, including the judicious selection of the mouse strains to be assessed.

The upside of the story

Pitted against the failures related to cancer treatments, the success in treating some forms of cancer has been remarkable. The 5-year survival for many forms of cancer has been impressive, and many drug treatments were accompanied by improved quality of life (Salas-Vega et al., 2017). Therapies for childhood leukemia, which only a few decades ago were very hard to eliminate, have achieved better than 90% success rates, although marked global inequities persist. The plant alkaloid taxol administered with other agents enhanced survival among breast cancer patients and can have positive effects in treating ovarian cancers. Gleevec, which inhibits tyrosine kinases, has produced wonders in the treatment of adult leukemia, chronic myeloid leukemia, acute lymphoblastic leukemia, and other forms of blood cancer. Likewise, the protein kinase inhibitor Cotellic used along with vemurafenib was effective for 30% of individuals with melanoma, whereas before their use the odds of survival were dismal.

The burgeoning development of a variety of immunotherapies (see Chapter 15) has been notable. Even though success rates for these treatments are still somewhat limited (20%–30%), they might be harbingers of still better treatments (or combination treatments). Other novel strategies take advantage of the body's propensity to get rid of unwanted or damaged proteins. Targeted protein degradation can come in several forms, including the development of methods (e.g., proteolysis-targeting chimeras; PROTACs) that can get at proteins that aren't readily accessible by standard drugs and hence were considered untreatable. It might be a bit early to celebrate the success of treatments that attempt to eliminate proteins that fuel cancer and other diseases, but for the moment the odds of new drug treatments coming into the field look promising (Scudellari, 2019).

Cancer-related mortality over the past 30 or so years has declined appreciably for breast, prostate, colorectal cancer, and even lung cancer. This was achieved through improved therapeutic methods as well as by earlier cancer detection, the timeliness of treatments, and being able to deal with side effects of therapies and that of cooccurring illnesses. Cancer mortality rates have also declined because of improved lifestyle factors being adopted, including a reduction in the number of people who smoke and increased focus on exercise and eating appropriately.

Cannabis and cancer

Cannabis (or its primary components, Δ^9-THC and cannabidiol) has been implicated in reducing the symptoms of several illnesses. Legal restrictions that had been in place for decades limited the experimental analysis of cannabis effects and we're now playing catch-up in determining its medical usefulness, as well as adverse effects that may arise. With the legalization of cannabis in many places, increasing attention has focused on evidence-based conclusions concerning the health benefits derived from cannabinoids, although obstacles to this line of research have not been fully eliminated.

Despite the overwhelming hype regarding the alleged therapeutic actions of cannabis, only a few health conditions were reliably found to be positively influenced by cannabinoids. These comprised treatment-resistant epilepsy and chronic pain, and the nausea resulting from cancer treatments (Parker, 2017). Given that chronic pain is experienced by more than 20% of adults, including about 35% of cancer survivors, finding treatments for this may be especially important. While some reports have supported cannabis use for cancer-related pain, its efficacy has not been overly impressive (Boland et al., 2020). Nonetheless, an increasingly greater number of older individuals have been using cannabis to ameliorate pain, insomnia, and anxiety, and there is certainly the need to determine the efficacy and safety of cannabis for older users. There is still a paucity of placebo-controlled, double-blind clinical studies concerning the influence of cannabis in pain control, and studies are needed to identify optimal doses that may be effective in this capacity.

Several reports suggested that cannabinoids might be effective in diminishing some cancer treatment side effects, such as chemotherapy-provoked neuropathy, sleep problems, and gastrointestinal disturbances, and with cancer cachexia and anorexia. However, the findings have not been entirely consistent and there have been concerns raised regarding negative interactions with some therapies. Studies in mice suggested that cannabis can affect cell proliferation and metastasis associated with multiple forms of cancer and could thus be useful as a component of cancer therapies (Nigro et al., 2021). The potential actions of cannabis on cancer could occur through actions on immune, inflammatory, oxidative stress processes, and inhibition of angiogenesis. In the absence of randomized controlled trials evaluating cannabis efficacy, it is premature to advocate for its use to treat cancer. In the end, the whole notion may well go up in smoke.

Problems in determining treatment efficacy

Evaluating the efficacy of cancer therapies comprises a multistep process to determine adverse effects of the treatment, whether it resulted in tumor regression, development of tolerance, and ultimately whether cancer-related mortality was reduced. This requires years of research and for patients desperate for effective treatments, the long delays are unacceptable. From the perspective of pharmaceutical companies, the substantial difficulties in receiving approval for treatments of various illnesses, have predictably produced efforts to have their products go through the pipeline more quickly. Regulatory agencies have typically been up to the task of ensuring the safety and efficacy of drug treatments, but some procedures adopted for approval of cancer therapies have been questioned. The certain need for new and better treatments, coupled with the persistence of pharmaceutical companies, has resulted in the relaxation of criteria for drugs to reach the market. The eagerness to develop new and effective treatments has allowed drug approvals to be based on *surrogate endpoints*, such as the duration of progression-free survival or the response rate (RR), which reflects the percentage of patients that exhibit tumor shrinkage based on a quasiarbitrary decision as to what is deemed a positive treatment response. Unfortunately, the RR was often relatively low (Chen et al., 2019), which is particularly germane as many drug therapies that initially show positive effects in shrinking tumor size, subsequently encounter resistance, so cancer cells continue to proliferate. Importantly, these surrogate end points do not uniformly speak to whether the treatment extended life or enhanced quality of life. In an analysis of 17 new therapies, 8 were not successful in extending life relative to placebo treatment, and only 3 could be considered as having positive effects in this regard (Braillon, 2016). Perhaps this will change with the development of appropriate biomarkers, although there are occasions in which the markers and targets for treatment are off base.

Complementary and alternative medicine

While the expression "desperate times call for desperate measures" may be a cliché, it does not diminish its truth. It is not unusual for distraught and anguished patients to seek alternatives to standard medical treatments. Patients may seek these treatments because certain cancers have historically not been treatable or because they're reluctant to endure the presumed hardships of radiotherapy and chemotherapy. Possibly knowing that their physician might frown on the use of alternative approaches, particularly the use of supplements, 29% of patients had not disclosed that they were using these compounds (Sanford et al., 2019).

The increasing movement to seek alternative treatments has been abetted by delays in receiving standard therapies, the possibility of experiencing adverse long-term effects, a loss of faith in legitimate practitioners, and at times, sheer gullibility born of desperation. The acceptance (normalization) of supplements as cures for virtually any illness has been helped along by the persistence of alternative treatments offered in gossip magazines ubiquitous at the checkout counters of supermarkets, on the internet, and in health food stores. Not surprisingly, they have been pushed as a component of naturopathic treatments. Of course the irony is inescapable, given that whereas many people distrust evidence-based treatments (witness the high rate of vaccine hesitancy) they are more than willing to use supplements to prevent or treat disease, despite the sheer lack of evidence supporting claims for their efficacy, and the potential dangers they can pose. Within the general population, including cancer patients and family caregivers, it is estimated that 39% believed that cancer could be cured through alternative therapies. The rationale for using alternative medicine treatments is understandable, but by no means should this be misconstrued as advisable.

Part of the problem has come from legislation that has been loose in prohibiting supplements from being pushed for therapeutic purposes. To be sure, manufacturers and marketers of supplements are responsible for assuring that the products do not produce adverse effects (which often remains questionable), and customers are misled into believing that "natural" products must be safe. So long as purveyors of supplements do not explicitly state that their compound acts against a specific disease, they are free to make statements that it possesses certain biological actions *consistent* with a positive effect in diminishing illnesses. Thus, we regularly see all manner of buzzwords (natural, antiinflammatory, antioxidant, immune enhancement) being thrown about supporting the efficacy of various supplements, typically without a shred of evidence from randomized controlled trials that could support the claims that they affect diseases.

The FDA has offered a list of treatments considered not to be of value, but the purveyors of supplements have kept peddling them, alluding to their beneficial actions, as well as in attenuating side effects that might develop owing to standard therapies. The alternative treatments have included apricot pits (or a compound present in the apricot kernels, variously referred to as vitamin B17, laetrile, and amygdalin), shark cartilage, green tea, baking soda, red wine extracts, herbal tonics, ginkgo biloba, garlic, and apple cider vinegar, none of which have any positive actions. Cesium chloride (a type of salt) did not do much for cancer patients other than produce some nasty side effects; colon detoxes are ineffective in treating cancers or anything else, and fortunately have gone off the radar as a fad procedure. Diets high in alkaline were promoted with the thought that cancer cells can't survive in acidic diets, and Gerson therapy comprising drinking 13 glasses of organic juices daily (together with vegetarian meals that do not contain any salt or spice, as well as supplements and regularly scheduled enemas) seems to be making a small comeback. Numerous cancer treatments (scientifically unsubstantiated) have been relentlessly endorsed, but in each case, they have failed and failed miserably.

Aside from these supplements, homeopathy continues as a favorite alternative treatment for whatever ails you, as are chiropractic treatments, reflexology, and acupuncture. A few others, such as cupping, and reiki therapy, have been more on the fringe, but they too have a following, and one of the latest magic treatments is that of biofield tuning (using tuning forks to clear the energetic blockage stemming from traumatic events) may yet develop an audience. Several other "science-like" treatments have come around that are not based on research findings or failed to be replicable. Nonetheless, they do have a customer base, and it has become increasingly common for crowd-funding efforts to be directed toward these "last hope" procedures.

To some degree, these alternative treatment procedures might be fine, so long as patients also continued with their physician-recommended treatments. The sad fact is that patients who opted for the alternative treatments were more likely to refuse standard cancer therapy and were twice as likely to die prematurely (Johnson et al., 2018). If this weren't a sufficiently great problem, some unauthorized treatments can produce negative outcomes, diminishing the actions of chemotherapy, or increase the risk for some cancers, as in the case of high levels of vitamin E or supplements containing selenium.

With the realization that buzzwords are important to gain a following, the term "alternative" treatments have been replaced by new catchy phrases, such as "complementary and alternative medicine" (CAM), "integrative medicine," as well as "functional medicine," all of which sound more alluring, but the fundamentals related to these new approaches are much the same as their predecessors. Aside from pushing herbals and supplements, proponents of CAM and functional medicine throw in concepts related to inflammation, hormones, and reactive oxygen species, genetics and epigenetics, and their impact on psychological disorders that accompany standard therapies. These alternative strategies also push for a patient-centered approach as opposed to a disease-centered approach (another of today's buzzwords), even though the two are certainly not exclusive of one another. Proponents of functional medicine maintain that it focuses on getting at the root of diseases, rather than simply treating symptoms, the implication being that physicians practicing usual medicine do not do this. The use of these approaches has been successful from the perspective of having become a multibillion-dollar business. One of many views offered by Caulfield (2017), a well-known and highly credible critic of sham "therapies" has commented on alternative treatments that have been burrowing their way into medical treatments. As he suggested, "doubt absolutely every single claim that suggests a significant breakthrough. Doubt everything…. If you adopt this approach you will be pleasantly surprised when something actually pans out. More important, this nothing-ever-works-as-promised strategy will be correct 99 percent of the time." Unfortunately, one can readily feel that scientists and physicians who have worked on legitimate therapies to treat cancer and to disseminate the importance of evidence-based medicine have been screaming into the ether.

One could go on at length discussing the flaws related to CAM and functional medicine, but we'll leave it here, particularly as Gorski (2019) has been exceptionally convincing in pointing to the shortcomings of these approaches. It can reasonably be accepted that some out-of-the-box treatment methods can influence disease processes, but these claims need to be supported by scientifically rigorous and reliable evaluations of their efficacy, including long-term prospective studies. Experimental trials have on occasion been undertaken to assess whether certain supplements can produce beneficial effects, even though there was no a priori justification to do so. Engaging in senseless trials without reason, as Gorski and Novella (2014) put it, is tantamount to "testing whether magic works." In some instances, certain treatments (e.g., homeopathy) have been tested on dozens of occasions, and large-scale studies have repeatedly indicated that no health benefits were obtained. So how many more repetitions of these trials will be necessary to convince the true believers that these treatments have no discernible benefits? The answer may be in the adage "that insanity comprises doing the same thing over and over again and expecting different results."

Fig. 13.1 shows some of the approaches that comprise CAM therapies. Some treatments, such as yoga, may produce positive effects in diminishing cancer-related fatigue and enhancing aspects of quality of life and may influence immune and inflammatory functioning (Danhauer et al., 2019). Likewise, some nonconventional treatments such as those that fall into the category of complementary interventions (e.g., acupuncture, tai chi, religious and spiritual practices) have, in some instances, been associated with lower levels of stress-related hormones and levels of proinflammatory cytokines. However, cancer progression was typically not diminished by these approaches and it similarly appeared that the effects on other serious, chronic conditions were inconsequential. Review papers concerning the effectiveness of CAM treatments are available regarding a wide range of illnesses, such as different types of heart disease, several autoimmune disorders, inflammatory bowel disease, to name just a few. In almost all cases the conclusions drawn were that "limited information was available" or that "further research was necessary" and that the "curative potential for CAM has not been documented consistently."

CAM centers have increasingly appeared within hospitals, which could be taken as acknowledgment that it is a legitimate approach to treat diseases. Of course, the position could be taken that while most CAM treatments might not be useful in diminishing cancer progression, it's far better that these be present within the context of legitimate treatment centers than having patients be left to the collusions and delusions provided through social media. At the very least, it ought to reduce the likelihood of patients abandoning standard treatments.

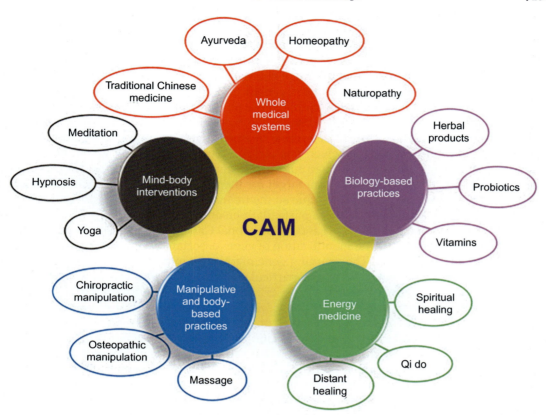

FIG. 13.1 Diverse approaches used in complementary and alternative medicine. *From Subramani, R., Lakshmanaswamy, R., 2017. Complementary and alternative medicine and breast cancer. Prog. Mol. Biol. Transl. Sci. 151, 231–274.*

Besides, even if these alternative approaches have little benefit in limiting cancer progression, they might allow individuals to feel as if they have a semblance of control over their lives, which is important in limiting distress and anxiety. To quote John Lennon, "Whatever gets you thru the night."

Precision treatment: Obstacles and challenges

Throughout this book we have returned to the theme that individual differences exist regarding the effects of cancer therapies, and that the adoption of a precision medicine approach might be desirable in selecting a therapeutic strategy for individual patients. Ideally, it would be possible to distinguish between features of cancer cells and then direct attacks accordingly. However, given the devices that cancer cells adopt to maintain their integrity, coupled with technical and strategic difficulties that have been encountered, reaching the point of being able to institute precision medicine broadly has been challenging.

Genetic and epigenetic factors in precision medicine

Precision medicine approaches have focused on the identification of patient-specific inherited mutations, as well as tumor-associated antigens (neoantigens) that arose owing to mutations associated with tumor formation (e.g., Schumacher and Schreiber, 2015). Most often the focus of precision medicine has concerned genetic and epigenetic marks that could predict the efficacy of treatments. In some instances, these might contribute to the prediction of disease occurrence and they may serve to select the most efficacious treatments. Capitalizing on genetic and epigenetic factors to determine treatment methods is, in theory, an eminently reasonable option. In practice, however, this can be exceedingly difficult given the multiple mutations and epigenetic changes that might contribute to the appearance of cancers and metastasis. Core genes might serve as a target for therapeutic agents, but each individual's cancer has been described as being unique so the treatment requires the identification of the specific *genetic profile* for the cancer in a particular person.

The Cancer Genome Atlas (TCGA) and International Cancer Genome Consortium (ICGC) were established to facilitate the identification of genes associated with tumors from patients with diverse forms of cancer. The resulting gene maps were intended to facilitate the identification of patients who would gain from particular treatments, but this commendable effort was only modestly successful, as fewer than 25% of patients with common cancers benefited from this approach. It was consequently proposed that a better method be adopted to enhance the selection of therapies. This gave rise to an approach to create the Cancer Dependency Map in which cancer vulnerabilities would be identified by perturbing genes (e.g., using CRISP genome editing) and proteins associated with multiple cancer types at different clinical stages and evaluating the responses that occur (Boehm et al., 2021). In its essence, this approach is meant to establish the factors that individual tumors depend on for their proliferation, and then defining the most effective ways of undermining these tumor "dependencies." This will require enormous resources but may well pay dividends over the long run.

Matching individuals to treatments

A major limitation to precision medicine approaches has concerned the difficulty of matching patients with particular gene profiles to specific treatments. Of 2600 patients enrolled in a gene sequencing program, only 6.4% could be paired with a targeted drug (Meric-Bernstam et al., 2015). A subsequent report from the University of Michigan Comprehensive Cancer Center's sequencing program revealed that there have been improvements in this regard. Of 500 patients assessed, 72% were eligible for a clinical trial, and 11% were eventually enrolled in such a trial, more than double that seen 4 years earlier. Using TARGET, a molecular profiling program based on next-generation sequencing of circulating tumor DNA (ctDNA), 11% of patients could be matched to treatments based on blood analyses (Rothwell et al., 2019). In the I-PREDICT trials, higher matching scores were obtained based on tumor DNA sequencing, and treatment success was achieved by combination therapies at higher than usual rates (Sicklick et al., 2019). Finally, when therapy was based on DNA sequencing or RNA expression within biopsied tissue (the WINTHER trial), 33% matching scores were obtained and treatments yielded longer progression-free survival (Rodon et al., 2019). Still more impressive matching scores were attained through an intensive effort based on a multidisciplinary approach that included investigators from the fields of bioinformatics, genetics, and physicians from multiple specialties. Indeed, 62% of 429 patients were successfully matched, and therapies based on several markers led to improved progression-free and overall mortality than in patients who had not been matched or who had received low matching scores (Kato et al., 2020).

Clearly, the effectiveness of matching genes to treatments has become more impressive. It is telling, however, that although a trial conducted by the National Cancer Institute's Molecular Analysis for Therapy Choice (NCI-MATCH) had impressive success using targeted treatments against cancer with specific mutations, implementation of treatment lagged owing to resource limitations (Flaherty et al., 2020). With so many individuals being affected by cancer, it is important to consider whether it is practically and economically feasible to undertake such approaches widely. And the view can be taken that too much attention is being devoted to precision medicine based on genetic markers, and insufficient attention is being paid to other markers.

Keeping hopes in check through CRISPR-Cas9

Even though numerous new approaches to cancer treatment have yielded encouraging outcomes, on occasion the initial excitement has needed to be tempered, only to be reinvigorated as variations of the approach emerged. CRISPR/Cas9 was an innovative genetic engineering method to disable, repair, or replace bits of DNA (Jinek et al., 2012).[c] Not long afterward the notion was advanced that gene-editing methods, such as CRISPR-Cas9, could be used to identify single and multiple genomic variations that contribute to cancer development and could potentially be adopted to prevent or treat some forms of cancer. However, the exuberance was beset by the concerns of off-target effects of the editing procedure and the possibility that cells that had been edited might promote cancer development. Concerns also existed that immune responses would be mounted against nucleases associated with CRISPR, but a method was developed to get around the problem of Cas9 being recognized as foreign and thus susceptible to targeting by the immune system. Also, anti-CRISPR methods were developed (kill switches) to limit the likelihood of off-target effects. Related approaches were established to increase the specificity of the procedure and to broaden the number of sites that can be modified.

Some new methods, such as meticulous integration Cas9, or miCas9, has the capacity to precisely place a DNA fragment into the genome without producing unintended insertions or deletions. It similarly appeared that using long single-stranded DNA (ssDNA) may be more effective in producing on-target changes while producing fewer off-target actions. Base editing (swapping one DNA nucleotide for another) can be adopted to limit off-target effects although this method experienced a few growing pains. A variant of CRISPR, that of "prime editing," can make small

---

[c] Jennifer Doudna and Emanuelle Charpentier won the 2021 Nobel Prize in Chemistry for their discovery and use of CRISPR-Cas9 in bacteria. The potential of CRISPR to cure diseases was certain, and consequently strong and extensive claims and counter-claims were made concerning the patent for the procedure. Several hearings were held pitting the Institutions that employed Doudna and Charpentier against that of Feng Zhang and his group who were the first to demonstrate the efficacy of CRISPR-Cas9 in eukaryotic cells (those that have a clearly defined nucleus). The courts seemed to provide a mixed verdict, essentially cutting the baby in half, which seemed fitting given that the method similarly could cut the genome in half.

edits to DNA, such as additions, deletion, or exchanges of single nucleotides without the risk of off-target effects. This procedure was shown to be capable of modifying genes linked to Tay Sachs disease and sickle cell anemia and could potentially be used to eliminate genes that contribute to negative outcomes associated with CAR T cancer therapy (Anzalone et al., 2019). Through another approach, that of using pools of guide RNAs, it was possible to identify and enhance the expression and presentation of genes that encode immunogenic antigens to deal with cancer cells (Wang et al., 2019). Despite the improvements of this technology, it needs to be kept in mind that many genes have been associated with cancers, and thus simply editing one of these may be insufficient to alter the course of an ongoing disease. Instead, concurrent targeting of several genes might offer a better approach to treatment. In fact, a small phase I clinical trial in three patients revealed that CRISPR could be used to edit 3 genes related to T cell effectiveness and that this approach was feasible in cancer immunotherapy, although several off-target gene changes were provoked (Stadtmauer et al., 2020).

To be sure, the many efforts involving CRISPR technologies are still at relatively early stages of development and evaluation but have nonetheless created enormous excitement. Once the limitations and safety of the procedures have been fully addressed, it may be possible to treat diseases that have been viewed as untreatable.

Beyond the genome

Precision medicine in a rudimentary form had been practiced for decades in cancer treatments and has been fundamental in determining therapeutic strategies. With the appreciation that hormones played a prominent role in cancer occurrence and progression, commensurately greater attention was given to the possibility that of therapeutic approaches to detect and treat neuroendocrine-related tumors (Hofland et al., 2018). As inflammatory factors can influence pathways that cause changes of the estrogen receptor, leading to endocrine resistance (Stender et al., 2017), identifying these could allow for the development of strategies to reprogram cytokine functioning, thereby eliminating the resistance that otherwise occurs. As discussed in a previous chapter, the abundance of certain microbiota can provide fundamental information about the features of tumors that might be present and there have certainly been indications that features of the microbiome can inform therapeutic efficacy.

In addition to liquid biopsies to detect specific types of cancer in blood, the presence of bladder cancer can also be determined through urine samples which may preclude the use of invasive procedures. With the increasing ability afforded through urinary metabolomics, it is also possible to detect lung, breast, bladder, prostate, and ovarian cancers (Burton and Ma, 2019). Analyses conducted in blood or urine allow for the monitoring of the tumor over time following treatment, which is critical given the high rate of cancer recurrence.

Having multiple markers related to cancer could provide a more accurate index of the therapeutic strategy that should be used relative to that based on a single marker. Yet being able to identify a constellation of markers, can provide predictive value above and beyond that of single genetic and epigenetic marks, and hence could be incorporated into the arsenal of information needed to attack cancers effectively. Ultimately, successful treatment of cancers will be facilitated by having adequate markers that predict the efficacy of treatments, but as described in Fig. 13.2, this may comprise predictors beyond those of the genome and will require broad teams and the availability of diverse therapeutic approaches.

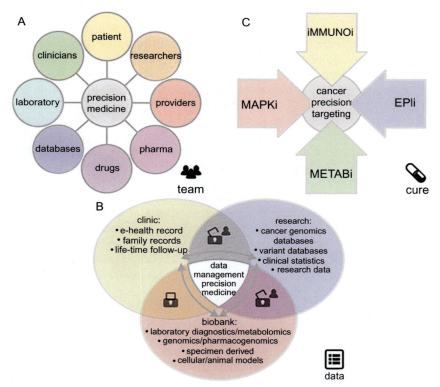

FIG. 13.2 Essential components in precision medicine and its application to cancer. The precision medicine infrastructure relies on a fruitful interplay between (A) a collaborative research team of clinicians and scientists, (B) personalized data allowing for fast and seamless interpretation, and (C) targeted pharmacological strategies. Precision disease management comprises targeted, personalized treatment aligned with the patient's genotype offering confidence to receive/provide care and hope for a cure. In malignant melanoma, rational, orthogonal combinations of immune checkpoint inhibitors (IMMUNOi), epigenetic inhibitors (EPIi), metabolic inhibitors (METABi), or inhibitors of specific signaling pathways such as the mitogen-activated protein kinase inhibitors (MAPKi) will provide the best standard of care. Balanced and effective data sharing is based on patient consent, secure data exchange, and synergistic, standardized data formats. *From Filipp, F.V., 2017. Precision medicine driven by cancer systems biology. Cancer Metastasis Rev. 36, 91–108.*

Therapies agnostic to cancer type

Several immune-related factors can determine the effectiveness of cancer treatments, such as the "foreignness" of the cancer cells, the sensitivity of tumors to immune challenges, the ability of immune cells to infiltrate tumors, the absence of features (checkpoints) that determine whether tumor cells will be attacked, the lack of inhibitory molecules, as well as other characteristics of the tumor. A "cancer immunogram" that considers multiple attributes of the tumor and the person can be created to decide the optimal strategy for individual patients (Blank et al., 2016). It might be still better if an approach agnostic to the type of cancer being treated were developed so that treatments are effective for a broad range of cancers irrespective of the site where the tumor started. This could, in theory, be accomplished through treatments that target cancer cells which share molecular features (Luoh and Flaherty, 2018). Several agents and approaches have already been adopted, such as those that involve either the programmed death 1 site (PD-1), a pan-tropomyosin receptor tyrosine kinase (TRK) inhibitor, and a pan-TRK, anaplastic lymphoma kinase (ALK) and ROS-1 inhibitor (Lavacchi et al., 2020). Unfortunately, these agents can encounter problems, including the development of treatment resistance, and they may not be equally effective for cancers that involve different tissues.

Aside from the agents that have so far received FDA approval as agnostic therapies, many other agents can potentially influence multiple forms of cancer. As we have discussed earlier, the p53 mutation is present within 40% of tumors, thus targeting this protein could be useful in treating several types of cancer. It will be recalled that the p53 protein is especially important in tumor suppression and is involved in sounding the alarm when DNA damage exists, so it can be repaired or pushed to cell suicide with the assistance of several heat shock proteins (Ikwegbue et al., 2019). Ordinarily, p53 serves to suppress inflammation whereas activation of the NFκB pathway is integral to the production of inflammation as well as varied tissue irritating factors (Gudkov and Komarova, 2016). An imbalance in the antagonistic actions between NFκB and p53 may affect the appearance and progression of some forms of cancer. Thus, by concurrently focusing on these factors it may be possible to affect several forms of cancer.

Aside from p53, it will be recalled that the KRAS mutation appears in about 25%–30% of solid tumors, and is particularly frequent in some forms of cancer. Thus, targeting cells with these mutations may be effective in treating these cancers, but doing so directly has proven to be difficult. One approach to overcome these difficulties involved attacking KRAS indirectly by targeting the cellular scaffold that supports the KRAS protein that regulates cell growth. By targeting a specific protein, connector enhancer of kinase suppressor of RAS (Cnksr1), it was possible to interfere with the growth of colon and lung cancer in mice, without disturbing healthy cells (Indarte et al., 2019). There are still other ways of dealing with cancers involving a KRAS mutation. Specifically, the KRAS mutation may be accompanied by other mutations (e.g., on a gene dubbed KEAP1) that altered metabolism ordinarily associated with KRAS. By targeting aspects of such co-mutations, it may be possible to treat KRAS-related cancer (Best et al., 2019). There have also been indications that KRAS-related cancers could be dealt with by targeting mutant KRAS-specific inhibitors or by binding a small molecule to a mutant form of KRAS, so a single amino acid glycine was substituted for cysteine, thereby preventing KRAS from entering an "on" state that allows for tumor progression. The upshot of this has been the development of new compounds that in combination with other agents, such as PD-1 inhibitors, may be able to deal with KRAS-related cancers (Canon et al., 2019). These manipulations have been assessed in humans, and for the moment the outcomes were encouraging.

Altering the functioning of p53 or KRAS hasn't been the only approach to dealing with multiple cancers. Manipulations that influence the BRAF mutation, which is also involved in several forms of cancer, could potentially be useful in their treatment (Bernocchi et al., 2021). As well, a compound (S63845) was developed that seemed to affect about 25% of tumors (acute myeloid leukemia, lymphoma, multiple myeloma, and solid cancers) by targeting MCL1, a member of the BCL2 family, which regulates apoptosis (cell death). MCL1 acts as a pro-survival protein that permits some types of cancer cells to evade apoptosis, and thus essentially remain immortal. Hence, when MCL1 is eliminated, cells regain their mortality and cancer should be slowed (Kotschy et al., 2016). Indeed, the use of this compound has been receiving increasingly more attention in dealing with diverse forms of cancer. Aside from these approaches, therapies have been developed to limit the actions of mutations that promote cancer progression by targeting epidermal growth factor receptor (EGFR), which has been implicated in several forms of cancer.

An old miracle drug?

It had long been recommended that baby aspirin (81 mg) be taken to prevent heart problems and stroke, especially if such events had already been encountered. Then we learned that aspirin neither prevents heart attack nor stroke, and whatever benefits that might be accrued might be offset by risks of serious intestinal bleeding. Based on such reports the American College of Cardiology no longer recommends aspirin for the prevention of heart disease or stroke. Despite these cautions, about half of those who are older than 70 have continued to take their daily aspirins.

Is it premature to abandon taking a daily dose of baby aspirin? There may be benefits of daily aspirin apart from actions on heart disease. Low-dose aspirin and other NSAIDs may contribute to cancer prevention, likely due to actions against the inflammatory factors associated with tumor development and growth. Several detailed reviews revealed that aspirin was associated with reduced occurrence of several gastrointestinal cancers, such as bowel, esophageal, stomach, gastric cardia, hepatobiliary, and pancreatic cancer (22%–39%). Likewise, regular low dose aspirin use was accompanied by reductions in lung, breast, endometrial, ovarian, and prostate cancers (Santucci et al., 2021), and effects such as these were apparent among women at elevated risk of breast cancer related to familial and genetic factors. Aside from observational and retrospective studies, the possible anticancer effects of aspirin were observed in a large-scale prospective study conducted over 10 years (Ajrouche et al., 2019), and the use of aspirin was associated with reduced occurrence of colorectal cancer assessed over 22 years (Figueiredo et al., 2021).

A systematic review indicated that aspirin was associated with reduced metastatic spread of several cancers and in a subsequent updated systematic review and meta-analysis of 118 studies that considered 18 forms of cancer, aspirin use that accompanied other treatments was associated with a 20% reduction in death related to various forms of cancer (Elwood et al., 2021). While the antiinflammatory effects of aspirin could potentially play a role in attenuating metastases, the effects obtained may also occur because aspirin prevents platelets from clumping, thereby limiting the ability of cancer cells to hide from immune attackers. In this regard, activation of platelet cyclooxygenase (COX-1) and the subsequent thromboxane A_2 release were primarily responsible for the inhibition of metastasis (Lucotti et al., 2019).

The benefits of aspirin were also apparent in association with immunotherapy, possibly owing to its inhibitory action on prostaglandin E, which ordinarily provokes immune activation. Other Cox-2 inhibitors, such as celecoxib (Celebrex) also have positive effects by inhibiting IDO1-induced immunosuppression, thereby enhancing the effectiveness of immunotherapy in eliminating cancers. Equally interesting was the finding that treatment with a Cox-2 inhibitor and the β-norepinephrine antagonist propranolol for 5 days before and after breast cancer surgery diminished the subsequent occurrence of metastases (Shaashua et al., 2017).

Biomarkers have been identified that indicated who might be best served by aspirin treatment for specific forms of cancer. Methylation of BRCA1 could predict mortality related to breast cancer among aspirin users, and a mutation of the PIK3CA gene that is associated with uncontrolled cell growth was identified in patients who did well in response to aspirin. As well, longer survival occurred among breast cancer patients who had used high doses of aspirin (325 mg) than among patients who used a low dose (81 mg) and was most useful in those carrying a PIK3CA mutation (Zhou et al., 2019). Similarly, responses to aspirin in breast cancer metastases might be linked to the PIK3CA gene, and among patients with head and neck cancer carrying the PIK3CA oncogene, the use of NSAIDs for at least 6 months improved survival, possibly through actions on prostaglandin E2 (Hedberg et al., 2019).

Despite the positive actions of aspirin, the effects of aspirin vary with an individual's age and may even have untoward effects on individuals who were 70 years of age or older. Specifically, continuous aspirin intake between the ages of 50 and 70 had preventive effects, but there were no benefits after the age of 70 (Guo et al., 2021). Importantly, in a follow-up study among more than 19,000 individuals aged around 65–70 years, daily use of 100 mg of aspirin (averaging 4.7 years) was associated with more cancer-related deaths than was observed in a group that took placebo (McNeil et al., 2018). Of course, it cannot be ignored that aspirin use can lead to intestinal damage and hence caution needs to be exercised. That said, specific microbiota, such as *Lactobacillus acidophilus* or *Bifidobacterium adolescentis*, could attenuate gut ulceration provoked by an NSAID, and oral intake of Bifidobacteria breve in humans can limit these negative actions (Mortensen et al., 2019).

Summary and conclusions

If cancer could be cured readily, then effective treatments would have been developed long ago. Cancer cells are wily and relentless. They change their characteristics to deal with threats and adopt various strategies to circumvent the best defenses that can be mustered to defeat them. Problems encountered in treating patients effectively have been tied to detecting cancer early, which entails developing detection procedures with high sensitivity and selectivity that can recognize the presence of cancer at a time when it was relatively manageable. Developing effective therapeutic strategies has also been beset by problems related to overdiagnosis and unintended consequences that have emerged. Making matters more problematic is that even when something seemed to be working in rodent models, it may not translate well to humans. Too often, treatments that appeared to work, run out of steam as resistance develops to their effects, and should the cancer have reached the point of metastasizing, the odds of effective treatment are much reduced.

Against the backdrop of this negativity, the field has seen notable successes. Many cancers can be treated more successfully, and even some cancers that previously were considered largely untreatable, now are more readily eliminated. In other instances, success could only be measured in small increments. A few months of life extension for the patients may not seem like much, but it may offer clues to scientists that they're on the right track, and baby steps will hopefully develop into great strides.

Whether it's dealing with the initial tumor, treatment resistance, or metastasis, a precision medicine approach holds considerable promise, especially if greater attention is focused on finding and targeting tumor vulnerabilities. At the same time, an enormous number of genetic and epigenetic processes that may be linked to cancer, and a constellation of processes indirectly related to cancer might also play into the progression and treatment of the illness.

There's no question that diet, exercise, sleep, and stressors can influence the cancer process, and could be relevant to the development of treatment strategies.

Time and history have made it clear that each new generation of cancer researchers has been resilient in the face of repeated failures. There is likely little hubris among cancer researchers who have repeatedly attempted to navigate the complex maze and multiple routes that seem to culminate in dead ends. Yet they persist (obsess even) in their search for effective therapies, which one day may offer true hope for defeating cancer.

References

Ajrouche, A., De Rycke, Y., Dalichampt, M., Messika Zeitoun, D., Hulot, J.S., et al., 2019. Reduced risk of cancer among low-dose aspirin users: data from French health care databases. Pharmacoepidemiol. Drug Saf. 28, 1258–1266.

Anzalone, A.V., Randolph, P.B., Davis, J.R., Sousa, A.A., Koblan, L.W., et al., 2019. Search-and-replace genome editing without double-strand breaks or donor DNA. Nature 576, 149–157.

Baker, M., Dolgin, E., 2017. Cancer reproducibility project releases first results. Nature 541, 269–270.

Begley, C.G., Ellis, L.M., 2012. Drug development: raise standards for preclinical cancer research. Nature 483, 531–533.

Bernocchi, O., Sirico, M., Corona, S.P., Strina, C., Milani, M., et al., 2021. Tumor type agnostic therapy carrying BRAF mutation: case reports and review of literature. Pharmaceuticals 14, 159.

Best, S.A., Ding, S., Kersbergen, A., Dong, X., Song, J.Y., et al., 2019. Distinct initiating events underpin the immune and metabolic heterogeneity of KRAS-mutant lung adenocarcinoma. Nat. Commun. 10, 4190.

Blank, C.U., Haanen, J.B., Ribas, A., Schumacher, T.N., 2016. Cancer immunology. The "cancer immunogram". Science 352, 658–660.

Boehm, J.S., Garnett, M.J., Adams, D.J., Francies, H.E., Golub, T.R., et al., 2021. Cancer research needs a better map. Nature 589, 514–516.

Boland, E.G., Bennett, M.I., Allgar, V., Boland, J.W., 2020. Cannabinoids for adult cancer-related pain: systematic review and meta-analysis. BMJ Support. Palliat. Care 10, 14–24.

Braillon, A., 2016. Lowering the high cost of cancer drugs—I. Mayo Clin. Proc. 91, 397.

Burton, C., Ma, Y., 2019. Current trends in cancer biomarker discovery using urinary metabolomics: achievements and new challenges. Curr. Med. Chem. 26, 5–28.

Canon, J., Rex, K., Saiki, A.Y., Mohr, C., Cooke, K., et al., 2019. The clinical KRAS(G12C) inhibitor AMG 510 drives anti-tumour immunity. Nature 575, 217–223.

Caulfield, T., 2017. Stem Cell Miracles Don't Live up to the Hype. https://healthydebate.ca/2017/03/topic/stem-cells/. Accessed January 2019.

Chen, E.Y., Raghunathan, V., Prasad, V., 2019. An overview of cancer drugs approved by the US food and drug administration based on the surrogate end point of response rate. JAMA Intern. Med. 179, 915–921.

Corsello, S.M., Nagari, R.T., Spangler, R.D., Rossen, J., Kocak, M., et al., 2020. Discovering the anti-cancer potential of non-oncology drugs by systematic viability profiling. Nat. Cancer 1, 235–248.

Danhauer, S.C., Addington, E.L., Cohen, L., Sohl, S.J., Van Puymbroeck, M., et al., 2019. Yoga for symptom management in oncology: a review of the evidence base and future directions for research. Cancer 125, 1979–1989.

Davis, C., Naci, H., Gurpinar, E., Poplavska, E., Pinto, A., et al., 2017. Availability of evidence of benefits on overall survival and quality of life of cancer drugs approved by European Medicines Agency: retrospective cohort study of drug approvals 2009-13. BMJ 359, j4530.

de Vries, R., Muller, M., van der Noort, V., Theelen, W., Schouten, R.D., et al., 2019. Prediction of response to anti-PD-1 therapy in patients with non-small-cell lung cancer by electronic nose analysis of exhaled breath. Ann. Oncol. 30, 1660–1666.

Elwood, P.C., Morgan, G., Delon, C., Protty, M., Galante, J., et al., 2021. Aspirin and cancer survival: a systematic review and meta-analyses of 118 observational studies of aspirin and 18 cancers. ecancermedicalscience 15, 1258.

Emerson, M.A., Golightly, Y.M., Aiello, A.E., Reeder-Hayes, K.E., Tan, X., et al., 2020. Breast cancer treatment delays by socioeconomic and health care access latent classes in Black and White women. Cancer 126, 4957–4966.

Figueiredo, J.C., Jacobs, E.J., Newton, C.C., Guinter, M.A., Cance, W.G., et al., 2021. Associations of aspirin and non-aspirin non-steroidal anti-inflammatory drugs with colorectal cancer mortality after diagnosis. J. Natl. Cancer Inst. 113, 833–840.

Flaherty, K.T., Gray, R., Chen, A., Li, S., Patton, D., et al., 2020. The molecular analysis for therapy choice (NCI-MATCH) trial: lessons for genomic trial design. J. Natl. Cancer Inst. 112, 1021–1029.

Fojo, T., Mailankody, S., Lo, A., 2014. Unintended consequences of expensive cancer therapeutics-the pursuit of marginal indications and a me-too mentality that stifles innovation and creativity: the John Conley Lecture. JAMA Otolaryngol. Head Neck Surg. 140, 1225–1236.

Galanzha, E.I., Menyaev, Y.A., Yadem, A.C., Sarimollaoglu, M., Juratli, M.A., et al., 2019. In vivo liquid biopsy using Cytophone platform for photoacoustic detection of circulating tumor cells in patients with melanoma. Sci. Transl. Med. 11, eaat5587.

Gorski, D., 2019. The Cleveland Clinic Publishes a Study Claiming to Show Benefits From Functional Medicine. It Doesn't. https://sciencebasedmedicine.org/cleveland-clinic-publishes-functional-medicine/. Accessed April 2021.

Gorski, D.H., Novella, S.P., 2014. Clinical trials of integrative medicine: testing whether magic works? Trends Mol. Med. 20, 473–476.

Gudkov, A.V., Komarova, E.A., 2016. p53 and the carcinogenicity of chronic inflammation. Cold Spring Harb. Perspect. Med. 6, A026161.

Guo, C.G., Ma, W., Drew, D.A., Cao, Y., Nguyen, L.H., et al., 2021. Aspirin use and risk of colorectal cancer among older adults. JAMA Oncol. 7, 428–435.

Hanna, T.P., King, W.D., Thibodeau, S., Jalink, M., Paulin, G.A., et al., 2020. Mortality due to cancer treatment delay: systematic review and meta-analysis. BMJ 371, m4807.

Hedberg, M.L., Peyser, N.D., Bauman, J.E., Gooding, W.E., Li, H., et al., 2019. Use of nonsteroidal anti-inflammatory drugs predicts improved patient survival for PIK3CA-altered head and neck cancer. J. Exp. Med. 216, 419–427.

Heiden, B.T., Eaton Jr., D.B., Engelhardt, K.E., Chang, S.H., Yan, Y., et al., 2021. Analysis of delayed surgical treatment and oncologic outcomes in clinical stage I non-small cell lung cancer. JAMA Netw. Open 4, e2111613.

Hoffmann, T.C., Del Mar, C., 2017. Clinicians' expectations of the benefits and harms of treatments, screening, and tests: a systematic review. JAMA Intern. Med. 177, 407–419.

Hofland, J., Zandee, W.T., de Herder, W.W., 2018. Role of biomarker tests for diagnosis of neuroendocrine tumours. Nat. Rev. Endocrinol. 14, 656–669.

Ikwegbue, P.C., Masamba, P., Mbatha, L.S., Oyinloye, B.E., Kappo, A.P., 2019. Interplay between heat shock proteins, inflammation and cancer: a potential cancer therapeutic target. Am. J. Cancer Res. 9, 242–249.

Indarte, M., Puentes, R., Maruggi, M., Ihle, N.T., Grandjean, G., et al., 2019. An inhibitor of the pleckstrin homology domain of CNK1 selectively blocks the growth of mutant KRAS cells and tumors. Cancer Res. 79, 3100–3111.

Jinek, M., Chylinski, K., Fonfara, I., Hauer, M., Doudna, J.A., et al., 2012. A programmable dual-RNA-guided DNA endonuclease in adaptive bacterial immunity. Science 337, 816–821.

Johnson, S.B., Park, H.S., Gross, C.P., James, B.Y., 2018. Complementary medicine, refusal of conventional cancer therapy, and survival among patients with curable cancers. JAMA Oncol. 4, 1375–1381.

Kato, S., Kim, K.H., Lim, H.J., Boichard, A., Nikanjam, M., et al., 2020. Real-world data from a molecular tumor board demonstrates improved outcomes with a precision N-of-One strategy. Nat. Commun. 11, 4965.

Klevorn, L.E., Teague, R.M., 2016. Adapting cancer immunotherapy models for the real world. Trends Immunol. 37, 354–363.

Kotschy, A., Szlavik, Z., Murray, J., Davidson, J., Maragno, A.L., et al., 2016. The MCL1 inhibitor S63845 is tolerable and effective in diverse cancer models. Nature 538, 477–482.

Lavacchi, D., Roviello, G., D'Angelo, A., 2020. Tumor-agnostic treatment for cancer: when how is better than where. Clin. Drug Investig. 40, 519–527.

Lu, C.Y., Zhang, F., Lakoma, M.D., Madden, J.M., Rusinak, D., 2014. Changes in antidepressant use by young people and suicidal behavior after FDA warnings and media coverage: quasi-experimental study. BMJ 348, g3596.

Lucotti, S., Cerutti, C., Soyer, M., Gil-Bernabe, A.M., Gomes, A.L., et al., 2019. Aspirin blocks formation of metastatic intravascular niches by inhibiting platelet-derived COX-1/thromboxane A2. J. Clin. Invest. 129, 1845–1862.

Luo, H., Zhao, Q., Wei, W., Zheng, L., Yi, S., et al., 2020. Circulating tumor DNA methylation profiles enable early diagnosis, prognosis prediction, and screening for colorectal cancer. Sci. Transl. Med. 12, eaax7533.

Luoh, S.W., Flaherty, K.T., 2018. When tissue is no longer the issue: tissue-agnostic cancer therapy comes of age. Ann. Intern. Med. 169, 233–239.

Mailankody, S., Prasad, V., 2017. Overall survival in cancer drug trials as a new surrogate end point for overall survival in the real world. JAMA Oncol. 3, 889–890.

McKinney, S.M., Sieniek, M., Godbole, V., Godwin, J., Antropova, N., et al., 2020. International evaluation of an AI system for breast cancer screening. Nature 577, 89–94.

McNeil, J.J., Nelson, M.R., Woods, R.L., Lockery, J.E., Wolfe, R., et al., 2018. Effect of aspirin on all-cause mortality in the healthy elderly. N. Engl. J. Med. 379, 1519–1528.

Menon, U., Gentry-Maharaj, A., Burnell, M., Singh, N., Ryan, A., et al., 2021. Ovarian cancer population screening and mortality after long-term follow-up in the UK Collaborative Trial of Ovarian Cancer Screening (UKCTOCS): a randomised controlled trial. Lancet 397, 2182–2193.

Meric-Bernstam, F., Brusco, L., Shaw, K., Horombe, C., Kopetz, S., et al., 2015. Feasibility of large-scale genomic testing to facilitate enrollment onto genomically matched clinical trials. J. Clin. Oncol. 33, 2753–2762.

Merton, R.K., 1936. The unanticipated consequences of purposive social action. Am. Sociol. Rev. 1, 894–904.

Mortensen, B., Murphy, C., O'Grady, J., Lucey, M., Elsafi, G., et al., 2019. Bifidobacteriumbreve Bif195 protects against small-intestinal damage caused by acetylsalicylic acid in healthy volunteers. Gastroenterology 157, 637–646 e634.

Moynihan, R., Henry, D., Moons, K.G., 2014. Using evidence to combat overdiagnosis and overtreatment: evaluating treatments, tests, and disease definitions in the time of too much. PLoS Med. 11, e1001655.

Nigro, E., Formato, M., Crescente, G., Daniele, A., 2021. Cancer initiation, progression and resistance: are phytocannabinoids from *Cannabis sativa* L. promising compounds? Molecules 26, 2668.

Nordström, T., Discacciati, A., Bergman, M., Clements, M., Aly, M., et al., 2021. Prostate cancer screening using a combination of risk-prediction, magnetic resonance imaging and targeted prostate biopsies: a randomised trial. Lancet Oncol. 22, 12401249.

Parker, L.A., 2017. Cannabinoids and the Brain. MIT Press, Cambridge, MA.

Prasad, V., Jena, A.B., 2014. The Peltzman effect and compensatory markers in medicine. Health 2, 170–172.

Prinz, F., Schlange, T., Asadullah, K., 2011. Believe it or not: how much can we rely on published data on potential drug targets? Nat. Rev. Drug Discov. 10, 712.

Rawla, P., 2019. Epidemiology of prostate cancer. World J. Oncol. 10, 63–89.

Rodon, J., Soria, J.C., Berger, R., Miller, W.H., Rubin, E., et al., 2019. Genomic and transcriptomic profiling expands precision cancer medicine: the WINTHER trial. Nat. Med. 25, 751–758.

Rothwell, D.G., Ayub, M., Cook, N., Thistlethwaite, F., Carter, L., et al., 2019. Utility of ctDNA to support patient selection for early phase clinical trials: the TARGET study. Nat. Med. 25, 738–743.

Salas-Vega, S., Iliopoulos, O., Mossialos, E., 2017. Assessment of overall survival, quality of life, and safety benefits associated with new cancer medicines. JAMA Oncol. 3, 382–390.

Sanford, N.N., Sher, D.J., Ahn, C., Aizer, A.A., Mahal, B.A., 2019. Prevalence and nondisclosure of complementary and alternative medicine use in patients with cancer and cancer survivors in the United States. JAMA Oncol. 5, 735–737.

Santucci, C., Gallus, S., Martinetti, M., La Vecchia, C., Bosetti, C., 2021. Aspirin and the risk of nondigestive tract cancers: an updated meta-analysis to 2019. Int. J. Cancer 148, 1372–1382.

Sartor, O., de Bono, J., Chi, K.N., Fizazi, K., Herrmann, K., et al., 2021. Lutetium-177–PSMA-617 for metastatic castration-resistant prostate cancer. NEJM 385, 1091–1103.

Schumacher, T.N., Schreiber, R.D., 2015. Neoantigens in cancer immunotherapy. Science 348, 69–74.

Scudellari, M., 2019. Protein-slaying drugs could be the next blockbuster therapies. Nature 567, 298–301.

Shaashua, L., Shabat-Simon, M., Haldar, R., Matzner, P., Zmora, O., et al., 2017. Perioperative COX-2 and beta-adrenergic blockade improves metastatic biomarkers in breast cancer patients in a phase-II randomized trial. Clin. Cancer Res. 23, 4651–4661.

Shen, C., Ferro, E.G., Xu, H., Kramer, D.B., Patell, R., Kazi, D.S., 2021. Underperformance of contemporary phase III oncology trials and strategies for improvement. J. Natl. Compr. Cancer Netw. 1, 1–7.

Sicklick, J.K., Kato, S., Okamura, R., Schwaederle, M., Hahn, M.E., et al., 2019. Molecular profiling of cancer patients enables personalized combination therapy: the I-PREDICT study. Nat. Med. 25, 744–750.

Siegel, D.A., Fedewa, S.A., Henley, S.J., Pollack, L.A., Jemal, A., 2021. Proportion of never smokers among men and women with lung cancer in 7 US states. JAMA Oncol. 7, 302–304.

Sotgia, F., Fiorillo, M., Lisanti, M.P., 2017. Mitochondrial markers predict recurrence, metastasis and tamoxifen-resistance in breast cancer patients: early detection of treatment failure with companion diagnostics. Oncotarget 8, 68730–68745.

Stadtmauer, E.A., Fraietta, J.A., Davis, M.M., Cohen, A.D., Weber, K.L., et al., 2020. CRISPR-engineered T cells in patients with refractory cancer. Science 367, eaba7365.

Stender, J.D., Nwachukwu, J.C., Kastrati, I., Kim, Y., Strid, T., et al., 2017. Structural and molecular mechanisms of cytokine-mediated endocrine resistance in human breast cancer cells. Mol. Cell 65, 1122–1135 e1125.

Sun, S., Yang, S., Hu, X., Zheng, C., Song, H., et al., 2020. Combination of immunomagnetic separation with aptamer-mediated double rolling circle amplification for highly sensitive circulating tumor cell detection. ACS Sens. 5, 3870–3878.

Wang, G., Chow, R.D., Bai, Z., Zhu, L., Errami, Y., et al., 2019. Multiplexed activation of endogenous genes by CRISPRa elicits potent antitumor immunity. Nat. Immunol. 20, 1494–1505.

Welch, H.G., Prorok, P.C., O'Malley, A.J., Kramer, B.S., 2016. Breast-cancer tumor size, overdiagnosis, and mammography screening effectiveness. N. Engl. J. Med. 375, 1438–1447.

Welch, H.G., Kramer, B.S., Black, W.C., 2019. Epidemiologic signatures in cancer. N. Engl. J. Med. 381, 1378–1386.

Zhou, Y., Simmons, J., Jordan, C.D., Sonbol, M.B., Maihle, N., et al., 2019. Aspirin treatment effect and association with PIK3CA mutation in breast cancer: a biomarker analysis. Clin. Breast Cancer 19, 354–362 e357.

CHAPTER 14

Traditional therapies and their moderation

OUTLINE

The role of serendipity	431	Nutrients related to treatment resistance	443
Chemotherapy and radiation therapy	432	Manipulating energy and metabolism	444
Chemotherapeutic approaches	433	Fasting diet in relation to cancer progression and treatment resistance	445
Radiation therapies	433		
Side effects of chemotherapy and radiation therapy	434	Selectively starving tumor cells	448
The challenge of treatment resistance	438	Microbiota in relation to chemotherapy	451
Mechanisms related to treatment resistance	438	Eubiosis and dysbiosis	452
Gene mutations	438	Summary and conclusions	455
Autophagy	441	References	455
Apoptosis and anastasis	442		

The role of serendipity

Anyone critical of progress in cancer therapy needs only to read the book *Emperor of All Maladies* (Mukherjee, 2011) to be dissuaded of this clearly incorrect perspective. Many hundreds of years of perverse treatments preceded relatively effective therapies being developed, but even in the 20th century, many approaches amounted to trial-and-error experiments in the hope that something better than the preceding failed therapies would evolve. Ironically, the advent of chemotherapy to save lives was prompted by effects observed in response to highly toxic and deadly mustard gas used during World War I. Although mustard gas was not used during World War II, mustard bombs were stockpiled on the S.S. John Harvey in the Italian port of Bari Harbor when it was bombed by Luftwaffe planes in September 1943. Survivors who jumped from the ship, drenched in the toxin, later experienced pronounced lymphopenia and myeloid cell suppression. Although the presence of mustard gas was covered up so that the Nazi government didn't become aware of its presence, thanks to the diligence of a chemical warfare specialist, Lieutenant Colonel Stewart Francis Alexander, it became apparent that mustard gas could be used to inhibit the multiplication of fast-dividing cells.

This observation encouraged further research into the possibility of chemical agents being used to deal with other fast-growing cells, notably cancer cells. Shortly thereafter Goodman and Gilman, together with several associates, assessed a variant of mustard gas (nitrogen mustard) to determine whether it could inhibit the progression of non-Hodgkin's lymphoma. The treatment had only transient positive effects but ushered in a novel way of approaching cancer therapies. At about the same time, Farber and his associates developed an interest in whether bone marrow functioning could be affected by folate, which came from green leafy vegetables. Eventually, this work led to the development of methotrexate in treating childhood leukemia, which was subsequently adopted as a chemotherapeutic agent to treat solid tumors. Despite the initial skepticism concerning the use of chemotherapy to treat cancers, often being derided as a poison, by the 1960s and 1970s it was accepted as a legitimate method of treating cancer, most often being used in conjunction with other treatments.

Over the ensuing years, more chemotherapeutic agents were discovered that came from naturally occurring products (e.g., an alkaloid of the Madagascar periwinkle) or those that were developed against cell division. As the therapeutic arsenal increased, approaches were adopted that entailed the concurrent use of several agents, each attacking a tumor through a different route.

The history of radiation to treat cancer has a somewhat longer history than that of chemotherapy. Very soon after Röntgen discovered X-rays in 1895, several reports emerged indicating that these rays could be used to treat skin malignancies as well as other forms of cancer. Radium therapies began to be used early as well, being applied through salts and inhalation. Regrettably, it also gained the attention of snake oil purveyors who peddled low doses that could be used as part of skin creams, fortified waters, inhalants, and the equivalent of bath salts, until it was found to be deadly. Radium was also used to illuminate watch dials. As we now know, the "radium girls" who painted radium on these dials, often bringing the brushes to their tongue and lips to produce a fine point, suffered miserably from radiation poisoning.

By the 1930s, higher doses of radiation were adopted to treat some cancers by administering over several days. In many respects, the procedures used today, albeit in a far more sophisticated and precise form, largely follow from procedures used 80 years earlier. Attacking cancer cells by radiation and chemotherapeutic agents, together with surgery, continue to be the mainstay of therapy. In this chapter, we will provide brief discussions of these cancer therapies and their varied manipulations, making sure to address how lifestyle factors moderate their effects.

Chemotherapy and radiation therapy

Chemotherapy influences the processes involved in cell division thereby destroying these fast-growing cells. Specifically, chemotherapeutic agents damage the cell's DNA and inhibit enzymes needed to repair the damage and may interfere with metabolites that sustain actively growing cancer cells. Chemotherapy can be undertaken using a single powerful agent or a cocktail of chemotherapeutic agents varying with the specific type of cancer being treated. Each of the agents comprising this cocktail might operate through different processes that affect diverse phases of cancer growth, and hence drug toxicity and the chances of resistance developing are diminished. However, this can be difficult to achieve, and treatment resistance remains a major obstacle to successful therapy. Not to mention the damage to healthy cells, and the excoriating side effects of chemotherapeutic agents.

Chemotherapeutic approaches

Chemotherapeutic agents may be used as the primary or sole method of treating cancer but are also used in conjunction with other cancer treatments, serving as a neoadjuvant therapy to diminish the size of tumors so that they can more readily be eliminated through surgery or radiation therapy. They are also used as adjuvant therapies after surgery to eliminate cancer cells that might remain. As chemotherapy can diminish the adverse effects of cancer cells when other treatments had been unsuccessful, it may be used in a palliative capacity to make patients more comfortable. The specific chemotherapeutic agents used and how they are delivered are dependent on the nature of the cancer being treated, its stage of development, the presence of specific features (e.g., the presence of specific hormone receptors), and what other treatments had previously been administered. Typically, chemotherapy is delivered intravenously, thereby affecting all parts of the body, but it can also be delivered to specific sites (e.g., through nanoparticles) to kill cancer cells that are localized.

As with so many other cancer-related processes, multiple balances may be in play in the progression of cancer after chemotherapy. To achieve optimal effects, the treatments must not be too low or too high ("the Goldilocks Zone"). For instance, chemotherapy may cause biphasic changes in the expression of p21 (which acts as a tumor suppressor), inhibiting cell division at higher doses, but can produce cancer cell proliferation at lower doses. In fact, moderate p21 expression present soon after chemotherapy can foster cancer cell proliferation (Hsu et al., 2019). It seems that the influence of p21 varies with the presence of other factors that can either enhance cell survival or cell death (genotoxic stress), such as the presence or the absence of the tumor protein p53 that regulates cell division. These interactive effects raise the question concerning what other factors influence the action of p21, especially as it plays a role in inflammatory processes, and in some conditions (e.g., sepsis) may contribute to macrophages switching from a proinflammatory state to one that is immunosuppressive. There is also the issue of whether the actions of chemotherapy are influenced by extrinsic factors (diet, exercise, stressors) that influence sterile inflammation. As we will see, numerous biological processes are responsible for the beneficial effects of chemotherapy, many of which are subject to lifestyle and psychosocial determinants.

Radiation therapies

Radiation therapy kills or slows down growth of cancer cells by damaging their DNA, and is generally adopted when the cancer is localized to a particular site. The use of radiation is a component of cancer treatment for about 50% of patients varying with the nature and stage of the cancer being treated, how close the tumor is to other radiation-sensitive tissues, and features of the patient (general health, age). As in the case of chemotherapy, the treatment may be used before surgery to shrink the size of a tumor to make it more manageable, or it may be adopted following chemotherapy or surgery to diminish cancer recurrence. Among individuals living with advanced cancer, palliative radiation therapy may be adopted to diminish symptoms (particularly pain), enhance quality of life, and briefly extend life, although its use needs to be weighed against possible side effects that could occur.

Radiation therapy most often entails the delivery of photon (e.g., gamma rays) or proton beams to specific sites. By applying treatments over days, damage to healthy cells can be minimized, whereas the likelihood of hitting cancer cells during that portion of the cell cycle when DNA damage is most likely to occur is maximized. To limit the side effects of radiation therapy, several approaches have been used in which the features of the radiation itself were altered. Typically, the tumor and nearby cells were exposed to radiation for 10–30 min over multiple sessions, but newer approaches, namely "stereotactic body radiation therapy" allowed for fewer higher radiation doses to be safely delivered to discrete regions. However, in the case of some cancers, such as small cell lung cancer, surgery generally produced more favorable outcomes than did stereotactic body radiation. Another new tactic has been developed in which the same dose of radiation can be delivered in less than a second. This *Flash radiation therapy* may result in diminished side effects, including reduced inflammation and cognitive disturbances (Montay-Gruel et al., 2021). It was similarly shown that following lumpectomy, a single treatment with targeted intraoperative radiotherapy (TARGIT-IORT) could produce effects comparable to external beam radiotherapy, but with fewer side effects (Vaidya et al., 2020).

As radiation therapy is designed to disturb the DNA of cancer cells, its influence could potentially be enhanced by agents that disturb DNA repair, such as PARP inhibitors. More commonly, internal radiation therapy may be adopted, depending on the nature of the cancer and its spread. Brachytherapy may be undertaken in which radioactive capsules, seeds, or ribbons are placed directly into specific sites within the body (and may or may not be removed after some time). This procedure, which has been used in the treatment of breast and prostate cancer, as well as cervical, uterine, lung, rectal, and eye cancer, is less likely to damage surrounding tissue relative to externally delivered radiotherapy although some side effects are not uncommon (e.g., swelling, bruising, urinary-related symptoms). Based on this approach, molecules have been created that can bind to protein receptors on cancer cells and simultaneously deliver a payload comprising radioactive particles that can destroy these cells while limiting side effects.

The effectiveness of radiation treatment can be limited by several factors, including the possibility of seriously damaging organs, as in the case of cancers that have spread to gastrointestinal sites. In addition, many tumors are considered "cold tumors" as they do not have immune cells surrounding them. But if a very small amount of radiation is delivered to these sites, it can essentially act as a stress signal which summons immune cells to the target site, thereby converting these tumors to a "hot" status. In experiments that combined this with an immunotherapeutic agent, the effectiveness of the treatment was elevated in mice and importantly, in a small number of patients enrolled in the study (Herrera et al., 2021).

Side effects of chemotherapy and radiation therapy

The cost of cancer being held in abeyance can be psychologically and physically taxing, and the choices made by some patients to seek alternative therapies might come from their fear of standard treatment as opposed to believing that these strategies are a better option. Often, the side effects aren't as bad as the patient had imagined or expected, but their expectancies can create a nocebo effect so that negative actions will be more likely to be elicited by the treatment.[a] The side effects of treatments can often be influenced by physical fitness (and exercise), sleep and circadian rhythms, and social support. While many of the adverse physical and psychological ramifications experienced by cancer patients stem from the treatments administered, the cancer itself (or the distress caused by diagnosis and/or anticipation of receiving cancer treatment) may also produce multiple biological changes that disturb well-being and create a physical and psychological milieu that fosters illness progression and may interact with the therapies received.

Physical effects

Cancer therapies may place a considerable burden on healthy cells so that biological aging, estimated on the basis of the accumulation of epigenetic changes, was advanced by as much as 4.9 years, often manifested by fatigue and elevated inflammation (Xiao et al., 2021). Chemotherapy was also associated with reduced methylation (indicative of diminished epigenetic changes) at sites associated with the production of cytokines, resulting in elevated levels of circulating IL-6 and TNF-α that were linked to fatigue (Smith et al., 2014).

The side effects of cancer therapies vary appreciably across individuals but commonly comprise nausea, vomiting, loss of appetite, fatigue, fever, and hair loss. Other longer-lasting and more serious effects comprise the development of autoimmune disorders and cardiac-related problems (e.g., cardiomyopathy, arrhythmia, heart attack, high blood pressure, blood clots, stroke). Of patients diagnosed with cancer, particularly breast, prostate, or bladder cancer, and to an extent melanoma and endometrial cancer, 11.3% died of cardiovascular problems, often during the first year after diagnosis. And for women with breast cancer who experienced a heart attack, there was an elevated risk of cancer recurrence, with 60% more likely to die of the disease relative to women absent a heart attack. In mice that had incurred an experimentally induced heart attack, cancer-related death was earlier and tied to epigenetic changes, including those related to increased abundance of immature myeloid-derived suppressor cells that inhibit attack on tumors by cytotoxic T cells (Koelwyn et al., 2020).

Among women treated for breast cancer, particularly those who received anthracyclines, cardiovascular disease caused more fatalities than the cancer itself, especially among women who smoked or had obesity. Some treatments, such as doxorubicin, a form of anthracycline, is particularly likely to damage heart cells, which could potentially be attenuated by treatment

---

[a] General descriptions of side effect severity might not be as informative as one would have liked. Terms such as "generally well tolerated" aren't precise, and even very powerful side effects are sometimes described as being "manageable" (Sacks et al., 2019). Such vague terms are unhelpful and even deceptive, and it would be better to have research reports describe the specific features that comprise side effects, as well as to specifically itemize those side effects that affect quality of life and daily functioning. With this information in hand, patients can make better decisions concerning what they're willing to endure in order to extend their life for some uncertain period of time. With the information being accessible on varied internet sites, and patients increasingly wanting to be partner in determining their destiny, clear communication has become increasingly important.

with either the growth factor VEGF-B, an angiotensin-converting-enzyme (ACE) inhibitor, statins, or manipulations that influence mitochondrial dynamics and function (e.g., Bhagat and Kleinerman, 2020).

As noted in earlier chapters, stress reduction procedures, sleeping well, exercising, and maintaining a proper diet, can enhance well-being and limit some of the side effects associated with cancer and its treatment. It was also maintained that the damaging effects of doxorubicin could be diminished by exercise preconditioning (Gibson et al., 2013). However, a degree of confusion still exists concerning how much exercise cancer patients should receive and what an ideal diet comprises. Should the diet include energy-rich foods or is intermittent fasting a better strategy? Likewise, although patients are often advised not to consume a high fiber diet owing to the provocation of bloating and diarrhea, fiber can limit radiation-induced inflammation that might have detrimental effects (Patel et al., 2020).

Cognitive disturbances

Given the neurobiological changes that can be induced by chemotherapy, as well as the distress created by a cancer diagnosis and the harsh treatments that patients undergo, it is not unusual for cancer patients to develop cognitive and behavioral disturbances. Some of these actions may come about through microbial, mitochondrial, and immune alterations, as well as hormonal or brain neurochemical changes. These psychological disturbances may depend on the individual's stressor history, their ability to deal with uncertainty, their coping resources, including social support networks, as well as a constellation of other psychosocial factors. Diminishing the neurobiological actions of chemotherapy is of paramount importance not only to maintain the patient's mental well-being but also because depression and PTSD that may develop could diminish the efficacy of cancer treatments.

Most often, cognitive disturbances produced by chemotherapy are transient, but in some instances, they can be sufficiently pronounced to undermine general quality of life. On occasion, the effects can be detected as long as 10 years later, possibly being tied to persistent sleep disturbances and impaired sleep efficiency. Pharmacological treatments were often ineffective in ameliorating cognitive disturbances associated with chemotherapy, whereas cognitive therapy was successful in some domains, such as verbal memory, attention, and processing speed, and in some individuals, cognitive disturbances could be attenuated by mindfulness-based stress reductions training. Interventions comprising physical activity and cognitive training were also effective in diminishing chemotherapy-related cognitive disturbances, although the heterogeneity across patients might require further analyses to determine best practices.

Chemo brain

One of the common cognitive disturbances stemming from chemotherapy is a phenomenon called chemo brain (or "chemo fog"), which consists of memory impairments, difficulty concentrating or in making decisions, and disturbed multitasking abilities (Paquet et al., 2013). The emergence of these problems may be linked to fatigue, infection, sleep disturbances, pain, nutritional issues, surgery, anesthetics, and other specific drugs used as part of the cancer treatment.

Aside from affecting aging of somatic cells, chemotherapy can increase aging of brain processes, and thus favor the appearance of chemo brain. These effects are more likely to be produced by some chemotherapeutic agents preferentially affecting certain brain regions. Like many other side effects of chemotherapy, chemo brain is more likely to be apparent in women than in men and has been more frequent in postmenopausal women, the presence of particular hormones, and nutritional deficiencies.

Cytokine-related epigenetic changes were implicated in the cognitive disturbances stemming from chemotherapy, as were reductions of the growth factor BDNF, a key neurotrophic factor associated with neuroplasticity, synaptogenesis, and memory processes. Moreover, disturbed neuronal myelination was proposed as contributing to the cognitive disturbances associated with methotrexate chemotherapy (Geraghty et al., 2019). Ultimately, determining why some individuals will develop chemo brain whereas others are spared may be an important component in decisions concerning cancer therapies and the development of strategies to minimize treatment side effects.

Limiting side effects of chemotherapy and radiation

Since chemotherapy and radiation treatments aren't selective in the sense that they not only attack cancer cells but also affect healthy cells, approaches were developed to circumvent these actions while still maintaining the beneficial actions of the primary therapy. It would be highly desirable, of course, to be able to identify those patients who would be at high risk for developing severe side effects. Using a genetic signature that comprised a network of genes that were primarily associated with cell division, particularly within the bone marrow, it was possible to determine which nonsmall-cell cancer patients would be most likely to exhibit severe side effects in response to the combination of gemcitabine and carboplatin (Björn et al., 2020). Thus, patients who were at low risk for these side effects could be treated with higher doses of gemcitabine and carboplatin, whereas those at high risk would need to receive alternative treatments.

Methods have been developed to limit some side effects of radiation and chemotherapy, such as antiemetic drugs to prevent or diminish nausea and vomiting, but many other negative consequences of these therapies remain prominent. To some extent, exercise can reduce some of the side effects of radiation and chemotherapy, such as diminishing fatigue, anemia, and sleep problems, and particularly effective outcomes could be obtained through personalized strategies that may need to be modified over the course of therapy based on the patient's medical status. A review of several reports among patients with hematological malignancies revealed that exercise enhanced muscular strength and endurance, diminished fatigue, and improved emotional well-being; this was accompanied by improved NK cell functioning and diminished inflammatory factors that were related to epigenetic changes. Additionally, in mice, exercise promoted a dose-dependent reduction of tumor burden accompanied by reduced VEGF and hence diminished angiogenesis, as well as enhanced immune functioning (Sitlinger et al., 2020).

Dietary factors are essential in maintaining well-being during and after cancer treatment. Anemia following chemotherapy contributes to fatigue and the propensity for severe bleeding can be diminished by foods that increase iron. However, the presence of cancer itself often produces anorexia and cachexia, and the distress and depression that may accompany cancer can also affect both exercise and food consumption. These problems are compounded by cancer therapy affecting various aspects of consummatory and digestive processes. During therapy, patients may reject certain types of foods because of disturbed taste and smell, appetite loss, promotion of dry and sore mouth, difficulties in swallowing, lactose intolerance, nausea, as well as diarrhea, or constipation, which may cause disturbed nutritional status. Over the course of chemotherapy, reduced intake of antioxidant nutrients can promote greater inflammation and oxidative stress, particularly among overweight women, pointing to the importance of healthy antiinflammatory diets being maintained.

Various organizations have offered recommendations concerning which foods to eat and which to avoid during cancer therapy. The recommendations described in Chapter 13 (Arends et al., 2017) may be particularly pertinent as it considers multiple factors, including the nature of the cancer, stage of the disease, and characteristics of the patients. The National Cancer Institute has also provided an informative review that provides important food-related information for cancer patients and their caregivers. However, as we will see later in this chapter, there continues to be the question of whether cancer therapy can best be facilitated by particular dietary approaches. For example, those that promote adequate energy for patients who otherwise are more in need of critical nutrients, versus those—such as intermittent fasting—that may diminish the survival of cancer cells.

The challenge of treatment resistance

Among the greatest problems encountered in attempting to deal with cancers, as mentioned in Chapter 1, is that resistance often develops in response to treatments that initially seemed to be effective. In fact, multidrug resistance may develop even to drugs structurally and functionally different from the compounds used in initial therapy.

Mechanisms related to treatment resistance

The development of treatment resistance can occur through diverse mechanisms. Cancer cells have developed a vast repertoire of ways to evade our immune defenders and getting around therapies that ought to eliminate them. Many of these processes, depicted in Fig. 14.1, are also involved in the development of treatment resistance. Beyond the factors described in the figure, treatment resistance may be related to processes associated with drug distribution, metabolism, and excretion that can vary with repeated treatments as well as through intrinsic (sex, weight, culture, ethnicity, and circadian rhythms) and extrinsic factors (e.g., diet, exercise, and interaction with other drugs). Cancer cells can also develop an overabundance of efflux pumps that rid them of toxins, and in so doing they also clear the chemotherapy drugs. Drug resistance may also be related to changes within the tumor microenvironment, including acidity, glucose availability, and oxygenation of tumor cells.

Gene mutations

As in the case of bacterial adjustments to antibiotics, cancer cells may find ways of evading efforts to destroy them by adopting specific evolutionary paths that allow them to survive and proliferate (Hata et al., 2016). The genetic instability characteristic of cancers and the disturbances of DNA repair processes may foster elevated heterogeneity of cancer cells that favor the development of treatment resistance. With time or upon cancer recurrence,

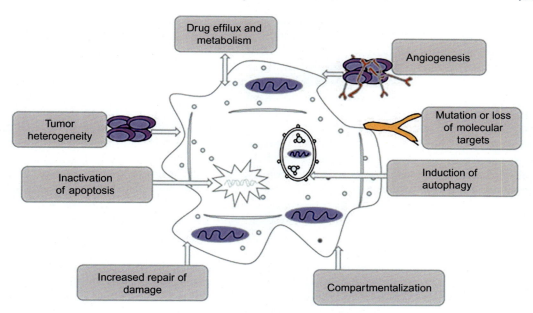

FIG. 14.1 A summary of the ways by which cancer cells become resistant to chemotherapy and various kinds of genotoxic or metabolic stresses. *From Sui, X., Chen, R., Wang, Z., Huang, Z., Kong, N., et al., 2013. Autophagy and chemotherapy resistance: a promising therapeutic target for cancer treatment. Cell Death Dis. 4, e838.*

and in response to therapeutics, cancer cells can morph and display new phenotypic characteristics, possibly owing to epigenetic changes or the occurrence of multiple mutations, so the effective treatment requires multitargeted therapies. Appreciation of the genomic correlates of the treatment response and the development of resistance may be fundamental for the development of effective therapeutic strategies. But the multiple mutations in tumor cells can differ appreciably between individuals diagnosed with the same type of cancer, and may also vary between tumors in each patient. Consequently, learning how to deal with drug resistance in one patient might not offer clues about resistance in a second patient.

The tumor microenvironment is harsh, in which nutrients are limited, oxygen levels are low, and acidity is high. As a result, premalignant cells may engage a Warburg phenotype that allows them to survive. These conditions, in essence, allow selection for those cells that can survive, and their clones will be equally capable of thriving. The survival of cancer cells within the tumor environment can concurrently be facilitated because it diminishes immune cells' access to the tumor and when cancer cells are attacked, they may "switch their states" and gene expression characteristics are altered, requiring a different anticancer drug treatment. It also appeared that the tumor microenvironment contains a considerable number of oxidized fat molecules (the LDL form of fat) that can affect cytotoxic T cells so that they are less able to destroy cancer cells. As it happens, because of their need for energy, these T cells increase the presence of a fat transporter CD36, but this has the effect of creating even more oxidized fat that further undermines their killing potential. Aside from these factors, the presence of certain proteins (e.g., PTIP, CHD4, and PARP1) can destabilize DNA copying, resulting in the cancer cells becoming resistant to chemotherapy (Ray Chaudhuri et al., 2016).

The presence of specific mutations (e.g., KRAS and BRAF) has been associated with several forms of cancer. These mutations might not only contribute to certain cancers occurring but also may signal that they might be particularly difficult to treat. Previously, we mentioned that the KRAS protein is a fundamental signaling mechanism that instructs cells to proliferate and function in specific ways. Mutations that promote KRAS overactivation may allow growth signals to persist in an "on" position so that cancer progression continues largely unabated even in the face of treatment. Colorectal cancer, for instance, is often associated with elevated growth factors and their receptors (e.g., EGFR), explaining why antibodies against these receptors have been used in treating these cancers. However, the presence of a BRAF or KRAS mutation predicts a poor treatment response to growth factor inhibitors, more so for the KRAS mutation. The presence of these mutations also signal whether individuals will be less (or more) responsive to chemotherapeutic agents and the development of treatment resistance.

Other processes related to cell growth may contribute to cancer progression and treatment resistance. The mTOR pathway is involved in the regulation of cell survival, metabolism, and growth, but when disturbed, various forms of cancer can arise along with increased treatment resistance. Furthermore, in response to targeted cancer therapies, the increased occurrence of cancer-promoting mutations may, in part, be regulated by mTOR, thereby affecting treatment resistance (Cipponi et al., 2020). It also appears that in some instances, such as in BRCA mutations in which DNA repair mechanisms are compromised, PARP inhibitors can disturb a secondary repair mechanism thereby limiting tumor progression. However, resistance frequently develops to these treatments so that cancers again begin to grow aggressively. Although several approaches have been attempted to enhance the effectiveness of PARP inhibitors, determining the mechanisms by which they have their primary effects—and how resistance develops—has been elusive.

While radiation and chemotherapy are frequently effective as cancer treatments, brain tumors, such as glioblastoma, have been notoriously difficult to tame. There are several reasons for this, including how their cellular plasticity endows them with treatment resistance. Yet for these cancer cells to behave as they do, two proteins—galectin 1 and HOXA—may be necessary since suppression of these proteins, at least within preclinical models, enhanced tumor sensitive to radiation. In essence, by using this approach it may be possible to attack the root cause of the illness rather than simply eliminating the features of the disease (Sharanek et al., 2021).

As well as the processes outline in predicting treatment resistance, the accumulated cellular damage that comes with aging, including diminished immune responses and elevated inflammation, may contribute to treatment resistance. Further, aging fibroblasts may promote a diminished response to DNA damage produced by reactive oxygen species, thereby permitting resistance to therapies (Kaur et al., 2016). Likewise, with age, melanoma cells are better able to take up lipids from the tumor microenvironment through upregulated functioning of the fatty acid transporter FATP2, thereby affording them better protection from cancer therapies. Therefore, by using agents that target the KRAS pathway and concurrently administering a FATP2 inhibitor, tumors in old mice disappeared and did not return as they might otherwise (Alicea et al., 2020).

Taking advantage of natural selection

The problems inherent in the development of treatment resistance gave rise to a different approach to cancer treatment that was, in a sense, based on the Darwinian theory of natural selection (Gatenby and Brown, 2020). It is generally thought that with continued treatment and the emergence of an increasing number of mutations, some cancer cells become progressively hardier, giving rise to stronger clones, allowing them to outcompete both healthy cells as well as less aggressive cancer cells, and concurrently develop greater resistance to therapeutic agents. Accordingly, the view was taken that therapeutic strategies necessarily would have to consider ways of dealing with the evolution of progressively hardier cancer cells. Rather than attacking cancer cells with high-intensity, high-dose treatments, the alternative was to adopt a less intense approach that would foster weaker treatment resistance, even if this meant that some cancer cells would remain viable (Enriquez-Navas et al., 2016). This would entail a variety of therapies being administered sequentially but altered before treatment resistance developed. A "first strike" against the cancer could be used to reduce the size and heterogeneity of the cancer cells, followed by a "second strike" that further diminished the cancer so that it could be controlled for an extended period (Gatenby et al., 2019). Essentially, cancer would be viewed as a chronic disease, but patients would remain alive and would have to learn to live with their disease.

An approach based on "game theory" that had become popular in the fields of economics and psychology, has been offered for cancer therapies (Stanková et al., 2019). The basic assumption is that, unlike the oncologist who can behave rationally and flexibly, cancer cells respond solely to the ongoing conditions, and thus do not adapt (evolve) to treatments that had not yet been applied—as ruthless as cancer cells might be, they lack clairvoyance or the ability to plan. Accordingly, oncologists can stay ahead of the cancer cells by modifying treatments before the cancer cells can adapt. This will become possible through better metrics to gauge the changes occurring within the tumor and its microenvironment. Based on a similar perspective, using a small number of prostate cancer patients a powerful agent administered intermittently (primarily during periods when the tumor appeared to be growing), yielded improved outcomes and resistance was less likely to develop (Zhang et al., 2017).

This method of cancer treatment was entirely antithetical to the common view that eradicating cancer was essential even if this entailed the therapy itself being hazardous to the patient. Predictably, this approach wasn't fervently embraced, but increasing interest continues to be shown in exploring its use in cancer treatment. Parhaps as certain types of cancer cells mutate and grow rapidly, the opportunity of using this approach may be limited to only some forms of cancer.

Autophagy

Autophagy is a mechanism by which the body can clear damaged or aged cells so that healthier cells can take over their functioning to maintain and restore effective biological performance. As autophagy is an important component that maintains the functioning and aggressiveness of cancer stem cells, it might be a good target to limit resistance. Like so many other biological processes, autophagy can operate in either a protective capacity or one in which tumor growth is stimulated by providing them with energy (Chmurska et al., 2021). While manipulations of autophagy could theoretically be used in cancer therapies and in reducing treatment resistance, this would require sophisticated balancing of different procedures. To date, few therapies focusing on autophagy have been effective in limiting treatment resistance, although combining the antimalarials chloroquine (CQ) and hydroxychloroquine (HCQ) with trastuzumab (Herceptin), or radiotherapy plus chemotherapy, yielded encouraging outcomes in HER2-positive breast cancer and glioblastoma. Potentially better effects could be provided by administering inhibitors of autophagy through nanoparticles so that the agent will accumulate in cancer cells.

Apoptosis and anastasis

There is another perspective that may be relevant for the recurrence of cancer cells and the development of resistance. While apoptosis is needed so that damaged cells neither use up essential resources nor turn into cancerous cells, there are conditions in which it might be advantageous for cells to survive (e.g., when a challenge to cells is transient). It turns out that a process exists, termed anastasis, wherein prosurvival processes are activated so that cells do not necessarily die once apoptosis has been initiated. In the case of cells being challenged by chemo- or radiation therapy, both apoptosis and anastasis may be induced. If the stress to the cell is prolonged, then apoptotic mechanisms may win out and the cell will die. But if the challenge is transient, then anastatic mechanisms may prevail and the cell will survive. If the body's objective was to prevent cell death, then anastasis would be a useful mechanism, but if the objective was to kill the offending (cancerous) cells, then their survival would backfire, as these cells may rejuvenate and multiply once treatment ceases. Accordingly, it would be beneficial to develop adjunct treatments to turn on apoptotic pathways, perhaps through activation of AKT1 that may be an important component of this process, so that targeted cells would die off (Sun et al., 2020).

While chemotherapy often leads to cell death, cancer cells may become "senescent" so that they stay alive but are inactive. In this senescent (hibernation) state, the cancer cells do not take up the chemotherapeutic agents and thus can reemerge after the threat has passed. When chemotherapy was used to eliminate acute myeloid leukemia, a subset of cancer cells went into a senescent state yet adopted an inflammatory profile that could subsequently encourage cancer growth. This dual state appeared to be dependent on a protein called ATR, in that treatment with an ATR inhibitor kept these cells from entering a senescent state, thereby making them vulnerable to destruction (Duy et al., 2021). In keeping with these findings, breast cancer cells in mice can survive treatment and remain in a quasidormant but interactive state. By affecting surrounding immune cells, they eventually turn on cytokines, which summon macrophages and other immune cells to deposit collagen and stimulate the cancer cells to flourish and grow once more. A particular chemokine, CCL5, may be critical to this process, potentially making it a possible target to prevent metastasis (Walens et al., 2019).

Nanoparticles in treating cancer and limiting treatment resistance

With improvements of nanotechnologies that allow delivery of small drug doses to specific target sites, the effectiveness of treatments may be enhanced, while the potential for tolerance would be reduced, as would side effects of the treatment. Using nanoparticles to reach the mitochondria (the site of energy production of the cell) together with near-infrared laser treatment, it was possible to temporarily affect efflux pumps, thereby allowing for chemotherapy to have maximal effects, at least for several days (Wang et al., 2018). In addition to nanoparticles being used to have chemotherapeutic agents attack cancer cells directly, biodegradable nanoparticles have been designed that can program immune cells so that they will be capable of identifying and then destroying cancer cells. Nanotherapies also can be used to turn cold tumors hot, thereby making them more responsive to chemotherapeutic agents and to immunotherapies.

Nanoparticles can be made more effective by the incorporation of targeted antibodies, peptides, or other biomolecules. Unfortunately, problems can still come up because the nanoparticles may be recognized as being foreign, thereby causing an immune response to be mounted against them. Accordingly, efforts have been made to use "biomimetic materials" that combine the actions of naturally occurring features of cells or their membranes with bioengineered substances to produce effective delivery vehicles (Yong et al., 2019). Since not every agent in the chemotherapeutic arsenal is effective in dealing with specific cancers, an approach was developed that comprised the use of an implantable device that allows many agents to be applied to a tumor (without each of the agents being able to interact with another) and then determining which drug was best based on their ability to kill tumor cells (Jonas et al., 2015).

Remarkably, microbots have also been created based on neutrophil characteristics (hence called neutrobots) that can pass the blood-brain barrier, home in on glioma cells (with the help of directed magnetic fields) where they deliver a drug payload (Zhang et al., 2021). For cancers that have irregular shapes and are difficult to access, such as brain glioblastoma, nanoparticles delivered directly to the tumor can be combined with a precision medicine approach (e.g., based on tumor's gene or cell surface markers) to improve outcomes. Nanoparticles can also be combined with immunotherapies to elicit potent actions against tumors distal from the initial treatment site.

Nutrients related to treatment resistance

The cancer cell's voracious appetite for glucose, and the ability to reprogram glucose metabolism, is fundamental for cancer growth and expansion; but it also may be instrumental in the promotion of resistance to treatment. It is possible to modify some of the downstream actions that stem from altered glucose metabolism by using metabolic inhibitors that decrease glucose availability to cancer cells, but other methods to achieve this or to enhance the effects of these treatments are available. This could include manipulations that affect circulating glucose levels, as well as IGF-1 or insulin-like growth factor-binding protein-1 (IGFBP-1). Dietary manipulations could also be used to diminish the development of chemoresistance (Nencioni et al., 2018). Indeed, several mechanisms presumed to influence cancer occurrence and the development of treatment resistance, such as mTOR functioning, are influenced by diet and by antidiabetic treatments, such as metformin (Amin et al., 2019). Likewise, prebiotic treatments, such as inulin or mucin added to the diet, can act on melanoma growth associated with a BRAF mutation, thereby delaying the development of cancer drug resistance (Li et al., 2020).

Types of foods or components of certain foods have been implicated as possibly being able to influence treatment resistance. Isothiocyanate and sulforaphane in some cruciferous vegetables, such as broccoli, cabbage, and cauliflower could potentially act in this capacity. The polyphenol, quercetin, which is present in cruciferous vegetables and many fruits, favored apoptosis and reduced viability of pancreatic ductal adenocarcinoma cells, but unlike the response to other therapies, resistance did not develop (Fan et al., 2016). Curcumin, that has antiinflammatory and antioxidant properties, may have beneficial effects in the treatment of breast, prostate, and colorectal cancer, and could have positive actions when combined with gemcitabine in the treatment of pancreatic cancer. But more than this, because curcumin can inhibit p-glycoprotein from pushing therapeutic agents out of the cell, it might be effective in preventing resistance to chemotherapy (Karthika and Sureshkumar, 2019). Given that many powerful pharmaceutical compounds have been ineffective in limiting chemoresistance, it would be reasonable to be suspicious of claims that foods or food components would be successful in acting in this capacity. If nothing else, however, the available data have been sufficient to encourage further randomized clinical trials to evaluate their efficacy.

Manipulating energy and metabolism

Because of their rapid multiplication, cancer cells need considerable energy in the form of sugars as well amino acids that form proteins and other biomolecules. By depriving cancer cells of nutrients, it is possible to promote their demise. Dietary manipulations can influence immune activity relevant to cancer therapies and may also diminish toxicity related to treatments, possibly owing to the moderation of the inflammatory response. In our earlier discussion of metformin, whose primary use in treating diabetes stems from its ability to reduce sugar formed by the liver, it was indicated that this drug may also be relevant to managing cancer since it could deprive cancer cells of an energy source as well as by altering microbial processes. Indeed, glucose deprivation can produce variations of calcium influx across the plasma membrane, leading to several sequential steps that culminate in cell death. Whatever the mechanism, metformin can augment the effects of other treatments, such as the proapoptotic agent venetoclax, in the treatment of certain types of leukemia and breast cancer (Haikala et al., 2019).

As much as manipulating energy provided to cancer cells could potentially exert beneficial effects, cancer cells find ways of getting around such defenses. A proposition based on a mathematical model suggested that when metastasis is prevented by cutting off energy supply through reduced glycolysis and oxygen availability, the cancer cells find alternative ways of getting their way, becoming more and persistently aggressive, even after they have been reoxygenated (Godet et al., 2019). With low oxygen levels, cancer cells are more likely to metastasize, and thus hypoxia in cancer cells has been associated with a poorer prognosis.

The plasticity of cancer cells in the face of obstacles makes them difficult to eradicate, thus requiring more creative ways of dealing with them. As discussed earlier, AMP-activated protein kinase (AMPK) and hypoxia-inducible factors (HIFs) are fundamental for oxidative phosphorylation and glycolysis. These proteins interact with metabolic pathways associated with glucose oxidation, glycolysis, and fatty acid oxidation, thereby influencing energy supplied to cancer cells. When the energy source of metastasizing cells is limited, both glycolysis and oxidative phosphorylation pathways are activated. By targeting both HIF and growth factors, such as VEGF (which also affects HIF), positive effects can be produced (Choueiri and Kaelin, 2020). In essence, because cancer cells have the capacity to switch from glycolysis to oxidative phosphorylation during cancer progression, blocking one pathway might not diminish cancer growth and metastasis, whereas interfering with both pathways may be more effective.

Another finding relevant to energy processes obtained by cancer cells comes from the observation that the nucleolus located within the nucleus, which is involved in the formation of energy-producing mitochondria, is enlarged in tumor cells and their number is increased. A causal connection between changes of nucleoli and cancer led to methods to diminish cancer progression (Carotenuto et al., 2019). The nucleolus grows by consuming guanosine triphosphate (GTP), a form of energy needed by some cancer cells. Thus, it was reasoned that by inhibiting inosine monophosphate dehydrogenase (IMPDH)—which fuels GTP production—cancer growth in a mouse model of glioblastoma could be limited. In fact, many existing therapies that operate through diverse pathways to affect cancer may share actions on nucleolus functioning.

Owing to the essential nature of mitochondria in energy production, targeting mitochondrial genes ought to be a good target for eliminating cancers. However, because drugs that affect mitochondria can be toxic, alternative methods of doing so have been sought. As it happens, mitochondrial DNA can be inhibited so that cancer cells are starved without producing toxic effects. For a period of time, therefore, cancer cells would be in a more vulnerable state so that they can be attacked directly by other agents (Bonekamp et al., 2020).

Fasting diet in relation to cancer progression and treatment resistance

While fasting or calorie restriction initially became popular as a way of reducing weight, it has since been maintained that it might also be beneficial in cancer treatment. Preclinical studies indicated that intermittent calorie restriction enhanced the effects of chemotherapeutic agents in dealing with transplanted or carcinogen-induced breast tumors. It seems that fasting-elicited alterations of glucose metabolism can limit the proliferation of tumor cells even though fasting may also diminish the action of immune cells that rely on glycolysis (Turbitt et al., 2019). In fact, in HER2-negative stage II/III breast cancer patients, a fasting-mimicking diet maintained for just 3 days before and during neoadjuvant chemotherapy was associated with tumor cell loss and reduced DNA damage in T cells that might otherwise be provoked by the chemotherapy (de Groot et al., 2020). It also appeared that cycles of a fasting-mimicking diet together with high doses of vitamin C could synergistically diminish cancers involving a KRAS mutation (Di Tano et al., 2020). It is particularly significant, as well, that fasting can act against some of the side effects elicited by the cancer treatments, and could potentially limit treatment resistance (Nencioni et al., 2018).

The actions of fasting diets may come about by influencing growth factors that affect tumorigenesis. In this respect, VEGF functioning can be altered through intermittent fasting, thereby affecting the formation of blood vessels and activation of antiinflammatory macrophages, causing more fat to be burned, hence acting against cancer. These diets can also create a metabolic milieu that can diminish the capacity for tumor survival in response to chemotherapeutic treatments and enhance the effects of hormone therapy (Caffa et al., 2020). Intermittent calorie restriction could similarly limit some adverse actions associated with chemotherapy by eliciting hematopoietic stem cell-based regeneration, while concurrently reducing IGF-1 and leptin, and increasing adiponectin (Chen et al., 2016). Aside from these processes, fasting or fasting-mimicking diets might promote health benefits through circadian rhythm alterations and gut microbial changes.

Several diets have been used to produce energy restrictions that could affect cancer progression, but the data have not been conclusive as to which procedure is most practical, beneficial, and safe. Regrettably, patient adherence to severe diets was often poor so that pathological complete responses (lack of all signs of cancer in tissue following treatment) were not apparent, making it essential to develop better approaches to enhance patient cooperation.

Ketogenic diets and cancer treatment

The use of ketogenic diets has been argued to contribute to the prevention or amelioration of features germane to a wide array of chronic illnesses, although support for this has been variable. A ketogenic diet comprises low carbohydrate and high-fat intake, and is meant to reduce glucose accumulation and push fats into ketone bodies (acetone, acetoacetic acid, beta-hydroxybutyric acid) that are typically not consumed by cancer cells. To an extent, the presumed benefits of a ketogenic diet in diminishing cancer occurrence or progression are somewhat puzzling given that the low fiber intake required with this diet diminishes beneficial gut bacteria that form SCFAs that have multiple health benefits owing to their antiinflammatory and antitumorigenic effects. In fact, it has been argued that with the exception of controlling seizure occurrence, the short-term benefits of a ketogenic diet are far outweighed by the risks of chronic diseases that could be produced (Crosby et al., 2021).

Based on the actions of a ketogenic diet, it was proposed that it might be useful in enhancing the effects of standard therapies that attenuate tumor progression. Among other things, such a diet may create an inhospitable environment for cancer cells by depriving them of glucose, glutamine, and carbohydrates, and might thus be useful as an add-on in the treatment of certain cancers (Weber et al., 2020). A ketogenic diet could also engender positive effects by reprogramming the energy metabolism of tumor cells as well as related processes within the tumor microenvironment. This diet could also enhance innate and adaptive immune responses (e.g., reduced dendritic cells, T cells, and NK cells), and increase the availability of metabolically protective $\gamma\delta$ T cells that limit inflammation (Goldberg et al., 2020). A ketogenic diet may also have its anticancer effects by reducing insulin levels. Specifically, the altered insulin could diminish the activity of proliferative signaling pathways involving PI3K and mTOR, while concurrently increasing oxidative stress within cancer cells. However, a ketogenic diet can cause a rebound effect in which insulin levels are increased, thereby reactivating the PI3K pathway that could promote tumor growth. It was especially interesting that in a mouse model of pancreatic cancer, a ketogenic diet that kept blood sugar spikes under control, PIK3 inhibitors were better able to diminish cancer progression (Hopkins et al., 2018).

A ketogenic diet augmented the effects of neoadjuvant therapy in patients with metastatic breast cancer and when a ketogenic diet was combined with other treatments, tumor progression could be inhibited, possibly acting through antiinflammatory, antiangiogenic, and proapoptotic processes. For instance, combining a ketogenic diet with a tumor-inhibiting antibiotic (6-diazo-5-oxo-L-norleucine) produced beneficial effects in glioblastoma, in which cancer stem cells and mesenchymal cells were more readily destroyed (Mukherjee et al., 2019).

Although the data supporting the benefits of a ketogenic diet have been encouraging, there may still be too few studies, and too few participants within each study, to form definitive conclusions concerning its usefulness in cancer therapies. Furthermore, maintaining a ketogenic diet can be difficult because of its palatability and side effects (e.g., constipation, loss of energy, leg cramps), and consequently, compliance has been relatively poor in patients undergoing cancer therapy. A less stringent form of this high-fat, low-carb diet in which ketone supplementation was provided, could be as useful as the standard ketogenic diet. There is certainly cause to be optimistic concerning the benefits of a ketogenic diet in conjunction with other therapies, but additional data based on individual patient characteristics are needed. In the context of other purported benefits of a ketogenic diet, such as in diminishing obesity and in taming type 2 diabetes, it was suggested that the hype may have outpaced the actual evidence supporting these claims. Whether this is accurate and whether it is also the case for the usefulness of a ketogenic diet in cancer treatment remains to be fully established.

While severe diets can prevent some diseases, several issues need to be considered concerning whether and when this procedure is appropriate for cancer patients. Cancer frequently promotes cachexia and anorexia, and even in the absence of these severe disturbances, cancer may lead to malnutrition and metabolic disturbances, as well as impaired repair of tissues damaged by radiation and chemotherapy. Even if diet restriction can produce some benefits, these need to be weighed against the adverse effects that might come about. As we described in Chapter 7, nutrition recommendations were offered in an extensive report that considered diverse dietary factors at different cancer stages and provided recommendations for cancer survivors (Arends et al., 2017). This set of recommendations is far too extensive to be described here, but foremost among the many recommendations was that screening for malnutrition ought to be a primary component of cancer care. Moreover, as with so many aspects of cancer treatment, a one-size-fits-all perspective was discouraged, and instead it was recommended that diet and energy manipulations should be considered on an individual basis varying with multiple characteristics (e.g., the presence of obesity, general health, age, sex) some of which may change over the course of the disease.

In general, it was recommended that patients obtain adequate liquids, energy-enriched foods, and foods high in protein—such as fish, poultry, lean red meat, eggs, low-fat dairy products, whole grains, nuts, dried beans, peas, lentils, and raw fruits and vegetables. Other foods should be avoided, particularly those high in sodium, added sugars or saturated fats, as well as the use of alcohol. Vitamin and mineral intake should not be used willy-nilly but should be restricted to reversing deficiencies that may have occurred. Contrary to the movement supporting fasting diets, it was explicitly recommended that cancer patients not adopt any diet that could diminish energy resources. Similar cautions were advanced because of the risks of malnutrition and sarcopenia that could emerge because of intense diets, and more preclinical data are needed to identify which cancers would benefit from fasting, at what stage this would be beneficial, and which diet restriction approaches would be optimal.

Revisiting angiogenesis

It will be recalled that an important characteristic of some cancer cells is that they promote the formation of blood vessels in tumors to facilitate the provision of oxygen and other nutrients. For a time, there had been considerable excitement (and some doubts) over the possibility that treatments that inhibit angiogenesis (e.g., andostatin and angiostatin or antivascular endothelial growth factor monoclonal antibodies) would limit tumor growth (Kerbel and Folkman, 2002). The initial hope that antiangiogenic treatments would be helpful was not fulfilled as survival was not extended for a variety of different cancers (Li et al., 2017). When positive effects were seen, these were restricted to certain cancers and only during specific stages, and treatment resistance often developed. Furthermore, inhibiting the formation of fine blood vessels that feed tumors also limits access of immune cells to engage in the battle, which might favor metastasis. This is not to say that angiogenesis inhibitors do not have a place in cancer treatment as they may be advantageous for a subset of ovarian cancer patients. Relevant markers will likely need to be identified, perhaps based on the presence of specific proteins that affect blood vessel formation, to determine which patients might benefit from this treatment.

Using specific dietary components to enhance the effects of antiangiogenic treatments has not been widely evaluated clinically. However, foods containing genistein (e.g., soybeans), lycopene (a natural pigment in the carotenoid family), resveratrol, curcumin, quercetin, and several other foods may have antiangiogenic actions and could potentially act against microscopic tumors. This view is consistent with the supposition that specific foods, as part of a broader diet, may prevent cancer occurrence and that the benefits of a Mediterranean diet may, in part, be due to antiangiogenic actions.

Selectively starving tumor cells

Starving oneself to destroy cancer cells may not be practical, despite the suggestion that intermittent fasting may have positive effects. In fact, in the absence of sufficient nutrients, immune cells might become exhausted and ineffectual in contending with cancer, thereby increasing the advantage of tumor cells over immune defenses. To counter this, it would be necessary to identify and alter the metabolic pathways that feed specific immune cells, thereby dealing with the disease more effectively.

It is feasible to starve tumor cells selectively by limiting the availability of glucose and amino acids, such as methionine, asparagine, and histidine, as shown in Fig. 14.2. This could also be achieved by preventing the transport of amino acids into cells by altering the functioning of the transporter enzyme ASCT2 by deleting the gene that encodes this enzyme or by drugs that inhibit this enzyme's functioning. In the case of relatively aggressive cancers (e.g., triple-negative breast cancer) that need fatty acids and the amino acid glutamine to grow prodigiously, inhibitors of both these components may have positive effects in slowing tumor progression. A glutamine antagonist not only limits tumor growth but also is effective in modifying T cell functioning so that the response to different forms of immunotherapy can be enhanced (Leone et al., 2019). To quench their appetite, tumors can use a pump, referred to as SLC1A5, so that glutamine is pumped into cancer cells. Hence, preventing the uptake of glutamine (e.g., through drugs, such as IMD-0345) might starve fast-growing cancer cells (Feng et al., 2021). In some instances where cancer cells are deprived of glutamine, they can find alternate energy sources and become more aggressive. In this regard, following glutamine starvation, a decline occurs in the level of one of its metabolites, alpha-ketoglutarate, which might contribute to cancer progression. When a form of alpha-ketoglutarate was supplied to animals, signs of intestinal tumors diminished (Tran et al., 2020), raising the possibility of a similar treatment being applied in conjunction with therapies in humans.

Aside from glutamine, restricting dietary methionine, serine, and glycine could also have positive actions on cancer progression and could affect treatment resistance. There is also reason to believe that histidine and asparagine could affect cancer progression as well as breast cancer metastasis, and histidine might contribute to the actions of methotrexate in attenuating cancer (Kanarek et al., 2018). Some cancer cells lack an enzyme fundamental in

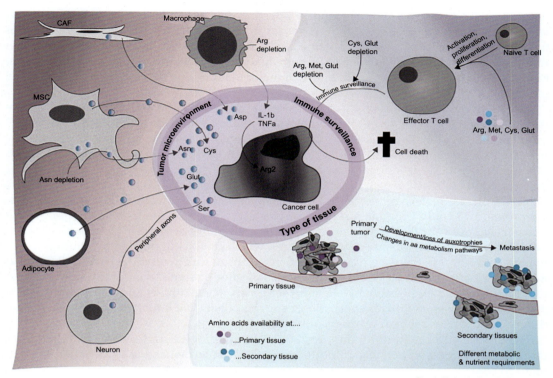

FIG. 14.2 The efficacy of therapeutic interventions using amino acid availability is determined by different external factors including the type of tissue, the tumor microenvironment, and immune surveillance. Tumor cells originating from the same cancer have different metabolic and nutrient requirements at their primary or metastatic sites. Furthermore, cells from the tumor microenvironment can supply cancer cells with depleted amino acids. To effectively fight cancer cells, T cells need to get activated, proliferate, and differentiate, which consumes high amounts of amino acids. As a consequence, immune surveillance is strongly dependent on amino acid availability. Abbreviations: *AA*, amino acids; *CAF*, cancer-associated fibroblast; *MSC*, mesenchymal stem cells. From Butler, M., van der Meer, L.T., van Leeuwen, F.N., 2021. Amino acid depletion therapies: starving cancer cells to death. Trends Endocrinol. Metab. 32, 367–381.

the production of methionine and thus rely on foods, such as red meat and eggs to obtain this amino acid. As such, anticancer effects could potentially be obtained through specific dietary restrictions or inhibition of enzymes that cancer cells need to obtain nutrients. Cancer cells display overactive methionine metabolism that is essential for their survival so that methionine deprivation can result in cancer cells being unable to grow and could augment the actions of cancer therapies (Wanders et al., 2020). In mouse models, restriction of dietary methionine could enhance the effects of treatments for a variety of cancers even if undertaken for only a week. This outcome was apparent in attacking treatment-resistant cancers, possibly through actions on KRAS and knockout of p53 (Gao et al., 2019).

Gut microbiota may contribute to the effectiveness of cancer therapy by affecting methionine. Specifically, bacteria of phyla such as *Firmicutes* and *Verrucomicrobia*, are related to the metabolism of dietary methionine, thereby affecting dendritic cell and T cell functioning as well as cancer cell metabolism so that responses to immunotherapy are affected. Whether or not beneficial effects on cancer are obtained by changes of specific microbiota that affect methionine or other amino acids, such as by combinations of prebiotics or narrow-spectrum antibiotics, remains to be determined, but such manipulations warrant experimental appraisal (Abid and Koh, 2019).

Like methionine, serine is required by cancer cells to proliferate and consequently diminishing the availability of serine through dietary manipulations could diminish cancer cell proliferation (Maddocks et al., 2017). However, serine deprivation causes a compensatory increase of serine synthesis that could undo the effects of the treatment. As expected, by inhibiting the synthesis of endogenous serine, the impact of dietary serine deprivation could be augmented, thereby enhancing the effects of anticancer agents (Tajan et al., 2021). It was also observed that in the face of metabolic stress, such as serine deprivation, the tumor suppressor protein p53 could promote a shift of aerobic glycolysis to glutathione synthesis so that cancer cells could survive. However, in p53 deficient tumors, serine and glycine deprivation can inhibit the growth of some types of cancer, which can be augmented by altering oxidative processes and perhaps by modifying microorganism functioning (Ke et al., 2020). It was mentioned earlier that fructose may be pivotal in promoting some forms of cancer (e.g., hepatocellular carcinoma), possibly by affecting serine-glycine synthesis and pentose phosphate pathways. Consistent with this suggestion, drug treatments that inhibit these pathways could diminish the cancer-promoting effects of fructose.

The SSRI sertraline inhibits the production of methionine and serine, more so when mitochondrial metabolism was inhibited, and thus could be used as an adjunctive treatment to control some forms of cancer (Geeraerts et al., 2021). In a mouse model of childhood sarcoma, treatment with the SSRI paroxetine reduced the number of insulin-like growth factor receptors (IGFR) on malignant cells, thereby inhibiting tumor growth. These effects may have stemmed from actions related to G protein-coupled receptors (GPCRs) and receptor tyrosine kinases (RTKs) (Crudden et al., 2021). That said, SSRIs can interact with other drugs so that negative effects can arise, such as a pronounced increase in breast cancer recurrence among women being treated with tamoxifen (Wedret et al., 2019).

The importance of essential amino acids in cancer promotion was documented in a timely and informative review, which indicated that dietary supplementation that affects amino acids and nutrients, such as histidine and mannose, might also be useful in affecting cancer therapies (Kanarek et al., 2020). Moreover, diminishing asparagine, arginine, and cystine, through diet or by pharmacological treatments, could enhance the response to cancer treatments. Glutamine, glutamate, and aspartate are also related to tumor growth and their modification by pharmacologic treatments might augment the impact of anticancer treatments. It is highly unlikely that a one-size-fits-all approach is feasible to enhance cancer therapy by altering amino acids. Diverse cancers vary with respect to the energy sources they require, and dietary manipulations do not affect nutrients to a comparable extent in all tissues or tumor microenvironments. Moreover, the actions of diets may interact in different ways with drug treatments. Thus, even if manipulations of amino acids could have beneficial effects, this will vary with numerous features of the tumor as well as those of the individual, necessitating strategies that take these differences into account.

The obesity paradox

It is curious that although obesity has been tied to cancer occurrence, enhanced responses to some therapies have been observed in patients with obesity. The source for this "obesity paradox" is uncertain particularly given the pro-tumorigenic actions ordinarily associated with obesity. It has been argued that this paradox may not be at all paradoxical as several biases may have crept into the findings. For instance, being at increased risk for various illnesses, individuals with obesity may see their physician more often than other people, making it possible that cancer would be detected relatively early and thus be more manageable. The influence of obesity may also be confounded with other variables that may skew the data to suggest better survival among those with obesity (Lennon et al., 2016). Moreover, consideration needs to be given to the inclusion and exclusion criteria of reported studies. For instance, were patients who smoked or who had a comorbid condition (e.g., heart disease) excluded from the studies so that the healthiest obese people were represented? Importantly, it is not only essential to assess how long individuals had been obese. As described earlier, those who had a long history of having obesity may be at elevated risk of poor treatment responses, whereas obesity that developed late may not lead to such adverse outcomes. Ultimately, studies assessing the relationship will need to be well structured. Finally, understanding the processes that govern the link between obesity and therapeutic efficacy will require a much deeper appreciation of the complex interactions that involve biological processes associated with obesity (including microbial changes).

Microbiota in relation to chemotherapy

Based on our earlier discussions, it should be clear that microbiota, including bacteria, fungi, viruses, and bacteriophages have been associated with several forms of cancer. While there is evidence suggesting that therapeutic benefits can be derived from specific microbiota, their translation to effective therapies has not been fully realized. Nonetheless, an impressive case has been made that we are presently at the cusp of being able to undertake large-scale human studies that may allow the development of personalized strategies to facilitate cancer therapies (Sepich-Poore et al., 2021).

Just as diverse microbes have been associated with cancer progression and occurrence, greater diversity of gut bacteria in mice was associated with numerous biological alterations that could influence the response to radiation and chemotherapy (Kovács et al., 2020). Specific bacteria could similarly either enhance or diminish the activity, efficacy, and toxicity of several therapies. The positive impact of microbiota might stem from the production of bacterial metabolites and changes of oncogenic toxins (Zitvogel et al., 2017), augmented bacterial translocation and the maturation of T_{17} cells within the intestine and effector lymph nodes, or by altering the tumor microenvironment (Iida et al., 2013). Microbiota can shape multiple immune and inflammatory responses elicited by chemotherapies, including monocytes/macrophages and NK cells that are capable of recognizing PAMPs and microbe-associated molecular patterns (MAMPs), and may even influence immune memory. Thus, by targeting the microbiome it might be possible to enhance the efficacy of several anticancer treatments, such as chemotherapy, checkpoint modulators, and adoptive T cell transfer therapy (Bashiardes et al., 2017). Furthermore, when the microecology of the gut is altered (e.g., by chemotherapy), some of the bacteria present may take on pathogenic characteristics (pathobionts) leading to illnesses. As such, targeted modification of microbiota or their metabolites could potentially attenuate some of the comorbid effects of cancer treatments. This is especially pertinent as the presence of microbial factors within the tumor microenvironment may alter the chemical structure of therapeutic agents, thereby undermining their functioning.

In contrast to the effects of enhancing the presence of specific microorganisms, depleting the gut microbiome by antibiotics can exacerbate cancer cell proliferation. In a mouse model of lung cancer, antibiotic treatment interfered with the effects of the therapy that could be partially attenuated by *Lactobacillus* (Gui et al., 2015). It was similarly observed that a cocktail of 11 specific bacterial strains promoted immune activity and could inhibit cancer progression in germ-free mice that had been genetically engineered to be vulnerable to melanoma (Li et al., 2019).

The effects of antibiotics on cancer occurrence were also documented in humans. In patients who received gram-positive antibiotic treatments, progression-free survival was diminished in response to cyclophosphamide for chronic lymphocytic leukemia (CLL) as well as cisplatin for relapsed lymphoma (Pflug et al., 2016). With the use of an antibiotic for over 6 months, cancer occurrence was elevated in the ascending colon by about 17%, often appearing within 5–10 years. A small increase in cancer risk was evident even after a single course of antibiotic use (Lu et al., 2021). Although antibiotics can enhance cancer within diverse tissues, it is somewhat unexpected that antibiotic use increased colon cancer but reduced rectal cancer (Zhang et al., 2019). The source for these differences is unclear but may be related to the nature and distribution of specific microbes throughout the gut.

Eubiosis and dysbiosis

Despite chemotherapy being a mainstay of cancer therapy, it can facilitate pronounced dysbiosis within the microbial community (Alexander et al., 2017), which can then feedback to influence the response to the chemotherapeutic agent. Relatively persistent gut microbiota disturbances can occur in association with cancer and chemotherapy that can disturb immune functioning, hence worsening disease progression. Likewise, some of the side effects of chemo- and radiotherapy, such as fatigue, were aligned with microbial changes that developed over time.

The presence of specific bacteria, notably *Mycoplasma hyorhinis*, *Fusobacterium nucleatum*, and *Gammaproteobacteria*, may undermine the efficacy of cancer treatments or promote resistance to therapy, including gemcitabine treatment of colon cancer (Geller et al., 2017). In line with these findings, the efficacy of the chemotherapeutic agent cisplatin was linked to microbiota that influenced reactive oxygen species (Iida et al., 2013). As described in Fig. 14.3, a healthy microbial state (eubiosis) can be established by diverse lifestyles, but in the face of dysbiosis varied diseases can emerge. As we will see, dysbiosis can ostensibly be attenuated by varied manipulations that involve pre- and probiotics and could thus be used as adjuvant cancer therapies, but there is debate as to whether such procedures are uniformly effective, particularly if the specific microbiota disturbances present have not been identified or acted upon.

Microbial processes may contribute to complications related to cancer or its treatment, potentially affecting bacterial infection that may occur during chemotherapy and can modulate

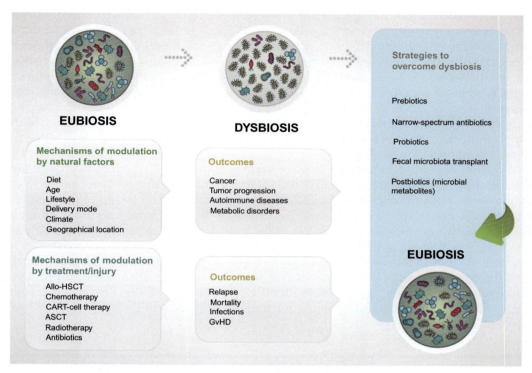

FIG. 14.3 Mechanisms of gut microbiota modulation, consequences, and potential solutions. A eubiotic state, in which a healthy microbial state is present, is affected by a variety of lifestyles. When this state is disturbed so that dysbiosis occurs, the risk of diseases is markedly increased (upper panels). Unfortunately, various cancer treatments can disturb gut homeostasis (eubiosis) increasing the likelihood of cancer relapse and mortality (lower panels). It is believed that methods can be adopted (right-hand side of the figure) to recreate a eubiotic state and could have beneficial actions in relation to cancer occurrence and its treatment. *From Uribe-Herranz, M., Klein-Gonzalez, N., Rodriguez-Lobato, L.G., Juan, M., de Larrea, C.F., 2021. Gut microbiota influence in hematological malignancies: from genesis to cure. Int. J. Mol. Sci. 22.*

the toxic side effects of therapeutic agents (Roy and Trinchieri, 2017). In the latter regard, microbial factors influence the anticancer actions of cyclophosphamide and oxaliplatin by modulating T cell functioning or by promoting the production of reactive oxygen species. Microbiota present within the tumor microenvironment could likewise affect treatments to eliminate cancer depending on the specific bacteria present and could also influence resistance to treatment (Geller et al., 2017). Being able to identify and target specific microbiota could potentially limit side effects linked to gastrointestinal disturbances, thereby allowing for sustained treatment and hence better outcomes of therapy.

Numerous processes associated with microbial variations can moderate the influence of chemotherapies, including those related to translocation, immunomodulation, metabolism, enzymatic degradation, reduced diversity and ecological variation (Alexander et al., 2017), and immune- and inflammatory-related processes within the tumor microenvironment (Iida et al., 2013). Certain microbes are also able to change the characteristics of cancer cells, such as modifying their receptors, influencing the activation of immune checkpoints, and altering immunosuppressive actions (Thiele Orberg et al., 2017). Additionally, some bacteria or their by-products can neutralize chemotherapeutic agents by altering their chemical structure or by affecting their metabolism. An analysis of 30 drugs used in cancer treatments revealed that certain bacteria enhanced the efficacy of 6 of these compounds, whereas the effectiveness of 10 other drugs was diminished by specific microbiota (Lehouritis et al., 2015). A broader investigation that included analysis of 76 human gut bacteria in the metabolism of 271 orally administered drugs indicated that many of these compounds were subject to chemically produced modifications by the microbes that were present (Zimmermann et al., 2019).

Microbiota can be manipulated through various approaches, which can enhance innate immune activity, immune cell cytotoxicity, gut barrier functioning, and prevention of colonization with harmful microbes. As such, these approaches could be implemented to augment the effects of anticancer treatments. However, the vast interindividual differences that exist within microbial communities make it difficult to determine in whom a particular drug's actions would be diminished. By assessing communities of microbes, it may be possible to determine in whom certain agents may have their best effects. Although we have emphasized the importance of the diversity of bacterial communities for cancer treatment efficacy, it may be possible to identify core microorganisms linked to therapeutic effectiveness and which could thereby contribute to an individualized approach to dealing with cancer.

Cancer-related pain and microbiota

Cancer-related pain or that produced by treatments (e.g., neuropathic pain stemming from chemotherapy) may be a considerable problem associated with cancer management. In general, gut microbiota can affect pain processes by directly or indirectly influencing primary nociceptive neurons and by affecting inflammation associated with CNS processes. Thus, microbiota alterations produced through diet and pharmabiotic manipulations could influence chronic pain conditions. Specifically, by altering microbiota, a diet low in saturated fat and added sugars can diminish oxidative stress, activation of Toll-like receptors that promote inflammatory responses and glial activation, and could potentially limit the influence of peripheral inflammation on brain processes, thereby diminishing cancer-related chronic pain (Nijs et al., 2020).

Studies in mice indicated that inflammatory cytokine and microbiota changes may be responsible for neuropathic pain provoked by some chemotherapeutic agents, such as paclitaxel and oxaliplatin. Accordingly, it may be possible to diminish pain related to chemotherapy (e.g., peripheral neuropathy) through selective or broad modulation of microbiota (Zhong et al., 2019). For example, elevated pain sensitivity produced by oxaliplatin was attenuated in germ-free mice or following microbiota depletion by an antibiotic and could be reinduced by microbial restoration (Shen et al., 2017).

Appreciation of the ties between different lifestyle factors and microbiota that could influence pain perception might encourage more attention to the serious problem of pain associated with cancer. This might give rise to randomized controlled trials that determine which specific lifestyle manipulations are most effective for pain management.

Summary and conclusions

Surgery, chemotherapy, and radiotherapy have been the primary ways of dealing with cancers, varying with the nature of the cancer and the stage at which it was discovered. Increasingly more potent chemotherapeutic agents have been developed and sophisticated combination therapies have led to improved outcomes. Even cancers that had at one time been seen as untreatable, have become highly manageable. This said, numerous obstacles have prevented these treatments from being more effective. In many cases, the side effects produced are sufficiently intense so that therapies need to be curtailed. Indeed, therapies could produce damage to healthy organs that lead to early mortality. Furthermore, while many cancers can be detected early so that they can be treated effectively, several forms of cancer cannot readily be detected at an early stage, making these procedures less effective. But even when the therapy initially appears to be effective, resistance to treatment may develop.

Although a wide assortment of tactics were developed to get around roadblocks to efforts that limited adverse effects but still made treatments effective, many of the problems inherent in cancer therapies persist. Aside from approaches that focused on the development of pharmacological agents to enhance cancer therapies, preclude treatment resistance, and diminish side effects, increasingly greater attention has focused on modification of lifestyles even though eating, physical abilities, and sleep are severely undermined by the illness itself and by the therapies used. Recent years have seen increasing attention being dedicated to manipulations of eating, energy processes, and microbiota as part of the effort to provide optimum benefits of chemotherapy and radiotherapy. To an appreciable extent, these approaches have been effective and have improved patient quality of life. Still, the limits of standard therapeutic strategies encouraged alternative approaches to deal with cancer and many of these novel therapies may gain from the endorsement of varied lifestyles.

References

Abid, M.B., Koh, C.J., 2019. Probiotics in health and disease: fooling Mother Nature? Infection 47, 911–917.

Alexander, J.L., Wilson, I.D., Teare, J., Marchesi, J.R., Nicholson, J.K., et al., 2017. Gut microbiota modulation of chemotherapy efficacy and toxicity. Nat. Rev. Gastroenterol. Hepatol. 14, 356–365.

Alicea, G.M., Rebecca, V.W., Goldman, A.R., Fane, M.E., Douglass, S.M., et al., 2020. Changes in aged fibroblast lipid metabolism induce age-dependent melanoma cell resistance to targeted therapy via the fatty acid transporter FATP2. Cancer Discov. 10, 1282–1295.

Amin, S., Lux, A., O'Callaghan, F., 2019. The journey of metformin from glycaemic control to mTOR inhibition and the suppression of tumour growth. Br. J. Clin. Pharmacol. 85, 37–46.

Arends, J., Bachmann, P., Baracos, V., Barthelemy, N., Bertz, H., et al., 2017. ESPEN guidelines on nutrition in cancer patients. Clin. Nutr. 36, 11–48.

Bashiardes, S., Tuganbaev, T., Federici, S., Elinav, E., 2017. The microbiome in anti-cancer therapy. Semin. Immunol. 32, 74–81.

Bhagat, A., Kleinerman, E.S., 2020. Anthracycline-induced cardiotoxicity: causes, mechanisms, and prevention. Adv. Exp. Med. Biol. 1257, 181–192.

Björn, N., Badam, T.V.S., Spalinskas, R., Branden, E., Koyi, H., et al., 2020. Whole-genome sequencing and gene network modules predict gemcitabine/carboplatin-induced myelosuppression in non-small cell lung cancer patients. NPJ Syst. Biol. Appl. 6, 25.

Bonekamp, N.A., Peter, B., Hillen, H.S., Felser, A., Bergbrede, T., 2020. Small-molecule inhibitors of human mitochondrial DNA transcription. Nature 588, 712–716.

Caffa, I., Spagnolo, V., Vernieri, C., Valdemarin, F., Becherini, P., et al., 2020. Fasting-mimicking diet and hormone therapy induce breast cancer regression. Nature 583, 620–624.

Carotenuto, P., Pecoraro, A., Palma, G., Russo, G., Russo, A., 2019. Therapeutic approaches targeting nucleolus in cancer. Cells 8, 1090.

Chen, Y., Ling, L., Su, G., Han, M., Fan, X., et al., 2016. Effect of intermittent versus chronic calorie restriction on tumor incidence: a systematic review and meta-analysis of animal studies. Sci. Rep. 6, 33739.

Chmurska, A., Matczak, K., Marczak, A., 2021. Two faces of autophagy in the struggle against cancer. Int. J. Mol. Sci. 22, 2981.

Choueiri, T.K., Kaelin Jr., W.G., 2020. Targeting the HIF2-VEGF axis in renal cell carcinoma. Nat. Med. 26, 1519–1530.

Cipponi, A., Goode, D.L., Bedo, J., McCabe, M.J., Pajic, M., et al., 2020. MTOR signaling orchestrates stress-induced mutagenesis, facilitating adaptive evolution in cancer. Science 368, 1127–1131.

Crosby, L., Davis, B., Joshi, S., Jardine, M., Paul, J., et al., 2021. Ketogenic diets and chronic disease: weighing the benefits against the risks. Front. Nutr. 8, 702802.

Crudden, C., Shibano, T., Song, D., Dragomir, M.P., Cismas, S., et al., 2021. Inhibition of G protein–coupled receptor kinase 2 promotes unbiased downregulation of igf1 receptor and restrains malignant cell growth. Cancer Res. 81, 501–514.

de Groot, S., Lugtenberg, R.T., Cohen, D., Welters, M.J.P., Ehsan, I., et al., 2020. Fasting mimicking diet as an adjunct to neoadjuvant chemotherapy for breast cancer in the multicentre randomized phase 2 DIRECT trial. Nat. Commun. 11, 3083.

Di Tano, M., Raucci, F., Vernieri, C., Caffa, I., Buono, R., et al., 2020. Synergistic effect of fasting-mimicking diet and vitamin C against KRAS mutated cancers. Nat. Commun. 11, 2332.

Duy, C., Li, M., Teater, M., Meydan, C., Garrett-Bakelman, F.E., et al., 2021. Chemotherapy induces senescence-like resilient cells capable of initiating AML recurrence. Cancer Discov. 11, 1542–1561.

Enriquez-Navas, P.M., Kam, Y., Das, T., Hassan, S., Silva, A., et al., 2016. Exploiting evolutionary principles to prolong tumor control in preclinical models of breast cancer. Sci. Transl. Med. 8, 327ra324.

Fan, P., Zhang, Y., Liu, L., Zhao, Z., Yin, Y., et al., 2016. Continuous exposure of pancreatic cancer cells to dietary bioactive agents does not induce drug resistance unlike chemotherapy. Cell Death Dis. 7, e2246.

Feng, Y., Pathria, G., Heynen-Genel, S., Jackson, M., James, B., et al., 2021. Identification and characterization of imd-0354 as a glutamine carrier protein inhibitor in melanoma. Mol. Cancer Ther. 20, 816–832.

Gao, X., Sanderson, S.M., Dai, Z., Reid, M.A., Cooper, D.E., et al., 2019. Dietary methionine influences therapy in mouse cancer models and alters human metabolism. Nature 572, 397–401.

Gatenby, R.A., Brown, J.S., 2020. The evolution and ecology of resistance in cancer therapy. Cold Spring Harb. Perspect. Med. 10, a040972.

Gatenby, R.A., Zhang, J., Brown, J.S., 2019. First strike-second strike strategies in metastatic cancer: lessons from the evolutionary dynamics of extinction. Cancer Res. 79, 3174–3177.

Geeraerts, S.L., Kampen, K.R., Rinaldi, G., Gupta, P., Planque, M., et al., 2021. Repurposing the antidepressant sertraline as shmt inhibitor to suppress serine/glycine synthesis–addicted breast tumor growth. Mol. Cancer Ther. 20, 50–63.

Geller, L.T., Barzily-Rokni, M., Danino, T., Jonas, O.H., Shental, N., et al., 2017. Potential role of intratumor bacteria in mediating tumor resistance to the chemotherapeutic drug gemcitabine. Science 357, 1156–1160.

Geraghty, A.C., Gibson, E.M., Ghanem, R.A., Greene, J.J., Ocampo, A., et al., 2019. Loss of adaptive myelination contributes to methotrexate chemotherapy-related cognitive impairment. Neuron 103, 250–265 e258.

Gibson, N.M., Greufe, S.E., Hydock, D.S., Hayward, R., 2013. Doxorubicin-induced vascular dysfunction and its attenuation by exercise preconditioning. J. Cardiovasc. Pharmacol. 62, 355–360.

Godet, I., Shin, Y.J., Ju, J.A., Ye, I.C., Wang, G., et al., 2019. Fate-mapping post-hypoxic tumor cells reveals a ROS-resistant phenotype that promotes metastasis. Nat. Commun. 10, 4862.

Goldberg, E.L., Shchukina, I., Asher, J.L., Sidorov, S., Artyomov, M.N., et al., 2020. Ketogenesis activates metabolically protective gammadelta T cells in visceral adipose tissue. Nat. Metab. 2, 50–61.

Gui, Q.F., Lu, H.F., Zhang, C.X., Xu, Z.R., Yang, Y.H., 2015. Well-balanced commensal microbiota contributes to anti-cancer response in a lung cancer mouse model. Genet. Mol. Res. 14, 5642–5651.

Haikala, H.M., Anttila, J.M., Marques, E., Raatikainen, T., Ilander, M., et al., 2019. Pharmacological reactivation of MYC-dependent apoptosis induces susceptibility to anti-PD-1 immunotherapy. Nat. Commun. 10, 620.

Hata, A.N., Niederst, M.J., Archibald, H.L., Gomez-Caraballo, M., Siddiqui, F.M., et al., 2016. Tumor cells can follow distinct evolutionary paths to become resistant to epidermal growth factor receptor inhibition. Nat. Med. 22, 262–269.

Herrera, F.G., Ronet, C., Ochoa de Olza, M., Barras, D., Crespo, I., et al., 2021. Low dose radiotherapy reverses tumor immune desertification and resistance to immunotherapy. Cancer Discov. https://doi.org/10.1158/2159-8290. In press.

Hopkins, B.D., Pauli, C., Du, X., Wang, D.G., Li, X., et al., 2018. Suppression of insulin feedback enhances the efficacy of PI3K inhibitors. Nature 560, 499–503.

Hsu, C.H., Altschuler, S.J., Wu, L.F., 2019. Patterns of early p21 dynamics determine proliferation-senescence cell fate after chemotherapy. Cell 178, 361–373 e312.

Iida, N., Dzutsev, A., Stewart, C.A., Smith, L., Bouladoux, N., et al., 2013. Commensal bacteria control cancer response to therapy by modulating the tumor microenvironment. Science 342, 967–970.

Jonas, O., Landry, H.M., Fuller, J.E., Santini Jr., J.T., Baselga, J., et al., 2015. An implantable microdevice to perform high-throughput in vivo drug sensitivity testing in tumors. Sci. Transl. Med. 7, 284ra257.

Kanarek, N., Keys, H.R., Cantor, J.R., Lewis, C.A., Chan, S.H., et al., 2018. Histidine catabolism is a major determinant of methotrexate sensitivity. Nature 559, 632–636.

Kanarek, N., Petrova, B., Sabatini, D.M., 2020. Dietary modifications for enhanced cancer therapy. Nature 579, 507–517.

Karthika, C., Sureshkumar, R., 2019. Can curcumin along with chemotherapeutic drug and lipid provide an effective treatment of metastatic colon cancer and alter multidrug resistance? Med. Hypotheses 132, 109325.

Kaur, A., Webster, M.R., Marchbank, K., Behera, R., Ndoye, A., et al., 2016. sFRP2 in the aged microenvironment drives melanoma metastasis and therapy resistance. Nature 532, 250–254.

Ke, W., Saba, J.A., Yao, C.H., Hilzendeger, M.A., Drangowska-Way, A., et al., 2020. Dietary serine-microbiota interaction enhances chemotherapeutic toxicity without altering drug conversion. Nat. Commun. 11, 2587.

Kerbel, R., Folkman, J., 2002. Clinical translation of angiogenesis inhibitors. Nat. Rev. Cancer 2, 727–739.

Koelwyn, G.J., Newman, A.A.C., Afonso, M.S., van Solingen, C., Corr, E.M., et al., 2020. Myocardial infarction accelerates breast cancer via innate immune reprogramming. Nat. Med. 26, 1452–1458.

Kovács, T., Mikó, E., Ujlaki, G., Sári, Z., Bai, P., 2020. The microbiome as a component of the tumor microenvironment. Adv. Exp. Med. Biol. 1225, 137–153.

Lehouritis, P., Cummins, J., Stanton, M., Murphy, C.T., McCarthy, F.O., et al., 2015. Local bacteria affect the efficacy of chemotherapeutic drugs. Sci. Rep. 5, 14554.

Lennon, H., Sperrin, M., Badrick, E., Renehan, A.G., 2016. The obesity paradox in cancer: a review. Curr. Oncol. Rep. 18, 56.

Leone, R.D., Zhao, L., Englert, J.M., Sun, I.M., Oh, M.H., et al., 2019. Glutamine blockade induces divergent metabolic programs to overcome tumor immune evasion. Science 366, 1013–1021.

Li, Q., Wu, T., Jing, L., Li, M.J., Tian, T., et al., 2017. Angiogenesis inhibitors for the treatment of small cell lung cancer (SCLC): a meta-analysis of 7 randomized controlled trials. Medicine (Baltimore) 96, e6412.

Li, Y., Tinoco, R., Elmen, L., Segota, I., Xian, Y., et al., 2019. Gut microbiota dependent anti-tumor immunity restricts melanoma growth in Rnf5(−/−) mice. Nat. Commun. 10, 1492.

Li, Y., Elmen, L., Segota, I., Xian, Y., Tinoco, R., et al., 2020. Prebiotic-induced anti-tumor immunity attenuates tumor growth. Cell Rep. 30, 1753–1766 e1756.

Lu, S.S.M., Mohammed, Z., Häggström, C., Myte, R., Lindquist, E., et al., 2021. Antibiotics use and subsequent risk of colorectal cancer: a Swedish nationwide population-based study. J. Natl. Cancer Inst., Djab125.

Maddocks, O.D.K., Athineos, D., Cheung, E.C., Lee, P., Zhang, T., et al., 2017. Modulating the therapeutic response of tumours to dietary serine and glycine starvation. Nature 544, 372–376.

Montay-Gruel, P., Acharya, M.M., Jorge, P.G., Petit, B., Petridis, I.G., et al., 2021. Hypofractionated FLASH-RT as an effective treatment against glioblastoma that reduces neurocognitive side effects in mice. Clin. Cancer Res. 27, 775–784.

Mukherjee, S., 2011. The Emperor of All Maladies: A Biography of Cancer. Simon and Schuster, NY.

Mukherjee, P., Augur, Z.M., Li, M., Hill, C., Greenwood, B., et al., 2019. Therapeutic benefit of combining calorie-restricted ketogenic diet and glutamine targeting in late-stage experimental glioblastoma. Commun. Biol. 2, 200.

Nencioni, A., Caffa, I., Cortellino, S., Longo, V.D., 2018. Fasting and cancer: molecular mechanisms and clinical application. Nat. Rev. Cancer 18, 707–719.

Nijs, J., Tumkaya Yilmaz, S., Elma, O., Tatta, J., Mullie, P., et al., 2020. Nutritional intervention in chronic pain: an innovative way of targeting central nervous system sensitization? Expert Opin. Ther. Targets 24, 793–803.

Paquet, L., Collins, B., Song, X., Chinneck, A., Bedard, M., et al., 2013. A pilot study of prospective memory functioning in early breast cancer survivors. Breast 22, 455–461.

Patel, P., Malipatlolla, D.K., Devarakonda, S., Bull, C., Rascon, A., et al., 2020. Dietary oat bran reduces systemic inflammation in mice subjected to pelvic irradiation. Nutrients 12, 2172.

Pflug, N., Kluth, S., Vehreschild, J.J., Bahlo, J., Tacke, D., et al., 2016. Efficacy of antineoplastic treatment is associated with the use of antibiotics that modulate intestinal microbiota. Oncoimmunology 5, e1150399.

Ray Chaudhuri, A., Callen, E., Ding, X., Gogola, E., Duarte, A.A., et al., 2016. Replication fork stability confers chemoresistance in BRCA-deficient cells. Nature 535, 382–387.

Roy, S., Trinchieri, G., 2017. Microbiota: a key orchestrator of cancer therapy. Nat. Rev. Cancer 17, 271–285.

Sacks, C.A., Miller, P.W., Longo, D.L., 2019. Talking about toxicity—"what we've got here is a failure to communicate". NEJM 381, 1406–1408.

Sepich-Poore, G.D., Zitvogel, L., Straussman, R., Hasty, J., Wargo, J.A., Knight, R., 2021. The microbiome and human cancer. Science 371, eabc455.

Sharanek, A., Burban, A., Hernandez-Corchado, A., Madrigal, A., Fatakdawala, I., et al., 2021. Transcriptional control of brain tumor stem cells by a carbohydrate binding protein. Cell Rep. 36, 109647.

Shen, S., Lim, G., You, Z., Ding, W., Huang, P., et al., 2017. Gut microbiota is critical for the induction of chemotherapy-induced pain. Nat. Neurosci. 20, 1213–1216.

Sitlinger, A., Brander, D.M., Bartlett, D.B., 2020. Impact of exercise on the immune system and outcomes in hematologic malignancies. Blood Adv. 4, 1801–1811.

Smith, A.K., Conneely, K.N., Pace, T.W., Mister, D., Felger, J.C., 2014. Epigenetic changes associated with inflammation in breast cancer patients treated with chemotherapy. Brain Behav. Immun. 38, 227–236.

Stanková, K., Brown, J.S., Dalton, W.S., Gatenby, R.A., 2019. Optimizing cancer treatment using game theory: a review. JAMA Oncol. 5, 96–103.

Sun, G., Ding, X.A., Argaw, Y., Guo, X., Montell, D.J., 2020. Akt1 and dCIZ1 promote cell survival from apoptotic caspase activation during regeneration and oncogenic overgrowth. Nat. Commun. 11, 5726.

Tajan, M., Hennequart, M., Cheung, E.C., Zani, F., Hock, A.K., et al., 2021. Serine synthesis pathway inhibition cooperates with dietary serine and glycine limitation for cancer therapy. Nat. Commun. 12, 366.

Thiele Orberg, E., Fan, H., Tam, A.J., Dejea, C.M., Destefano Shields, C.E., et al., 2017. The myeloid immune signature of enterotoxigenic *Bacteroides fragilis*-induced murine colon tumorigenesis. Mucosal Immunol. 10, 421–433.

Tran, T.Q., Hanse, E.A., Habowski, A.N., Li, H., Gabra, M.B.I., et al., 2020. alpha-Ketoglutarate attenuates Wnt signaling and drives differentiation in colorectal cancer. Nat. Cancer 1, 345–358.

Turbitt, W.J., Demark-Wahnfried, W., Peterson, C.M., Norian, L.A., 2019. Targeting glucose metabolism to enhance immunotherapy: emerging evidence on intermittent fasting and calorie restriction mimetics. Front. Immunol. 10, 1402.

Vaidya, J.S., Bulsara, M., Baum, M., Wenz, F., Massarut, S., et al., 2020. Long term survival and local control outcomes from single dose targeted intraoperative radiotherapy during lumpectomy (TARGIT-IORT) for early breast cancer: TARGIT-A randomised clinical trial. BMJ 370, m2836.

Walens, A., DiMarco, A.V., Lupo, R., Kroger, B.R., Damrauer, J.S., et al., 2019. CCL5 promotes breast cancer recurrence through macrophage recruitment in residual tumors. eLife 8.

Wanders, D., Hobson, K., Ji, X., 2020. Methionine restriction and cancer biology. Nutrients 12, 684.

Wang, H., Gao, Z., Liu, X., Agarwal, P., Zhao, S., et al., 2018. Targeted production of reactive oxygen species in mitochondria to overcome cancer drug resistance. Nat. Commun. 9, 562.

Weber, D.D., Aminzadeh-Gohari, S., Tulipan, J., Catalano, L., Feichtinger, R.G., et al., 2020. Ketogenic diet in the treatment of cancer—where do we stand? Mol. Metab. 33, 102–121.

Wedret, J.J., Tu, T.G., Paul, D., Rousseau, C., Bonta, A., Bota, R.G., 2019. Interactions between antidepressants, sleep aids and selected breast cancer therapy. Ment. Illn. 11, 8115.

Xiao, C., Miller, A.H., Peng, G., Levine, M.E., Conneely, K.N., et al., 2021. Association of epigenetic age acceleration with risk factors, survival, and quality of life in patients with head and neck cancer. Int. J. Radiat. Oncol. Biol. Phys. 111, 157–167.

Yong, T., Zhang, X., Bie, N., Zhang, H., Zhang, X., et al., 2019. Tumor exosome-based nanoparticles are efficient drug carriers for chemotherapy. Nat. Commun. 10, 3838.

Zhang, J., Cunningham, J.J., Brown, J.S., Gatenby, R.A., 2017. Integrating evolutionary dynamics into treatment of metastatic castrate-resistant prostate cancer. Nat. Commun. 8, 1816.

Zhang, J., Haines, C., Watson, A.J.M., Hart, A.R., Platt, M.J., et al., 2019. Oral antibiotic use and risk of colorectal cancer in the United Kingdom, 1989-2012: a matched case-control study. Gut 68, 1971–1978.

Zhang, H., Li, Z., Gao, C., Fan, X., Pang, Y., et al., 2021. Dual-responsive biohybrid neutrobots for active target delivery. Sci. Robot. 6.

Zhong, S., Zhou, Z., Liang, Y., Cheng, X., Li, Y., et al., 2019. Targeting strategies for chemotherapy-induced peripheral neuropathy: does gut microbiota play a role? Crit. Rev. Microbiol. 45, 369–393.

Zimmermann, M., Zimmermann-Kogadeeva, M., Wegmann, R., Goodman, A.L., 2019. Mapping human microbiome drug metabolism by gut bacteria and their genes. Nature 570, 462–467.

Zitvogel, L., Daillere, R., Roberti, M.P., Routy, B., Kroemer, G., 2017. Anticancer effects of the microbiome and its products. Nat. Rev. Microbiol. 15, 465–478.

CHAPTER 15

Immunotherapies and their moderation

OUTLINE

Immunotherapeutic approaches	462
Stem cell therapy	464
Induced pluripotent stem cells (iPSCs)	464
Graft-vs-host disease	465
Microbial factors in stem cell therapy	465
Nonspecific immunotherapies	467
Cytokine-based therapy	468
Role of indoleamine-2,3-dioxygenase in cancer processes	469
Treatment vaccines	471
Virally related cancer	471
Cancer vaccines	472
Oncolytic viral therapy	472
Antibody-based therapies	474
Checkpoint inhibitors	475
CAR T-cell therapy	478
Broader use of CAR T therapy	479
Enhancing immunotherapy's effectiveness	479
Combination therapies	480
Moderating the actions of checkpoint inhibitors through lifestyles	481
Side effects of checkpoint inhibitors and CAR T therapy	482
Modifying the side effects of CAR T therapy	485
Use of biomarkers	485
Genetic characteristics related to the individual (systemic host factors)	486
Markers that reflect features of the tumor	486
Features of the tumor microenvironment	488
Biomarkers for immunotherapy side effects	488
Influence of microbiota on treatment responses to cancer therapies	489
Microbiota and immunotherapy	489
Prebiotics and probiotics in cancer treatment	495
Summary and conclusions	497
References	497

Immunotherapeutic approaches

Immunotherapy is often considered a relatively new approach to tackle diseases, dating back about two decades. But its advent can arguably be attributed to approaches used centuries ago to prevent the development of smallpox. Fluid or dried scab from smallpox pustules was rubbed into scratches that had been made on an individual's skin, which resulted in the development of immunity to the virus. This procedure, dubbed "variolation," produced a mild form of the disease, thereby protecting the person from subsequently developing a more severe illness that was deadly in about 30% of infected people. Unfortunately, in some instances, severe illness developed and led to death in about 1%–2% of those people. Moreover, even mildly infected individuals could spread the disease to others, not unlike asymptomatic COVID-19 infected people can do so today. This forerunner of inoculation and vaccinology, as crude and often counterproductive as it was, had indicated that through some biological process the procedure activated factors that fought off illness.

Much later it was noted that bacterial infections could result in tumor regression, and in the 1890s it was observed that improvements in cancer patients occurred if they developed an infectious illness, such as influenza. This occurred for reasons entirely different from those associated with variolation. But having noticed this relationship, William Coley injected his patients with cultures of streptococcal bacteria (including directly into the tumor), which led to tumor shrinkage in some patients. The thinking had been that this outcome stemmed from the infection causing immune cell production that turned against the cancer cells. Coley's toxin, as it came to be known, was used to treat hundreds of patients. Although this treatment protocol went by the wayside, having been supplanted by chemotherapy, it influenced some forms of immunotherapies as we now know them. With an increased understanding of how the immune system operates, further strategies evolved to enhance immune functioning, culminating in an array of techniques that could deal with numerous forms of cancer.

While chemotherapy and radiotherapy are aimed at directly destroying cancer cells, the objective of immunotherapy has been that of enhancing immune processes that could attack and destroy these rogue cells or eliminate the defenses of cancer cells so that they could more readily be dispatched. Several forms of immunotherapy have been developed to treat a variety of cancers (see Fig. 15.1), with the positive effects of immunotherapies leading to the belief that an entirely new chapter of cancer treatments had been reached, despite these approaches benefiting only a minority of patients. Still, the promise for still more effective treatment coming about has encouraged further efforts to develop variations of these procedures.

To reach these goals, it may be important to consider multiple elements that contribute to immune functioning, including the influence of intrinsic and numerous extrinsic factors (e.g., features of the host as well as both experiential and environmental factors). These may be combined so that a "set point" is reached wherein anticancer actions exceed pro-cancer factors, thereby enhancing the response to immunotherapies (Chen and Mellman, 2017). Defining this set point requires complex computational formulae that include genetic (both DNA and RNA) and epigenetic data, proteomics, and earlier clinical findings regarding the conditions where treatments were or were not effective. To an extent, this is currently impractical. Therefore, immunotherapy may be reserved for patients with certain forms of cancer that had not responded positively to other treatments, patients with biomarkers that are directly relevant to the actions of the immunotherapy (e.g., being positive for the presence of

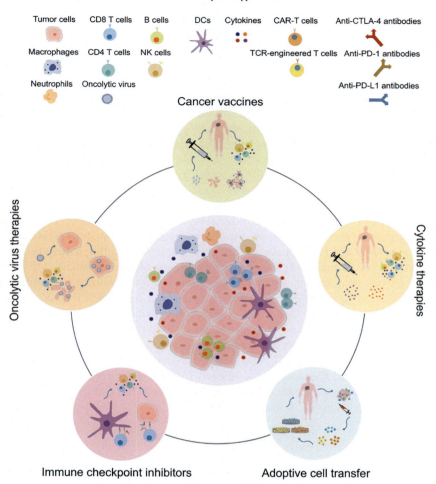

FIG. 15.1 The major categories of immunotherapy. Different forms of cancer immunotherapy, including oncolytic virus therapies, cancer vaccines, cytokine therapies, adoptive cell transfer, and immune checkpoint inhibitors, have evolved and shown promise in clinical practice. The basic principles of each strategy and the corresponding cellular and molecular underpinnings involved in each step are depicted. *CAT-T*, chimeric antigen receptor T-cell; *DCs*, dendritic cells; *NK*, natural killer; *TCR*, T-cell receptor. *Figure and caption from Zhang, Y., Zhang, Z., 2020. The history and advances in cancer immunotherapy: understanding the characteristics of tumor-infiltrating immune cells and their therapeutic implications. Cell. Mol. Immunol. 17, 807–821.*

certain checkpoints), and those with tumors that are high in genetic hypermutability (i.e., a predisposition to mutations). Of particular significance, there has been a push to be better able to predict who would best be served by these approaches (e.g., through the identification of specific genetic or epigenetic biomarkers). In this chapter, we will introduce some of the common immunotherapeutic approaches that have been adopted. Our purpose is not to provide a detailed review of these procedures but instead to point to some of the variables, particularly lifestyles (primarily diet and microbiota) that can affect the efficacy of these therapies.

Stem cell therapy

Stem cells comprise undifferentiated cells that could divide and differentiate to become one of the many cell types that make up the human body, including critical immune cells. However, mutations within stem cells may contribute to the development of cancers, relapse, and metastasis, as well as multidrug resistance. Stem cell therapies are meant to bolster the immune arsenal to deal with cancers and can augment the actions of immune-based therapies. As such, it certainly is not a "traditional therapy" and might fit more comfortably as an immunotherapy, but technically it hasn't been considered as an immunotherapy. Nonetheless, it is included in this chapter since stem cells serve as the progenitor cellular system that will strengthen efforts by the immune system to attack and destroy cancer cells.

Stem cell transplants are performed when endogenous stem cells have not been functioning as they should, as in the case of cells that are damaged in some forms of cancer. The objective of hematopoietic stem cell therapy is to eliminate unhealthy blood-forming cells and replace them with those that are healthy. To this end, healthy stem cells can be obtained from the bone marrow of a healthy donor, but embryonic stem cells can also be obtained from umbilical cord blood. A small number of stem cells are also present in adult blood and fat tissue. Stem cell transplantation comprises the patient first receiving high-dose chemotherapy and/or radiation to eliminate the unhealthy cells, after which the stem cells obtained from a matching bone marrow donor are transfused into the patient. Stem cells can also be obtained from the blood of a healthy matching donor who had received treatment (filgrastim) to increase the production of stem cells. The transplanted stem cells reach the bone marrow and then produce healthy blood cells so that the restored immune cells will effectively attack the cancer cells. Stem cell therapies have shown promise in a variety of cancer types, such as small-cell lung cancer, colon cancer, and multiple myeloma, but they have primarily been used to treat leukemia and lymphoma.

Induced pluripotent stem cells (iPSCs)

The use of embryonic stem cells was not widely adopted, largely owing to the controversy surrounding the harvesting of these cells from unused embryos following in vitro fertilization procedures, which resulted in the destruction of a human embryo. Thus, a form of stem cell therapy came about that entailed the use of induced pluripotent stem cells (iPSCs) (Takahashi and Yamanaka, 2006).[a] This involves the use of adult somatic cells that were genetically reprogrammed, converting them into an embryonic-like state, and so can form any type of cell and keep dividing. In this way, they can form the basis of therapy that could potentially attack a variety of cancers. As promising as it might be, this approach can produce immune complications, especially mutations that occur with continuous cell multiplication. Furthermore, mutations

---

[a] The 2012 Nobel Prize in Physiology and Medicine was jointly awarded to John B. Gurdon and Shinya Yamanaka. Their work was fundamental in showing that the dogma of cell maturation being unidirectional wasn't entirely correct and the developmental clock could in fact be turned backward. Mature cells could be manipulated so that they returned to an immature pluripotent stage, after which they could be coaxed to become any type of cell in the human body.

may occur within mitochondrial DNA that lack effective repair mechanisms. In addition, immune responses can be instigated that can act against the regeneration of tissues as intended.

The effectiveness of stem cell therapy could be enhanced by targeting specific cancer stem cells that act in a tumor-initiating capacity. Biomarkers could potentially be identified (e.g., genetic and epigenetic signatures) that are characteristic of particular stem cells fundamental to the cancer process. This might entail the determination of cancer stem cells involved in developmental pathways that serve to maintain stemlike states (e.g., Wnt, Notch, or Hedgehog) or those stem cells that are most often enriched in cancers, such as those with CD44 markers. Other promising markers comprise extracellular factors, including vascular niches, hypoxia, tumor-associated macrophages, and cancer-associated fibroblasts (Yang et al., 2020). Likewise, structural proteins, such as plectin, may also fit the bill. It is typically expressed intracellularly, but its presence on the cell surface might be indicative of tumor invasion and metastasis (Raymond et al., 2019).

Graft-vs-host disease

Following stem cell transplantation, when immune cells are knocked out and new immune cells are replenished, patients may be highly vulnerable to diseases. To avoid bacterial infection, they may be treated with antibiotics that eliminate gut bacteria, including those that may be beneficial. A further difficulty may arise following *allogeneic* transplant in which stem cells come from another person (i.e., Person X cells harvested and given as donor cells to Person Y) as opposed to *autologous* transplants where the cells come from the patient (Person Y cells harvested and returned back to Person Y). The patient and donor are matched based on major tissue types, but they may still differ on other features. As a result, graft-versus-host disease (GvHD) may occur in which the donor cells attack the host leading to damage, particularly of the skin, the gastrointestinal tract, and liver. As described in Fig. 15.2, the tissue damage produced by chemotherapy and/or radiotherapy administered prior to stem cell transplantation causes proinflammatory cytokine release by antigen presenting cells (APCs). The inflammatory signals together with antigen presentation by APCs promote an alloreactive T-cell response. Thus, the donor T cells migrated to the target tissues, including skin, intestine and liver, where inflammatory responses cause cell death. The risk for GvHD depends on the source of the graft (blood versus bone marrow) and the age of the donor. It may comprise a chronic condition that requires the use of immunosuppressive agents, and if it involves the gastrointestinal tract, then particular dietary protocols need to be followed that comprise lots of fluids, small meals, and foods consistent with those that resemble a gluten-free diet.

Microbial factors in stem cell therapy

Microbial factors may contribute to the effects promoted by hematopoietic stem cell transplantation, and there is reason to believe that the loss of bacteria stemming from antibiotic treatments might have contributed to the development of GvHD (Shono and van den Brink, 2018). Patients who received allogeneic transplants displayed marked loss of microbiota diversity and altered levels of some beneficial microbiota, notably *Lactobacillus* (Rashidi et al., 2020). Mortality rates following bone marrow transplants were greater among patients who exhibited the most pronounced loss of microbiota diversity (Peled et al., 2020), and these relations were tied to variations of circulating neutrophils, monocytes, and lymphocytes.

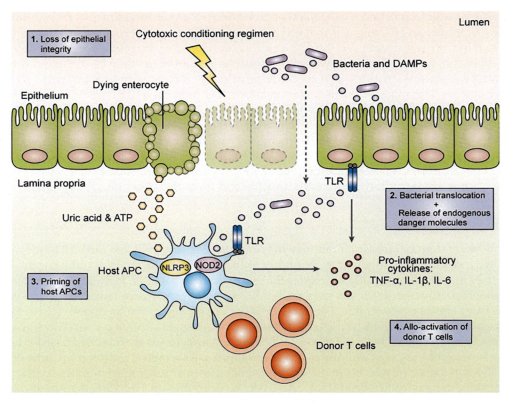

FIG. 15.2 Schematic overview of the initiation phase of acute graft-versus-host disease. During the toxic conditioning regimen with total-body irradiation and/or chemotherapy, the destruction of intestinal epithelial cells leads to the loss of epithelial barrier function. The subsequent translocation of luminal bacteria as well as the release of endogenous danger molecules such as adenosine triphosphate (ATP) and uric acid result in the production of proinflammatory cytokines. Activated host and/or donor antigen-presenting cells then prime alloreactive donor T cells, which perpetuate acute GVHD. *APC*, antigen-presenting cell; *DAMP*, danger-associated molecular pattern; *IL*, interleukin; *NOD2*, nucleotide-binding oligomerization domain; *NLRP3*, NACHT, LRR, and PYD domains-containing protein 3; *TLR*, toll-like receptor; *TNF*, tumor necrosis factor. *Figure and caption from Heidegger, S., van den Brink, M.R., Haas, T., Poeck, H., 2014. The role of pattern-recognition receptors in graft-versus-host disease and graft-versus-leukemia after allogeneic stem cell transplantation. Front. Immunol. 5, 337.*

It has not been examined extensively, but it seems that microbial manipulations can be used to affect allogeneic hematopoietic stem cell transplantation, especially concerning the development of GvHD, as well as posttreatment infection and cancer relapse (Shono and van den Brink, 2018). The adverse effects of stem cell transplantation were exacerbated by treatments that diminished *Lactobacillus*, which could be reversed by the reintroduction of these microbes. Similarly, a combination of *Clostridial* strains that produce high levels of butyrate had positive effects in mice with GvHD (Mathewson et al., 2016). The damaging effects associated with hematopoietic stem cell transplantation could also be diminished by diet changes that favored bacterial diversity and elevated SCFAs, as well as intravenous feeding that included glucose, salts, amino acids, lipids, vitamins, and minerals (total parenteral nutrition).

The occurrence of GvHD could likewise be abated by fecal microbiota transplantation, but more such trials are needed to assure its reliability.

It would be ideal if the gut bacteria of individuals were evaluated and then treatments offered to increase or decrease those microbial components that were out of kilter. Of course, some low abundance microbes can have considerable effects on outcomes, and many unknown bacteria may also be pivotal in affecting the response to stem cell transplants. Ultimately, however, these approaches can prove to be difficult if patients require antibiotics to stem bacterial illnesses such as bacteremia or sepsis.

The use of iPSC-based approaches has gone well beyond that of cancer therapy and could potentially be used in modeling neurodegenerative diseases, such as Alzheimer's and Parkinson's disease, as well as in new drug discovery, drug screening, and toxicity tests. To be sure, there are still many kinks that must be worked out before iPSCs can be broadly applied to multiple diseases, but there is ample reason for optimism.

From hope to hype

"The flame that burns twice as bright, burns half as long." This statement, which has been attributed to the Chinese Philosopher Lao Tzu, is apt in the context of stem cells being used commercially to profit from the misery of others. Numerous rogue clinics have opened across many countries, more than 700 within the United States, to offer stem cell therapies, ostensibly for treatment of diabetes, epilepsy, stroke, spinal cord injuries, brain injury, Parkinson's disease, Alzheimer's disease, liver diseases, multiple sclerosis, autism, and varied cancers. These procedures may amount to taking cells from one part of the body, concentrating them (by centrifugation), and then administering them to another part of the body. The FDA has not authorized stem cell use for these conditions, necessitating oversight to protect sick and distressed people from those who are keen to take advantage of them. Remarkably, an investigation of 166 clinics revealed that fewer than 50% of physicians working there had been adequately trained to deal with the illnesses that they were alleged to be treating (Fu et al., 2019). Despite limits that may be placed on these clinics, patients can find their way to countries that are lax in regulating these clinics, perhaps even welcoming "stem cell tourism."

Nonspecific immunotherapies

Nonspecific immunotherapies comprise treatments that do not target specific cancer cells but instead promote immune activity that could act against tumors. These treatments can be administered on their own or as adjuvants to enhance the influence of other immunotherapeutic approaches. One method of doing so is to promote the functioning of antigen-presenting cells so that B cells will produce antibodies, while T_h cells will promote the recruitment of macrophages to engulf tumor cells.

Based on the view that T_{reg} cells within tumors protect cancer cells by inhibiting the actions of effector T cells, cancer progression can be slowed by inhibiting the actions of some forms of T_{reg} cells. However, this could favor the development of autoimmune disorders. As it happens, the CD36 receptor and high levels of genes involved in lipid uptake and metabolism are present in intratumoral T_{reg} cells so in mice engineered to lack CD36, the tumor burden was reduced. Moreover, by administering CD36 antibodies to affect T_{reg} cells within the tumors in conjunction with immunotherapy, cancer progression was abated (Wang et al., 2020).

In some instances, the actions of T_{reg} cells can be modified by affecting markers that are present on the cell surface. When immune cells lose the surface marker neuropilin-1 (Nrp 1), the integrity of T_{reg} cells is lost, thereby allowing cytotoxic T cells to kill tumor cells. Thus, by inhibiting Nrp 1, the efficacy of immunotherapy may be enhanced while minimizing collateral damage to healthy cells and avoiding the occurrence of autoimmune disorders (Overacre-Delgoffe et al., 2017). Similarly, in the presence of inflammation, the downregulation of a key transcription factor (Helios) may promote the conversion of T_{reg} cells into effector T cells. So not only was the inhibition of effector T cells removed, but newly created T cells joined the battle against the tumor (Nakagawa et al., 2016). While not diminishing the potential efficacy and use of these approaches, several immunotherapies were developed that on face value seemed to have greater potential to deal with diverse cancers, so these approaches took center stage.

Cytokine-based therapy

The administration of cytokines can enhance immune functioning leading to attacks on tumor cells and have thus been used as nonspecific cancer immunotherapies. Therapy that comprised IL-2 infusion was effective in treating cancer in a small percentage of individuals (15%), but because of its short half-life, patients had to be treated with high doses every few hours over several days. Severe side effects developed in response to the therapy, and eventually its use was curtailed in favor of other treatments. However, interest in the potential benefits of IL-2 has reemerged and may yet prove to be an effective therapeutic strategy if its half-life could be extended and the toxic side effects could be diminished.

One of the alternatives to IL-2 therapy involved the use of IFN-α. Virtually all mammalian cells express receptors for type I interferons (e.g., IFN-α, IFN-β, IFN-ε) so the cytokines that act on these receptors can diminish infection stemming from viruses, bacteria, parasites, and fungi, and IFN-α could be used against the progression of some forms of cancer. The type II interferon IFN-γ shares many actions with type I IFNs, serving as an antiviral factor, and concurrently promotes immunoregulatory actions that aid in the prevention of collateral damage associated with excessive immune activation. IFNs, which are produced by diverse types of immune cells, can have dual actions during infections. When present acutely, they can promote antiviral responses, but when they are present chronically, as in the case of persistent infection, harmful inflammatory actions can be engendered (McNab et al., 2015).

Differences in the functioning of IFN-α and IFN-γ exist, and while IFN-γ has been used in cancer therapy, the IFN-α form had primarily been used to this end, including in the treatment of melanoma, multiple myeloma, renal cell cancer, non-Hodgkin lymphoma, and some forms of leukemia. The influence of IFN-α may come about by affecting immune activity (primarily increasing T and NK cell functioning) and the tumor vasculature (Borden, 2019). Owing to serious side effects of IFN-α, interest in cytokine therapy for cancer seemed to linger. A resurgence of interest in cytokine therapies arose with the enhancement of safety and efficacy of these agents, as well as with their augmented half-life and improved targeting of their actions within the tumor microenvironment. Newer IFN-based therapeutics have also been used experimentally in conjunction with other therapeutic agents to yield better outcomes.

Factors complicating cytokine-based immunotherapies: Depression and indoleamine-2,3-dioxygenase (IDO)

Although IFN-α was effective in the treatment of some patients with cancer or hepatitis C, marked side effects were frequently produced, including severe depression (Miller and Raison, 2016) together with multiple other cognitive disturbances. Patients sometimes exhibited concentration and memory impairments, as well as a host of neurocognitive disturbances, such as confusion, disorientation, irritability, anxiety, disturbed vigilance, diminished alertness, and psychotic-like features. Thus, this treatment might provoke a nonspecific state (perhaps reflecting general toxicity) that favors the appearance of multiple symptoms. In a considerable proportion of cases, the severity of these cognitive disturbances was sufficiently robust to necessitate treatment discontinuation (Pape et al., 2019).

The development of depression in response to IFN-α treatment could be predicted based on many of the same features that have been tied to the emergence of depression in other contexts. This included the presence of pretreatment depressive symptoms and relatively marked ACTH and cortisol elevations (Capuron and Miller, 2011). Moreover, stressful experiences may act synergistically with IFN-α therapy in the provocation of the negative actions that emerge (Anisman et al., 2007), which may be especially relevant for cancer patients who face considerable distress.

The depressive actions of IFN-α therapy may come about because it promotes a reduction of brain serotonin levels owing to the degradation of the 5-HT precursor tryptophan. It is more likely, however, that depressive illness develops owing to elevated conversion of tryptophan to kynurenine through the enzyme indoleamine 2,3-dioxygenase (IDO). The formation of oxidative metabolites, 3-hydroxykynurenine and quinolinic acid, promotes neurotoxic actions that can disturb anterior cingulate and prefrontal cortex neural circuits, thereby leading to depression (Miller and Raison, 2016). Concurrently, innate immune functioning involved in the detection of pathogenic microorganisms can influence caspase-1 and related factors which instigate inflammation. Together with elevated activity within oxidative and nitrosative (overproduction of nitric oxide) pathways, DNA damage may be incurred that fosters affective symptoms (Moylan et al., 2014). The inflammation and altered oxygen and nitrosative functioning that promote depression could be exacerbated by psychosocial stressor experiences together with lifestyle factors (e.g., proinflammatory diet, poor sleep, sedentary behaviors, obesity) as well as altered gut functioning (Moylan et al., 2014).

Our focus on depression related to cytokine therapy is not only predicated on this being a serious side effect that limits the use of this therapy. Depression, as we've repeatedly seen, has been associated with the emergence of cancer and may directly or indirectly undermine the efficacy of cancer therapies. Whether providing patients with cognitive therapy or mindfulness training prior to cancer therapy is effective in enhancing cancer outcomes is not certain, but it may preempt or limit the misery of depression that would otherwise develop.

Role of indoleamine-2,3-dioxygenase in cancer processes

Besides contributing to the production of depression following cytokine therapies, IDO and tryptophan-2,3-dioxygenase (TDO), which are involved in tryptophan catabolism to form kynurenine, may contribute to cancer promotion. The relevance to cancer has come from the finding that kynurenic acid, a metabolite of the amino acid L-tryptophan, which is activated by IDO, influences antiinflammatory gene expression within adipose tissue and has been linked to various diseases. Furthermore, the baseline ratio between kynurenine and tryptophan in humans was related to the presence of several forms of cancer and was predictive of overall and progression-free survival among nonsmall-cell lung cancer patients being treated with a checkpoint inhibitor (Botticelli et al., 2020).

The potential significance of IDO and TDO has also come from reports that these enzymes were elevated in tumor microenvironments and tumor-draining lymph nodes and predicted poorer cancer prognosis in preclinical models. In this regard, when the tryptophan-kynurenine pathway is excessively activated, negative inflammatory changes can be created that promote pathological outcomes. Essentially, IDO contributes to the regulation of immune processes, thereby enhancing antitumorigenic actions and limiting the occurrence of autoimmune disorders, but a tipping point can be reached in which the pro-tumorigenic actions of IDO will predominate (Munn and Mellor, 2016). When elevated for extended times, widespread immune-related disturbances can be provoked by IDO and TDO, comprising suppression of NK cells, local CD8+ T effector cells, CD4+ T cells, and myeloid-derived suppressor cells. In addition to acting on immune tolerance, IDO can also affect checkpoint molecules that act on T-cell functioning. Together, the immune cell variations effectively create a tumor microenvironment that is conducive to tumor growth and such effects are especially prominent when inflammation stemming from the presence of IDO and TDO is not resolved.

As the presence of high IDO levels may act against the effectiveness of therapies, it was proposed that inhibiting IDO and TDO functioning could influence cancer progression and metastasis. Despite the potential of IDO-1 inhibitors in cancer treatments, the outcomes reported were not as impressive as had been hoped. Many factors could play into this outcome, including the specific combination therapies that were used as well as individual difference factors related to the ratio between kynurenine and tryptophan. Accordingly, patient stratification based on IDO-1 and TDO-2 expression could have promoted better results (Opitz et al., 2020). Additionally, kynurenine-related processes could be engaged as an effective add-on treatment among patients receiving cancer chemotherapy, and as described later in this chapter could significantly enhance the actions of other immunotherapies.

Stress and lifestyles influence indoleamine-2,3-dioxygenase

As both intrinsic and extrinsic factors, including gut microbiome processes, can influence the availability and the metabolism of tryptophan, it is reasonable to expect that lifestyles, including diet and exercise, could potentially influence the efficacy of IDO inhibitors, which can then enhance the effectiveness of immunotherapy. Given that exercise causes skeletal muscles to increase the conversion of kynurenine to kynurenic acid (Agudelo et al., 2019), it could also have effects on diseases involving this pathway and may have implications for fatigue resistance and stressor-provoked depressive illness.

In addition to promoting metabolic changes, dietary factors that lead to adiposity and impaired glucose homeostasis can influence IDO activity, which may thus affect cancer progression and the efficacy of treatments. In this regard, diverse nutrients influence metabolites of tryptophan-kynurenine activity, which might act to integrate energy metabolism involving adipocytes, immune cells, and muscle fibers. Moreover, through altered immune functioning and IDO activation, the increase of kynurenine synthesis together with tryptophan catabolites was related not only to depressive disorders but also to comorbid obesity, which over the long term can influence cancer occurrence.

Beyond these actions, inflammation tied to obesity, as well as the inflammation associated with kynurenine changes, may contribute to the comorbidity between cancer and depression and may offer the opportunity of enhancing cancer treatments. In the presence of obesity, IDO-elicited reductions of tryptophan can inhibit microbial conversion of tryptophan to beneficial metabolites and the production of IL-22, which regulates mucosal barrier functions together with intestinal immunity. By inhibiting IDO, insulin sensitivity may be enhanced, inflammation reduced, the gut mucosal barrier maintained, and lipid metabolism regulated appropriately, thereby enhancing metabolic health (Laurans et al., 2018).

Treatment vaccines

Virally related cancer

A small number of viruses (oncoviruses) were identified that could promote cancers, likely by altering genomic stability, disturbing DNA repair processes, increasing cell proliferation, and by stimulating inflammation.[b] It was estimated that 12%–20% of cancers stem from viral infection. Human T-cell leukemia virus-1 was the first virus identified that could promote cancer, notably leukemia, and was associated with the development of rheumatic illness and neurologic disease. The virus can be transmitted through blood transfusion, sexual behaviors, and can also be transferred from a mother to her infant through breastfeeding. Still other viruses, including hepatitis B and hepatitis C, have been linked to liver cancer and non-Hodgkin's lymphoma. Cancers have also been linked to Kaposi's sarcoma-associated herpesvirus (Kaposi's sarcoma) and human T-cell lymphotropic virus.

The best known of the oncogenic viruses is likely human papillomavirus (HPV), which has been associated with cervical carcinoma, as well as cancer of the vagina, anus, penis, and throat. It seems that HPV can suppress immune functioning and cancer is consequently more likely to develop, and this could occur through DNA methylation effects brought about by the virus. HPV has received appreciable attention because young women and men could be immunized against this virus (parental objections to their kids being vaccinated notwithstanding), and vaccines might also be used therapeutically (Garbuglia et al., 2020).

Epstein-Barr virus, a form of herpesvirus, is present within most of the population, but its presence for an extended period has been associated with several different types of cancer, such as Burkitt's lymphoma, Hodgkin's lymphoma, posttransplant lymphoproliferative disease, nasopharyngeal carcinoma, and some forms of stomach cancer. Strains of the Epstein-Barr virus may produce mutations so certain forms of cancer develop, presumably having these effects predominantly among individuals with other immune disturbances that increase their cancer vulnerability, or because this virus can interact with genetic, environmental, and

---

[b] In the mid-1960s, the notion began to take hold that certain cancers, such as some forms of leukemia, might be of viral origin. By the 1970s, then President Richard Nixon became convinced that the "War on Cancer" should include the analysis of viruses in the provocation of cancer. If polio could be beaten with a vaccine, so could cancer! Unfortunately, after considerable time and resources had been expended, it was clear that only modest progress had been made. Yet with all its shortcomings, this program may have spurred cancer research, as well as cancer prevention efforts. It was also found that viruses are associated with cancer occurrence, some of which can be prevented by vaccines.

dietary influences, although these have not been identified (Bakkalci et al., 2020). Vaccines have been developed to prevent cancers related to both HPV and hepatitis, but not for other viruses. Ideally, the prevention of other oncolytic viruses could occur with individuals taking appropriate behavioral measures. The reality, however, is that preventing transmission of a virus, such as the Epstein-Barr virus, is virtually impossible, which is why 90%–95% of individuals carry it.

Cancer vaccines

Treating cancers with the equivalent of a vaccine has spread beyond that of viruses that could instigate cancer development. Being unique to already existent cancer cells, neoantigens ought to be an ideal target for vaccines tailored to match these features. As a case in point, in a small sample of melanoma patients who had received a personalized vaccine, their T cells recognized these markers and destroyed the related cancers, although cancer recurred in some patients (Ott et al., 2017). Such approaches were also useful in brain cancer models, modestly extending the life of glioblastoma patients, and might be useful in treating breast and ovarian cancers. A vaccine-like treatment for some forms of lung cancer similarly added several months to life, and still better outcomes were obtained when it was combined with other types of anticancer agents or localized radiation. In this regard, by blocking norepinephrine receptors along with vaccination, immune effector functioning was enhanced, and tumor cell growth was diminished. Other combination treatments that included vaccination have similarly provided encouraging results, and when administered directly to the tumor, this procedure may have distinct advantages over other therapeutic methods. Among other things, this permits the presentation of tumor antigens directly to T cells, alters an immunosuppressive tumor microenvironment to one that is immunostimulatory, and produces limited side effects. Microgram amounts of two immune-stimulating agents administered directly into a solid tumor in mice similarly led to the cancer being eliminated. The vaccine in this instance comprised a short stretch of DNA (a CpG oligonucleotide, specifically a toll-like receptor 9 ligand), which together with adjacent immune cells amplified the expression of the OX40 receptor present on the surface of T cells, thereby leading to an attack on the cancer cells. Using a PET tracer to image noninvasive OX40, it was possible to predict the tumor response within days following vaccine treatment. Significantly, some activated T cells left the region to hunt down cancer cells present elsewhere in the body, thereby limiting metastasis (Sagiv-Barfi et al., 2018).

Customizing vaccines for individual patients based on genomic information obtained from a tumor could be combined with other immunotherapies so that a vaccine therapy would be especially effective. With new technologies, identification of a wider array of biomarkers may be possible, including the presence of multiple mutations. The application of machine learning tools, which includes greater consideration of the tumor microenvironment, high-throughput single-cell sequencing, and circulating tumor DNA detection, could result in a better personalized vaccine approach to cancer treatment (Sahin and Türeci, 2018).

Oncolytic viral therapy

Viral therapies come in several forms, including those in which a naturally occurring virus or a genetically engineered virus attacks cancer cells, or through the induction of antitumor immunity that involves innate and adaptive immune mechanisms. For example, when coxsackievirus was administered to patients with acute myeloid leukemia or multiple myeloma, immune effector cells were activated, including elevated NK cell activity and cytotoxic T lymphocytes that were specific for tumor-associated antigens, leading to attack on the cancer cells. In some instances, however, an immune response developed to the virus that had been administered, thereby undermining the treatment effectiveness. Thus, to be useful, it is necessary to modify the virus so that it avoids an attack by the immune system.

Another effective approach entailed the use of viruses that were engineered to attack only cancerous cells while concurrently boosting immune functioning or removing viral defense mechanisms. It was also demonstrated that injecting an inactivated virus (e.g., seasonal influenza vaccine) into lung tumors could result in these tumors becoming "hot," as reflected by increased infiltration of dendritic cells and CD8+ T cells that can kill the cancer cells. This increased the susceptibility of tumors previously resistant to immunotherapy to checkpoint blockade, a process in which factors that normally strengthen a tumor cell's survival against immune attack are inhibited (Newman et al., 2020). Similarly, the infusion of a modified form of a virus (e.g., polio or herpesvirus) directly into the tumor site can directly cause the immune system to become excited and ready to dispatch the tumor. A prime-boost process was also developed wherein two viruses encoded a common antigen, and specific bacteria could likewise be combined with an oncolytic virus to promote potent tumor attacks (Martin and Bell, 2018).

The demonstration that RNA-based vaccines could be used to prevent COVID-19 infection has obviously been a remarkable advance in preventing infectious disease. Earlier attempts to deal with infectious illnesses, such as rabies and influenza, had already been shown to engender some positive effects, although the procedures weren't fully effective. This approach had also been explored as a means of taming some forms of cancer (Pardi et al., 2018), and a small trial of a personalized RNA-based approach that targeted multiple cancer mutations provided encouraging results (Sahin et al., 2017). Since these initial demonstrations of vaccine efficacy in cancer, several methods were developed to enhance the effectiveness of mRNA vaccines. This entailed improvement to the structural features of the mRNA molecule together with enhanced stability as well as more selective mRNA targeting (Van Hoecke et al., 2021). The success of an mRNA-based vaccine to treat cancer, reinforced by its effectiveness in the development of COVID-19 vaccines, will no doubt encourage the use of this approach in producing further viable cancer treatments.

In addition to viruses, bacteria can be enlisted to diminish tumors. This approach has not been widely adopted but recent advances have been made to use bacteria to interfere with the progression of solid tumors. Bacteria can be used to target specific types of tumors where they release toxins that destroy cancer cells. Moreover, engineered bacteria aimed at cancer cells with specific genetic or epigenetic features can be used to provide selective payloads to tumors while sparing healthy cells. By engineering nonpathogenic bacteria to release a nanobody antagonist of the cancer cell marker CD47 within the tumor's microenvironment, inflammation and angiogenesis can be affected, making it possible to promote tumor regression and prevention of metastasis (Chowdhury et al., 2019).

Oncolytic viral therapies have enormous potential but are encumbered by several obstacles. These comprise multiple physical barriers so that the viruses encounter difficulty penetrating solid cancers, which may be exacerbated by the characteristics of the tumor microenvironment. Furthermore, the immune system can neutralize and clear oncolytic viruses, thereby limiting the amount of virus that reaches the tumor. Varied approaches are being developed to overcome the challenges encountered using oncolytic viral therapies and combination therapies (e.g., with immune checkpoint inhibitors) have been having some success. As well, to overcome barriers that largely prevent access to tumors, bacterial carriers were designed that could migrate and localize to their intended target.

Given that cancer cells require enormous amounts of energy, varied manipulations that influence metabolic processes can affect cancer growth and the response to therapeutic agents. It can reasonably be expected that the influence of viral therapies on cancer progression would be modifiable by nutritional factors and by the presence of microbiota or their metabolites. As viral therapy is still a relatively new treatment strategy, the influence of diet and other lifestyles in moderating its effectiveness has not been extensively examined. Nonetheless, glioblastoma cells that were grown in a medium free of nutrients were more affected by herpes simplex virus therapy than were nondeprived cells (Esaki et al., 2016). Furthermore, food restriction (24 h before treatment and an additional 96 h afterward) enhanced the benefits of the oncolytic measles vaccine virus (MeV-GFP) in treating human colorectal carcinoma cell lines without affecting normal colon cells (Scheubeck et al., 2019). The presence of metabolic hormones, such as leptin, can affect the ability of T cells to operate efficiently, such as when oncolytic viruses engineered to express leptin were shown to diminish tumor cells (Rivadeneira et al., 2019). While relevant to the question of dietary influences, these procedures do not necessarily speak to the efficacy of nutrient manipulations in enhancing viral therapies in vivo.

It is important to underscore yet again that the success of immunotherapies is dependent on the effective functioning of immune cells. This can be problematic, as tumor cells may diminish microenvironment nutrients and concurrently create an abundance of toxic by-products that can negatively affect T cell functioning. This means that manipulations that affect the nutritional capacity of immune cells within the tumor microenvironment can profoundly influence their effects on cancer cells. Indeed, drugs that affect nutrient uptake, such as the diabetic medication metformin, may positively influence local immune functioning and negatively affect cancer growth. As discussed later, the effectiveness of several immunotherapies can similarly be influenced by diets and microbiota. Admittedly, it is surprising that more data are not available concerning their actions in moderating the impact of oncolytic viral therapies.

Antibody-based therapies

Monoclonal antibodies have been manufactured that flag cancer cells for destruction and trigger an immune response to attack tumors directly. Antibodies can also be used to indirectly affect tumors, such as by inhibiting energy supplies to cancer cells, including inhibition of intra-tumor angiogenesis (blood vessel growth). Chemical antibodies, aptamers, have also been developed for endogenous substrates, and such single-stranded nucleic acid oligonucleotides (i.e., having a small number of nucleotides) could be designed to attach to cancer cells selectively. It may also be possible to target intracellular targets (as opposed to those that are on cell surfaces) that have been implicated in cancer development. Antibody therapies have also been the forerunner of other immunotherapies, notably the use of checkpoint inhibitors.

Checkpoint inhibitors

Hardwired inhibitory mechanisms are essential for self-tolerance and for regulating immune responses so that damage to healthy tissue does not occur. Immune cells have checkpoint molecules on their surface whose functions can be turned down by healthy cells, thereby preventing them from attacking normal tissues. Cancer cells capitalize on this by cloaking themselves with molecules suggestive of being a normal healthy cell. Consequently, they can avoid attack by immune cells. The therapeutic challenge then is to inhibit checkpoints on cancer cells and expose them for the frauds that they are and unleash the actions of immune cells.

CTLA-4 immunotherapy

This approach to inhibit checkpoints initially produced exceptionally encouraging outcomes in the treatment of solid tumors, although only a minority of patients (20%–30%) showed a positive response (Demaria et al., 2019). Several stimulatory checkpoints were targeted for treatment purposes, with cytotoxic T-lymphocyte-associated antigen-4 (CTLA-4) located on T lymphocytes being among the first used in cancer therapy. When CTLA-4 is bound to a particular protein (dubbed B7) early in an immune response, T cells are restrained so that they do not kill these healthy cells, but cancer cells may also have this protein on their surface, thereby evading attack by T cells. By inhibiting the binding of CTLA-4 to B7, the brakes are taken off immune cells so that they can attack the cancer cells. One of the difficulties with anti-CTLA-4 drugs was that they behaved relatively indiscriminately so the side effects emerged, including effects on multiple organs and systems. The gastrointestinal tract and skin are most often affected but adverse events may also occur in the liver, kidney, and pancreas, as well as the endocrine and nervous systems.

PD-1/PD-L1 immunotherapy

Other checkpoint inhibitors were developed that were expected to have more selective actions and thus would produce greater treatment success with fewer side effects (Havel et al., 2019). Ordinarily, the inhibitory checkpoint molecule Programmed Death-1 (PD-1) present on T cells, B cells, and myeloid cells can serve as an off-switch so they do not attack healthy cells within the body. The programmed death-ligand 1 (PD-L1) present on normal cells can bind to the PD-1 checkpoint, providing the inhibitory signal to suppress the adaptive arm of the immune system. Once again, some cancer cells also carry the PD-L1 protein, giving them a way to avoid immune attacks. Checkpoint inhibitors in the form of PD-1 antagonists (antibodies) or agents that interfere with PD-L1 allow the immune system to dispatch these tumor cells as described in Fig. 15.3.

The PD-1 inhibitors were more effective in a wider range of cancers than the anti-CTLA-4 agents, and toxicity levels were lower. Even a single dose of a neoadjuvant anti-PD-1 inhibitor was sufficient to produce an antitumor response in 8 of 27 melanoma patients, which could be predicted based on variations of CD8+ T cells in the tumor microenvironment and in circulation (Huang et al., 2019a). The early response to checkpoint inhibitors (within 3 weeks of commencing immunotherapy), including T cell turnover, T cell receptor repertoire dynamics, and the expansion of a subset of peripheral effector T cells, could predict the effectiveness of the procedure (Valpione et al., 2020). Indeed, among patients who received immunotherapy to treat metastatic melanoma, those with the greatest number of CD8+ T cell clones 3 weeks after treatment were most likely to fare relatively well over 6 months (Fairfax et al., 2020).

Checkpoint inhibitors have been effective for a wide range of solid tumors, but as already indicated, only a modest percentage of patients show a positive treatment response. Among many other obstacles, T cells encounter difficulties in moving through the fibrous mass of tumors so they are less able to attack cancer cells. By genetically altering the features of T cells, their capacity to get around these obstacles can augment the therapeutic response (Tabdanov et al., 2021). The limitations of checkpoint inhibitors may also be related to the diverse genetic features of patients, disturbances in the communication between immune and tumor cells, and the immunosuppressive and hypoxic microenvironment surrounding the tumor. These issues are further complicated by the heterogeneity that exists concerning antigen processing and presentation of targets by tumors, and not all neoantigens will be recognized by immune cells. For that matter, there remains uncertainty regarding the features of T cells that are needed to produce the most effective therapeutic responses. Among other things, immunotherapy might not be effective unless the tumor expressed neoantigens that were recognized by both CD4 and CD8 T cells. It is also possible that the effectiveness of

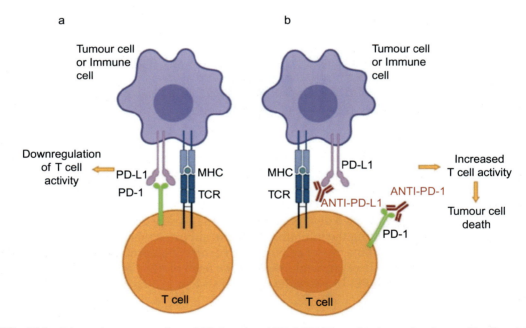

FIG. 15.3 Schematic representation of PD-1 and anti-PD-1/PD-L1 mechanisms of action on T cell activity. Activated T cells at secondary lymphoid organs/tumor tissue (A) will upregulate the expression of coinhibitory cell surface receptor PD-1. Binding of PD-1 to its ligands, PD-L1 or PD-L2, found on the surface of several immune cells as well as tumor cells, will inhibit signaling downstream of the TCR, thus downregulating T cell activity. (B) Targeting PD-1 or PD-L1 with antibody therapeutics can reinvigorate exhausted T cells at the tumor site, increase the activity, consequently allowing T cell-mediated tumor cell killing. *Figure and caption from Alard, E., Butnariu, A.B., Grillo, M., Kirkham, C., Zinovkin, D.A., et al., 2020. Advances in anti-cancer immunotherapy: CAR-T cell, checkpoint inhibitors, dendritic cell vaccines, and oncolytic viruses, and emerging cellular and molecular targets. Cancers (Basel) 12, 1826.*

immunotherapeutic agents is diminished in the presence of weak or limited tumor-related inflammatory responses. This difficulty might be overcome by reprogramming T_{reg} cells so that they selectively release IFN-γ in tumor cells, thereby enhancing infiltration by NK and CD8+ T cells. In this way, the effectiveness of a PD-1 inhibitor could be enhanced without promoting an autoimmune response (Di Pilato et al., 2019). The position has been taken that it might be advantageous to target T_{reg} cells in the tumor microenvironment and concurrently act against chronic inflammation provoked by macrophages, thereby attenuating oxidative stress and augmenting the regulation of metabolic processes. A related strategy to enhance immunotherapy was to enhance the killing potential of T cells, which could be accomplished by epigenetically silencing the actions of genes that otherwise limit the power of T cells.

Some tumors do not produce PD-L1 so checkpoint inhibitors do not have a target to block and thereby arrest cancer progression. Alternatively, excessive levels of PD-L1 are produced by some cancer cells, and instead of being displayed on the cell's surface, the PD-L1 is transported to lymph nodes through exosomes that serve as molecular freighters sailing through the bloodstream or lymphatic system. Having reached the lymph nodes, the PD-L1 proteins undermine immune functioning so that the tumors are not targeted. Accordingly, therapy that combined inhibition of exosome formation and delivery of PD-L1 inhibitors yielded superior anticancer effects. It is still some time before these approaches can be used in humans, especially as the use of PD-1/PD-L1 antagonists may still encounter problems such as the development of treatment resistance. Nonetheless, these findings might be relevant for therapy that is otherwise not influenced by immunotherapy alone.

As we have stated previously, chronic systemic inflammation can have negative consequences, and when the inflammation occurs within tumors, this is accompanied by elevated levels of infiltrating T cells and IFN-γ signaling. At this point, the opportunity for checkpoint immunotherapy can be enhanced. Yet, as we've seen repeatedly, cancer cells are often capable of activating pathways that promote suppression of immune functioning. For example, among melanoma and kidney cancer patients treated with the PD-1 antibody nivolumab, the conversion of tryptophan to kynurenine may be increased, leading to poorer treatment response, possibly because of the immunosuppressive actions of kynurenine. Indeed, median survival among those with a greater than 50% increase of the kynurenine to tryptophan ratio was half that of patients with low levels of this ratio (Li et al., 2019). It was disappointing, however, that inhibiting IDO-1 did not enhance the effects of immunotherapy among melanoma patients. It seems that antitumor treatments can cause the formation of aberrant presentation of specific peptides on the surface of cancer (melanoma) cells, thereby protecting them. Hence, by altering the presence of these peptides, the effects of immunotherapy might be enhanced (Bartok et al., 2021).

Given that checkpoint inhibitors involving T cells aren't universally effective in taming cancers, variants of this procedure have been sought, such as targeting multiple epitopes, or developing strategies that offer greater immune activation, but without more profound adverse reactions. While the focus of research on checkpoint inhibitors had primarily concerned T cells, it was considered that better outcomes might be obtained by including actions on other immune cells. As tumor-associated macrophages also express PD-1, disrupting the binding between PD-1 and PD-L1 may have its anticancer effects by affecting these immune cells (Gordon et al., 2017). Furthermore, checkpoints also constrain NK cells; hence, focusing on the inhibition of PD-1 on NK cells could enhance the therapeutic response, especially if this was combined with other treatments (Hsu et al., 2018). The effectiveness of immune checkpoint inhibitors was also aligned with the presence of B cells situated in immune cell clusters present within the tumor, and the possibility was raised that B cells might be a viable target to promote immunotherapy (Helmink et al., 2020).

TIM-3 immunotherapy

In addition to CTLA-4 and PD-1/PD-L1 checkpoint inhibitors, other tumor-associated antigens are expressed on the surface of some cancer cells that may contribute to their unregulated proliferation. T cell immunoglobulin and mucin domain-containing protein 3 (TIM-3), which is present on IFN-γ producing CD4+ and CD8+ T cells, may act in this capacity. Thus, it was suggested that blocking both TIM-3 and PD-1 can promote actions superior to that of PD-1 blockade alone (Wolf et al., 2020). It was similarly indicated that cancer cells produce relatively high levels of CD24 protein that can be sensed by macrophages through a particular receptor, SIGLEC-10. Blocking the interaction between CD24 and SIGLEC-10 resulted in macrophages becoming voracious eaters of cancer cells in vitro and in mice with transplanted human ovarian tumors or triple-negative breast cancer cells (Barkal et al., 2019).

CAR T-cell therapy

Adoptive cell transfer (ACT) represented a critical advance within the immunotherapeutic arsenal, primarily being used to treat some hematological malignancies, including lymphoma, acute lymphoblastic leukemia, chronic lymphocytic leukemia, and multiple myeloma. One form of ACT, that of CAR T therapy, entails blood being taken from a patient, after which the purified T cells are engineered to produce chimeric (artificial) antigen receptors (CARs) that target a CD19 antigen found on cancer cells. These cells are then expanded (grown) and readministered to the patient. With the transfer of these T cells to patients, they multiply further and after receiving a message to do so, CAR T cells home in on cancer cells and then dispatch them (June and Sadelain, 2018). Ideally, the reprogrammed T cells would be selective for specific antigens, but several peptide antigens can react to a single T-cell receptor, making selective attacks more difficult, and off-target recognition might promote immune toxicity.

This treatment had shown sufficiently impressive effects for it to receive a breakthrough designation from the FDA in 2014. In a trial of 30 acute lymphoblastic leukemia (ALL) patients who had not responded well to other treatments, an impressively high remission rate was obtained (Maude et al., 2014). Remarkably, in some instances, positive outcomes were achieved within weeks of treatment (Davila et al., 2014). This procedure was effective in more than half the patients treated for B-cell lymphoma or non-Hodgkin lymphoma, whereas less success was achieved in treating chronic lymphocytic leukemia (CLL). While numerous factors might contribute to the limited effects among CLL patients, it is significant that responders exhibited greater pretreatment expression of genes whose transcription was more readily elicited by STAT3 and IL-6, possibly promoting greater CAR T cell multiplication. Responders also had more T cells present that expressed the IL-6 receptor, as well as lymphocytes with memory-like features, and hence could help sustain remission. The nonresponders, in contrast, exhibited upregulation of genes linked to glycolysis, exhaustion, and apoptosis (Fraietta et al., 2018).

Several ways have been devised to enhance the effectiveness of CAR T therapy and to diminish cancer recurrence. These have ranged from dual targeting of antigens, as well as elevated expansion and persistence of T cells. However, most cells used to promote killing by CAR T cells have receptors, such as Fas that binds with FasL, which is frequently present in the tumor microenvironment, leading to downregulated immune functioning and diminished effectiveness of the therapy. Through engineering T cells in which Fas variants do not bind with FasL, the influence of CAR T therapy can be improved (Yamamoto et al., 2019). From the perspective of our discussions regarding the influence of lifestyles, it is significant that an obesity-promoting diet in rodents was able to drive production of chemokines (e.g., CXCL1) that augmented Fas-promoted resistance to immunotherapy.

Not all cells within a tumor contain markers that are targeted by CAR T cells, so they might escape destruction. However, CAR T cells not only attack and kill those cancer cells that express the FasL marker, but they may also attack adjacent cancer cells. Among patients with non-Hodgkin's lymphoma, for instance, elevated Fas levels were associated with greater persistence of the positive effects of therapy (Upadhyay et al., 2021). Thus, increased "bystander killing" and longer survival could be attained by providing immunotherapy that augments Fas-FasL signaling.

Broader use of CAR T therapy

The effectiveness of CAR T therapy was initially restricted to liquid cancers but increasing attention focused on using this procedure to attack solid tumors. As described earlier, T cells have a difficult time infiltrating solid tumors, and even after arriving at their intended destination, the CAR T cells might not bind with the tumor cells present. Being hypoxic, acidic, and immunosuppressive, the tumor microenvironment is not hospitable for the CAR T cells. If CAR T cells can be administered regionally, they could be effective with fewer side effects developing (Hou et al., 2021).

Problematic effects could potentially be overcome in other ways. For example, myeloid-derived suppressor cells may be present that protect solid tumors from attack by the immune system. By using an antibody to eliminate these suppressor cells, access to the tumor can be facilitated, thereby allowing for better outcomes in response to CAR T therapy. An innovative approach to destroy solid tumors has been that of generating a "tandem CAR" that creates a specific antigen on the cell surface after immunization so that enhanced T cell proliferation occurs when the CAR Ts that had migrated to the lymph node microenvironment were again stimulated (Ma et al., 2019). Yet another approach to enhance the effectiveness of CAR T therapy has been through a naturally occurring RNA (RN7SL1) that can boost T cells so that they search for cancer cells that had evaded attack and can do so for longer periods as well as being able to infiltrate tumors more readily.

Enhancing immunotherapy's effectiveness

The effectiveness of CAR T therapy may be hampered by the development of T cell exhaustion that occurs in more than half of the patients. The source for the exhaustion might be due to a family of proteins, Nr4A, that contribute to gene regulation. As expected, tumors were reduced and CAR T cell performance was augmented among mice lacking Nr4A transcription factors (Chen et al., 2019). In other studies, the focus was on the AP-1 transcription factor that plays a fundamental role in gene regulation associated with growth factors and cytokines as well as cell differentiation, proliferation, and apoptosis. Differential gene expression patterns were detected between exhausted and nonexhausted cells that were mediated by c-Jun expression, which is a component of AP-1. Hence, when CAR T cells were modified they overexpressed the oncogenic transcription factor c-Jun, CAR T exhaustion was diminished, and the therapy was augmented in mice (Lynn et al., 2019).

As with checkpoint inhibitors, CAR T therapy could also be applied in the development of reprogrammed macrophages and preclinical studies pointed to positive outcomes using NK cells (Mehta and Rezvani, 2018). As NK cells can be derived from induced pluripotent stem cells, they could potentially be stored in biobanks and then used for allogeneic "universal" cell therapy. It is also exciting that a T cell has been discovered that could potentially find and destroy a variety of cancer cell types. This T cell has receptors that could bind to the MR1 molecule present in cancer cells with disturbed metabolism . Such T cells, which are grown from peripheral blood mononuclear cells, responded to a variety of cancer cell lines in multiple tissues, killing them effectively without damaging healthy cells (Crowther et al., 2020). While these are all viable approaches to enhance the effectiveness and the breadth of CAR T therapy, they are experimental in nature, using animal models, and therefore await proof of efficacy in human cancers.

Nonetheless, despite the hurdles encountered by immunotherapies, it can be expected that the next generation of treatments will likely be effective and less toxic for a broader range of cancers. This will likely be dependent on appropriately identifying the features of different tumor microenvironments, the genetic nature of the tumor cells, and the multiple subclasses of immune cells and inflammatory cytokines that support the growth and invasion of these cells (Binnewies et al., 2018). Being able to engineer T cells (e.g., through CRISPR) they are better able to navigate the tumor microenvironment and are capable of greater functionality and may thereby enhance the effectiveness of CAR T therapies. It will also be advantageous to develop strategies to provide CAR T cells with their metabolic demands. Finally, as we keep emphasizing, the effectiveness of therapies could potentially be enhanced through appropriate individualized treatment approaches. At the very least, this might inform the value of one therapeutic approach over another so that patients would be more likely to receive therapies that might prove effective and not undergo distressing procedures that are apt to be ineffective.

Combination therapies

Combination chemotherapies, as already indicated, often have effects that are superior to those provided by monotherapies. The superior outcomes of combination therapies might reflect additive or synergistic effects, and when used at lower doses of each therapeutic agent, treatment resistance might be less likely to develop. As anti-CTLA-4 and anti-PD-1 treatments act on different checkpoints, together they might have positive effects beyond that of either treatment alone. This was evident in nonsmall-cell lung cancer and among melanoma patients in whom a PD-1 inhibitor combined with a CTLA-4 checkpoint inhibitor prior to surgery had more positive effects than single treatments. As expected, the enhanced outcomes realized with combined checkpoint inhibitors may also be accompanied by increased side effects. Pretreating rodents with a TNF-α inhibitor that reduced inflammation successfully limited the side effects of a combination of a PD-1 and a CTLA-4 inhibitor, thereby permitting treatment that might not otherwise be feasible (Perez-Ruiz et al., 2019). Likewise, an NLRP3 inhibitor that diminishes inflammation (by affecting the inflammasome) might enhance the effects of immunotherapy when treatment resistance had developed (Tengesdal et al., 2021).

The effectiveness of PD-1/PD-L1 immunotherapy can be enhanced when combined with other treatments, including cancer vaccines and chemotherapy. In a small trial, in situ vaccination enhanced the effects of PD-1 blockade so the tumors shrank in 8 of 11 patients and in some cases was effective for several years (Hammerich et al., 2019). Blockade of the PD-1 checkpoint combined with virotherapy also yielded enhanced responses, and checkpoint inhibitors combined with stem cell transplants could be effective in treating non-Hodgkin's lymphoma in which relapse occurs frequently.

Preclinical studies have also pointed to superior outcomes in treating nonsmall-cell lung cancer when checkpoint inhibitors were combined with PARP inhibitors that ordinarily enhance DNA repair. Moreover, consistent with the antitumor actions of β-norepinephrine antagonists, pronounced benefits were apparent in mice when it was combined with PD-1 immunotherapy in treating malignant melanoma. Benefits could also be accrued by combining checkpoint inhibition with agents that act against angiogenesis and the effects of PD-L1 immunotherapy could be augmented by epigenetic modulation that influences T cell trafficking into the tumor microenvironment. Stemlike T cells are also present within lymph nodes located near tumors, some of which may find their way to the tumor to enhance the ability of exhausted T cells. If the actions of these stemlike cells could be harnessed, they might enhance the actions of cancer therapies. The combination treatments described here are only a few of the very many that have been developed and which continue to be assessed, although the full potential of checkpoint inhibitors in cancer treatments has not yet been fully realized.

Owing to the promising effects attained through different immunotherapies, an increasingly greater number of clinical trials continue to be undertaken to find the most effective therapies. To an extent, immunotherapies have been a victim of their success. With so many ongoing clinical trials, there may be an insufficient number of patients to allow for rapid testing of all the new compounds (Kaiser, 2018). Despite the potential of combination therapies, the zeal to find the best treatments may have resulted in the evaluation of cancer therapies moving too quickly, and it may be time to reconsider the best approaches concerning how immune therapies are tested. New machine learning methods have been capable of predicting the efficacy of drug combinations and the very same approach could ostensibly be used to predict the occurrence of side effects. If nothing else, this might be included as an important step in deciding the specific therapies that ought to be administered to cancers in individual patients.

Moderating the actions of checkpoint inhibitors through lifestyles

A veritable armada of agents is in development with the hope that tumors will be attacked more effectively. This has in many instances produced promising outcomes, but too often these benefits are limited. Perhaps the time has come to determine whether lifestyle factors that can affect metabolic and immune functioning will influence the efficacy of therapies, including the actions of checkpoint inhibitors (Deshpande et al., 2020). Some features that influence the impact of checkpoint inhibitors are not modifiable (aging, sex, race) but many others are flexible and might permit improved therapeutic outcomes. By virtue of their actions on metabolic processes, microbiota, immune function, and inflammation, factors such as diet, exercise, stressful experiences, and sleep-related processes could potentially affect therapeutic outcomes (e.g., Baiden-Amissah and Tuyaerts, 2019). Based on findings pointing to the benefits of short-term fasting diets for other therapies, it had been considered that diet manipulations might be a route to enhance the effects of immunotherapy. As described earlier, several studies have pointed to this possibility, although this can be difficult for patients with advanced cancers. Foods with antiinflammatory actions can limit the occurrence of some forms of cancer and could potentially augment the effects of therapies. There is also reason to believe that food components, such as high levels of salt, which have typically been viewed as having negative effects in relation to some illnesses and are typically considered as undesirable for cancer patients, may actually have some benefits. In mice maintained on a high salt diet, NK cell activation was provoked, apparently by increasing intratumoral *Bifidobacterium* presence, and enhanced the effectiveness of PD-1 immunotherapy. In fact, fecal transplants from mice maintained on a high salt diet to normally fed mice resulted in the response to immunotherapy being enhanced, again pointing to the importance of dietary salt in in mediating these actions (Rizvi et al., 2021).

Preclinical studies have likewise pointed to exercise being able to modestly enhance the effects of anti-PD-1 immunotherapy and in reducing side effects of the treatment (Martín-Ruiz et al., 2020). For instance, in a mouse model of triple-negative breast cancer, exercise diminished the accumulation of myeloid-derived suppressor cells that ordinarily suppress immune activity, while increasing the activation of NK and CD8 T cells, and concurrently enhancing the actions of PD-1 blockade and radiotherapy in diminishing tumor progression (Wennerberg et al., 2020). In mice with induced breast cancer, exercise enhanced the effects of checkpoint inhibitors by reprogramming the tumor microenvironment and increasing CD8+ T cell infiltration by affecting the chemokine receptor CXCR3 (Gomes-Santos et al., 2021). In a review of the pertinent research, several other mechanisms were offered by which exercise could enhance the effects of immunotherapy (Zhang et al., 2019). Specifically, in addition to enhancing innate and adaptive immunity, exercise could enhance immune cell trafficking into a tumor, modify the tumor microenvironment, reduce tumor hypoxia, and influence the availability of nutrients to tumors.

Because of the marked effects of sustained exercise regimens on various aspects of immune activity, it was maintained that these routines could enhance the benefits of CAR T therapy (Gustafson et al., 2021). But even if tumor progression was not slowed beyond that provided by the primary therapy, patients were more likely to report improved quality of life and returned to work sooner than individuals who had not exercised.

Side effects of checkpoint inhibitors and CAR T therapy

Generally, PD-1/PD-L1 inhibitors were better tolerated than chemotherapy, although in some individuals checkpoint inhibitors promoted pronounced immune-related adverse events, such as autoimmune disorders involving multiple organs. As depicted in Fig. 15.4, to varying degrees, CTLA-4 and PD-1 inhibitors can promote a wide range of side effects. Given that single immunotherapies may be accompanied by adverse reactions, it is not surprising

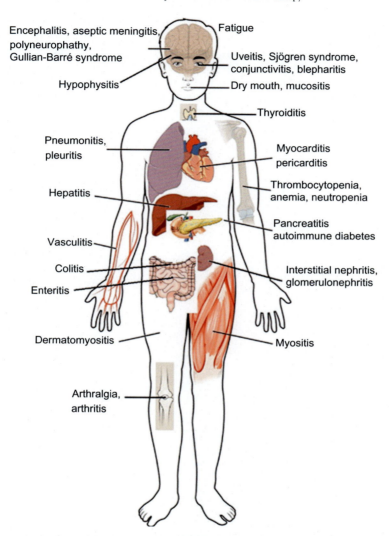

FIG. 15.4 Common immune-related adverse events associated with immune checkpoint blockade. Immune checkpoint blockade can result in the inflammation of most organs. *Figure and caption from Morgado, M., Plácido, A., Morgado, S., Roque, F., 2020. Management of the adverse effects of immune checkpoint inhibitors. Vaccines, 8 575.*

that combined checkpoint inhibitor therapy would elicit stronger side effects, and similar actions would be provoked when other combinations were used.

The side effects of CAR T therapy are more profound than those elicited by checkpoint inhibitors. A few days following infusion of CAR T cells, dose-dependent signs of neurotoxicity may develop, reflected by headaches, memory loss, hallucinations, seizure, reduced consciousness, delirium, and in some cases cerebral bleeding. An especially troubling side effect of CAR T therapy is the development of a "cytokine storm" (cytokine-release syndrome), stemming from the conjoint release of several proinflammatory cytokines. Additionally, "macrophage activation syndrome" and "tumor lysis syndrome" may occur in which the contents of cells are dumped into the bloodstream. Ordinarily, when T cells attack foreign cells, these pathogens die off slowly and the debris is removed. However, when CAR T cells attack, the cancer cells swell and eventually burst, causing nearby immune cells to release excessive amounts of cytokines (Liu et al., 2020). The appearance of a cytokine storm may initially begin within a particular organ but systemic inflammation and multiorgan failure can follow (Paucek et al., 2019). As described in Fig. 15.5, still other severe outcomes can occur, including vascular leakage together

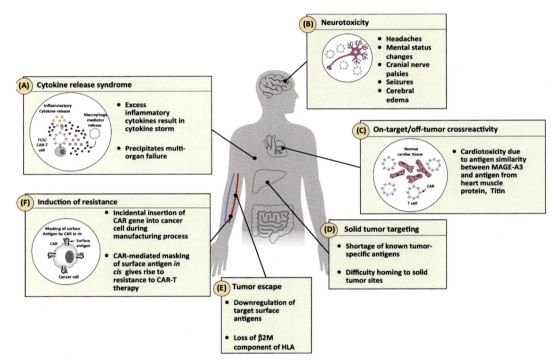

FIG. 15.5 Overview of toxicities and limitations of T cell receptor (TCR)/chimeric antigen receptor T cell (CAR T) therapy. (A) Cytokine release syndrome (CRS) is a syndrome of toxicities related to a progressive systemic inflammatory response. CRS is maintained by infused engineered CAR T cells, which are activated in vivo upon binding target antigen, and which then release superfluous inflammatory cytokines. (B) Symptoms of neurotoxicity range from mild headaches and confusion to medical emergencies such as cerebral edema. (C) On-target/off-tumor cross-reactivity occurs when TCR/CAR T cells target antigens present in healthy tissues. Cardiotoxicity is one example in which TCR-T cells targeting the MAGE-A3 antigen in melanoma were cross-reactive with antigens in heart muscle. (D) Solid tumor treatment with TCR/CAR T faces challenges in tumor homing and infiltration. Furthermore, there are limited amounts of known specific solid tumor antigens to target in order to obtain positive therapeutic effects. (E) Mechanisms of tumor escape from TCR/CAR T cells often include the downregulation of target antigens in cancer cells, as well as loss of β2M expression. (F) Induction of resistance to CAR T cell therapy was observed in a case study where the accidental introduction of a CAR gene into a cancer cell during the cell manufacturing process created resistance to CAR T cell therapy owing to masking of antigen by CAR in *cis* signaling. *Figure and caption from Paucek, R.D., Baltimore, D., Li, G., 2019. The cellular immunotherapy revolution: arming the immune system for precision therapy. Trends Immunol. 40, 292–309.*

with hypotension, hypoxia, and multiorgan toxicity. Clearly, this therapy can be exceptionally distressing, and it has been remarked that CAR T therapy can elicit symptoms sufficiently broad and entrenched as to require a small army of specialists to keep patients alive.

Modifying the side effects of CAR T therapy

Extensive efforts were made to get around the difficulties produced by CAR T therapy. The possibility was explored that manipulations of the biochemicals that cause tumor cell destruction, such as perforin/granzyme B, could be altered to limit cytokine release syndrome (Liu et al., 2020). It also seemed that IL-6 and IL-1 receptor antagonists, as well as glucocorticoids, could act against many of the negative consequences of CAR T therapy without affecting tumor remission (Norelli et al., 2018). A novel CAR T molecule has also been developed (CD19-BBz) that is effective in killing cancer cells but produces fewer cytokine changes, thus limiting the occurrence of a cytokine storm (Ying et al., 2019). Based on a computationally designed approach, it was found that concurrently manipulating two molecules that affected T cell activation could switch off the CAR T actions. This STOP-CAR T manipulation has obvious implications to diminish the adverse effects of the standard CAR T therapy and may even allow for tweaking of the treatment to yield better outcomes (Giordano-Attianese et al., 2020).[c]

As nutritional status and the availability of micronutrients may be important in limiting inflammation and oxidative stress, foods that have the most favorable antiinflammatory effects could potentially limit cytokine storm generation. The influence of these nutrients would certainly have profound effects on highly vulnerable individuals, especially those with obesity in whom TNF-α, IL-6, and MCP-1 would be most elevated, and eating-related hormones, such as leptin that foster an inflammatory milieu, might also contribute to the impact of CAR T therapy. Once more, microbiota can play a role in the cytokine storm given their role in inflammatory processes and in determining lung health through bidirectional communications that occur between the lung and gut microbiota.

Use of biomarkers

Despite the considerable gains made in cancer therapies, there's no way around the fact that these treatments are still only effective in a subset of patients, and in many cases, progression-free survival is often extended by less than 6 months. There is the hope that treatments agnostic to the type of cancer might be available at some time, possibly using CRISPR to facilitate the recognition of cancer cells (Crowther et al., 2020). Short of this, it is essential to define which markers or sets of biomarkers predict whether a particular form of cancer is likely to appear. Furthermore, it is important to determine the optimal therapeutic strategy that ought to be used, and which will have the fewest and least intense side effects. A

---

[c] The features of a cytokine storm associated with CAR T are very similar to those associated with COVID-19, and it seems that treatments to control the CAR T cytokine storm (e.g., an IL-6 receptor blocking antibody) can also diminish those associated with COVID-19. In essence, what had been learned from experiences with cytokine storm elicited by CAR-T may have benefited individuals who developed this syndrome in association with COVID-19 infection.

thorough review described a range of biomarkers that have been used to predict treatment response (Havel et al., 2019), including markers related to characteristics of the individual and those related to cancer. We will briefly turn to a few of the varied biomarkers that can be used to predict the effectiveness of cancer therapies and the side effects that can be encountered.

Genetic characteristics related to the individual (systemic host factors)

Numerous features of the individual have been linked to the development and progression of cancers, including the presence of assorted polymorphisms, epigenetic factors, genome-wide markers (e.g., DNA damage and repair pathways, MAPK pathway, IDO-1), and gut microbiome constituents. Generally, biomarkers related to the individual fall into several groups that comprise genes that reflect DNA repair disturbances, the presence of specific oncogenes, mutations of genes that contribute to tumor suppression, acquired mutations that promote cancer as well as germline mutations that serve in this capacity. For instance, the TP53 gene that is responsible for the p53 protein serves as a transcription factor to suppress tumor cell formation, so the mutations of this gene are associated with more than 50% of cancers. The BRCA1 and BRCA2 genes are involved in tumor suppression, so the mutation of this gene, which can be inherited, allows cancer cells to proliferate uncontrollably. Similarly, the immune receptor NLRX1, which serves as a tumor suppressor, was very low in models of colorectal cancer and could thus serve as a marker (and perhaps a target) for those with this form of cancer (Koblansky et al., 2016). Finally, overexpression of the MYC gene, which regulates cellular metabolism, has been tied to immune disturbances and the development of several types of cancer. The presence of these gene mutations not only signal the potential for cancer occurrence, but when these genes are inherited, individuals might have the opportunity to take actions to prevent disease occurrence as we discussed in our discussion of the BRCA1 and BRCA2 mutations.

Markers that reflect features of the tumor

The presence of numerous biomarkers can predict later tumor development, which might allow for the adoption of interventions so that precancerous cells or disease-promoting processes can be intercepted. These might entail genetic and epigenetic markers, or they might comprise microbiota known to trigger specific forms of cancer. Several of the mutations that have been identified have broad actions in that they are associated with a variety of different types of cancer. This is the case for the KRAS mutation that has been associated with about 25% of cancer types, and in some instances its presence can predict a poor response to therapies as in the case of low survival for nonsmall-cell lung cancer. By targeting specific sites on the KRAS gene that allowed for neoantigens to be expressed on the tumor surface, it may be possible to use T cell therapy to attack the cancer cells (Bear et al., 2021).

Yet another protein-coding gene, AIM2, can serve as a sensor to detect altered or damaged DNA as well as the presence of foreign DNA from viruses, bacteria, and parasites that cause activation of the inflammasome. Thus, AIM2 can be used as a cancer biomarker and dysregulation of AIM2 might contribute to a variety of diseases, including cancer. Targeting a factor that enhances AIM2 can limit colon cancer growth and may be effective among individuals lacking this protein. Moreover, it may be an attractive target to counter resistance that develops to gemcitabine in the treatment of pancreatic ductal adenocarcinoma (Wei et al., 2021).

Peripheral blood biomarkers can also be used to predict the efficacy of cancer treatments. Analyses of single blood cells using high-dimensional single-cell mass cytometry allowed for the determination of up to 50 different proteins in blood mononuclear cells that could predict the response to PD-1 immunotherapy (Krieg et al., 2018). Analyses of markers in blood have already been developed that allow for the detection of five different cancers as early as 4 years prior to the predicted onset of disease (Chen et al., 2020). There have also been indications that genetic markers associated with immune-related processes can predict the effectiveness of therapies. These comprise a wide assortment of markers, such as those reflecting T cell functionality, checkpoint regulators, chromatin modifiers, JAK-STAT pathway activation, and HLA variability (Keenan et al., 2019). Even if these markers do not allow for effective prevention strategies to be adopted, they can be instrumental in detecting cancers early and possibly making them more amenable to therapy.

While some markers may predict several types of cancer, others may be selective for certain tissue-specific cancers. For instance, a marker that may prove useful for lung cancer treatment was based on the finding that the characteristics of tissue-resident memory cells are related to the strength of the cytotoxic T cell response (Ganesan et al., 2017). The diversity of specific biomarkers can provide clues concerning optimal treatments in cancers that are difficult to treat. In the case of triple-negative breast cancer, for instance, a set of six proteins (kinases) were identified that signaled a particularly poor prognosis and high rate of cancer recurrence. Being able to identify which of these proteins was present permitted the determination of drug combinations that yielded excellent results in mice and might turn out to be useful within clinical settings (Zagorac et al., 2018).

Phenotypes comprising mutational load have been used to predict the effectiveness of immunotherapy. These features commonly include tumor cell antigens and neoantigens, the expression of viral genes and driver gene mutations, and indices of an overall mutational burden. Moreover, tumors with greater microsatellite instability (in which predisposition to mutations exists owing to impaired DNA repair) can serve as a marker in predicting the response to PD-1 immunotherapy (Mandal et al., 2019).

There are instances in which genetic mutations prevent tumors from either recognizing or receiving communication from T cells that instruct them to "stop growing," thereby undermining the efficacy of some anti-PD-1 agents. Accordingly, the presence of mutations related to signaling processes could serve as markers to determine the potential efficacy of checkpoint inhibitors. The expression of PD-L1 within tumor tissue is also predictive of the treatment response. Those tumors that produce high levels of PD-L1 respond relatively favorably to checkpoint inhibitors and could serve as a biomarker to decide whether a checkpoint inhibitor ought to be used for treatment. It also appeared that in some lung cancer patients the levels of a newly identified tumor suppressor, A20, was diminished allowing for the cancer to grow unimpeded. When A20 is low in cancer cells, they do not express sufficient PD-L1, making it less likely that immunotherapy involving PD-1/PD-L1 will be of value, and hence can be eliminated as a potential therapy (Breitenecker et al., 2021).

With so many potential tumor markers being present, it can be difficult to identify which are relevant to cancer occurrence and progression. Indeed, these markers can be present even in the absence of cancer, and conversely the presence of cancer may not always be accompanied by specific markers. Furthermore, as cancer cells can mutate over time, repeated analysis is necessary through liquid biopsies to determine optimal therapeutic strategies. In the end, problems related to sensitivity and specificity are ever-present, as is the possibility of overdiagnosis. Still, with increasing methods of determining markers related to protein structure and function, together with their expression pattern (proteomics), it will be possible to get a better handle on the treatments that will be best for specific cancers. Furthermore, combining proteomics with genome analyses (proteogenomics) allows for early cancer cell detection as well as differentiation of slow-growing and aggressive rapidly dividing cancer cells.

Features of the tumor microenvironment

Markers related to immune functioning, such as which immune cells have gathered in the tumor's immediate neighborhood, may provide essential information regarding the optimal immunotherapy to combat the disease. The presence of a T cell-inflamed tumor microenvironment (e.g., involving IFN-γ-related mRNA) has been linked to a beneficial outcome in response to PD-1 blockade (Cristescu et al., 2018). Relatedly, the abundance of tumor-infiltrating immune cell subpopulations (e.g., those characterized as FOXP3+ T_{reg} cells) might be indicative of the prognosis related to forms of cancer, such as nonsmall-cell lung cancer.

The presence of tumor-associated macrophages (TAMs) was likewise linked to numerous solid tumors and predicted cancer progression and resistance to treatment that occurred owing to their metabolic actions, so modifying metabolic processes can influence the efficacy of immunotherapy (Bader et al., 2020). In mice, the presence of cancer cells that expressed integrin molecules ($\alpha\nu\beta3$), a form of membrane glycoprotein, was more likely to become treatment resistant. By using an antibody that recognizes $\alpha\nu\beta3$, TAMs could be redirected to selectively attack tumors that had this marker present, thereby enhancing the effects of therapies. Conjoint analysis of the presence of TAMs and immune checkpoints may have still greater prognostic value than single markers. Of course, many other biomarkers have been identified within the tumor microenvironment, varying with different types of tumors, and may ultimately further enhance strategies to deal with diverse cancers.

Biomarkers for immunotherapy side effects

To a significant extent, the usefulness of anticancer treatments comes down to destroying cancer cells while leaving healthy cells unaffected and side effects limited. Accordingly, the success of the next round of immune-based therapies may, to a significant extent, be dependent on bringing the side effects of the treatment under control. Given the individual differences in the development of side effects, it would be desirable to identify biomarkers that predict vulnerability to treatment side effects. Overall, being female and having low levels of IL-6 independently predicted more severe adverse effects of checkpoint inhibitors, and pretreatment cytokine levels could also be used to predict the development of posttreatment autoimmune disorders. Several other markers that predicted side effects of immunotherapy have supported the contention that diminishing inflammation might effectively reduce adverse outcomes (Havel et al., 2019). Unfortunately, treatments that reduce inflammation may concurrently cause immune disturbances that leave individuals vulnerable to other illnesses. A form of microRNA, specifically miR-199, inhibits cyclin-dependent kinase 2 (CDK2) and is thus able to diminish the migration of neutrophils and hence limit inflammation, but without compromising immune functioning (Hsu et al., 2019). It is still too early to determine how effective this treatment may ultimately become, but such findings point to the possibility of better approaches to tame inflammation being developed as part of an effective treatment regimen.

While it is certainly important to diminish the side effects of treatments, the adverse events in patients who received PD-1/PD-L1 immunotherapy were directly related to progression-free and overall survival. This was apparent irrespective of cancer grade and has been observed across a range of cancer types (Matsuoka et al., 2020). Findings such as these raise the possibility that identifying the mechanisms underlying the development of treatment-related adverse events may provide important clues to promote enhanced efficacy of therapeutic outcomes.

Influence of microbiota on treatment responses to cancer therapies

Microbiota and immunotherapy

Just as the impact of chemotherapy could be influenced by microbial abundance and diversity, these factors markedly influence the actions of cancer immunotherapy. This is brought into sharper focus by knowledge that varied lifestyle factors modify the composition of gut microbiota. Increasing evidence has supported the view that commensal microbiota within the gut or specific cancer tissues might be reliable markers of the efficacy of cancer therapies, as might the microbes within the tumor microenvironment. As depicted in Fig. 15.6, a case has been made concerning the link between microbiota and the efficacy of immune checkpoint inhibitors in the treatment of gastrointestinal cancers, renal cell carcinoma, non-small-cell lung cancer, epithelial tumors, and malignant melanoma.

Consistent with a causal role for diverse microbiota in cancer treatment, fecal transplants from colorectal cancer patients to mice resulted in the local immune environment being altered, polyp formation increased, and pro-carcinogenic signals augmented. Conversely, fecal transplants from patients who responded positively to therapy yielded an improved response to therapy in mice (Gopalakrishnan et al., 2018). Among germ-free mice, bacterial reconstitution with fecal material from patients who responded to treatment led to enhanced T cell responses, better tumor control, and superior efficacy of anti-PD-L1 therapy (Matson et al., 2018). This did not occur, however, when fecal material was transferred from patients who had not responded to PD-1 antagonism but could if mice received oral supplementation with *Akkermansia muciniphila* (Routy et al., 2018b).

Selective fecal microbiota transplantation could be adopted to promote a diverse microbiome, thereby restoring effective epithelial defenses, and could diminish the side effects of immunotherapy; accordingly, this might be used as an adjunctive treatment among patients receiving immunotherapy. It was cautioned, however, that it is still uncertain whether the transfer of microbes mighty be accompanied by other factors that can have harmful effects. This said, it is possible to screen for the absence of infectious agents before transferring them to cancer patients. When this was done, transferring fecal microbiota from patients who responded to immunotherapy to melanoma patients who had not responded to immunotherapy (pembrolizumab or nivolumab), 6 of 15 showed tumor reduction or long-term disease stabilization. The patients who were responders to treatment displayed elevated abundance

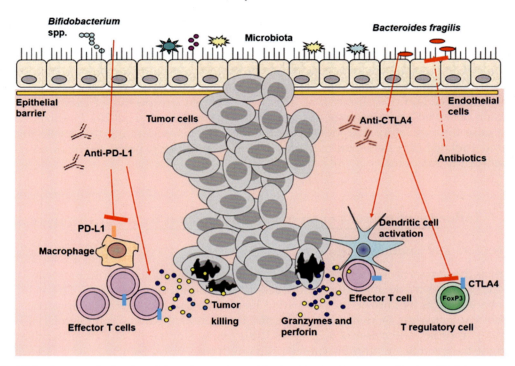

FIG. 15.6 Immune checkpoint inhibitor interactions with microbiota and immune cells. Anti-PD-L1 treatment cooperates with resident *Bifidobacterium* spp., leading to activation of dendritic cells, promoting the activation, expansion, and function of T effector cells. Anti-CTLA4 promotes the enrichment of resident *Bacteroides* spp. and enhances dendritic and effector T cell activation while suppressing T regulatory cell function. All of these mechanisms improve antitumor efficacy, in contrast to antibiotics that might decrease it. *Red arrows* represent stimulation, whereas *red arrows with red rectangle* represent inhibition. Figure and caption from Velikova, T., Krastev, B., Lozenov, S., Gencheva, R., Peshevska-Sekulovska, M., Nikolaev, G., Peruhova, M., 2021. Antibiotic-related changes in microbiome: the hidden villain behind colorectal carcinoma immunotherapy failure. Int. J. Mol. Sci. 22, 1754.

of microbes that had previously been found to enhance T cell functioning, increased CD8+ T cell activation, and lowered occurrence of IL-8 expressing myeloid cells (Davar et al., 2021). A phase I clinical trial evaluated the efficacy of fecal microbiota transplantation and re-induction of anti-PD-1 immunotherapy in patients who exhibited anti-PD-1-refractory metastatic melanoma. Of 10 patients, three exhibited partial or complete clinical responses comprising enhanced immune cell infiltrates and gene expression profiles within the thin layer of connective tissue that makes up the gut mucous membrane (lamina propria) and the tumor microenvironment (Baruch et al., 2021).

Just as increasing the abundance and diversity of some microbiota may have positive effects on cancer therapies, in mice treated with antibiotics to deplete their microbiota the effectiveness of immunotherapeutic agents was impaired (Iida et al., 2013). Such effects were likewise observed in patients treated for solid tumors, including nonsmall-cell lung cancer, renal cell cancer, and malignant melanoma. A meta-analysis that included 2740 patients confirmed that antibiotics limited the efficacy of immune checkpoint inhibitors, so the overall and progression-free survival was abbreviated (Huang et al., 2019b). Although it should be noted that the negative actions of antibiotics were apparent in patients who had received broad-spectrum antibiotic treatment before cancer therapy. Such an effect was not evident among patients who received the antibiotic treatment at the same time as the immune checkpoint inhibitors (Pinato et al., 2019). Fig. 15.7 provides a general summary of the effects of

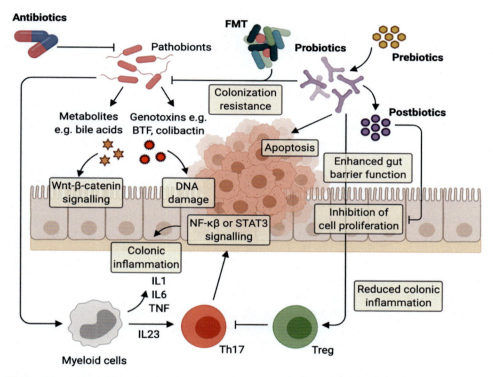

FIG. 15.7 The gut microbiota can influence colorectal carcinogenesis via a variety of mechanisms, including microbial-derived factors such as metabolites or genotoxins. Skewed host-microbe interactions contribute to the activation of pro-carcinogenic inflammatory pathways that ultimately lead to the progression of CRC. Antibiotic usage is effective in eradicating pathobionts, but its nonselective antimicrobial actions can affect gut homeostasis by also killing health-promoting bacteria and therefore reducing its application in CRC management. Prebiotic function fosters probiotic growth. Probiotics act through different anticancerogenic mechanisms: (i) probiotics can inhibit the colonization of pathogenic bacteria, (ii) they can enhance barrier function increasing mucin production and tight junction protein expression, (iii) they promote homeostatic immune responses, contributing to the expansion of antiinflammatory responses by T_{reg} cells and the modulation of proinflammatory cytokine release, (iv) they promote apoptosis on cancer cells. Postbiotics induce selective cytotoxicity against tumor cells as well as the control of tumor cell proliferation by inhibiting NFATc3 activation. Finally, fecal microbiota transplantation (FMT) could be used in CRC management to restore microbiome normobiosis and therefore induce homeostatic immune responses; nevertheless, potential complications associated with FMT include the risk of introducing new pathobionts and the spread of disease-associated genes. *Figure and caption from Perillo, F., Amoroso, C., Strati, F., Giuffrè, M.R., Díaz-Basabe, A., Lattanzi, G., Facciotti, F., 2020. Gut microbiota manipulation as a tool for colorectal cancer management: recent advances in its use for therapeutic purposes. Int. J. Mol. Sci. 21, 5389.*

antibiotics and immune-potentiating microorganisms as well as fecal microbiota transplantation on the response to therapy.

Enhancing PD-1/PD-L1 immunotherapy through microbial changes

Consistent with the position that diminished nutrient availability can enhance the effects of therapies, short-term starvation enhanced the effects of PD-1 blockade on cancer progression and prevented metastasis, apparently reflecting actions of reduced IGF-1 and down-regulated sensitivity of its receptor (Ajona et al., 2020). Protein restriction similarly affected tumor-associated macrophages and tumor progression, and there is ample reason to believe that dietary manipulations can have beneficial effects through changes of gut microbiota.

Although the view has often been endorsed that the balance between varied microbiota is essential in determining gut health, specific microbes and their combinations may have especially pertinent actions. Owing to the very great number of bacterial species that exist, it can be difficult to point to specific microbiota or sets of microorganisms relevant in predicting efficacious treatments, although several taxa and species have been suspected of enhancing positive effects in response to immunotherapies, as described in Fig. 15.8. Yet an abundance of select microbiota, such as *A. muciniphila, Bacteroides fragilis, Bifidobacterium*, and *Faecalibacterium*, was associated with improved responses to CTLA-4 or PD-1/PD-L1 therapy (Routy et al., 2018a). Depending on the specific microorganism constituency, a pro- or an antiinflammatory environment can be created, which could influence the efficacy of immunotherapies and moderate the toxicity that could emerge.

Supporting the causal connection between microbiota and the efficacy of cancer treatments, germ-free mice or those that had been treated with antibiotics did not respond to immunotherapy, but when given *B. fragilis* the therapeutic response to CTLA-4 inhibitors was promoted (Vétizou et al., 2015). Likewise, feeding mice a diet containing several *Bifidobacterium* species could augment the efficacy of a PD-L1 antibody in diminishing cancer growth (Longhi et al., 2020) as could *A. muciniphila* (Routy et al., 2018b). It seems that *Bifidobacterium* enhanced PD-L1 blockade efficacy by augmenting dendritic cell function, enhancing CD8+ T cell priming and their accumulation within the tumor's microenvironment (Roy and Trinchieri, 2017). The benefits of this bacterium might come from its ability to migrate to tumors where they activate the immune signaling STING pathway, which promotes the activation of immune cells. Therefore, mice that otherwise might not respond to immunotherapy can be transformed into treatment responders by dint of specific transfer of *Bifidobacterium* bacteria.

Several other bacterial species were related to enhancements of immunotherapy in humans, and the baseline presence of some of these predicted the response to treatment. The richness of the microbiota and the presence of specific bacteria were associated with longer progression-free survival in melanoma patients treated with immunotherapy, as well as in the treatment of nonsmall-cell lung cancer and renal cell carcinoma. The therapeutic efficacy of immunotherapy in the treatment of highly lethal pancreatic ductal adenocarcinoma could similarly be enhanced by modifications of microbiota within the tumor microenvironment. Paralleling these findings, in multiple myeloma patients who exhibited minimal residual disease following treatment, the abundance of *Eubacterium hallii* was greater than in patients who had not shown a positive treatment response (Pianko et al., 2019). Once more, these reports point to metabolic components and gut microbiota being tied to tumor progression and could potentially be used to predict the response to treatment and might also be instrumental in causing cold tumors to turn hot.

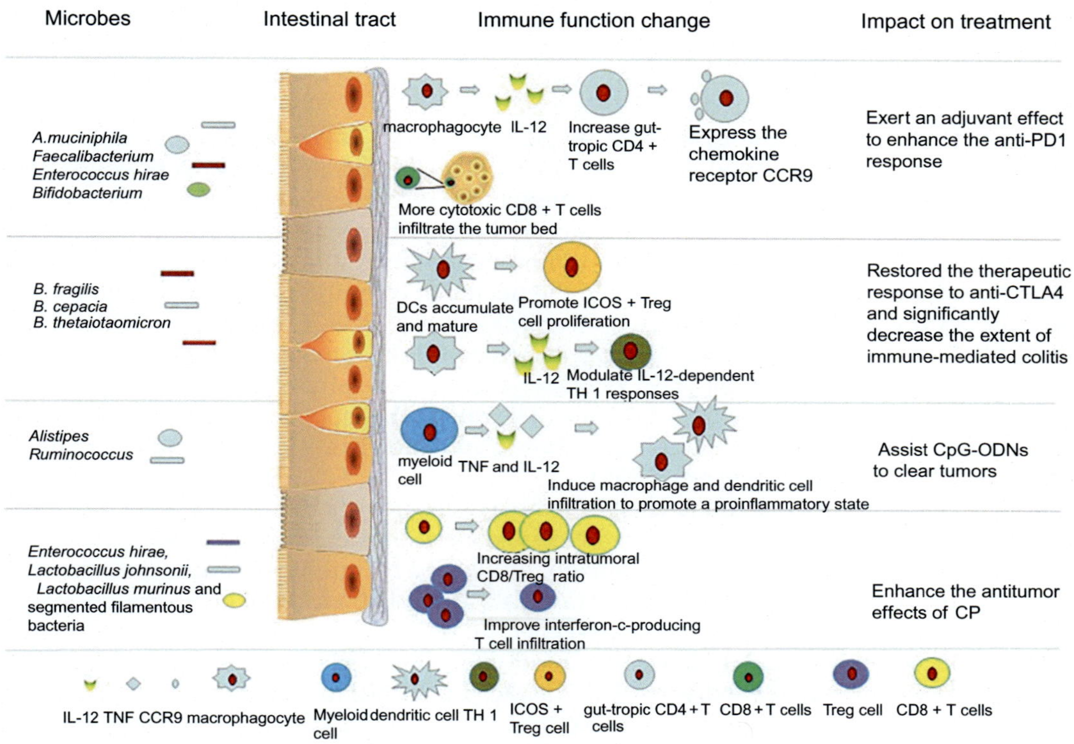

FIG. 15.8 Intestinal microbiota influence the therapeutic effect of drugs on tumors by regulating the immune system. Gut microbiota enhance some cancer immunotherapies (e.g., anti-PD-L1 efficacy) by reactivating dendritic cells. Dendritic cells can boost CD8+ T cell responses and increase the number of gut-tropic CD4+ T cells to defeat tumors. Gut microbiota can also trigger dendritic cell maturation and modulate IL-12–dependent T_h1 responses to restore the therapeutic response to another form of immunotherapy, that of anti-CTLA4. The microbiota are associated with side effects of immunotherapy. When germ-free mice receiving anti-CTLA-4 monoclonal antibodies were monopolized with *B. fragilis*, plasmacytoid DCs accumulated and matured in mesenteric lymph nodes. Dendritic cells can promote T_{reg} cells to proliferate in the lamina propria. The gut microbiota help CpG oligonucleotides to promote myeloid cells to secrete proinflammatory cytokines such as TNF and IL-12. These cytokines induce macrophage and dendritic cell infiltration to promote a proinflammatory state. The body develops an antigen-specific adaptive T cell antitumor immunity to clear tumors in this proinflammatory microenvironment. Some of the intestinal microbiota affect the antitumor efficacy of cyclophosphamide. *E. hirae* translocation could improve the intratumoral $CD8/T_{reg}$ ratio. In addition, the Gram-negative *Barnesiella intestinihominis* improved interferon-c–producing T cell infiltration in cancer lesions to enhance the antitumor effects of cyclophosphamide. *Figure and caption from Ma, W., Mao, Q., Xia, W., Dong, G., Yu, C., et al., 2019. Gut microbiota shapes the efficiency of cancer therapy. Front. Microbiol. 10, 1050.*

The combination of specific commensal microbes might behave cooperatively during cancer progression and in response to immunotherapies. Among patients who exhibited a positive response to immunotherapy, the levels of *Bifidobacterium longum*, *Collinsella aerofaciens*, and *Enterococcus faecium* were appreciably elevated (Matson et al., 2018). Likewise, melanoma patients who responded to anti-PD-1 therapy exhibited greater bacterial diversity and greater abundance of *Clostridial*, whereas the poor treatment responders displayed greater abundance of *Bacteroidales*. It is thought that *B. fragilis* and *Burkholderia* may act in tandem to protect healthy tissue from damage that could be accrued because of cancer therapy, and *B. fragilis* together with *Bifidobacterium* may enhance the effect of checkpoint blockade in their antitumor actions.

Several other microorganisms, such as *A. muciniphila*, *Ruminococcus*, *Eubacterium*, and *Alistipes*, were overrepresented among patients who exhibited a positive treatment response to a PD-1 inhibitor, whereas underrepresentation of *Bifidobacterium* and *Parabacteroides* were seen in nonresponders. Elevated abundance and diversity of microbiota were similarly associated with a superior response to PD-1 immunotherapy in the treatment of nonsmall-cell lung cancer through interactions between the microbial community present within the lung and that of the gut (the gut-lung axis) (Nagasaka et al., 2020). Consistent with these findings, PD-1 immunotherapy was effective in hepatocellular carcinoma, preferentially influencing individuals with greater taxa richness, and the diversity of microbiota became still more pronounced over the course of therapy. The differential appearance of microbiota could be detected within 3–6 weeks following treatment initiation and could be used to predict the efficacy of the therapy (Zheng et al., 2019).

Much as resistance develops to chemotherapy, resistance to immune checkpoint blockade also occurs, possibly being related to the composition of the gut microbiota. Among cancer patients who were or were not positively influenced by immunotherapy, gut microbiota could be distinguished from one another (Yi et al., 2019). Patients with greater abundance of bacteria from the *Ruminococcaceae* family generally fared better with PD-1 immunotherapy than did those with low abundance, but more than this, the response to immunotherapy was highly dependent on the diversity of gut bacteria (Gopalakrishnan et al., 2018). For that matter, the pronounced differences in response to therapy in humans, including treatment resistance, might be linked to the marked differences of microbiota populations that exist across individuals.

The full gamut of mechanisms by which microbiota could affect the response to immunotherapies is uncertain, although it does seem that toxigenic bacteria favored cancer development by altering inflammation or activation of immune cells. Aside from these actions, the link to cancer treatment response could occur through interactions between PAMPs and both antigen-presenting cells and innate effectors. Moreover, cytokine production by antigen-presenting cells or lymphocytes can promote systemic changes that elicit both local and distal tumorigenic consequences (Helmink et al., 2019).

Overall, it seems that specific microbiota or microbial combinations can influence the effects of immunotherapy through various immune routes that may additively or synergistically promote positive outcomes. Further, microbial factors could potentially serve as biomarkers to predict which patients will benefit from immunotherapy, whether the effects of the treatments would be long-lasting, and whether the side effects can be tolerated.

Advancing microbial treatments in modulating the effects of checkpoint inhibitors

Strong evidence has supported microbial involvement in triggering the progression of several forms of cancer and in the modulation of various forms of cancer therapies. It was indicated, however, that limited convincing data have been provided showing that microbiota in humans cause the development of cancers (Sepich-Poore et al., 2021). In the absence of prospective studies, it is uncertain whether microbial factors cause cancer occurrence rather than reflecting the actions of cancer on the microbiome, or simply being an associated but noncausal constitutional element of cancer patients.

To determine more fully the influence of microbiota in moderating the effects of cancer therapies, further detailed analyses are necessary to define and evaluate the influence of specific microbiota in conjunction with immunotherapy. The marked individual differences that exist in gut microbiota may result in individuals being differentially responsive to the specific microbial adjunctive treatments. This is especially the case given that the effects of these treatments may also vary with different types of cancers. As the clinical response to immunotherapy may also be dependent on the individual's genetic signature, it is reasonable to assume that further complex interactions would need to be considered for the combined use of microbial treatments and immunotherapy. Issues would also need to be addressed concerning the dose of microbial treatments, the frequency of their administration, as well as the timing of these treatments relative to immunotherapy. The complexity of such analyses is exponentially greater given that microorganisms, such as specific fungi and viruses, interact with bacteria and could affect immune responses. Of course, analyses would necessarily need to consider the influence of dietary factors, epigenetic alterations that come about with diet, as well as individual difference factors, such as age and obesity (Baiden-Amissah and Tuyaerts, 2019). Even different geographical regions within a country—as well as specific cultural and ethnic communities and dietary practices—can influence the presence of common microbial species that are present across individuals and can thus have very different implications for microbiota related to disease occurrence and the response to therapies.

Enhancing CAR T therapy through microbial changes

Unlike the large number of studies that revealed positive relations between microbiota and immune checkpoint inhibitors, the data linking microbiota to the influence of CAR T therapy have been lacking. It was observed, however, that the abundance of microbiota and the presence of certain microbes in CAR T treatment responders differed from that of nonresponders (Smith et al., 2019). Specifically, *Oscillospiraceae*, *Ruminococcaceae*, and *Lachnospiraceae* were more abundant in treatment responders, whereas *Peptostreptococcaceae* was more abundant in nonresponders. Unfortunately, the high abundance of *Lachnospiraceae* was also associated with greater toxicity. In addition to these specific microbes, the SCFA butyrate increased the cytotoxic actions of CD8+ T cells and similarly enhanced the action of CAR T cells (Luu et al., 2021). Despite the limited data, given the beneficial actions of specific microbiota in moderating the actions of checkpoint inhibitors, it would be important to determine whether similar actions are apparent in conjunction with CAR T therapy. As it stands, an impressive case has been made supporting the contention that microbiota might be influential in moderating the influence of CAR T therapy (Abid et al., 2019), but firm conclusions will need to await studies that prospectively assess the influence of microbiota in this form of treatment.

Prebiotics and probiotics in cancer treatment

Since microbiota can affect immunotherapeutic responses, microbial manipulations through the ingestion of pre- or probiotic supplements have been proposed to enhance cancer treatments and in the management of treatment-related toxicity. Research in animals, as well as several studies in humans, supported the suggestion that pre- and probiotic supplements can influence cancer development and modify cancer progression. Reviews of the pertinent clinical trials of colorectal patients also indicated that preoperative probiotics and synbiotics enhanced gut barrier functioning, reduced inflammatory factors, diminished the side effects of chemotherapy, and lessened postoperative infectious complications (Darbandi et al., 2020). There are, to be sure, several ways by which microbiota supplements could affect the response to cancer treatments, such as by influencing the metabolism or toxicity of drugs, suppress the development and growth of pathogens, regulate gut enzymatic activity, and enhance immune functioning.

In some quarters, considerable optimism had been expressed concerning the potential benefits of microbial supplements in limiting cancer occurrence and enhancing treatments, but the data were not overly convincing. Most studies were conducted in rodents, and insufficient data have been available from human trials. While several reports indicated that microbial supplements could diminish colon inflammation, such effects weren't necessarily accompanied by diminished colorectal cancer. It has been maintained that the observed positive actions largely comprised reduced side effects, and the data were not sufficiently strong to form definitive conclusions regarding treatment effects on cancer progression. The complexity regarding microbial factors and the response to cancer treatments was apparent when the actions of probiotics were evaluated in melanoma patients treated through immunotherapy. Some probiotic supplements seemed to have beneficial effects, but negative outcomes could also emerge, as they did in a small study in which patients treated with a probiotic exhibited a poorer response to immunotherapy.

The provision of probiotic supplements did not reduce stay in the ICU or mortality in hospitalized patients, although ICU-acquired pneumonia was reduced. Unfortunately, treatment with probiotics among patients in the ICU substantially increased the risk for *Lactobacillus* bacteremia, a relatively infrequent condition that favors diminished host defenses and increased illness vulnerability (Yelin et al., 2019). Furthermore, probiotics did not consistently promote microbiota reconstitution following antibiotic treatment, and in some instances even disturbed the appearance of important microorganisms, which contrasts with the effects of fecal microbiota transplants that were effective in reconstituting the microbiota (Suez et al., 2018). For the moment, it does not appear that probiotic supplementation among seriously ill patients is beneficial, but it remains possible that they might have positive actions based on a personalized approach to the probiotics consumed (Zmora et al., 2019).

This brings us back to the question raised earlier as to what constitutes a good microbial mix, and what consequences can be expected when this balance is disturbed. Given the exceptional complexity of the microbiome and its interactions with other systems, it shouldn't be surprising that our understanding of whether or how pre- and probiotics might come to influence cancer has yet to be deduced (Helmink et al., 2019). The effectiveness of treatment might be related to a deficiency or excess of specific microbes or could be related to broad dysbiosis. Accordingly, the development of a profile of microbiota metabolites could potentially be used to predict the fate of anticancer drugs.

With the enormous interest in the potential therapeutic effects that could be obtained by manipulation of microbiota, heavy investments have been made in developing related treatments for cancer and a variety of other chronic diseases. Taking advantage of the links between microbiota and immune functioning may well be possible, but as we've repeatedly seen, getting to that place on the road may be limited by numerous obstacles. It is unfortunate that like so many other over-the-counter supplements, probiotic formulations are being sold to treat all manner of diseases, despite not having been systematically assessed for safety or efficacy. In fact, under conditions where microbiota abundance and diversity are not impaired, it is debatable whether probiotic supplements are effective in colonizing the gut, and there is the question of whether these compounds are even safe, let alone effective. So as difficult as it might be, it would be propitious to adopt an approach that identifies specific microbiota or metabolite deficiencies and administer probiotics on this basis (Zmora et al., 2018).

Summary and conclusions

Despite what might appear to be slow progress, our understanding of cancer processes and the development of effective cancer therapies has been advancing at an inexorable pace. With this has come a greater appreciation for the importance of sustaining individual mental wellness and what steps need to be taken to ensure this pursuant to cancer therapy. Of particular significance is that efforts in developing new therapies have been matched by greater emphasis on having people adopt prevention measures, which is of utmost importance given that cancer has been viewed as being preventable in more than 40% of cases. Numerous lifestyle factors, particularly diet and exercise, might not only serve in a preventive capacity but can also augment the efficacy of cancer treatments. Obtaining proper sleep and diminishing distress can also serve in a similar manner, although the available data are less extensive.

Diet and exercise (and avoidance of sedentary behaviors) can have positive effects on cancer occurrence and its treatment, including action against excessive inflammatory processes. These lifestyles also affect microbiota that can act against the initiation and progression of numerous types of cancer, may serve as biomarkers relevant to treatments, and can affect the response to several forms of therapy. These outcomes might occur through complex interactions: between an array of microbes within the gut and at other sites, microbiota within the tumor microenvironment, and with various host factors. In the final analysis, to fully understand the dynamics involved in the actions of microbiota in the context of therapeutics, a systems biology approach will be required that incorporates microbial ecology, immunology, cancer cell biology, and computational biology (Xavier et al., 2020).

References

Abid, M.B., Shah, N.N., Maatman, T.C., Hari, N., 2019. Gut microbiome and CAR-T therapy. Exp. Hematol. Oncol. 8, 31.

Agudelo, L.Z., Ferreira, D.M.S., Dadvar, S., Cervenka, I., Ketscher, L., et al., 2019. Skeletal muscle PGC-1alpha1 reroutes kynurenine metabolism to increase energy efficiency and fatigue-resistance. Nat. Commun. 10, 2767.

Ajona, D., Ortiz-Espinosa, S., Lozano, T., Exposito, F., Calvo, A., et al., 2020. Short-term starvation reduces IGF-1 levels to sensitize lung tumors to PD-1 immune checkpoint blockade. Nat. Cancer 1, 75–85.

Anisman, H., Poulter, M.O., Gandhi, R., Merali, Z., Hayley, S., 2007. Interferon-alpha effects are exaggerated when administered on a psychosocial stressor backdrop: cytokine, corticosterone and brain monoamine variations. J. Neuroimmunol. 186, 45–53.

Bader, J.E., Voss, K., Rathmell, J.C., 2020. Targeting metabolism to improve the tumor microenvironment for cancer immunotherapy. Mol. Cell 78, 1019–1033.

Baiden-Amissah, R.E.M., Tuyaerts, S., 2019. Contribution of aging, obesity, and microbiota on tumor immunotherapy efficacy and toxicity. Int. J. Mol. Sci. 20, 3586.

Bakkalci, D., Jia, Y., Winter, J.R., Lewis, J.E., Taylor, G.S., et al., 2020. Risk factors for Epstein Barr virus-associated cancers: a systematic review, critical appraisal, and mapping of the epidemiological evidence. J. Glob. Health 10, 010405.

Barkal, A.A., Brewer, R.E., Markovic, M., Kowarsky, M., Barkal, S.A., et al., 2019. CD24 signalling through macrophage Siglec-10 is a target for cancer immunotherapy. Nature 572, 392–396.

Bartok, O., Pataskar, A., Nagel, R., Laos, M., Goldfarb, E., et al., 2021. Anti-tumour immunity induces aberrant peptide presentation in melanoma. Nature 590, 332–337.

Baruch, E.N., Youngster, I., Ben-Betzalel, G., Ortenberg, R., Lahat, A., et al., 2021. Fecal microbiota transplant promotes response in immunotherapy-refractory melanoma patients. Science 371, 602–609.

Bear, A.S., Blanchard, T., Cesare, J., Ford, M.J., Richman, L.P., et al., 2021. Biochemical and functional characterization of mutant KRAS epitopes validates this oncoprotein for immunological targeting. Nat. Commun. 12, 4365.

Binnewies, M., Roberts, E.W., Kersten, K., Chan, V., Fearon, D.F., et al., 2018. Understanding the tumor immune microenvironment (TIME) for effective therapy. Nat. Med. 24, 541–550.

Borden, E.C., 2019. Interferons alpha and beta in cancer: therapeutic opportunities from new insights. Nat. Rev. Drug Discov. 18, 219–234.

Botticelli, A., Vernocchi, P., Marini, F., Quagliariello, A., Cerbelli, B., et al., 2020. Gut metabolomics profiling of non-small cell lung cancer (NSCLC) patients under immunotherapy treatment. J. Transl. Med. 18, 49.

Breitenecker, K., Homolya, M., Luca, A.C., Lang, V., Trenk, C., et al., 2021. Down-regulation of A20 promotes immune escape of lung adenocarcinomas. Sci. Transl. Med. 13, eabc3911.

Capuron, L., Miller, A.H., 2011. Immune system to brain signaling: neuropsychopharmacological implications. Pharmacol. Ther. 130, 226–238.

Chen, J., Lopez-Moyado, I.F., Seo, H., Lio, C.J., Hempleman, L.J., et al., 2019. NR4A transcription factors limit CAR T cell function in solid tumours. Nature 567, 530–534.

Chen, X., Gole, J., Gore, A., He, Q., Lu, M., et al., 2020. Non-invasive early detection of cancer four years before conventional diagnosis using a blood test. Nat. Commun. 11, 3475.

Chen, D.S., Mellman, I., 2017. Elements of cancer immunity and the cancer-immune set point. Nature 541, 321–330.

Chowdhury, S., Castro, S., Coker, C., Hinchliffe, T.E., Arpaia, N., et al., 2019. Programmable bacteria induce durable tumor regression and systemic antitumor immunity. Nat. Med. 25, 1057–1063.

Cristescu, R., Mogg, R., Ayers, M., Albright, A., Murphy, E., et al., 2018. Pan-tumor genomic biomarkers for PD-1 checkpoint blockade-based immunotherapy. Science 362, eaar 3593.

Crowther, M.D., Dolton, G., Legut, M., Caillaud, M.E., Lloyd, A., et al., 2020. Genome-wide CRISPR-Cas9 screening reveals ubiquitous T cell cancer targeting via the monomorphic MHC class I-related protein MR1. Nat. Immunol. 21, 178–185.

Darbandi, A., Mirshekar, M., Shariati, A., Moghadam, M.T., Lohrasbi, V., et al., 2020. The effects of probiotics on reducing the colorectal cancer surgery complications: a periodic review during 2007-2017. Clin. Nutr. 39, 2358–2367.

Davar, D., Dzutsev, A.K., McCulloch, J.A., Rodrigues, R.R., Chauvin, J.M., et al., 2021. Fecal microbiota transplant overcomes resistance to anti-PD-1 therapy in melanoma patients. Science 371, 595–602.

Davila, M.L., Riviere, I., Wang, X., Bartido, S., Park, J., et al., 2014. Efficacy and toxicity management of 19-28z CAR T cell therapy in B cell acute lymphoblastic leukemia. Sci. Transl. Med. 6, 224ra225.

Demaria, O., Cornen, S., Daeron, M., Morel, Y., Medzhitov, R., et al., 2019. Harnessing innate immunity in cancer therapy. Nature 574, 45–56.

Deshpande, R.P., Sharma, S., Watabe, K., 2020. The confounders of cancer immunotherapy: roles of lifestyle, metabolic disorders and sociological factors. Cancer 12, 2983.

Di Pilato, M., Kim, E.Y., Cadilha, B.L., Prussmann, J.N., Nasrallah, M.N., et al., 2019. Targeting the CBM complex causes Treg cells to prime tumours for immune checkpoint therapy. Nature 570, 112–116.

Esaki, S., Rabkin, S.D., Martuza, R.L., Wakimoto, H., 2016. Transient fasting enhances replication of oncolytic herpes simplex virus in glioblastoma. Am. J. Cancer Res. 6, 300–311.

Fairfax, B.P., Taylor, C.A., Watson, R.A., Nassiri, I., Danielli, S., et al., 2020. Peripheral CD8(+) T cell characteristics associated with durable responses to immune checkpoint blockade in patients with metastatic melanoma. Nat. Med. 26, 193–199.

Fraietta, J.A., Lacey, S.F., Orlando, E.J., Pruteanu-Malinici, I., Gohil, M., et al., 2018. Determinants of response and resistance to CD19 chimeric antigen receptor (CAR) T cell therapy of chronic lymphocytic leukemia. Nat. Med. 24, 563–571.
Fu, W., Smith, C., Turner, L., Fojtik, J., Pacyna, J.E., et al., 2019. Characteristics and scope of training of clinicians participating in the us direct-to-consumer marketplace for unproven stem cell interventions. JAMA 321, 2463–2464.
Ganesan, A.P., Clarke, J., Wood, O., Garrido-Martin, E.M., Chee, S.J., et al., 2017. Tissue-resident memory features are linked to the magnitude of cytotoxic T cell responses in human lung cancer. Nat. Immunol. 18, 940–950.
Garbuglia, A.R., Lapa, D., Sias, C., Capobianchi, M.R., Del Porto, P., 2020. The use of both therapeutic and prophylactic vaccines in the therapy of papillomavirus disease. Front. Immunol. 11, 188.
Giordano-Attianese, G., Gainza, P., Gray-Gaillard, E., Cribioli, E., Shui, S., et al., 2020. A computationally designed chimeric antigen receptor provides a small-molecule safety switch for T-cell therapy. Nat. Biotechnol. 38, 426–432.
Gomes-Santos, I.L., Amoozgar, Z., Kumar, A.S., Ho, W.W., Roh, K., et al., 2021. Exercise training improves tumor control by increasing CD8+ T-cell infiltration via CXCR3 Signaling and sensitizes breast cancer to immune checkpoint blockade. Cancer Immunol. Res. 9, 765–778.
Gopalakrishnan, V., Spencer, C.N., Nezi, L., Reuben, A., Andrews, M.C., et al., 2018. Gut microbiome modulates response to anti-PD-1 immunotherapy in melanoma patients. Science 359, 97–103.
Gordon, S.R., Maute, R.L., Dulken, B.W., Hutter, G., George, B.M., et al., 2017. PD-1 expression by tumour-associated macrophages inhibits phagocytosis and tumour immunity. Nature 545, 495–499.
Gustafson, M.P., Wheatley-Guy, C.M., Rosenthal, A.C., Gastineau, D.A., Katsanis, E., et al., 2021. Exercise and the immune system: taking steps to improve responses to cancer immunotherapy. J. Immunother. Cancer 9, e001872.
Hammerich, L., Marron, T.U., Upadhyay, R., Svensson-Arvelund, J., Dhainaut, M., et al., 2019. Systemic clinical tumor regressions and potentiation of PD1 blockade with in situ vaccination. Nat. Med. 25, 814–824.
Havel, J.J., Chowell, D., Chan, T.A., 2019. The evolving landscape of biomarkers for checkpoint inhibitor immunotherapy. Nat. Rev. Cancer 19, 133–150.
Helmink, B.A., Khan, M.A.W., Hermann, A., Gopalakrishnan, V., Wargo, J.A., 2019. The microbiome, cancer, and cancer therapy. Nat. Med. 25, 377–388.
Helmink, B.A., Reddy, S.M., Gao, J., Zhang, S., Basar, R., et al., 2020. B cells and tertiary lymphoid structures promote immunotherapy response. Nature 577, 549–555.
Hou, A.J., Chen, L.C., Chen, Y.Y., 2021. Navigating CAR-T cells through the solid-tumour microenvironment. Nat. Rev. Drug Discov. 20, 531–550.
Hsu, J., Hodgins, J.J., Marathe, M., Nicolai, C.J., Bourgeois-Daigneault, M.C., et al., 2018. Contribution of NK cells to immunotherapy mediated by PD-1/PD-L1 blockade. J. Clin. Invest. 128, 4654–4668.
Hsu, A.Y., Wang, D., Liu, S., Lu, J., Syahirah, R., et al., 2019. Phenotypical microRNA screen reveals a noncanonical role of CDK2 in regulating neutrophil migration. Proc. Natl. Acad. Sci. 116, 18561–18570.
Huang, A.C., Orlowski, R.J., Xu, X., Mick, R., George, S.M., et al., 2019a. A single dose of neoadjuvant PD-1 blockade predicts clinical outcomes in resectable melanoma. Nat. Med. 25, 454–461.
Huang, X.Z., Gao, P., Song, Y.X., Xu, Y., Sun, J.X., et al., 2019b. Antibiotic use and the efficacy of immune checkpoint inhibitors in cancer patients: a pooled analysis of 2740 cancer patients. Oncoimmunology 8, e1665973.
Iida, N., Dzutsev, A., Stewart, C.A., Smith, L., Bouladoux, N., et al., 2013. Commensal bacteria control cancer response to therapy by modulating the tumor microenvironment. Science 342, 967–970.
June, C.H., Sadelain, M., 2018. Chimeric antigen receptor therapy. N. Engl. J. Med. 379, 64–73.
Kaiser, J., 2018. Too much of a good thing? Science 359, 1346–1347.
Keenan, T.E., Burke, K.P., Van Allen, E.M., 2019. Genomic correlates of response to immune checkpoint blockade. Nat. Med. 25, 389–402.
Koblansky, A.A., Truax, A.D., Liu, R., Montgomery, S.A., Ding, S., et al., 2016. The innate immune receptor nlrx1 functions as a tumor suppressor by reducing colon tumorigenesis and key tumor-promoting signals. Cell Rep. 14, 2562–2575.
Krieg, C., Nowicka, M., Guglietta, S., Schindler, S., Hartmann, F.J., et al., 2018. High-dimensional single-cell analysis predicts response to anti-PD-1 immunotherapy. Nat. Med. 24, 144–153.
Laurans, L., Venteclef, N., Haddad, Y., Chajadine, M., Alzaid, F., et al., 2018. Genetic deficiency of indoleamine 2,3-dioxygenase promotes gut microbiota-mediated metabolic health. Nat. Med. 24, 1113–1120.
Li, H., Bullock, K., Gurjao, C., Braun, D., Shukla, S.A., et al., 2019. Metabolomic adaptations and correlates of survival to immune checkpoint blockade. Nat. Commun. 10, 4346.

Liu, Y., Fang, Y., Chen, X., Wang, Z., Liang, X., et al., 2020. Gasdermin E-mediated target cell pyroptosis by CAR T cells triggers cytokine release syndrome. Sci. Immunol. 5.

Longhi, G., van Sinderen, D., Ventura, M., Turroni, F., 2020. Microbiota and cancer: the emerging beneficial role of bifidobacteria in cancer immunotherapy. Front. Microbiol. 11, 575072.

Luu, M., Riester, Z., Baldrich, A., Reichardt, N., Yuille, S., et al., 2021. Microbial short-chain fatty acids modulate CD8(+) T cell responses and improve adoptive immunotherapy for cancer. Nat. Commun. 12, 4077.

Lynn, R.C., Weber, E.W., Sotillo, E., Gennert, D., Xu, P., et al., 2019. c-Jun overexpression in CAR T cells induces exhaustion resistance. Nature 576, 293–300.

Ma, L., Dichwalkar, T., Chang, J.Y.H., Cossette, B., Garafola, D., et al., 2019. Enhanced CAR-T cell activity against solid tumors by vaccine boosting through the chimeric receptor. Science 365, 162–168.

Mandal, R., Samstein, R.M., Lee, K.W., Havel, J.J., Wang, H., et al., 2019. Genetic diversity of tumors with mismatch repair deficiency influences anti-PD-1 immunotherapy response. Science 364, 485–491.

Martin, N.T., Bell, J.C., 2018. Oncolytic virus combination therapy: killing one bird with two stones. Mol. Ther. 26, 1414–1422.

Martín-Ruiz, A., Fiuza-Luces, C., Rincon-Castanedo, C., Fernandez-Moreno, D., Galvez, B.G., et al., 2020. Benefits of exercise and immunotherapy in a murine model of human non-small-cell lung carcinoma. Exerc. Immunol. Rev. 26, 100–115.

Mathewson, N.D., Jenq, R., Mathew, A.V., Koenigsknecht, M., Hanash, A., et al., 2016. Gut microbiome-derived metabolites modulate intestinal epithelial cell damage and mitigate graft-versus-host disease. Nat. Immunol. 17, 505–513.

Matson, V., Fessler, J., Bao, R., Chongsuwat, T., Zha, Y., et al., 2018. The commensal microbiome is associated with anti-PD-1 efficacy in metastatic melanoma patients. Science 359, 104–108.

Matsuoka, H., Hayashi, T., Takigami, K., Imaizumi, K., Shiroki, R., et al., 2020. Correlation between immune-related adverse events and prognosis in patients with various cancers treated with anti PD-1 antibody. BMC Cancer 20, 656.

Maude, S.L., Frey, N., Shaw, P.A., Aplenc, R., Barrett, D.M., et al., 2014. Chimeric antigen receptor T cells for sustained remissions in leukemia. N. Engl. J. Med. 371, 1507–1517.

McNab, F., Mayer-Barber, K., Sher, A., Wack, A., O'Garra, A., 2015. Type I interferons in infectious disease. Nat. Rev. Immunol. 15, 87–103.

Mehta, R.S., Rezvani, K., 2018. Chimeric antigen receptor expressing natural killer cells for the immunotherapy of cancer. Front. Immunol. 9, 283.

Miller, A.H., Raison, C.L., 2016. The role of inflammation in depression: from evolutionary imperative to modern treatment target. Nat. Rev. Immunol. 16, 22–34.

Moylan, S., Berk, M., Dean, O.M., Samuni, Y., Williams, L.J., et al., 2014. Oxidative & nitrosative stress in depression: why so much stress? Neurosci. Biobehav. Rev. 45, 46–62.

Munn, D.H., Mellor, A.L., 2016. IDO in the tumor microenvironment: inflammation, counter-regulation, and tolerance. Trends Immunol. 37, 193–207.

Nagasaka, M., Sexton, R., Alhasan, R., Rahman, S., Azmi, A.S., et al., 2020. Gut microbiome and response to checkpoint inhibitors in non-small cell lung cancer—a review. Crit. Rev. Oncol. Hematol. 145, 102841.

Nakagawa, H., Sido, J.M., Reyes, E.E., Kiers, V., Cantor, H., et al., 2016. Instability of Helios-deficient Tregs is associated with conversion to a T-effector phenotype and enhanced antitumor immunity. Proc. Natl. Acad. Sci. 113, 6248–6253.

Newman, J.H., Chesson, C.B., Herzog, N.L., Bommareddy, P.K., Aspromonte, S.M., et al., 2020. Intratumoral injection of the seasonal flu shot converts immunologically cold tumors to hot and serves as an immunotherapy for cancer. Proc. Natl. Acad. Sci. 117, 1119–1128.

Norelli, M., Camisa, B., Barbiera, G., Falcone, L., Purevdorj, A., et al., 2018. Monocyte-derived IL-1 and IL-6 are differentially required for cytokine-release syndrome and neurotoxicity due to CAR T cells. Nat. Med. 24, 739–748.

Opitz, C.A., Somarribas Patterson, L.F., Mohapatra, S.R., Dewi, D.L., Sadik, A., et al., 2020. The therapeutic potential of targeting tryptophan catabolism in cancer. Br. J. Cancer 122, 30–44.

Ott, P.A., Hu, Z., Keskin, D.B., Shukla, S.A., Sun, J., et al., 2017. An immunogenic personal neoantigen vaccine for patients with melanoma. Nature 547, 217–221.

Overacre-Delgoffe, A.E., Chikina, M., Dadey, R.E., Yano, H., Brunazzi, E.A., et al., 2017. Interferon-gamma drives Treg fragility to promote anti-tumor immunity. Cell 169, 1130–1141 e1111.

Pape, K., Tamouza, R., Leboyer, M., Zipp, F., 2019. Immunoneuropsychiatry—novel perspectives on brain disorders. Nat. Rev. Neurol. 15, 317–328.

Pardi, N., Hogan, M.J., Porter, F.W., Weissman, D., 2018. mRNA vaccines—a new era in vaccinology. Nat. Rev. Drug Discov. 17, 261–279.

Paucek, R.D., Baltimore, D., Li, G., 2019. The cellular immunotherapy revolution: arming the immune system for precision therapy. Trends Immunol 40, 292–309.

Peled, J.U., Gomes, A.L.C., Devlin, S.M., Littmann, E.R., Taur, Y., et al., 2020. Microbiota as predictor of mortality in allogeneic hematopoietic-cell transplantation. N. Engl. J. Med. 382, 822–834.

Perez-Ruiz, E., Minute, L., Otano, I., Alvarez, M., Ochoa, M.C., et al., 2019. Prophylactic TNF blockade uncouples efficacy and toxicity in dual CTLA-4 and PD-1 immunotherapy. Nature 569, 428–432.

Pianko, M.J., Devlin, S.M., Littmann, E.R., Chansakul, A., Mastey, D., et al., 2019. Minimal residual disease negativity in multiple myeloma is associated with intestinal microbiota composition. Blood Adv. 3, 2040–2044.

Pinato, D.J., Howlett, S., Ottaviani, D., Urus, H., Patel, A., et al., 2019. Association of prior antibiotic treatment with survival and response to immune checkpoint inhibitor therapy in patients with cancer. JAMA Oncol. 5, 1774–1778.

Rashidi, A., Kaiser, T., Graiziger, C., Holtan, S.G., Rehman, T.U., et al., 2020. Gut dysbiosis during antileukemia chemotherapy versus allogeneic hematopoietic cell transplantation. Cancer 126, 1434–1447.

Raymond, A.C., Gao, B., Girard, L., Minna, J.D., Gomika Udugamasooriya, D., 2019. Unbiased peptoid combinatorial cell screen identifies plectin protein as a potential biomarker for lung cancer stem cells. Sci. Rep. 9, 14954.

Rivadeneira, D.B., DePeaux, K., Wang, Y., Kulkarni, A., Tabib, T., et al., 2019. Oncolytic viruses engineered to enforce leptin expression reprogram tumor-infiltrating t cell metabolism and promote tumor clearance. Immunity 51, 548–560 e544.

Rizvi, Z.A., Dalal, R., Sadhu, S., Kumar, Y., Kumar, S., et al., 2021. High-salt diet mediates interplay between NK cells and gut microbiota to induce potent tumor immunity. Sci. Adv. 7, eabg5016.

Routy, B., Gopalakrishnan, V., Daillere, R., Zitvogel, L., Wargo, J.A., et al., 2018a. The gut microbiota influences anticancer immunosurveillance and general health. Nat. Rev. Clin. Oncol. 15, 382–396.

Routy, B., Le Chatelier, E., Derosa, L., Duong, C.P.M., Alou, M.T., et al., 2018b. Gut microbiome influences efficacy of PD-1-based immunotherapy against epithelial tumors. Science 359, 91–97.

Roy, S., Trinchieri, G., 2017. Microbiota: a key orchestrator of cancer therapy. Nat. Rev. Cancer 17, 271–285.

Sagiv-Barfi, I., Czerwinski, D.K., Levy, S., Alam, I.S., Mayer, A.T., et al., 2018. Eradication of spontaneous malignancy by local immunotherapy. Sci. Transl. Med. 10, eaan4488.

Sahin, U., Türeci, O., 2018. Personalized vaccines for cancer immunotherapy. Science 359, 1355–1360.

Sahin, U., Derhovanessian, E., Miller, M., Kloke, B.P., Simon, P., et al., 2017. Personalized RNA mutanome vaccines mobilize poly-specific therapeutic immunity against cancer. Nature 547, 222–226.

Scheubeck, G., Berchtold, S., Smirnow, I., Schenk, A., Beil, J., et al., 2019. Starvation-induced differential virotherapy using an oncolytic measles vaccine virus. Viruses 11, 614.

Sepich-Poore, G.D., Zitvogel, L., Straussman, R., Hasty, J., Wargo, J.A., et al., 2021. The microbiome and human cancer. Science 371.

Shono, Y., van den Brink, M.R.M., 2018. Gut microbiota injury in allogeneic haematopoietic stem cell transplantation. Nat. Rev. Cancer 18, 283–295.

Smith, M., Littmann, E., Slingerland, J., Clurman, A., Slingerland, A.E., et al., 2019. Intestinal microbiome analyses identify biomarkers for patient response to CAR T cell therapy. Biol. Blood Marrow Transplant. 25, S177.

Suez, J., Zmora, N., Zilberman-Schapira, G., Mor, U., Dori-Bachash, M., et al., 2018. Post-antibiotic gut mucosal microbiome reconstitution is impaired by probiotics and improved by autologous fMT. Cell 174, 1406–1423 e1416.

Tabdanov, E.D., Rodriguez-Merced, N.J., Cartagena-Rivera, A.X., Puram, V.V., Callaway, M.K., et al., 2021. Engineering T cells to enhance 3D migration through structurally and mechanically complex tumor microenvironments. Nat. Commun. 12, 2815.

Takahashi, K., Yamanaka, S., 2006. Induction of pluripotent stem cells from mouse embryonic and adult fibroblast cultures by defined factors. Cell 126, 663–676.

Tengesdal, I.W., Menon, D.R., Osborne, D.G., Neff, C.P., Powers, N.E., 2021. Targeting tumor-derived NLRP3 reduces melanoma progression by limiting MDSCs expansion. Proc. Natl. Acad. Sci. 118. e2000915118.

Upadhyay, R., Boiarsky, J.A., Pantsulaia, G., Svensson-Arvelund, J., Lin, M.J., et al., 2021. A critical role for fas-mediated off-target tumor killing in t-cell immunotherapy. Cancer Discov. 11, 599–613.

Valpione, S., Galvani, E., Tweedy, J., Mundra, P.A., Banyard, A., et al., 2020. Immune-awakening revealed by peripheral T cell dynamics after one cycle of immunotherapy. Nat. Cancer 1, 210–221.

Van Hoecke, L., Verbeke, R., Dewitte, H., Lentacker, I., Vermaelen, K., et al., 2021. mRNA in cancer immunotherapy: beyond a source of antigen. Mol. Cancer 20, 48.

Vétizou, M., Pitt, J.M., Daillere, R., Lepage, P., Waldschmitt, N., et al., 2015. Anticancer immunotherapy by CTLA-4 blockade relies on the gut microbiota. Science 350, 1079–1084.

Wang, H., Franco, F., Tsui, Y.C., Xie, X., Trefny, M.P., et al., 2020. CD36-mediated metabolic adaptation supports regulatory T cell survival and function in tumors. Nat. Immunol. 21, 298–308.

Wei, L., Lin, Q., Lu, Y., Li, G., Huang, L., et al., 2021. Cancer-associated fibroblasts-mediated ATF4 expression promotes malignancy and gemcitabine resistance in pancreatic cancer via the TGF-beta1/SMAD2/3 pathway and ABCC1 transactivation. Cell Death Dis. 12, 334.

Wennerberg, E., Lhuillier, C., Rybstein, M.D., Dannenberg, K., Rudqvist, N.P., et al., 2020. Exercise reduces immune suppression and breast cancer progression in a preclinical model. Oncotarget 11, 452–461.

Wolf, Y., Anderson, A.C., Kuchroo, V.K., 2020. TIM3 comes of age as an inhibitory receptor. Nat. Rev. Immunol. 20, 173–185.

Xavier, J.B., Young, V.B., Skufca, J., Ginty, F., Testerman, T., et al., 2020. The cancer microbiome: distinguishing direct and indirect effects requires a systemic view. Trends Cancer 6, 192–204.

Yamamoto, T.N., Lee, P.H., Vodnala, S.K., Gurusamy, D., Kishton, R.J., et al., 2019. T cells genetically engineered to overcome death signaling enhance adoptive cancer immunotherapy. J. Clin. Invest. 129, 1551–1565.

Yang, L., Shi, P., Zhao, G., Xu, J., Peng, W., et al., 2020. Targeting cancer stem cell pathways for cancer therapy. Signal Transduct. Target. Ther. 5, 8.

Yelin, I., Flett, K.B., Merakou, C., Mehrotra, P., Stam, J., et al., 2019. Genomic and epidemiological evidence of bacterial transmission from probiotic capsule to blood in ICU patients. Nat. Med. 25, 1728–1732.

Yi, M., Jiao, D., Qin, S., Chu, Q., Li, A., et al., 2019. Manipulating gut microbiota composition to enhance the therapeutic effect of cancer immunotherapy. Integr. Cancer Ther. 18. 1534735419876351.

Ying, Z., Huang, X.F., Xiang, X., Liu, Y., Kang, X., et al., 2019. A safe and potent anti-CD19 CAR T cell therapy. Nat. Med. 25, 947–953.

Zagorac, I., Fernandez-Gaitero, S., Penning, R., Post, H., Bueno, M.J., et al., 2018. In vivo phosphoproteomics reveals kinase activity profiles that predict treatment outcome in triple-negative breast cancer. Nat. Commun. 9, 3501.

Zhang, X., Ashcraft, K.A., Betof Warner, A., Nair, S.K., Dewhirst, M.W., 2019. Can exercise-induced modulation of the tumor physiologic microenvironment improve antitumor immunity? Cancer Res. 79, 2447–2456.

Zheng, Y., Wang, T., Tu, X., Huang, Y., Zhang, H., et al., 2019. Gut microbiome affects the response to anti-PD-1 immunotherapy in patients with hepatocellular carcinoma. J. Immunother. Cancer 7, 193.

Zmora, N., Zilberman-Schapira, G., Suez, J., Mor, U., Dori-Bachash, M., et al., 2018. Personalized gut mucosal colonization resistance to empiric probiotics is associated with unique host and microbiome features. Cell 174, 1388–1405 e1321.

Zmora, N., Suez, J., Elinav, E., 2019. You are what you eat: diet, health and the gut microbiota. Nat. Rev. Gastroenterol. Hepatol. 16, 35–56.

CHAPTER 16

Moving forward—The science and the patient

OUTLINE

Where we stand	503	Living with dignity	509
Moving forward	504	Relation between the patient and the health provider	510
Genomic advances	504		
Precision medicine through artificial intelligence	505	Dying with dignity	512
		Summary and conclusions	514
Prevention vs treatment	506	References	515
One planet, one health, one cure	508		

Where we stand

Each time a novel remedy to treat cancer was announced, journalists tripped over each other to report the information. When the hyped-up newspaper articles appear, some patients phone researchers or cancer centers, asking when the treatment will be available, and ask whether they can enroll in a trial to test certain treatments. Their chances of getting into a trial might be reasonably good, although not necessarily the trial they would like, and the outcomes will likely be less dramatic than the newspaper articles suggested. The vision of a cure too often simply turned out to be an illusion.

There have been many skirmishes in this war, and scientists have won several key battles with some types of cancer. Without doubt, some of the therapeutic strategies that have been developed over the past few decades have been remarkable in their effectiveness, particularly leukemia and some forms of breast cancer, and even melanoma has seen an increase in survival rates. With the new immunotherapies and the possibility of developing more profitable markers, incremental gains can be expected, and perhaps breakthroughs will be achieved.

The expense of the research to develop novel treatments, coupled with the costs of promoting new therapies have resulted in extraordinary costs for drug treatments, in some instances running as high as hundreds of thousands of dollars. This obviously is a burden for government health programs and insurance companies, and simply unattainable for those who are uninsured. Pharmaceutical companies have argued that the price of therapeutics is fair given the expense of developing them, but it certainly isn't fair if the treatment has no actual benefit. Perhaps the cost of the drug should be tied to the clinical benefits obtained (Del Paggio et al., 2017) as the drug company Novartis did in offering its $475,000 drug Kymriah with a money-back guarantee should the drug fail to do its job.

Moving forward

Considerable attention has been devoted to determine the drivers for treatment resistance and metastasis, and impressive advances have been made in defining the genetic characteristics of diverse tumors, features of the microenvironment that can be targeted to treat the disease, or aspects of the individual that predict outcomes. Despite the potential of these advances, it has been difficult to translate them into therapeutic benefits. It has become certain that a focus on the identification of biomarkers that predict cancer occurrence may be instrumental in enhancing therapeutic outcomes, especially if these biomarkers can be detected years before the disease was first detected, thereby allowing treatments to be initiated before frank cancers are apparent. Given the slow pace of cancer cures being developed, it can be argued that the approaches that have been adopted are too narrow, and it might be preferable to adopt a broader more holistic method. This might include the contribution of numerous biological processes that are influenced by lifestyle factors, particularly diet, exercise, sleep, and methods of reducing the impact of stressors.

Genomic advances

Technological advances were made to identify features of cancer cells, such as through single-cell RNA sequencing. And discoveries are being made in nanotechnologies, counting cancer cells as they pass by electrodes to determine whether cancer treatments are having the desired effect. These innovations have considerable benefits, and deep learning algorithms can provide still better predictions of cancer occurrence and progression. Indeed, the genetic analyses provided by the Pan-Cancer Analysis of Whole Genomes (PCAWG) Consortium revealed that deep learning analyses based on patterns of somatic passenger mutations using whole-genome sequencing of 24 cancer types were able to classify metastatic tumors to a much better extent than could experienced pathologists (Jiao et al., 2020). The benefits of advanced technologies will be exponentially greater when these are paired with genome editing, cellular reprogramming, and tissue engineering.

Yet despite this rosy scenario, in several respects we may be getting ahead of ourselves. When genetic markers can identify specific targets for treatment, these may not be "druggable," being elusive to pharmacotherapy. In considering the microbiota as a tool for developing cancer therapies, what baseline should we be working from? Essentially, what is "normal"? The very same question can be addressed concerning genes related to precision medicine for cancer treatment (Manrai et al., 2018). Normality might differ across diverse ethnic groups, and so even if an appropriate reference condition was available for genome analyses, would this be applicable for all ethnicities and cultures. After all, certain polymorphisms are more common in Asian people than in other groups, which may be significant in making decisions to adopt the most appropriate therapies (Bachtiar et al., 2019). And in defining normal, it is important to consider epigenetic changes that might have occurred owing to environmental or social triggers, or those that accumulated with age or with stressors, as well as those transmitted across generations.

To get around some of these issues, a variant of the precision medicine approach was adopted that seems to hold promise. Instead of considering individual gene mutations, it was thought that many processes may converge on particular proteins that influence multiple cancers. Using different methodological tools to focus on core factors based on individual cells from 36 different tumors, it was possible to identify clusters of tumors based on multiple biological factors that could predict optimal treatments (Garofano et al., 2021). Similarly, through a complex computational tool, it was possible to identify hundreds of "Master Regulators" that fell into 24 distinct Master Regulator Blocks (or clusters). By targeting distinct blocks or sets of blocks, it was possible to determine the types of cancers that developed and might thus inform treatment strategies, including whether treatment resistance was apt to develop. In some instances, drugs that could affect these blocks were already available, whereas in other cases adequate drug therapies have yet to be developed (Paull et al., 2021).

Markers related to experiences and environmental factors

The development and progression of cancer is known to be affected by numerous hormones, growth factors, and microorganisms. Some of these may be associated with specific, identifiable genetic factors, whereas others may be related to diet, sedentary behaviors, or stressor experiences that might be tied to epigenetic changes. Likewise, some of these actions may be associated with prenatal or early life influences. As much as genetic markers may be fundamental to the development of a precision medicine approach to determine optimal therapeutic strategies, the incorporation of additional data relevant to lifestyles may provide added value to enhance the efficacy of individualized treatments. However, with each additional variable included in such analyses the practical and performative difficulties become much greater without effective ways of translating the information.

Precision medicine through artificial intelligence

The idea of adopting precision medicine in cancer therapies has been especially exciting, and in the case of several cancers, this has already made a difference for therapies. But what about the treatments for all those other cancers that haven't yet gained appreciably from the new approaches that have been developed? One can only hope that advances will be made even for the most "untreatable" types of cancer. These cancers would need to be identified very early; once they have reached the point of being detectable, they might already have escaped any available means to thwart their progression. When critical biomarkers can be identified very early and their meaning deciphered, these may be linked to effective therapies.

The future of precision medicine, as Topol (2019) suggested, lies in the progression of artificial intelligence. To a significant extent everyone's biology is unique, and it is certain that one-size-fits-all approaches should have long gone by the wayside. Arguably, treating cancers on an individual basis will be challenging and could strain financial resources. Still, with the integration of data through large data networks that permit sharing of information concerning the multiple markers that predict cancer occurrence, the efficacy of treating cancer may be enhanced as might the ability to predict therapies that are less likely to become treatment resistant.

The enormous amount of data obtained through whole-genome and epigenome analyses has required artificial intelligence to make sense of the data on accurate predictions relating to various facets of cancer detection, progression, and treatment. Even at this early point, AI has been effective in helping to interpret information obtained through diverse imaging techniques that can help with cancer identification, and the ability to determine and assess morphometric characteristics that are fundamental in the analysis of cancer processes (Bera et al., 2019). In fact, a combination of optical imaging and deep learning was readily and quickly able to determine the major histopathologic classes of brain tumors during surgery (Hollon et al., 2020). A deep learning method was also developed to predict early indications of cancer occurrence that are usually discovered too late for treatments to be effective, such as pancreatic cysts that will likely develop into cancer. Various approaches were likewise developed to classify cells based on the tumor's microenvironment, as well as on genomic, transcriptomic, and proteomic data. Related methods have been used to detect epigenetic changes to predict cervical and renal carcinoma, and algorithms were developed to diagnose melanoma, while machine learning approaches were developed to stratify ovarian cancer based on blood biomarkers to attain optimum treatment outcomes.

Through machine learning, it was possible to identify genes that are the main drivers of cancer as well as genes that are more distal from the cancer-driving gene, such as those that influence energy regulation and are altered by epigenetic processes rather than frank gene mutations. And through a combination of proteomic and genetic analyses that were interpreted through advanced AI, it was possible to track the sequence of changes that occurred in the development of different forms of cancer (Schulte-Sasse et al., 2021).

These are only a few of the many procedures that have been developed and will continue to evolve. Given the influence of lifestyle factors on cancer evolution and response to treatments, it is reasonable to suppose that the inclusion of these influences within machine learning approaches will enhance the ability to detect and treat diverse forms of cancer and other chronic illnesses.

Prevention vs treatment

Scientists and physicians who are more pessimistic (or realistic) might feel that a "cure" for many cancers will be some time off. This provides even more reason for promoting prevention strategies even if this might be an uphill battle especially as cancers may occur because of genetic and epigenetic factors, random mutations, or mutations that result from carcinogens. And in many instances, individuals have little control over toxicants or diverse events that can affect them (carcinogens in the atmosphere). Yet, about 40%–45% of cancers are thought to be preventable. Some forms of cancer can be prevented by vaccines (e.g., HPV-provoked cancers), others can be limited by eating the right foods, diminishing excess weight, remaining active, as well as staying out of the sun and away from tobacco products, cutting back on alcohol, and as former US President Obama asserted, just "don't do dumb shit."

Despite the obvious remedies related to prevention, within Western countries, more than 20% of people continue to smoke, and within Eastern European countries and Russia, smoking and alcohol consumption is much more prevalent. But best not to wag our finger at these countries, when in North America, Britain, and the EU countries the occurrence of people being overweight or having obesity is stunning, and is becoming progressively worse even among children. It's probably fair to conclude that smokers and those who overeat and don't exercise, are failing to attend to their long-term health prospects; and even if it crossed their mind, the response might well be "it won't happen to me." Should they be struck by a disease such as cancer, these same individuals may well be part of the chorus of those who sadly wonder, "Why me?"

The adage "an ounce of prevention is better than a pound of cure," didn't arise for no reason, and it's not an issue that is much debated. To be sure, many people are unaware of the factors that can cause diseases or why they got sick; but the fundamental problem encountered in prevention is not simply a lack of knowledge concerning risk factors, but rather a failure to actually comply with health recommendations. Knowing the hazards of smoking or alcohol consumption might not lead to appropriate behavioral change, and knowing about bad eating habits might not alter the tendency to add an order of fries to that already overflowing tray of comfort foods. These behaviors are not even curtailed when individuals are aware that they're in a genetically high-risk group for a particular illness (Hollands et al., 2016). The answer to why a person gets sick can be complex, but at times, one can simply point to behavior.

If individuals are ready to engage in measures to prevent disease development and believe that they're capable of doing so, then methods are available to help them achieve their individual goals, and these could be accomplished through individualized prevention programs. The future of health care, as it is related to preventive and precision medicine will necessarily involve broader systematic and creative approaches. One strategy in this respect, which is not exclusive of other methods, is the P4 medicine approach. This entails four components; (a) *predictive*: identification of early predictors of illness, (b) *preventive*: developing ideal approaches to limit illness occurrence, (c) *personalized*: identifying specific markers that are linked to effective treatment methods, and (d) *participatory*: patients, physicians, and health workers have a voice in treatments thereby allowing for better policy and implementation strategies (Hood and Auffray, 2013).

Given the vast number of people who need to pay more attention to illness prevention, one-on-one therapies might not be practical. Certain health problems have, to some extent, emerged because of multiple challenges, such as industrialization, pollutants, poverty, poor housing, and overcrowding. These conditions, as well as new emerging diseases and those that have gained the ability to evade treatments, need to be dealt with through community-level approaches, or those that involve whole societies. The influence of these variables ought to be factored into a precision medicine approach (Khoury et al., 2016).

Although cancer prevention research has frequently focused on genetic biomarkers, too often the role of poverty and related factors have received short shrift despite the numerous reports attesting to their role in cancer detection, occurrence, and progression.

As discussed in an earlier chapter, many prenatal and early postnatal experiences that are tied to maternal disadvantages, have pronounced and lasting implications for health risks encountered in adulthood. The recognition of the various psychosocial processes and early experiences associated with chronic illness ought to be fundamental in altering some of the basic strategies related to prevention and treatment (Wise, 2019). It was suggested that in the absence of attention to multiple factors, including diverse lifestyle factors and changes in individual characteristics over time, personalized medicine will be less successful in meeting health needs. There is certainly reason for optimism in the search for best practices, but this needs to be tempered by the complexity of the issues faced, the practical limits of adopting the relevant strategies, and the feasibility of governments, industry, research funding agencies, and communities to be willing and able to implement appropriate and durable health-management programs (McCarthy and Birney, 2021).

One planet, one health, one cure

Recent years have seen increasing adoption of a "One Health" transdisciplinary approach that advocates for the strengthening of programs to foster well-being and prevent diseases. The essential perspective is that disease prevention and the health of the planet more generally will come from a better understanding of the intersections between humans, animals, plants, and the environment. Being able to achieve these goals necessarily encompasses varied global as well as regional issues, such as environmental contamination, conflicts related to the misuse of natural habitats, degradation of ecosystems, food and water safety, conservation of natural resources, antimicrobial resistance, the emergence of new diseases, and disaster preparedness. A fundamental component of a One Health approach includes basic research and implementation science as well as the development of educational and policy programs. It is anticipated that this might not only be instrumental in the prevention of chronic and infectious diseases but also could facilitate the development of novel therapeutic strategies.

The One Health notion isn't new but was only formalized about a decade ago, but its value to clinical practices has slowly been gaining traction. Studies regarding diets, exercise, and environmental contaminants have long pointed to health risks on the one side and health benefits on the other. Once the old and sterile concept of the immune system and the brain being independent of one another was discarded, greater appreciation of processes linked to cancer and the alleviation of side effects of therapies administered became possible. Similarly, the understanding that experiential and environmental factors could influence the expression of genes (epigenetic variations) was instrumental in altering the general view that inherited genes are not alone in determining disease occurrence. By straying from conventional approaches to cancer processes and treatments, the role for microbiota has been evolving, and this too holds promise for improved auxiliary treatments. To maximize the health benefits further, distinct silos across multiple disciplines ought to be dismantled, and then reintegrated so that cross-fertilization occurs between different fields, including those that focus on the tenets of a One Health approach. The past decades have seen increasingly improved technologies to deal with diseases, but with this comes the responsibility of assuring that stewardship of people, animals, plants, and environments (natural and built) is well managed (Amuasi et al., 2020).

Living with dignity

The life expectancy of men and women in the developed countries has increased dramatically over the past century, thanks in part to improved access to medical treatments and enhanced lifestyles, although the gains have not been realized equally in all ethnic and geographical groups. In 1920 about 6.5% of infants died before their first birthday, whereas in 2020 infant mortality was about 0.2%. For individuals born about 1920, life expectancy in most developed countries was about 60 years; those born in 1970 were expected to live to about 72 years of age, and it was estimated that individuals born in 2020 can be expected to live until about 85–90 years of age. The latter estimates may be a bit optimistic and are, of course, predicated on catastrophic events not occurring, including the potentially disastrous effects stemming from climate change or pandemics worse than COVID-19. It is, of course, gratifying that we're living longer, but to what extent has this been met by increased health span? Globally, as well as in the developed countries, disability-free years have lagged increased life span, more so in less developed countries. To a significant extent, this occurred owing to the impact of noncommunicable diseases, pointing to the need for improved intervention strategies that might enhance living well.

Some older people might feel that they have lived a good life, accomplished all that they had hoped for, and when they pass, they will leave behind a satisfying legacy. Others, however, might have a different view of their life and the course of aging. They may have outlived all their friends, and they may have lost their independence. For them, simply existing can be a lonely experience, and despite not having any severe illnesses, these individuals might simply feel that each day is too long, and staying alive is simply something that must be endured. For older people who are beset by poor health and diminished quality of life, their ordeal can be far more taxing. A fundamental question for older folks is "how long do they want to live"? In a remarkably astute article in *The Atlantic*, Emanuel (2014) made an impressive case for not wanting to live past 75. Much better to go peacefully and be remembered well, than becoming a burden. Maybe 75 is a bit short, and 80 might be a better age at which to pass into that good night, but that would depend on the individual's physical and mental health. If the zest and zing is still there at 80 or 85, why stop? But many are not so lucky. What a patient is prepared to live through to gain a few more months or even years can vary from one person to another. One only needs to read Gawande's (2014) descriptions to appreciate, even second hand, what life is like for aging individuals, the travails encountered, the frequent false hopes experienced, and the indignities suffered. Essential to the well-being of older patients is that they receive optimal treatments beyond that provided by the myriad technologies that have helped extend life. Given the importance of psychosocial factors in preserving the mental and physical health of patients, these elements need to be broadly addressed.

Aside from his views on the contribution that AI will have on the future of medicine, Topol (2015) has been a staunch advocate of bringing the physician closer to the patient, and for patient rights to be acknowledged and acted upon in a potentially patient unfriendly system. In an insightful book published some years ago, Shem (1978) pointed to the importance of patients not being dehumanized by a medical system, and that for this to occur, physicians need to connect and be empathetic with the patient. They are, after all, a major source of social support and hope for patients. Headway has been made on this front, but with so many tribulations encountered by physicians, such as completing endless electronic forms, the care of patients may be falling by the wayside. The modern health system is in danger of taking the humanity out of the physician, and in the process damaging the patient. The irony is inescapable.

Relation between the patient and the health provider

The relationship between a patient and their health providers (physicians, nurse-practitioners, nurses, physician assistants) can have a significant impact on the patient's psychological and physical well-being. Among other things, they can offer needed social support to patients, and by keeping patients informed about all aspects of their treatments, health providers can diminish the unpredictability and uncertainty being faced. Consequently, patients may feel enhanced agency and self-efficacy and might be better prepared to deal with the hardships that may follow.

For best results to be obtained there ought to be a good match between the values and desires of the patient and the perceptions of the physician in this regard. Some patients might prefer an empathetic physician who is kind and patient and generally beloved. Others are less inclined to be affected by the attitudes and bedside manner of the physician and simply prefer the most competent physician even if s/he appears cold and aloof. Another consideration is fairly important in considering the match between the patient and the physician. Just as group members with whom we share an identity may be especially useful as support resources in dealing with stressors, patients may be more positive or receptive when the advice comes from doctors of their own group. For instance, African American patients have expressed greater comfort when dealing with physicians who are similarly African American (Cabral and Smith, 2011). However, only 5% of physicians within the United States are African American, clearly being underrepresented relative to the general population, resulting in patients not receiving treatment from their preferred doctor. Considering the importance of the doctor-patient alliance, training doctors from diverse groups is necessary not simply to meet equity and inclusion goals, but to meet patient desires and needs.

Physician-patient communication: Shared decision-making

The pervasive attitude years ago that "doctor knows best" has since been replaced—to a considerable extent, by greater attention to the patients' expressed needs and goals. In the context of serious illnesses, patients and their family members are most often consulted about the treatment options available and are encouraged to participate in decision-making. Among some patients, sharing in decisions concerning their treatment is a top priority and this empowerment may have contributed to enhanced patient satisfaction with clinical treatment, and compliance with treatments was improved (Náfrádi et al., 2017). Other patients, in contrast, particularly as they typically have little knowledge about the relevant issues, prefer to leave decisions to the physician. Even so, being invited to be involved in critical decisions might leave them feeling that they are not simply a victim of the situation and instead have a degree of agency over their destiny.

Transparency and open communication are essential for the physician-patient relationship. Failure to establish proper communication between physicians and patients (or surrogates, such as family members) increases the likelihood of patients not obtaining the care they desire. Likewise, when patients believed that their concerns had not been adequately considered or if trust in the physician was diminished, treatment compliance frequently suffered, symptoms could worsen, and the benefits of treatments were not fully obtained.

As much as a match between treatment preference expressed by patients and the treatment offered is essential, at times the physicians' perceptions might not be congruent with those expressed by the patient (Melzer et al., 2020). Physicians typically make great efforts to understand patient fears and goals, and they will likely consider the patient's concerns regarding the trade-offs that exist between getting well and possible impaired functioning stemming from treatments. However, physicians might not always fully appreciate the patient's priorities. The physician might consider treatment effectiveness as the primary goal, along with side effects that could emerge. The patient, however, may be more concerned with the aftermath of the treatment, including whether they will experience pain or disability, and the possible loss of independence and role status. When communication between the patient and physician is faulty, patients may not receive the treatment they wanted or expected ("silent misdiagnosis") (Mulley et al., 2012).

Although patients are the final arbiters as to the treatments they receive, the physician is not obliged to administer treatments simply because a patient desires them. There are times when patients might be inclined to make some very bad decisions, such as opting for unproven alternative medicines. Through verbal and nonverbal cues, the physician might indicate their discomfort with the patient's wishes, but in the end, patients have the right to make their own decisions. Patients might desire opioid-based pain killers, but the physician might see this as counterproductive and will not comply with the requested medication. Likewise, a patient might request surgery to remove a tumor even though the available data indicate that this will be of no value and may even be harmful (particularly if cancer has already metastasized broadly). Here again, the physician might not be inclined to undertake the specific procedures that the patient requests. Of course, the patient is free to request other opinions and may find a surgeon willing to comply with the patient's wishes.

Decision regret

Simply being involved in the decision-making process doesn't necessarily result in patients being happy with the decisions they made. Some patients who had undergone surgery as a component of cancer therapy subsequently expressed regret for this decision, particularly when incongruence existed between the patients' preferred and actual decision-making roles (Wang et al., 2018). Even when the decisions that had been made were entirely reasonable, decisional regret may have stemmed from diminished physical or psychological health together with low quality of life subsequently experienced. Some patients indicated that having had the opportunity to engage in decisions might have limited decision regret, but others regretted the decision that they had made as they had been in a shocked and distressed state when the decision had been made (Connolly and Reb, 2005). Equally concerning is that patients might have unrealistic expectations concerning what can be done for them, essentially trusting that their physician will cure them. When the expected outcome doesn't materialize, some patients might hold their physician responsible. In essence, the benefits of trust in their physician may turn to blame when positive outcomes are not realized (Taha et al., 2011).

It is not unusual for physicians to be in the unenviable position of having to juggle maintenance of their integrity, helping to maintain patient spirits, and all the while still providing them with accurate information that will allow for appropriate decisions. More often than not physicians expressed willingness to inform patients about a serious illness with a good prognosis, but more than half were understandably reluctant to communicate a diagnosis with a poor prognosis. On many occasions, the prognosis is not cut and dried, which might require that some ambiguity be maintained about the future, but this too should be explained to patients. This would seem rather obvious, but in some countries the attitude persists that if the patient does not need to hear bad news, then it's best to keep it from them, thereby allowing patients to live with less anxiety and greater hope for the future (see Karimi Rozveh et al., 2017). As patients might soon enough become aware of their dire situation, these initial acts to spare patients from grief may foster uncertainty, anxiety, and dissatisfaction, as well as promoting dissolution of the patient-doctor alliance, which may influence the quality of terminal care.

Dying with dignity

There are occasions when physicians treating frail, older patients may ask themselves "when do we stop" treatment. Should they be going through extraordinary lengths to keep patients alive who would, even under the best circumstances, be left in exceedingly poor condition? These issues won't be going away soon, particularly with yesterday's baby boomers having reached an older age. Perhaps the value of treatment for the patient ought to be considered in the context of the probability of positive outcomes being obtained, the physical and psychological costs incurred, and the individual's likely quality of life following treatment. Guidelines have been created to facilitate decisions related to whether the benefits of treatment will outweigh the risks (Hurria et al., 2014), but abstract guidelines don't necessarily translate well into the realities that are faced.

For cancer patients and their families, several very difficult decisions need to be made, sometimes when they are least able to plan, weighed down by the gravity of the situation and its attending emotions. Decisions might involve how aggressively to battle cancer, or should procedures be limited to relieve pain and discomfort? Are patients better off living at home or should they enter a hospice care facility? And if patients have no hope of recovery, should a "Do Not Resuscitate" order be issued on behalf of the family, with the assurance that this instruction will be diligently followed.

Relatedly, there has been the question of when a person should stop making efforts to beat an illness. Most people believe that plans concerning end-of-life care should be discussed with family members, but it's an uncomfortable topic and consequently, a minority of older individuals have these conversations. It is certainly important for patients and family members to consult with their family physician to develop future plans and to be assured that the patient's wishes will be honored as much as this will be possible. This might entail consideration of the patient's values and care preference, illness prognosis, and fears that exist. But for whatever reasons, these are frequently not discussed with the family physician.

Understandably, when an individual learns that they have a progressive, incurable illness that may require harsh treatments for them to survive, sometimes being accompanied by intense pain or loss of self, they might be inclined not to live longer. Much too often, however, they might feel pressure from family members or friends who encourage them to "courageously fight the disease" and "not to give up." This might well be commendable—but only to a point. Once the battle has gone on too long, however that's defined, or when the person has suffered a long and painful illness, then they should be given the opportunity to leave.

Debate on the issue of assisted dying has had a long history, but laws allowing this have recently been enacted in relatively progressive countries. Opponents of assisted dying have raised multiple objections to this practice. Those with a religious orientation have viewed suicide and facilitating death as a sin, hence the term "committing suicide" rather than more appropriate terminology, such as physician-assisted death or medical assistance in dying (MAiD).[a] Others have argued that assisted dying counters the Hippocratic oath (do no harm), or they resort to the tired cliché that this might be a slippery slope that could lead to abuse of the most vulnerable (e.g., leading to euthanasia without patient consent). The contention was also advanced that legislation to permit euthanasia could hinder the development of more compassionate approaches to enhance care for those who are dying, and concerns were expressed that assisted dying might lead to a contagion of people seeking to die well before it would be advisable to do so.

On the other side of the debate are those who have advocated that *respect for life* also means being able to die with dignity. None of us ought to die in slow painful increments, piece by piece. None of us should have to be maintained in a drugged stupor because of intractable pain or withering away in a comatose state. Most assuredly, none of us ought to be subjected to the misery and humiliation of being unable to control virtually all aspects of our lives, even bowel movements. It is more than a little disturbing that despite the reluctance to actively help patients die, there is much less resistance to allow patients to linger for extended periods. In some strange logical confusion, a distinction has been drawn between assisted dying and that of withholding or withdrawing treatment, even though this differentiation is entirely artificial and lacking any moral significance or justification.

Physicians are often pushed by family members who are intent that their loved one survive as long as possible. Many physicians and nurses witness "futile care" being delivered, resulting in patients living for slightly longer, accompanied by numerous indignities. In these circumstances what would doctors and nurses want for themselves or their loved ones? Knowing what lays ahead, doctors often prefer not to have any extreme measures taken, and thus are likely to sign do not resuscitate (DNR) orders. It was suggested that doctors routinely subject their patients to treatments that they wouldn't have chosen for themselves (O'Mahony, 2018) and nurses more often chose less aggressive treatments for themselves than those delivered to their patients.

Two decades ago, the majority of the US physicians felt that assisted dying was morally permissible under specific conditions, and this position was also expressed within the United Kingdom. Understandably, being in favor of medical assistance in dying is less often matched by physicians being willing to engage in this practice. Fortunately, with legislation approving physician-assisted dying, the feared "slippery slope" view has not materialized, and requests for assisted dying have been infrequent. After legalization, the frequency of physician-assisted deaths increased somewhat, particularly among cancer patients, but overall widespread abuse of these measures was not apparent (Emanuel et al., 2016).

---

[a] Words matter. A 2013 US Gallup poll revealed that most people (70%) were in favor of allowing a doctor to "end the [terminally ill] patient's life by some painless means," but far fewer people (51%) were in favor of doctors helping patients "commit suicide."

There is another element concerning patient rights and opportunities that have received much less attention but has important implications. When extremely ill patients have received multiple treatments without success, they ought to have the *right-to-try* medications that have not undergone the full gamut of trials to assure safety and effectiveness (e.g., a drug may only be at Phase II of clinical trials). This had been done before in other contexts (e.g., in people who had contracted Ebola virus) and compassionate use of experimental drugs was used in other instances. Surely, if patients had the right-to-die, then they would also have the right-to-try. A law to allow this was enacted in the United States in 2018, provided that it was made clear to the patient that there was only a slight chance that the treatment would have much effect and that the treatment could worsen the condition or create side effects and other unintended consequences. Of course, the drug maker would have to be prepared to provide the drug on a compassionate care basis, and the patient would have to agree not to hold anyone responsible if things went awry.

Summary and conclusions

Numerous writers have ventured into the unknown attempting to predict where we would be in the treatment of illnesses at some time in the future (50 years seems to be a favorite). In the main, these predictions are based on technological advances that can be expected. Those of us who were present 50 years ago could hardly have imagined some of the approaches currently being used. For that matter, we didn't even know what we didn't know, let alone predict what might be coming down the pipeline. With the exponential growth in science and technology related to medicine, we might do better focusing on what can be expected 10 years down the road. Arguably, even making accurate predictions of highly predictable events might not move policy makers to make the right decisions. The lack of preparedness for a pandemic—despite repeated warnings—has most certainly taught us that.

A couple of generations from now we can count on people looking back at the treatments offered during the early portion of this century and wonder how primitive we were. We can also count on our descendants wondering about the failure of societies to adopt even the most basic strategies to prevent chronic illnesses. Individual efforts have already been instrumental in reducing cancer occurrence and deaths within the developed countries, but so much more can be accomplished by effectively advocating for lifestyle changes and by people actually adopting these simple measures. This will be facilitated by government agencies taking on a greater role in prevention efforts. In fact, even the incidence of cancers that are genetically determined could be diminished by the adoption of healthy lifestyles. But knowing what to do and then having this translate to people engaging in healthy behaviors is another thing. Implementation is most certainly essential for individuals to gain from the considerable investments that have been made in cancer research. More must be done than simply hope for the best. What will be critical are better methods to move in this direction through individual, community, and large-scale government efforts, including a considerable increase in financial investments, if we are to reach these lifesaving goals.

References

Amuasi, J.H., Lucas, T., Horton, R., Winkler, A.S., 2020. Reconnecting for our future: the Lancet One Health Commission. Lancet 395, 1469–1471.

Bachtiar, M., Ooi, B.N.S., Wang, J., Jin, Y., Tan, T.W., et al., 2019. Towards precision medicine: interrogating the human genome to identify drug pathways associated with potentially functional, population-differentiated polymorphisms. Pharmacogenomics J. 19, 516–527.

Bera, K., Schalper, K.A., Rimm, D.L., Velcheti, V., Madabhushi, A., 2019. Artificial intelligence in digital pathology—new tools for diagnosis and precision oncology. Nat. Rev. Clin. Oncol. 16, 703–715.

Cabral, R.R., Smith, T.B., 2011. Racial/ethnic matching of clients and therapists in mental health services: a meta-analytic review of preferences, perceptions, and outcomes. J. Couns. Psychol. 58, 537–554.

Connolly, T., Reb, J., 2005. Regret in cancer-related decisions. Health Psychol. 24, S29–S34.

Del Paggio, J.C., Sullivan, R., Schrag, D., Hopman, W.M., Azariah, B., et al., 2017. Delivery of meaningful cancer care: a retrospective cohort study assessing cost and benefit with the ASCO and ESMO frameworks. Lancet Oncol. 18, 887–894.

Emanuel, E.J., 2014. Why I Hope to Die at 75. [Press release]. Retrieved from https://www.theatlantic.com/magazine/archive/2014/10/why-i-hope-to-die-at-75/379329/.

Emanuel, E.J., Onwuteaka-Philipsen, B.D., Urwin, J.W., Cohen, J., 2016. Attitudes and practices of euthanasia and physician-assisted suicide in the United States, Canada, and Europe. JAMA 316, 79–90.

Garofano, L., Migliozzi, S., Oh, Y.T., D'Angelo, F., Najac, R.D., et al., 2021. Pathway-based classification of glioblastoma uncovers a mitochondrial subtype with therapeutic vulnerabilities. Nat. Cancer 2, 141–156.

Gawande, A., 2014. Being Mortal: Medicine and What Matters in the End. Metropolitan Books.

Hollands, G.J., French, D.P., Griffin, S.J., Prevost, A.T., Sutton, S., et al., 2016. The impact of communicating genetic risks of disease on risk-reducing health behaviour: systematic review with meta-analysis. BMJ 352, i1102.

Hollon, T.C., Pandian, B., Adapa, A.R., Urias, E., Save, A.V., et al., 2020. Near real-time intraoperative brain tumor diagnosis using stimulated Raman histology and deep neural networks. Nat. Med. 26, 52–58.

Hood, L., Auffray, C., 2013. Participatory medicine: a driving force for revolutionizing healthcare. Genome Med. 5, 110.

Hurria, A., Wildes, T., Blair, S.L., Browner, I.S., Cohen, H.J., et al., 2014. Senior adult oncology, version 2.2014: clinical practice guidelines in oncology. J. Natl. Compr. Cancer Netw. 12, 82–126.

Jiao, W., Atwal, G., Polak, P., Karlic, R., Cuppen, E., et al., 2020. A deep learning system accurately classifies primary and metastatic cancers using passenger mutation patterns. Nat. Commun. 11, 728.

Karimi Rozveh, A., Nabi Amjad, R., Karimi Rozveh, J., Rasouli, D., 2017. Attitudes toward telling the truth to cancer patients in Iran: a review article. Int. J. Hematol. Oncol. Stem Cell Res. 11, 178–184.

Khoury, M.J., Iademarco, M.F., Riley, W.T., 2016. Precision public health for the era of precision medicine. Am. J. Prev. Med. 50, 398–401.

Manrai, A.K., Patel, C.J., Ioannidis, J.P.A., 2018. In the era of precision medicine and big data, who is normal? JAMA 319, 1981–1982.

McCarthy, M., Birney, E., 2021. Personalized profiles for disease risk must capture all facets of health. Nature 597, 175–177.

Melzer, A.C., Golden, S.E., Ono, S.S., Datta, S., Crothers, K., et al., 2020. What exactly is shared decision-making? A qualitative study of shared decision-making in lung cancer screening. J. Gen. Intern. Med. 35, 546–553.

Mulley, A.G., Trimble, C., Elwyn, G., 2012. Stop the silent misdiagnosis: patients' preferences matter. BMJ 345, e6572.

Náfrádi, L., Nakamoto, K., Schulz, P.J., 2017. Is patient empowerment the key to promote adherence? A systematic review of the relationship between self-efficacy, health locus of control and medication adherence. PLoS One 12, e0186458.

O'Mahony, S., 2018. Do doctors die better than philosophers? Lancet 391, 1474–1475.

Paull, E.O., Aytes, A., Jones, S.J., Subramaniam, P.S., Giorgi, F.M., et al., 2021. A modular master regulator landscape controls cancer transcriptional identity. Cell 184, 334–351 e320.

Schulte-Sasse, R., Budach, S., Hnisz, D., Marsico, A., 2021. Integration of multiomics data with graph convolutional networks to identify new cancer genes and their associated molecular mechanisms. Nat. Mach. Intell. 3, 513–526.

Shem, S., 1978. The House of God: A Novel. R. Marek Publishers, NY.

Taha, S.A., Matheson, K., Paquet, L., Verma, S., Anisman, H., 2011. Trust in physician in relation to blame, regret, and depressive symptoms among women with a breast cancer experience. J. Psychosoc. Oncol. 29, 415–429.

Topol, E.J., 2015. The Patient Will See You Now. Basic Books, NY.
Topol, E.J., 2019. Deep Medicine: How Artificial Intelligence Can Make Healthcare Human Again. Basic Books, NY.
Wang, A.W., Chang, S.M., Chang, C.S., Chen, S.T., Chen, D.R., et al., 2018. Regret about surgical decisions among early-stage breast cancer patients: effects of the congruence between patients' preferred and actual decision-making roles. Psychooncology 27, 508–514.
Wise, J., 2019. Childhood poverty is linked to adult cancer risk, says report. BMJ 353, i2808.

Index

Note: Page numbers followed by *f* indicate figures, *t* indicate tables, and *b* indicate boxes.

A

Acute lymphoblastic leukemia (ALL), 478
Acute myeloid leukemia (AML), 125–126
Adaptive (acquired) immunity
   antibody molecule, 56, 57*f*
   antigen neutralization, 56–57
   B lymphocytes
      antigen neutralization, 56–57
      IgA, 55–56
      IgE, 55–56
      IgG, 55
      immunoglobulins, 55
      plasma cells, 55
      structure, immunoglobulin, 56, 57*f*
   discrimination between self and nonself, 54
   diversity, 54
   memory, 54
   proliferation and elimination, 54
   self-limitation, 54
   specificity, 54
   T lymphocytes
      antigen presenting cells (APCs), 58
      immune functioning, 58–59
      immune steps, cancer progression, 58, 59*f*
      major histocompatibility complex (MHC), 57–58
Adoptive cell transfer (ACT), 478
Allostasis and allostatic overload, 138, 138*f*
Angiogenesis, 62, 447*b*
Antibody-based therapies
   CAR T-cell therapy
      acute lymphoblastic leukemia (ALL), 478
      adoptive cell transfer (ACT), 478
      chronic lymphocytic leukemia (CLL), 478
      Fas variants, 479
      use of, 479
   checkpoint inhibitors
      CTLA-4 immunotherapy, 475
      PD-1/PD-L1 immunotherapy, 475–478, 476*f*
      TIM-3 immunotherapy, 478
Antigen presenting cells (APCs), 58, 465
Applied behavior analysis (ABA), 386
Artificial intelligence, 506
Autophagy, 441–442

B

BBB. *See* Blood-brain-barrier (BBB)
Biomarkers
   immunotherapy
      AIM2, 486
      genetic characteristics, 486
      greater microsatellite instability, 487
      KRAS mutation, 486
      PD-1/PD-L1, 487
      peripheral blood biomarkers, 487
      side effects of, 488–489
      tumor-associated macrophages (TAMs), 488
   overweight/obesity, 258
   predict cancer occurrence, 504
Blood-brain-barrier (BBB), 85, 179–180
B lymphocytes
   antigen neutralization, 56–57
   IgA, 55–56
   IgE, 55–56
   IgG, 55
   immunoglobulins, 55
   plasma cells, 55
   structure, immunoglobulin, 56, 57*f*
Brain and immune system interactions
   blood-brain-barrier (BBB), 179–180
   brain parenchyma, 180
   claudin-5 (cldn5), 179–180
   cytokine molecules, 179–180
   glioblastoma development, 180
   intra- and intersystem communication, 179
   microglia, 180, 181*f*
   proinflammatory cytokine, 180–182
   psychological problems, 178–179
   sensitization effects, 180–182

C

Cachexia
   eating and nutrition, 247–248
   exercise and, 325, 332*b*
Cancer
   acquired resistance, 20
   active defenses
      exosomes, 18–19

Cancer (Continued)
    integrin CD11b, 18
    macrophages, 18
    myeloid cells, 18
    tumor-associated macrophages (TAMs), 18
  average annual percent change (AAPC), 6, 7f
  awareness for, 6–8
  being unneighborly, 26–27
  in canines, 12–13
  in cats, 13
  cellular aging, 4
  circumventing defenses
    glioblastomas, 17
    mammalian target of rapamycin (mTOR), 16
    neoepitopes, 16
    programmed death-ligand 1 (PD-L1) protein, 16–17
  classification, 11–13, 11t
  development and progression
    apoptosis/programmed cell death, 9
    benign tumor, 9
    biomechanical features, 9–10
    characteristics of, 9, 10f
    malignant tumor, 9
    metastasis/secondary tumor, 9
    tumor microenvironment, 10
  exercise and
    beneficial effects, 319–320, 321f
    cachexia, 325
    depression, 326–327
    epigenetic changes, 321–322
    fatigue, 323–324
    features and side effects, 322–327
    high-intensity, 319
    insulin-like growth factor-1 (IGF-1), 328
    molecular mechanisms, 316, 317f
    neuropathy, 326
    physical activity, 316
    principles of, 327–328, 328f
    progression, 318–319
    prophylactic actions, 316
    quality of life, 323–325
    sleep disturbances, 325–326
    supervised exercise, 318
  extrinsic factors, 3–4, 3f
  five-year survival rates, USA, 5–6, 5f
  hormones and hormone receptors
    colorectal cancer, 29
    estrogen/testosterone functioning, 28
    hormone replacement therapy (HRT), 28
    metformin, 29
    triple-negative breast cancer, 28
  immune surveillance
    cancer immunoediting process, 14, 15f
    elimination phase, 14, 15f
    equilibrium phase, 14, 15f
    escape phase, 14, 15f
    inflammatory factors, 14
    tumor dormancy, 13–14
  incidence of, 4
  inflammatory factors
    consequences, 30
    inflammation-related carcinogenesis, 30, 31f
    innate immune cells, 30
    mechanisms of, 31–33
  intrinsic resistance, 20
  intrinsic vs. extrinsic contributions, 33–34
  lifestyle modifications, 4
  "loser" cells, 19–20
  metaanalyses, 8
  metastases
    APOE gene, 21
    disseminated cancer cells, 22
    glucose and the Warburg effect, 24–27, 25f, 27f
    integrins, 23
    neutrophil extracellular traps (NETs), 22–23
    nutritional and energy support, 23–24
  microbiota
    antibiotic-mediated microbiota depletion, 285, 285f
    antibiotic use, 286
    breast cancer, 296
    colorectal cancer, 293–296
    diet and, 283
    dysbiosis, 293
    epigenetic involvement, 283
    Escherichia coli, 282–283
    fecal transplants, 286–287
    female reproductive system cancers, 296
    germ-free mice, 284–285
    lateral transmission, 295b
    lung cancer, 297–298
    pancreatic adenocarcinoma, 283
    parasites and, 298
  side effects, standard cytotoxic treatments, 6
  sleep and
    circadian disturbances, 355
    circadian misalignment, 359–360
    health risks, 360–361
    hormones, 355
    IL-6 levels, 352
    night shift work, 358–359
    The Nurses' Health Studies, 359–360
    polymorphisms, 357
    sleep duration, 352–353
  stress and
    cortisol, 198–200, 201f
    estrogen, 200
    immune processes, 195–196, 197f
    norepinephrine, 196–198, 199f
    oxytocin, 200–202
  "supercompetitors", 19–20

Index

survivors, 2
systematic reviews, 8
treatment, 2
"winner" cells, 19–20
Cancer screening
  breast cancer, 403–404
  caveats
    appraisal delay, 408
    behavioral delay, 408
    delay in surgical procedures, 410
    delays in seeking help, 410
    illness delay, 408
    overdiagnosis, 408, 409*t*
  colon cancer, 404–405
  complementary and alternative medicine, 416–419, 419*f*
  drug repurposing, 412
  early detection, 402, 403*t*
  lung cancer, 405–406
  precision medicine
    applications, 422, 423*f*
    aspirin, 425*b*
    "cancer immunogram", 423–424
    components in, 422, 423*f*
    CRISPR-Cas9, 421*b*
    genetic and epigenetic factors, 420
    KRAS mutation, 424
    matching individuals, 420–422
    MCL1, 424–425
    obstacles and challenges, 419–426
    p53 protein, 424
  prostate cancer, 404
  recommendations for, 406*b*
  risk compensation theory, 411
  sensitivity *vs.* specificity, 403
  treatment development
    cannabis and cancer, 415*b*
    difficulties, 412–413
    drug treatments, 414
    drug trials, 413
    efficacy of, 415–416
    mortality rates, 414–415
    preclinical studies, 413–414
    Reproducibility Project, 413
  unintended consequences, 411
Carcinogens
  cancer development, 110
  climate change and cancer, 110*b*
  DNA-repair, 109
CAR T-cell therapy
  acute lymphoblastic leukemia (ALL), 478
  adoptive cell transfer (ACT), 478
  checkpoint inhibitors, 481–482
  chronic lymphocytic leukemia (CLL), 478
  combination therapies, 480–481
  effectiveness of, 479–480
  Fas variants, 479
  side effects, 482–485, 483–484*f*
  use of, 479
CBT. *See* Cognitive behavioral therapy (CBT)
CD36 antibodies, 467–468
Cellular stress responses
  endoplasmic reticulum (ER), 165
  heat shock proteins (HSPs), 165
  reactive oxygen species (ROS), 164–165
  unfolded protein response (UPR), 165
Chemokines, 62–63
Chemotherapy
  in cell division, 432–433
  chemotherapeutic agents, 433
  p21 expression, 433
  side effects of, 434–438
Cholesterol
  functions, 259
  high-density lipoproteins (HDL), 259
  27-hydroxycholesterol (27-HC), 260
  low-density lipoproteins (LDL), 259
  N-glycolylneuraminic acid (Neu5Gc), 260
  trimethylamine oxide (TMAO), 260
Chromosome-associated regulatory RNAs (carRNAs), 98
Chronic lymphocytic leukemia (CLL), 478
Chronotherapy, 363
Circadian clock network, 347–349, 348*f*
Circadian rhythms
  chronic challenges, 347
  circadian clock network, 347–349, 348*f*
  hormones, 347
  immune and cytokine changes, 349–350, 350–351*f*
  implications of, 363–364
  melatonin, 347
Cognitive behavioral therapy (CBT)
  benefits of, 387
  disadvantages, 388
  internet-delivered, 387–388
  psychotherapy, 387
  supportive group therapy, 388
Common-sense model, 379–380
Corticotropin releasing hormone (CRH)
  amygdala activation, 150, 151*f*
  food selection, 153
  prototypical stress system, 150–151, 151*f*
Cortisol, 150–153, 152*b*
Counterproductive behaviors, 369–370
CRH. *See* Corticotropin releasing hormone (CRH)
Cytokines
  angiogenesis, 62
  antiinflammatory, 60–62
  based therapy
    administration of, 468
    depression, 469

Cytokines *(Continued)*
    indoleamine-2,3-dioxygenase (IDO), 469–471
    interferons, 468
  chemokines, 62–63
  exercise
    calorie-restricted diet, 314–315
    C-reactive protein, 314–315
    inflammatory effects, 314
    myokines, 315–316, 315*f*
  immunosuppressive effects, 62
  JAK-STAT pathways, 62
  messenger proteins, 60–62
  monocyte chemoattractant protein-1 (MCP-1), 62
  nuclear factor kappa-light-chain-enhancer of activated B cells (NFκB), 62
  types and functions, 60–62, 60–61*t*
  variations and stress
    antiinflammatory actions, 188
    early-life experiences, 189–191
    humans, stressor-elicited, 188–189, 190*f*
    IL-17, 187
    IL-18, 187
    proinflammatory actions, 188

D

Danger-associated molecular patterns (DAMPs)
  adenosine triphosphate (ATP), 50
  high-mobility group box 1 (HMGB1) protein, 50
  necrotic cells, 50
Depression
  antidepressants, 215
  chemotherapeutic agents, 215
  diagnosis and treatment, 214
  magic mushrooms, 216*b*
  periodic tests, 213–214
  postsurgical symptoms, 214–215
  predictor of, 212–213
  symptoms, 213
Diet and weight loss
  famine preparedness mode, 235–236
  microbiota, 235
  periodic fasting, 238–239
  polyphenols, 235
  prebiotics, 234–235
  probiotics, 234–235
  right choice of, 236*b*
Digestive process
  amylase, 227–228
  bile acids, 227–228
  cholecystokinin, 227–228
  functions, 228*f*
  organs involved in, 227, 228*f*
  pepsin, 227–228
  short-chain fatty acids (SCFAs), 228
  taurine, 227–228

Diverse life stressors
  chronic illnesses, 209–210
  job-related stress, 207–208
  loneliness, 208–209
  surgery
    cognitive decline, 210
    cyclooxygenase 2 (COX-2), 212
    general anesthetics, 211
    metastasis, 210
    preoperative diet, 211
Do not resuscitate (DNR) orders, 512–513
Dying with dignity, 513
Dysbiosis
  microbiota, 70, 87–88, 293
  traditional therapy, 452–454, 453*f*

E

Early-life stressors
  being bullied, 169
  downstream consequences, 169–170, 170*f*
  influence of, 205
  mortality, 169
Eating and nutrition
  brain neuronal functioning, 227
  cancer occurrence
    alcohol consumption, 239–240
    anticancer effects, diet ingredients, 240*b*
    antioxidants, 244*b*
    cachexia, 247–248
    fats and meat proteins, 245–247
    fiber, 242–244
    food constituents and, 241–242, 243*f*
    Mediterranean diet, 244–245
    nonmeat proteins, 244–245
    nutrition needs, 247–248
  diets and weight loss
    famine preparedness mode, 235–236
    microbiota, 235
    periodic fasting, 238–239
    polyphenols, 235
    prebiotics, 234–235
    probiotics, 234–235
    right choice of, 236*b*
  digestive process
    amylase, 227–228
    bile acids, 227–228
    cholecystokinin, 227–228
    functions, 228*f*
    organs involved in, 227, 228*f*
    pepsin, 227–228
    short-chain fatty acids (SCFAs), 228
    taurine, 227–228
  food marketing, 226
  hormonal and brain processes
    central nervous system (CNS), 229

fat, 232–233
fiber content, 233
ghrelin, 229–230, 232*f*
hypothalamic mechanism, 229
incretins, 231–234
insulin, 230–231
leptin, 229, 232*f*
polyunsaturated fats (PUFAs), 233
processed foods, 234
saturated fats, 233
trans fats, 233
ultra-processed foods, 234
unprocessed foods, 234
unsaturated fats, 233
innate and learned food preferences, 226*b*
obesity, 226
selection pressures, 225–226
stress hormone output, 225–226
e-cigarettes, 371*b*
Epigenetics
aging, 112–114, 113*b*
definition, 110
DNA methylation, 111
environmental challenges, 111
intergenerational and transgenerational changes, 115–118
limitations of, 118–119
microRNA variations, 111
prenatal challenges, 111
silencing genes, 112
targeting epigenetic modifications, 114–115, 115*f*
telomeres, 113*b*
Eubiosis, 70, 279
Euthanasia, 513
Exercise
aerobic exercise, 312
anaerobic exercise, 312
cancer and
beneficial effects, 319–320, 321*f*
cachexia, 325
depression, 326–327
epigenetic changes, 321–322
fatigue, 323–324
features and side effects, 322–327
high-intensity, 319
insulin-like growth factor-1 (IGF-1), 328
molecular mechanisms, 316, 317*f*
neuropathy, 326
physical activity, 316
principles of, 327–328, 328*f*
progression, 318–319
prophylactic actions, 316
quality of life, 323–325
sleep disturbances, 325–326
supervised exercise, 318

cognitive benefits, 311–312
cytokine changes
calorie-restricted diet, 314–315
C-reactive protein, 314–315
inflammatory effects, 314
myokines, 315–316, 315*f*
disadvantages, 334–335
epinephrine activation, 312–313
"Exercise as Medicine", 312
microbiota
in athletes, 330–331
cachexia, 332*b*
cancer development and progression, 331–332, 331*f*
dietary and environmental factors, 329, 330*f*
fecal transplants, 331
in humans, 329–330
in rodents, 329
moderate *vs.* strenuous, 313–314
morning exercise, 320*b*
mortality rate, 312
sedentary behaviors, 332–334
social identity, 334
want *vs.* need, 333*b*
Exposome, 3–4, 3*f*
Exposure therapy and systematic desensitization, 385–386

F

Fatty acids
docosahexaenoic acid (DHA), 268
eicosapentaenoic acid (EPA), 268
omega-3 fatty acids, 268–270, 269*f*
polymorphonuclear myeloid-derived suppressor cells (PMN-MDSCs), 270
Flash radiation therapy, 434

G

Genetics
cancer process and variables, 119–120, 120*f*
carcinogens
cancer development, 110
climate change and cancer, 110*b*
DNA-repair, 109
climate change and cancer, 110*b*
driver genes, 105*b*
epigenetic processes
aging, 112–114, 113*b*
definition, 110
DNA methylation, 111
environmental challenges, 111
intergenerational and transgenerational changes, 115–118
limitations of, 118–119
microRNA variations, 111
prenatal challenges, 111
silencing genes, 112

Genetics *(Continued)*
    targeting epigenetic modifications, 114–115, 115*f*
    telomeres, 113*b*
    Mendelian genetics to molecular biology
        blood types, Punnett square, 95, 96*f*
        CRISPR-Cas9, 97
        cystic fibrosis, Punnett square, 94–95, 95*f*
        epistatic interaction, 95
        gene's expressivity, 95
        heritability, 96
        heterozygous offspring, 94
        homozygous offspring, 94
        Huntington's disease, 94–95
        intermediate inheritance, 95
        monozygotic and dizygotic twins, 96
        penetrance, 95
        poly-genic effects, 95
        segregating generations, 96–97
    natural selection and, 93–94
    nucleic acids
        chromosome-associated regulatory RNAs (carRNAs), 98
        DNA, 98, 99*f*
        gene signatures, 105–106
        "junk DNA", 98–99
        multihit hypothesis, 106–108
        mutations of cancer genes, 102–108
        nucleotides, 98, 99*f*
        posttranslational modification, 98
        protein production, 98
        transcription, 98, 99*f*
        translation, 98, 99*f*
    population-level genetics
        ethnic and racial variations, 123–125
        sex variations, 121–123, 122*f*, 124*f*
    precision (personalized) medicine
        acute myeloid leukemia (AML), 125–126
        cancer-related phenotypes, 129
        epigenetic marker, 127
        gene expression profiles, 125–126
        microRNA methylation, 126
        occurrence and progression, cancer, 128–129, 128*f*
        pleiotropy, 127
        Research Domain Criteria (RDoC), 127–128
        symptoms of illness, 127
    stressors, 108
Genomic technological advances
    genetic markers, 505
    "Master Regulators", 505
    Pan-Cancer Analysis of Whole Genomes (PCAWG) Consortium, 504
    precision medicine, 504–505
Growth factors
    colony-stimulating factors (CSFs), 162–163
    epidermal growth factor receptor (EGFR) gene, 161–162
    fibroblast growth factor (FGF), 160–161
    influence of, 163–164
    insulin-like growth factor (IGF-1), 162
    neurotrophins and brain functioning, 158
    role of, 157–158, 161*f*
    transforming growth factor-β (TGF-β), 162
    vascular endothelial growth factor (VEGF), 162

H

Hormone replacement therapy (HRT), 28
Health behaviors
    alcohol use, 371–372
    antivaccine attitudes, 375*b*
    behavior change, 374–376
    behavior gap, 376–377
    cigarette smoking, 370–371
    counterproductive behaviors, 369–370
    default option, 377*b*
    e-cigarettes, 371*b*
    intervention approaches
        annual medical check-up, 373–374
        risky behaviors, 372
        selective prevention approaches, 373
        universal prevention approaches, 373
    lung cancer, 370–371
    mouth and lip cancer, 370–371
    nudge theory, 377–378
    psychosocial and cognitive approaches
        applied behavior analysis (ABA), 386
        behavior change methods, 385
        cognitive behavioral therapy (CBT), 386–388
        common-sense model, 379–380
        exposure therapy and systematic desensitization, 385–386
        Health Belief Model, 380–381, 381*f*
        illness perceptions, 378–379
        mindfulness meditation, 388–389
        positive emotions, animals, 391*b*
        positive psychology, 391–392
        Protection Motivation Theory, 381
        Self-efficacy Theory, 382
        Theory of Planned Behavior, 383
        Theory of Reasoned Action, 382, 383*f*
        Transtheoretical Model (TTM), 383–385, 384*f*
    scrotal cancer, 370
    social support
        limitations, 395
        social cure, 393
        social homeostasis, 392
        social identity, 394
        unmarried *vs.* married individuals, 393
Health Belief Model, 380–381, 381*f*
Hormonal and brain process, eating

central nervous system (CNS), 229
fat, 232–233
fiber content, 233
ghrelin, 229–230, 232f
hypothalamic mechanism, 229
incretins, 231–234
insulin, 230–231
leptin, 229, 232f
polyunsaturated fats (PUFAs), 233
processed foods, 234
saturated fats, 233
trans fats, 233
ultra-processed foods, 234
unprocessed foods, 234
unsaturated fats, 233
Human gut
  bacterial interactions, 71
  bacteriophages, 71–72
  challenges, 71
  dysbiosis, 70
  eubiosis, 70
  risk factors, 70–71
  virome, 71–72

I

Immune surveillance
  cancer immunoediting process, 14, 15f
  elimination phase, 14, 15f
  equilibrium phase, 14, 15f
  escape phase, 14, 15f
  inflammatory factors, 14
  tumor dormancy, 13–14
Immune tolerance
  B cells, 65–66
  B$_{reg}$ cells, 65–66
  CD8+ memory T cells, 65
  limiting attacks, 66
  negative selection, 65
  T$_{reg}$ cells, 65
Immunity and cancer
  adaptive (acquired) immunity
    antibody molecule, 56, 57f
    antigen neutralization, 56–57
    B lymphocytes, 55–57
    discrimination between self and nonself, 54
    diversity, 54
    memory, 54
    proliferation and elimination, 54
    self-limitation, 54
    specificity, 54
    T lymphocytes, 57–59, 59f
  antibiotics, 39
  cytokines
    angiogenesis, 62

    antiinflammatory, 60–62
    chemokines, 62–63
    immunosuppressive effects, 62
    JAK-STAT pathways, 62
    messenger proteins, 60–62
    monocyte chemoattractant protein-1 (MCP-1), 62
    nuclear factor kappa-light-chain-enhancer of activated B cells (NFκB), 62
    types and functions, 60–62, 60–61t
  evolutionary role
    chronic inflammation, 42
    N-acetylneuraminic acid (Neu5Ac), 42
    natural selection, 43
    N-glycolylneuraminic acid (Neu5Gc), 42
    sialic acid-binding immunoglobulin-type lectins (Siglecs), 42–43
  global issues, 40
  immune tolerance
    B cells, 65–66
    Breg cells, 65–66
    CD8+ memory T cells, 65
    limiting attacks, 66
    negative selection, 65
    T$_{reg}$ cells, 65
  increased life span, 39
  inflammasome
    fibroblasts, 64
    IL-1β, 63–64
    IL-18, 63–64
    NOD-like receptors (NLRs), 63–64
    types, 63
  innate immunity
    danger-associated molecular patterns (DAMPs), 49–52, 51f
    granulocyte-macrophage colony-stimulating factor (GM-CSF), 46
    inflammation, 45
    leukocytes, 43–44, 44f
    lymphoid cells, 44–45
    macrophages, 45–46
    myeloid cells, 44–45
    neutrophils, 46–47
    NK cells, 47
    pathogen-associated membrane patterns (PAMPs), 47–48
    pattern recognition receptors (PRRs), 47–48
    peritoneal macrophages, 45–46
    phagocytosis, 44–45
    reactive oxygen species (ROS), 46–47
    resident tissue macrophages, 45–46
    respiratory bursts, 44–45
    sterile inflammation, 48–52
    tissue injury, 45
    toll-like receptors (TLRs), 48, 49f

Immunity and cancer (Continued)
  lymphoid system
    afferent lymph, 53
    cell trafficking, 52–53
    dendritic cells (DCs), 53
    efferent lymph, 53
    lymph nodes, 52–53, 52f
    spleen, 53
    T and B cell, 53
  mammalian immune responses, 40
  overpopulation, 40
  red blood cells (erythrocytes), 41
  stress
    acute vs. chronic challenges, 183–184, 183f
    features of, 182–183
    macrophages and dendritic cell functioning, 182
  stressor-elicited immune changes, 185–186
  stressor-provoked responses
    in humans, 186–187
    in rodents, 184–187
    sex differences, 185
  vaccines, 39
  white blood cells (leukocytes), 41
Immunotherapy
  antibody-based therapies
    CAR T-cell therapy, 478–479
    checkpoint inhibitors, 475–478
  biomarkers
    AIM2, 486
    genetic characteristics, 486
    greater microsatellite instability, 487
    KRAS mutation, 486
    PD-1/PD-L1, 487
    peripheral blood biomarkers, 487
    side effects of, 488–489
    tumor-associated macrophages (TAMs), 488
  CAR T therapy
    effectiveness of, 479–482
    side effects, 482–485, 483–484f
  categories of, 462, 463f
  Coley's toxin, 462
  combination therapies, 480–481
  cytokine-based therapy
    administration of, 468
    depression, 469
    indoleamine-2,3-dioxygenase (IDO), 469–471
  interferons, 468
  intrinsic and extrinsic factors, 462–463
  microbiota and
    CAR T therapy, 495
    colorectal carcinogenesis, 490–492, 491f
    fecal transplantation, 489–490
    immune checkpoint inhibitor interactions, 489, 490f
    microbial adjunctive treatments, 495
    PD-1/PD-L1 therapy, 492–494, 493f
    role for, 489
  nonspecific, 467–471
  prebiotics and probiotics
    in cancer treatment, 495–496
    in ICU patients, 496
    microbiota metabolites, 496–497
  stem cell therapy
    graft-versus-host disease (GvHD), 465–467, 466f
    hematopoietic, 464
    induced pluripotent stem cells (iPSCs), 464–465, 467
    limitations, 467
    microbial factors, 465–467
    mutations, 464
    progenitor cellular system, 464
    transplantation, 464
  treatment vaccines
    cancer vaccines, 472
    Epstein-Barr virus, 471–472
    human papillomavirus (HPV), 471
    human T-cell leukemia virus-1, 471
    oncolytic viral therapy, 472–474
  variolation, 462
Inflammasome
  fibroblasts, 64
  IL-1β, 63–64
  IL-18, 63–64
  NOD-like receptors (NLRs), 63–64
  types, 63
Innate immunity
  danger-associated molecular patterns (DAMPs)
    adenosine triphosphate (ATP), 50
    high-mobility group box 1 (HMGB1) protein, 50
    necrotic cells, 50
  granulocyte-macrophage colony-stimulating factor (GM-CSF), 46
  inflammation, 45
  leukocytes, 43–44, 44f
  lymphoid cells, 44–45
  macrophages, 45–46
  myeloid cells, 44–45
  neutrophils, 46–47
  NK cells, 47
  pathogen-associated membrane patterns (PAMPs), 47–48
  pattern recognition receptors (PRRs), 47–48
  peritoneal macrophages, 45–46
  phagocytosis, 44–45
  reactive oxygen species (ROS), 46–47
  resident tissue macrophages, 45–46
  respiratory bursts, 44–45
  sterile inflammation, 48–52
  tissue injury, 45
  toll-like receptors (TLRs), 48, 49f

K

Ketogenic diets, 446*b*

L

Life expectancy, 509
Lung cancer, 370–371
Lymphoid system
    afferent lymph, 53
    cell trafficking, 52–53
    dendritic cells (DCs), 53
    efferent lymph, 53
    lymph nodes, 52–53, 52*f*
    spleen, 53
    T and B cell, 53

M

Machine learning, 506
Mammalian target of rapamycin (mTOR), 16, 264, 440
"Master Regulators", 505
Medical assistance in dying (MAiD), 513
Mendelian genetics to molecular biology
    blood types, Punnett square, 95, 96*f*
    CRISPR-Cas9, 97
    cystic fibrosis, Punnett square, 94–95, 95*f*
    epistatic interaction, 95
    gene's expressivity, 95
    heritability, 96
    heterozygous offspring, 94
    homozygous offspring, 94
    Huntington's disease, 94–95
    intermediate inheritance, 95
    monozygotic and dizygotic twins, 96
    penetrance, 95
    poly-genic effects, 95
    segregating generations, 96–97
Metastasis
    APOE gene, 21
    cancer, development and progression, 9
    disseminated cancer cells, 22
    diverse life stressors, 210
    glucose and the Warburg effect, 24–27, 25*f*, 27*f*
    integrins, 23
    neutrophil extracellular traps (NETs), 22–23
    nutritional and energy support, 23–24
Microbial factors
    breastfeeding
        bacterial profile, 78
        cross-fostering paradigm, 77–78
        mycobiota, 78
        nutrients, 78–79
        oxytocin, 78–79
        prebiotics, 78–79
    vaginal birth, 79–80
Microbiota
    bacterial species, 280
    beneficial and nonbeneficial bacteria, 282
    cancer
        antibiotics and germ-free conditions, 284–286, 285*f*
        breast and female reproductive system cancers, 296
        colorectal cancer, 293–296
        diet and, 283
        dysbiosis, 293
        epigenetic involvement, 283
        Escherichia coli, 282–283
        fecal transplants, 286–287
        lung cancer, 297–298
        pancreatic adenocarcinoma, 283
        parasites and, 298
    CAR T therapy, 495
    colorectal carcinogenesis, 490–492, 491*f*
    dysbiotic state, 279
    eubiotic state, 279
    exercise and
        in athletes, 330–331
        cachexia, 332*b*
        cancer development and progression, 331–332, 331*f*
        dietary and environmental factors, 329, 330*f*
        fecal transplants, 331
        in humans, 329–330
        in rodents, 329
    fecal transplantation, 489–490
    food components and actions, 281, 281*f*
    gut immune functioning, 280
    immune checkpoint inhibitor interactions, 489, 490*f*
    health and
        anxiety, 84
        autotoxins, 69–70
        bacterial interactions, 71
        bacteriophages, 71–72
        blood-brain-barrier (BBB), 85
        breastfeeding, 77–80
        challenges, 71
        communication pathways, microbiota-gut-brain axis, 85, 86*f*
        depression, 84
        dysbiosis, 87–88
        early postnatal influences, 76–77
        environmental toxicants, 88–89
        Escherichia coli, 74
        eubiosis, 70
        good and bad microbes, 87–88
        human intestine, 72–73, 72*f*
        humoral immunity, 69
        illness comorbidities, 83–87
        immune functioning, 80–83, 82*f*
        inflammation, 81–83
        intrinsic and extrinsic factors, 73, 74*f*

Microbiota (Continued)
    magic bullets, 69–70
    norepinephrine, 85
    prenatal influences, 75–76
    probiotic formulations, 87–88
    risk factors, 70–71
    Salmonella, 74
    serotonin, 85
    short-chain fatty acids (SCFA), 85
    vaginal birth, 79–80
    virome, 71–72
Mediterranean diet, 281
microbial adjunctive treatments, 495
PD-1/PD-L1 therapy
    checkpoint blockade, 493–494
    hepatocellular carcinoma, 494
    therapeutic effects, 492, 493f
probiotics, 280
role for, 489
short-chain fatty acids (SCFA)
    free fatty acids (FFA), 290–291
    G protein-coupled receptors (GPCRs), 290–291
    gut barrier functioning, 288, 290f
    gut-brain communication pathways, 288, 289f
    probiotic diets, 288
stress and
    adolescent and adult, 193–195
    in centenarians, 194b
    challenges, 191
    HPA activity, 192
    posttraumatic stress disorder (PTSD), 193
    role of, 192–193
traditional therapy
    antibiotics, 452
    cancer-related pain and, 454b
    dysbiosis, 452–454, 453f
    eubiosis, 452–454, 453f
    gut bacteria, 451–452
Mindfulness meditation, 388–389
Monocyte chemoattractant protein-1 (MCP-1), 62
Mouth and lip cancer, 370–371
mTOR. *See* Mammalian target of rapamycin (mTOR)
Myokines, 315–316, 315f

N

N-acetylneuraminic acid (Neu5Ac), 42
Neuropilin-1 (Nrp 1), 468
Neuropsychiatric and neurodegenerative disorders
    Alzheimer's disease, 217
    Parkinson's disease, 218
    schizophrenia, 217
Neutrophil extracellular traps (NETs), 22–23
N-glycolylneuraminic acid (Neu5Gc), 42
Nonspecific immunotherapy, 467–471

Nucleic acids
    chromosome-associated regulatory RNAs (carRNAs), 98
    DNA, 98, 99f
    gene signatures, 105–106
    "junk DNA", 98–99
    multihit hypothesis, 106–108
    mutations of cancer genes
        BRCA1 and BRCA2, 103–104
        extrachromosomal DNA (ecDNA), 104
        inherited, 101–102
        PTEN gene, 103–104
        random mutations, 99–101
        TP53, 103
    nucleotides, 98, 99f
    posttranslational modification, 98
    protein production, 98
    transcription, 98, 99f
    translation, 98, 99f

O

"One Health" transdisciplinary approach, 508
Overweight/obesity
    awareness of, 255
    β-carotene supplementation, 273
    biomarkers, 258
    cancer occurrence, 254
    cholesterol
        27-hydroxycholesterol (27-HC), 260
        functions, 259
        high-density lipoproteins (HDL), 259
        low-density lipoproteins (LDL), 259
        N-glycolylneuraminic acid (Neu5Gc), 260
        trimethylamine oxide (TMAO), 260
    dietary amino acids, 271
    elevated body-mass index (BMI), 254
    fat accumulation, 256
    fat and lipid metabolism, 266–268
    fat-soluble vitamins, 272
    fatty acids
        docosahexaenoic acid (DHA), 268
        eicosapentaenoic acid (EPA), 268
        omega-3 fatty acids, 268–270, 269f
        polymorphonuclear myeloid-derived suppressor cells (PMN-MDSCs), 270
    genetic differences, 256–257
    genetic influences, 260–261
    ghrelin, 259
    glucose
        artificial sweeteners, 263
        fructose, 265–266
        hyperglycemia, 263
        insulin, 262
        mammalian target of rapamycin (mTOR) pathway, 264
        metformin, 263–264

PI3K activation, 264–265
serotonin production, 262
sugar industry, 263
Warburg effect, 262
leptin, 258
low-fat diet, 256
"metabolically healthy obesity", 254
metabolic regulation, 258
mortality rate, 253–254
obese offspring, 255
polyphenols, 270
risk factor, 255
sex differences, 256–257
shorter life span, 253–254
sugar, 263
tumor progression, 257–258, 257f
vitamin A, 273
vitamin C, 272–273
vitamin D, 272
water-soluble vitamins, 272
Women's Health Initiative, 256

P

P4 medicine approach, 507
Pan-Cancer Analysis of Whole Genomes (PCAWG) Consortium, 504
Pathogen-associated membrane patterns (PAMPs), 47–48
Pattern recognition receptors (PRRs), 47–48
Physician-assisted death, 513
Physician-patient communication
decision regret, 511–512
do not resuscitate (DNR) orders, 512–513
dying with dignity, 513
medical assistance in dying (MAiD), 513
right-to-try medications, 514
shared decision-making, 510
transparency and open communication, 510
treatment preference, 511
Population-level genetics
ethnic and racial variations, 123–125
sex variations
dimorphism, 121–122
drug trials, 121
ethnic and racial variations, 123–125
genetic factors, 121
incidence and mortality rates, 121, 122f
microbiota, 123
Positive psychology, 391–392
Posttraumatic stress disorder (PTSD), 193, 207
Precision medicine
acute myeloid leukemia (AML), 125–126
cancer-related phenotypes, 129
cancer screening
applications, 422, 423f
aspirin, 425b
"cancer immunogram", 423–424
components in, 422, 423f
CRISPR-Cas9, 421b
genetic and epigenetic factors, 420
KRAS mutation, 424
matching individuals, 420–422
MCL1, 424–425
obstacles and challenges, 419–426
p53 protein, 424
epigenetic marker, 127
gene expression profiles, 125–126
genomic technological advances, 504–505
microRNA methylation, 126
occurrence and progression, cancer, 128–129, 128f
pleiotropy, 127
Research Domain Criteria (RDoC), 127–128
symptoms of illness, 127
Prenatal stress
epigenetic changes, 166–168
hormonal consequences, 166
premature birth, 165–166
prenatal infection, 168
stressors, 205–206
Protection Motivation Theory, 381
PRRs. *See* Pattern recognition receptors (PRRs)
Psychological well-being
cancer treatment, 213–214
depression
antidepressants, 215
chemotherapeutic agents, 215
diagnosis and treatment, 214
magic mushrooms, 216b
periodic tests, 213–214
postsurgical symptoms, 214–215
predictor of, 212–213
symptoms, 213
Psychosocial and cognitive approaches
applied behavior analysis (ABA), 386
behavior change methods, 385
cognitive behavioral therapy (CBT), 386–388
common-sense model, 379–380
exposure therapy and systematic desensitization, 385–386
Health Belief Model, 380–381, 381f
illness perceptions, 378–379
mindfulness meditation, 388–389
positive emotions, animals, 391b
positive psychology, 391–392
Protection Motivation Theory, 381
Self-efficacy Theory, 382
Theory of Planned Behavior, 383
Theory of Reasoned Action, 382, 383f
Transtheoretical Model (TTM), 383–385, 384f

Psychosocial stressors
 mouse model, 202
 prospective analyses, humans, 203–204
 retrospective analyses, humans, 202–203
PTSD. *See* Posttraumatic stress disorder (PTSD)

R

Radiation therapy
 brachytherapy, 434
 flash radiation therapy, 434
 side effects of, 434–438
 stereotactic body radiation therapy, 434
Resilience and vulnerability, 145–146, 146f

S

SCFA. *See* Short-chain fatty acids (SCFA)
Scrotal cancer, 370
Selective fecal microbiota transplantation, 489
Self-efficacy Theory, 382
Short-chain fatty acids (SCFA)
 free fatty acids (FFA), 290–291
 G protein-coupled receptors (GPCRs), 290–291
 gut barrier functioning, 288, 290f
 gut-brain communication pathways, 288, 289f
 probiotic diets, 288
Sialic acid-binding immunoglobulin-type lectins (Siglecs), 42–43
Sleep
 blood flow, importance of, 341–342
 cancer and
  circadian disturbances, 355
  circadian misalignment, 359–360
  health risks, 360–361
  hormones, 355
  IL-6 levels, 352
  night shift work, 358–359
  The Nurses' Health Studies, 359–360
  polymorphisms, 357
  sleep duration, 352–353
 circadian disruptions and diurnal variations
  gut microbiota, 361–363
  innate lymphoid cells (ILC3), 361
  microbial regulation, 361, 362f
 circadian rhythms
  chronic challenges, 347
  circadian clock network, 347–349, 348f
  hormones, 347
  immune and cytokine changes, 349–350, 350–351f
  implications of, 363–364
  melatonin, 347
 disturbances, 343
 dreams, 341
 eating and, 346
 exercise and, 345–346
 functions of
  glymphatic system, 343–344
  non-rapid eye movement (NREM), 345
  rapid eye movement (REM), 345
  sleep loss, consequences of, 344–345
 importance of, 342
 internal forces and humors, 341–342
 narcolepsy, 354
 recuperative powers, 342
 restless leg syndrome, 354
 serotonin, 343
 sleep apnea, 353–354
 sleep disturbances, 343
Smoking and alcohol consumption, 507
Stem cell therapy
 graft-*versus*-host disease (GvHD), 465–467, 466f
 hematopoietic, 464
 induced pluripotent stem cells (iPSCs), 464–465, 467
 limitations, 467
 microbial factors, 465–467
 mutations, 464
 progenitor cellular system, 464
 transplantation, 464
Stereotactic body radiation therapy, 434
Stress
 brain and immune system interactions
  blood-brain-barrier (BBB), 179–180
  brain parenchyma, 180
  claudin-5 (cldn5), 179–180
  cytokine molecules, 179–180
  glioblastoma development, 180
  intra- and intersystem communication, 179
  microglia, 180, 181f
  proinflammatory cytokine, 180–182
  psychological problems, 178–179
  sensitization effects, 180–182
 cancer
  cortisol, 198–200, 201f
  estrogen, 200
  immune processes, 195–196, 197f
  norepinephrine, 196–198, 199f
  oxytocin, 200–202
 cancer occurrence/progression, 178
 children, cancer occurrence, 206–207
 cytokine variations
  antiinflammatory actions, 188
  early-life experiences, 189–191
  humans, stressor-elicited, 188–189, 190f
  IL-17, 187
  IL-18, 187
  proinflammatory actions, 188

diverse life stressors
   chronic illnesses, 209–210
   job-related stress, 207–208
   loneliness, 208–209
   surgery, 210–212
early-life stressors, 205
immunity
   acute *vs.* chronic challenges, 183–184, 183*f*
   features of, 182–183
   macrophages and dendritic cell functioning, 182
   stressor-elicited immune changes, 185–186
   stressor-provoked responses, 184–187
microbiota and
   adolescent and adult, 193–195
   in centenarians, 194*b*
   challenges, 191
   HPA activity, 192
   posttraumatic stress disorder (PTSD), 193
   role of, 192–193
neuropsychiatric and neurodegenerative disorders
   Alzheimer's disease, 217
   Parkinson's disease, 218
   schizophrenia, 217
prenatal stressors, 205–206
psychological well-being
   cancer treatment, 213–214
   depression as, 212–216, 213*b*, 216*b*
psychosocial stressors
   mouse model, 202
   prospective analyses, humans, 203–204
   retrospective analyses, humans, 202–203
posttraumatic stress disorder (PTSD), 207
Stressors
   allostasis and allostatic overload, 138, 138*f*
   appraisals
      characteristics, 139, 140*f*
      Dunning-Kruger effect, 139–140
      illusory superiority, 139–140
   availability heuristics, 141
   breast cancer diagnosis, 137
   cellular stress responses
      endoplasmic reticulum (ER), 165
      heat shock proteins (HSPs), 165
      reactive oxygen species (ROS), 164–165
      unfolded protein response (UPR), 165
   chronic illnesses and pain, 137–138
   cognitive and behavioral control, 136–137
   coping method
      cognitive flexibility, 145
      cognitive restructuring, 144–145
      emotion-focused, 144
      problem-solving, 144
   decisional control, 136–137
   decision-making, 141
   early-life stressors
      being bullied, 169
      downstream consequences, 169–170, 170*f*
      mortality, 169
   effects on illness, 135–136
   growth factors
      colony-stimulating factors (CSFs), 162–163
      epidermal growth factor receptor (EGFR) gene, 161–162
      fibroblast growth factor (FGF), 160–161
      influence of, 163–164
      insulin-like growth factor (IGF-1), 162
      neurotrophins and brain functioning, 158
      role of, 157–158, 161*f*
      transforming growth factor-β (TGF-β), 162
      vascular endothelial growth factor (VEGF), 162
   heuristics, 140–141
   information control, 136–137
   learned helplessness, 136–137
   misappraisals, 140–142
   negative consequences, 136
   neurobiological actions
      acetylcholine (ACh), 150
      aging, 152–153
      corticotropin releasing hormone (CRH), 150–153, 151*f*
      cortisol, 150–153, 152*b*
      endocannabinoid (eCB) system, 156–157
      estrogen, 154
      neuropeptide Y (NPY), 155–156
      neurotransmitter alterations, 148
      oxytocin, 155
      peripheral biological changes, 147–148
      peripheral norepinephrine, 149–150
      prolactin, 154–155
      psychological challenges, 147
   prenatal stress
      epigenetic changes, 166–168
      hormonal consequences, 166
      premature birth, 165–166
      prenatal infection, 168
   resilience and vulnerability, 145–146, 146*f*
   stress sensitization, 139
   systemic challenges, 136

T

TAMs. *See* Tumor-associated macrophages (TAMs)
Theory of Planned Behavior, 383
Theory of Reasoned Action, 382, 383*f*
T lymphocytes
   antigen presenting cells (APCs), 58
   immune functioning, 58–59
   immune steps, cancer progression, 58, 59*f*
   major histocompatibility complex (MHC), 57–58

Traditional therapy
  angiogenesis, 447b
  autophagy, 441–442
  chemotherapy
    in cell division, 432–433
    chemotherapeutic agents, 433
    p21 expression, 433
    side effects of, 434–438
  energy and metabolism
    angiogenesis p21, 447b
    energy-producing mitochondria, 445
    fasting diets, 445–448
    glycolysis, 444–445
    guanosine triphosphate (GTP), 445
    ketogenic diets, 446b
    metformin, 444
    obesity paradox, 451b
    oxidative phosphorylation, 444–445
    selectively starving tumor cells, 448–451, 449f
  methotrexate development, 432
  microbiota
    antibiotics, 452
    cancer-related pain and, 454b
    eubiosis and dysbiosis, 452–454, 453f
    gut bacteria, 451–452
  mustard gas (nitrogen mustard), 431–432
  obesity paradox, 451b
  radiation therapy
    brachytherapy, 434
    flash radiation therapy, 434
    side effects of, 434–438
    stereotactic body radiation therapy, 434
  radium therapies, 432
  starving tumor cells, 448–451
  treatment resistance
    apoptosis and anastasis, 442–443
    autophagy, 441–442
    development mechanism, 438, 439f
    KRAS mutation, 440
    mammalian target of rapamycin (mTOR) pathway, 440
    nanoparticles, 442b
    natural selection, 441b
    nutrients, 443–444
    tumor microenvironment, 439–440
Transtheoretical Model (TTM), 383–385, 384f
Tumor-associated macrophages (TAMs), 488